Before We Are Born

ESSENTIALS OF EMBRYOLOGY AND BIRTH DEFECTS

Sixth Edition

Keith L. Moore, PhD, FIAC, FRSM

Professor Emeritus, Division of Anatomy, Department of Surgery
Faculty of Medicine, University of Toronto, Toronto, Ontario, Canada

Former Professor and Head, Department of Anatomy,
University of Manitoba
and former Professor and Chairman
Department of Anatomy and Cell Biology
University of Toronto

T.V.N. Persaud, MD, PhD, DSc, FRCPath (Lond.)

Professor Emeritus and Former Head
Department of Human Anatomy and Cell Science
Professor of Pediatrics and Child Health
Professor of Obstetrics, Gynecology, and Reproductive Sciences
Faculty of Medicine, University of Manitoba,

Consultant in Pathology and Clinical Genetics
Health Sciences Centre
Winnipeg, Manitoba, Canada

Visiting Professor, St. George's University Medical School
Grenada, West Indies

Before We Are Born

ESSENTIALS OF EMBRYOLOGY AND BIRTH DEFECTS

SAUNDERS
An Imprint of Elsevier Science

SAUNDERS
An Imprint of Elsevier Science

The Curtis Center
Independence Square West
Philadelphia, Pennsylvania 19106-3399

BEFORE WE ARE BORN: ESSENTIALS OF EMBRYOLOGY
AND BIRTH DEFECTS SIXTH EDITION ISBN 0-7216-9408-X

NOTICE

Medical Assisting is an ever-changing field. Standard safety precautions must be followed but as new research and clinical experience broaden our knowledge, changes in treatment and drug therapy may become necessary or appropriate. Readers are advised to check the most current product information provided by the manufacturer of each drug to be administered to verify recommended dose, the method and duration of administration, and contraindications. It is the responsibility of the treating physician, relying on experience and knowledge of the patient, to determine dosages and the best treatment for each individual patient. Neither the Publisher nor the author assume any liability for any injury and/or damage to persons or property arising from this publication.

The Publisher

Library of Congress Cataloging-in-Publication Data

Moore, Keith L.
Before we are born: essentials of embryology and birth defects/Keith L. Moore, T.V.N Persaud.—6th ed.
p.; cm.
Includes bibliographical references and index.
ISBN 0-7216-9408-X
1. Embryology, Human. 2. Abnormalities, Human.
I. Persaud, T.V.N. II. Title.
[DNLM: 1. Embryology. 2. Abnormalities. QS 604 M822b 2003]

QM601.M757 2003
612.6′4—dc21

2002021796

Listed here are the latest translated editions of this book together with the languages for the translations and the publishers.

Portugese *(6th Edition)* — Editora Guanabara Koogan S.A.
Spanish *(6th Edition)* — McGraw-Hill Interamericana S.A.

Acquisitions Editor: Jason Malley

RDC/DNP

Printed in China

Last digit is the print number: 9 8 7 6 5 4 3 2 1

To our wives, Marion and Gisela, our children, and our grandchildren: Melissa, Kristin, Alicia, Lauren, Mitchel, Caitlin, Jayme, Courtney, and Brooke (KLM) and Brian Lucas (TVNP)

Preface

This sixth edition of *Before We Are Born* presents the essentials of normal and abnormal development for students of medicine and the associated health sciences. *Before We Are Born* is now specially designed for shorter, one-semester courses and for students reviewing for professional examinations. This work is a digest of our larger book, *The Developing Human: Clinically Oriented Embryology*, 7th edition. Therefore, readers wishing more information on any of the subjects should consult the more comprehensive text.

An important feature of this book is the section entitled *Clinically Oriented Questions* that appears at the end of each chapter. Over the years we have been asked these and other questions by students and lay people. Because of misconceptions derived from newspaper articles and discussions on television and radio, we have done our best to give answers that are supported by current research and medical practice.

This sixth edition contains more clinically oriented material than previous editions. These sections are highlighted in color to set them apart from the rest of the text. This edition also contains more color photographs of embryos (normal and abnormal). Many of the illustrations have been revised and improved using *three-dimensional renderings* and utilizing color more effectively. There are also additional diagnostic (ultrasound and MRI) images of embryos and fetuses, and scanning electron micrographs to illustrate three-dimensional aspects of embryos.

The teratology content has increased because the study of abnormal development is so helpful in understanding risk estimation, the causes of anomalies, and how birth defects may be prevented. Molecular aspects of developmental biology have been highlighted throughout the book, especially in those areas that appear promising for clinical medicine and future research. Successful completion of the national board examinations now requires knowledge of this aspect of development.

We have continued our attempts to present an *easy-to-read account of human development before birth*. Every chapter has been revised thoroughly to reflect new research findings and their clinical significance. The chapters are organized to present a systematic and logical approach to the development of embryos and fetuses. In addition to being updated, the text also includes more information that is useful in clinical practice and material that will be most helpful to those taking problem-solving courses.

Keith L. Moore
Vid Persaud

Acknowledgments

Many of our colleagues (listed alphabetically) have helped with the preparation of this edition. It is a pleasure to record our indebtedness to them: Dr. Anne M.R. Agur, Associate Professor of Anatomy, Division of Anatomy, Department of Surgery, University of Toronto, Toronto, Ontario. *Dr. Steve Ahing*, Associate Professor, Faculty of Dentistry, University of Manitoba, Winnipeg; *Dr. Judy Anderson*, Professor of Anatomy and Cell Science, University of Manitoba, Winnipeg; *Dr. Kunwar Bhatnagar*, Professor of Anatomy, School of Medicine, University of Louisville, Louisville, KY; *Dr. Albert Chudley*, Professor of Pediatrics and Child Health, Director of Clinical Genetics, Health Sciences Centre, University of Manitoba, Winnipeg; *Dr. Blaine Cleghorn*, Associate Professor, Faculty of Dentistry, University of Manitoba, Winnipeg; *Dr. Angelika Dawson*, Director, Cytogenetics Laboratory, Health Sciences Centre and Associate Professor, Department of Biochemistry and Medical Genetics, University of Manitoba, Winnipeg; *Dr. Marc Del Bigio*, Professor of Pathology, University of Manitoba, Winnipeg; *Dr. Raymond Gasser*, Louisiana State University School of Medicine, New Orleans, LA; *Dr. Barry Grayson*, New York University Medical Center, Institute of Reconstructive Plastic Surgery, New York, NY; *Dr. Christopher Harman*, Department of Obstetrics, Gynecology and Reproductive Sciences, University of Maryland, Baltimore, MD; *Dr. John A. Jane, Sr.*, David D. Weaver Professor of Neurosurgery, University of Virginia Health System, Charlottesville, VA; *Dr. Dagmar Kalousek*, Program Head, Cytogenetics/Embryopathology Laboratory and Professor of Pathology, University of British Columbia, Vancouver BC; *Dr. Peeyush Lala*, Professor Emeritus, Faculty of Medicine, University of Western Ontario, London, Ontario; *Dr. Edward A. Lyons*, Professor of Radiology and Obstetrics and Gynecology, University of Manitoba, Winnipeg; *Dr. Moshe Matilsky*, Director of the IVF Laboratory, Reproductive Medicine Unit, Department of Obstetrics and Gynecology, Poriya Government Hospital, M.P. Hagalil Hatachton, Israel; *Dr. John B. Mulliken*, Associate Professor of Surgery and Director, Craniofacial Center, The Children's Hospital, Harvard Medical School, Boston, MA; *Professor T.S. Ranganathan*, St. George's University, School of Medicine, Grenada; *Dr. Gregory Reid*, Assistant Professor, Department of Obstetrics, Gynecology and Reproductive Sciences, University of Manitoba, Winnipeg; *Dr. Prem Sahni*, Department of Radiology, Children's Hospital, Winnipeg, Manitoba; *Dr. Kohei Shiota*, Professor and Chairman of the Department of Anatomy and Developmental Biology and Director of the Congenital Anomaly Research Center, Kyoto University, Japan; *Dr. Joseph Siebert*, Research Associate Professor, Children's Hospital and Regional Medical Center, Seattle, WA; *Dr. Gerald S. Smyser*, Altru Health System, Grand Forks, ND; *Dr. Pierre Soucy*, Professor and Chief, Division of Paediatric General Surgery, Children's Hospital of Eastern Ontario, University of Ottawa, Ottawa; *Dr. Mark Torchia*, Associate Professor, Department of Surgery, University of Manitoba, Winnipeg; Dr. Brunno L. Vendittelli, New York University Medical

Center, Institute of Reconstructive Plastic Surgery, New York, NY; and *Dr. Michael Wiley*, Professor and Chair, Division of Anatomy, Department of Surgery, University of Toronto, Ontario.

The new illustrations were prepared by Hans Neuhart, President of the Electronic Illustrators Group in Fountain Hills, AZ. Marion Moore in Toronto and Barbara Clune in Winnipeg did the word processing and helped with review of the manuscript, as did Gisela Persaud in Winnipeg. William Schmitt, Publishing Director, Medical Textbooks; Jason Malley, Medical Editor; Joan Sinclair, Supervisor, Full-Service Production, Saunders, Elsevier Science; Ellen Sklar, Production Editor, P.M. Gordon Associates, and their colleagues have been most helpful with our work. To all these people, we extend our sincere thanks. Last, but not least, we thank our wives, Marion and Gisela, for their continued understanding and support.

Keith L. Moore
Vid Persaud

Contents

Preface vii

Acknowledgments ix

1 Introduction to Human Embryology 1
Embryological Terminology 2
Importance of and Advances in Embryology 2
Descriptive Terms 7
Clinically Oriented Questions 7

2 Human Reproduction 9
Reproductive Organs 10
Gametogenesis 13
Female Reproductive Cycles 19
Transportation of Gametes 24
Maturation of Sperms 24
Viability of Gametes 26
Summary of Reproduction 26
Clinically Oriented Questions 26

3 The First Week of Human Development 27
Fertilization 28
Cleavage of the Zygote 31
Formation of the Blastocyst 31
Summary of the First Week of Development 35
Clinically Oriented Questions 35

4 The Second Week of Human Development 37
Continuation of Embryonic Development 38
Development of the Chorionic Sac 41

Implantation Sites of the Blastocyst 42
Summary of Implantation of the Blastocyst 43
Summary of the Second Week of Development 44
Clinically Oriented Questions 44

5 The Third Week of Human Development 45

Gastrulation: Formation of Germ Layers 46
Neurulation: Formation of Neural Tube 53
Development of Somites 53
Development of the Intraembryonic Coelom 56
Early Development of the Cardiovascular System 56
Development of Chorionic Villi 57
Summary of the Third Week of Development 58
Clinically Oriented Questions 60

6 Organogenetic Period: Human Development During Weeks Four to Eight 61

Folding of the Embryo 62
Germ Layer Derivatives 62
Control of Embryonic Development 66
Highlights of Development in Weeks Four to Eight 67
Estimation of Embryonic Age 75
Summary of Development During Weeks Four to Eight 75
Clinically Oriented Questions 77

7 Fetal Period: Ninth Week to Birth 77

Estimation of Fetal Age 78
Highlights of the Fetal Period 79
Expected Date of Delivery 83
Factors Influencing Fetal Growth 83
Procedures for Assessing Fetal Status 85
Summary of the Fetal Period 87
Clinically Oriented Questions 88

8 The Placenta and Fetal Membranes 89

The Placenta 90
Parturition (Childbirth) 99
Amnion and Amniotic Fluid 105
Yolk Sac 108
Allantois 110
Multiple Pregnancies 110
Summary of the Placenta and Fetal Membranes 114
Clinically Oriented Questions 115

9 Human Birth Defects 117
Teratology: The Study of Abnormal Development 118
Anomalies Caused by Genetic Factors 119
Anomalies Caused by Environmental Factors 127
Anomalies Caused by Multifactorial Inheritance 137
Summary of Human Birth Defects 137
Clinically Oriented Questions 138

10 Body Cavities, Mesenteries, and Diaphragm 139
The Embryonic Body Cavity 140
Development of the Diaphragm 145
Congenital Diaphragmatic Hernia 147
Summary of Development of the Body Cavities 149
Clinically Oriented Questions 150

11 The Pharyngeal Apparatus 151
Pharyngeal Arches 152
Pharyngeal Pouches 159
Pharyngeal Grooves 161
Pharyngeal Membranes 162
Development of the Thyroid Gland 166
Development of the Tongue 167
Development of the Salivary Glands 171
Development of the Face 171
Development of the Nasal Cavities 176
Development of the Palate 178
Summary of the Pharyngeal Apparatus 186
Clinically Oriented Questions 186

12 The Respiratory System 189
Development of the Larynx 190
Development of the Trachea 190
Development of the Bronchi and Lungs 193
Summary of the Respiratory System 198
Clinically Oriented Questions 199

13 The Digestive System 201
The Foregut 202
Development of the Spleen 210
The Midgut 212
The Hindgut 222
Summary of the Digestive System 226
Clinically Oriented Questions 227

14 The Urogenital System 229
Development of the Urinary System 230
Development of the Suprarenal Glands 243
Development of the Genital System 243
Development of the Inguinal Canals 256
Summary of the Urogenital System 260
Clinically Oriented Questions 261

15 The Cardiovascular System 263
Early Development of the Heart and Vessels 264
Further Development of the Heart 268
Anomalies of the Heart and Great Vessels 284
Aortic Arch Derivatives 291
Aortic Arch Anomalies 293
Fetal and Neonatal Circulations 295
Development of the Lymphatic System 299
Summary of the Cardiovascular System 303
Clinically Oriented Questions 304

16 The Skeletal System 305
Development of Bone and Cartilage 306
Development of Joints 308
Development of the Axial Skeleton 310
Development of the Appendicular Skeleton 317
Summary of the Skeletal System 320
Clinically Oriented Questions 320

17 The Muscular System 323
Development of Skeletal Muscle 324
Development of Smooth Muscle 327
Development of Cardiac Muscle 327
Summary of the Muscular System 327
Clinically Oriented Questions 327

18 The Limbs 329
Early Stages of Limb Development 330
Final Stages of Limb Development 330
Cutaneous Innervation of the Limbs 332
Blood Supply to the Limbs 334
Anomalies of Limbs 334
Summary of Limb Development 340
Clinically Oriented Questions 340

19 The Nervous System 343
Origin of the Nervous System 344
Development of the Spinal Cord 344
Congenital Anomalies of the Spinal Cord 350
Development of the Brain 352
Congenital Anomalies of the Brain 362
Development of the Peripheral Nervous System 366
Development of the Autonomic Nervous System 369
Summary of the Nervous System 370
Clinically Oriented Questions 370

20 The Eye and Ear 371
Development of the Eye and Related Structures 372
Development of the Ear 380
Summary of Development of the Eye 385
Summary of Development of the Ear 386
Clinically Oriented Questions 386

21 The Integumentary System 387
Development of the Skin 388
Development of the Hair 389
Development of the Nails 391
Development of the Mammary Glands 392
Development of the Teeth 393
Summary of the Integumentary System 399
Clinically Oriented Questions 400

References and Suggested Readings 401

Answers to Clinically Oriented Questions 409

Index 421

1

Introduction to Human Embryology

Human development begins when an **oocyte** (ovum) from a female is fertilized by a **sperm** (spermatozoon) from a male. Development involves many changes that transform a single cell, the **zygote**, into a multicellular human being. **Human embryology** is concerned with the origin and development of a human being from a zygote to the birth of an infant. The stages of development that occur before birth are illustrated in the *Timetables of Human Prenatal Development* (Figs. 1 – 1 and 1 – 2).

Embryological Terminology

Most embryological terms have Latin (L.) or Greek (Gr.) origins. An understanding of the origin of terms can serve as a memory key. The term *zygote*, for example, is derived from the Greek word *zygtos*, meaning yoked, which indicates that the sperm and oocyte unite to form a new cell, the zygote.

Oocyte (L. ovum, ***egg).*** This term refers to the female germ or sex cell produced in the *ovaries*. When mature, the oocyte is called a *secondary oocyte*, or mature ovum.

Sperm (***spermatozoon***). This term refers to the male germ or sex cell produced in the *testes* (testicles). Numerous sperms (spermatozoa) are expelled from the male urethra during ejaculation.

Zygote. This cell, formed by the union of an oocyte and a sperm, is the beginning of a new human being (i.e., an embryo). The expression *fertilized ovum* refers to a secondary oocyte (ovum) that has been impregnated by a sperm; when fertilization is complete, the oocyte becomes a zygote.

Fertilization Age. It is difficult to determine exactly when fertilization (conception) occurs because the process cannot be observed in vivo (within the living body). Physicians calculate the age of the embryo or fetus from the first day of the last normal menstrual period (LNMP). This is the *gestational age*, which is about 2 weeks longer than the *fertilization age* because the oocyte is not fertilized until about 2 weeks after the preceding menstruation (see Fig. 1 – 1). Consequently, when a physician states the age of an embryo or fetus, 2 weeks must be deducted to determine the actual or fertilization age.

Cleavage. Mitotic cell division, or cleavage of the zygote, forms embryonic cells called *blastomeres*. The size of the early embryo remains the same because the blastomeres become smaller with each succeeding cell division.

Morula. When 12 to about 32 blastomeres (cells) have formed, a *morula* is created, which is the ball of cells resulting from cleavage of the zygote. It is so called because the cluster of cells resembles the fruit known as a mulberry or blackberry (L. *morus*). The morula stage is reached about 3 days after fertilization, just as the developing human enters the uterus from the uterine tube (fallopian tube).

Blastocyst. After the morula passes from the uterine tube into the uterus, a fluid-filled cavity — *the blastocystic cavity* — forms inside it. This change converts the morula into a blastocyst, which contains an *inner cell mass* or embryoblast that will form the embryo.

Embryo. This term refers to the developing human during its early stages of development. The *embryonic period* extends to the end of the eighth week, by which time the beginnings of all major structures are present.

Conceptus. This term refers to the entire products of conception from fertilization onward (the embryo or fetus) and its membranes (e.g., placenta).

Primordium. This term refers to the beginning or first discernible indication of an organ or structure. The term *anlage*, from the German, is a synonym.

Fetus. After the embryonic period (8 weeks), the developing human is called a fetus. During the *fetal period* (ninth week to birth), differentiation and growth of the tissues and organs formed during the embryonic period occur. Functional maturation of the organs and the rate of body growth are remarkable, especially during the third and the fourth months (see Fig. 1 – 2), and weight gain is phenomenal during the terminal months.

Trimester. This is a period consisting of *3 calendar months*. Obstetricians commonly divide the 9-month period of gestation into three trimesters. The most critical stages of development occur during the first trimester, when embryonic and early fetal development is occurring.

Abortion (***L.*** aboriri, ***to miscarry***). This term refers to the expulsion from the uterus of an embryo or fetus before it is *viable* (i.e., mature enough to survive outside the uterus).

Importance of and Advances in Embryology

The study of prenatal stages of development, especially those occurring during the embryonic period, helps us to understand the normal relationships of adult body structures and the causes of congenital anomalies. Much of the modern practice of obstetrics involves applied or **clinical embryology**. Because some children may have disorders, such as spina bifida and congenital heart disease, resulting from maldevelopment, the significance of embryology is readily apparent to pediatricians. Advances in surgery, especially in procedures involving the prenatal and pediatric age groups, have made a knowledge of human development more clinically significant. A thorough understanding and correction of most congenital anomalies (e.g., cleft palate and cardiac defects) depend on an understanding of normal development and the deviations that have occurred.

Rapid advances in the field of molecular biology have led to the use of sophisticated techniques (e.g., *recombinant DNA technology*, chimeric models, transgenic mice, and stem cell manipulation) in research laboratories to explore such diverse issues as the genetic regulation of morphogenesis, the temporal and regional expression of specific genes, and the mechanisms by which cells are committed to form the various parts of the embryo. For the first time, researchers are beginning to understand how, when, and where selected genes are activated and expressed in the embryo during normal and abnormal development. For example, *endogenous retinoic acid* has been identified as an important regulatory substance in embryonic development. Apparently, it acts as a transcriptional activator for specific genes that are involved in embryonic patterning.

The critical role of **homeobox-containing (HOX) genes** and other molecular factors in regulating early embryonic development is rapidly being delineated. In 1995, Edward B. Lewis, Christiane Nüsslein-Volhard, and Eric F. Wieschaus were awarded the *Nobel prize for physiology or medicine* for their discovery of genes that control embryonic development. Such discoveries are contributing to a better understanding of the causes of spontaneous abortion and congenital anomalies.

In 1997, Ian Wilmut and colleagues were the first to produce a mammal (a sheep dubbed *Dolly*) by cloning using the technique of somatic cell nuclear transfer. Since then, other animals (mice, cows, and pigs) have been cloned successfully from cultured differentiated adult cells. Interest in human cloning has generated considerable debate because of social, ethical, and legal implications. Moreover, there is concern that cloning may result in an increase in the number of infants born with birth defects and serious diseases.

Human embryonic stem cells are pluripotential and capable of developing into diverse cell types. The isolation and culture of human embryonic stem cells hold great promise for the development of molecular therapies as a result of the sequencing of the human genome.

Descriptive Terms

In anatomy and embryology, special terms of position and direction are used, and various planes of the body are referred to in sections. Descriptions of the adult are based on the anatomical position, the position in which the body is erect, the upper limbs are at the sides, and the palms are directed anteriorly. The descriptive terms of position, direction, and planes used for embryos are shown in Figure 1-3.

Clinically Oriented Questions

1. Should you be able to reproduce the timetable of human development and know the characteristic features of each stage?
2. What is the difference between the terms *conceptus* and *embryo*? What are the products of conception?
3. Why do we study human embryology? Does it have any practical value in medicine and other health sciences?
4. Some say that animal and human embryos look alike. Is this true?
5. Physicians date a pregnancy from the first day of the LNMP, as explained earlier, but the embryo does not start to develop until about 2 weeks later. Why do they do this?

The answers to these questions are at the back of the book.

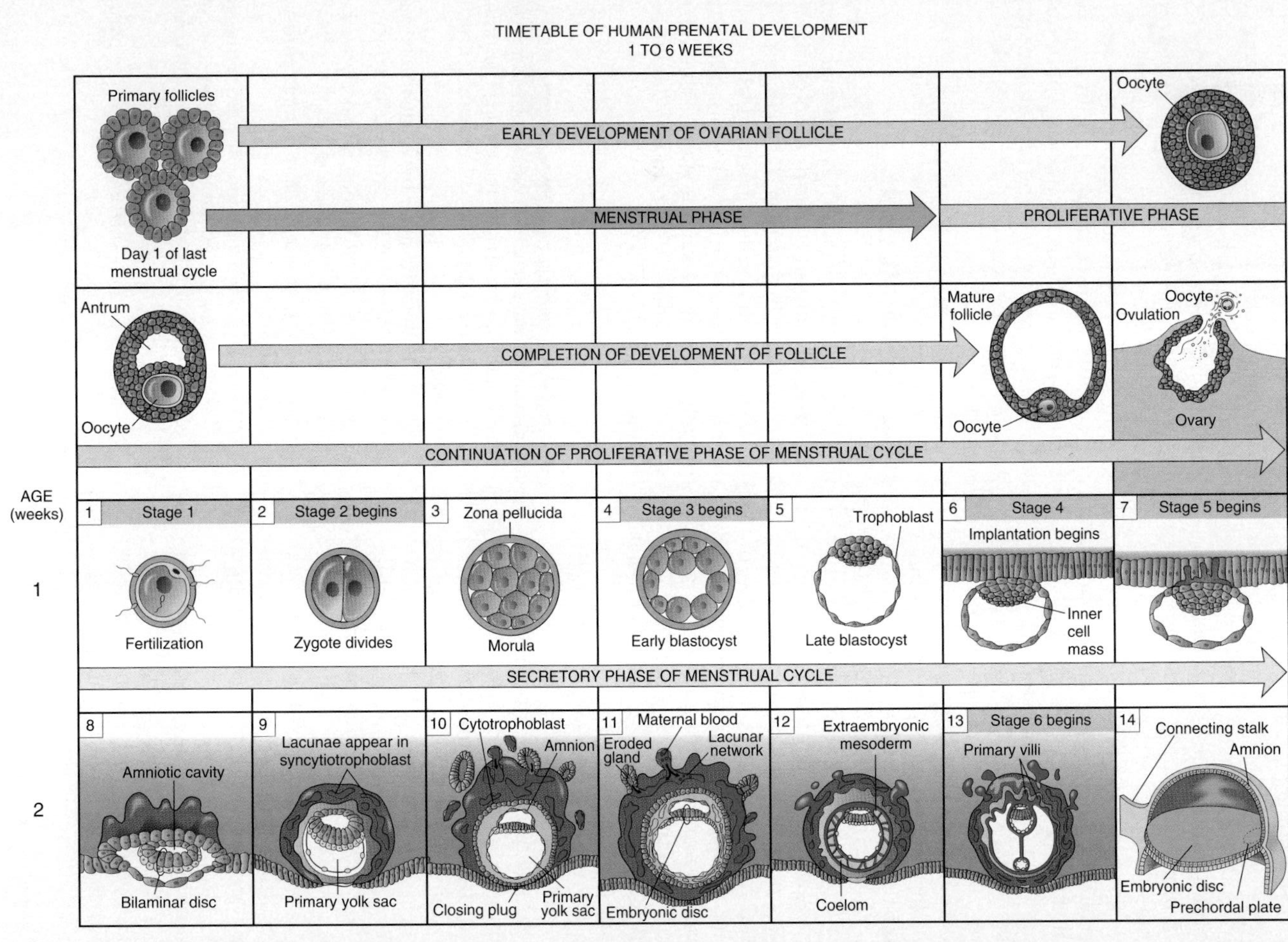

Figure 1 - 1. Early stages of human embryonic development. Development of an ovarian follicle containing an oocyte, ovulation, and the phases of the menstrual cycle are illustrated. Development begins at fertilization, about 14 days after the onset of the last menstrual period. Cleavage of the zygote in the uterine tube, implantation of the blastocyst, and early development of the embryo are also shown.

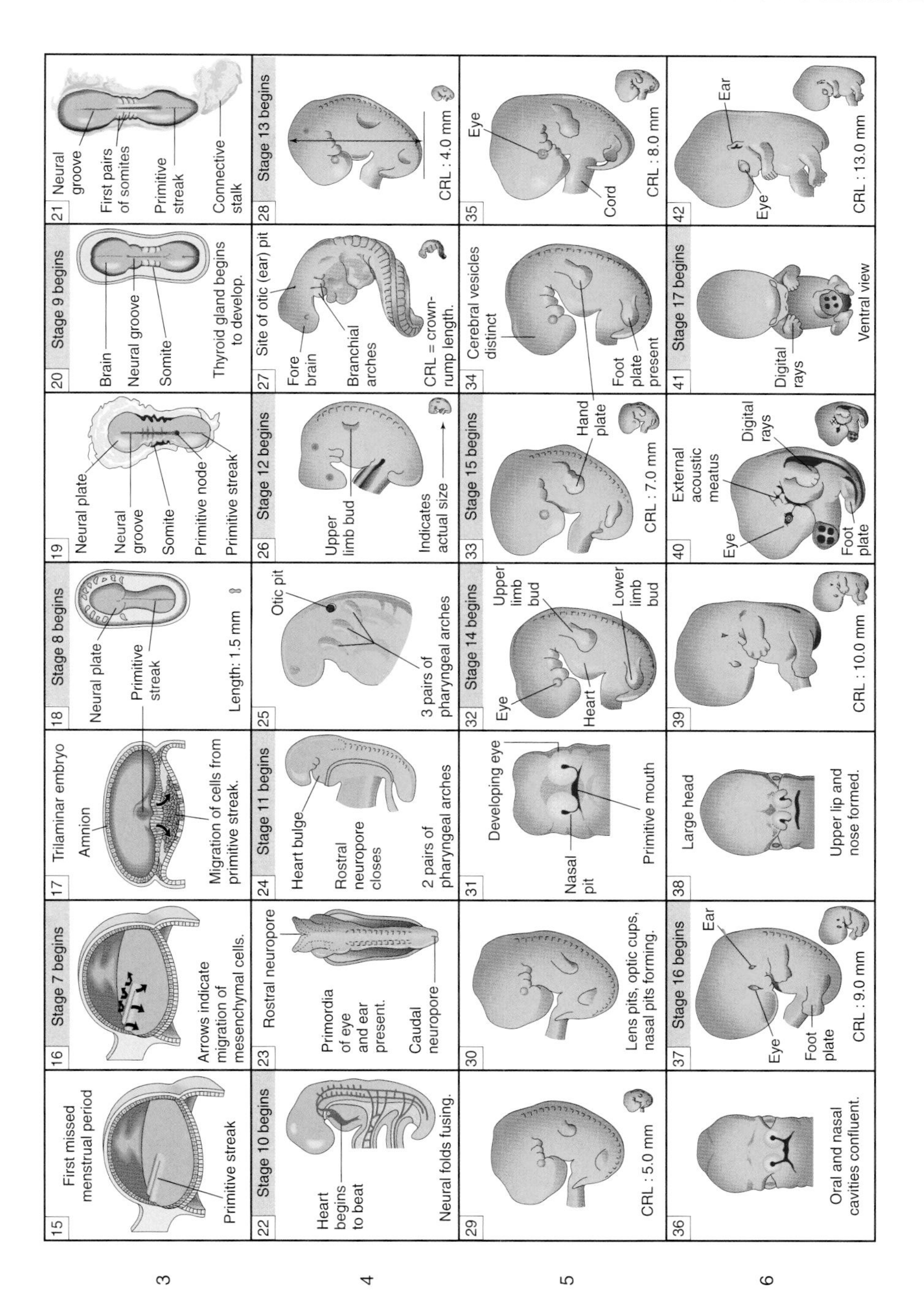
3
4
5
6
15
First missed menstrual period
Primitive streak
16
Stage 7 begins
Arrows indicate migration of mesenchymal cells.
17
Trilaminar embryo
Amnion
Migration of cells from primitive streak.
18
Stage 8 begins
Neural plate
Primitive streak
Length: 1.5 mm
19
Neural plate
Neural groove
Somite
Primitive node
Primitive streak
20
Stage 9 begins
Brain
Neural groove
Somite
Thyroid gland begins to develop.
21
Neural groove
First pairs of somites
Primitive streak
Connective stalk
22
Stage 10 begins
Heart begins to beat
Neural folds fusing.
23
Rostral neuropore
Primordia of eye and ear present.
Caudal neuropore
24
Stage 11 begins
Heart bulge
Rostral neuropore closes
2 pairs of pharyngeal arches
25
Otic pit
3 pairs of pharyngeal arches
26
Stage 12 begins
Upper limb bud
Indicates actual size
27
Site of otic (ear) pit
Fore brain
Branchial arches
CRL = crown-rump length.
28
Stage 13 begins
CRL : 4.0 mm
29
CRL : 5.0 mm
30
Lens pits, optic cups, nasal pits forming.
31
Developing eye
Nasal pit
Primitive mouth
32
Stage 14 begins
Eye
Upper limb bud
Heart
Lower limb bud
33
Stage 15 begins
Hand plate
CRL : 7.0 mm
34
Cerebral vesicles distinct
Foot plate present
35
Eye
Cord
CRL : 8.0 mm
36
Oral and nasal cavities confluent.
37
Stage 16 begins
Ear
Eye
Foot plate
CRL : 9.0 mm
38
Large head
Upper lip and nose formed.
39
CRL : 10.0 mm
40
External acoustic meatus
Eye
Digital rays
Foot plate
41
Stage 17 begins
Digital rays
Ventral view
42
Ear
Eye
CRL : 13.0 mm

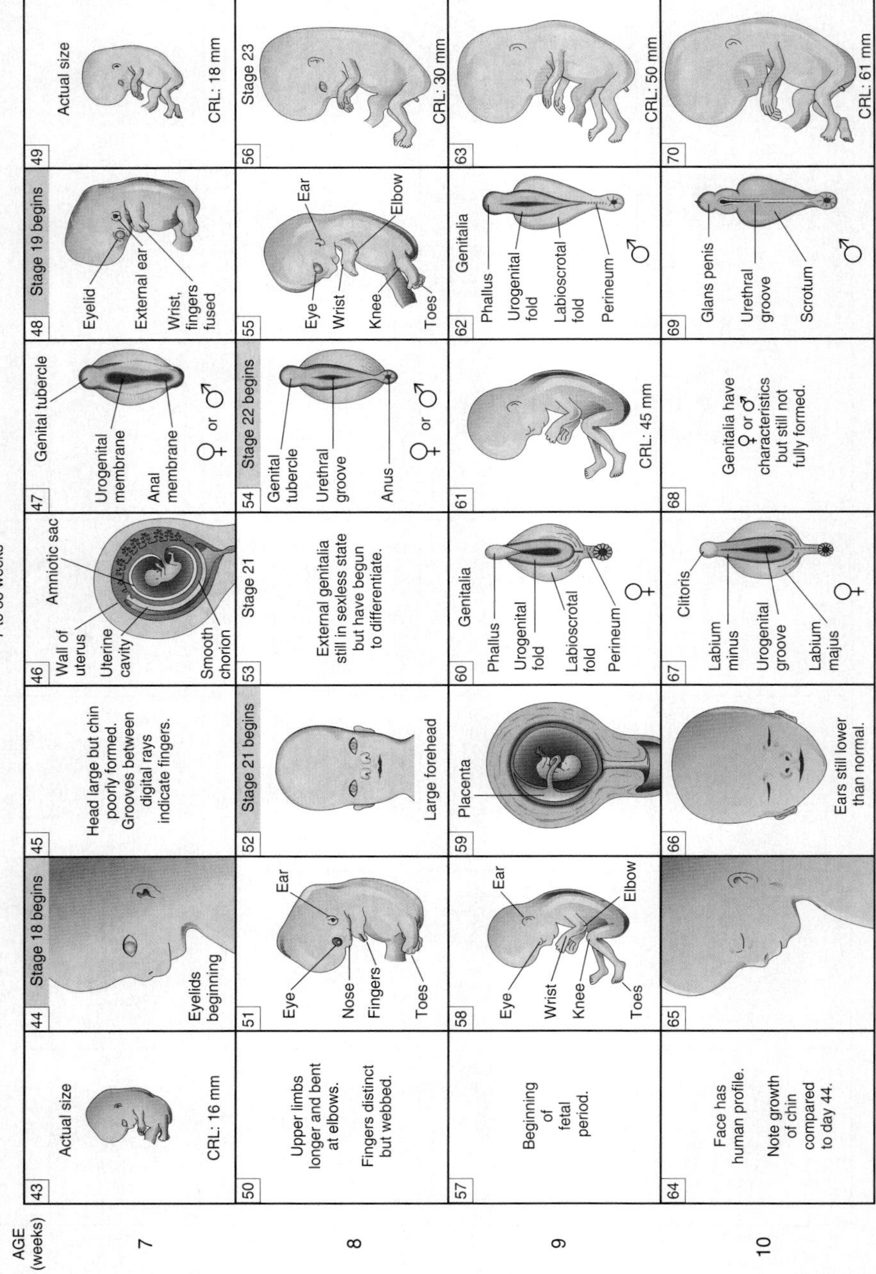

TIMETABLE OF HUMAN PRENATAL DEVELOPMENT
7 to 38 weeks
AGE (weeks)
7
8
9
10
43
Actual size
CRL: 16 mm
44
Stage 18 begins
Eyelids beginning
45
Head large but chin poorly formed. Grooves between digital rays indicate fingers.
46
Wall of uterus
Uterine cavity
Amniotic sac
Smooth chorion
47
Genital tubercle
Urogenital membrane
Anal membrane
♀ or ♂
48
Stage 19 begins
Eyelid
External ear
Wrist, fingers fused
49
Actual size
CRL: 18 mm
50
Upper limbs longer and bent at elbows.
Fingers distinct but webbed.
51
Eye
Ear
Nose
Fingers
Toes
52
Stage 21 begins
Large forehead
53
Stage 21
External genitalia still in sexless state but have begun to differentiate.
54
Stage 22 begins
Genital tubercle
Urethral groove
Anus
♀ or ♂
55
Eye
Ear
Wrist
Elbow
Knee
Toes
56
Stage 23
CRL: 30 mm
57
Beginning of fetal period.
58
Eye
Ear
Wrist
Elbow
Knee
Toes
59
Placenta
60
Genitalia
Phallus
Urogenital fold
Labioscrotal fold
Perineum
♀
61
CRL: 45 mm
62
Genitalia
Phallus
Urogenital fold
Labioscrotal fold
Perineum
♂
63
CRL: 50 mm
64
Face has human profile.
Note growth of chin compared to day 44.
65
66
Ears still lower than normal.
67
Clitoris
Labium minus
Urogenital groove
Labium majus
♀
68
Genitalia have ♀ or ♂ characteristics but still not fully formed.
69
Glans penis
Urethral groove
Scrotum
♂
70
CRL: 61 mm

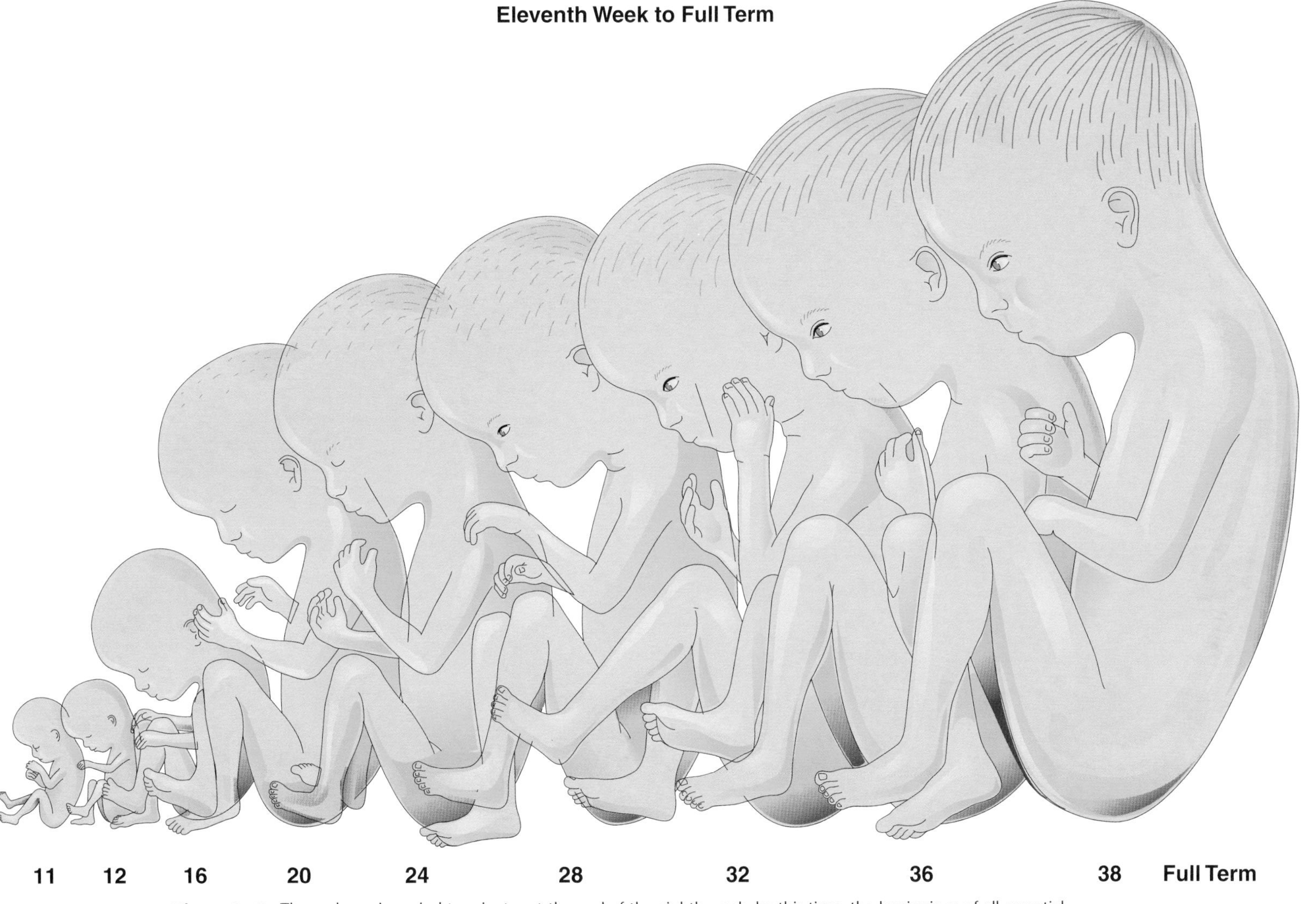

Figure 1–2. The embryonic period terminates at the end of the eighth week; by this time, the beginnings of all essential structures are present. The fetal period, extending from the ninth week to birth, is characterized by growth and elaboration of structures. Sex is clearly distinguishable by 12 weeks. Fetuses are viable 22 weeks after fertilization, but their chances of survival are not good until they are several weeks older. The 11- to 38-week fetuses pictured are shown at about half their actual sizes.

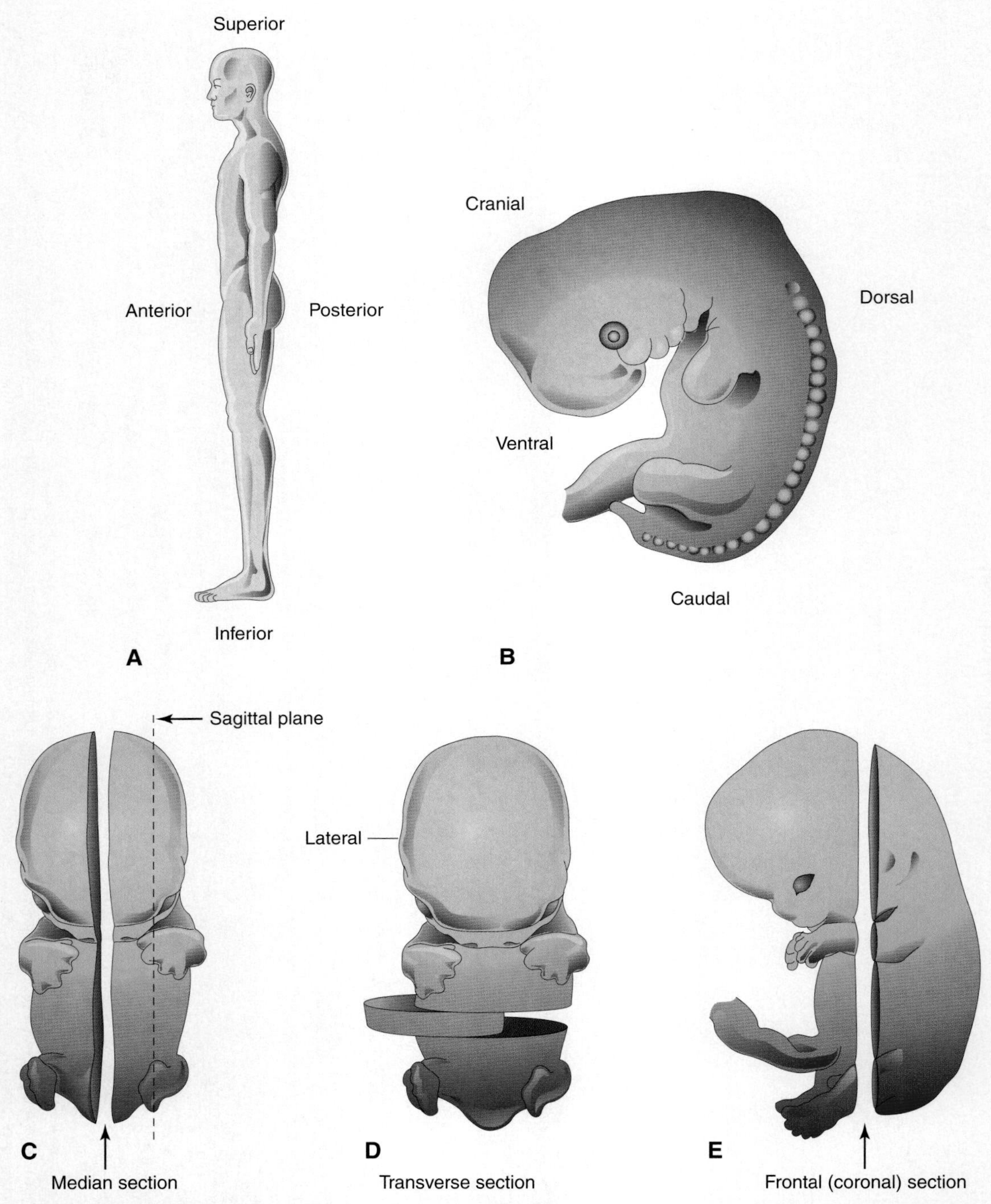

Figure 1-3. Drawings illustrating descriptive terms of position, direction, and planes of the body. *A*, Lateral view of an adult in the anatomical position. *B*, Lateral view of a 5-week embryo. *C* and *D*, Ventral views of 6-week embryos. The *median plane* is an imaginary vertical plane of section that passes longitudinally through the body, dividing it into right and left halves. A *transverse plane* refers to any plane that is at right angles to both the median and frontal planes. *E*, Lateral view of a 7-week embryo. A *frontal (coronal) plane* is any vertical plane that intersects the median plane at a right angle and divides the body into front (anterior or ventral) and back (posterior or dorsal) parts. In describing development, it is necessary to use words denoting the position of one part to another, or to the body as a whole. For example, the vertebral column (backbone) develops in the dorsal part of the embryo, and the sternum (breast bone) develops ventral to it in the ventral part of the embryo.

2 *Human Reproduction*

Human reproduction involves the union of an **oocyte** (ovum) from a female and a **sperm** (spermatozoon) from a male. Each cell brings a half share of genetic information to the union so that the new cell, a **zygote**, receives the genetic information required for directing the development of a new human being. The reproductive system in both sexes is designed to ensure the successful union of the sperm and oocyte, a process termed — *fertilization*.

Puberty begins when secondary sex characteristics (pubic hair, for example) first appear. It is the period — usually between the ages of 12 and 15 years in females and 13 and 16 years in males — when an individual becomes capable of sexual reproduction. Although the most obvious changes are in the reproductive system, puberty affects the whole body (consider, for example, the attendant increase in growth rate, termed the *pubertal growth spurt*). **Menarche** (first menstruation) may occur in 8- to 11-year-old girls. Puberty is largely completed by age 16. *Puberty in males* begins later (13 to 16 years of age); however, signs of sexual maturity may appear in 12-year-old boys. Puberty ends when mature sperms are formed.

Reproductive Organs

Each sex has reproductive or *sex organs* that produce and transport *gametes* (sex cells) from the *sex glands*, or gonads, to the site of fertilization in the uterine tube (Fig. 2–1). The penis, the male sex organ, deposits sperms (spermatozoa), produced by the *testes*, in the vagina of the female genital tract during sexual intercourse.

Female Reproductive Organs

The **vagina** (see Fig. 2–1*A*) serves as the excretory passage for menstrual fluid, receives the penis during sexual intercourse, and forms the inferior part of the birth canal. The size and appearance of the **vaginal orifice** vary with the condition of the *hymen*, a thin fold of mucous membrane that surrounds the vaginal orifice (see Fig. 2–3).

Oocytes are produced by the **ovaries**, which are located in the pelvic cavity, one on each side of the uterus (see Fig. 2–1*A*). When released from the ovary at *ovulation*, the secondary oocyte (ovum) passes into one of two trumpet-shaped *uterine tubes* (fallopian tubes). These tubes open into the *uterus* (L., womb), which protects and nourishes the embryo and fetus until birth.

Uterus

The uterus is a thick-walled, pear-shaped organ (Fig. 2–2). It varies considerably in size. The uterus consists of two main parts:

- **Body**, the expanded superior two thirds
- **Cervix**, the cylindrical inferior third

The **fundus** is the rounded part of the uterine body that lies superior to the orifices of the uterine tubes. The body of the uterus narrows from the fundus to the **isthmus**, the constricted region between the body and cervix. The lumen of the cervix, the **cervical canal**, has a constricted opening or *os* (ostium) at each end. The **internal os** communicates with the cavity of the body of the uterus, whereas the **external os** communicates with the vagina. The walls of the body of the uterus consist of three layers:

- **Perimetrium**, a thin external layer of peritoneum
- **Myometrium**, a thick smooth muscle layer
- **Endometrium**, a thin internal mucous membrane

At the peak of its development, the endometrium is 4 to 5 mm thick. During the luteal (secretory) phase of the menstrual cycle (see Fig. 2–8), *three layers of the endometrium* can be distinguished microscopically (see Fig. 2–2*C*):

- **Compact layer**, consisting of densely packed connective tissue around the neck of the uterine glands
- **Spongy layer**, composed of edematous connective tissue containing the dilated, tortuous bodies of the uterine glands
- **Basal layer**, containing the blind ends of the uterine glands

The basal layer of the endometrium has its own blood supply and is not cast off during menstruation. The compact and spongy layers, known collectively as the *functional layer*, disintegrate and are shed at menstruation and after *parturition* (delivery of a baby).

Uterine Tubes

The uterine tubes, measuring 10 to 12 cm long and 1 cm in diameter, extend laterally from the **horns** (L. *cornua*) of the uterus (see Fig. 2–2*A*). The tubes carry oocytes from the ovaries and sperms entering from the uterus to reach the fertilization site in the **ampulla of the uterine tube** (see Fig. 2–2*B*). The uterine tube also conveys the dividing zygote to the uterine cavity. Each tube opens into a horn of the uterus at its proximal end and into the peritoneal cavity at its distal end. The uterine tube is divided into four parts: the infundibulum, ampulla, isthmus, and uterine part.

Ovaries

The ovaries are almond-shaped glands located on each side of the uterus (see Fig. 2–2*B*). The ovaries produce estrogen and progesterone, the hormones responsible for the development of secondary sex characteristics and regulation of pregnancy. The ovaries also produce oocytes.

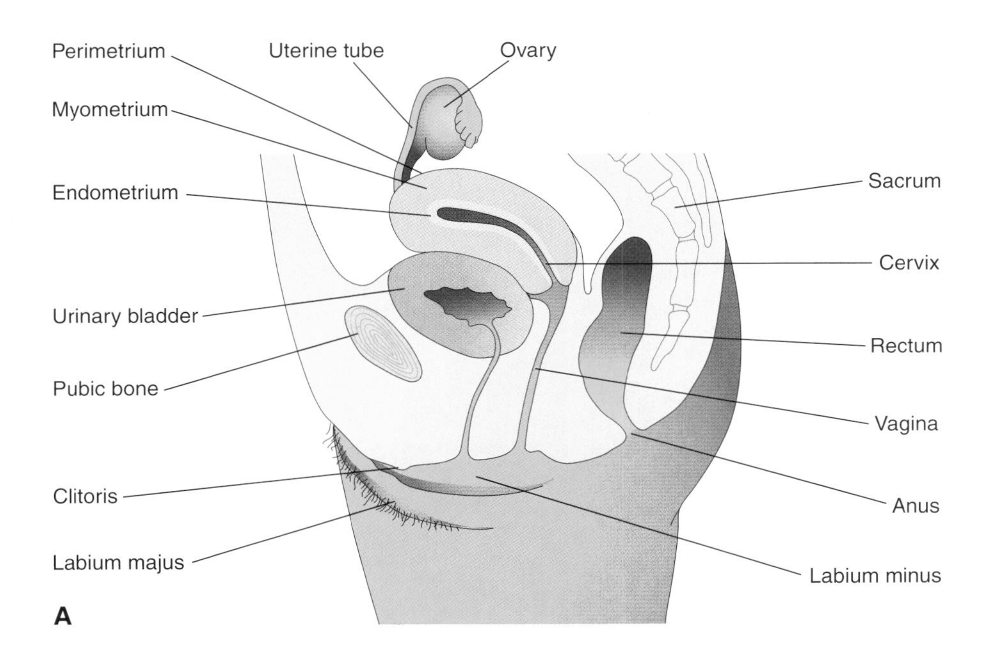

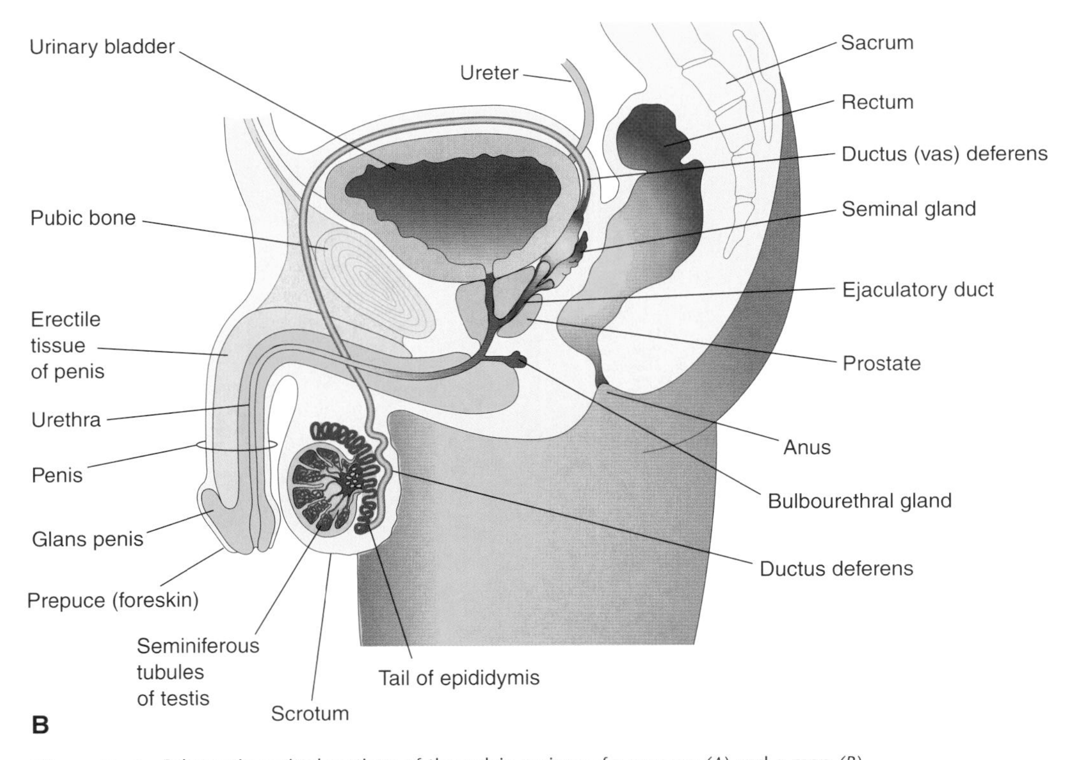

Figure 2 - 1. Schematic sagittal sections of the pelvic regions of a woman *(A)* and a man *(B)*.

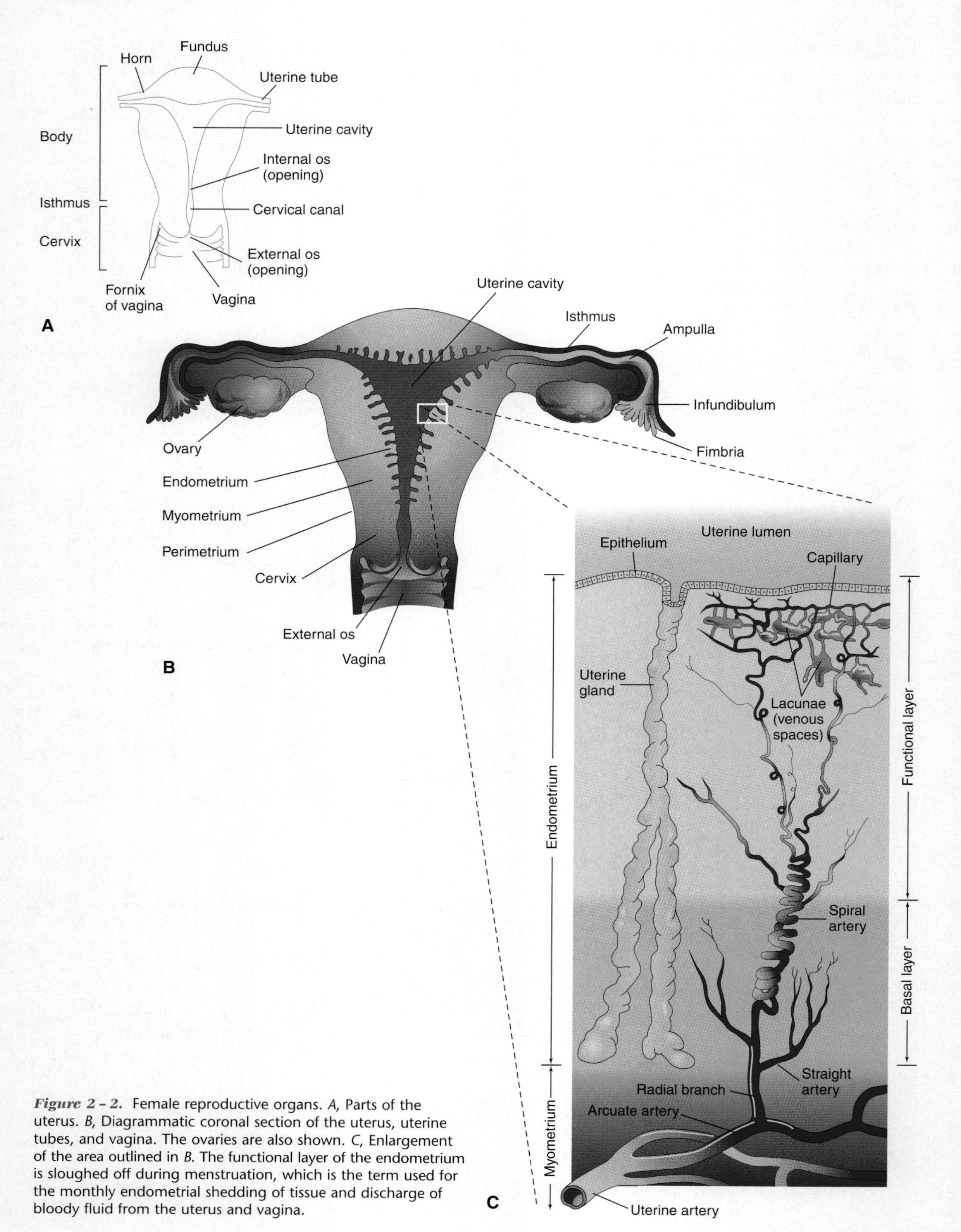

Figure 2 - 2. Female reproductive organs. *A*, Parts of the uterus. *B*, Diagrammatic coronal section of the uterus, uterine tubes, and vagina. The ovaries are also shown. *C*, Enlargement of the area outlined in *B*. The functional layer of the endometrium is sloughed off during menstruation, which is the term used for the monthly endometrial shedding of tissue and discharge of bloody fluid from the uterus and vagina.

Female External Sex Organs

The female external sex organs are known collectively as the **vulva**, or pudendum (Fig. 2 – 3). The **labia majora** (L., major lips), fatty external folds of skin, conceal the vaginal orifice, the opening of the vagina. Inside these labia are two smaller folds of mucous membrane, the **labia minora** (L., minor lips). The **clitoris**, a small erectile organ, is situated at the superior junction of these folds. The clitoris, the morphologic equivalent of the penis, is very important in the sexual excitement of a female. The vagina and urethra open into a cavity, the **vestibule of the vulva** (the cleft between the labia minora). The size and appearance of the **vaginal orifice** varies with the condition of the **hymen**, a fold of mucous membrane that surrounds the vaginal orifice. During early fetal life the hymen covers the vaginal orifice. The hymen usually ruptures during the perinatal period.

Male Reproductive Organs

The male reproductive system (see Fig. 2 – 1*B*) includes the testes, epididymis, ductus deferens (vas deferens), prostate, seminal glands (vesicles), bulbourethral glands, ejaculatory ducts, and urethra. *Sperms* are produced in the *testes,* two oval-shaped glands (gonads) that are suspended in the scrotum.

Each **testis** consists of many highly coiled seminiferous tubules that produce sperms. Immature sperms pass from the testis into a single, complexly coiled tube, the **epididymis**, where they are stored. It takes many days for sperms to mature in the epididymis. From the epididymis, the **ductus deferens** carries the sperms to the ejaculatory duct. The ductus deferens passes from the scrotum through the inguinal canal into the abdominal cavity. The ductus then descends into the pelvis, where it fuses with the duct of the seminal gland to form the **ejaculatory duct**, which enters the urethra.

The **urethra** is a tube leading from the urinary bladder to the outside of the body; its spongy part runs through the **penis** (see Fig. 2 – 1*B*). Within the penis, three columns of **erectile tissue** surround the urethra. During sexual excitement, this tissue fills with blood under increased pressure. This causes the penis to become erect and allowing it to be more easily inserted into the vagina during sexual intercourse. Ejaculation of semen — sperms mixed with seminal fluid produced by a number of glands: including the *seminal glands*, *bulbourethral glands*, and *prostate* — occurs when the penis is further stimulated. Consequently, the male urethra transports both urine and semen, but not at the same time.

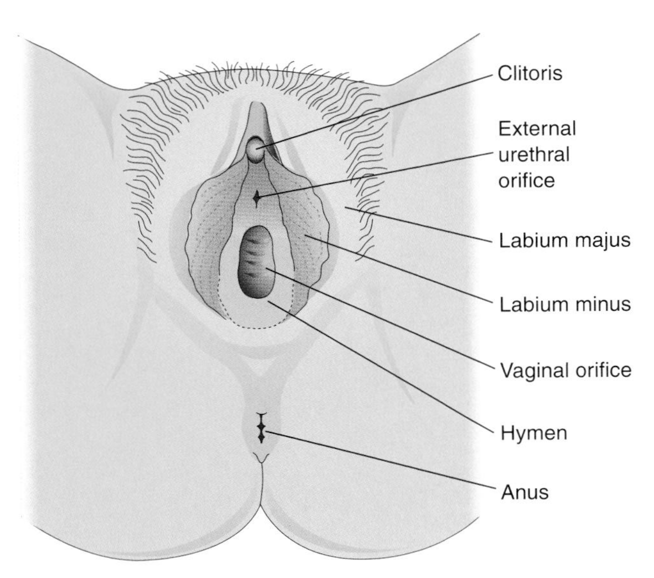

Figure 2 - 3. External female genitalia. The labia are spread apart to show the external urethral and vaginal orifices.

Gametogenesis

The *sperm and oocyte are highly specialized sex cells* (Fig. 2 – 4). These gametes each contain half the number of required chromosomes (i.e., 23 instead of 46). The number of chromosomes is reduced during a special type of cell division called **meiosis**. This type of cell division occurs during the formation of gametes, a process termed — **spermatogenesis** in males and **oogenesis** (ovogenesis) in females.

Gametogenesis (gamete formation) is the process by which specialized generative cells — **gametes**, or germ cells (or, more particularly, oocytes in females and sperms in males) — are formed (Fig. 2 – 5). This process, which involves the chromosomes and cytoplasm of the gametes, prepares these sex cells for their role in *fertilization*.

Meiosis

Meiosis consists of two meiotic cell divisions (Fig. 2 – 6), during which the chromosome number of the germ cells is reduced to half (23, the *haploid* number) the number present in other cells in the body (46, the *diploid* number).

During the **first meiotic division**, the chromosome number is reduced from diploid to haploid. *Homologous chromosomes* (one from each parent) pair during prophase and then separate during anaphase, with one representative of each pair going to each pole of the meiotic spindle. Homologous chromosomes, or homologs, are pairs of chromosomes of one type, one inherited from each parent. At this stage, they are *double chromatid chromosomes*. The X and Y chromosomes are not homologs; however, they have homologous segments at

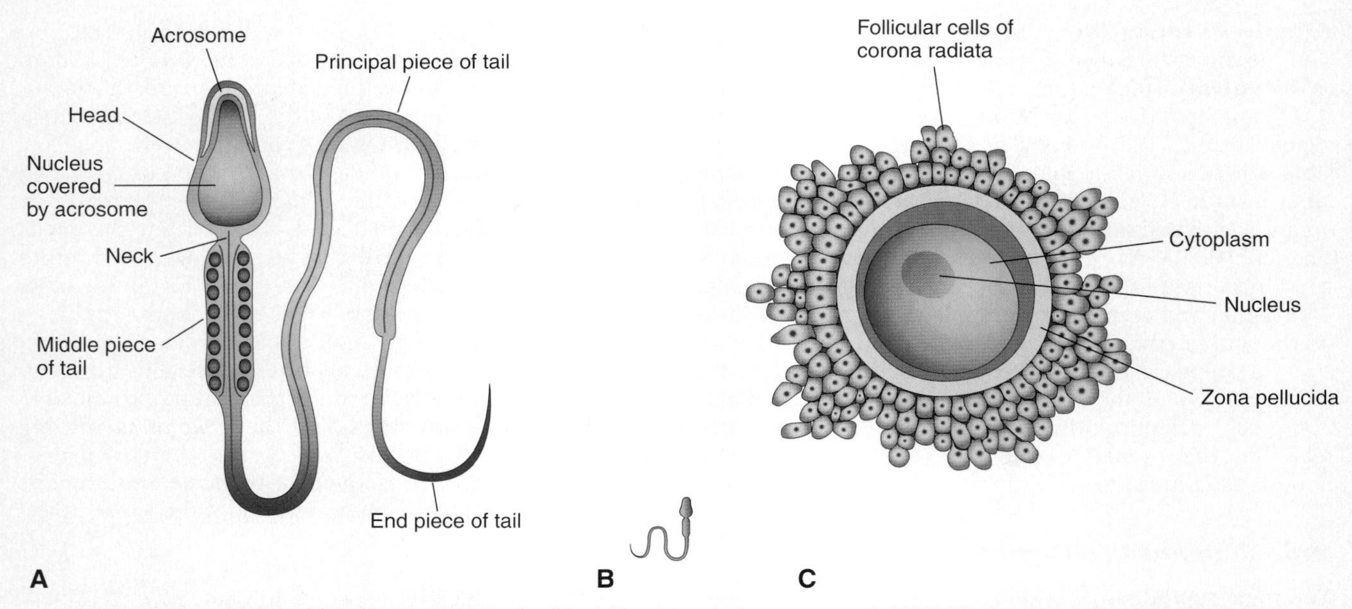

Figure 2-4. Male and female gametes (sexual cells). *A*, The parts of a human sperm (×1250). The head, composed mostly of the nucleus, is partly covered by the acrosome, an organelle containing enzymes. The tail of the sperm consists of three regions: middle piece, principal piece, and end piece. *B*, A sperm drawn to about the same scale as the oocyte in *C*. *C*, A human secondary oocyte (× 200), shown surrounded by the zona pellucida and corona radiata.

the tips of their short arms. They pair in these regions only. By the end of the first meiotic division, each new cell formed (secondary spermatocyte or secondary oocyte) has the *haploid chromosome number* (double chromatid chromosomes); that is, each new cell contains half the number of chromosomes of the preceding cell (primary spermatocyte or primary oocyte). This separation or disjunction of paired homologous chromosomes is the *physical basis of segregation* or separation of allelic genes during meiosis.

The **second meiotic division** follows the first division without a normal interphase (i.e., without an intervening step of DNA replication). Each chromosome divides, and each half, or *chromatid*, is drawn to a different pole of the meiotic spindle; thus, the haploid number of chromosomes (23) is retained, and each daughter cell formed by meiosis has the reduced haploid number of chromosomes, with one representative of each chromosome pair (now a single chromatid chromosome). The second meiotic division is similar to ordinary mitosis except that the chromosome number of the cell entering the second meiotic division is haploid.

Meiosis is a significant process of cell division because:

- It provides for *constancy of the chromosome number* from generation to generation by reducing the chromosome number from diploid to haploid, thereby producing haploid gametes.
- It allows random *assortment of maternal and paternal chromosomes* between the gametes.
- It relocates segments of maternal and paternal chromosomes by *crossing-over of chromosome segments*, which "shuffles" the genes and produces a recombination of genetic material.

Spermatogenesis

Spermatogonia are transformed into mature germ cells, or sperms. This maturation process begins at *puberty* (13 to 16 years of age) and continues into old age (see Fig. 2-5). Prior to maturation, these spermatogonia remain dormant in the seminiferous tubules of the testes from the late fetal period until puberty, at which time they begin to increase in number. After several mitotic cell divisions, the spermatogonia grow and undergo gradual changes that transform them into **primary spermatocytes**—the largest germ cells in the seminiferous tubules. Each primary spermatocyte subsequently undergoes a reduction division — the *first meiotic division* — to form two haploid **secondary spermatocytes**, which are about half the size of primary spermatocytes. Subsequently, the secondary spermatocytes undergo a *second meiotic division* to form four haploid **spermatids**, which are about half the size of secondary spermatocytes. During this division, no further reduction occurs in the number of chromosomes. The spermatids are gradually transformed into four mature sperms during a process known

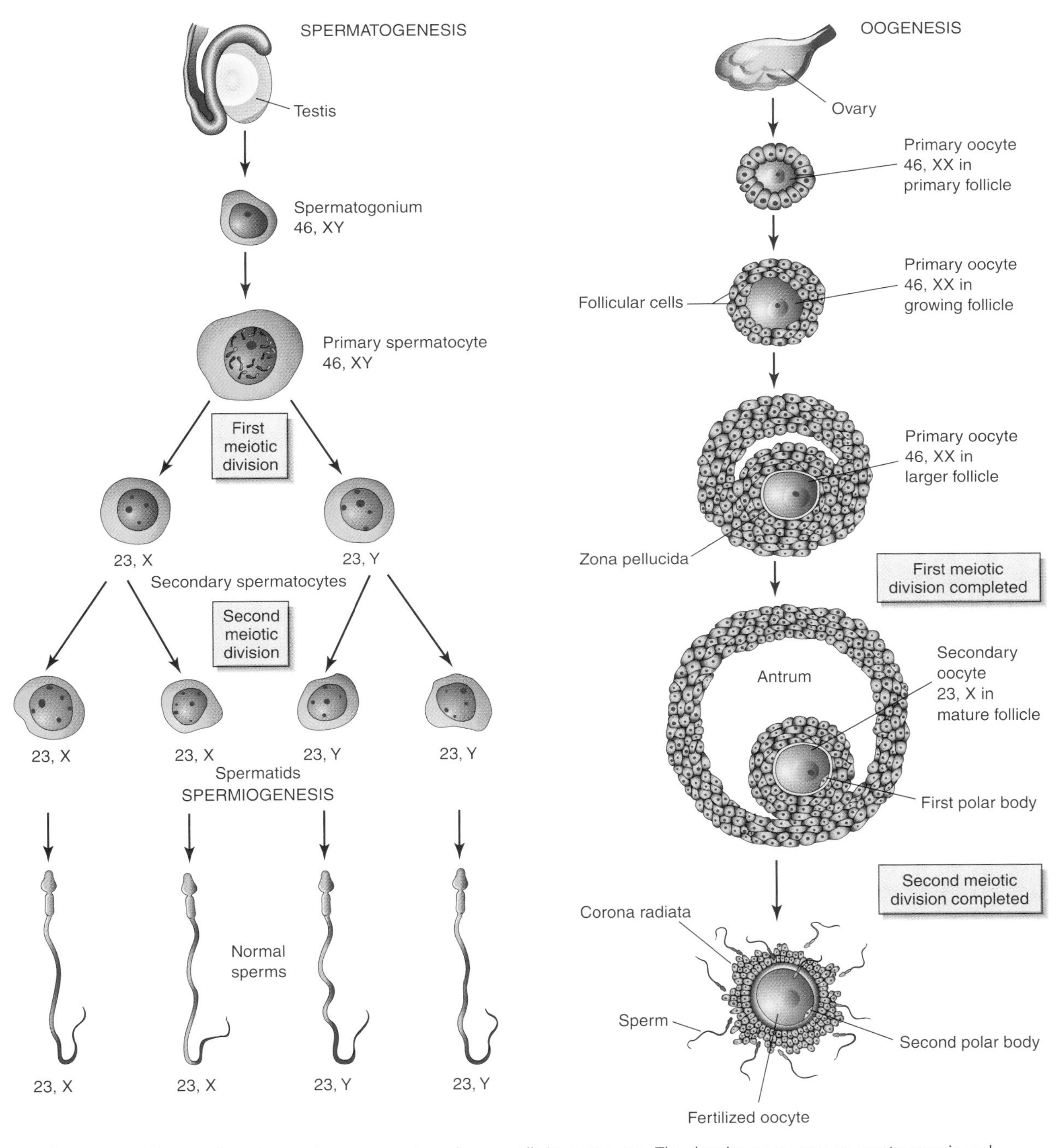

Figure 2-5. Normal Gametogenesis, or conversion of germ cells into gametes. The drawings compare spermatogenesis and oogenesis. Oogonia are not shown in this figure because they differentiate into primary oocytes before birth. The chromosome complement of the germ cells is shown at each stage. The number designates the total number of chromosomes, including the sex chromosome(s) (shown after the comma). *Note:* (1) Following the two meiotic divisions, the diploid number of chromosomes, 46, is reduced to the haploid number, 23; (2) four sperms form from one primary spermatocyte, whereas only one mature oocyte results from maturation of a primary oocyte; (3) the cytoplasm is conserved during oogenesis to form one large cell, the mature oocyte. The polar bodies are small nonfunctional cells that eventually degenerate.

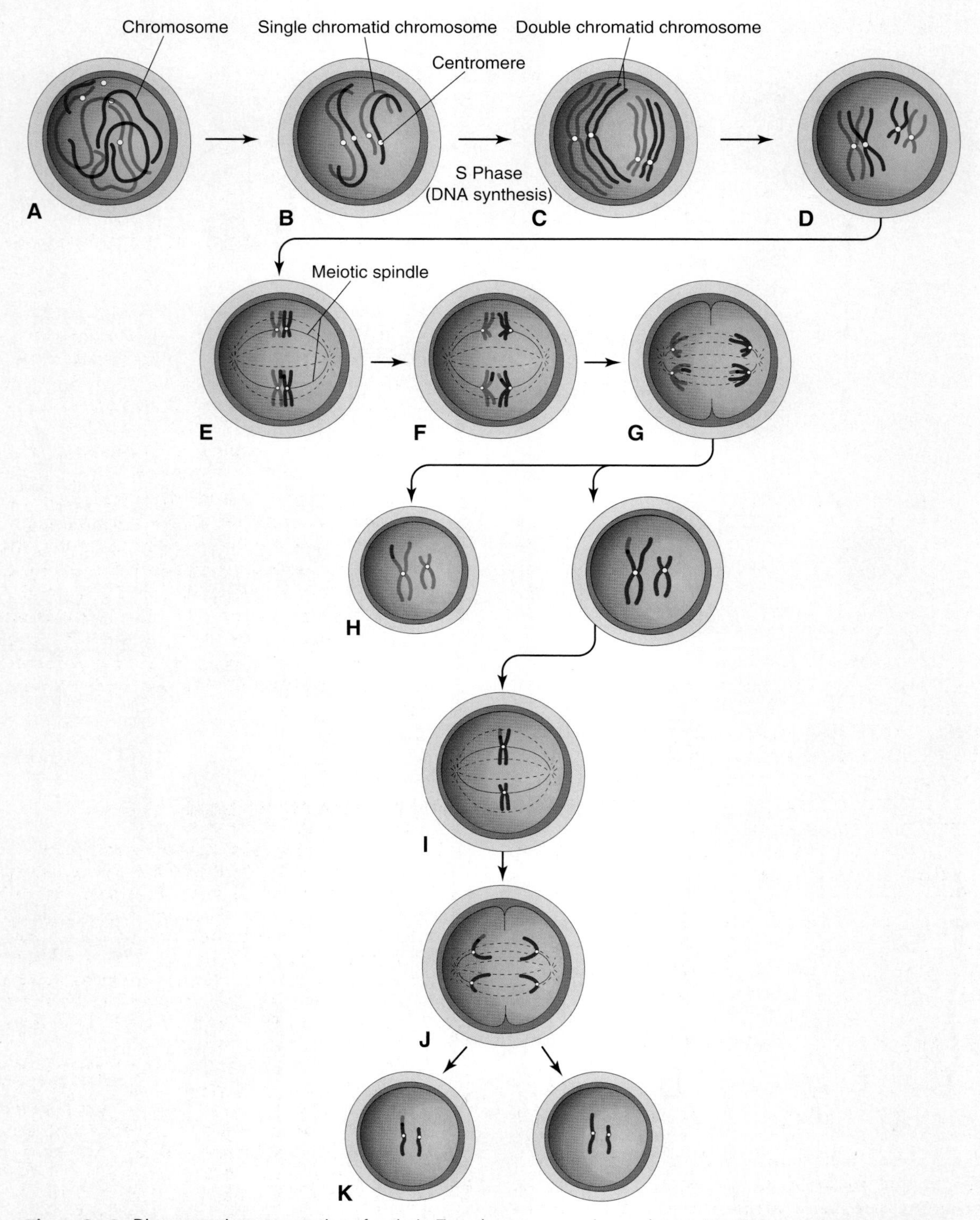

Figure 2-6. Diagrammatic representation of meiosis. Two chromosome pairs are shown. *A* to *D*, Stages of prophase of the first meiotic division. The homologous chromosomes approach each other and pair; each member of the pair consists of two chromatids. Observe the single crossover in one pair of chromosomes, resulting in the interchange of chromatid segments. *E*, Metaphase. The two members of each pair become oriented on the meiotic spindle. *F*, Anaphase. *G*, Telophase. The chromosomes migrate to opposite poles. *H*, Distribution of parental chromosome pairs at the end of the first meiotic division. *I* to *K*, Second meiotic division, which is similar to mitosis except that the cells are haploid.

as **spermiogenesis**. During this *metamorphosis* (change in form), the nucleus condenses, the **acrosome** forms, and most of the cytoplasm is shed. Spermatogenesis, including spermiogenesis, requires about 2 months for completion and normally continues throughout the reproductive life of a male. When spermiogenesis is complete, sperms enter the lumina of the seminiferous tubules. The sperms then move to the **epididymis** (see Fig. 2 – 1*B*), where they are stored and become functionally mature.

When ejaculated, the **mature sperm** is a free-swimming, actively motile cell consisting of a head and tail (see Fig. 2 – 4*A*). The **neck of the sperm** is the junction between the head and tail. The **head of the sperm**, forming most of the bulk of the sperm, contains the nucleus of the cell, which has 23 chromosomes. The anterior two thirds of the head is covered by the **acrosome**, a caplike organelle containing enzymes that facilitate sperm penetration during fertilization. The **tail of the sperm** provides the motility of the sperm, assisting with its transport to the site of fertilization in the ampulla of the uterine tube. It consists of three segments: the *middle piece*, *principal piece*, and *end piece*. The *middle piece of the tail* contains the energy-producing cytoplasmic and mitochondrial apparatus, which produces the lashing movements of the tail.

Oogenesis

Oogenesis refers to the sequence of events by which **oogonia** are transformed into **oocytes** (see Fig. 2 – 5). This maturation process begins during the fetal period but is not completed until after *puberty* (12 to 15 years of age). Except during pregnancy, this cyclic process occurs monthly during the reproductive life of females. During early fetal life, primordial ova — oogonia — proliferate by mitotic cell division. These oogonia enlarge to form **primary oocytes** before birth. By the time of birth, all primary oocytes have completed the prophase of the first meiotic division. These oocytes remain in prophase until puberty. Shortly before ovulation, a primary oocyte completes the *first meiotic division* (see Fig. 2 – 5). Unlike the corresponding stage of spermatogenesis, however, the division of cytoplasm is unequal. The **secondary oocyte** receives almost all the cytoplasm, whereas the **first polar body** receives very little, causing this small, nonfunctional cell to degenerate after a short time. At ovulation, the nucleus of the secondary oocyte begins the *second meiotic division* but progresses only to metaphase, at which point division is arrested.

If the secondary oocyte is fertilized by a sperm, the second meiotic division is completed. Again, most cytoplasm is retained by one cell, the *fertilized oocyte* (see Fig. 2 – 5). The other nonfunctional cell, the **second polar body**, is very small and soon degenerates. The secondary oocyte released at ovulation is surrounded by a covering of amorphous material known as the **zona pellucida** and a layer of follicular cells called the **corona radiata** (see Fig. 2 – 4*C*). Compared with ordinary cells, the secondary oocyte is large, being just visible to the unaided eye as a tiny speck. Up to 2 million primary oocytes are usually present in the ovaries of a newborn female infant. Most of these oocytes regress during childhood so that, by puberty, no more than 40,000 remain. Of these, only about 400 mature and are expelled at ovulation during the reproductive period.

Comparison of Male and Female Gametes

The sperm and secondary oocyte are dissimilar in several ways because of their adaptation for specialized roles in reproduction. Compared with the sperm, the oocyte is massive and immotile (see Fig. 2 – 4), whereas the microscopic sperm is highly motile. The mature oocyte also has an abundance of cytoplasm, whereas the sperm has very little. The sperm bears little resemblance to an oocyte or any other cell because of its sparse cytoplasm and specialization for motility.

In terms of sex chromosome constitution, **there are two kinds of normal sperm** (see Fig. 2 – 5): 22 autosomes plus an X chromosome (i.e., 23, X); and 22 autosomes plus a Y chromosome (23, Y). **There is only one kind of normal secondary ovum**: 22 autosomes plus an X chromosome (i.e, 23, X). *The difference in sex chromosome complement forms the basis of primary sex determination.*

Abnormal Gametogenesis

Disturbances of meiosis during gametogenesis, such as **nondisjunction** (Fig. 2 – 7), result in the formation of chromosomally abnormal gametes. If involved in fertilization, these gametes with numerical chromosome abnormalities cause abnormal development, as occurs in infants with **Down syndrome** (see Chapter 9). The likelihood of chromosomal abnormalities in the embryo increases significantly after the mother is 35. The older parents are at the time of conception, the more likely they are to have accumulated mutations that the embryo might inherit.

During meiosis, homologous chromosomes sometimes fail to separate and go to opposite poles of the germ cell. As a result of this error of cell division — **nondisjunction** — some gametes have 24 chromosomes and others only 22 (see Fig. 2 – 7). If a gamete with 24 chromosomes unites with a normal one with 23 chromosomes during fertilization, a zygote with 47 chromosomes forms (see Fig. 9 – 2). This condition is called **trisomy** because of the presence of three representatives of a particular chromosome instead of the usual two. If a gamete with only 22 chromosomes unites with a normal one, a zygote with 45 chromosomes forms.

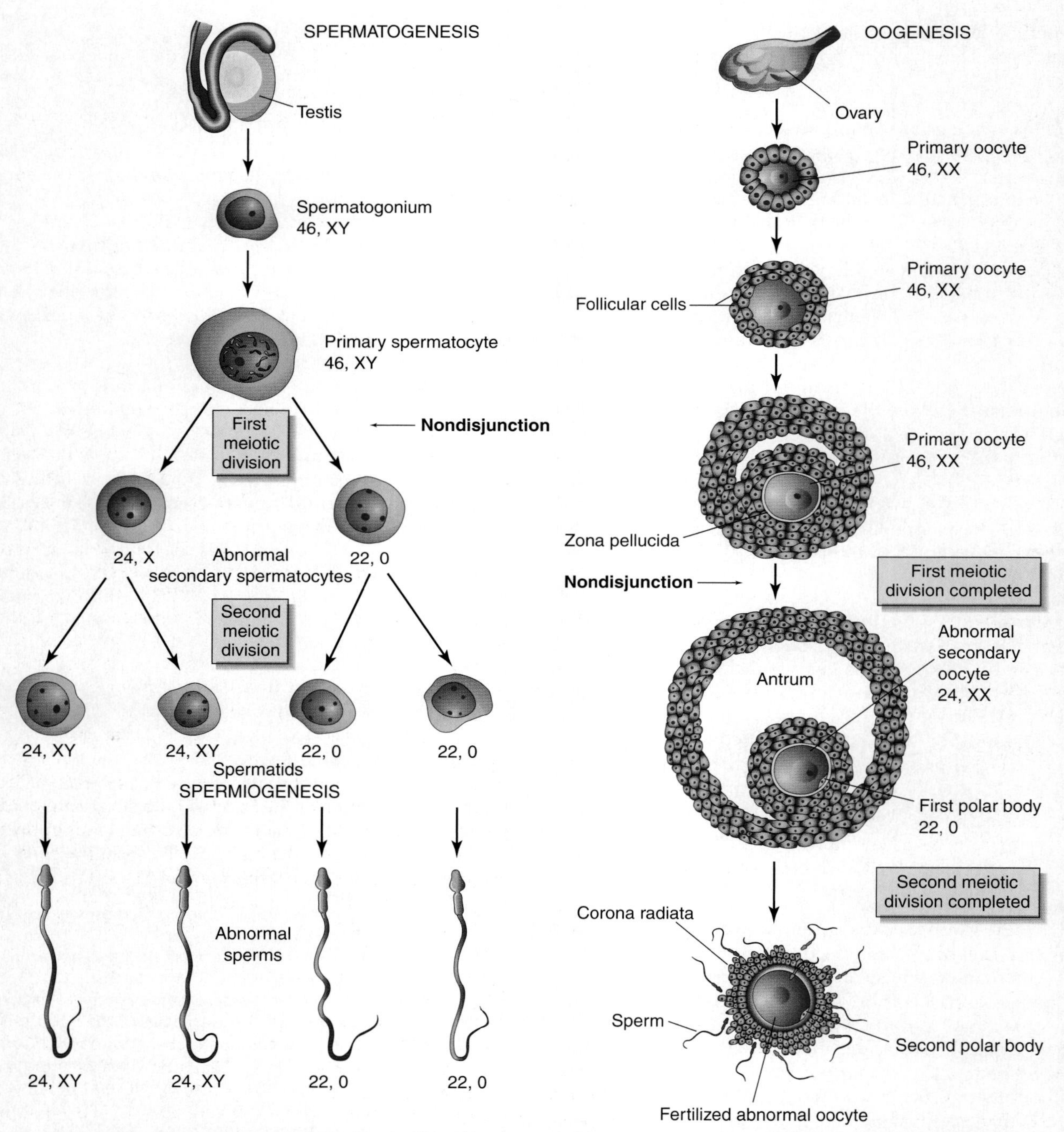

Figure 2 – 7. Abnormal gametogenesis. The drawings show how nondisjunction, an error in cell division, results in an abnormal chromosome distribution in germ cells. Although nondisjunction of sex chromosomes is illustrated, a similar defect may occur during the division of autosomes. When nondisjunction occurs during the first meiotic division of spermatogenesis, one secondary spermatocyte contains 22 autosomes plus an X and a Y chromosome, whereas the other one contains 22 autosomes and no sex chromosome. Similarly, nondisjunction during oogenesis may give rise to an oocyte with 22 autosomes and two X chromosomes (as shown) or one with 22 autosomes and no sex chromosome.

This condition is known as **monosomy** because only one representative of the particular chromosomal pair is present. For clinical conditions associated with numerical disorders of chromosomes, see Chapter 9.

Up to 10% of sperms in an ejaculate may be grossly abnormal (e.g., having two heads), but it is generally believed that these abnormal sperms do not fertilize oocytes owing to their lack of normal motility. Most morphologically abnormal sperms are unable to pass through the mucus in the cervical canal. Irradiation, severe allergic reactions, environmental pollutants, and certain antispermatogenic agents have been reported to increase the percentage of abnormally shaped sperms. However, such sperms are not believed to affect fertility unless their number exceeds 20%.

Female Reproductive Cycles

Commencing at puberty and normally continuing throughout the reproductive years, female humans undergo monthly reproductive cycles regulated by the **hypothalamus, pituitary gland, ovaries, uterus, uterine tubes, vagina,** and **mammary glands** (Fig. 2 – 8). These monthly cycles prepare the reproductive system for pregnancy. *Gonadotropin-releasing hormone* (GnRH) is synthesized by neurosecretory cells in the hypothalamus and is carried by the *hypophysial portal system* to the anterior lobe of the pituitary gland. GnRH stimulates the release of two hormones produced by this gland, which act on the ovaries:

- *Follicle-stimulating hormone* (FSH) stimulates the development of ovarian follicles and the production of **estrogen** by its follicular cells.
- *Luteinizing hormone* (LH) serves as the "trigger" for ovulation (release of a secondary oocyte) and stimulates the follicular cells and corpus luteum to produce **progesterone**.

These hormones also produce growth of the ovarian follicles and endometrium.

Ovarian Cycle

FSH and LH produce cyclic changes in the ovaries (development of follicles, ovulation, and formation of the corpus luteum) collectively known as the **ovarian cycle**. During each cycle, FSH promotes growth of several primary (ovarian) follicles (see Fig. 2 – 8); however only one of them usually develops into a mature follicle and ruptures through the surface of the ovary, expelling its oocyte. Hence, 4 to 11 follicles degenerate each month.

Follicular Development

Development of an ovarian follicle (Figs. 2 – 8 and 2 – 9) is characterized by:

- Growth and differentiation of a primary oocyte
- Proliferation of follicular cells
- Formation of the zona pellucida
- Development of a connective tissue capsule, the theca folliculi (Gr. *theke*, box)

The **theca folliculi** differentiates into two layers: an internal vascular and glandular layer — called the *theca interna* — and a capsulelike layer known as the *theca externa*. Thecal cells are thought to produce an *angiogenesis factor* that promotes growth of blood vessels in the theca interna (see Fig. 2 – 9*B*); these vessels provide nutritive support for follicular development.

Ovulation

The follicular cells divide actively, producing a stratified layer around the oocyte (see Fig. 2 – 9*A*). The ovarian follicle soon becomes oval and the oocyte eccentric in position because proliferation of the follicular cells occurs more rapidly on one side. Subsequently, fluid-filled spaces appear around the cells; these spaces coalesce to form a single cavity, the **antrum**, containing **follicular fluid**. After the antrum forms, the ovarian follicle is called a vesicular or **secondary follicle** (see Fig. 2 – 9*B*). The primary oocyte is pushed to one side of the follicle where it is surrounded by a mound of follicular cells, known as the **cumulus oophorus**, which projects into the enlarged antrum. The follicle continues to enlarge until it reaches maturity and forms a bulge on the surface of the ovary (Fig. 2 – 10*A*). At that point, it is a **mature ovarian follicle**. Around midcycle (14 days in an average 28-day menstrual cycle), the ovarian follicle, under the influence of FSH and LH, undergoes a sudden *growth spurt*, producing a cystic swelling (bulge) on the surface of the ovary. A small, oval, avascular spot, the **stigma**, soon appears on this swelling (see Fig. 2 – 10*A*). Prior to ovulation, the secondary oocyte and some cells of the cumulus oophorus detach from the interior of the distended follicle (see Fig. 2 – 10*B*).

Ovulation is triggered by a surge of LH production. Ovulation usually follows the LH peak by 12 to 24 hours. The **LH surge**, elicited by the high estrogen level in the blood (see Fig. 2 – 11), appears to cause the stigma to balloon out, forming a vesicle. The stigma then ruptures, expelling the secondary oocyte with the follicular fluid (see Fig. 2 – 10*D*). Expulsion of the oocyte is the result of intrafollicular pressure and, possibly, contraction of smooth muscle in the theca externa owing to stimulation by prostaglandins. *Enzymatic digestion of the follicular wall* seems to be one of the principal mechanisms leading to ovulation.

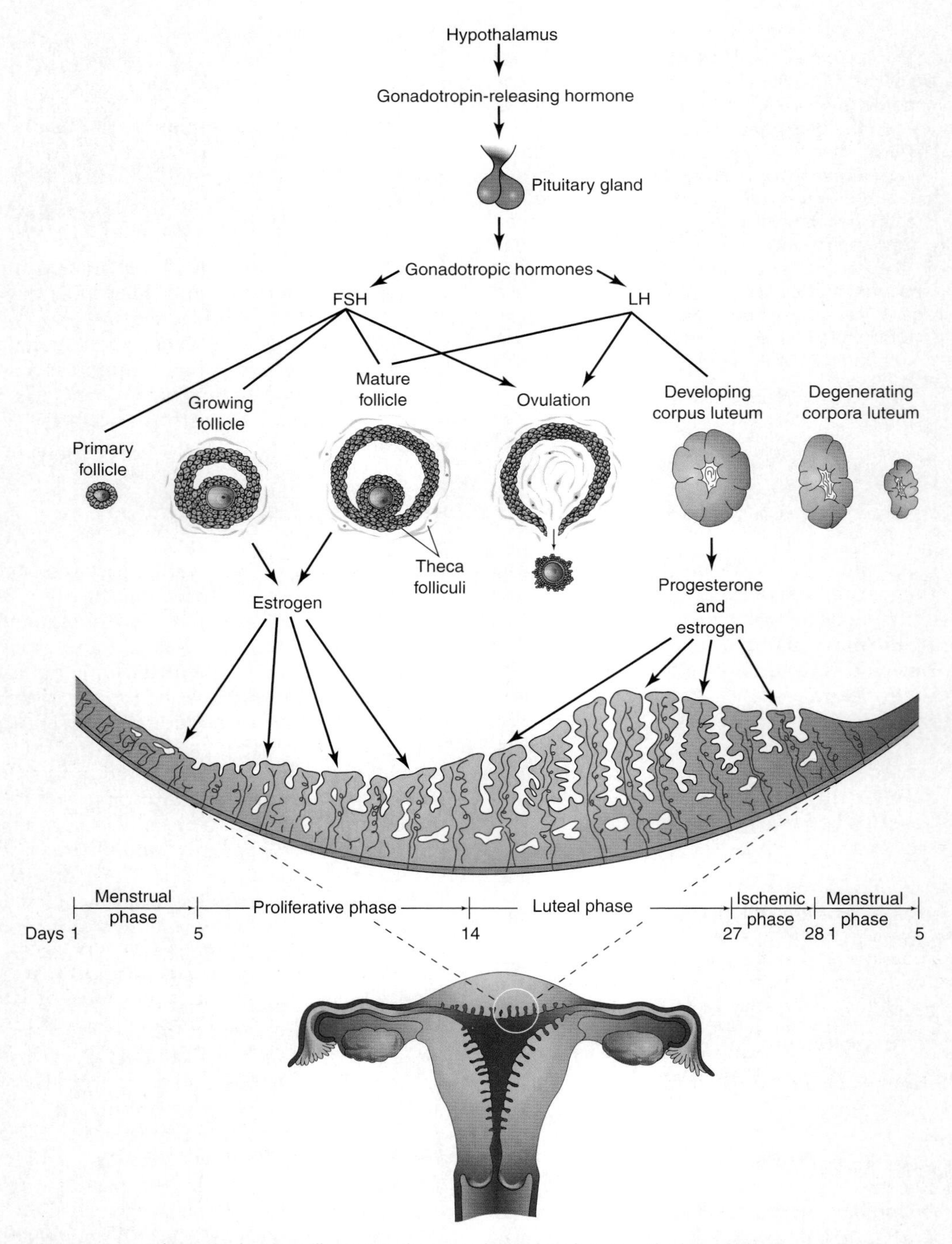

Figure 2-8. Schematic drawings illustrating the interrelationships between the hypothalamus, pituitary gland, ovaries, and endometrium. One complete menstrual cycle and the beginning of another are shown. Changes in the ovaries — referred to as the *ovarian cycle* — are induced by the gonadotropic hormones (follicle-stimulating hormone [FSH] and luteinizing hormone [LH]). Hormones from the ovaries (estrogens and progesterone) then promote cyclic changes in the structure and function of the endometrium, comprising the *menstrual cycle*. Thus, the cyclic activity of the ovary is intimately linked with changes in the uterus. The ovarian cycles are governed by the rhythmic endocrine control of the adenohypophysis of the pituitary gland, which, in turn, is controlled by gonadotropin-releasing hormone (GnRH) that is produced by neurosecretory cells in the hypothalamus.

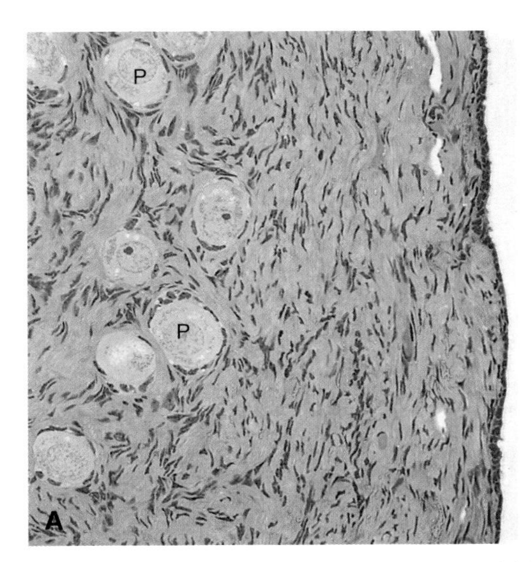

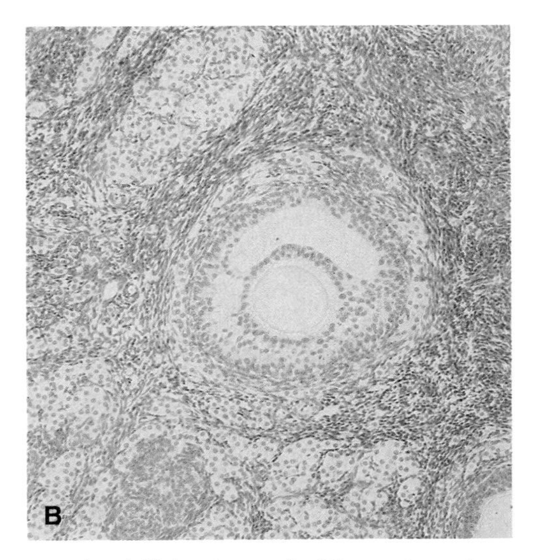

Figure 2 – 9. Photomicrographs of sections from adult human ovaries. *A*, Light micrograph of the ovarian cortex demonstrating mostly primordial follicles (P), which are primary oocytes surrounded by follicular cells (X270). *B*, Light micrograph of a secondary follicle. Observe the primary oocyte and the follicular fluid surrounded by membrana granulosa (X132). (From Garttner, LP, Hiatt JL: *Color Textbook of Histology*, 2nd ed. Philadelphia, WB Saunders, 2001.)

The expelled secondary oocyte is surrounded by the zona pellucida and one or more layers of follicular cells, which are radially arranged to form the corona radiata and cumulus oophorus (Figs. 2 – 9*C* and 2 – 10*C*); together, these form the oocyte-cumulus complex. The LH surge also seems to induce resumption of the first meiotic division of the primary oocyte. Hence, mature ovarian follicles contain secondary oocytes.

Mittelschmerz and Ovulation

A variable amount of abdominal pain, called mittelschmerz (Ger. *mittel*, mid; *schmerz*, pain), accompanies ovulation in some women. Mittelschmerz may be used as a symptom of ovulation; however, there are better indicators, such as the *basal body temperature*, which usually follows a slight drop followed by a sustained rise after ovulation.

Anovulation and Hormones

Some women do not ovulate because of an inadequate release of gonadotropins; as a result, they are unable to become pregnant in the usual way. In some of these women, *ovulation can be induced* by the administration of gonadotropins or an ovulatory agent. Such agents stimulate the release of pituitary gonadotropins (FSH and LH), resulting in maturation of several ovarian follicles and multiple ovulations. The incidence of multiple pregnancy increases when ovulation is induced.

Corpus Luteum

Shortly after ovulation, the walls of the ovarian follicle and theca folliculi collapse and are thrown into folds (Fig. 2 – 10*D*). Under the influence of LH, the walls of the follicle develop into a glandular structure, the **corpus luteum**, which secretes primarily *progesterone* but some estrogen as well. These hormones, particularly progesterone, cause the endometrial glands to secrete a thick mucoid and also prepare the endometrium for implantation of the blastocyst.

If the oocyte is fertilized, the corpus luteum enlarges to form a *corpus luteum of pregnancy* and increases its hormone production. When pregnancy occurs, degeneration of the corpus luteum is prevented by *human chorionic gonadotropin* (hCG), a hormone secreted by the syncytiotrophoblast of the chorion (see Chapter 3), which is rich in LH. The corpus luteum of pregnancy remains functionally active throughout the first 20 weeks of pregnancy. After that time, the placenta assumes the task of producing the estrogen and progesterone that are necessary for the maintenance of pregnancy (see Chapter 8).

If the oocyte is not fertilized, the corpus luteum will begin to involute and degenerate about 10 to 12 days after ovulation. It is then called a *corpus luteum of menstruation*. The corpus luteum is subsequently transformed into white scar tissue in the ovary, forming *corpus albicans* (atretic or degenerating corpus luteum). Except during pregnancy, ovarian cycles normally persist throughout the reproductive life of women, terminating at **menopause** — (permanent cessation of menstruation).

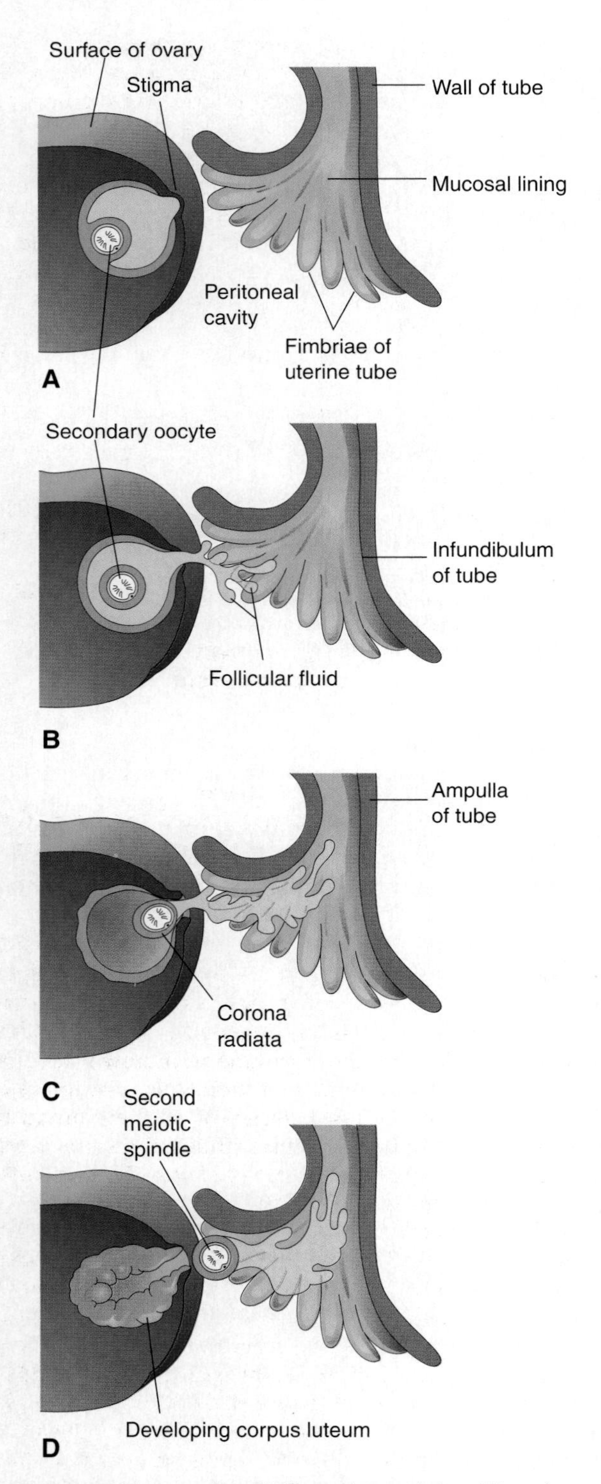

Figure 2 – 10. Diagrams (*A* to *D*) illustrating ovulation. When the stigma ruptures, the secondary oocyte is expelled from the ovarian follicle with the follicular fluid. After ovulation, the wall of the follicle collapses and is thrown into folds. The follicular wall is then transformed into a glandular structure called the corpus luteum.

Menstrual Cycle

The menstrual cycle is the period during which the oocyte matures, is ovulated, and enters the uterine tube (Fig. 2 – 11). The hormones produced by the ovarian

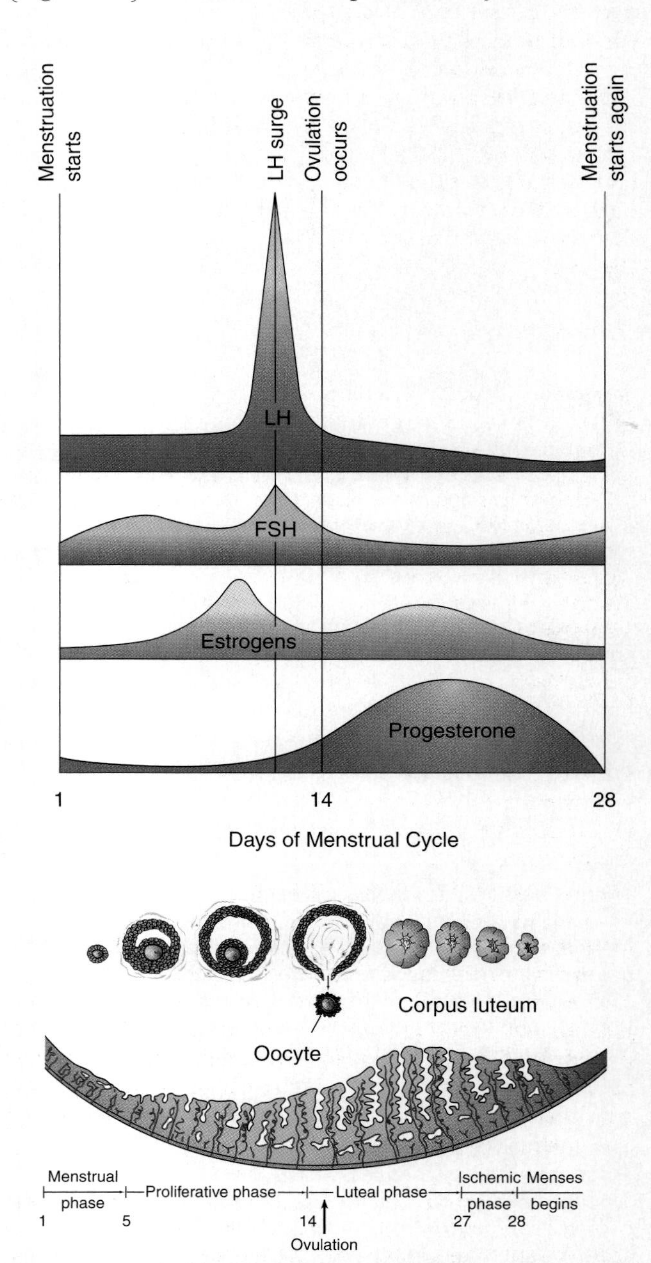

Figure 2 – 11. The blood levels of various hormones during the menstrual cycle. Follicle-stimulating hormone (FSH) stimulates the ovarian follicles to develop and produce estrogens. The level of estrogens rises to a peak just before the luteinizing hormone (LH) surge induces ovulation. Ovulation normally occurs 24 to 36 hours after the LH surge. If fertilization does not occur, the blood levels of circulating estrogens and progesterone fall. This hormone withdrawal causes the endometrium to regress and menstruation to start again.

follicles and corpus luteum (estrogen and progesterone) produce cyclic changes in the endometrium of the uterus. These monthly changes in the uterine lining constitute the **endometrial cycle**, commonly referred to as the **menstrual cycle** (period) because *menstruation* (flow of blood from the uterus) is an obvious event. *The normal endometrium is a "mirror" of the ovarian cycle* because it responds in a consistent manner to the fluctuating concentrations of ovarian hormones. The average menstrual cycle is 28 days, with day 1 of the cycle corresponding to the day on which menstrual flow begins. Menstrual cycles can vary in length by several days in normal women. In 90% of women, the length of the cycles ranges between 23 and 35 days. Almost all these variations result from alterations in the duration of the proliferative phase of the cycle.

Anovulatory Menstrual Cycles

In *anovulatory cycles,* the endometrial changes are minimal; the proliferative endometrium develops as usual, but no ovulation occurs and no corpus luteum forms. Consequently, the endometrium does not progress to the luteal phase; it remains in the proliferative phase until menstruation begins. *Suppression of ovulation is the basis for the success of birth control pills.* Estrogen in *birth control pills,* with or without progesterone, acts on the hypothalamus and pituitary gland, resulting in inhibition of secretion of GnRH, FSH, and LH, the secretion of which is essential for ovulation to occur.

Phases of the Menstrual Cycle

The menstrual cycle is divided into three main phases for descriptive purposes only (see Fig. 2-11). In actuality, *the menstrual cycle is a continuous process*; each phase gradually passes into the next one.

Menstrual Phase. The first day of menstruation is the beginning of the menstrual cycle. The functional layer of the uterine wall is sloughed off and discarded with the menstrual flow, which usually lasts 4 to 5 days. The menstrual flow, or **menses**, discharged through the vagina consists of varying amounts of blood combined with small pieces of endometrial tissue. After menstruation, the eroded endometrium is thin.

Proliferative Phase. The proliferative phase, lasting about 9 days, coincides with growth of ovarian follicles and is controlled by estrogen secreted by these follicles. There is a two- to three-fold increase in the thickness of the endometrium and in its water content during this *phase of repair and proliferation.* Early during this phase, the surface epithelium reforms and covers the endometrium. The glands increase in number and length, and the spiral arteries elongate.

Luteal Phase. The luteal (secretory) phase, lasting about 13 days, coincides with the formation, function, and growth of the corpus luteum. The progesterone produced by the corpus luteum stimulates the glandular epithelium to secrete a glycogen-rich, mucoid material. The glands become wide, tortuous, and saccular, and the endometrium thickens because of the influence of progesterone and estrogen from the corpus luteum and the increase in fluid in the connective tissue. As the spiral arteries grow into the superficial compact layer, they become increasingly coiled (see Fig. 2-2*C*). The venous network becomes complex and shows large *lacunae* (spaces).

If fertilization occurs:

- Cleavage of the zygote and blastogenesis (formation of the blastocyst) occur.
- The blastocyst begins to implant in the endometrium on about the sixth day of the luteal phase (day 20 of a 28-day cycle).
- hCG, a hormone produced by the syncytiotrophoblast of the chorion (see Chapter 4), keeps the corpus luteum secreting estrogens and progesterone.
- The luteal (secretory) phase continues and menstruation does not occur.

If fertilization does not occur:

- The corpus luteum degenerates.
- Estrogen and progesterone levels decrease, and the secretory endometrium enters an *ischemic phase* during the last day of the luteal phase.
- Menstruation occurs.

Ischemia (reduced blood supply) occurs as the spiral arteries constrict. This constriction results from the decrease in the secretion of hormones, primarily progesterone, by the degenerating corpus luteum. In addition to vascular changes, the hormone withdrawal results in the stoppage of glandular secretion, a loss of interstitial fluid, and a marked shrinking of the endometrium. Toward the end of the **ischemic phase,** the spiral arteries become constricted for longer periods. This results in *venous stasis* and patchy ischemic necrosis (death) in the superficial tissues. Eventually, rupture of damaged vessel walls follows, and blood seeps into the surrounding connective tissue. Small pools of blood form and break through the endometrial surface, resulting in bleeding into the uterine lumen and from the vagina.

As small pieces of the endometrium detach and pass into the uterine cavity, the torn ends of the arteries bleed into the uterine cavity, resulting in a loss of 20 to 80 ml of blood. Eventually, over 3 to 5 days, the entire compact layer and most of the spongy layer of the endometrium are discarded in the *menses.* Remnants of the spongy and basal layers remain to undergo regeneration during the subsequent proliferative phase of the endometrium.

If pregnancy occurs, the menstrual cycles cease and the endometrium passes into a *pregnancy phase*. With the termination of pregnancy, the ovarian and menstrual cycles resume after a variable period (usually 6 to 10 weeks if the woman is not breastfeeding her baby). If pregnancy does not occur, the reproductive cycles normally continue until the end of a woman's reproductive life — *menopause*, the permanent cessation of the menses — usually between the ages of 48 and 55 years. The syndrome of endocrine, somatic (body), and psychological changes occurring at the termination of the reproductive period is called the *climacteric* (climacterium).

Transportation of Gametes

Transportation of the gametes refers to the way the oocyte and sperms meet each other in the ampulla of the uterine tube, the usual site of fertilization (see Fig. 2-12).

Oocyte Transport

The secondary oocyte is expelled at ovulation from the ovarian follicle along with the escaping follicular fluid (see Fig. 2-10*D*). During ovulation, the fimbriated (fringed) end of the uterine tube comes in close proximity to the ovary. The fingerlike processes of the tube, the *fimbriae*, move back and forth over the ovary. The sweeping action of the fimbriae and the fluid currents produced by the cilia of the mucosal cells of the fimbriae "sweep" the secondary oocyte into the funnel-shaped infundibulum of the uterine tube. The oocyte then passes into the ampulla of the tube, primarily as a result of *waves of peristalsis* — movements of the wall of the tube characterized by alternate contraction and relaxation — that move in the direction of the uterus.

Sperm Transport

From their storage site in the epididymis (mainly in its tail), sperms are rapidly transported to the urethra by peristaltic contractions of the thick muscular coat of the ductus deferens (Fig. 2-12). As sperms pass by the accessory sex glands — *seminal glands (vesicles), prostate*, and *bulbourethral glands* — their secretions are added to sperm-containing fluid in the ductus deferens and urethra. The number of sperms deposited in the fornix of the vagina during sexual intercourse ranges from 200 to 600 million. The sperms pass slowly through the cervical canal simply by movements of their tails. The enzyme *vesiculase*, produced by the seminal glands, coagulates some of the *semen* (seminal fluid containing sperms) and forms a vaginal plug that may prevent backflow of semen into the vagina. At the time of ovulation, the cervical mucus increases in amount and becomes less viscid, making it more favorable for sperm transport. Passage of sperms through the uterus and uterine tubes results mainly from muscular contractions of the walls of these organs. *Prostaglandins* in the semen stimulate uterine motility and help to move the sperms through the uterus and tubes to the site of fertilization. *Fructose* in the semen, secreted by the seminal glands, provides an energy source for the sperms.

The volume of the **ejaculate** (sperms suspended in secretions from accessory sex glands) averages 3.5 ml, with a range of 2 to 6 ml. Sperms move 2 to 3 mm per minute but the speed varies with the pH of the environment. They are nonmotile during storage in the epididymis but become motile in the ejaculate. They move slowly in the acid environment of the vagina but more rapidly in the alkaline environment of the uterus. Only about 250 sperms reach the fertilization site. Most sperms degenerate during their passage through the female genital tract.

Maturation of Sperms

Freshly ejaculated sperms are unable to fertilize oocytes. They must first undergo a *period of conditioning* — **capacitation** — lasting about 7 hours. During this period, a glycoprotein coat and seminal proteins are removed from the surface of the sperm's acrosome. The membrane components, as well as the membrane potential, are altered. Capacitated sperms show no morphological changes, but they exhibit increased activity. Sperms are usually capacitated in the uterus or uterine tubes by substances secreted by these parts of the female genital tract. The *acrosome reaction* of sperms must be completed before a sperm can fuse with the oocyte. ZP3 (a protein present in the zona pellucida), prostaglandin E, progesterone, and calcium ions play a critical role in the acrosome reaction. When capacitated sperms come into contact with the corona radiata surrounding a secondary oocyte, they undergo changes that result in the development of perforations in the acrosome (see Fig. 3-1). Multiple point fusions of the plasma membrane of the sperm and the external acrosomal membrane occur. Breakdown of the membranes at these sites produces apertures. The changes induced by the acrosome reaction are associated with the release from the acrosome of enzymes, including *hyaluronidase* and *acrosin*, that facilitate fertilization.

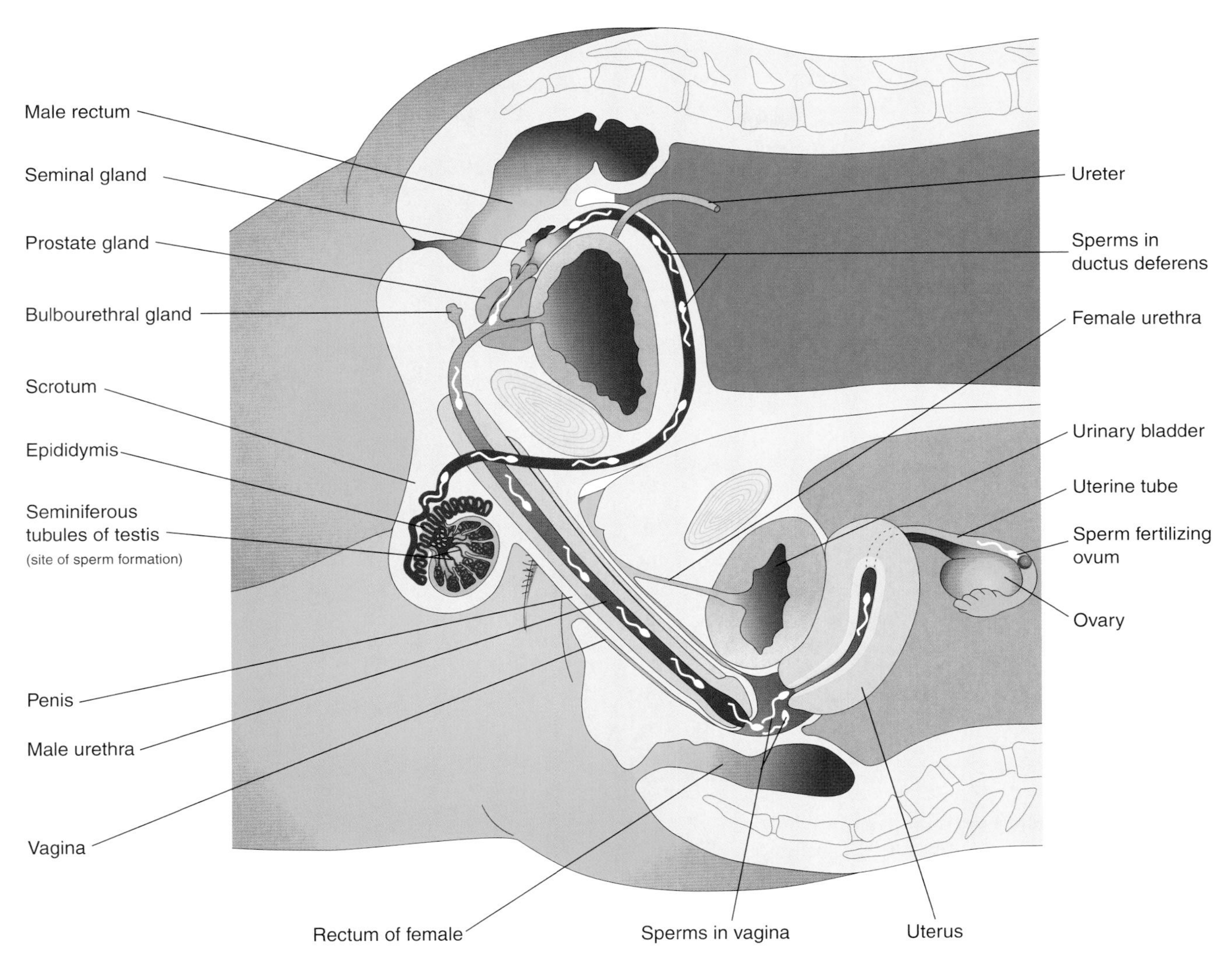

Figure 2 - 12. Schematic sagittal section of male and female pelves. Sperms are produced in the seminiferous tubules of the testis and stored in the epididymis. During ejaculation, the sperms pass along the ductus deferens and ejaculatory duct and enter the urethra, where they mix with secretions from the seminal glands, prostate, and bulbourethral glands. This mixture, called semen, is deposited in the superior part of the vagina close to the external os of the uterus. The sperms then pass through the cervix and the cavity of the uterus into the uterine tubes, where fertilization usually occurs. Sexual intercourse that is interrupted before the male ejaculates is called coitus interruptus.

Sperm Counts

During evaluation of male fertility, an analysis of semen is made. Sperms account for less than 10% of the semen. The remainder of the ejaculate consists of the secretions of the accessory sex glands: seminal glands (60%), prostate (30%), and bulbourethral glands (10%). In the ejaculate of normal males, there are usually more than 100 million sperms per milliliter of semen. Although there is much variation in individual cases, men whose semen contains 20 million sperms per milliliter, or 50 million in the total specimen, are probably fertile. A man with less than 10 million sperms per milliliter of semen is likely to be sterile, especially when the specimen contains immotile and abnormal sperms. Thus, in assessing fertility potential, the total number and motility of sperms in the ejaculate are taken into consideration. For potential fertility, at least 40% of sperms should be motile after 2 hours, and some should be motile after 24 hours. It is believed that male infertility is the cause of one third to one half of unintentionally childless marriages. Male infertility may result from endocrine disorders, abnormal spermatogenesis, or obstruction of a genital duct (e.g., the ductus deferens).

Vasectomy

The most effective method of contraception in males is *deferentectomy* (excision of a segment of the ductus deferens). This surgical procedure, often called a **vasectomy**, is reversible in at least 50% of cases. Following vasectomy, there are no sperms in the ejaculate, but the amount of seminal fluid is the same as before the procedure.

Viability of Gametes

Human oocytes are usually fertilized within 12 hours after ovulation. In vitro observations have shown that oocytes cannot be fertilized after 24 hours, and they degenerate shortly thereafter. *Most human sperms probably do not survive for more than 48 hours in the female genital tract.* Some sperms are stored in folds of the mucosa of the cervix and are gradually released into the cervical canal, and then pass from the uterus into the uterine tubes. Semen that has been frozen to low temperatures may be kept for many years. Children have been born to women who have been artificially inseminated with semen stored for several years.

Summary of Reproduction

Fertilization involves the union of an oocyte from a female and sperm from a male. The reproductive system in both sexes is designed to produce gametes and ensure their union. The secondary oocyte develops in the ovary and is expelled from it at ovulation. It is carried into the infundibulum of the uterine tube by sweeping motions of the fimbriae of the tube. Peristaltic waves in the tube move the oocyte to the fertilization site in the ampulla of the tube. Sperms are produced in the seminiferous tubules of the testes and stored in the epididymis. During ejaculation, which usually occurs during sexual intercourse, semen is deposited in the vagina. Although there are several million sperms in the semen, only a few thousand pass through the cervical canal and uterine cavity and along the uterine tube. Only about 250 sperms reach the ampulla where fertilization occurs if a secondary oocyte is present.

Clinically Oriented Questions

1. Does a ruptured hymen indicate that a woman is not a virgin?
2. Some say that a woman can have an erection. Is this true?
3. There have been reports of a woman who claimed that she menstruated throughout her pregnancy. How could this happen?
4. If a woman forgets to take a birth control pill and then takes two, is she likely to become pregnant?
5. What is coitus interruptus? Some people believe that it is a safe method of birth control. Is this true?
6. What is the difference between spermatogenesis and spermiogenesis?
7. Some say that an IUD (intrauterine device) is a contraceptive. Is this correct?
8. What is the difference between the terms *menopause* and *climacteric*?

The answers to these questions are at the back of the book.

The First Week of Human Development

A zygote is a highly specialized, totipotent cell resulting from union of a sperm and an oocyte. The zygote contains chromosomes and genes that are derived from the mother and father. The zygote divides many times and is progressively transformed into a multicellular human being through cell division, migration, growth, and differentiation. The stages and duration of pregnancy described in clinical medicine are calculated from the commencement of the mother's *last normal menstrual period* (LNMP), which is about 14 days before conception occurs (see Fig. 1 – 1). LNMP indicates the *gestational age*, which overestimates the actual fertilization or embryonic age by 2 weeks.

Fertilization

The usual site of fertilization is the ampulla, the longest and widest part of the uterine tube (see Fig. 2 – 2*B*). If the oocyte is not fertilized here, it slowly passes along the tube to the uterus, where it degenerates and is resorbed. Fertilization is a complex sequence of "coordinated molecular events" that begins with contact between a sperm and an oocyte (Fig. 3 – 1) and ends with the intermingling of maternal and paternal chromosomes at metaphase of the first mitotic division of the **zygote** (see Fig. 3 – 2*E*). Carbohydrate- and protein-binding molecules on the surface of the gametes are involved in sperm chemotaxis and gamete recognition, as well as in the process of fertilization. Fertilin-α and fertilin-β, sperm membrane proteins, have been implicated in sperm-egg interactions.

Phases of Fertilization

The phases of fertilization are as follows (see Figs. 3 – 1 and 3 – 2):

- **Passage of a sperm through the corona radiata of the oocyte**. Dispersal of the follicular cells of the corona radiata results mainly from the action of the enzyme *hyaluronidase, which is* released from the acrosome of the sperm. *Tubal mucosal enzymes* also appear to assist hyaluronidase. Additionally, movements of the tail of the sperm are important in achieving penetration of the corona radiata.
- **Penetration of the zona pellucida**. The formation of a pathway for the sperm through the zona pellucida results from the action of enzymes released from the acrosome. The enzymes — *esterases, acrosin,* and *neuraminidase* — appear to cause lysis (Gr., dissolution) of the zona pellucida, thereby forming a path for the sperm to follow to the oocyte. The most important of these enzymes is *acrosin,* a proteolytic enzyme. Once the sperm penetrates the zona pellucida, a **zona reaction** (a change in its properties) occurs, which makes the zona pellucida impermeable to other sperms.
- **Fusion of the plasma membranes of oocyte and sperm**. The plasma (cell) membranes of the oocyte and sperm fuse and break down at the area of fusion. The head and tail of the sperm then enter the cytoplasm of the oocyte, but the sperm's plasma membrane remains behind (see Fig. 3 – 1*B*).
- **Completion of the second meiotic division of the oocyte and formation of the female pronucleus**. After entry of the sperm, the oocyte, which has been arrested in metaphase of the second meiotic division, completes the division and forms a mature oocyte and a second polar body (see Fig. 3 – 2*B*). Following decondensation of the maternal chromosomes, the nucleus of the mature oocyte becomes the female pronucleus.
- **Formation of the male pronucleus**. Within the cytoplasm of the oocyte, the nucleus of the sperm enlarges to form the male pronucleus, and the tail of the sperm degenerates (see Fig. 3 – 2*C*). During growth, the male and female pronuclei replicate their DNA — 1 n (haploid), 2 *c* (two chromatids).
- **Breakdown of the pronuclear membranes, condensation of the chromosomes, and arrangement of the chromosomes for mitotic cell division**, the first cleavage division (see Fig. 3 – 4*A*). The combination of 23 chromosomes in each pronucleus results in a zygote with 46 chromosomes.

Fertilization is completed within 24 hours of ovulation. An immunosuppressant protein — the *early pregnancy factor* (EPF) — is secreted by the trophoblastic cells and appears in the maternal serum within 24 to 48 hours after fertilization. EPF forms the basis of a pregnancy test during the first 10 days of development.

Dispermy and Triploidy

Although several sperms begin to penetrate the zona pellucida, usually only one sperm enters the oocyte and fertilizes it. Two sperms may participate in fertilization during an abnormal process known as *dispermy,* resulting in three haploid sets of chromosomes (triploidy). Triploid conceptions account for about 20% of spontaneous abortions occurring as a result of chromosomal abnormalities. The resulting **triploid embryos** (69 chromosomes) appear quite normal, but they are nearly always aborts. Aborted fetuses with triploidy have severe *intrauterine growth retardation* with a disproportionately small trunks and many other anomalies (e.g., of the central nervous system). A few triploid infants have been born, but all have died shortly after birth. Livebirths of triploid infants are uncommon, occurring in fewer than 1 in 2500 such pregnancies.

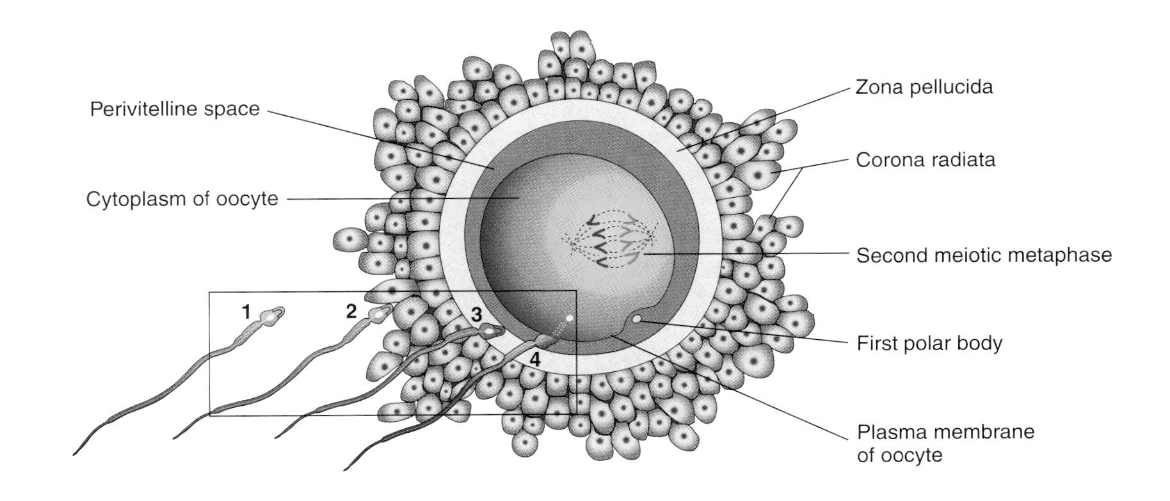

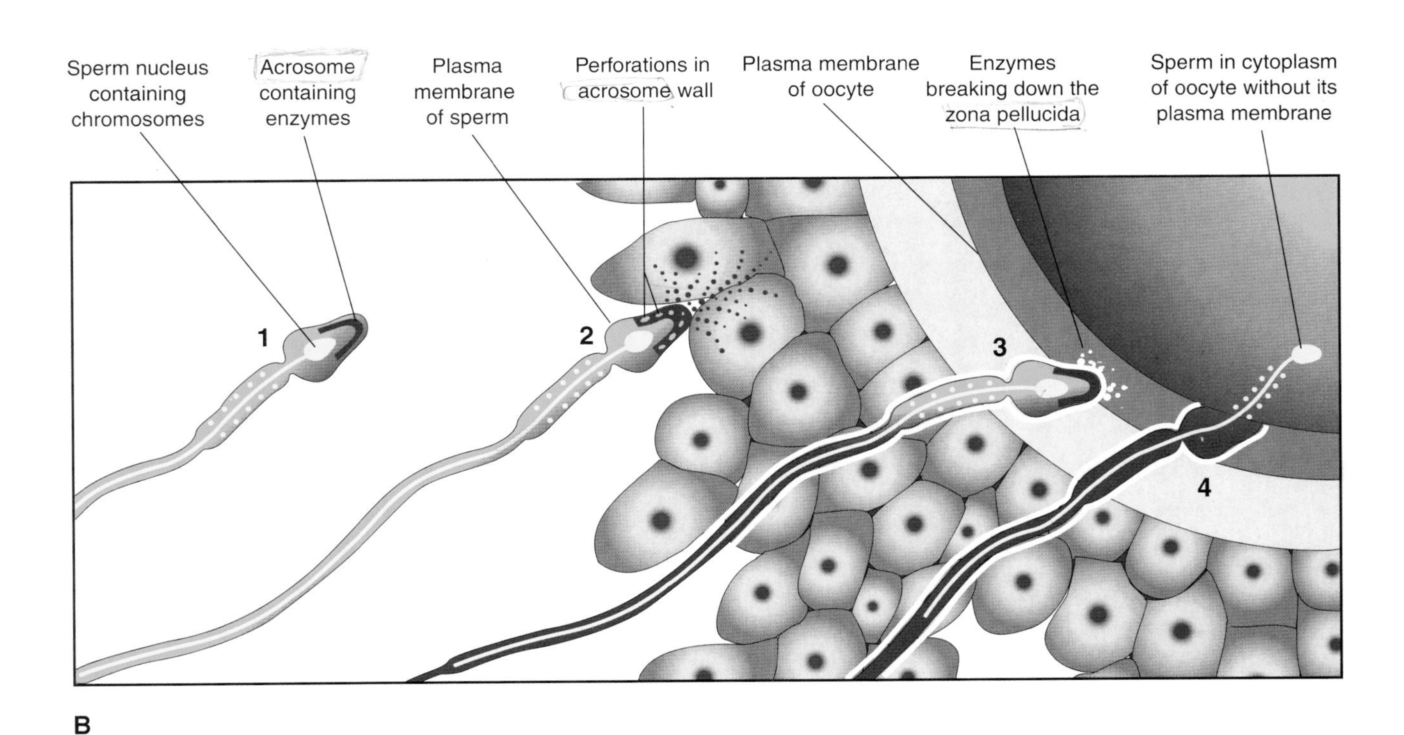

Figure 3-1. Acrosome reaction and a sperm penetrating an oocyte. The area outlined in *A* is detailed in *B*. *1*, Sperm during capacitation, a period of conditioning that occurs in the female reproductive tract. *2*, Sperm undergoing the acrosome reaction during which perforations form in the acrosome. *3*, Sperm digesting a path through the zona pellucida by the action of enzymes released from the acrosome. *4*, Sperm after entering the cytoplasm of the oocyte. Note that the plasma membranes of the sperm and oocyte have fused and that the head and tail of the sperm enter the oocyte.

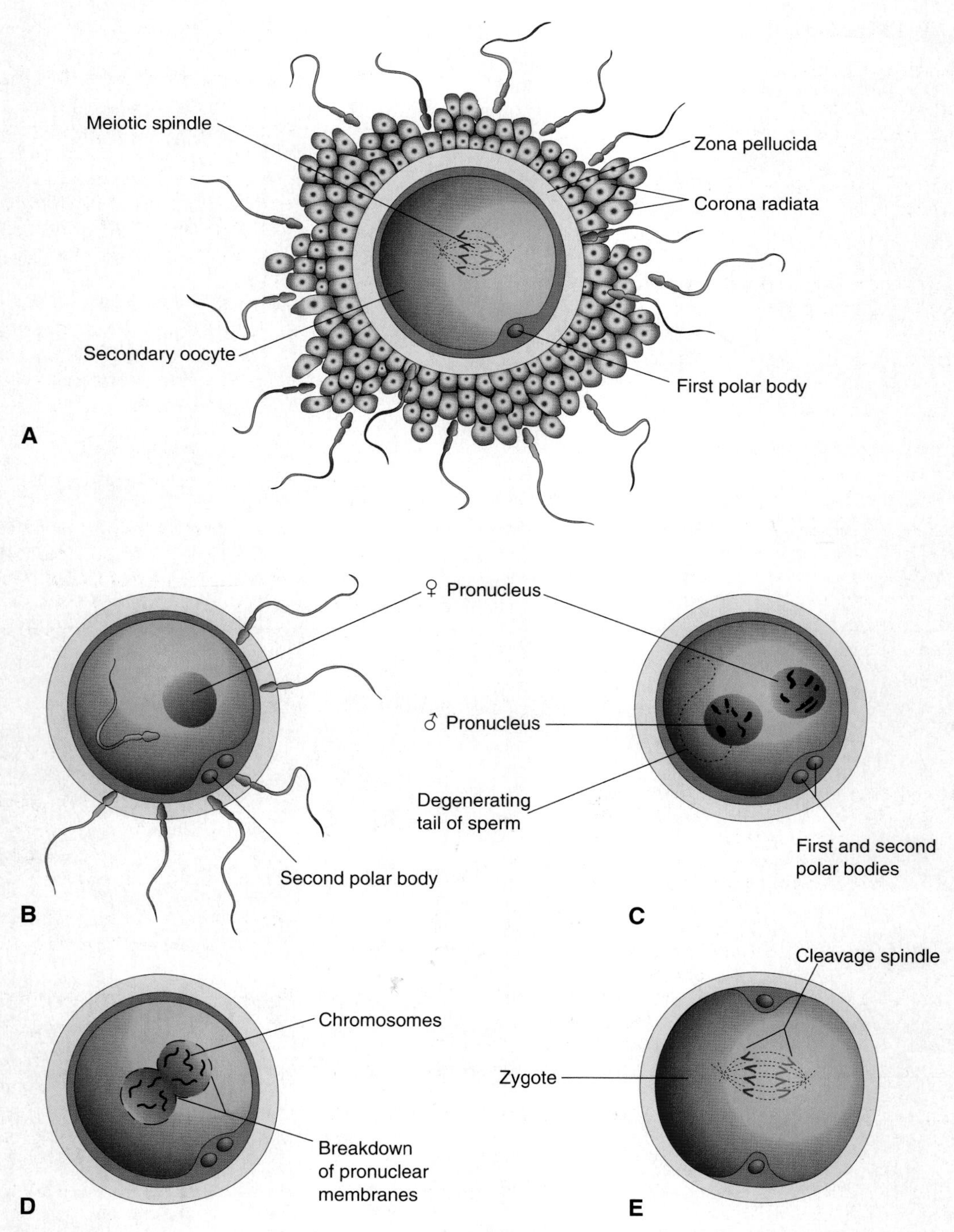

Figure 3-2. Diagrams illustrating fertilization, the process that begins when the sperm comes in contact with the secondary oocyte's plasma membrane and that ends with the intermingling of maternal and paternal chromosomes at metaphase of the first mitotic division of the zygote. *A,* The secondary oocyte is surrounded by several sperms. (Only 4 of the 23 chromosome pairs are shown.) *B,* The corona radiata has disappeared, a sperm has entered the oocyte, and the second meiotic division has occurred, resulting in the formation of a mature oocyte. The nucleus of the ovum is now the female pronucleus. *C,* The sperm head has enlarged to form the male pronucleus. *D,* The pronuclei are fusing. *E,* The zygote has formed; it contains 46 chromosomes, the diploid number.

Results of Fertilization

When fertilization occurs, it

- stimulates the secondary oocyte to complete the second meiotic division, producing the second polar body.
- restores the normal diploid number of chromosomes (46) in the zygote.
- results in variation of the human species through mingling of maternal and paternal chromosomes.
- determines the chromosomal sex of the embryo; an X-bearing sperm produces a female embryo and a Y-bearing sperm produces a male embryo.
- causes metabolic activation of the oocyte (ootid — the nearly mature oocyte), which initiates cleavage (cell division of zygote).

The zygote is genetically unique because half of its chromosomes come from the mother and half are derived from the father. The zygote contains a new combination of chromosomes that is different from that in the cells of either of the parents. *This mechanism forms the basis of biparental inheritance and variation of the human species.* Meiosis allows independent assortment of maternal and paternal chromosomes among the germ cells (see Fig. 2 – 6). *Crossing-over of chromosomes*, by relocating segments of the maternal and paternal chromosomes, "shuffles" the genes, thereby producing a recombination of genetic material.

In Vitro Fertilization and Embryo Transfer

The process of in vitro fertilization (IVF) of oocytes and transfer of the dividing (cleaving) zygotes (embryos) into the uterus has provided an opportunity for many women who are sterile (e.g., because of tubal occlusion) to bear children. The first of these IVF babies was born in 1978. The steps involved during IVF and embryo transfer are summarized in Figure. 3 – 3. The incidence of multiple pregnancies is higher with IVF than when pregnancy results from normal ovulation and passage of the morula into the uterus via the uterine tube. The incidence of spontaneous abortion of transferred embryos is also higher than normal in women undergoing IVF.

The technique of **intracytoplasmic sperm injection** (ICSI) involves injecting a sperm directly into the cytoplasm of the mature oocyte. This procedure is invaluable in cases of infertility resulting from blocked uterine tubes or oligospermia (reduced number of sperms).

Cleavage of the Zygote

Cleavage consists of repeated mitotic divisions of the zygote, resulting in a rapid increase in the number of cells. These cells — called **blastomeres** — become smaller with each cleavage division (Fig. 3 – 4). First the zygote divides into two blastomeres; these cells then divide into four blastomeres, then eight blastomeres, and so on. Cleavage normally occurs as the zygote passes through the uterine tube toward the uterus (see Fig. 3 – 6). During cleavage, the zygote is still contained within the zona pellucida (pellucid zone). Division of the zygote into blastomeres begins about 30 hours after fertilization. Subsequent divisions follow, forming progressively smaller blastomeres.

After the eight-cell stage, the blastomeres change their shape and tightly align themselves against each other to form a compact ball of cells. This phenomenon, termed — **compaction** — is probably mediated by cell surface adhesion glycoproteins. Compaction permits greater cell-to-cell interaction and is a prerequisite for segregation of the internal cells that form the inner cell mass or **embryoblast** of the blastocyst (see Fig. 3 – 4*E*). When there are 12 to 15 blastomeres, the developing human is called a **morula** (L., *morus*, mulberry) because of its resemblance to the fruit of the mulberry tree. The internal cells of the morula — the **inner cell mass or embryoblast** — are surrounded by a layer of flattened cells that form the outer cell mass, or **trophoblast**.

Formation of the Blastocyst

Shortly after the morula enters the uterus (about 4 days after fertilization), fluid passes from the uterine cavity through the zona pellucida to form a fluid-filled space — called the **blastocystic cavity** — inside the morula (see Fig. 3 – 4*E*). As fluid increases in the cavity, the blastomeres are separated into two parts:

- the **trophoblast** (Gr., *trophe*, nutrition), a thin outer cell that gives rise to the embryonic part of the placenta
- the **embryoblast**, a group of centrally located blastomeres that gives rise to the embryo

At this stage of development, the conceptus is called a **blastocyst**. The embryoblast now projects into the blastocystic cavity, and the trophoblast forms the wall of the blastocyst (see Fig. 3 – 4*F*). After the blastocyst has floated in the uterine secretions for about 2 days, the zona pellucida degenerates and disappears. *Shedding of the zona pellucida* or "hatching of the blastocyst" has been observed in vitro. Shedding of the zona pellucida permits the blastocyst to increase rapidly in size. While floating freely in the uterine cavity, the embryo derives nourishment from secretions of the uterine glands.

About 6 days after fertilization (day 20 of a 28-day menstrual cycle), the blastocyst attaches to the endome-

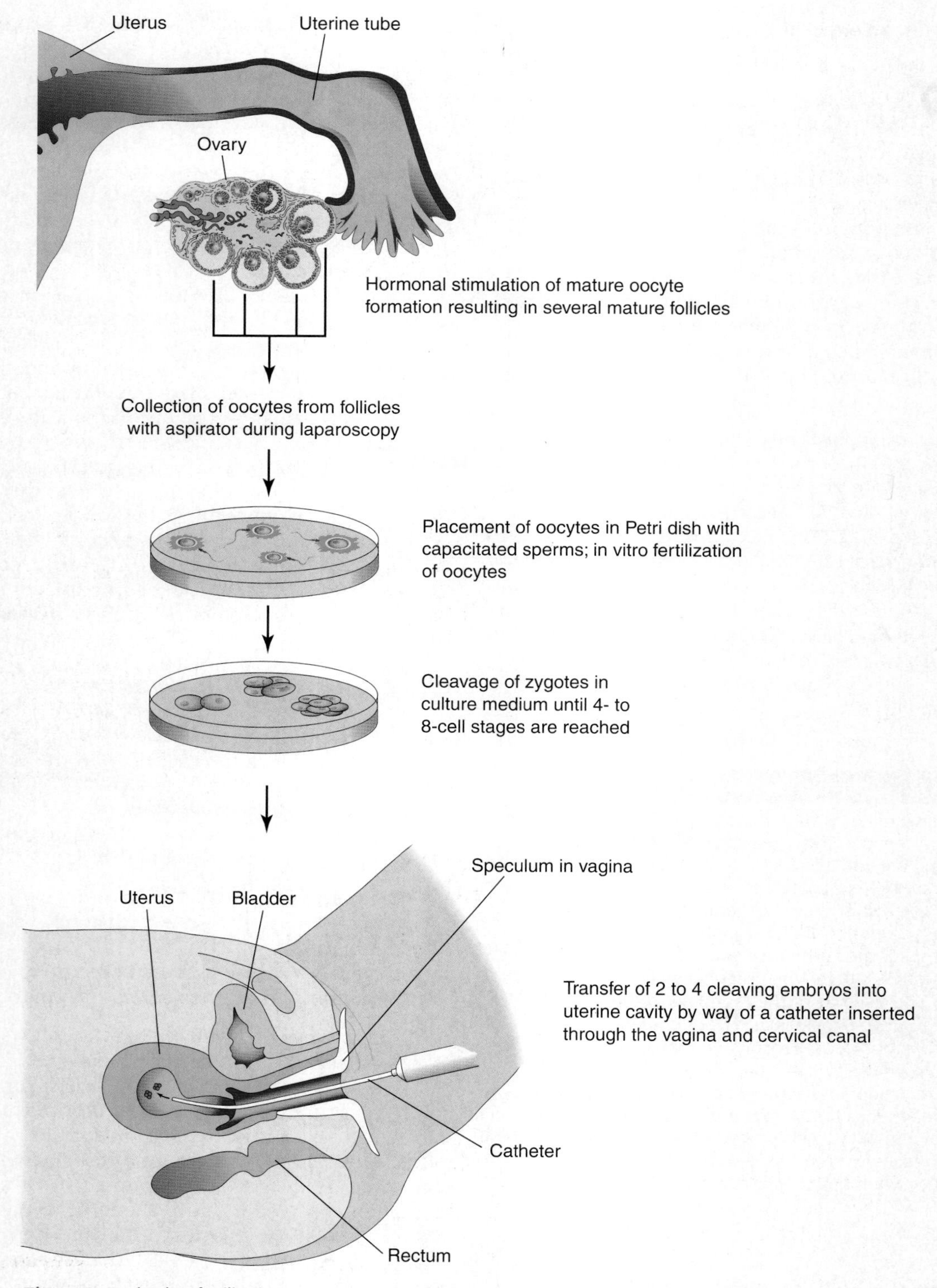

Figure 3-3. In vitro fertilization and embryo transfer procedures.

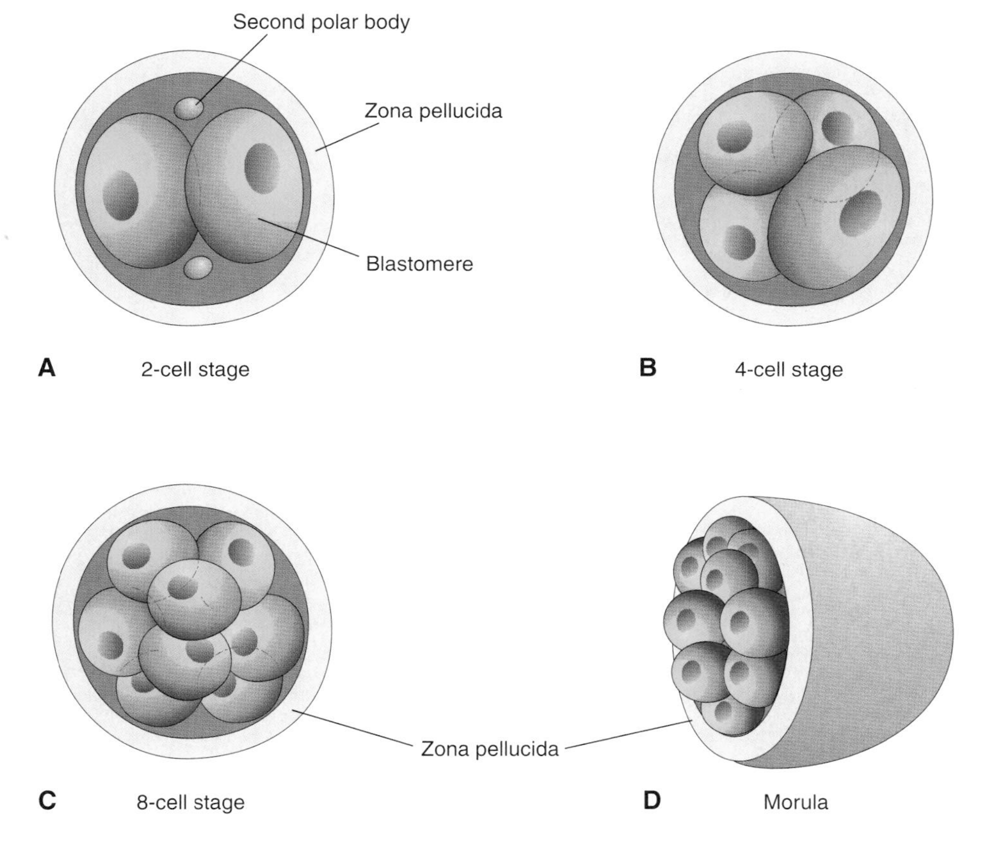

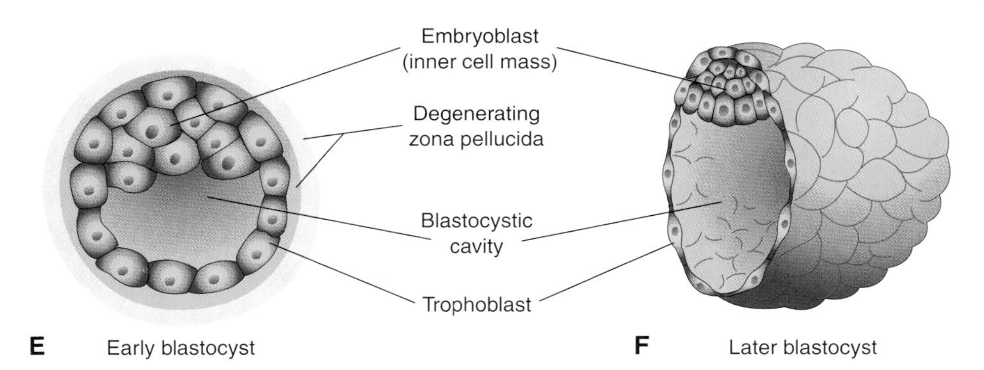

Figure 3 – 4. Drawings illustrating cleavage of the zygote and formation of the blastocyst. *A* to *D* show various stages of cleavage. The period of the morula begins at the 12- to 16-cell stage and ends when the blastocyst forms. *E* and *F* are sections of blastocysts. The zona pellucida disappears by the late blastocyst stage (5 days). The polar bodies shown in *A* are small, nonfunctional cells that soon degenerate. Cleavage and formation of the morula occur as the dividing zygote passes along the uterine tube. Blastocyst formation normally occurs in the uterus. Although cleavage increases the number of blastomeres, note that each of the daughter cells is smaller than the parent cells. As a result, there is no increase in the size of the developing embryo until the zona pellucida degenerates. The blastocyst then enlarges considerably. The embryoblast gives rise to the tissues and organs of the embryo.

trial epithelium, usually adjacent to the embryoblast (Fig. 3 – 5*A*). As soon as it attaches to the endometrial epithelium, the trophoblast starts to proliferate rapidly and differentiate into two layers (see Fig. 3 – 5*B*):

- the **cytotrophoblast** (cellular trophoblast), the inner layer of cells
- the **syncytiotrophoblast** (syncytial trophoblast), the outer syncytial layer consisting of a multinucleate protoplasmic mass formed by the fusion of cells without distinguishable cell boundaries

The fingerlike processes of the syncytiotrophoblast extend through the endometrial epithelium and invade the endometrial connective tissue. By the end of the first week, the blastocyst is superficially implanted in the compact layer of the endometrium and is deriving

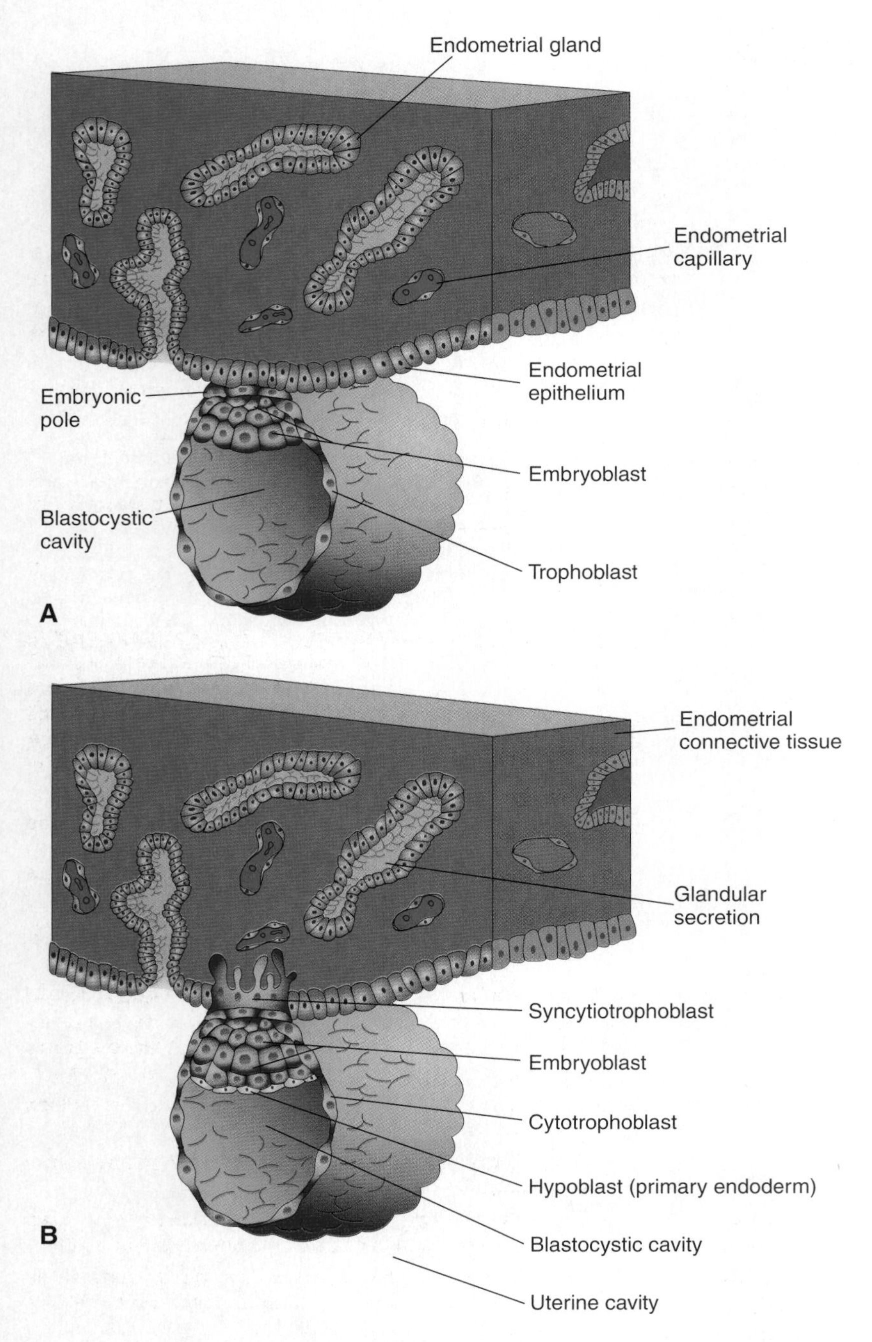

Figure 3–5. Attachment of the blastocyst to the endometrial epithelium during the early stages of its implantation. *A,* Six days; the trophoblast is attached to the endometrial epithelium at the embryonic pole of the blastocyst. *B,* Seven days; the syncytiotrophoblast has penetrated the epithelium and has started to invade the endometrial connective tissue. Some students have difficulty interpreting illustrations such as these because in histologic studies it is conventional to draw the endometrial epithelium upward, whereas in embryologic studies the embryo is usually shown with its dorsal surface upward. Because the embryo implants on its future dorsal surface, it would appear upside-down if the histologic convention were followed. In this book, the histologic convention is followed when the endometrium is the dominant consideration (see Fig. 2–2), and the embryologic convention is used when the embryo is the center of interest, as in these illustrations.

its nourishment from the eroded maternal tissues. The highly invasive syncytiotrophoblast rapidly expands adjacent to the embryoblast — the **embryonic pole.** The syncytiotrophoblast produces proteolytic enzymes that erode the maternal tissues, enabling the blastocyst to burrow into the endometrium. *At about 7 days,* a cuboidal layer of cells, called the **hypoblast**, appears on the surface of the embryoblast, facing the blastocystic cavity (see Fig. 3–5*B*). Comparative embryological data suggest that the hypoblast arises by delamination (separation) from the embryoblast.

Preimplantation Diagnosis of Genetic Disorders

Using currently available techniques of micromanipulation and DNA application, a cleaving zygote known to be at risk for a specific genetic disorder may be diagnosed before implantation. The sex of the embryo can be determined from a blastomere taken from a six- to eight-cell zygote and analyzed by DNA amplification of sequences from the Y chromosome. This procedure has been used to detect female embryos during IVF in cases in which a male embryo would be at risk for a serious X-linked disorder.

Abnormal Embryos and Spontaneous Abortions

Many early embryos are spontaneously aborted. The early implantation stages of the blastocyst are critical periods of development that may fail to occur because of inadequate production of progesterone and estrogen by the corpus luteum (see Fig. 2–8). Clinicians occasionally see a patient whose last menstrual period was delayed by several days and whose last menstrual flow was unusually profuse. Very likely, such patients have had an early spontaneous abortion; thus, the overall *early spontaneous abortion rate* is thought to be about 45%. *Early spontaneous abortions occur for a variety of reasons,* an important one being the presence of **chromosomal abnormalities**. Thus, the early loss of embryos represents the disposal of abnormal conceptuses that could not have developed normally.

Summary of the First Week of Development

When an **oocyte** comes into contact with a **sperm**, it completes the second meiotic division. As a result, a mature oocyte and a *female pronucleus* are formed. After the sperm enters the oocyte, the head of the sperm separates from the tail and enlarges to become the *male pronucleus*. Fertilization is complete when the pronuclei unite and the maternal and paternal chromosomes intermingle during metaphase of the first mitotic division of the **zygote** (Fig. 3–6). As it passes along the uterine tube toward the uterus, the zygote undergoes **cleavage** (a series of mitotic cell divisions), which results in the formation of smaller cells called **blastomeres**. About 3 days after fertilization, a ball of blastomeres, called the **morula**, enters the uterus. A *blastocystic cavity* soon forms in the morula, converting it into a **blastocyst** consisting of

- the *embryoblast*
- the *blastocystic cavity*, a fluid-filled space
- the *trophoblast*, a thin outer layer of cells

The **trophoblast** encloses the embryoblast and blastocystic cavity and later forms extraembryonic structures and the embryonic part of the placenta. After fertilization, the zona pellucida is shed and the trophoblast adjacent to the embryoblast attaches to the endometrial epithelium. The trophoblast adjacent to the embryonic pole differentiates into two layers, an outer **syncytiotrophoblast** (a multinucleated mass without distinct cell boundaries) and an inner cytotrophoblast (a mononucleated layer of cells). The syncytiotrophoblast invades the endometrial epithelium and underlying connective tissue. Concurrently, a cuboidal layer of **hypoblast** forms adjacent to the blastocystic cavity. By the end of the first week, the blastocyst is superficially implanted in the endometrium.

Clinically Oriented Questions

1. Although women do not commonly become pregnant after they are 48 years old, very elderly men can still impregnate women. Why is this? Is there an increased risk of Down syndrome or other congenital anomalies in the child when the father is older than 50 years of age?
2. Are there contraceptive pills for men? If not, what is the reason?
3. Is a polar body ever fertilized? If so, does the fertilized polar body give rise to a viable embryo?
4. What is the most common cause of spontaneous abortions during the first week of development?
5. Some clinicians have speculated that a woman could have dissimilar twins as a result of one oocyte being fertilized by a sperm from one man and another one being fertilized by a sperm from another man. Is this possible?
6. Do the terms *impregnation*, *conception*, and *fertilization* differ in meaning? If so, how?
7. Do the terms *cleavage* and *mitosis* of the zygote mean the same thing?
8. How is the dividing zygote (cleaving embryo) nourished during the first week? Do the blastomeres contain yolk?
9. Can one determine the sex of a cleaving embryo developing in vitro? If so, what medical reasons would there be for doing so?

The answers to these questions are at the back of the book.

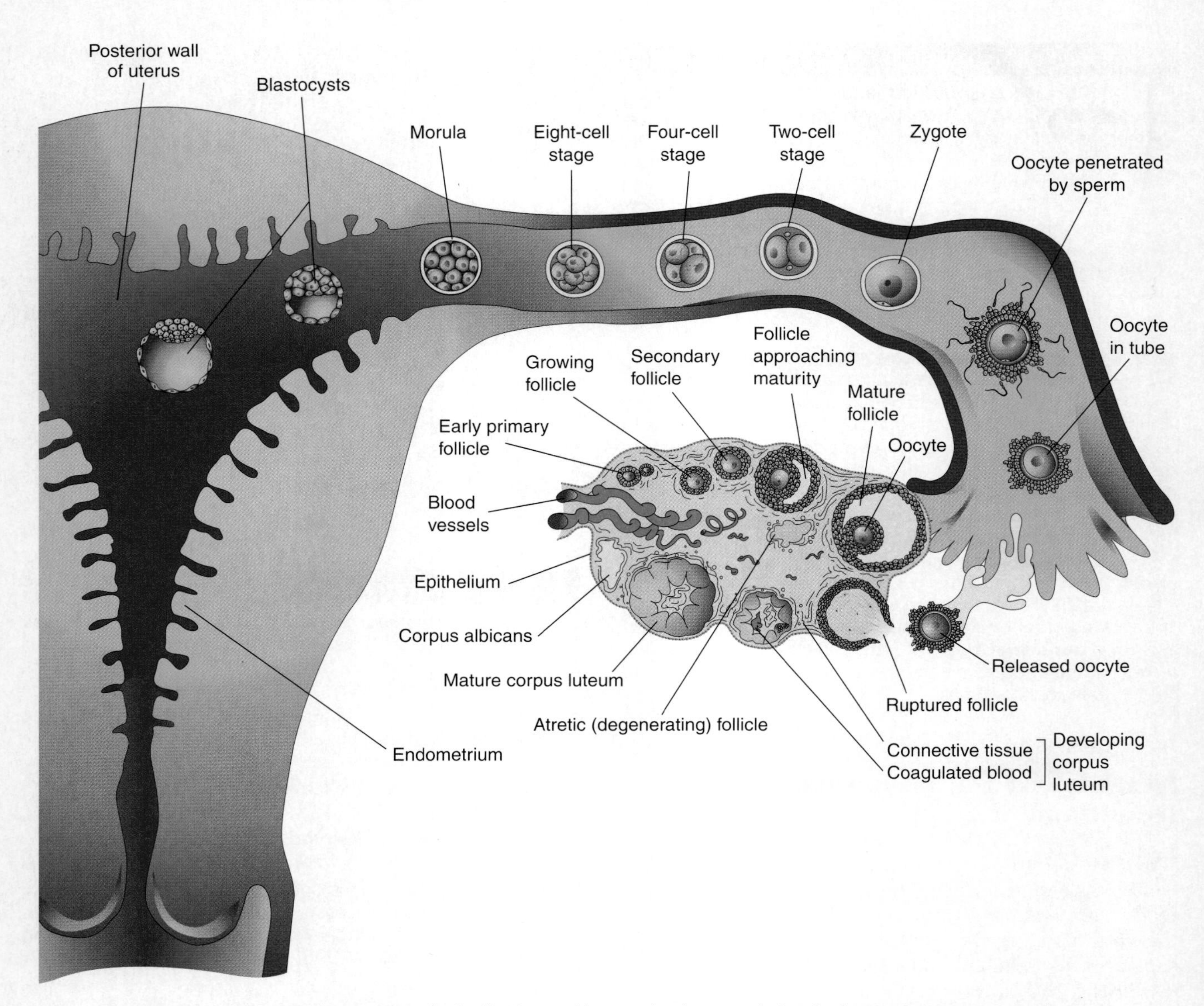

Figure 3-6. Summary of the ovarian cycle, fertilization, and human development during the first week. Stage 1 of development begins with fertilization and ends when the zygote forms. In stage 2 (days 2 to 3) the early stages of cleavage occur (from 2 to about 32 cells — [morula]). In stage 3 (days 4 to 5) the blastocyst becomes free. Stage 4 (days 5 to 6) is characterized by the blastocyst attaching to the wall of the uterus. The blastocysts are sectioned to show their internal structure.

4

The Second Week of Human Development

Implantation of the blastocyst is completed during the second week of embryonic development. As this crucial process occurs, changes occur in the embryoblast, producing a bilaminar embryonic disc composed of two layers, the epiblast and hypoblast (Fig. 4–1). The **embryonic disc** gives rise to the germ layers that form all the tissues and organs of the embryo. Extraembryonic structures forming during the second week include the amniotic cavity, amnion, yolk sac, connecting stalk, and chorionic sac.

Continuation of Embryonic Development

Implantation of the blastocyst commences at the end of the first embryonic week and is completed by the end of the second week. The actively erosive **syncytiotrophoblast** invades the endometrial connective tissue framework that supports the capillaries and glands. As this occurs, the blastocyst slowly embeds itself in the endometrium. The blastocyst implants in the endometrial layer at its embryonic pole (site of the embryoblast). Syncytiotrophoblastic cells from this region displace endometrial cells in the central part of the implantation site. The endometrial cells undergo *apoptosis* (programmed cell death), which facilitates the invasion of the maternal endometrium during implantation. Proteolytic enzymes produced by the syncytiotrophoblast and, probably, Fas ligand that is present in the syncytiotrophoblast, are involved in this process. The connective tissue cells around the implantation site become loaded with glycogen and lipids and assume a polyhedral appearance. Some of these cells — **decidual cells** — degenerate adjacent to the penetrating syncytiotrophoblast. The syncytiotrophoblast engulfs these degenerating cells, providing a rich source of *embryonic nutrition*. As the blastocyst implants, more trophoblast contacts the endometrium and differentiates into two layers (see Fig. 4–1*A*):

- the **cytotrophoblast**, a layer of mononucleated cells that is mitotically active and that forms new trophoblastic cells that migrate into the increasing mass of syncytiotrophoblast, where they fuse and lose their cell membranes
- the **syncytiotrophoblast**, a rapidly expanding, multinucleated mass in which no cell boundaries are discernible

The syncytiotrophoblast produces a hormone, *human chorionic gonadotropin* (hCG), which enters the maternal blood in the lacunae (L., hollow cavities) in the syncytiotrophoblast (see Fig. 4–1*C*). hCG maintains the endocrine activity of the corpus luteum in the ovary during pregnancy and forms the basis for *pregnancy tests*. Highly sensitive radioimmunoassays are available

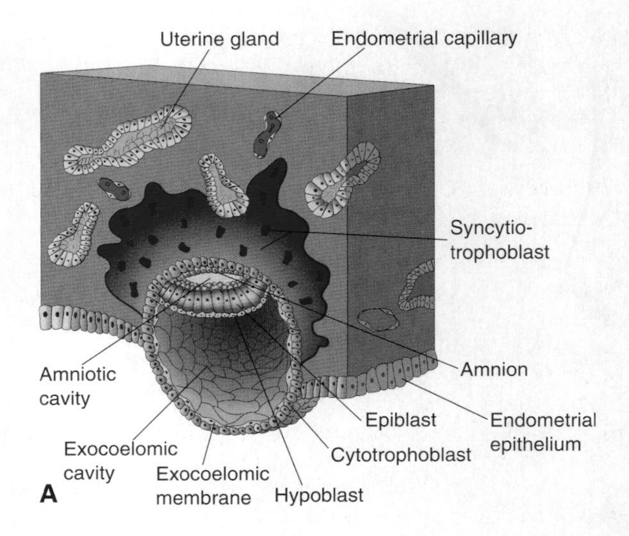

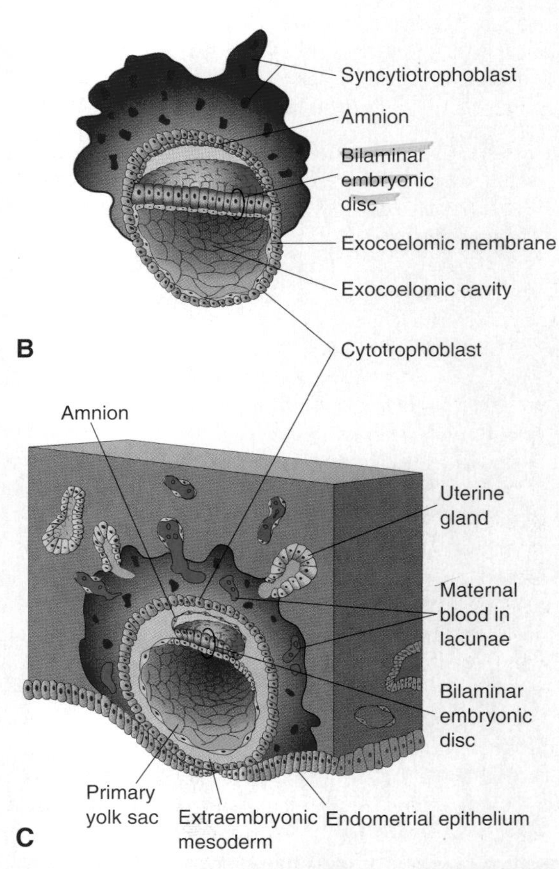

Figure 4–1. Implantation of a blastocyst. The actual size of the conceptus is about 0.1 mm. *A*, Drawing of a section of a partially implanted blastocyst (at about 8 days after fertilization). Note the slitlike amniotic cavity. *B*, An enlarged, three-dimensional sketch of a slightly older blastocyst after removal from the endometrium. Note the extensive syncytiotrophoblast. *C*, Section through a blastocyst at about 9 days. Note the lacunae appearing in the syncytiotrophoblast.

for detecting hCG. Enough hCG is produced by the syncytiotrophoblast at the end of the second week to yield a positive pregnancy test, even though the woman is probably unaware that she is pregnant.

Formation of the Amniotic Cavity, Embryonic Disc, and Yolk Sac

As implantation of the blastocyst progresses, a small cavity appears in the embryoblast, which is the primordium of the **amniotic cavity** (see Fig. 4 - 1*A*). Soon, amniogenic (amnion-forming) cells called *amnioblasts* separate from the epiblast and organize to form a thin membrane, the **amnion**, which encloses the amniotic cavity (see Fig. 4 - 1,*B* and *C*). Concurrently, changes occurring in the embryoblast result in the formation of a flattened, almost circular, bilaminar plate of cells — the **embryonic disc** — consisting of two layers (Fig. 4 - 2*B*):

- the **epiblast**, the thicker layer, consisting of high columnar cells related to the amniotic cavity
- the **hypoblast**, consisting of small cuboidal cells adjacent to the exocoelomic cavity

The epiblast forms the floor of the amniotic cavity and is continuous peripherally with the amnion. The hypoblast forms the roof of the **exocoelomic cavity** and is continuous with the thin wall of this cavity (see Fig. 4 - 1*B*). The cells that migrated from the hypoblast to form the **exocoelomic membrane** surround the blastocystic cavity and line the internal surface of the cytotrophoblast. The exocoelomic membrane and cavity soon become modified to form the **primary yolk sac**. The embryonic disc then lies between the amniotic cavity and the primary yolk sac (see Fig. 4 - 1*C*). Cells from the yolk sac endoderm form a layer of loosely arranged connective tissue, the **extraembryonic mesoderm**, which surrounds the amnion and yolk sac. Later, extraembryonic mesoderm is formed by cells that arise from the primitive streak (see Chapter 5).

As the amnion, embryonic disc, and primary yolk sac form, isolated cavities called **lacunae** appear in the syncytiotrophoblast (see Figs. 4 - 1*C* and 4 - 2). The lacunae soon become filled with a mixture of maternal blood from ruptured endometrial capillaries and cellular debris from eroded uterine glands. The maternal blood also contains hCG (produced by the syncytiotrophoblast), which maintains the *corpus luteum*, an endocrine glandular structure that secretes estrogen and progesterone for maintenance of the pregnancy (see Chapter 3). The fluid in the lacunar spaces, sometimes called *embryotroph* (*Gr. trophe*, nourishment), passes to the embryonic disc by diffusion. The communication of the eroded uterine vessels with the lacunae represents *the beginning of uteroplacental circulation*. When maternal blood flows into the lacunae, oxygen and nutritive substances become available to the extraembryonic tissues over the large surface of the syncytiotrophoblast. Because both arterial and venous branches of maternal blood vessels communicate with the lacunae, a primordial circulation of blood is established. *Oxygenated blood* passes into the lacunae from the *spiral endometrial arteries* in the endometrium;

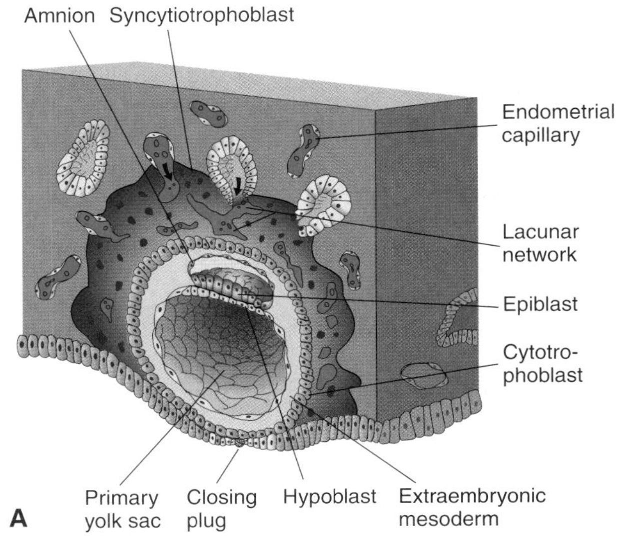

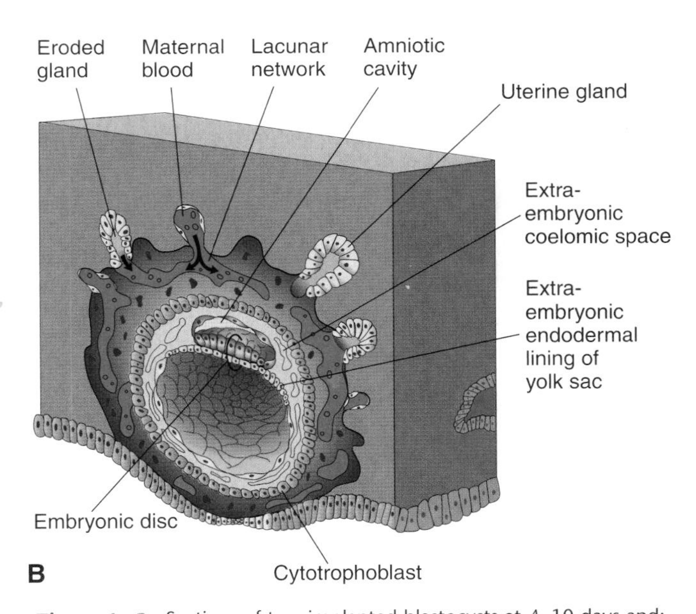

Figure 4 - 2. Sections of two implanted blastocysts at *A*, 10 days and; *B*, 12 days. These stages of development are characterized by the formation of lacunar networks — the communication of the blood-filled lacunae. In *B*, note that spaces have appeared in the extraembryonic mesoderm, indicating the beginning of the formation of the extraembryonic coelom.

deoxygenated blood is removed from them through endometrial veins (see Chapter 2).

The 10-day human conceptus (embryo and extraembryonic membranes) is completely embedded in the endometrium (see Fig. 4-2*A*). For about 2 days, there is a defect in the endometrial epithelium that is filled by a **closing plug**, a fibrinous coagulum of blood. By day 12, an almost completely regenerated uterine epithelium covers the closing plug (see Fig. 4-2*B*). As the conceptus implants, the endometrial connective tissue cells undergo a transformation known as the **decidual reaction**. After the cells swell because of the accumulation of glycogen and lipid in their cytoplasm, they are known as **decidual cells**. The primary function of the decidual reaction is to provide an immunologically privileged site for the conceptus.

In a 12-day embryo, adjacent syncytiotrophoblastic lacunae have fused to form **lacunar networks** (see Fig. 4-2*B*), giving the syncytiotrophoblast a spongelike appearance. The lacunar networks, which are particularly prominent around the embryonic pole, are the *primordia of the intervillous space of the placenta* (see Chapter 8). The endometrial capillaries around the implanted embryo first become congested and dilated to form *sinusoids*, which are thin-walled terminal vessels that are larger than ordinary capillaries. The syncytiotrophoblast then erodes the sinusoids, and maternal blood flows into the lacunar networks. Maternal blood flows in and out of the networks, establishing the *uteroplacental circulation*. The degenerated endometrial stromal cells and glands, together with the maternal blood, provide a rich source of material for embryonic nutrition. Close examination of Figures 4-1 and 4-2 shows that growth of the bilaminar embryonic disc (embryo) is slow compared to growth of the trophoblast.

As changes occur in the trophoblast and endometrium, the extraembryonic mesoderm increases and isolated extraembryonic coelomic spaces appear within it (see Fig. 4-2). These spaces rapidly fuse to form a large, isolated cavity, the **extraembryonic coelom** (Fig. 4-3A). This fluid-filled cavity surrounds the amnion and yolk sac except where they are attached to the chorion by the **connecting stalk.** As the extraembryonic coelom forms, the primary yolk sac decreases in size and a smaller **secondary yolk sac** forms (see Fig. 4-3*B*). This smaller yolk sac is formed by extraembryonic endodermal cells that migrate inside the primary yolk sac from the hypoblast of the embryonic disc (Fig. 4-4). During formation of the secondary yolk sac, a large part of the primary yolk sac is pinched off. The yolk sac contains fluid but no yolk. It may have a role in the selective transfer of nutritive materials to the embryonic disc. The trophoblast absorbs nutritive fluid from the lacunar networks in the syncytiotrophoblast; the fluid is then transferred to the embryo.

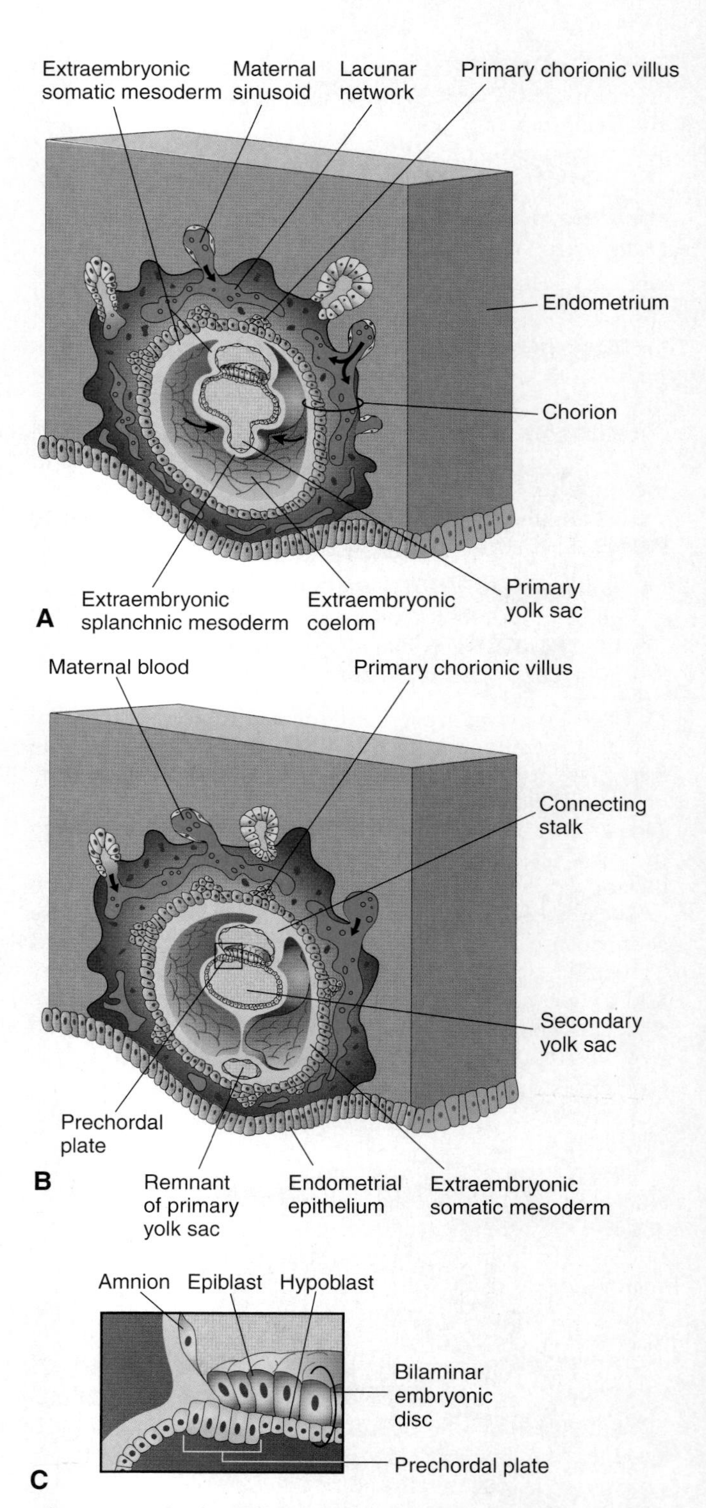

Figure 4-3. Sections of implanted human embryos. *A,* Thirteen days. Note the decrease in relative size of the primary yolk sac and the early appearance of primary chorionic villi. *B,* Fourteen days. Note the newly formed secondary yolk sac and the location of the prechordal plate. *C,* Detailed view of the prechordal plate area outlined in *B.*

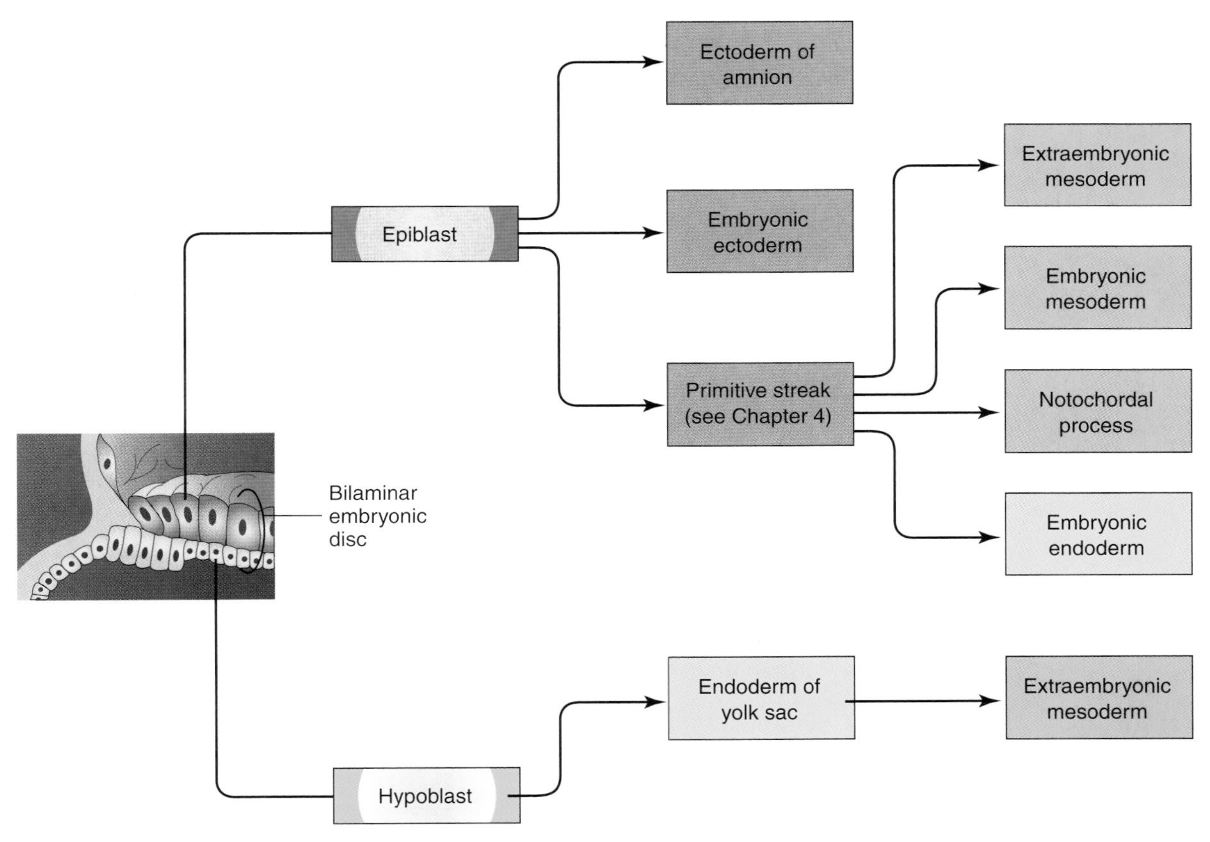

Figure 4 - 4. Origin of embryonic tissues. The colors in the boxes correspond to those used in drawings of sections of conceptuses.

Development of the Chorionic Sac

The end of the second week is characterized by the appearance of **primary chorionic villi** (see Figs. 4 - 3 and 4 - 5). Proliferation of cytotrophoblastic cells produces cellular extensions that grow into the overlying syncytiotrophoblast. The growth of the cytotrophoblastic extensions is thought to be induced by the underlying **extraembryonic somatic mesoderm**. The cellular projections form primary chorionic villi, the first stage in the development of the chorionic villi of the placenta. The extraembryonic coelom splits the extraembryonic mesoderm into two layers (see Fig. 4 - 3, *A* and *B*):

- the *extraembryonic somatic mesoderm*, which lines the trophoblast and covers the amnion
- the *extraembryonic splanchnic mesoderm*, which surrounds the yolk sac

The extraembryonic somatic mesoderm and the two layers of trophoblast form the **chorion**. *The chorion forms the wall of the chorionic sac* (gestational sac), within which the embryo and its amniotic and yolk sacs are suspended by the connecting stalk. The extraembryonic coelom is now called the **chorionic cavity**. The amniotic sac (with the embryonic epiblast forming its "floor") and the yolk sac (with the embryonic hypoblast forming its "roof") are analogous to two balloons pressed together (at the site of the embryonic disc) and suspended by a cord (connecting stalk) from the inside of a larger balloon (chorionic sac).

Transvaginal ultrasonography (endovaginal sonography) is used to measure chorionic (gestational) sac diameter. This measurement is valuable for evaluating early embryonic development and pregnancy outcome. The **14-day embryo** still has the form of a flat, bilaminar, embryonic disc, but endodermal cells in a localized area are now columnar and form a thickened circular area called the **prechordal plate** (see Fig. 4 - 3, *B* and *C*). This plate indicates the future site of the mouth and is an *important organizer of the head region.*

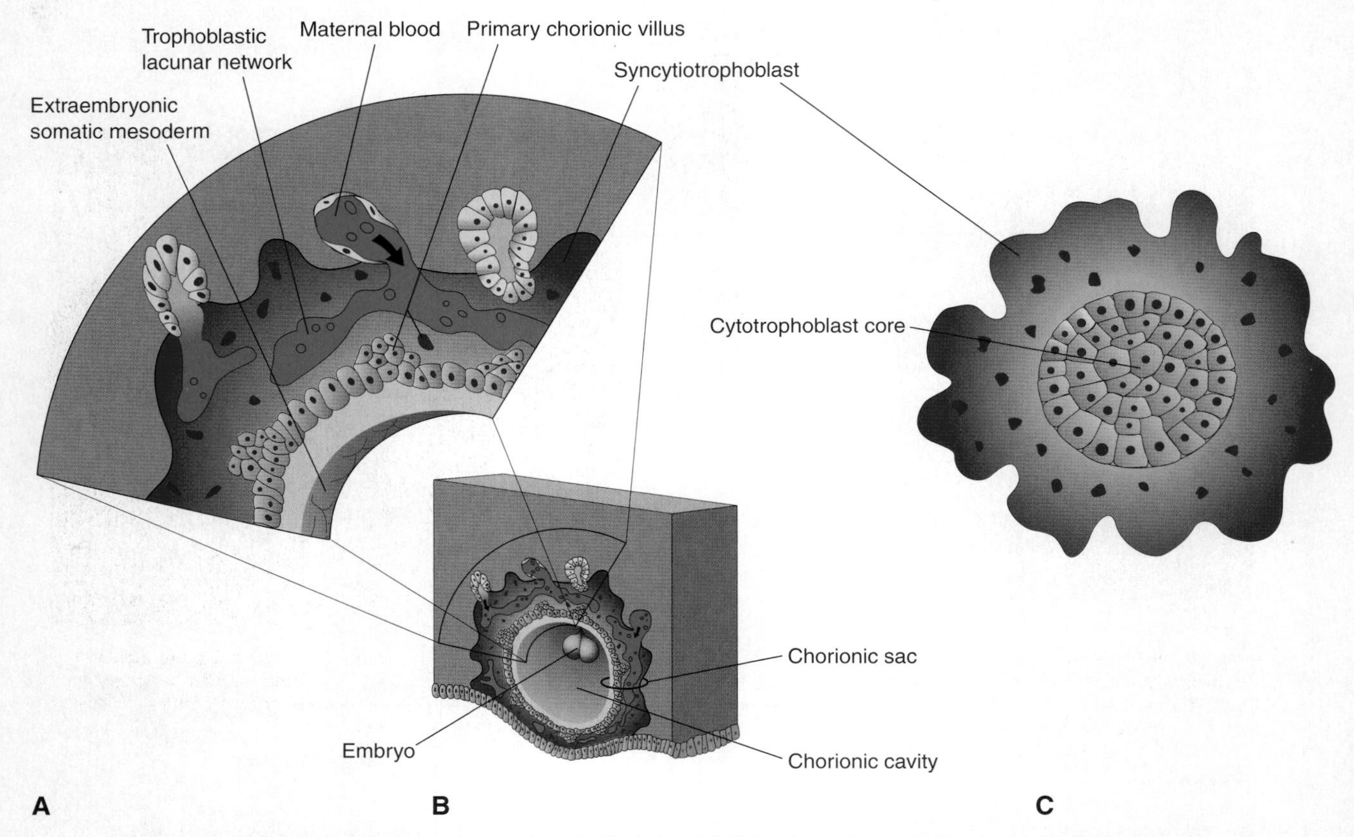

Figure 4–5. *A,* Detailed view of a section (outlined in *B*) of the wall of the chorionic sac. *B,* Sketch of a 14-day conceptus illustrating the chorionic sac and the primary chorionic villi (×6). *C,* Transverse section through a primary chorionic villus (×400).

Implantation Sites of the Blastocyst

Implantation of the blastocyst begins at the end of the first embryonic week and most often occurs in the endometrium of the uterus, usually superiorly in the body of the uterus and slightly more often on the posterior than on the anterior wall. Implantation of a blastocyst can be detected by ultrasonography and highly sensitive hCG radioimmunoassays during the second week.

Placenta Previa

Implantation of a blastocyst near the internal os (opening) of the uterus results in *placenta previa,* a placenta that partially or completely covers the os. Placenta previa may cause bleeding because of premature separation of the placenta during pregnancy.

Extrauterine Implantation Sites

Blastocysts may implant outside the uterus. Extrauterine implantations result in **ectopic pregnancies**; 95 to 97% of ectopic implantations occur in the uterine tube. *Most ectopic pregnancies are in the ampulla and isthmus of the uterine tube* (Fig. 4–6). Ectopic tubal pregnancy occurs in about 1 in 200 pregnancies in North America. A woman with a **tubal pregnancy** has the usual signs and symptoms of pregnancy (e.g., a missed menstrual period), but she may also experience abdominal pain and tenderness because of distention of the uterine tube, abnormal bleeding, and irritation of the pelvic peritoneum. The pain may be confused with *appendicitis* if the pregnancy is in the right uterine tube.

There are *several causes of tubal pregnancy;* however, they are often related to factors that delay or prevent transport of the cleaving zygote to the uterus (e.g., blockage of the uterine tube caused by scarring secondary to infection in the abdominopelvic cavity). Ectopic tubal pregnancies usually result in rupture of the uterine tube and hemorrhage into the peritoneal cavity during the first 8 weeks, followed by

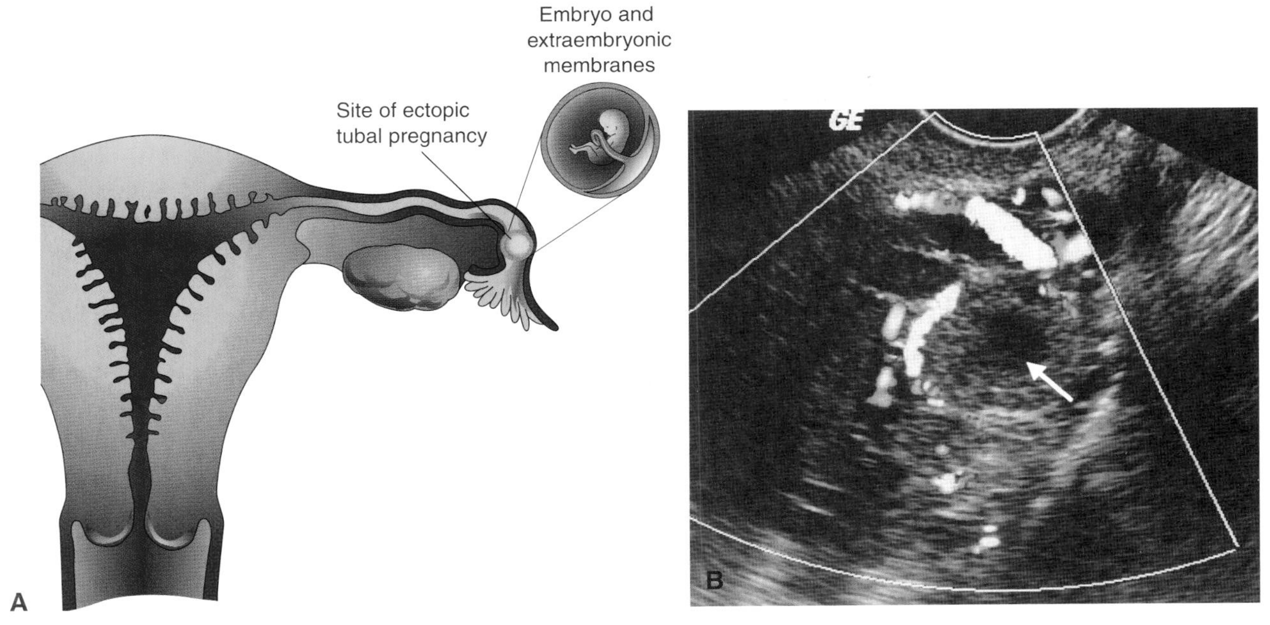

Figure 4-6. *A,* Coronal section of the uterus and uterine tube illustrating an ectopic tubal pregnancy. *B,* Ectopic tubal pregnancy (6 weeks). The sonogram of the uterine tube *(left)* shows a small gestational sac *(arrow)* with prominent vascularity in its periphery. This is characteristic of an ectopic pregnancy. (Courtesy of Dr. E. A. Lyons, Department of Radiology, Health Sciences Centre, University of Manitoba, Winnipeg, Canada.)

death of the embryo. Tubal rupture and hemorrhage constitute a threat to the mother's life. In exceptional cases, an **abdominal pregnancy** may continue to full term and the fetus may be delivered alive through an abdominal incision. Usually, however, an abdominal pregnancy creates a serious condition because the placenta attaches to abdominal organs, which causes considerable intraperitoneal bleeding.

Spontaneous Abortion of Embryos

Abortion is commonly defined as the termination of pregnancy before 20 weeks' gestation, before the period of viability of the embryo or fetus. Most abortions of embryos during the first 3 weeks occur spontaneously. *Sporadic* and *recurrent spontaneous abortions* are two of the most common gynecologic problems. The frequency of early abortions is difficult to establish because early abortions often occur before women are aware that they are pregnant. An abortion occurring just after the first missed period is very likely to be mistaken for delayed menstruation. Bleeding associated with an early abortion is *implantation bleeding* — not *menstrual bleeding* (menses). More than 50% of all known spontaneous abortions result from chromosomal abnormalities. The increased incidence of early abortions in older women probably results from the increasing frequency of *nondisjunction of chromosomes* during oogenesis (see Chapter 2).

Summary of Implantation of the Blastocyst

Implantation begins at the end of the first week and is completed by the end of the second week. Implantation may be summarized as follows:

- The zona pellucida surrounding the oocyte degenerates (day 5). Its disappearance results from enlargement of the blastocyst and degeneration caused by enzymatic lysis.
- The blastocyst adheres to the endometrial epithelium (day 6).
- The trophoblast begins to differentiate into two layers: the syncytiotrophoblast and the cytotrophoblast (day 7).
- The syncytiotrophoblast erodes endometrial tissues and the blastocyst starts to embed in the endometrium (day 8).
- Blood-filled lacunae appear in the syncytiotrophoblast (day 9).
- The blastocyst sinks beneath the endometrial epithelium, and the defect in the endometrial epithelium is filled by a closing plug (day 10).
- Lacunar networks form by fusion of adjacent lacunae (days 10 and 11).

- The syncytiotrophoblast erodes endometrial blood vessels, allowing maternal blood to seep in and out of lacunar networks, thereby establishing a *primordial uteroplacental circulation* (days 11 and 12).
- The defect in the endometrial epithelium gradually disappears as the endometrial epithelium is repaired (days 12 and 13).
- Primary chorionic villi develop (days 13 and 14).

Inhibition of Implantation

The administration of relatively large doses of estrogen ("morning-after pills") for several days, beginning shortly after unprotected sexual intercourse, usually does not prevent fertilization, but often prevents implantation. Normally, the endometrium progresses to the luteal (secretory) phase of the menstrual cycle as the zygote forms, undergoes cleavage, and enters the uterus. The large amount of estrogen, however, disturbs the normal balance between estrogen and progesterone that is necessary to prepare the endometrium for implantation. The "abortion pill" *RU486* destroys the conceptus by interrupting implantation through interference with the hormonal environment of the implanting embryo. An *intrauterine device* (IUD), which is inserted into the uterus through the vagina and cervix, usually interferes with implantation by causing a local inflammatory reaction. Some IUDs contain slow-release progesterone, which interferes with the development of the endometrium so that implantation does not usually occur.

Summary of the Second Week of Development

Rapid proliferation and differentiation of the trophoblast are important features of the second week. The various endometrial changes resulting from the adaptation of these tissues to implantation are known as the **decidual reaction**. Concurrently, the *primary yolk sac* forms, and extraembryonic mesoderm arises from the endoderm of the yolk sac. The **extraembryonic coelom** forms from cavities that develop in the *extraembryonic mesoderm*. The extraembryonic coelom later becomes the **chorionic cavity**. The primary yolk sac becomes smaller and gradually disappears as the secondary yolk sac develops. As these changes occur,

- The **amniotic cavity** appears as a space between the cytotrophoblast and the embryoblast.
- The embryoblast differentiates into a **bilaminar embryonic disc** consisting of the *epiblast*, related to the amniotic cavity, and the *hypoblast*, which is adjacent to the blastocystic cavity.
- The **prechordal plate** develops as a localized thickening of the hypoblast; this plate indicates the future site of the mouth and is also an important organizer of the head region.

Clinically Oriented Questions

1. What is meant by the term *implantation bleeding*? Is this the same as menses (menstrual fluid)?
2. Can a drug taken during the first 2 weeks of pregnancy cause abortion of the embryo?
3. Can an ectopic pregnancy occur in a woman who has an IUD?
4. Can a blastocyst that implants in the abdomen develop into a living, full-term fetus?
5. Can combined intrauterine and ectopic pregnancy occur?

The answers to these questions are at the back of the book.

5

The Third Week of Human Development

Rapid development of the embryo from the embryonic disc during the early part of the third week is characterized by the

- appearance of the primitive streak
- development of the notochord
- differentiation of the three germ layers from which all embryonic tissues and organs develop

The third week of embryonic development occurs during the week following the first missed menstrual period, that is, 5 weeks after the onset of the last normal menstrual period (LNMP). *Cessation of menstruation is usually the first indication that a woman may be pregnant.* However, missing a menstrual period is not always a certain sign of pregnancy; for example, delay of menstruation may result from illness.

Pregnancy Tests

Relatively simple tests are now available for detecting pregnancy. Most tests depend on the presence of an *early pregnancy factor* (EPF) in the maternal serum and *human chorionic gonadotropin* (hCG), a hormone produced by the syncytiotrophoblast and excreted in the mother's urine. EPF can be detected 24 to 48 hours after fertilization, and the production of hCG is sufficient to give a positive indication of pregnancy early in the second week of development. About 3 weeks after conception (Fig. 5 - 1) a normal pregnancy can be detected with ultrasonography.

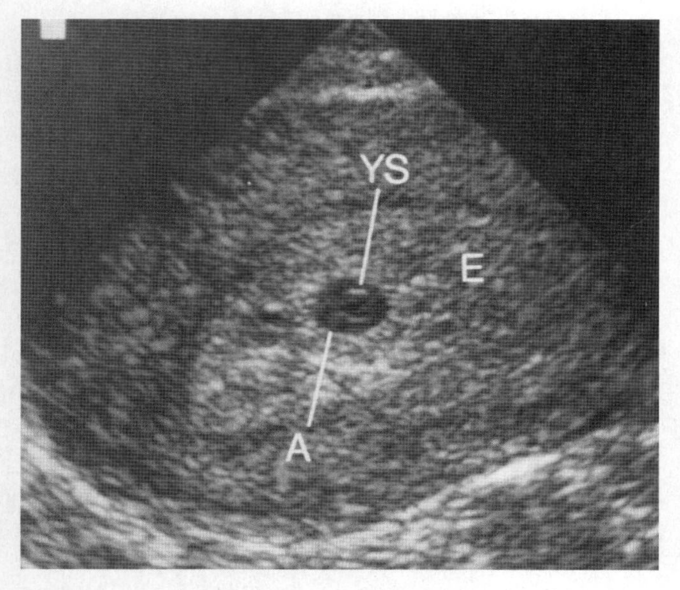

Figure 5 - 1. Endovaginal sonogram of a conceptus about 3 weeks after conception showing the amnion (A) and yolk sac (YS). As well, the endometrium (E) surrounding the conceptus is visible. (From Filly RA: Ultrasound evaluation during the first trimester of pregnancy. *In* Callen PW (ed): *Ultrasonography in Obstetrics and Gynecology*, 4th ed. Philadelphia, WB Saunders, 2000.)

Gastrulation: Formation of Germ Layers

Gastrulation is the process by which the bilaminar embryonic disc (Fig. 5 - 2*A*) is converted into a trilaminar embryonic disc. *Gastrulation is the beginning of* ***morphogenesis*** (the development of the form and structure of various organs and parts of the body) and is the significant event occurring during the third week. Gastrulation begins with formation of the **primitive streak** (see Fig. 5 - 2*B*). Each of the three germ layers (ectoderm, endoderm, and mesoderm) of the embryonic disc gives rise to specific tissues and organs.

- The *ectoderm* gives rise to the epidermis, the central and peripheral nervous systems, the retina of the eye, and various other structures (see Chapter 6).
- The *endoderm* is the source of the epithelial linings of the respiratory passages and gastrointestinal (GI) tract, including the glands opening into the GI tract and the glandular cells of associated organs, such as the liver and pancreas.
- The *mesoderm* gives rise to smooth muscular coats, connective tissues, and vessels associated with the tissues and organs. The mesoderm also forms most of the cardiovascular system and is the source of blood cells and bone marrow, the skeleton, striated muscles, and the reproductive and excretory organs.

Formation of the primitive streak and notochord are important processes occurring during gastrulation.

Primitive Streak

At the beginning of the third week, a thickened, linear band of epiblast, the **primitive streak**, appears caudally in the median plane of the dorsal aspect of the embryonic disc (see Figs. 5 - 2 and 5 - 3). The primitive streak results from the proliferation and migration of cells of the epiblast to the median plane of the embryonic disc. The precursor cells of the epiblast that give rise to the primitive streak are HNK-1 (cell surface molecule, a sulfated form of glucuronic acid) positive, and the formation of the primitive streak involves the expression of different regulatory genes. As the primitive streak elongates by addition of cells to its caudal end, its cranial end proliferates to form a **primitive node** (see Figs. 5 - 2, *B* and *C*). Concurrently, a narrow **primitive groove** develops in the primitive streak; that is continuous with a small depression in the primitive node, the **primitive pit**. As soon as the primitive streak appears, it is possible to identify the embryo's craniocaudal axis, its cranial and caudal ends, its dorsal and ventral surfaces, and its right and left sides. The primitive groove and pit result from the invagination (an inward movement) of

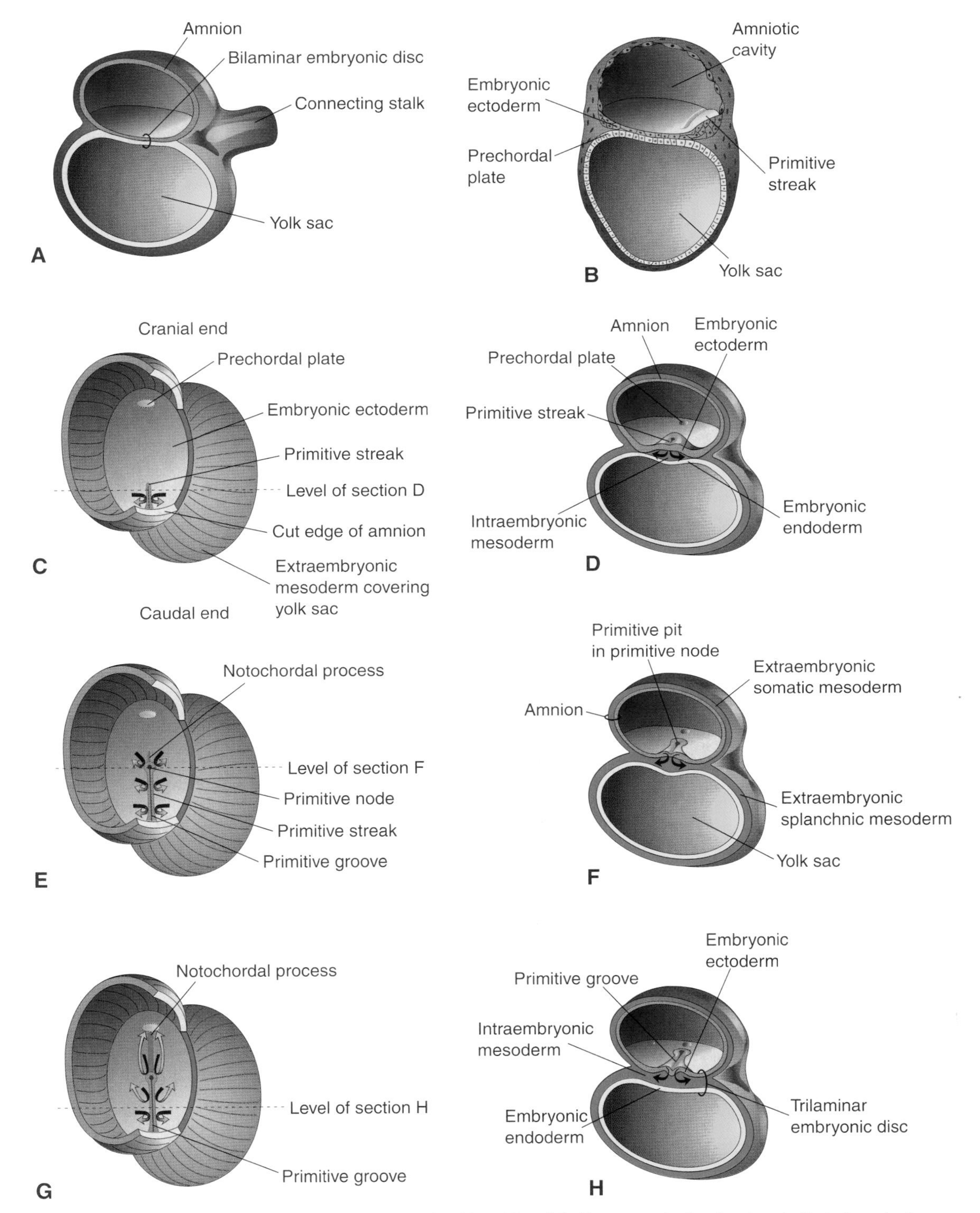

Figure 5-2. Formation of the trilaminar embryonic disc (days 15 to 16). The arrows in the drawings indicate invagination and migration of mesenchymal cells between the ectoderm and endoderm. *A, C, E,* and *G,* Dorsal views of the embryonic disc early in the third week, exposed by removal of the amnion. *B, D, F,* and *H,* Transverse sections through the embryonic disc at the levels indicated. The prechordal plate, indicating the head region in *C,* is depicted by light blue oval because it is a thickening of endoderm that cannot be seen from the dorsal surface.

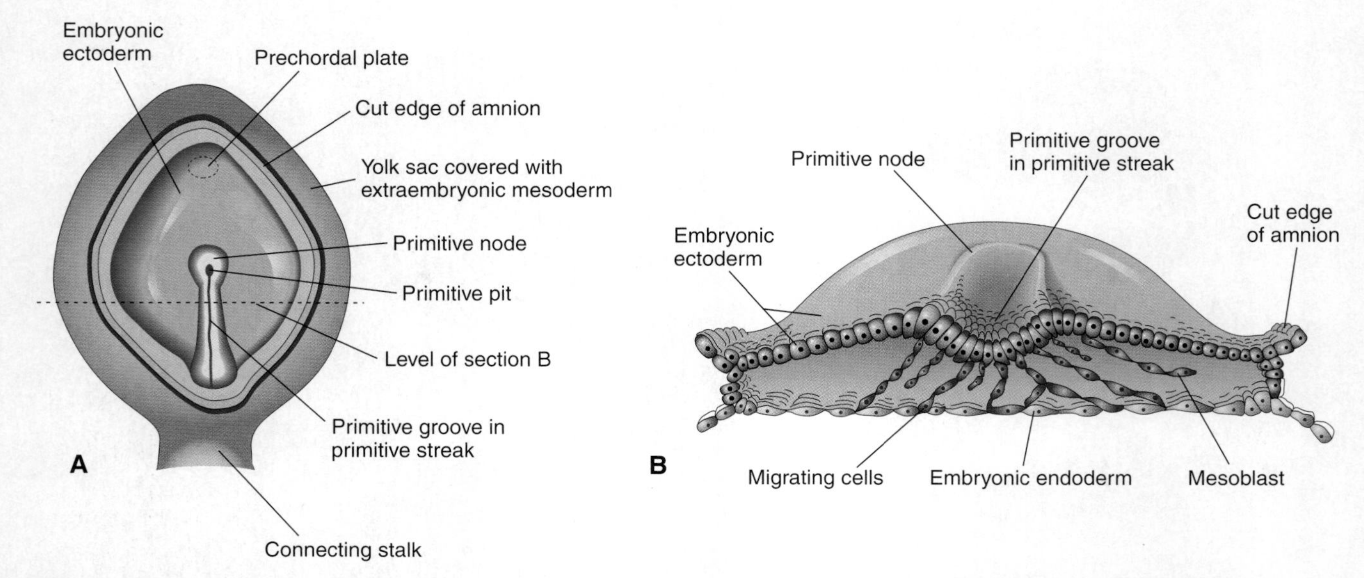

Figure 5-3. *A,* Dorsal view of a 16-day embryo. The amnion has been removed to expose the embryonic disc. *B,* Drawing of the cranial half of the embryonic disc during the third week. The disc has been cut transversely to show the migration of mesenchymal cells from the primitive streak to form mesoblast that soon organizes to form the intraembryonic mesoderm.

epiblastic cells, which is indicated by arrows in Figure 5-2*E*. Shortly after the primitive streak appears (see Fig. 5-3*A*), cells leave its deep surface and form **mesoblast,** a loose network of embryonic connective tissue (see Fig. 5-3*B*), that forms the supporting tissues of the embryo. Some mesenchyme forms a layer known as **intraembryonic**, or **embryonic mesoderm** (see Fig. 5-2*D*). Some cells of the epiblast of the primitive streak also displace the hypoblast, forming the **intraembryonic** or **embryonic endoderm** in the roof of the yolk sac. Cells remaining in the epiblast form the **intraembryonic** or **embryonic ectoderm**. Under the influence of various *embryonic growth factors*, mesenchymal cells migrate widely from the primitive streak. These cells have the potential to proliferate and differentiate into diverse types of cells, such as fibroblasts, chondroblasts, and osteoblasts. In summary, cells of the epiblast, through the process of gastrulation, give rise to all three germ layers in the embryo, which are the primordia of all its tissues and organs.

The primitive streak actively forms mesoderm until the early part of the fourth week (Fig. 5-4, *A* to *C*); thereafter, its production slows down. The streak diminishes in relative size and becomes an insignificant structure in the sacrococcygeal region of the embryo (see Fig. 5-4*D*).

Sacrococcygeal Teratoma

Remnants of the primitive streak may persist and give rise to a large tumor known as a *sacrococcygeal teratoma* (Fig. 5-5). Because they are derived from pleuripotent primitive streak cells, these tumors contain various types of tissues with elements of the three germ layers in incomplete stages of differentiation. Sacrococcygeal teratomas are the most common tumors in newborn infants and have an incidence of about 1 in 35,000 neonates. These tumors are usually surgically excised promptly, and the prognosis is good.

The Notochordal Process and Notochord

Some mesenchymal cells migrate cranially from the primitive node and pit, forming a median cellular cord, the **notochordal process** (Fig. 5-6, *A* to *C*). This process soon acquires a lumen, the **notochordal canal**. The notochordal process grows cranially between the ectoderm and endoderm until it reaches the **prechordal plate**, a small circular area of columnar endodermal cells. The rodlike notochordal process can extend no farther because the prechordal plate is firmly attached to the overlying ectoderm. These fused layers form the **oropharyngeal membrane** (Fig. 5-7*C*) located at the future site of the oral cavity (mouth).

Some mesenchymal cells from the primitive streak and notochordal process migrate laterally and cranially

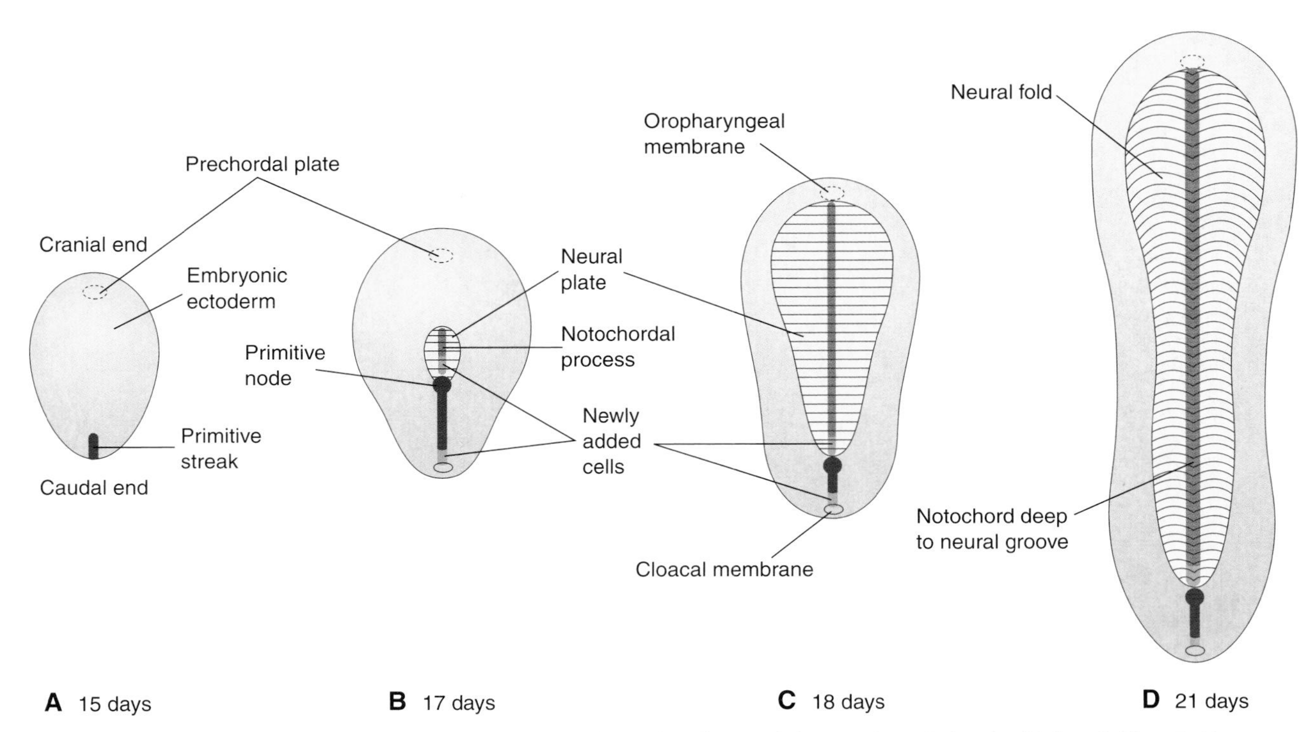

Figure 5-4. Dorsal views of the embryonic disc showing how it lengthens and changes shape during the third week. The primitive streak lengthens by addition of cells at its caudal end; the notochordal process lengthens by migration of cells from the primitive node. The notochordal process and adjacent mesoderm induce the overlying embryonic ectoderm to form the neural plate, the primordium of the central nervous system (CNS). Observe that, as the notochordal process elongates, the primitive streak shortens. At the end of the third week, the notochordal process is transformed into the notochord. Note that the embryonic disc is originally egg-shaped but soon becomes pear-shaped and then slipperlike as the notochord develops.

between the ectoderm and mesoderm until they reach the margins of the embryonic disc. There, these cells are continuous with the extraembryonic mesoderm covering the amnion and yolk sac (see Fig. 5-2*C*). This mesoderm is derived from the endoderm of the yolk sac (see Chapter 4). Some cells from the primitive streak migrate cranially on each side of the notochordal process and around the prechordal plate. They meet cranially to form the cardiogenic mesoderm in the **cardiogenic area**, where the primordium of the heart begins to develop at the end of the third week. Caudal to the primitive streak there is a circular area — the **cloacal membrane** — which indicates the future site of the anus (see Fig. 5-6*E*). The embryonic disc remains bilaminar here and at the **oropharyngeal membrane** because the embryonic ectoderm and endoderm are fused at these sites, thereby preventing migration of mesenchymal cells between them (Fig. 5-7, *A* to *C*).

The **notochord** is a cellular rod that

- defines the primordial axis of the embryo and gives it some rigidity
- serves as the basis for the development of the axial skeleton (bones of the head and vertebral column)
- indicates the future site of the vertebral bodies

The notochord develops as follows:

- The notochordal process elongates by invagination of cells from the primitive pit (see Fig. 5-4, *B* and *C*).
- The primitive pit extends into the notochordal process, forming the *notochordal canal* (see Fig. 5-6, *B* to *E*).
- The notochordal process is now a cellular tube that extends cranially from the primitive node to the prechordal plate.
- The floor of the notochordal process fuses with the underlying intraembryonic endoderm of the yolk sac.
- The fused layers gradually degenerate, resulting in the formation of openings in the floor of the notochordal process, which brings the notochordal canal into communication with the yolk sac (see Fig. 5-7*B*).

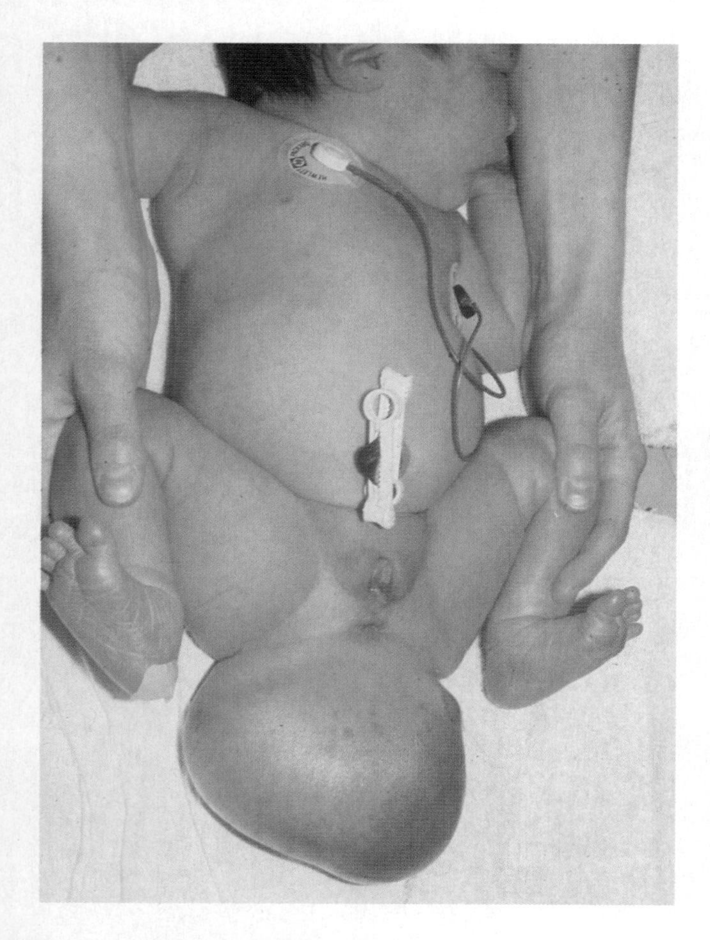

Figure 5-5. Female infant with a large sacrococcygeal teratoma that developed from remnants of the primitive streak. The tumor was surgically removed. (Courtesy of Dr. A. E. Chudley, Section of Genetics and Metabolism, Department of Pediatrics and Child Health, Children's Hospital, University of Manitoba, Winnipeg, Manitoba, Canada.)

- The openings rapidly become confluent, and the floor of the notochordal canal disappears (see Fig. 5-7*C*). The remains of the notochordal process form the flattened *notochordal plate* (see Fig. 5-*D*).
- Beginning at the cranial end of the embryo, the notochordal cells proliferate, and the notochordal plate begins infoldings to form the rod-shaped notochord (see Fig. 5-7, *F* and *G*).
- The proximal part of the notochordal canal persists temporarily as the *neurenteric canal* (see Fig. 5-7, *C* and *E*), which forms a transitory communication between the amniotic and yolk sac cavities. When development of the notochord is complete, the neurenteric canal is normally obliterated.
- The notochord becomes detached from the endoderm of the yolk sac, which again becomes a continuous layer (see Fig. 5-7*G*).

The notochord is an intricate structure around which the vertebral column forms (see Chapter 16). It extends from the oropharyngeal membrane to the primitive node. The notochord degenerates and disappears as the bodies of the vertebrae form, but it persists as the *nucleus pulposus* of each intervertebral disc. *The notochord functions as the primary inductor in the early embryo.* The developing notochord induces the overlying embryonic ectoderm to thicken and form the **neural plate** (see Fig. 5-7*C*), the primordium of the central nervous system (CNS).

Allantois

The **allantois** (Gr., *allas*, sausage) appears on about day 16 as a small, sausage-shaped diverticulum (outpouching) that extends from the caudal wall of the yolk sac into the connecting stalk (see Fig. 5-6, *B*, *C*, and *E*). The allantois is a large, saclike structure in embryos of reptiles, birds, and some mammals that has a respiratory function and/or acts as a reservoir for urine during embryonic life. The allantois remains very small in human embryos because the placenta and amniotic sac take over its functions. The allantois is involved with early blood formation in the human embryo and is associated with development of the urinary bladder (see Chapter 14). As the bladder enlarges, the allantois

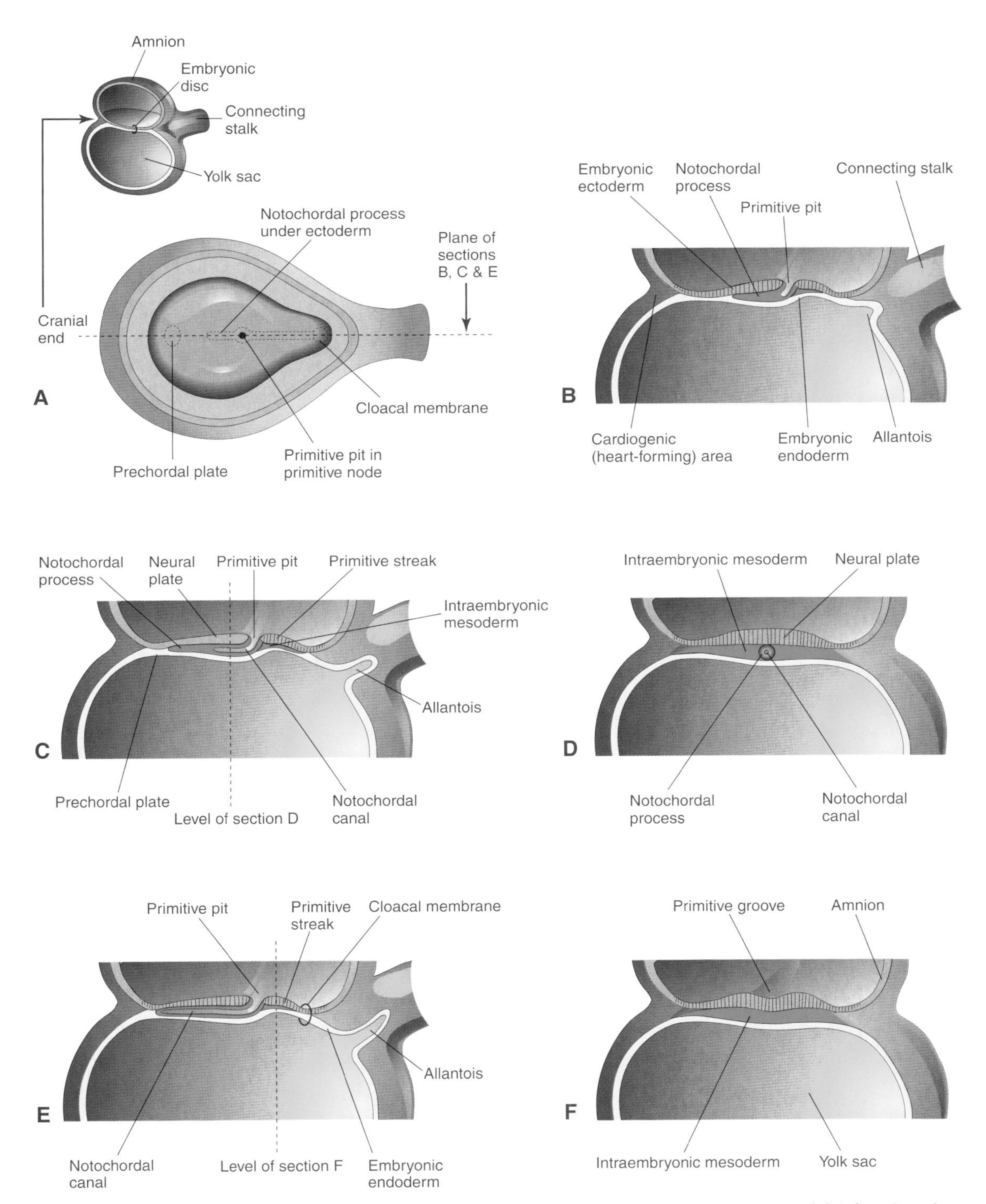

Figure 5 - 6. Drawings illustrating development of the notochordal process. The small sketch at the upper left is for orientation; the short arrow indicates the dorsal aspect of the embryonic disc. *A*, Dorsal view of the embryonic disc (at about 16 days), exposed by removal of the amnion. The notochordal process is shown as if it were visible through the embryonic ectoderm. *B*, *C*, and *E*, Median sections, at the same plane as shown in *A*, illustrating successive stages in the development of the notochordal process and canal. The stages shown in *C* and *E* occur at about 18 days. *D* and *F*, Transverse sections through the embryonic disc at the levels shown in *C* and *E*.

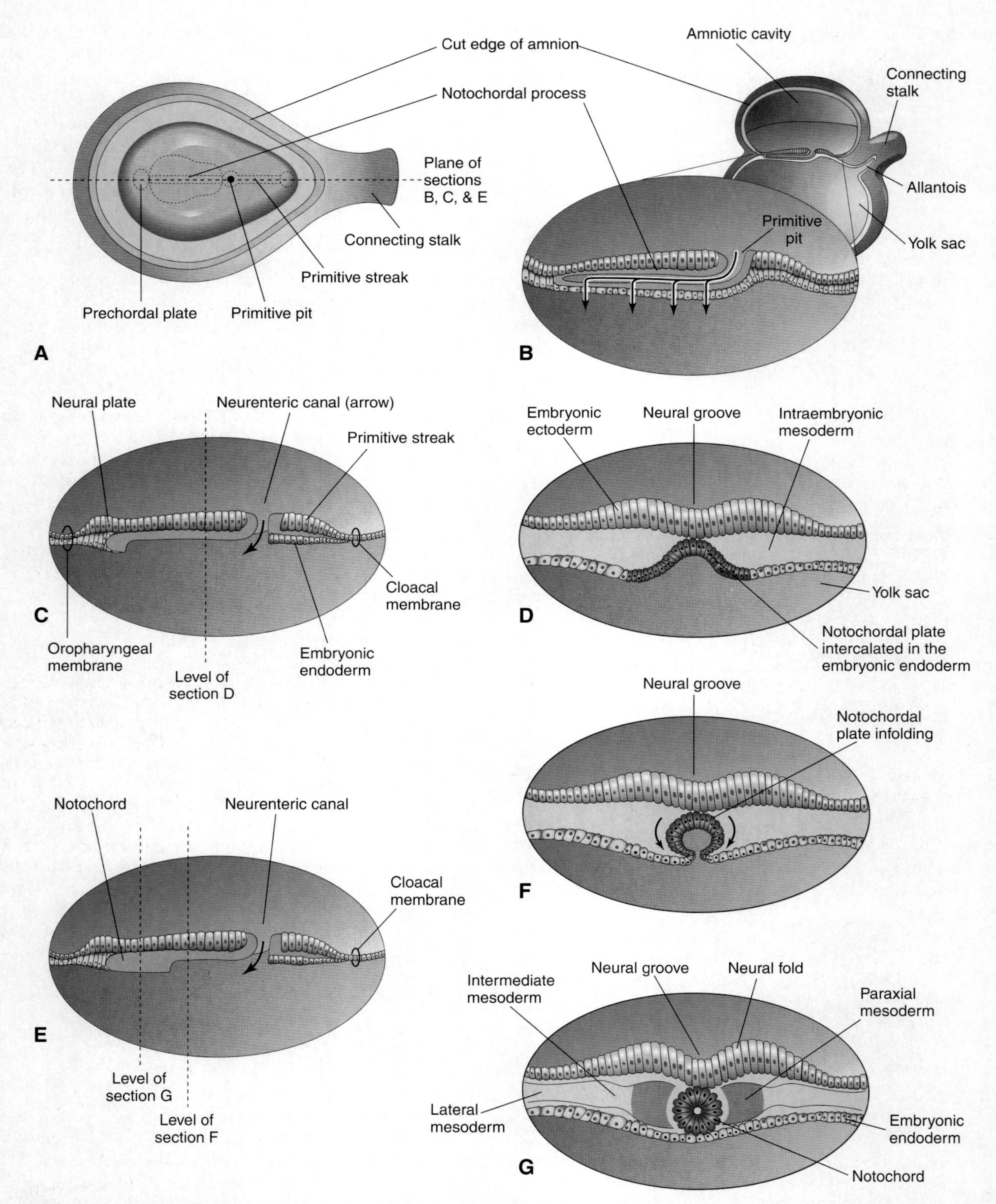

Figure 5-7. Further development of the notochord by transformation of the notochordal process. *A*, Dorsal view of the embryonic disc (at about 18 days), exposed by removing the amnion. *B*, Three-dimensional median section of the embryo. *C* and *E*, Similar sections of slightly older embryos. *D*, *F*, and *G*, Transverse sections of the trilaminar embryonic disc shown in *C* and *E*.

becomes the *urachus*, which is represented in adults by the *median umbilical ligament*. The blood vessels of the allantois become the umbilical arteries and veins (see Fig. 15 - 2).

Neurulation: Formation of Neural Tube

The processes involved in the formation of the neural plate and neural folds, and closure of these folds to form the neural tube, constitute neurulation. These processes are completed by the end of the fourth week.

Neural Plate and Neural Tube

As the notochord develops, the embryonic ectoderm over it thickens to form an elongated, slipperlike plate of thickened neuroepithelial cells — called the **neural plate**. *Neural plate formation is induced by the developing notochord*. The ectoderm of the neural plate (neuroectoderm) gives rise to the **CNS** — comprising the brain and spinal cord. Neuroectoderm also gives rise to various other structures, such as the retina. At first, the elongated neural plate corresponds precisely in length to the underlying notochord. It appears cranial to the primitive node and dorsal to the notochord and the mesoderm adjacent to it (see Fig. 5 - 4*B*). As the notochord elongates, the neural plate broadens and eventually extends cranially as far as the oropharyngeal membrane (see Figs. 5 - 4*C* and 5 - 7*C*). Eventually, the neural plate extends beyond the notochord. On about day 18, the neural plate invaginates along its central axis to form a median longitudinal **neural groove**, which has neural folds on each side (see Fig. 5 - 7*G*). The **neural folds** are particularly prominent at the cranial end of the embryo and are *the first signs of brain development*. By the end of the third week, the neural folds have begun to move together and fuse, converting the neural plate into a **neural tube** (Figs. 5 - 8 and 5 - 9).

Neural tube formation (neural induction) is a complex cellular and multifactorial process involving genes and extrinsic factors. The neural tube soon separates from the surface ectoderm. The free edges of the ectoderm fuse so that this layer becomes continuous over the neural tube and back of the embryo (see Fig. 5 - 9*E*). Subsequently, the surface ectoderm differentiates into the epidermis of the skin. Neurulation is completed during the fourth week (see Chapter 6). The molecular mechanism of neural tube formation in humans is unclear.

Neural Crest Formation

As the neural folds fuse to form the neural tube, some neuroectodermal cells lying along the crest of each neural fold lose their epithelial affinities and attachments to neighboring cells (see Fig. 5 - 9). As the neural tube separates from the surface ectoderm, **neural crest cells** migrate dorsolaterally on each side of the neural tube. They form a flattened, irregular mass, the **neural crest**, between the neural tube and the overlying surface ectoderm (see Fig. 5 - 9, *D* and *E*). The neural crest soon separates into right and left parts that migrate in a wave to the dorsolateral aspects of the neural tube. Many neural crest cells migrate in various directions and disperse widely within the mesenchyme. *Neural crest cells* are neuroectodermal cells that differentiate into various cell types, including the spinal ganglia (dorsal root ganglia) and the ganglia of the autonomic nervous system. The ganglia of cranial nerves V, VII, IX, and X are also partly derived from neural crest cells. In addition to forming ganglion cells, neural crest cells form the sheaths of peripheral nerves. They also form the meningeal coverings of the brain and spinal cord (at least the pia mater and arachnoid). Neural crest cells also contribute to the formation of pigment cells, the suprarenal (adrenal) medulla, and several skeletal and muscular components in the head (see Chapter 11). Laboratory studies indicate that BMP, Wnt, Notch, and FGF are involved in the signaling systems of neural crest formation and in the migration and differentiation of neural crest cells.

Abnormal Neurulation

Because the neural plate, primordium of the CNS, appears during the third week and gives rise to the neural folds, disturbance of neurulation may result in severe abnormalities of the brain and spinal cord (see Chapter 19). **Neural tube defects** (NTDs) are among the most common congenital anomalies. *Meroanencephaly* or anencephaly — partial absence of the brain — is the most severe defect. Although the term anencephaly (Gr. *an*, without; *enkephalos*, brain) is commonly used, it is a misnomer because the brain is not completely absent. Available evidence suggests that the primary disturbance (e.g., a teratogenic drug; see Chapter 9) affects the neuroectoderm, resulting in failure of the neural folds to fuse and form the neural tube in the brain region. This results in meroanencephaly and *spina bifida cystica* (see Chapter 19).

Development of Somites

As the notochord and the neural tube form, the intraembryonic mesoderm on each side of them proliferates to form a thick, longitudinal column of **paraxial mesoderm** (see Figs. 5 - 7*G* and 5 - 8*B*). Each column is continuous laterally with the **intermediate mesoderm**, which gradually thins into a layer of lateral mesoderm. The **lateral mesoderm** is continuous with the extraembryonic mesoderm covering the yolk sac and amnion. Toward

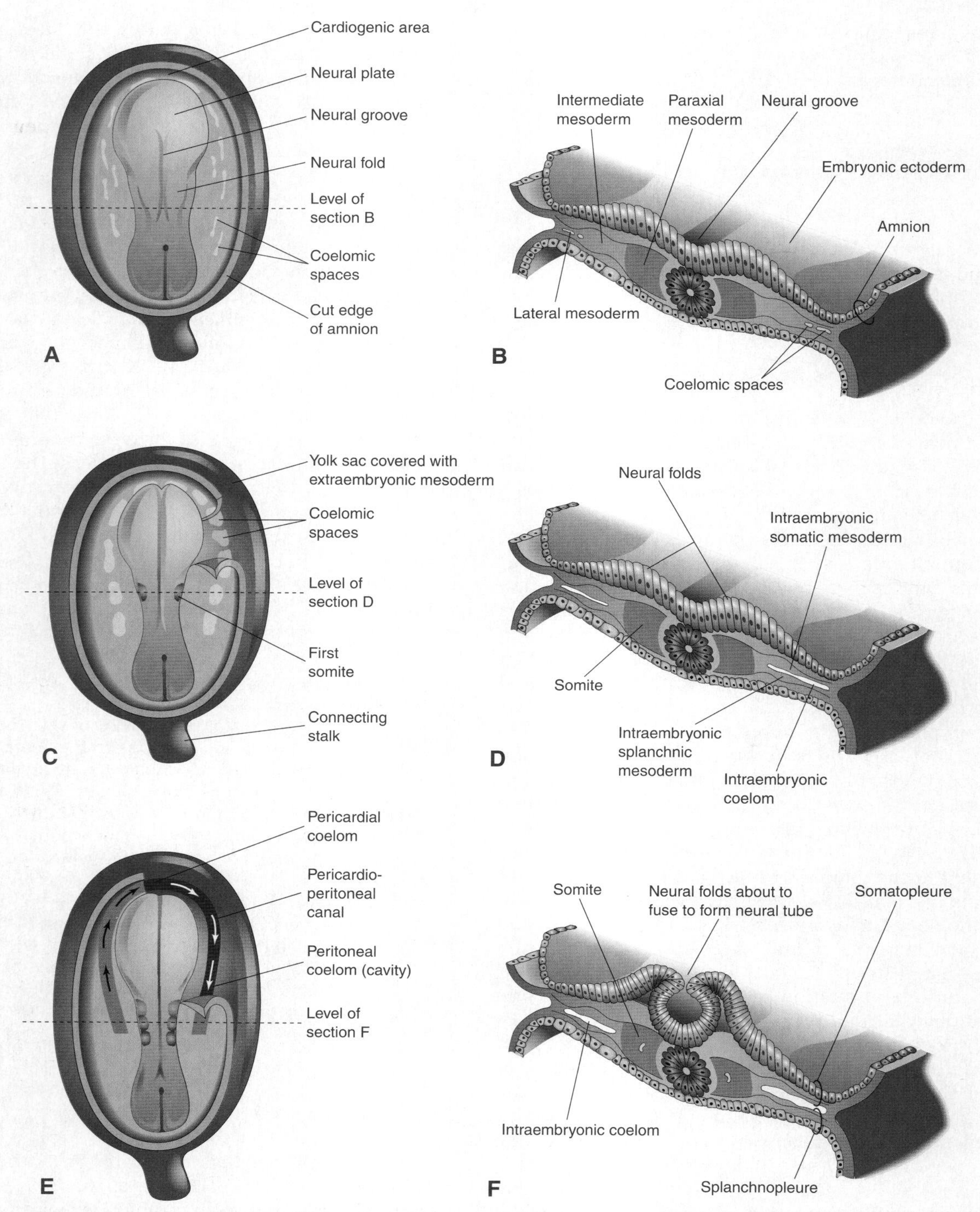

Figure 5-8. Drawings of embryos 19 to 21 days old, illustrating development of the somites and intraembryonic coelom. *A, C,* and *E,* Dorsal view of the embryo, exposed by removal of the amnion. *B, D,* and *F,* Transverse sections through the embryonic disc at the levels shown. *A,* Presomite embryo of about 18 days. *C,* An embryo of about 20 days, showing the first pair of somites. A portion of the somatopleure on the right has been removed to show the isolated coelomic spaces in the lateral mesoderm. *E, A* three-somite embryo (about 21 days old), showing the horseshoe-shaped intraembryonic coelom, exposed on the right by removal of part of the somatopleure.

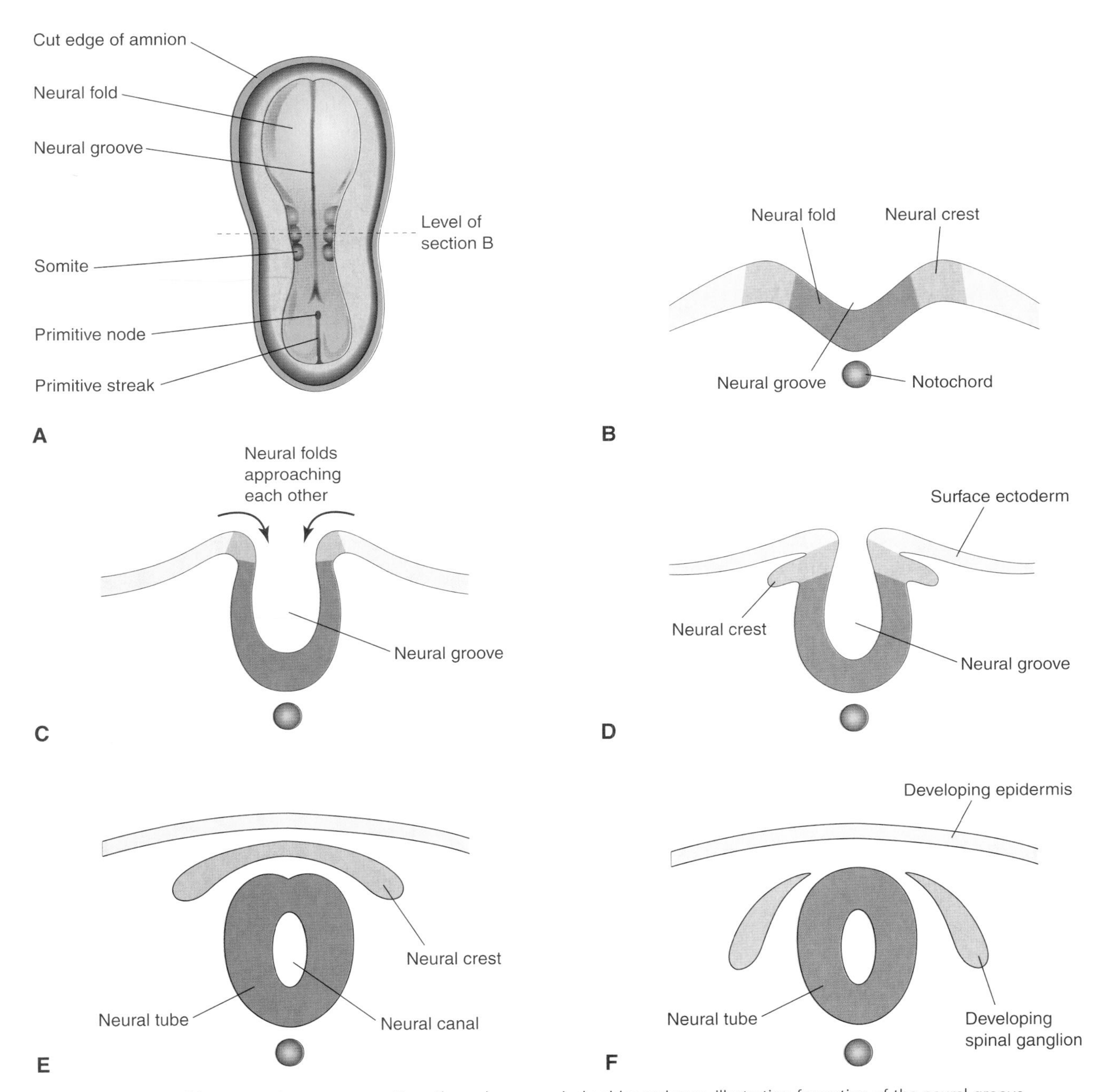

Figure 5-9. Diagrammatic transverse sections through progressively older embryos, illustrating formation of the neural groove, neural tube, and neural crest up to the end of the fourth week.

the end of the third week, the paraxial mesoderm differentiates and begins to divide into paired cuboidal bodies — called **somites** — on each side of the developing neural tube (see Fig. 5-8, *C* to *E*). The somites form distinct surface elevations on the embryo and appear somewhat triangular in transverse section (see Fig. 5-8, *C* to *F*). Because the somites are so prominent during the fourth and fifth weeks, they are used as one of the criteria for determining an embryo's age (see Chapter 6, Table 6-1). The somites first appear in the future occipital region of the embryo but soon extend craniocaudally, and give rise to most of the *axial skeleton* (bones of the head and vertebral column) and associated musculature, as well as to the adjacent dermis of the skin (see Chapters 16 and 21). The first pair of somites appears at the end of the third week (see Fig. 5-8*C*) near the cranial end of the notochord. Subsequent pairs form in a craniocaudal sequence. Experimental studies indicate that formation of somites from the paraxial mesoderm involves the expression of Notch pathway genes (Notch Signaling), Hox genes, and other signaling factors.

Development of the Intraembryonic Coelom

The intraembryonic coelom (body cavity) first appears as small, isolated, *coelomic spaces* in the lateral mesoderm and cardiogenic (heart-forming) mesoderm (see Fig. 5-8, *A* to *D*). These spaces coalesce to form a single, horseshoe-shaped cavity — the **intraembryonic coelom** (see Fig. 5-8*D*) — which divides the lateral mesoderm into two layers:

- a somatic or *parietal layer* that is continuous with the extraembryonic mesoderm covering the amnion
- a splanchnic or *visceral layer* that is continuous with the extraembryonic mesoderm covering the yolk sac

The *somatic mesoderm* and overlying embryonic ectoderm form the body wall, or **somatopleure** (see Fig. 5-8*F*), whereas the *splanchnic mesoderm* and underlying embryonic endoderm form the gut wall, or **splanchnopleure**. During the second month, the intraembryonic coelom is divided into three body cavities:

- *pericardial cavity*
- *pleural cavities*
- *peritoneal cavity*

For a description of these divisions of the intraembryonic coelom, see Chapter 10.

Early Development of the Cardiovascular System

At the beginning of the third week, *angiogenesis* (blood vessel formation) begins in the extraembryonic mesoderm of the yolk sac, connecting stalk, and chorion (Fig. 5-10). Embryonic blood vessels begin to develop about 2 days later. The early formation of the cardiovascular system correlates with the absence of a significant amount of yolk in the ovum and yolk sac and the consequent urgent need for blood vessels to bring oxygen and nourishment to the embryo from the maternal circulation through the placenta. At the end of the second week, embryonic nutrition is obtained from the maternal blood by diffusion through the extraembryonic coelom and yolk sac. During the third week a primitive uteroplacental circulation develops (Fig. 5-11).

Vasculogenesis and Angiogenesis

The formation of the embryonic vascular system involves two processes: *vasculogenesis* and *angiogenesis*. Blood vessel formation (vasculogenesis) in the embryo and extraembryonic membranes during the third week may be summarized as follows (see Fig. 5-10):

- Mesenchymal cells differentiate into endothelial cell precursors — **angioblasts** (vessel-forming cells) — that aggregate to form isolated angiogenic cell clusters — or blood **islands**.
- Small cavities appear within the blood islands by confluence of intercellular clefts.
- Angioblasts flatten to form endothelial cells that arrange themselves around the cavities in the blood islands to form the primitive endothelium.
- These endothelium-lined cavities soon fuse to form networks of endothelial channels (vasculogenesis).
- Vessels sprout into adjacent areas by endothelial budding and fuse with other vessels (angiogenesis).

Blood cells develop from the endothelial cells of vessels (*hemangioblasts*) as they develop on the yolk sac and allantois at the end of the third week (see Fig. 5-10, *E* and *F*). Blood formation (hematogenesis) does not begin in the embryo until the fifth week. It occurs first in various parts of the embryonic mesenchyme, chiefly the liver, and later in the spleen, bone marrow, and lymph nodes. Fetal and adult erythrocytes are derived from different hematopoietic progenitor cells (hemangioblasts). The mesenchymal cells surrounding the primordial endothelial blood vessels differentiate into the muscular and connective tissue elements of the vessels.

The **heart** and **great vessels** form from mesenchymal cells in the cardiogenic area (see Fig. 5-10*B*). Paired, endothelium-lined channels, termed — **endocardial heart tubes** — develop during the third week and fuse to form a **primordial heart tube**. The tubular heart joins with blood vessels in the embryo, connecting stalk, chorion, and yolk sac to form a primordial cardiovascular system (see Fig. 15-2). By the end of the third week, the blood is circulating, and the heart begins to beat on day 21 or 22 (about 5 weeks after the LNMP). The cardiovascular system is thus the first organ system to reach a functional state. The embryonic heart beat can be detected ultrasonographically using a Doppler probe during the fifth week, about 7 weeks after the LNMP (see Fig. 5-11).

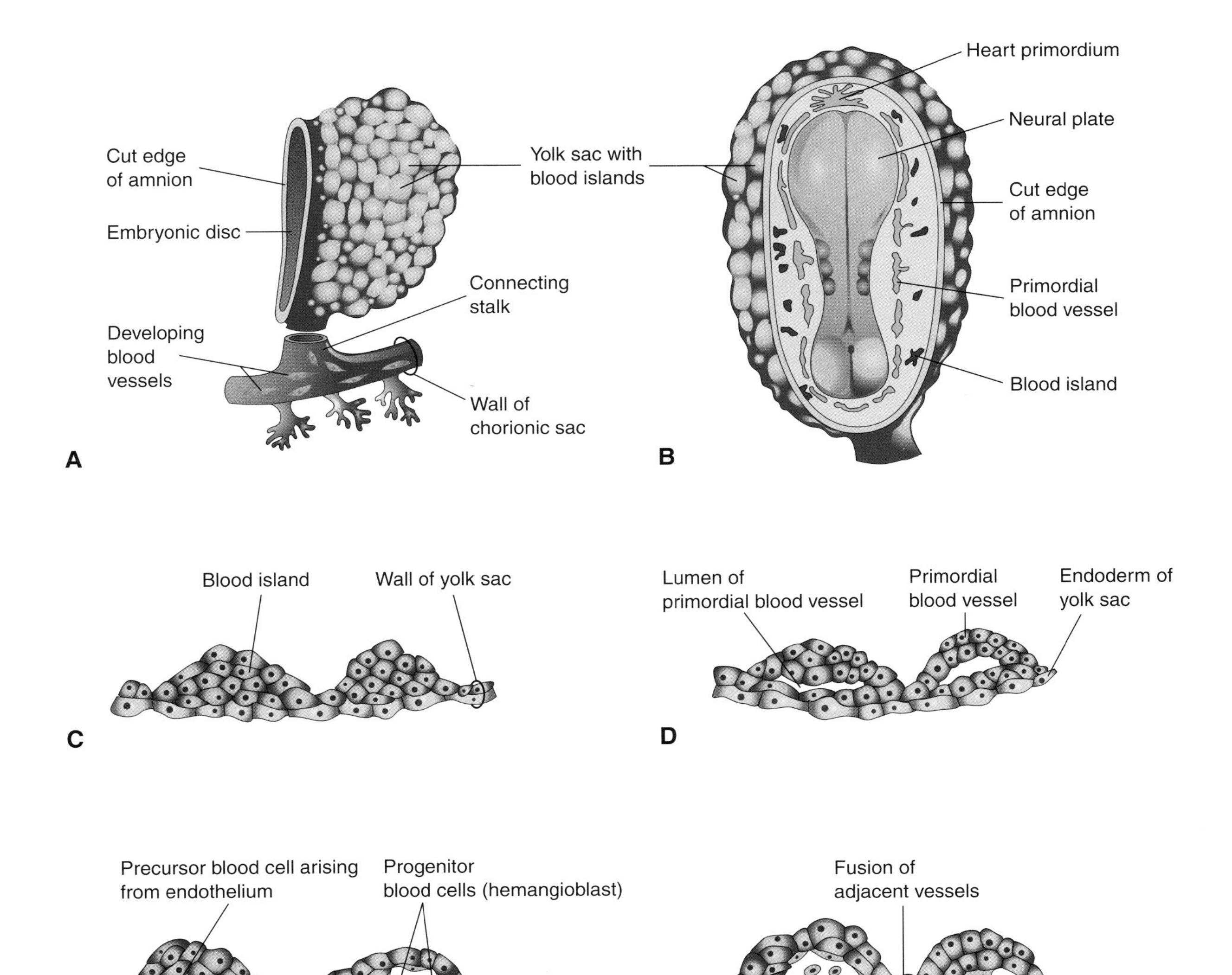

Figure 5-10. Successive stages in the development of blood and blood vessels. *A*, The yolk sac and a portion of the chorionic sac (at about 18 days). *B*, Dorsal view of the embryo exposed by removing the amnion. *C* to *F*, Sections of blood islands showing progressive stages in the development of blood and blood vessels.

Development of Chorionic Villi

Shortly after the **primary chorionic villi** appear at the end of the second week, they begin to branch. Early in the third week, mesenchyme grows into the primary villi, forming a core of loose mesenchymal (connective) tissue. The villi at this stage, known as — **secondary chorionic villi** — cover the entire surface of the chorionic sac (Fig. 5-12, *A* and *B*). Some mesenchymal cells in the villi soon differentiate into capillaries and blood cells (see Fig. 5-12, *B* and *C*). When capillaries are visible in the villi, they are called **tertiary chorionic villi** (see Figs. 15-2 and 5-12*D*). The capillaries in the chorionic villi fuse to form **arteriocapillary networks,** which soon become connected with the embryonic heart through vessels that differentiate in the mesenchyme of the chorion and connecting stalk (see

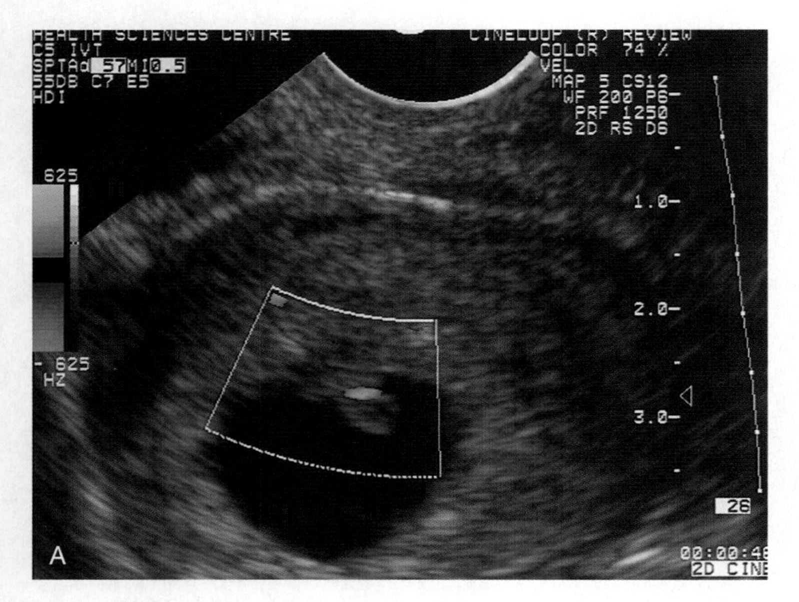

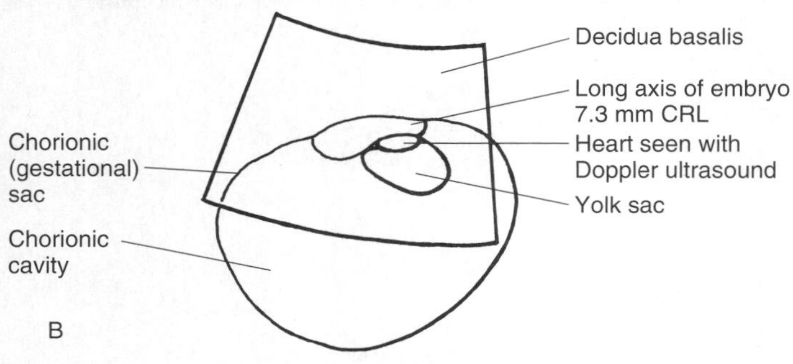

Figure 5 - 11. *A*, Sonogram of a 5-week embryo (7.2 mm) and its attached yolk sac within its chorionic (gestational) sac. The pulsating heart of the embryo was visualized using Doppler ultrasound. *B*, Sketch of the sonogram, provided for orientation and identification of structures. CRL, crown-rump length. (From Moore KL, Persaud TVN, Shiota K: *Color Atlas of Clinical Embryology*, 2nd ed. Philadelphia, WB Saunders, 2000. Courtesy of Dr. E. A. Lyons, Professor of Radiology and Obstetrics and Gynecology, Health Sciences Centre, University of Manitoba, Winnipeg, Manitoba, Canada.)

Fig. 15 - 2). By the end of the third week, embryonic blood begins to flow slowly through the capillaries in the chorionic villi. Oxygen and nutrients in the maternal blood in the intervillous space diffuse through the walls of the villi and enter the embryo's blood (see Fig. 5 - 12, *C* and *D*). Carbon dioxide and waste products diffuse from blood in the fetal capillaries through the wall of the villi into the maternal blood. Concurrently, cytotrophoblastic cells of the chorionic villi proliferate and extend through the syncytiotrophoblast to form a **cytotrophoblastic shell** (see Fig. 5 - 12*C*), which gradually surrounds the chorionic sac and attaches it to the endometrium. Villi that attach to the maternal tissues through the cytotrophoblastic shell are **stem chorionic villi** (anchoring villi). The villi that grow from the sides of the stem villi are called **branch chorionic villi** (terminal villi). It is through the walls of branch villi that the main exchange of material between the blood of the mother and the embryo takes place. The branch villi are bathed in continually changing maternal blood in the intervillous space.

Abnormal Growth of Trophoblast

Sometimes the embryo dies and the chorionic villi do not complete their development; that is, they do not become vascularized to form tertiary villi. These degenerating villi may form cystic swellings, called *hydatidiform moles*, which resemble a bunch of grapes. These moles exhibit variable degrees of trophoblastic proliferation and produce excessive amounts of hCG. In 3% to 5% of such cases, these moles develop into malignant trophoblastic lesions called — **choriocarcinomas**. Choriocarcinomas invariably metastasize (spread) by way of the blood to various sites, such as the lungs, vagina, liver, bone, intestine, and brain.

Summary of the Third Week of Development

Major changes occur in the embryo as the bilaminar embryonic disc is converted into a trilaminar embryonic disc during **gastrulation**. These changes begin with the appearance of the primitive streak.

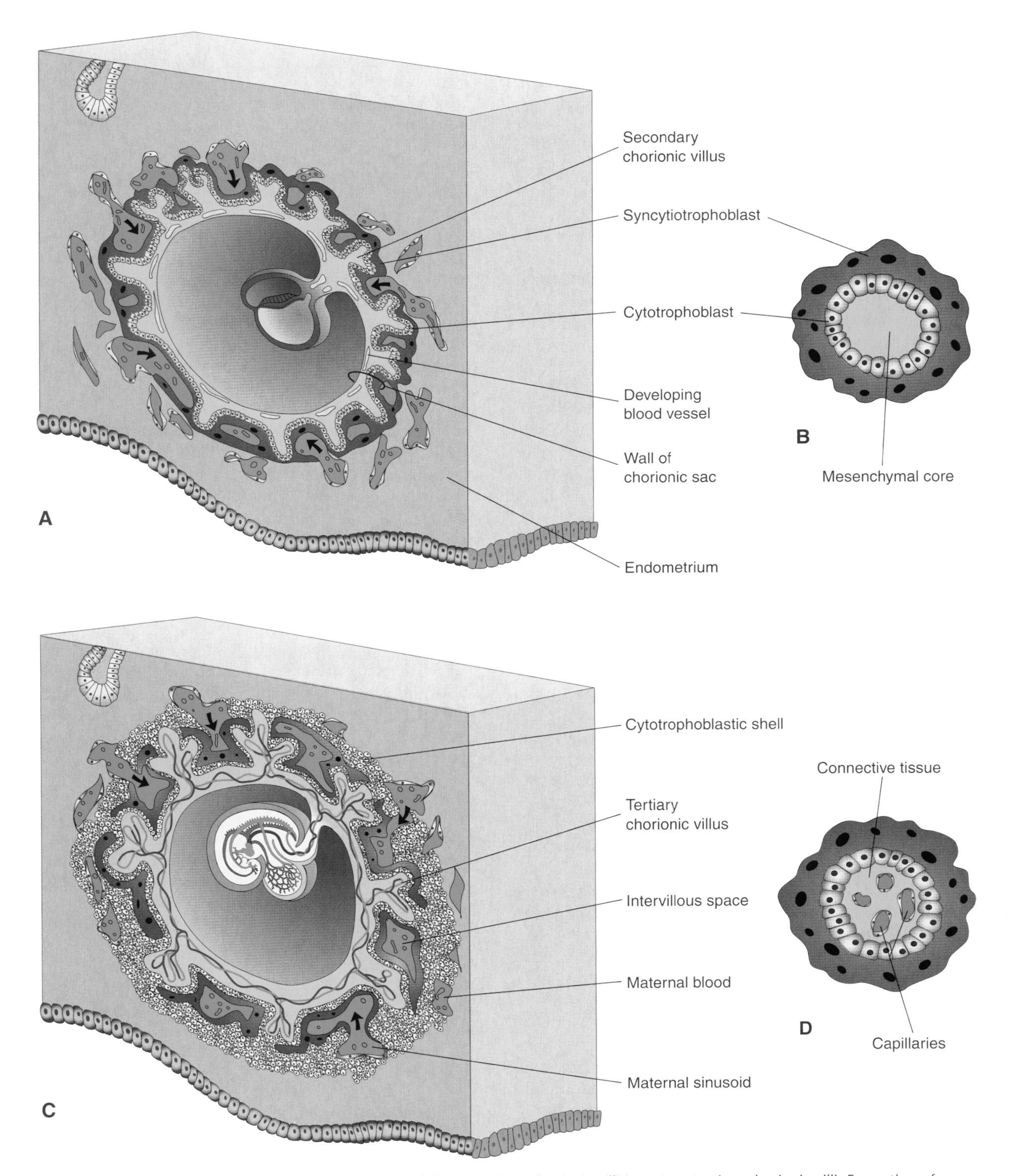

Figure 5 - 12. Diagrams illustrating development of the secondary chorionic villi into stem tertiary chorionic villi. Formation of the placenta is also shown. *A*, Sagittal section of an embryo (at about 16 days). *B*, Section of a secondary chorionic villus. *C*, Section of an implanted embryo (at about 21 days). *D*, Section of a tertiary chorionic villus.

The **primitive streak** first becomes evident at the beginning of the third week as a localized thickening of the epiblast. Invagination of epiblastic cells from the primitive streak gives rise to mesenchymal cells that migrate ventrally, laterally, and cranially between the epiblast and the hypoblast. As soon as the primitive streak begins to produce mesenchymal cells, the epiblastic layer becomes known as the embryonic ectoderm. Some cells of the epiblast displace the hypoblast and form the embryonic endoderm. Mesenchymal cells produced by the primitive streak soon organize into a third germ layer, the *intraembryonic mesoderm*. Cells from the primitive streak migrate to the edges of the embryonic disc, where they join the *extraembryonic mesoderm* covering the amnion and yolk sac. By the end of the third week, mesoderm exists between the ectoderm and the endoderm everywhere except at the oropharyngeal membrane, in the median plane occupied by the notochord, and at the cloacal membrane. Early in the third week, mesenchymal cells arising from the *primitive node* of the primitive streak form the *notochordal process*, which extends cranially from the primitive node as a rod of cells between the embryonic ectoderm and endoderm. The primitive pit extends into the notochordal process and forms a *notochordal canal*. When fully developed, the notochordal process extends from the primitive node to the prechordal plate. Openings develop in the floor of the notochordal canal that soon coalesce, leaving a *notochordal plate*. The notochordal plate soon begins infoldings to form the notochord — the primitive axis of the embryo around which the axial skeleton forms.

The **neural plate** appears as a thickening of the embryonic ectoderm, cranial to the primitive node. The neural plate is induced to form by the developing notochord. A longitudinal *neural groove* develops in the neural plate; the neural groove is flanked by *neural folds*. Fusion of the folds forms the neural tube, the primordium of the CNS. The process of neural plate formation and its infolding to form the neural tube is called *neurulation*. As the neural folds fuse to form the neural tube, neuroectodermal cells migrate dorsolaterally to form a neural crest between the surface ectoderm and the neural tube. The neural crest soon divides into two cell masses that give rise to the sensory ganglia of the cranial and spinal nerves. Other neural crest cells migrate from the neural tube and give rise to various other structures, such as the retina.

The mesoderm on each side of the notochord thickens to form longitudinal columns of paraxial mesoderm. Division of these paraxial columns into pairs of *somites* begins cranially by the end of the third week. The somites are compact aggregates of mesenchymal cells from which cells migrate to form the vertebrae, ribs, and axial musculature. During the third week, the number of somites present is a reliable indicator of the age of the embryo. The *coelom* within the embryo first appears as a number of isolated spaces in the lateral mesoderm and cardiogenic mesoderm. The coelomic vesicles subsequently coalesce to form a single, horseshoe-shaped cavity that eventually gives rise to the body cavities, such as the peritoneal cavity.

Blood vessels first appear on the yolk sac and allantois, as well as in the chorion. They develop within the embryo shortly thereafter. Spaces appearing within aggregations of mesenchyme are known as *blood islands*. The spaces soon become lined with endothelium derived from the mesenchymal cells. These primordial tubules sprout and unite with other vessels to form a primordial cardiovascular system. Toward the end of the third week, the heart is represented by paired endothelial heart tubes that are joined to blood vessels in the embryo and in the extraembryonic membranes (yolk sac, umbilical cord, and chorionic sac). By the end of the third week, the endothelial heart tubes have fused to form a tubular heart, which is joined to vessels in the embryo, yolk sac, chorion, and connecting stalk to form a *primordial cardiovascular system*. The primitive blood cells (hemangioblasts) are derived mainly from the endothelial cells of blood vessels in the walls of the yolk sac and allantois.

Primary chorionic villi become secondary chorionic villi as they acquire mesenchymal cores. Before the end of the third week, capillaries develop in the villi, transforming them into *tertiary chorionic villi*. Cytotrophoblastic extensions from these stem villi join to form a *cytotrophoblastic shell* that anchors the chorionic sac to the endometrium. The rapid development of chorionic villi during the third week greatly increases the surface area of the chorion for the exchange of nutrients and other substances between the maternal and embryonic circulations.

Clinically Oriented Questions

1. Do women who have been taking contraceptive pills for many years have more early spontaneous abortions than women who have used other contraceptive methods?
2. What is meant by the term *menstrual extraction*? Is this the same as an early induced abortion?
3. Can drugs and other agents cause congenital anomalies of the embryo if they are present in the mother's blood during the third week? If so, what organs would be most susceptible?
4. Are there increased risks for the embryo associated with pregnancies in women older than 40 of age? If so, what are they?

The answers to these questions are at the back of the book.

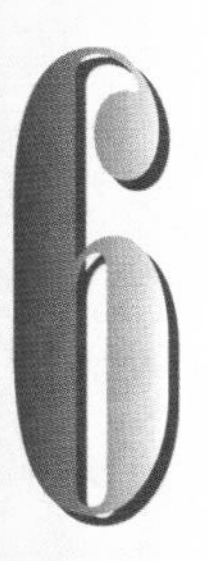

Organogenetic Period: Human Development During Weeks Four to Eight

The fourth to eighth weeks of development constitute most of the embryonic period; however, critical developmental events also occur during the first 3 weeks, such as *cleavage of the zygote, blastogenesis*, and early development of the nervous and cardiovascular systems. All major external and internal structures are established during the fourth to eighth weeks. By the end of this **organogenetic period**, all the main organ systems have begun to develop. As the tissues and organs form, the shape of the embryo changes so that, by the eighth week, it has a distinctly human appearance. Because the organ systems develop during the fourth to eighth weeks, exposure of embryos to teratogens during this period may cause major congenital anomalies. **Teratogens** are agents, such as drugs and viruses, that produce or raise the incidence of congenital anomalies (see Chapter 9). Teratogens act during the stage of active differentiation of a tissue or organ.

Folding of the Embryo

A significant event in the establishment of body form is folding of the flat trilaminar embryonic disc into a somewhat cylindrical embryo (Fig. 6 - 1). Folding occurs in both the median and horizontal planes and results from rapid growth of the embryo, particularly of its brain and spinal cord. Folding at the cranial and caudal ends and sides of the embryo occurs simultaneously. Concurrently, a relative constriction occurs at the junction of the embryo and yolk sac. Folding of the ends of the embryo ventrally produces head and tail folds that result in the cranial and caudal regions moving ventrally as the embryo elongates cranially and caudally (see Fig. 6 - 1A_2 to D_2).

Head and Tail Folds

By the beginning of the fourth week, the neural folds in the cranial region form the primordium of the brain. Later, the developing forebrain grows cranially beyond the oropharyngeal membrane and overhangs the developing heart. Concomitantly, the **septum transversum**, primordial heart, pericardial coelom, and oropharyngeal membrane move onto the ventral surface of the embryo (Fig. 6 - 2). During longitudinal folding, part of the endoderm of the yolk sac is incorporated into the embryo as the **foregut**. The foregut lies between the brain and heart, and the **oropharyngeal membrane** separates the foregut from the **stomodeum** (see Fig. 6 - 2*C*). After folding of the head, the septum transversum lies caudal to the heart, where it subsequently develops into the *central tendon of the diaphragm* (see Chapter 10). The head fold also affects the arrangement of the embryonic coelom (primordium of body cavities). Before folding, the coelom is a flattened, horseshoe-shaped cavity (see Fig. 6 - 1A_1). After folding, the pericardial coelom lies ventral to the heart and cranial to the septum transversum (see Fig. 6 - 2*C*). At this stage, the intraembryonic coelom communicates widely on each side with the extraembryonic coelom (see Fig. 6 - 1A_3).

Folding of the caudal end of the embryo results primarily from growth of the distal part of the neural tube, the primordium of the spinal cord (Fig. 6 - 3). As the embryo grows, the tail region projects over the **cloacal membrane** (future site of the anus). During folding, part of the endodermal germ layer is incorporated into the embryo as the **hindgut**. The terminal part of the hindgut soon dilates slightly to form the **cloaca**. Before folding, the primitive streak lies cranial to the cloacal membrane (see Fig. 6 - 3*B*); after folding, it lies caudal to it (Fig. 6 - 3*C*). The connecting stalk (primordium of the umbilical cord) is now attached to the ventral surface of the embryo, and the allantois — an endodermal diverticulum of the yolk sac — is partially incorporated into the embryo.

Lateral Folds

Folding of the sides of the embryo results from rapid growth of the spinal cord and somites, which produces right and left **lateral folds** (see Fig. 6 - 1A_3 to D_3). The lateral body wall folds toward the median plane, rolling the edges of the embryonic disc ventrally and forming a roughly cylindrical embryo. As the abdominal walls form, part of the endoderm germ layer is incorporated into the embryo as the **midgut**. Initially, there is a wide connection between the midgut and yolk sac (see Fig. 6 - 1A_2). After lateral folding, the connection is reduced to a *yolk stalk*, or vitelline duct (see Fig. 6 - 1C_2). The region of attachment of the amnion to the ventral surface of the embryo is also reduced to a relatively narrow umbilical region (see Figs. 6 - 1D_2 and D_3). As the **umbilical cord** forms from the connecting stalk, ventral fusion of the lateral folds reduces the region of communication between the intraembryonic and extraembryonic coelomic cavities to a narrow communication (see Fig. 6 - 1C_2). As the amniotic cavity expands and obliterates most of the extraembryonic coelom, the amnion forms the epithelial covering of the umbilical cord (see Fig. 6 - 1D_2).

Germ Layer Derivatives

The three germ layers (ectoderm, mesoderm, and endoderm) formed during gastrulation (see Chapter 5) give rise to the primordia of all tissues and organs. The specificity of the germ layers, however, is not rigidly fixed. The cells of each germ layer divide, migrate, aggregate, and differentiate in rather precise patterns

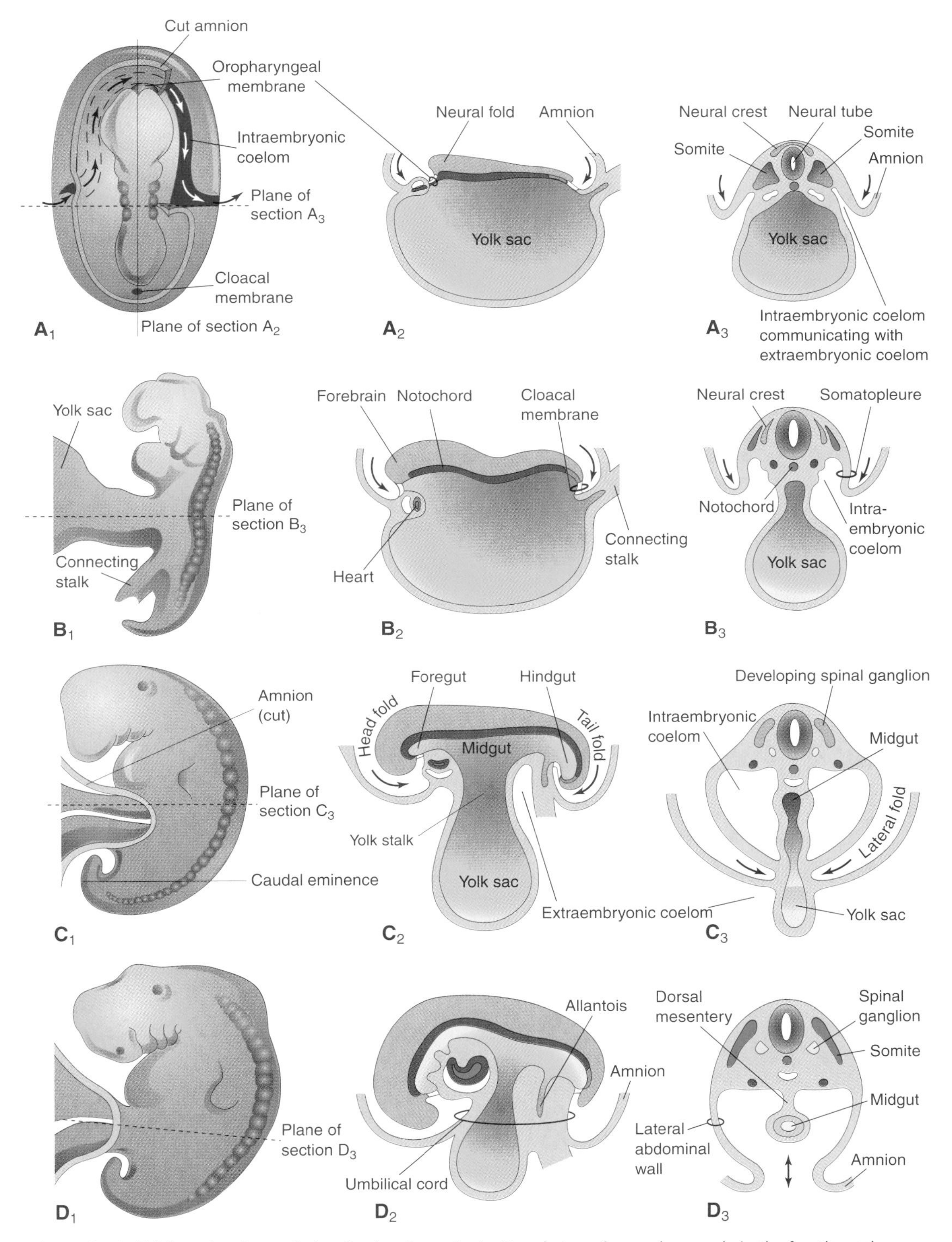

Figure 6 - 1. Folding of embryos during the fourth week. A_1, Dorsal view of an embryo early in the fourth week. Three pairs of somites are visible. The continuity of the intraembryonic coelom and extraembryonic coelom is illustrated on the right side by removal of a part of the embryonic ectoderm and mesoderm. B_1, C_1, and D_1, Lateral views of embryos at 22, 26, and 28 days, respectively. A_2, B_2, C_2, and D_2, Sagittal sections at the plane shown in A_1. A_3, to B_3, C_3, and D_3, Transverse sections at the levels indicated in A_1 to D_1.

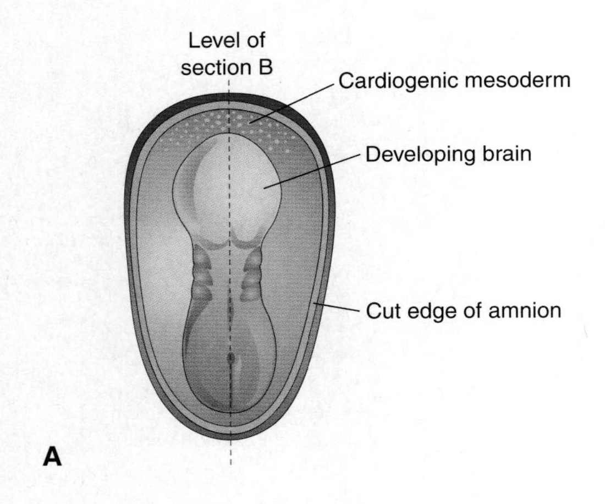

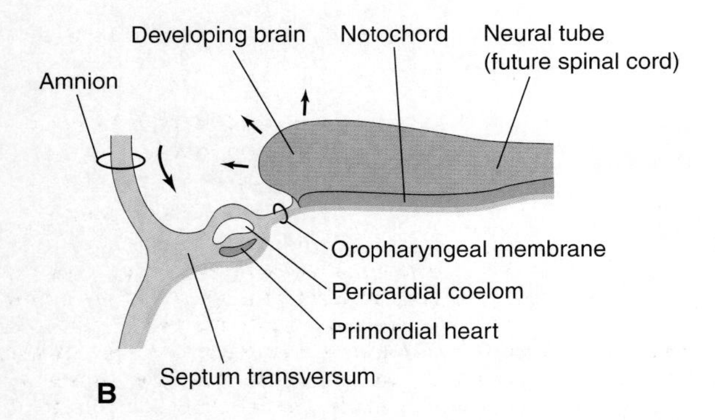

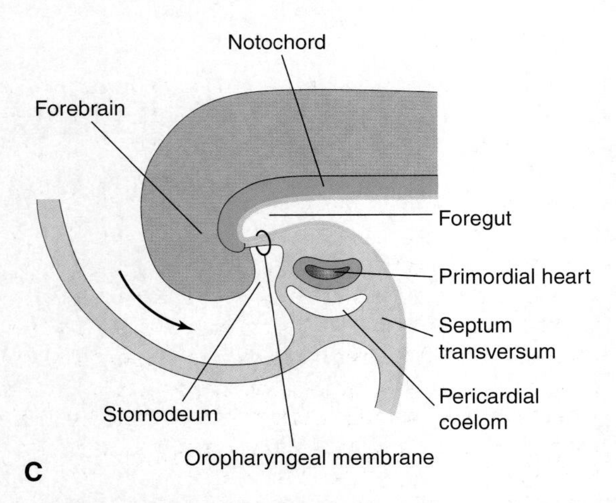

Figure 6-2. Folding of the cranial end of the embryo. *A,* Dorsal view of an embryo at 21 days. *B,* Sagittal section of the cranial part of the embryo at the plane shown in *A.* Observe the ventral movement of the heart. *C,* Sagittal section of an embryo at 26 days. Note that the septum transversum, heart, pericardial coelom, and oropharyngeal membrane have moved onto the ventral surface of the embryo. Observe also that part of the yolk sac is incorporated into the embryo as the foregut.

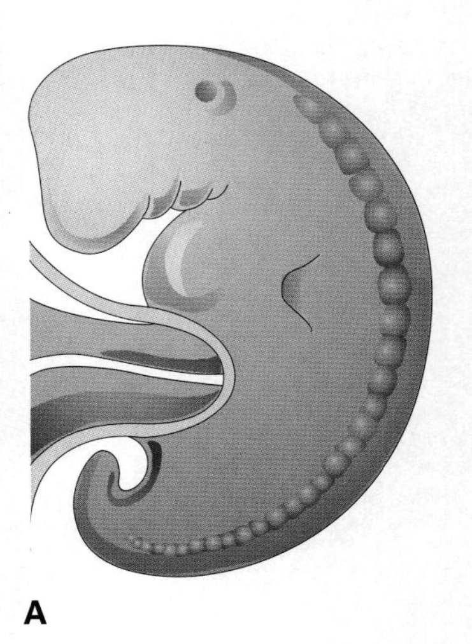

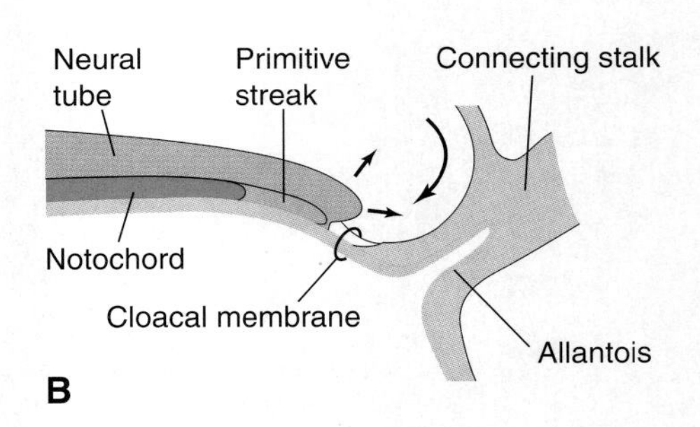

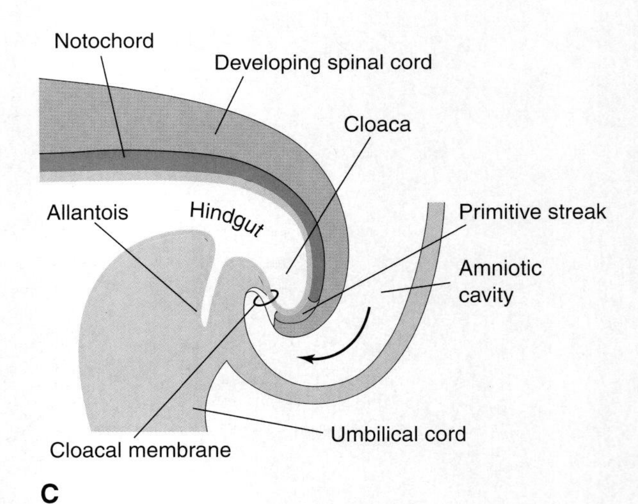

Figure 6-3. Folding of the caudal end of the embryo. *A,* Lateral view of a 4-week embryo. *B,* Sagittal section of the caudal part of the embryo at the beginning of the fourth week. *C,* Similar section at the end of the fourth week. Note that part of the yolk sac is incorporated into the embryo as the hindgut and that the terminal part of the hindgut has dilated to form the cloaca. Observe also the change in position of the primitive streak, allantois, cloacal membrane, and connecting stalk.

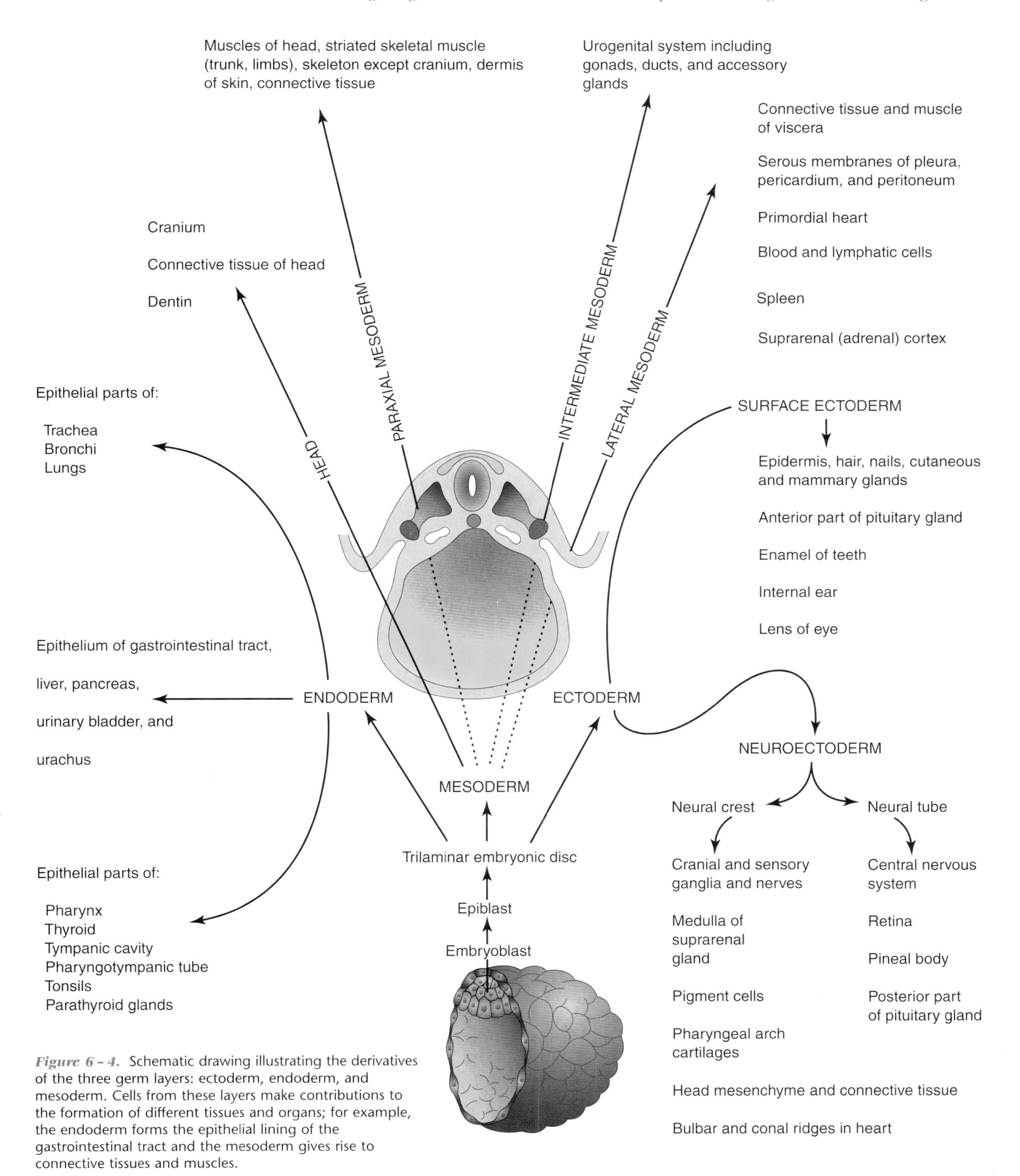

Figure 6 - 4. Schematic drawing illustrating the derivatives of the three germ layers: ectoderm, endoderm, and mesoderm. Cells from these layers make contributions to the formation of different tissues and organs; for example, the endoderm forms the epithelial lining of the gastrointestinal tract and the mesoderm gives rise to connective tissues and muscles.

as they form the various organ systems (*organogenesis*). The main germ layer derivatives are as follows (Fig. 6–4):

- The **ectoderm** gives rise to the central nervous system (CNS; brain and spinal cord); peripheral nervous system; sensory epithelia of eye, ear, and nose; epidermis and its appendages (hair and nails); mammary glands; pituitary gland; subcutaneous glands; and enamel of teeth. **Neural crest cells**, derived from the neuroectoderm, give rise to the cells of the spinal, cranial ganglia of cranial nerves V, VII, IX, and X, and autonomic ganglia; ensheathing cells of the peripheral nervous system; pigment cells of the dermis; muscle and connective tissues; bones of pharyngeal arch origin (see Chapter 11); suprarenal (adrenal) medulla; and meninges (coverings) of the brain and spinal cord.
- The **mesoderm** gives rise to connective tissue, cartilage, bone, striated and smooth muscles, heart, blood and lymphatic vessels, kidneys, ovaries and testes, genital ducts, serous membranes lining the body cavities (pericardial, pleural, and peritoneal), spleen, and the cortex of the suprarenal glands.
- The **endoderm** gives rise to the epithelial lining of the gastrointestinal and respiratory tracts; cells of the tonsils, thyroid and parathyroid glands, thymus, liver, and pancreas; the epithelial lining of the urinary bladder and most of the urethra; and the epithelial lining of the tympanic cavity, tympanic antrum, and pharyngotympanic (auditory) tube.

Control of Embryonic Development*

Development results from genetic plans in the chromosomes. Knowledge of the genes or hereditary units that control human development is increasing. Most information about developmental processes has come from studies in other organisms, especially *Drosophila* and mice, because of ethical problems associated with the use of human embryos for laboratory studies. Most developmental processes depend upon a precisely coordinated interaction of genetic and environmental factors. Several control mechanisms guide differentiation and ensure synchronized development, such as tissue interactions, regulated migration of cells and cell colonies, controlled proliferation, and programmed cell death. Each system of the body has its own developmental pattern, but most processes of morphogenesis are similar and are relatively simple. Underlying all these changes are basic regulating mechanisms.

Embryonic development is essentially a process of growth and increasing complexity of structure and function. Growth is achieved by mitosis (the somatic reproduction of cells), together with the production of extracellular matrices, whereas complexity is achieved through morphogenesis and differentiation. The cells that make up the tissues of very early embryos are pluripotential; that is, depending on circumstances, they are able to follow more than one pathway of development. This broad developmental potential becomes progressively restricted as tissues acquire the specialized features necessary for increased sophistication of structure and function. Such restriction presumes that choices must be made in order to achieve tissue diversification. At present, most evidence indicates that these choices are determined not as a consequence of cell lineage, but rather in response to cues from the immediate surroundings, including the adjacent tissues. As a result, the architectural precision and coordination that are often required for the normal function of an organ appear to be achieved by the interaction of its constituent parts during development.

The interaction of tissues during development is a recurring theme in embryology. The interactions that lead to a change in the course of development of at least one of the interactants are termed **inductions**. Numerous examples of such inductive interactions can be found in the literature; for example, during development of the eye, the optic vesicle is believed to induce the development of the lens from the surface ectoderm of the head. When the optic vesicle is absent, the eye fails to develop. Moreover, if the optic vesicle is removed and placed in association with surface ectoderm that is not usually involved in eye development, lens formation can be induced. Clearly then, development of a lens depends on the ectoderm acquiring an association with a second tissue. In the presence of the neuroectoderm of the optic vesicle, the surface ectoderm of the head follows a pathway of development that it would not otherwise have taken. In a similar fashion, many of the morphogenetic tissue movements that play such important roles in shaping the embryo also provide for the changing tissue associations that are fundamental to inductive tissue interactions.

The fact that one tissue can influence the developmental pathway adopted by another tissue presumes that a signal passes between the two interactants. Analysis of the molecular defects in mutant strains that show

* The authors are grateful to Dr. Michael Wiley, Associate Professor of Anatomy and Cell Biology, Department of Surgery, Faculty of Medicine, University of Toronto, for his assistance in preparing this section.

abnormal tissue interactions during embryonic development, as well as studies of the development of embryos with targeted gene mutation, have begun to reveal the molecular mechanisms of induction. The mechanism of signal transfer appears to vary with the specific tissues involved. In some cases, the signal appears to take the form of a diffusible molecule (e.g., sonic hedgehog) that passes from the inductor to the reacting tissue. In other instances, the message appears to be mediated through a nondiffusible, extracellular matrix that is secreted by the inductor and that comes into contact with the reacting tissue. In still other cases, the signal appears to require physical contact between the inducing and responding tissues. Regardless of the mechanism of intercellular transfer involved, the signal is translated into an intracellular message that influences the genetic activity of the responding cells.

To be competent to respond to an inducing stimulus, the cells of the reacting system must express the appropriate receptor for the specific inducing signal molecule, the components of the particular intracellular signal transduction pathway, and the transcription factors that will mediate the particular response. Experimental evidence suggests that the acquisition of competence by the responding tissue is often dependent on its previous interactions with other tissues. For example, the lens-forming response of head ectoderm to the stimulus provided by the optic vesicle appears to be dependent on a previous association of the head ectoderm with the anterior neural plate.

Highlights of Development During Weeks Four to Eight

The following descriptions summarize the main developmental events and changes in external form of the embryo during the fourth to eighth weeks. Criteria for estimating developmental stages in human embryos are listed in Table 6 - 1.

Fourth Week

At the beginning of the fourth week, the embryo is almost straight and has 4 to 12 *somites* that produce conspicuous surface elevations (Fig. 6 - 5*A*). The *neural tube* is formed opposite the somites, but it is widely open at the rostral and caudal neuropores (see Figs. 6 - 5*B* and 6 - 6). By 24 days' gestation, the first *pharyngeal arches* have appeared. The first (mandibular) arch and the second (hyoid) arch are distinct (see Figs. 6 - 5*C* and 6 - 7). The major part of the first pharyngeal arch gives rise to the mandible (lower jaw), and a rostral extension of the arch — the maxillary prominence — contributes to the maxilla (upper jaw). The embryo is now slightly curved because of the head and tail folds. The heart produces a large ventral prominence and pumps blood. Three pairs of **pharyngeal arches** are visible by 26 days' gestation (Fig. 6 - 8; see also Fig. 6 - 5*D*), and the rostral neuropore is closed. The **forebrain** produces a prominent elevation of the head, and folding of the embryo imparts a characteristic C-shaped curvature to the embryo. A long curved **caudal eminence** (tail-like structure) is present. **Upper limb buds** are recognizable by day 26 or 27 as small swellings on the ventrolateral body walls (see Fig. 6 - 5*D* and *E*). The **otic pits**, the primordia of the internal ears, are also visible. Ectodermal thickenings, called **lens placodes**, indicating the future lenses of the eyes, are visible on the sides of the head. The fourth pair of pharyngeal arches and the **lower limb buds** are visible by the end of the fourth week (see Fig. 6 - 5*E*). Toward the end of the fourth week, the tail-like **caudal eminence** is a characteristic feature (Fig. 6 - 9). Rudiments of many of the organ systems, especially the *cardiovascular system*, are established.

Fifth Week

Changes in body form are minor during the fifth week compared with those that occurred during the fourth week. Growth of the head exceeds that of other regions (Fig. 6 - 10). Enlargement of the head is caused mainly by the rapid development of the brain and facial prominences. The face soon contacts the heart prominence. The rapidly growing second pharyngeal arch overgrows the third and fourth arches, forming a lateral ectodermal depression on each side — the **cervical sinus**. The upper limb buds are paddle-shaped, whereas the lower limb buds are flipper-like. The *mesonephric ridges* indicate the site of the mesonephric kidneys, which are interim organs in humans.

Sixth Week

The upper limbs begin to show regional differentiation as the elbows and large **hand plates** develop (Fig. 6 - 11). The primordia of the digits — the **digital rays** — begin to develop in the hand plates, indicating the formation of digits (fingers). It has been reported that embryos in the sixth week show spontaneous movements, such as twitching of the trunk and limbs. Development of the lower limbs occurs somewhat later than that of the upper limbs. Several small swellings — **auricular hillocks** — develop around the pharyngeal groove or cleft between the first two pharyngeal arches. This groove becomes the **external acoustic meatus** (external auditory canal), and the hillocks fuse to form the *auricle*, the shell-shaped part of the external ear. Largely

Table 6 - 1. Criteria for Estimating Developmental Stages in Human Embryos

Age (Days)	Figure Reference	Carnegie Stage	No. of Somites	Length (mm)*	Main External Characteristics†
20–21	6–1A_1 6–2*A*	9	1–3	1.5–3.0	*Flat embryonic disc. Deep neural groove and prominent neural folds.* One to three pairs of somites present. Head fold evident.
22–23	6–5*A* 6–6*A*, *C*	10	4–12	2.0–3.5	*Embryo straight or slightly curved.* Neural tube forming or formed opposite somites, but widely open at rostral and caudal neuropores. First and second pairs of pharyngeal arches visible.
24–25	6–5*C* 6–7*A*	11	13–20	2.5–4.5	*Embryo curved owing to head and tail folds.* Rostral neuropore closing. Otic placodes present. Optic vesicles formed.
26–27	6–5*D* 6–8*A*	12	21-29	3.0–5.0	*Upper limb buds appear.* Rostral neuropore closed. Caudal neuropore closing. Three pairs of pharyngeal arches visible. Heart prominence distinct. Otic pits present.
28–30	6–5*E* 6–9*A*	13	30–35	4.0–6.0	*Embryo has* C*-shaped curve.* Caudal neuropore closed. Upper limb buds are flipperlike. Four pairs of pharyngeal arches visible. Lower limb buds appear. *Otic vesicles present.* Lens placodes distinct. Tail-like caudal eminence present.
31–32	6–10*A*	14	‡	5.0–7.0	*Upper limbs are paddle-shaped.* Lens pits and nasal pits visible. Optic cups present.
33–36		15		7.0–9.0	*Hand plates formed; digital rays present.* Lens vesicles present. Nasal pits prominent. *Lower limbs are paddle-shaped.* Cervical sinuses visible.
37–40		16		8.0–11.0	*Foot plates formed.* Pigment visible in retina. Auricular hillocks developing.
41–43	6–11*A*	17		11.0–14.0	*Digital rays clearly visible in hand plates.* Auricular hillocks outline future auricle of external ear. Trunk beginning to straighten. Cerebral vesicles prominent.
44–46		18		13.0–17.0	*Digital rays clearly evident in foot plates.* Elbow region visible. Eyelids forming. Notches between the digital rays in the hands. Nipples visible.
47–48		19		16.0–18.0	*Limbs extend ventrally.* Trunk elongating and straightening. Midgut herniation prominent.
49–51		20		18.0–22.0	*Upper limbs longer and bent at elbows. Fingers distinct but webbed.* Notches between the digital rays in the feet. Scalp vascular plexus appears.
52–53		21		22.0–24.0	*Hands and feet approach each other. Fingers are free and longer. Toes distinct but webbed.* Stubby tail present.
54–55		22		23.0–28.0	*Toes free and longer.* Eyelids and auricles of external ears more developed.
56	6–12	23		27.0–31.0	*Head more rounded and shows human characteristics.* External genitalia still have sexless appearance. Distinct bulge still present in umbilical cord, caused by herniation of intestines. *Caudal eminence ("tail") has disappeared.*

*The embryonic lengths indicate the usual range. In stages 9 and 10, the measurement is greatest length (*GL*); in subsequent stages crown-rump (*CR*) measurements are given.

† Based mainly on O'Rahilly R, Müller F: *Developmental Stages in Human Embryos.* Washington, Carnegie Institute of Washington, 1987.

‡At this and subsequent stages, the number of somites is difficult to determine and so is not a useful criterion. Refer to Moore KL, Persaud TVN, Shiota K: *Color Atlas of Clinical Embryology,* 2nd ed. Philadelphia, WB Saunders. 2000 for more color photographs of embryos.

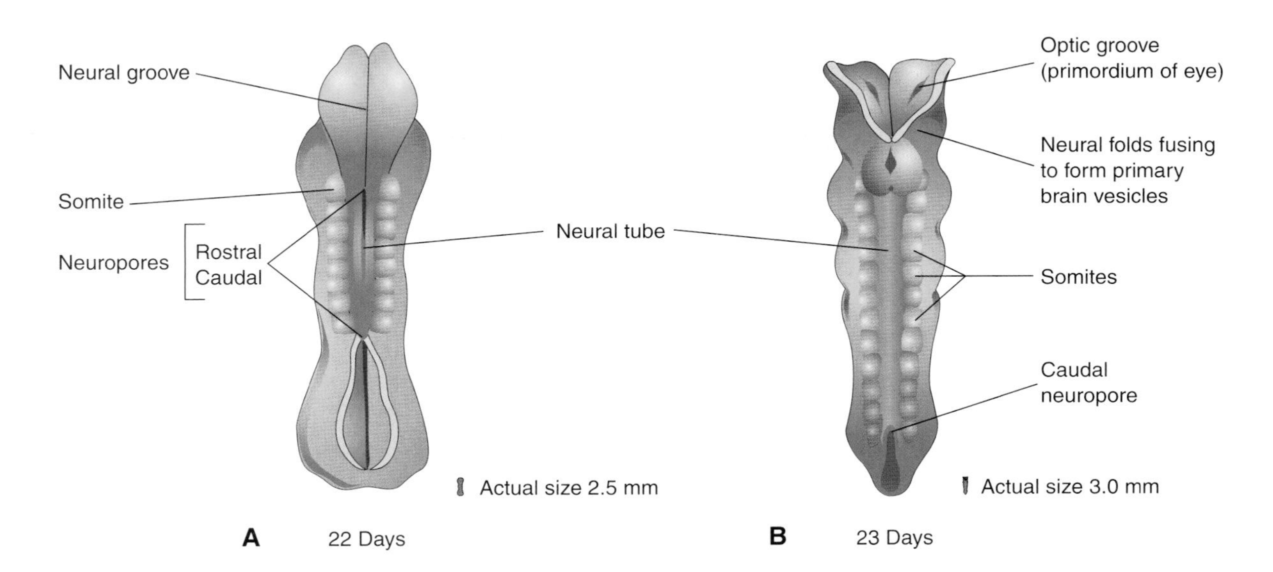

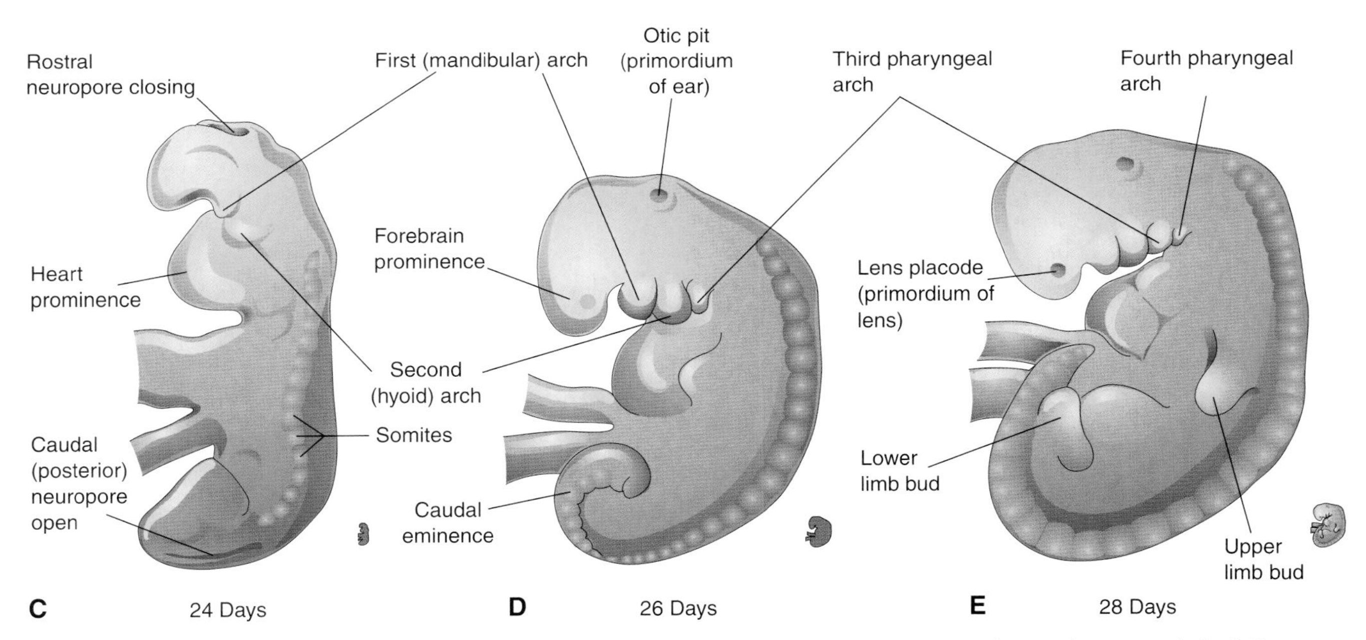

Figure 6-5. *A* and *B*, Drawings of dorsal views of embryos early in the fourth week showing 8 and 12 somites, respectively. *C*, *D*, and *E*, Lateral views of older embryos showing 16, 27, and 33 somites, respectively. The rostral neuropore is normally closed by 25 to 26 days, and the caudal neuropore is usually closed by the end of the fourth week.

because retinal pigment has formed, the eye is now obvious. The head is now much larger relative to the trunk and is bent over the large **heart prominence**. This head position results from bending in the cervical (neck) region. By this time, the trunk and neck have begun to straighten. It has been reported that embryos during the sixth week show reflex responses to touch.

Seventh Week

The limbs undergo considerable change during the seventh week. Notches appear between the digital rays in the hand plates, partially separating the future digits. Communication between the primordial gut and yolk sac is now reduced to a relatively slender duct, the *yolk stalk*. The intestines enter the extraembryonic coelom in

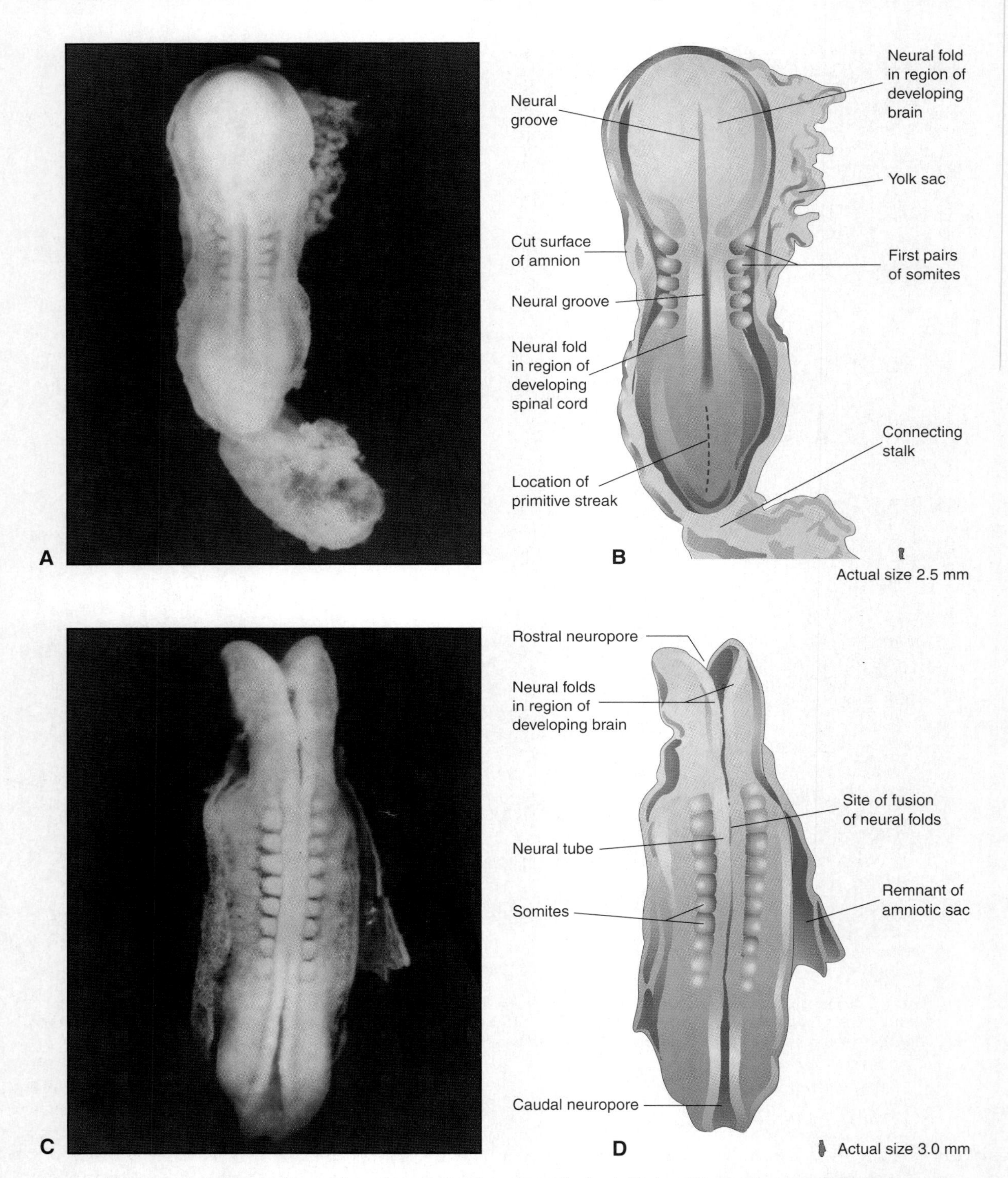

Figure 6 - 6. *A*, Dorsal view of a five-somite embryo at Carnegie stage 10, about 22 days. Observe the neural folds and neural groove. The neural folds in the cranial region have thickened to form the primordium of the brain. *B*, Drawing indicating the structures shown in *A*. Most of the amniotic and chorionic sacs have been cut away to expose the embryo. *C*, Dorsal view of a 10-somite embryo at Carnegie stage 10, about 23 days. The neural folds have fused opposite the somites to form the neural tube (primordium of the spinal cord in this region). The neural tube is in open communication with the amniotic cavity at the cranial and caudal ends through the rostral and caudal neuropores, respectively. *D*, Diagram indicating the structures shown in *C*.

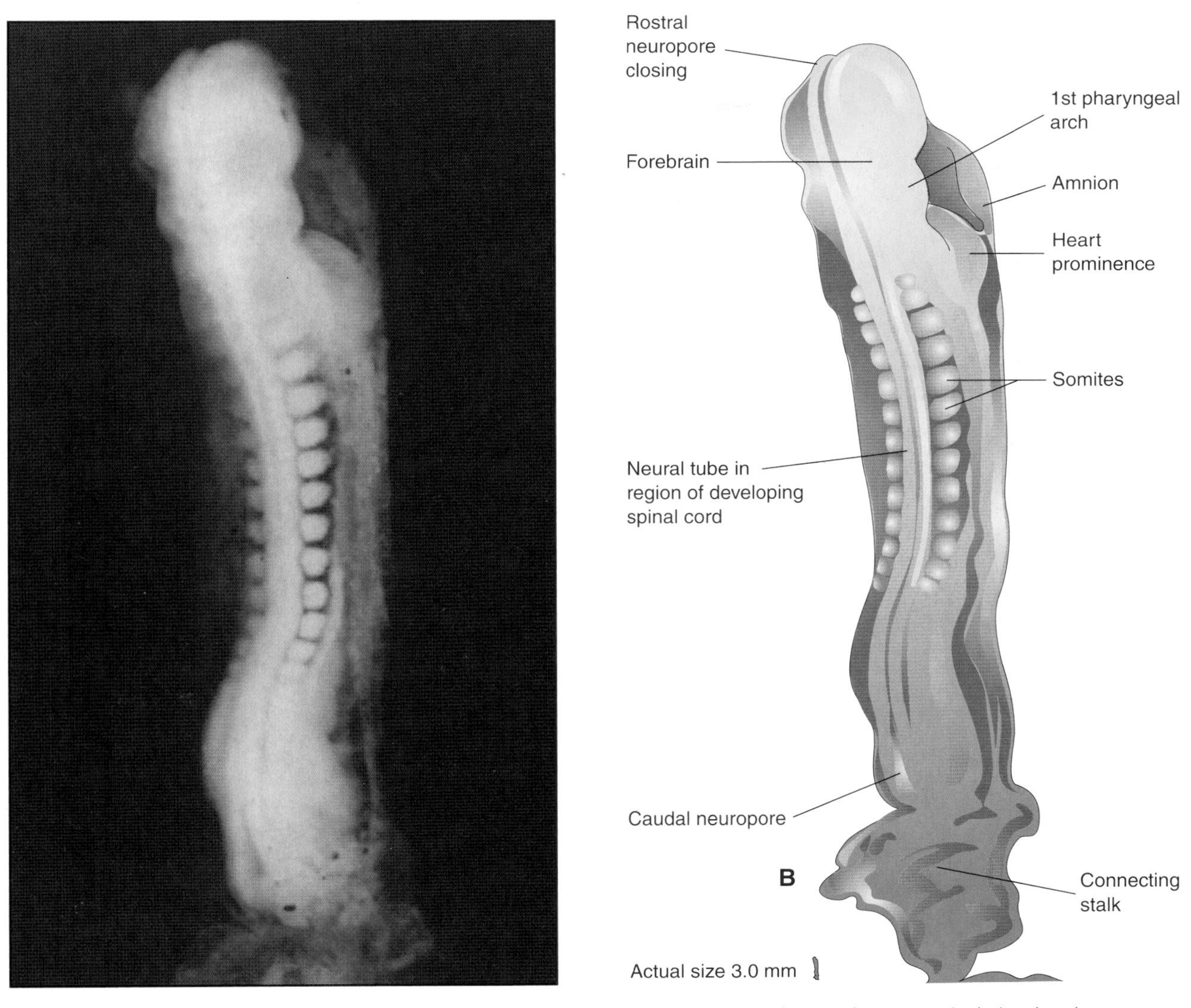

Figure 6 – 7. *A,* Dorsal view of a 13-somite embryo at Carnegie stage 11, about 24 days. The rostral neuropore is closing, but the caudal neuropore is wide open. *B,* Drawing indicating the structures shown in *A*. The embryo is curved because of folding at the cranial and caudal ends.

the proximal part of the umbilical cord. This **umbilical herniation** is a normal event in the embryo, occurring because the abdominal cavity is too small at this stage to accommodate the rapidly growing intestines.

Eighth Week

At the beginning of this final week of the embryonic period, the digits of the hand are separated but noticeably webbed. Notches are now clearly visible between the digital rays of the fan-shaped feet. The tail-like caudal eminence is still present, but is stubby. The **scalp vascular plexus** has appeared and forms a characteristic band around the head. By the end of the eighth week, all regions of the limbs are apparent, and the digits have lengthened and are completely separated (Fig. 6–12). *Purposeful limb movements first occur during this week.* Ossification begins in the lower limbs in the eighth week and is first recognizable in the femur. All evidence of the tail-like caudal eminence has disappeared by the end of the eighth week. The *scalp vascular plexus* now forms a slender band near the vertex (crown) of the head. The hands and feet approach each other ventrally. At the end of the eighth week, the embryo has distinct human characteristics; however, the head is still disproportionately large, constituting almost half of the embryo.

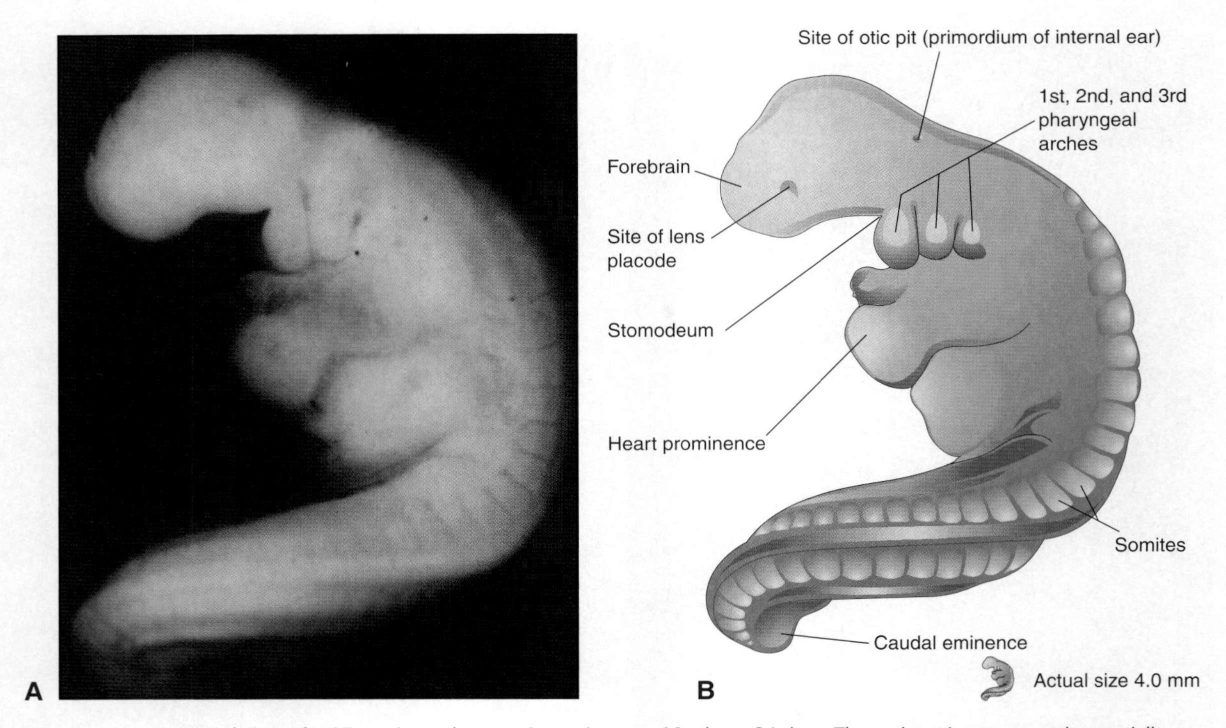

Figure 6–8. *A*, Lateral view of a 27-somite embryo at Carnegie stage 12, about 26 days. The embryo is very curved, especially its long tail-like caudal eminence. Observe the lens placode (primordium of the lens of the eye) and the otic pit, indicating early development of the internal ear. *B*, Drawing indicating the structures shown in *A*. The rostral neuropore is closed, and three pairs of pharyngeal arches are present. (*A*, From Nishimura H, Semba H, Tanimura T, Tanaka O: *Prenatal Development of the Human with Special Reference to Craniofacial Structures: An Atlas*. Washington, DC, National Institutes of Health, 1977.)

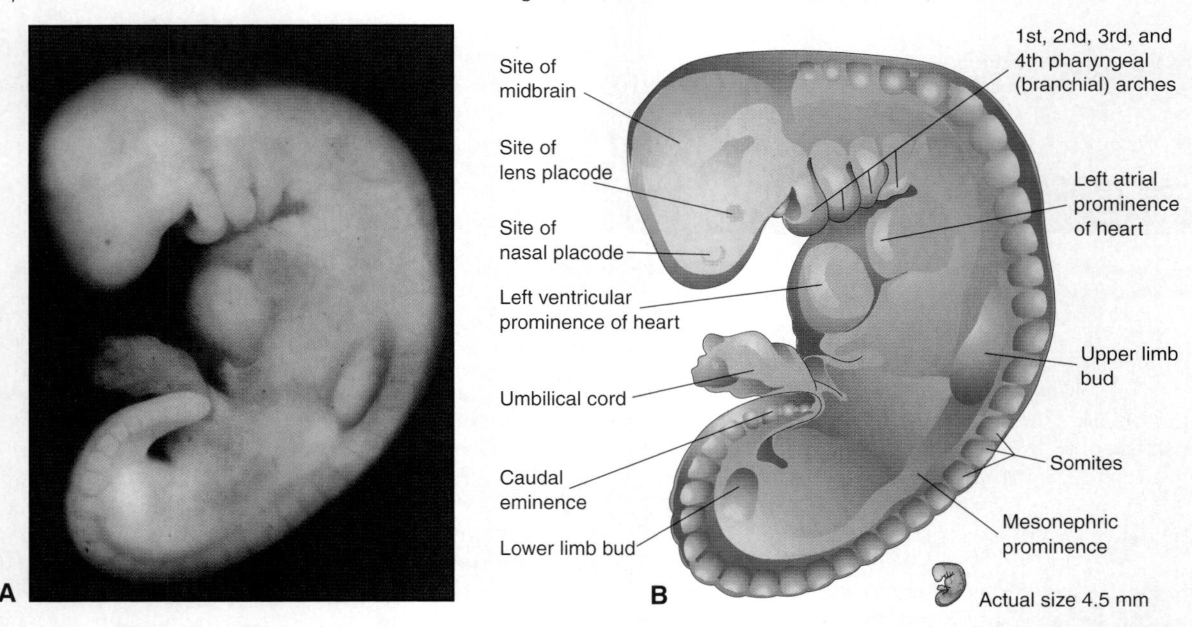

Figure 6–9. *A*, Lateral view of an embryo at Carnegie stage 13, about 28 days. The primordial heart is large, and its division into a primordial atrium and ventricle is visible. The rostral and caudal neuropores are closed. *B*, Drawing indicating the structures shown in *A*. The embryo has a characteristic C-shaped curvature, four pharyngeal arches, and upper and lower limb buds. (*A*, From Nishimura H, Semba H, Tanimura T, Tanaka O: *Prenatal Development of the Human with Special Reference to Craniofacial Structures: An Atlas*. Washington, DC, National Institutes of Health, 1977.)

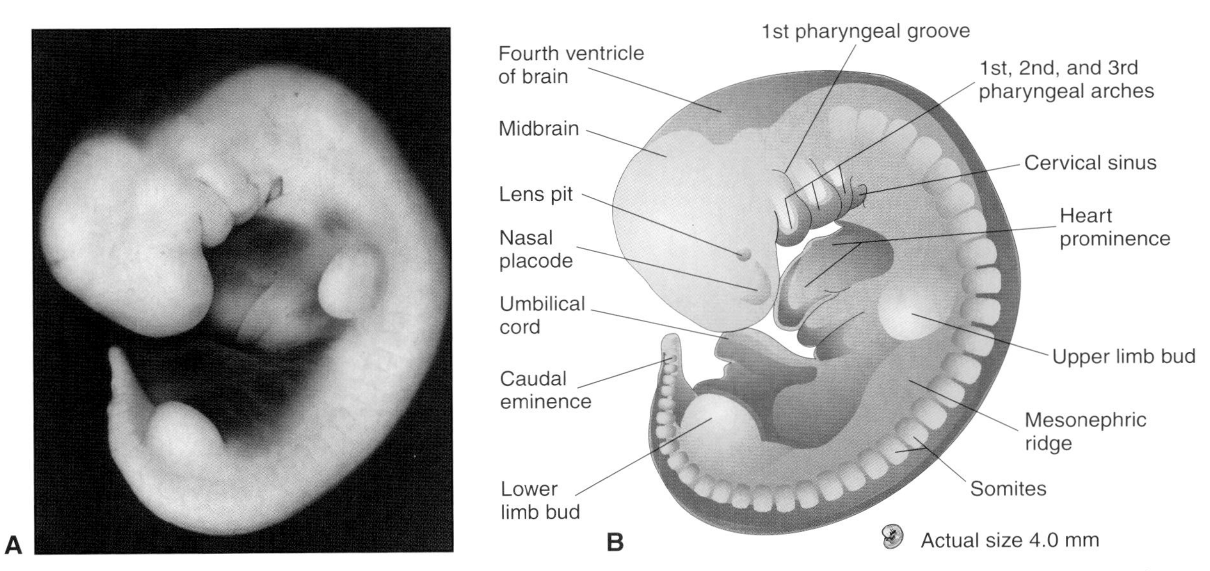

Figure 6 - 10. *A*, Lateral view of an embryo at Carnegie stage 14, about 32 days. The second pharyngeal arch has overgrown the third arch, forming a depression known as the cervical sinus. The mesonephric ridge indicates the site of the mesonephric kidney, an interim kidney (see Chapter 14). *B*, Drawing indicating the structures shown in *A*. The upper limb buds are paddle-shaped, whereas the lower limb buds are flipperlike. (*A*, From Nishimura H, Semba H, Tanimura T, Tanaka O: *Prenatal Development of the Human with Special Reference to Craniofacial Structures: An Atlas*. Washington, DC, National Institutes of Health, 1977.)

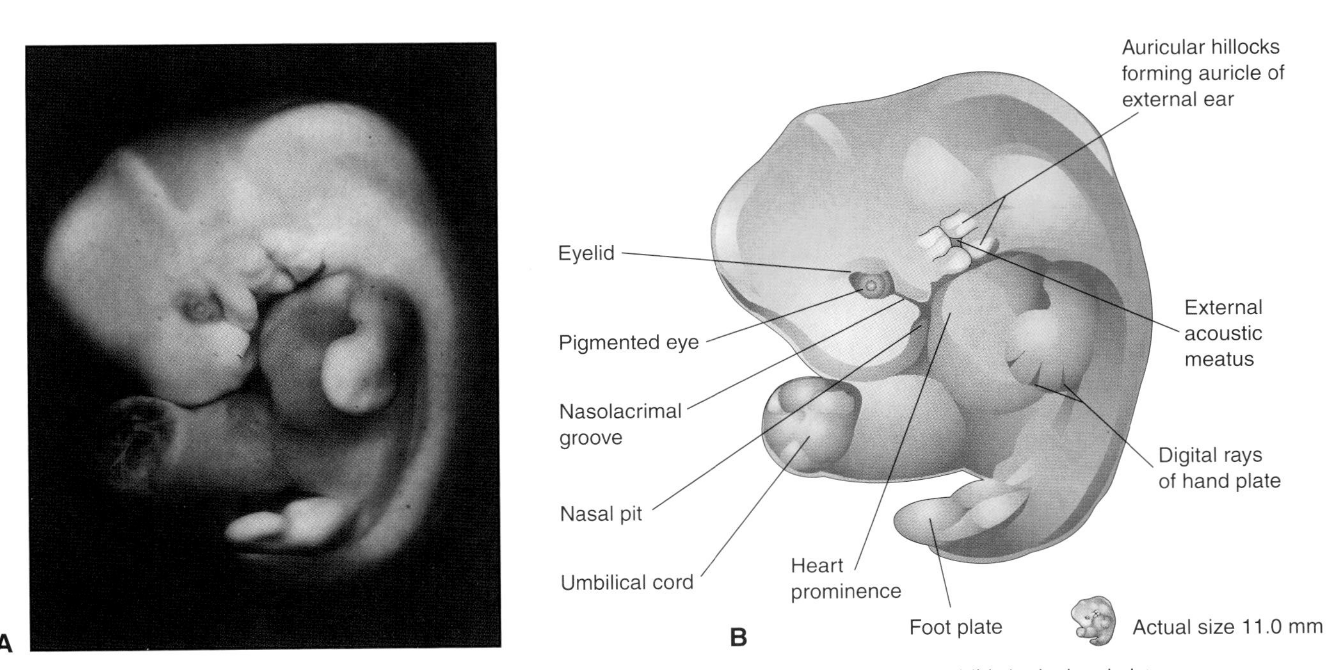

Figure 6 - 11. *A*, Lateral view of an embryo at Carnegie stage 17, about 42 days. Digital rays are visible in the hand plate, indicating the future site of the digits (fingers). *B*, Drawing indicating the structures shown in *A*. The eye, auricular hillocks, and external acoustic meatus are now clearly discernible.

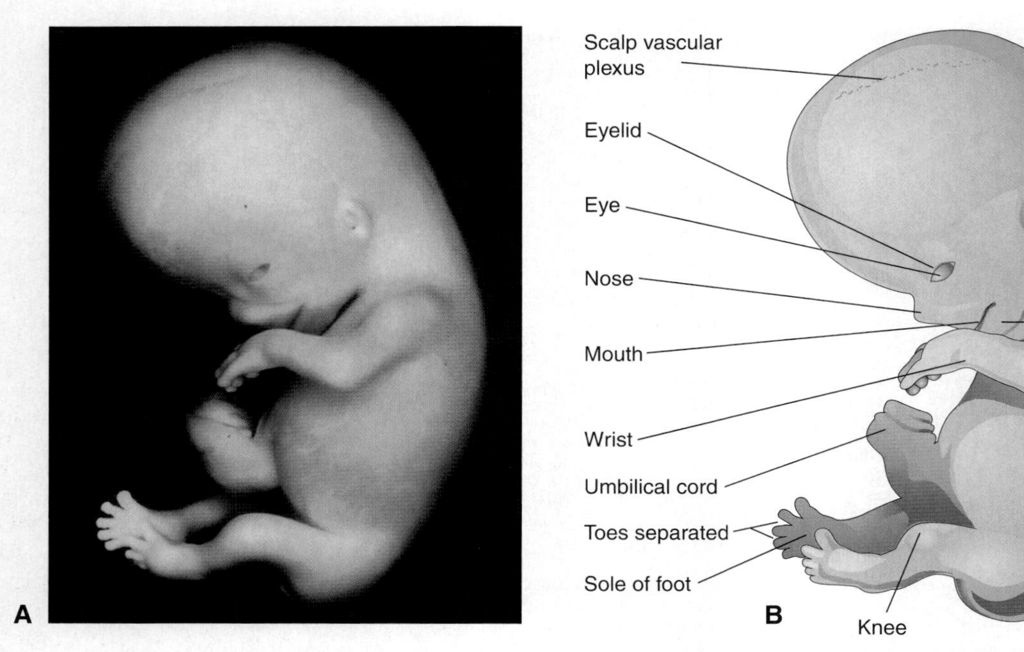

Figure 6-12. *A*, Lateral view of an embryo at Carnegie stage 23, about 56 days. The embryo now has a distinctly human appearance. *B*, Drawing indicating the structures shown in *A*. The scalp vascular plexus is reduced and the caudal eminence has disappeared. (*A*, From Nishimura H, Semba H, Tanimura T, Tanaka O: *Prenatal Development of the Human with Special Reference to Craniofacial Structures: An Atlas*. Washington, DC, National Institutes of Health, 1977.)

The neck region is established, and the eyelids are more obvious. The eyelids are closing, and by the end of the eighth week, they begin to unite by epithelial fusion. The intestines are still in the proximal portion of the umbilical cord. The auricles of the external ears begin to assume their final shape, but they are still low-set on the head. Although sex differences exist in the appearance of the external genitalia, they are not distinctive enough to permit accurate sexual identification (see Chapter 14).

Estimation of Gestational Age

Determination of the starting date of a pregnancy may be difficult in some instances, partly because it depends on the mother's memory of an event that occurred several weeks before she realized she was pregnant. Two reference points are commonly used for estimating age:

- onset of the last normal menstrual period (LNMP)
- probable time of fertilization (conception)

The LNMP is commonly used by clinicians to estimate the age of embryos and is a reliable criterion in most cases. *Ultrasonographic assessment of the size of the chorionic (gestational) sac* and its contents (Fig. 6-13) enables clinicians to obtain an accurate estimate of the date of conception.

The zygote does not form until about 2 weeks after the LNMP; consequently, 14 ± 2 days must be deducted from the gestational (menstrual) age to obtain the fertilization (conceptional) age of an embryo. Because it may be important to know the fertilization age of an embryo for determining its sensitivity to teratogenic agents (see Chapter 9), all statements about age should indicate the reference point used (i.e., days after the LNMP or days after the estimated time of fertilization).

Estimation of Embryonic Age

Estimates of the age of recovered embryos (e.g., after spontaneous abortion) are determined from their external characteristics and measurements of their length (see Table 6-1). Size alone may be an unreliable criterion because some embryos undergo a progressively slower rate of growth prior to death. The appearance of the developing limbs is a very helpful criterion for estimating embryonic age. Because embryos of the third and early fourth weeks are straight (Fig. 6-14*A*), measurements of them indicate the *greatest length* (GL). The sitting height, or *crown-rump length* (CRL), is used for older embryos (see Fig. 6-14*B* and *C*). Standing height, or crown-heel length (CHL), is sometimes measured for

8-week embryos. The *Carnegie Embryonic Staging System* is used internationally (see Table 6 – 1); its usage enables comparisons to be made between the findings of one person and those of another.

Ultrasonographic Examination of Embryos

Most women seeking obstetric care have at least one ultrasonographic examination during their pregnancy for one or more of the following reasons:

- estimation of gestational age for confirmation of clinical dating
- evaluation of embryonic growth when intrauterine growth retardation is suspected
- guidance during chorionic villus sampling (see Chapter 7)
- suspected ectopic pregnancy
- possible uterine abnormality
- detection of congenital anomalies

No biologic effects on embryos or fetuses from the use of diagnostic ultrasonographic evaluation have been confirmed. The size of an embryo in a pregnant woman can be estimated using ultrasonographic measurements. *Transvaginal* or *endovaginal sonography* permits accurate measurement of CRL in early pregnancy.

Summary of Development During Weeks Four to Eight

During weeks four to eight, which represents most of the embryonic period, *all major organs and systems of the body form from the three germ layers*. At the beginning of the fourth week, folding of the embryo in the median and horizontal planes converts the flat, trilaminar, embryonic disc into a C-shaped, cylindrical embryo. The formation of head, caudal eminence, and lateral folds is a continuous sequence of events that result in a constriction between the embryo and the yolk sac. During folding, part of the yolk sac is incorporated into the embryo and gives rise to the primordial gut. As the head region folds ventrally, part of the endodermal layer is incorporated into the developing embryonic head as the *foregut*. Folding of the head region also results in the oropharyngeal membrane and heart being carried ventrally, and the developing brain becoming the most cranial part of the embryo. As the caudal eminence folds ventrally, part of the endodermal layer is incorporated into the caudal end of the embryo as the *hindgut*. The terminal part of the hindgut expands to form the *cloaca*. Folding of the caudal (tail) region also results in the cloacal membrane,

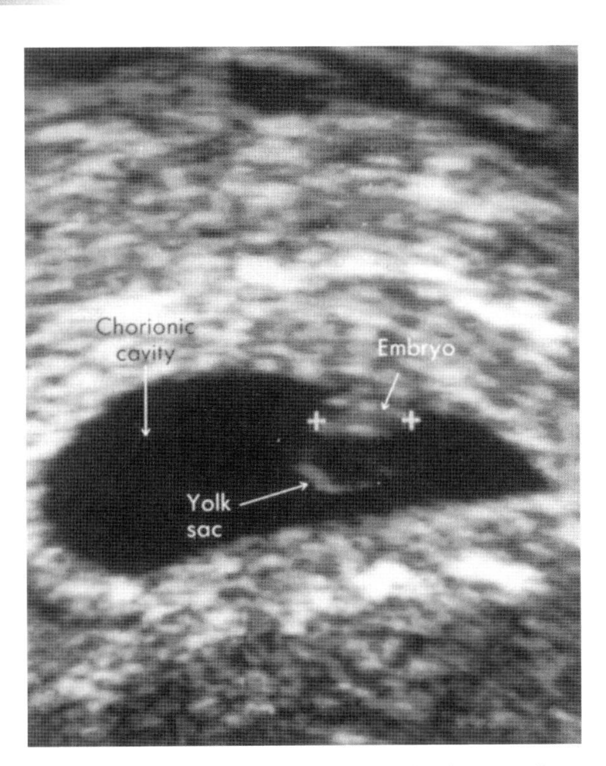

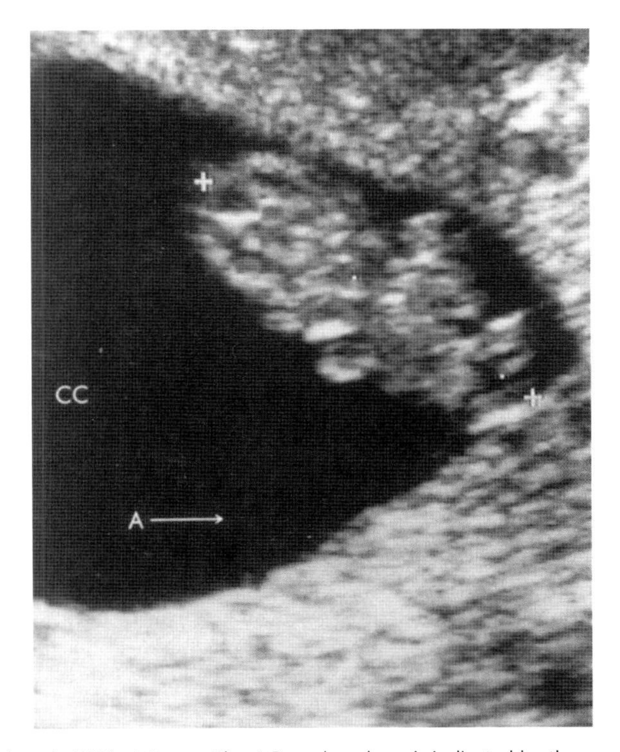

Figure 6 – 13. Ultrasonographic images of embryos. *A*, Crown-rump length (CRL), 4.8 mm. The 4.5-week embryo is indicated by the measurement cursors (+). Ventral to the embryo is the yolk sac. The chorionic cavity appears black. *B*, Coronal scan of a 6.5-week embryo (CRL, 2.09 cm). The upper limbs are visible. The embryo is surrounded by a thin amnion (A). The fluid in the chorionic sac (CC) is more particulate than the amniotic fluid. (Courtesy of Dr. E. A. Lyons, Professor of Radiology and Obstetrics and Gynecology, Health Sciences Centre, University of Manitoba, Winnipeg, Manitoba, Canada.)

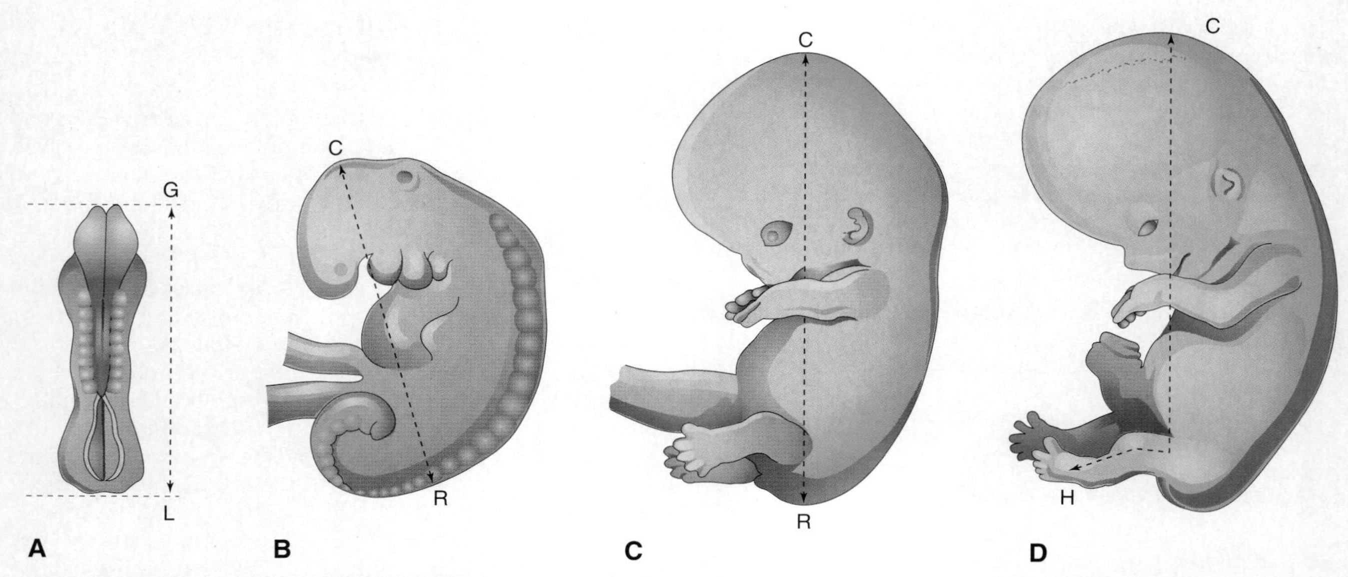

Figure 6-14. Sketches showing the methods used to measure the length of embryos. *A*, Greatest length. *B* and *C*, Crown-rump length. *D*, Crown-heel length.

allantois, and connecting stalk being carried to the ventral surface of the embryo. Folding of the embryo in the horizontal plane incorporates part of the endoderm into the embryo as the *midgut*. The yolk sac remains attached to the midgut by a narrow *yolk stalk*. During folding in the horizontal plane, the primordia of the lateral and ventral body walls are formed. As the amnion expands, it envelops the connecting stalk, yolk stalk, and allantois, thereby forming an epithelial covering for the *umbilical cord*.

The external appearance of the embryo is greatly affected by the formation of the brain, heart, liver, somites, limbs, ears, nose, and eyes. As these structures develop, the appearance of the embryo changes so that it has unquestionably human characteristics by the end of the eighth week. Because the beginnings of most essential external and internal structures are formed during the fourth to eighth weeks, this is the most critical period of development. Developmental disturbances during this period may give rise to major congenital anomalies of the embryo (see Chapter 9).

Clinically Oriented Questions

1. There is little apparent difference between an 8-week embryo and a 9-week fetus. Why do embryologists give them different names?
2. When does the embryo become a human being? What guidelines can be used to distinguish this stage?
3. Can the sex of embryos be determined by ultrasonograpic study? What other methods can be used to diagnose sex?

The answers to these questions are at the back of the book.

7

Fetal Period: Ninth Week to Birth

The transformation of an embryo to a fetus is gradual. Development during the fetal period is concerned primarily with rapid body growth and differentiation of tissues, organs, and systems. The rate of body growth during the fetal period is very rapid, and fetal weight gain is phenomenal during the terminal weeks (Table 7 – 1).

Viability of Fetuses

Viability is defined as the ability of fetuses to survive in the extrauterine environment (i.e., after premature birth). Fetuses weighing less than 500 gm at birth usually do not survive. However, if given expert postnatal care, some fetuses weighing less than this may survive; they are referred to as *extremely low birth weight* (*ELBW*) or *immature infants*. Many full-term, *low birth weight* (*LBW*) babies result from intrauterine growth retardation (*IUGR*). Most fetuses weighing between 1500 and 2500 gm survive but complications may occur; they are *premature infants*. Prematurity is one of the most common causes of morbidity and perinatal death.

Estimation of Fetal Age

If doubt arises about the age of a fetus in a woman with an uncertain medical history, ultrasonographic measurements of the crown-rump length (CRL) can be taken to determine its size and probable age, and to provide a reliable prediction of the *expected date of confinement* (EDC) for delivery of the fetus. Fetal head measurements and femur length may also be used to evaluate the age of the fetus. The intrauterine period may be divided into days, weeks, or months (Table 7 – 2), but confusion arises if it is not stated whether the age is calculated from the onset of the last normal menstrual period (LNMP) — the *gestational* or *menstrual age* — or from the estimated day of fertilization or conception — the *fertilization age*. Unless otherwise stated, fetal age in this book is calculated from the estimated time of fertilization, and months refer to calendar months. It is best to express fetal age in weeks and to state whether the beginning or end of a week is meant, because statements such as "in the 10th week" are nonspecific.

Table 7 – 1. Criteria for Estimating Fertilization Age During the Fetal Period

Age (weeks)	CR Length (mm)*	Foot Length (mm)*	Fetal Weight (gm)†	Main External Characteristics
Previable Fetuses				
9	50	7	8	*Eyelids closing or closed.* Head round. External genitalia still not distinguishable as male or female. Intestines in umbilical cord.
10	61	9	14	*Intestine in abdomen.* Early fingernail development.
12	87	14	45	*Sex distinguishable externally.* Well-defined neck.
14	120	20	110	*Head erect.* Lower limbs well developed. Early toenail development.
16	140	27	200	*Ears stand out from head.*
18	160	33	320	*Vernix caseosa covers skin.* Quickening (signs of life felt by mothers).
20	190	39	460	*Head and body hair (lanugo) visible.*
Viable Fetuses‡				
22	210	45	630	*Skin wrinkled* and red.
24	230	50	820	*Fingernails present.* Lean body.
26	250	55	1000	*Eyes partially open.* Eyelashes present.
28	270	59	1300	*Eyes open.* Good head of hair. Skin slightly wrinkled.
30	280	63	1700	*Toenails present.* Body filling out. Testes descending.
32	300	68	2100	*Fingernails extend to fingertips.* Skin smooth.
36	340	79	2900	*Body usually plump.* Lanugo almost absent. Toenails extend to toe tips. Flexed limbs; firm grasp.
38	360	83	3400	*Prominent chest*; breasts protrude. Testes in scrotum or palpable in inguinal canals. Fingernails extend beyond fingertips.

*These measurements are averages and so may not apply to specific cases; dimensional variations increase with age.
†These weights refer to fetuses that have been fixed for about 2 weeks in 10% formalin. Fresh specimens usually weigh about 5% less.
‡There is no sharp limit of development, age, or weight at which a fetus automatically becomes viable or beyond which survival is ensured, but experience has shown that it is uncommon for a baby to survive if its weight is less than 500 gm or its fertilization age or developmental age is less than 22 weeks. Even fetuses born during the 26- to 28-week period have difficulty surviving, mainly because the respiratory and central nervous systems are not completely differentiated. The term *abortion* refers to all pregnancies that terminate before the period of viability.

Table 7-2. Comparison of Gestational Time Units

Reference Point	Days	Weeks	Calendar Months	Lunar Months
Fertilization*	266	38	$8^3/_4$	$9^1/_2$
LNMP	280	40	$9^1/_4$	10

*The date of birth is calculated as 266 days after the estimated day of fertilization, or 280 days after the onset of the last normal menstrual period (*LNMP*). From fertilization to the end of the embryonic period (8 weeks), age is best expressed in days; thereafter, age is often given in weeks.

Clinically, the gestational period is divided into three **trimesters**, each lasting 3 months. At the end of the *first trimester*, all major systems are developed. At the end of the *second trimester*, the fetus may survive if born prematurely. The fetus matures during the *third trimester* and reaches a major developmental landmark at 35 weeks of gestation. It weighs about 2500 gm, which indicates the usual level for fetal maturity.

Various measurements and external characteristics are useful for estimating fetal age (see Table 7-1). The CRL is the method of choice for estimating fetal age until the end of the first trimester because there is very little variability in fetal size during this period. In the second and third trimesters, several structures can be identified and measured ultrasonographically, but the basic measurements used for age estimation are the following:

- *biparietal diameter* (BPD) — the diameter of the head measured between the two parietal eminences
- head circumference
- abdominal circumference
- femur length
- foot length

Foot length correlates well with CRL and is particularly useful for estimating the age of incomplete or macerated fetuses. *Fetal weight* is often a useful criterion for estimating age, but a discrepancy may exist between the age and the weight of a fetus, particularly if the mother had experienced metabolic disturbances, such as diabetes mellitus, during pregnancy. Cheek-to-cheek and transverse cerebellar measurements have also been used to assess fetal growth and gestational age, respectively. Determination of the size of a fetus, especially of its head, is of great value to the obstetrician for the management of certain patients, such as women with small pelves or fetuses with IUGR or congenital anomalies, or both.

Highlights of the Fetal Period

There is no formal staging system for the fetal period. However, it is helpful to consider the changes that occur in terms of periods of 4 to 5 weeks.

Nine to Twelve Weeks

At the beginning of the ninth week, the head constitutes half the CRL of the fetus (Figs. 7-1 and 7-2). Subsequently, growth in body length accelerates rapidly, and by the end of 12 weeks, the CRL has more than doubled (see Table 7-1). At 9 weeks, the face is broad, the eyes are widely separated, the ears are low-set, and the eyelids are fused. By the end of 12 weeks, *primary ossification centers* appear in the skeleton, especially in the cranium and long bones. The eyelids are fused throughout this period. Early in the ninth week, the legs are short and the thighs are relatively small. By the end of 12 weeks, the upper limbs have almost reached their final relative lengths, but the lower limbs are still not so well developed and are slightly shorter than their final relative lengths. *The external genitalia of males and females appear similar until the end of the ninth week.* Their mature fetal form is not established until the 12th week. Intestinal coils are clearly visible in the proximal end of the umbilical cord until the middle of

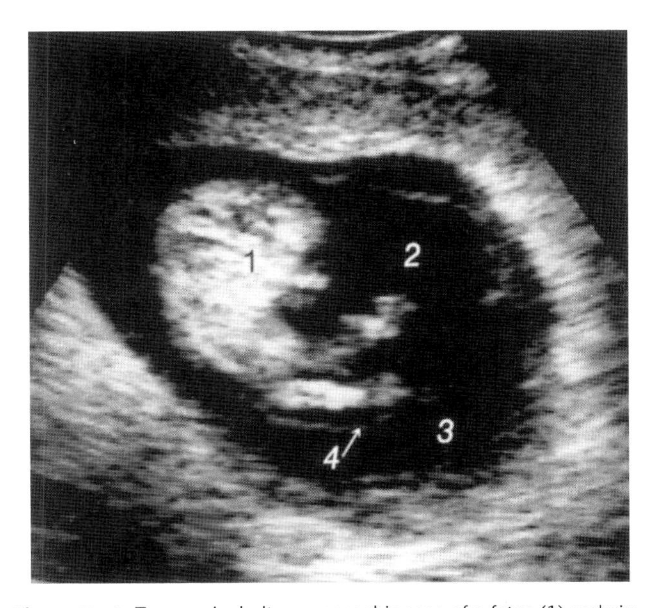

Figure 7-1. Transvaginal ultrasonographic scan of a fetus (1) early in the ninth week. The relationship of the fetus to the amniotic cavity (2), the chorionic cavity (3), and the amnion (4) is demonstrated. (From Wathen NC, Cass PL, Kitan MJ, Chard T: Human chorionic gonadotrophin and alpha-fetoprotein levels in matched samples of amniotic fluid, extra-embryonic coelomic fluid, and maternal serum in the first trimester of pregnancy. *Prenat Diagn* 11:145, 1991.)

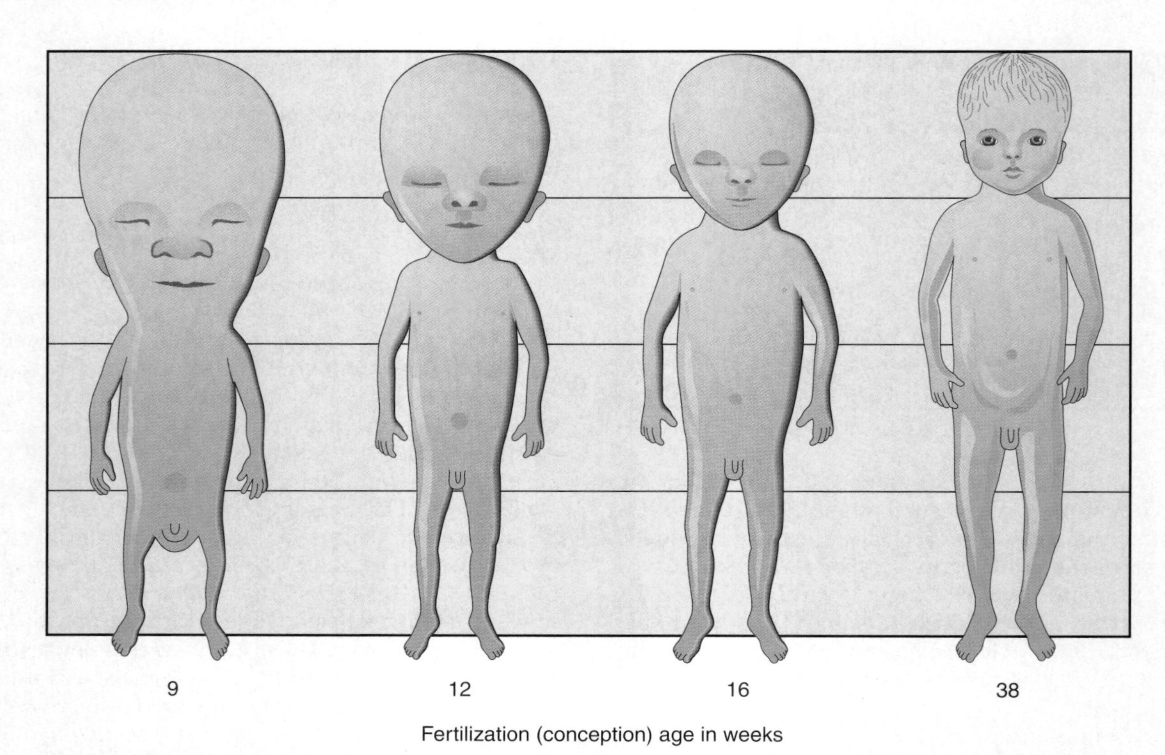

Figure 7 - 2. Diagram illustrating the changing proportions of the body during the fetal period. At 9 weeks, the head is nearly half the crown-heel length of the fetus. By 36 weeks, the circumferences of the head and the abdomen are approximately equal. All stages are drawn to the same total height.

the 10th week. By the 11th week, the intestines have returned to the abdomen (Fig. 7 - 3).

At the beginning of the fetal period, the liver is the major site of *erythropoiesis* (formation of red blood cells). By the end of the 12th week, this activity has decreased in the liver and has begun in the spleen. *Urine formation* begins between the 9th and 12th weeks, and urine is discharged into the amniotic fluid. The fetus reabsorbs some of this fluid after swallowing it. Fetal waste products are transferred to the maternal circulation by passing across the placental membrane (see Chapter 8).

Thirteen to Sixteen Weeks

Growth is very rapid during the 13th to 16th weeks of development (Figs. 7 - 4 and 7 - 5; see Table 7 - 1). By 16 weeks, the head is relatively small compared with that of the 12-week fetus, and the lower limbs have lengthened. Limb movements, which first occur at the end of the embryonic period (8 weeks), become coordinated by the 14th week, but are too slight to be felt by the mother. These movements are visible during ultrasonographic examinations. *Ossification of the fetal skeleton* is progressing during this period, and by the beginning of the 16th week, the bones are clearly visible on ultrasound images of the mother's abdomen. *Slow eye movements occur at 14 weeks* (16 weeks after the LNMP). Scalp hair patterning is also determined during this period. By 16 weeks, the ovaries in female fetuses are differentiated and contain primordial follicles that have oogonia. The external genitalia can be recognized by 14 weeks, and by 16 weeks, the appearance of the fetus is even more human because its eyes face anteriorly rather than anterolaterally.

Seventeen to Twenty Weeks

Growth slows down during weeks 17 to 20, but the fetus still increases its CRL by about 50 mm (see Figs. 7 - 4 and 7 - 6; Table 7 - 1). The limbs reach their final relative proportions, and fetal movements — known as **quickening** — are commonly felt by the mother. The skin is now covered with a greasy material, called **vernix caseosa**, which consists of a fatty secretion from the fetal sebaceous glands and dead epidermal cells. The vernix caseosa protects the delicate fetal skin from abrasions, chapping, and hardening that could result from exposure to the

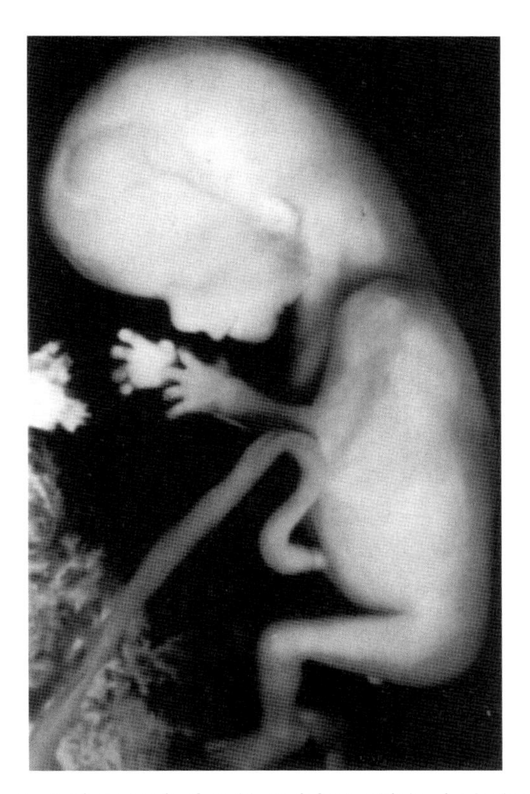

Figure 7-3. Photograph of an 11-week fetus with its chorionic and amniotic sacs removed (×1.5). Note that the head is relatively large. (Courtesy of Professor Jean Hay [Retired], Department of Human Anatomy and Cell Science, University of Manitoba, Winnipeg, Manitoba, Canada.)

amniotic fluid. Eyebrows and head hair are also visible at 20 weeks. The bodies of 20-week fetuses are usually completely covered with fine downy hair called *lanugo*, which helps to hold the vernix caseosa on the skin. *Brown fat* forms during weeks 17 through 20 and is the site of heat production, particularly in the newborn infant. This specialized adipose tissue produces heat by oxidizing fatty acids. By 18 weeks, the uterus is formed in female fetuses, and canalization of the vagina has begun. By this time, many *primordial ovarian follicles* containing oogonia have formed. In male 20-week fetuses, the *testes* have begun to descend, but they are still located on the posterior abdominal wall, as are the *ovaries* in female fetuses.

Twenty-One to Twenty-Five Weeks

Substantial weight gain occurs during weeks 21 to 25. Although still somewhat lean, the fetus is better proportioned. At 21 weeks, rapid eye movements begin; *blink-startle responses* have been reported at 22 to 23 weeks following application of a vibroacoustic noise source to the mother's abdomen. By 24 weeks, the secretory epithelial cells (type II pneumocytes) in the interalveolar walls of the lung have begun to secrete *surfactant*, a surface-active lipid that maintains the patency of the developing alveoli of the lungs. *Fingernails* are also present by 24 weeks. Although a 22-

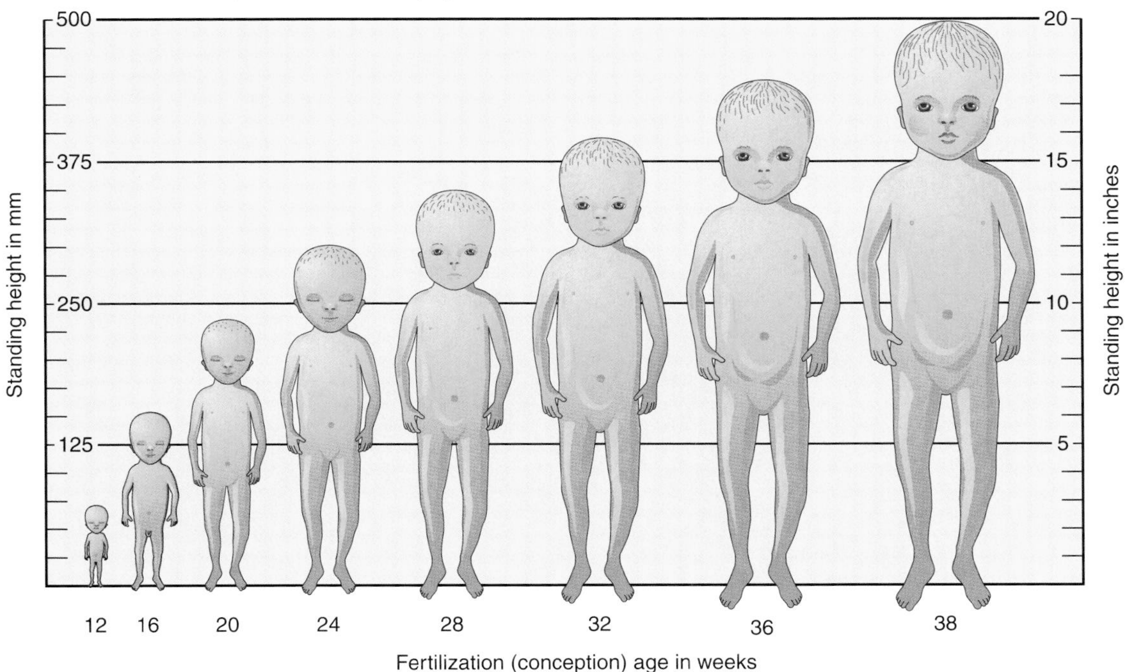

Figure 7-4. Diagram, drawn to scale, illustrating the progressive changes in the size of the human fetus.

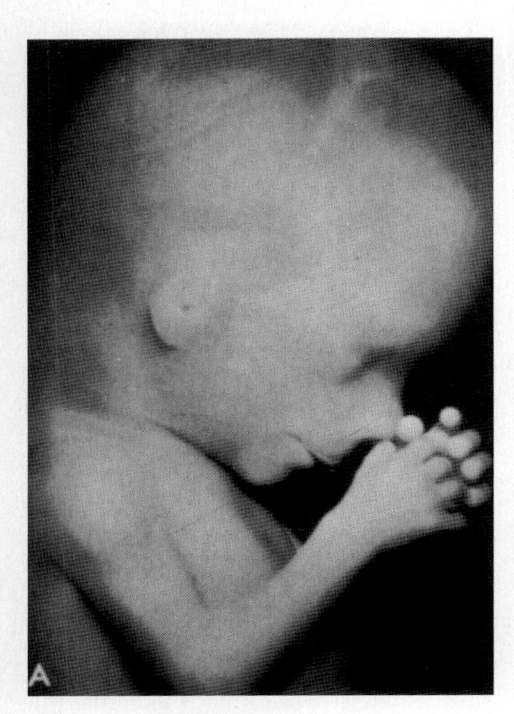

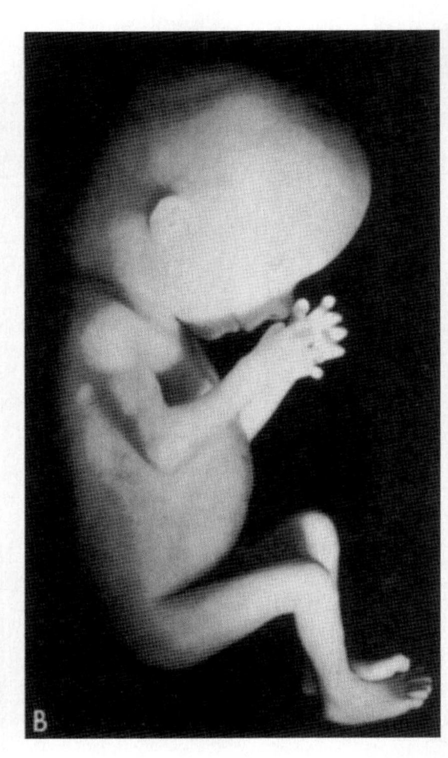

Figure 7 - 5. Photographs of a 13-week fetus. *A,* An enlarged photograph of the head and shoulders (×2). *B,* Actual size. (Courtesy of Professor Jean Hay [Retired], Department of Human Anatomy and Cell Science, University of Manitoba, Winnipeg, Manitoba, Canada.)

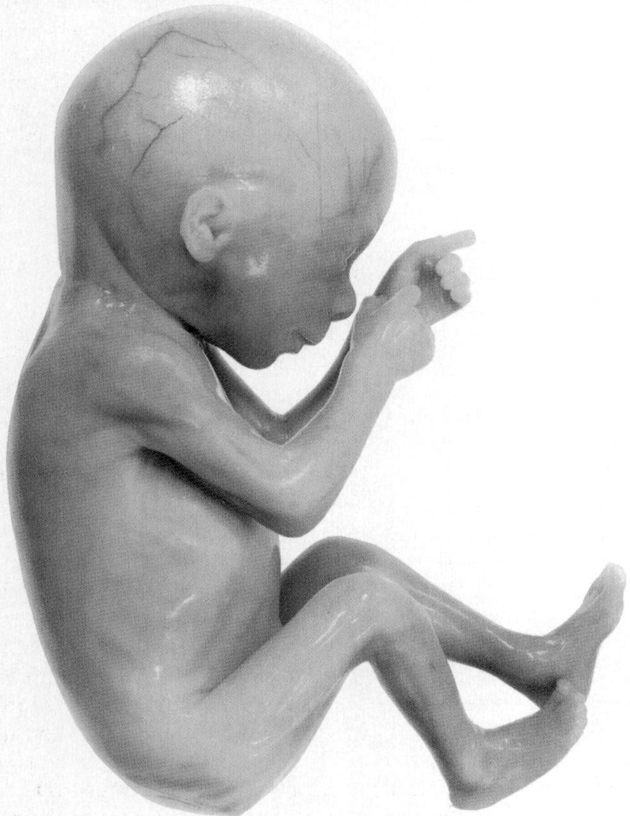

Figure 7 - 6. Photograph of a 17-week fetus (actual size). Fetuses at this age are unable to survive if born prematurely, mainly because their respiratory system is immature. (From Moore KL, Persaud TVN, Shiota K: *Color Atlas of Clinical Embryology*, 2nd ed. Philadelphia, WB Saunders, 2000.)

to 25-week fetus born prematurely may survive if given intensive care, it may die during early infancy because its respiratory system is still immature.

Twenty-Six to Twenty-Nine Weeks

At 26 to 29 weeks, a fetus can often survive if born prematurely and given intensive care because its *lungs are now capable of breathing air*. The lungs and pulmonary vasculature have developed sufficiently to provide adequate gas exchange. In addition, the central nervous system has matured to the stage at which it can direct rhythmic breathing movements and control body temperature. The greatest neonatal losses occur in LBW infants (weighing 2500 gm or less) and very low birth weight (VLBW) infants (weighing 1500 gm or less). *The eyelids are open at 26 weeks*, and lanugo and head hair are well developed. Toenails are visible and considerable subcutaneous fat is now present, smoothing out many of the wrinkles. *The fetal spleen is now an important site of hematopoiesis*, the process of formation and development of various types of blood cells and other formed elements. Erythropoiesis in the spleen ends by 28 weeks, by which time bone marrow has become the major site of this process.

Thirty to Thirty-Four Weeks

The pupillary light reflex of the eyes can be elicited by 30 weeks. Usually by the end of this period, the skin is pink and smooth, and the upper and lower limbs have a chubby appearance. Fetuses 32 weeks and older usually survive if born prematurely. If a normal-weight fetus is born during this period, it is "premature by date" as opposed to being "premature by weight."

Thirty-Five to Thirty-Eight Weeks

Fetuses at 35 weeks have a firm grasp and exhibit a spontaneous orientation to light. As term approaches (37 to 38 weeks), the nervous system is sufficiently mature to carry out some integrative functions. Most fetuses during this "finishing period" are plump (Fig. 7-7). By 36 weeks, the circumferences of the head and abdomen are approximately equal. Growth slows as the time of birth approaches (Fig. 7-8).

Normal fetuses usually have a CRL of 360 mm and weigh about 3400 gm. By full term, the amount of white fat is about 16% of body weight. A fetus adds about 14 gm of fat a day during these last weeks of gestation. The chest is prominent and the breasts often protrude slightly in both sexes. The testes are usually in the scrotum in full-term male infants; premature male infants commonly have undescended testes. Although the head is smaller at full term in relation to the rest of the body than it was earlier in fetal life, it still is one of the largest regions of the fetus. This is an important consideration related to its passage through the birth canal.

Expected Date of Delivery

The expected date of delivery (EDD) of a fetus is 266 days, or 38 weeks, after fertilization (i.e., 280 days, or 40 weeks, after the LNMP) (see Table 7-1). About 12% of babies, however, are born 1 to 2 weeks after the expected time of birth.

Determination of Delivery Date

The common delivery date method (*Nägele's rule*) for determining the EDD or the EDC is to count back 3 months from the first day of the LNMP and add 1 year and 7 days. For example:

- First day of the LNMP = January 4, 2003
- Subtract 3 months = October 4, 2002
- Add a year and 7 days = October 11, 2003 (the EDD)

In women with regular menstrual cycles, this method gives a reasonably accurate EDD. However, if the woman's cycles are irregular, miscalculations of 2 to 3 weeks may occur. In addition, *implantation bleeding* occurs in some pregnant women at the time of the first missed period (about 2 weeks after fertilization). Should the woman interpret this bleeding as a normal menstruation, the estimated time of birth could be miscalculated by 2 or more weeks. As a result, ultrasonographic examinations of the fetus — in particular, CRL measurements during the first trimester — are commonly used to predict the EDD more reliably.

Factors Influencing Fetal Growth

The fetus requires substrates for growth and production of energy. Gases and nutrients pass freely to the fetus from the mother through the placental membrane (see Chapter 8). **Glucose** is a primary source of energy for fetal metabolism and growth; *amino acids* are also required. These substances pass from the mother's blood to the fetus through the placental membrane. **Insulin**, which is required for the metabolism of glucose, is secreted by the fetal pancreas. Insulin, human growth hormone, and some small polypeptides (such as insulin-like growth factor I) are believed to stimulate fetal growth.

Many factors — maternal, fetal, and environmental — may affect prenatal growth. In general, factors operating throughout pregnancy, such as *cigarette smoking* and *consumption of alcohol*, tend to produce IUGR and small infants, whereas factors operating during the last trimester (e.g., maternal malnutrition) usually produce underweight infants with normal length and head size.

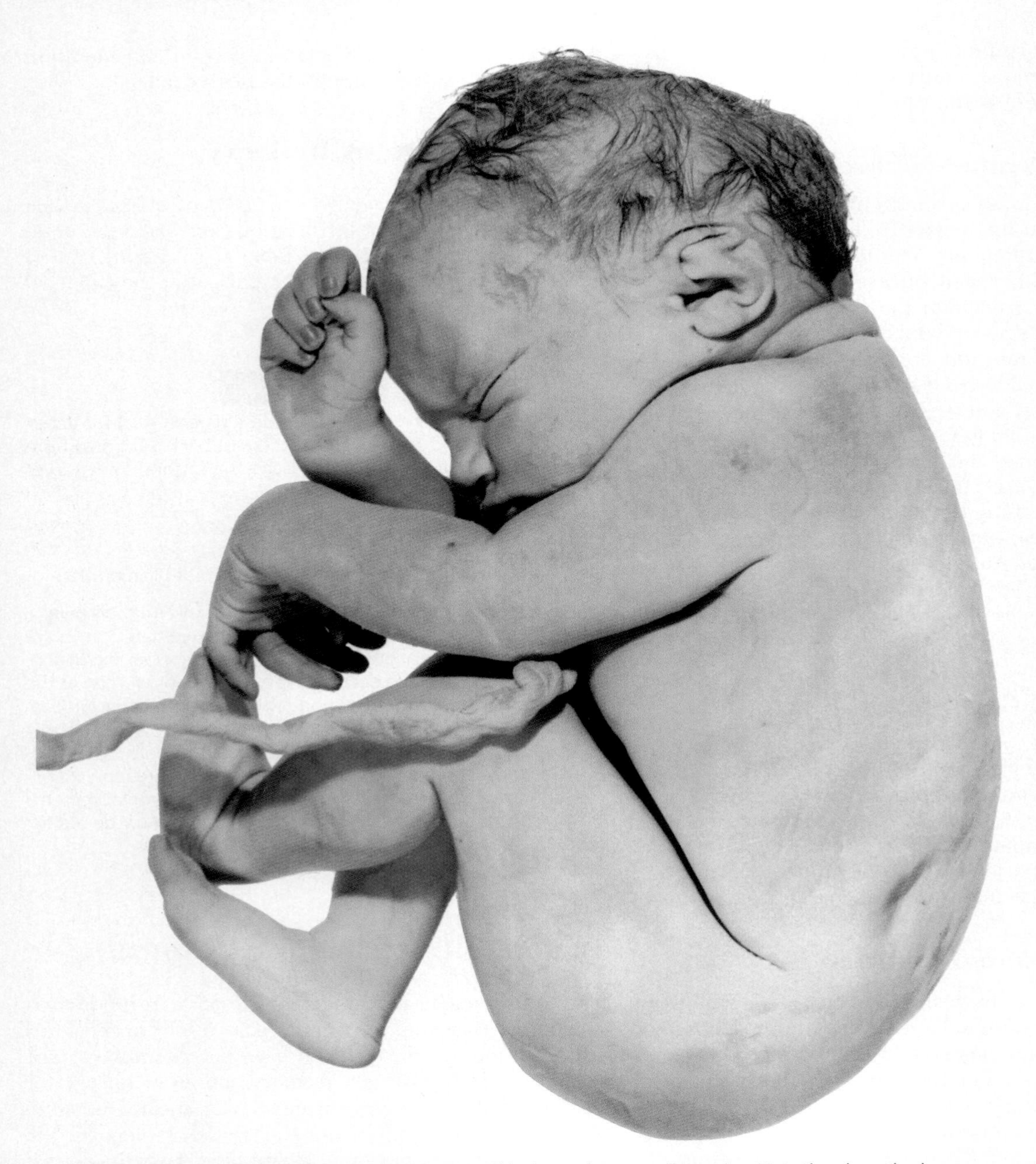

Figure 7-7. Photograph of a 36-week fetus. Fetuses at this size and age usually survive. Note the plump body resulting from the deposition of subcutaneous fat. This fetus' mother was killed in an automobile accident, and the fetus died before it could be delivered by cesarean section. (From Moore KL, Persaud TVN, Shiota K: *Color Atlas of Clinical Embryology*, 2nd ed. Philadelphia, WB Saunders, 2000.)

IUGR is usually defined as infant weight within the lowest 10th percentile for gestational age. Severe malnutrition resulting from a poor-quality diet is known to cause reduced fetal growth (see Fig. 7-8). Poor nutrition and faulty food habits are common during pregnancy and are not restricted to mothers living in poverty.

Cigarette Smoking

Smoking is a well-established cause of IUGR. The growth rate for fetuses of mothers who smoke cigarettes is less than normal during the last 6 to 8 weeks of pregnancy (see Fig. 7-8). On average, the birth weight of infants

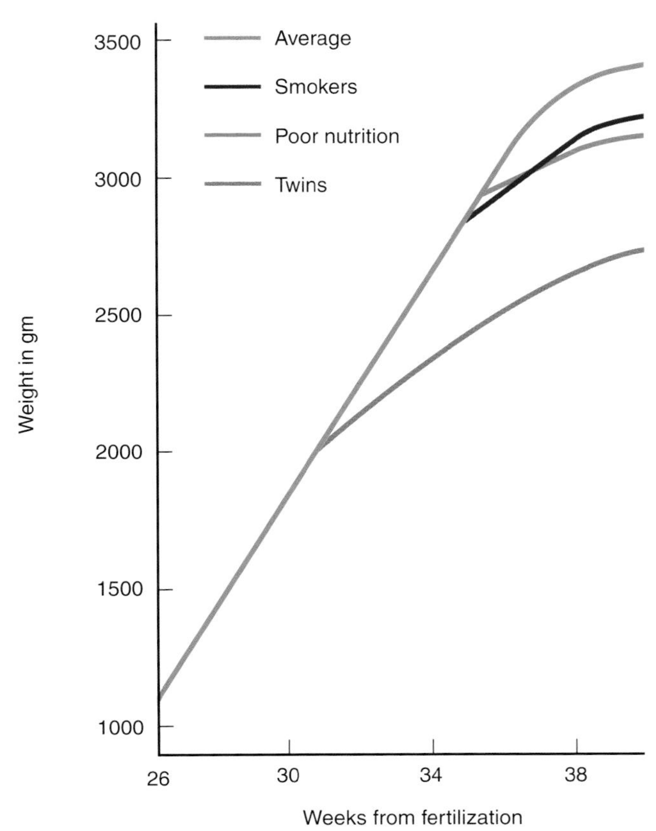

Figure 7 - 8. Graph showing the rate of fetal growth during the last trimester. The "average" refers to the rate of growth of babies born in the United States. After 36 weeks, the growth rate deviates from the straight line. The decline, particularly after full term (38 weeks), probably reflects inadequate fetal nutrition caused by placental changes. (Adapted from Gruenwald P: Growth of the human fetus. I. Normal growth and its variation. *Am J Obstet Gynecol* 94:1112, 1966.)

whose mothers smoke heavily during pregnancy is 200 gm less than normal, and *perinatal morbidity* is increased when adequate medical care is unavailable.

Multiple Pregnancy

Individuals of twin, triplet, and other multiple births usually weigh considerably less than infants resulting from a single pregnancy (see Fig. 7-8). It is evident that the total requirements of two or more fetuses exceed the nutritional supply available from the placenta during the third trimester.

Social Drugs

Infants born to alcoholic mothers often exhibit IUGR as part of the *fetal alcohol syndrome*. Similarly, the use of *marijuana* and other illicit drugs (e.g., *cocaine*) can cause IUGR and other obstetric complications.

Impaired Uteroplacental Blood Flow

Maternal-placental circulation may be reduced by conditions that decrease uterine blood flow (e.g., small chorionic or umbilical vessels, severe hypotension, and renal disease). Chronic reduction of uterine blood flow can cause *fetal starvation*, resulting in IUGR. Placental dysfunction or defects (e.g., infarction; see Chapter 8) can also cause IUGR. The net effect of these placental abnormalities is a reduction of the total area for exchange of nutrients between the fetal and maternal bloodstreams.

Genetic Factors and Growth Retardation

It is well established that genetic factors can cause IUGR. Repeated cases of IUGR in one family indicate that recessive genes may be the cause of the abnormal growth. In recent years, structural and numerical chromosomal aberrations have also been shown to be associated with cases of retarded fetal growth. IUGR is pronounced in infants with Down syndrome and is also characteristic of fetuses with trisomy 18 syndrome.

Procedures for Assessing Fetal Status

Perinatology is the branch of medicine that is concerned with the well-being of the fetus and newborn infant, generally encompassing the period from about 26 weeks after fertilization to 4 weeks after birth. The subspecialty of *perinatal medicine* combines aspects of obstetrics and pediatrics. A third-trimester fetus is commonly regarded as an *unborn patient* on whom diagnostic and therapeutic procedures may be performed. Several techniques are now available for assessing the status of the fetus and providing prenatal treatment if required. Fetal activity felt by the mother or palpated by the physician was the first identified clue to fetal well-being. Then the fetal heartbeat was detected, first by auscultation and later by electronic monitors. These techniques indicated when fetal stress and distress were occurring. Later, gonadotropic hormones were detected in maternal blood. Many new procedures for assessing the status of the fetus have been developed in the last two decades. It is now possible to treat many fetuses whose lives are in jeopardy.

Diagnostic Amniocentesis

Diagnostic amniocentesis is a common invasive prenatal diagnostic procedure. For prenatal diagnosis, amniotic fluid is sampled by inserting a hollow needle through the mother's anterior abdominal and uterine walls and into the amniotic cavity, piercing the chorion and amnion (Fig. 7 - 9*A*). A syringe is then attached to the needle, and amniotic fluid is withdrawn. Because relatively little amniotic fluid is present prior to the 14th week after the LNMP, amniocentesis is difficult to perform prior to this time. The amniotic fluid volume is approximately 200 ml, and 20 to 30 ml can be withdrawn safely. The procedure is relatively devoid of risk, especially when the procedure is performed by an experienced physician who uses ultrasonography as a guide for outlining the position of the fetus and placenta.

Alpha-Fetoprotein Assay

Alpha-fetoprotein (AFP), a glycoprotein that is synthesized in the fetal liver and yolk sac, escapes from the fetal circulation into the amniotic fluid in fetuses with open neural tube defects (NTDs), such as spina bifida with myeloschisis (see Chapter 19). Open NTDs refer to lesions that are not covered with skin. AFP can also enter the amniotic fluid from open ventral wall defects (VWDs), as occurs with gastroschisis and omphalocele (see Chapter 13).

Chorionic Villus Sampling

Biopsies of chorionic villi (mostly trophoblasts) may be performed by inserting a needle, guided by ultrasonography, through the mother's abdominal and uterine walls and into the uterine cavity (Fig. 7 - 9*B*). Chorionic villus sampling (CVS) may also be performed transcervically using real-time ultrasound guidance.

Diagnostic Value of Chorionic Villus Sampling

Biopsies of chorionic villi are performed to detect chromosomal abnormalities, inborn errors of metabolism, and X-linked disorders. CVS can be performed as early as the ninth week of gestation (7 weeks after fertilization). The rate of fetal loss is about 1%, slightly more than the risk from amniocentesis. The major advantage of CVS over amniocentesis is that it allows the results of chromosomal analysis to be available several weeks earlier.

Cell Cultures

Fetal sex and chromosomal aberrations can also be determined by studying the sex chromosomes in cultured fetal cells obtained during amniocentesis. These cultures are commonly done when an autosomal abnormality such as occurs in Down syndrome is suspected. Fluorescent in situ hybridization (**FISH**) analysis, using DNA-specific probes, is now widely used for the rapid detection of chromosomal aneuploidies. Inborn errors of metabolism in fetuses can also be detected by studying cell cultures. Enzyme deficiencies can be determined by incubating cells recovered from amniotic fluid and then detecting the specific enzyme deficiency in the cells.

Intrauterine Fetal Transfusion

Some fetuses with *hemolytic disease of the newborn* (HDN) can be saved by receiving intrauterine blood transfusions. The blood is injected through a needle inserted into the fetal peritoneal cavity. Over a period of 5 to 6 days, most of the injected blood cells pass into the fetal circulation through the diaphragmatic lymphatics. With recent advances in *percutaneous umbilical cord puncture*, blood can be transfused directly into the fetal cardiovascular system. The need for fetal blood transfusions has been reduced as a result of the treatment of Rh-negative mothers of Rh-positive fetuses with anti-Rh immunoglobulin. Consequently, *HDN is relatively uncommon now* because Rh immunoglobulin usually prevents development of this disease of the Rh system.

Percutaneous Umbilical Cord Blood Sampling

Fetal blood samples may be obtained from the umbilical vessels for chromosomal analysis by percutaneous umbilical cord blood sampling (PUBS). Ultrasonographic scanning facilitates the procedure by outlining the location of the vessels. PUBS is often performed about 20 weeks after the LNMP to obtain samples for chromosomal analysis when ultrasonographic or other examinations have shown characteristics of fetal anomalies, such as trisomy 13 (see Chapter 9).

Ultrasonography

Ultrasonography is the primary imaging modality in the evaluation of the fetus because of its wide availability, low cost, and lack of known adverse effects. The chorionic (gestational) sac and its contents may be visualized by ultrasonography during the embryonic and fetal periods (Fig. 7 - 10). Placental and fetal size, multiple births, abnormalities of placental shape, and abnormal presentations can also be determined. *Ultrasonographic scans* also give accurate measurements of the BPD of the fetal cranium, from which close estimates of fetal age and length can be made. Some developmental defects can also be detected prenatally by ultrasonography.

Computed Tomography and Magnetic Resonance Imaging

When planning fetal treatment, such as surgery, computed tomography (CT) and magnetic resonance imaging (MRI) may be used. These studies to can provide additional information about an abnormality detected ultrasonographically.

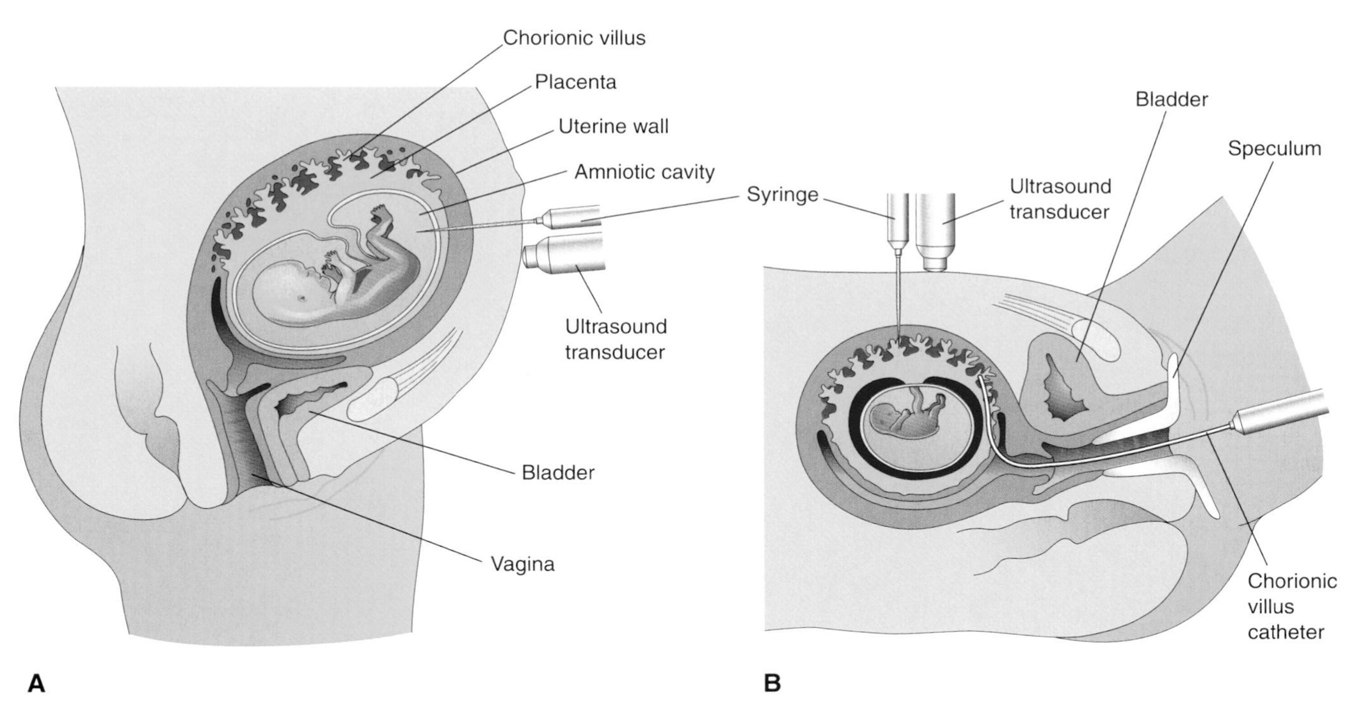

Figure 7-9. *A*, Drawing illustrating the technique of amniocentesis. Under ultrasonographic guidance, a needle is inserted through the mother's abdominal wall and uterine wall into the amniotic cavity. A syringe is attached, and amniotic fluid is withdrawn for diagnostic purposes. *B*, Drawing illustrating chorionic villus sampling (CVS). This technique is usually performed at about the ninth week after the last normal menstrual period (LNMP). Two sampling approaches are illustrated,: one through the maternal anterior abdominal wall using a spinal needle, and one through the vagina and cervical canal using a malleable catheter.

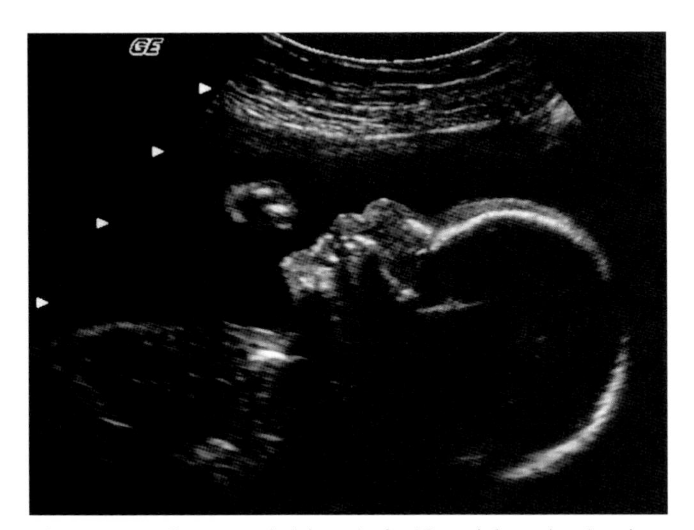

Figure 7-10. Sonogram (axial scan) of a 22-week fetus showing the facial profile of the fetus. The nose, lips, and mandible are visible. (Courtesy of Dr. EA Lyons, Department of Radiology, University of Manitoba, Health Sciences Centre, Winnipeg, Manitoba, Canada. From Moore KL, Persaud TVN, Shiota K: *Color Atlas of Clinical Embryology*, 2nd ed. Philadelphia, WB Saunders, 2000.)

Fetal Monitoring

Continuous fetal heart rate monitoring in high-risk pregnancies is routine and provides information about the oxygenation of the fetus. *Fetal distress*, as indicated by an abnormal heart rate or rhythm, suggests that the fetus is in jeopardy.

Summary of the Fetal Period

The fetal period begins 9 weeks after fertilization (11 weeks after the LNMP) and ends at birth. It is characterized by rapid body growth and differentiation of tissues and organ systems. An obvious change in the fetal period is the relative slowing of head growth compared with that of the rest of the body. By the beginning of the 20th week, lanugo and head hair appear, and the skin is coated with vernix caseosa. The eyelids are closed during most of the fetal period but begin to reopen at about 26 weeks. At this time, the fetus is usually capable of extrauterine existence, mainly because of the maturity of its respiratory system. Until about 30 weeks, the fetus appears reddish and wizened because of the thinness of its skin and the relative absence of subcutaneous fat. Fat usually develops rapidly during the last 6 to 8 weeks, giving the fetus a smooth, plump appearance. This terminal ("finishing") period is devoted mainly to building up tissues and preparing systems involved in the transition from intrauterine to extrauterine environments, primarily the respiratory and cardiovascular systems.

Various techniques are available for assessing the status of the fetus and for diagnosing certain diseases and developmental anomalies before birth. The physician can now determine whether or not a fetus has a particular disease or a congenital anomaly by using various diagnostic techniques, such as amniocentesis and ultrasonography. A prenatal diagnosis can now be established early enough to allow early termination of the pregnancy if desired, as in cases in which serious anomalies incompatible with postnatal life are diagnosed. In some cases, the fetus can be treated prenatally, for example, by administering drugs to correct cardiac arrhythmia or thyroid disorders. Surgical correction of selected life-threatening anomalies in utero is also possible, such as ureterostomies performed on fetuses that have ureters that do not open into the bladder.

Clinically Oriented Questions

1. Some say that the mature embryo twitches and that a first-trimester fetus moves its limbs. Is this true? If so, can the mother feel her baby kicking at this time?
2. Some reports suggest that vitamin supplementation around the time of conception will prevent neural tube defects (NTDs), such as spina bifida. Is there scientific proof for this statement?
3. Can the fetus be injured by the needle during amniocentesis? Is there a risk of inducing a miscarriage or causing maternal or fetal infection?

The answers to these questions are at the back of the book.

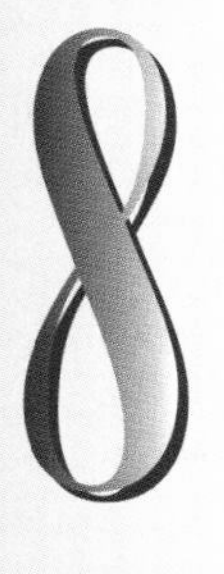

The Placenta and Fetal Membranes

The fetal part of the placenta and the fetal membranes separate the embryo or fetus from the endometrium of the uterus. An interchange of substances (e.g., nutrients and oxygen) occurs between the maternal and fetal blood through the *placenta*. The vessels in the umbilical cord connect the placental circulation with the fetal circulation. *The chorion, amnion, yolk sac, and allantois constitute the fetal membranes.* The placenta is a membrane-like organ that develops in higher mammals, such as humans. The fetal membranes develop from the zygote but do not participate in the formation of the embryo or fetus, except for parts of the yolk sac and allantois. Part of the yolk sac is incorporated into the embryo as the primordium of the gut. The allantois becomes a fibrous cord that is known as the urachus in the fetus and the median umbilical ligament in the adult.

The Placenta

The placenta, the primary site of nutrient and gas exchange between the mother and fetus, is a **fetomaternal organ** that has two components:

- a **fetal part** that develops from part of the chorionic sac
- a **maternal part** that is derived from the endometrium

The placenta and umbilical cord function as a *transport system* for substances passing between the mother and fetus. Nutrients and oxygen pass from the maternal blood through the placenta to the fetal blood, and waste materials and carbon dioxide pass from the fetal blood through the placenta to the maternal blood. The placenta and fetal membranes perform the following functions and activities:

- protection
- nutrition
- respiration
- excretion
- hormone production

Shortly after the birth of a baby, the placenta and fetal membranes are expelled from the uterus as the *afterbirth*.

Decidua

The decidua (L., *deciduus*, a falling off) refers to the *gravid endometrium*, the functional layer of the endometrium in a pregnant woman. The term *decidua* is appropriate because this part of the endometrium separates ("falls away") from the remainder of the uterus after *parturition* (childbirth).

Three regions of the decidua are named according to their relation to the implantation site (Fig. 8 – 1):

- **decidua basalis** — the part of the decidua deep to the conceptus that forms the maternal part of the placenta
- **decidua capsularis** — the superficial part of the decidua overlying the conceptus
- **decidua parietalis** — all the remaining parts of the decidua

In response to increasing progesterone levels in the maternal blood, the connective tissue cells of the decidua enlarge to form pale-staining **decidual cells**. These cells enlarge as glycogen and lipid accumulate in their cytoplasm. The cellular and vascular changes in the decidua resulting from pregnancy are referred to as the **decidual reaction**. Many decidual cells degenerate near the chorionic sac in the region of the *syncytiotrophoblast* and, together with maternal blood and uterine secretions, provide a rich source of nutrition for the embryo. Decidual regions clearly recognizable during *ultrasonography* are important in diagnosing early pregnancy.

Development of the Placenta

As described previously, early placental development is characterized by the rapid proliferation of the trophoblast and development of the chorionic sac and chorionic villi (see Chapters 4 and 5). By the end of the third week of development, the anatomical arrangements necessary for physiological exchanges between the mother and embryo have been established. By the end of the fourth week, a complex vascular network develops in the placenta, allowing maternal-embryonic exchanges of gases, nutrients, and metabolic waste products. Chorionic villi cover the entire chorionic sac until the beginning of the eighth week (see Figs. 8 – 1*C* and 8 – 2). As this sac grows, the villi associated with the decidua capsularis are compressed, reducing the blood supply to them. These villi soon degenerate (see Fig. 8 – 1*D*), producing a relatively avascular bare area, the **smooth chorion**. As the villi disappear, those associated with the decidua basalis rapidly increase in number, branch profusely, and enlarge (Fig. 8 – 3). This bushy part of the chorionic sac is known as the **villous chorion**.

Ultrasonography of the Chorionic Sac

The size of the chorionic sac is useful in determining the *gestational age* of embryos in pregnant women with uncertain menstrual histories. The early chorionic sac is filled with *chorionic fluid* because the amniotic sac containing the embryo and the yolk sac are relatively small. Growth of the chorionic sac is extremely rapid between the 5th and 10th weeks of development. Modern ultrasound devices, in particular,

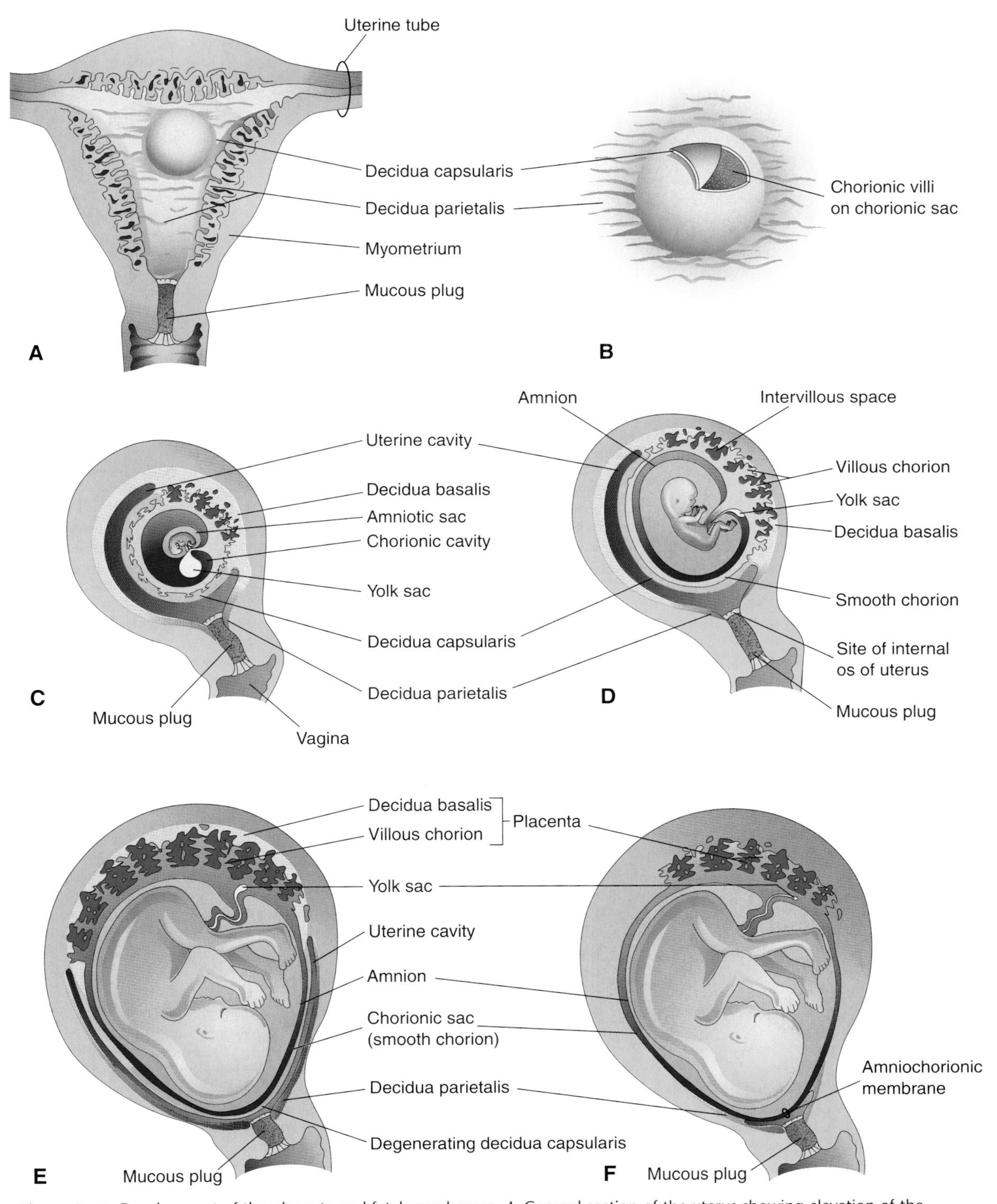

Figure 8 - 1. Development of the placenta and fetal membranes. *A*, Coronal section of the uterus showing elevation of the decidua capsularis by the expanding chorionic sac in a 4-week embryo. *B*, Enlarged drawing of the implantation site. The chorionic villi were exposed by cutting an opening in the decidua capsularis. *C* to *F*, Sagittal sections of the gravid uterus from the 5th to 22nd weeks, showing the changing relationship of the fetal membranes to the decidua. In *F*, the amnion and chorion are fused with each other and with the decidua parietalis, thereby obliterating the uterine cavity.

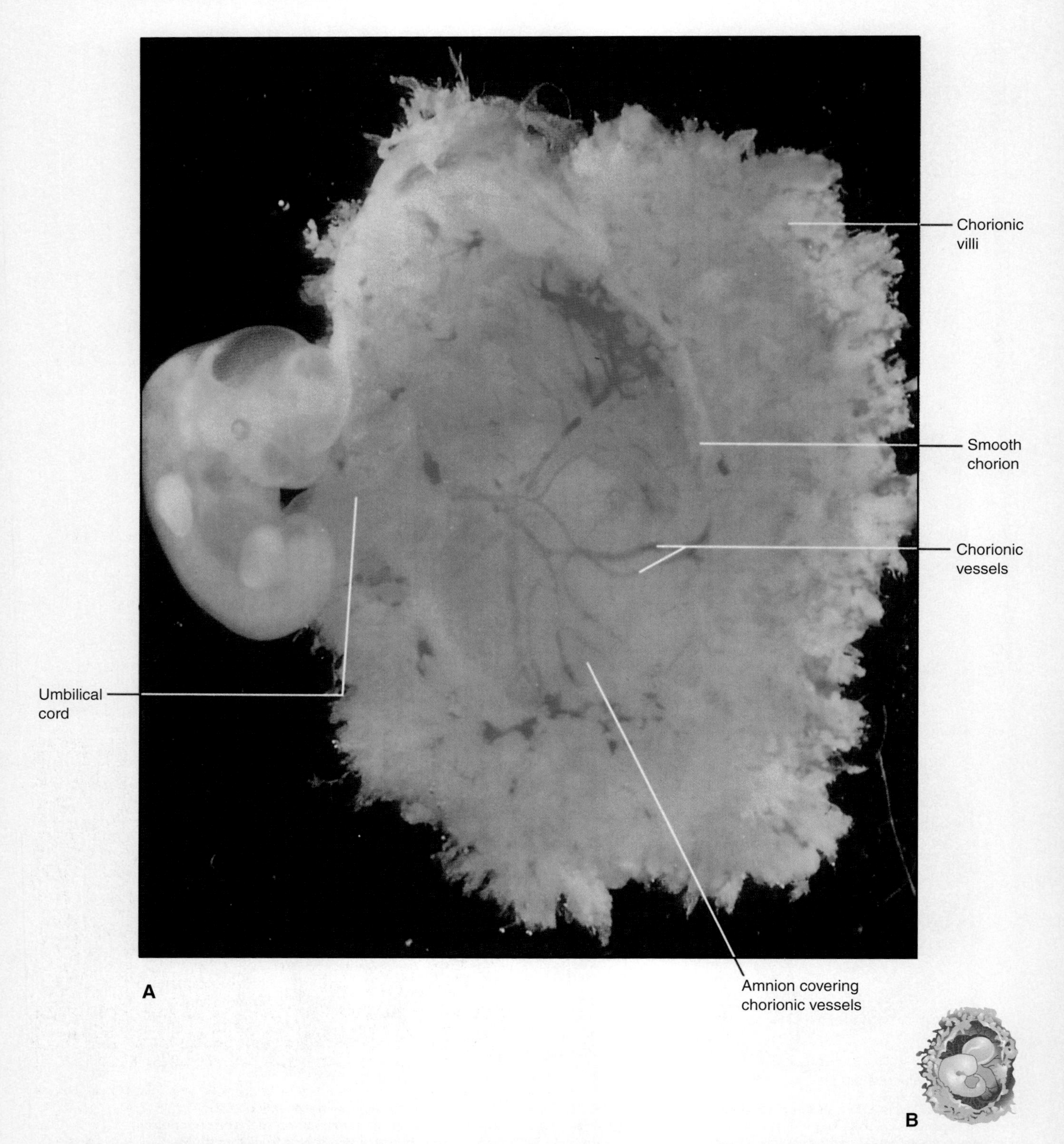

Figure 8 - 2. *A*, Lateral view of a spontaneously aborted embryo at Carnegie stage 14, about 32 days. The chorionic and amniotic sacs have been opened to show the embryo. *B*, Actual size of the embryo and its membranes.

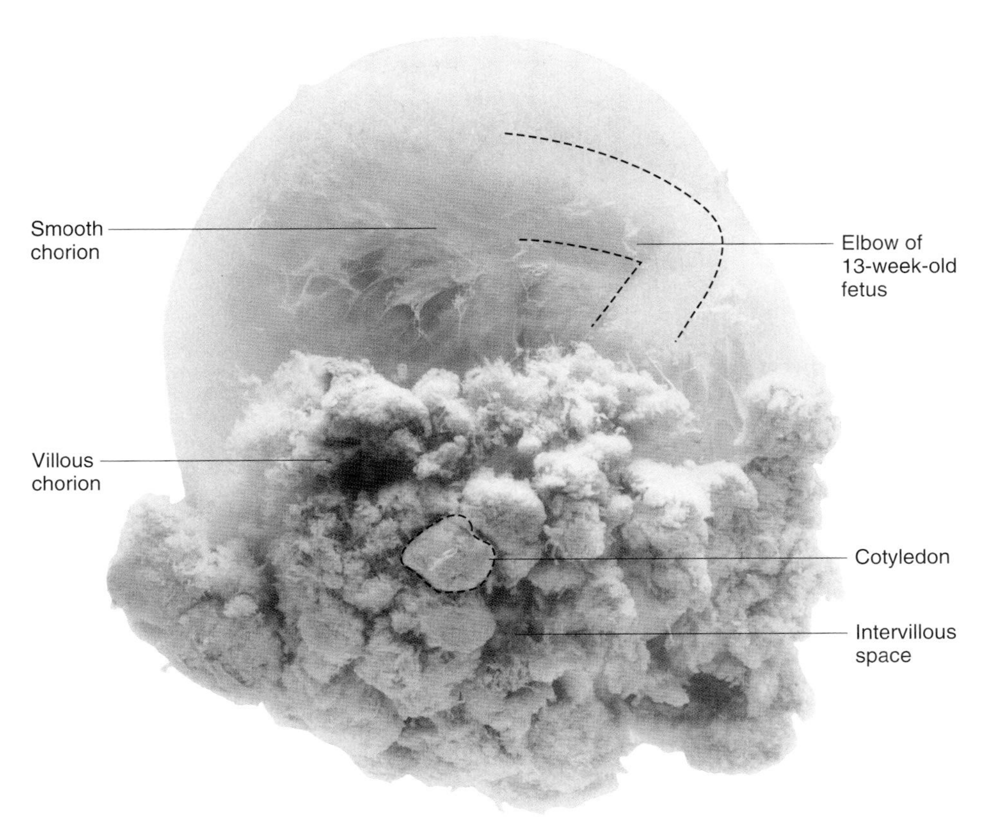

Figure 8 - 3. Photograph of a human chorionic sac containing a 13-week fetus. The villous chorion (chorion frondosum) is where chorionic villi persist and form the fetal part of the placenta. In situ, the cotyledons were attached to the decidua basalis, and the intervillous space was filled with maternal blood.

instruments equipped with intravaginal transducers, enable ultrasonographers to detect the chorionic (gestational) sac when it has a *median sac diameter* of 2 to 3 mm (Fig. 8 – 4). Chorionic sacs with this diameter indicate a gestational age of about 4 weeks and 3 to 4 days (i.e., about 18 days after fertilization).

The uterus, chorionic sac, and placenta enlarge as the embryo and fetus grow. The placenta has two parts (see Figs. 8 – 1*E* and 8 – 5):

- **The fetal part of the placenta** is formed by the *villous chorion*. The chorionic villi that arise from it project into the intervillous space, which contains maternal blood.
- **The maternal part of the placenta** is formed by the *decidua basalis*, the part of the decidua related to the fetal component of the placenta.

Fetomaternal Junction

The fetal part of the placenta (villous chorion) is attached to the maternal part of the placenta (decidua basalis) by the **cytotrophoblastic shell**, the external layer of trophoblastic cells on the maternal surface of the placenta (see Fig. 8 – 6). The *chorionic villi*, which are

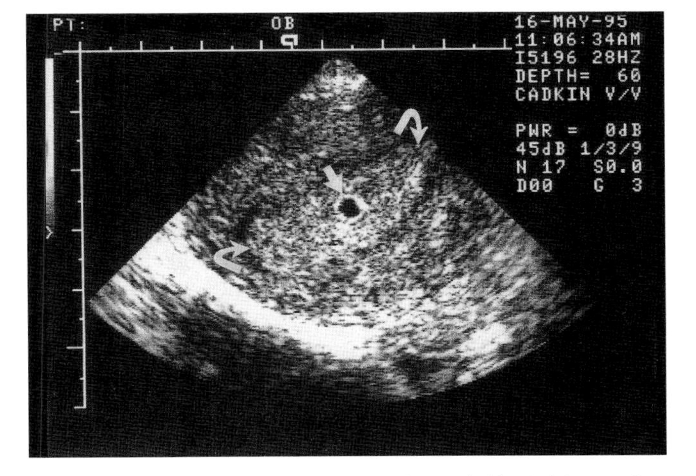

Figure 8 - 4. Transverse endovaginal scan of an early (day 34) intrauterine chorionic (gestational) sac. A cross-section of the chorionic sac (straight arrow) demonstrates a thick bright (echogenic) ring of chorionic villi surrounding the black chorionic fluid. The ring is surrounded by the less echogenic decidua parietalis (lateral borders indicated by curved arrows). (Courtesy of Dr. Alan V. Cadkin, M.D., Fort Myers, Florida.)

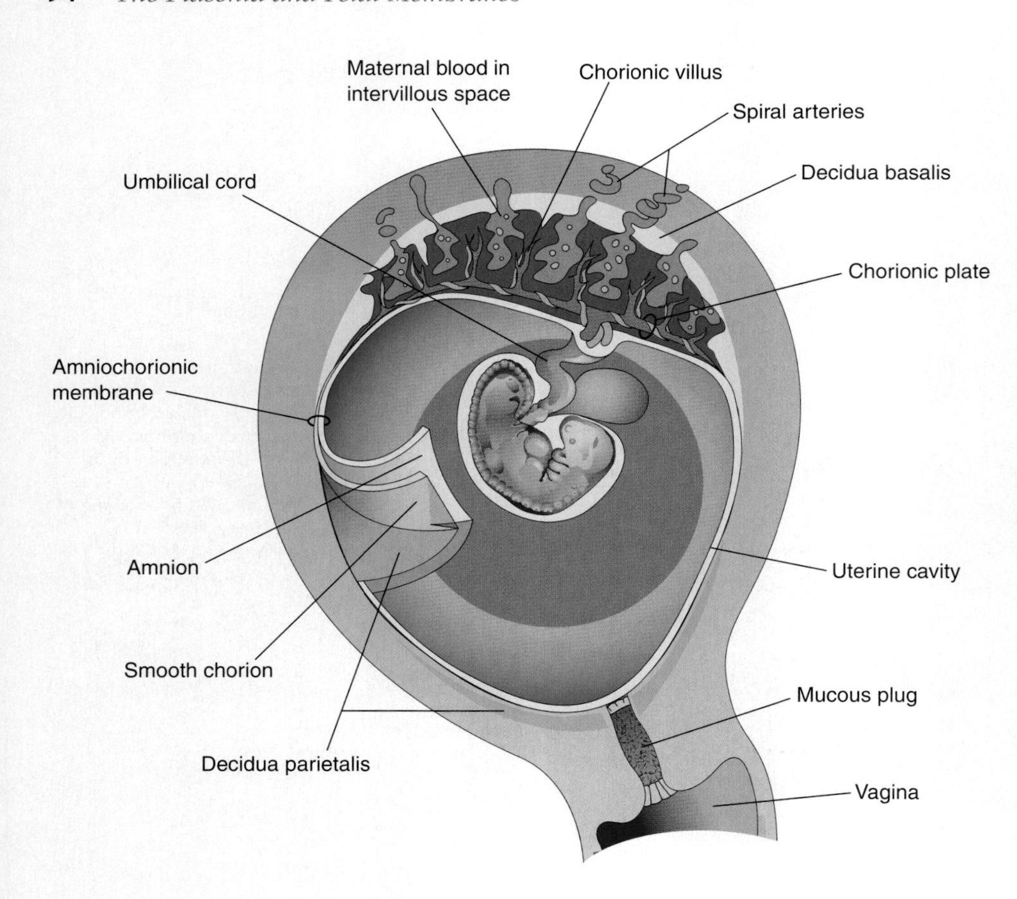

Figure 8-5. Drawing of a sagittal section of a gravid uterus at 4 weeks showing the relationship of the fetal membranes to each other and to the decidua and embryo. The amnion and smooth chorion have been cut and reflected to show their relationship to each other and to the decidua parietalis.

attached firmly to the decidua basalis through the cytotrophoblastic shell, anchor the *chorionic sac* to the decidua basalis. Endometrial arteries and veins pass freely through gaps in the cytotrophoblastic shell and open into the intervillous space. The **shape of the placenta** is determined by the shape of the persistent area of chorionic villi (see Fig. 8-1*F*). Usually, this is a circular area, giving the placenta a discoid shape. As the chorionic villi invade the decidua basalis during placental formation, decidual tissue is eroded to enlarge the intervillous space. This erosion produces several wedge-shaped areas of decidua — **placental septa** — that project toward the **chorionic plate** (see Fig. 8-6). The placental septa divide the fetal part of the placenta into irregular convex areas called **cotyledons** (see Fig. 8-3). Each cotyledon consists of two or more stem villi and their many branch villi.

The **decidua capsularis**, the layer of decidua overlying the implanted chorionic sac, forms a capsule over the external surface of the sac (see Fig. 8-1*A* to *D*). As the conceptus enlarges, the decidua capsularis bulges into the uterine cavity and becomes greatly attenuated. Eventually, the decidua capsularis makes contact and fuses with the decidua parietalis, thereby slowly obliterating the uterine cavity (see Fig. 8-1*E* and *F*). By 22 to 24 weeks, reduced blood supply to the decidua capsularis causes it to degenerate and disappear. After disappearance of the decidua capsularis, the smooth part of the chorionic sac fuses with the decidua parietalis.

Intervillous Space

The intervillous space of the placenta contains maternal blood, which is derived from the lacunae that developed in the syncytiotrophoblast during the second week of development (see Fig. 4-1*C*). This large, blood-filled space results from the coalescence and enlargement of the lacunar networks. The intervillous space of the placenta is divided into compartments by the **placental septa**; however, free communication occurs between the compartments because the septa do not reach the **chorionic plate** (see Fig. 8-6) — the part of the chorionic membrane associated with the placenta. Maternal blood enters the intervillous space from the **spiral arteries** in the decidua basalis; these endometrial arteries pass through gaps in the cytotrophoblastic shell and discharge blood into the intervillous space. This large space is drained by endometrial veins that also penetrate the cytotrophoblastic shell. The

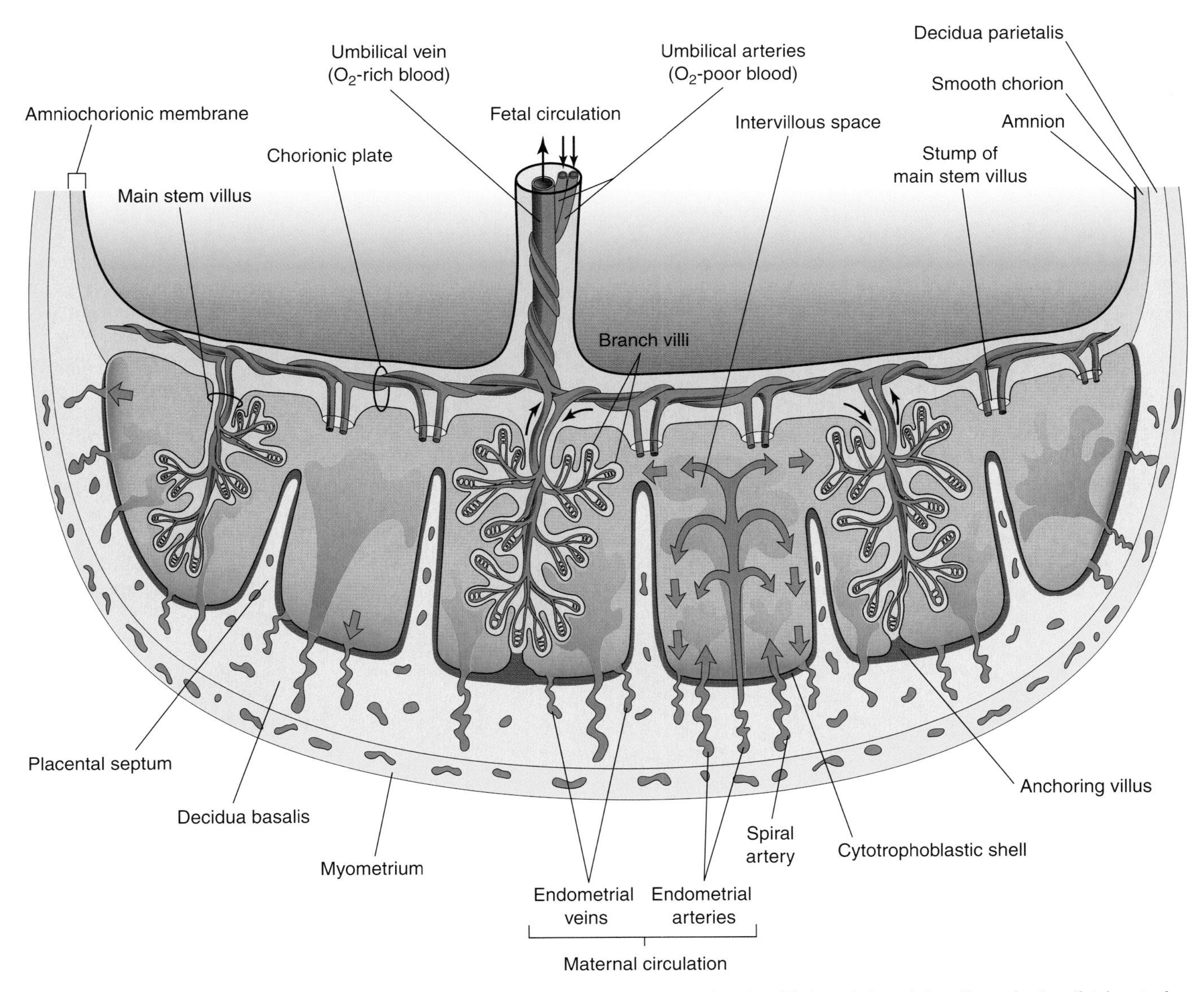

Figure 8 - 6. Schematic drawing of a transverse section through a full-term placenta, showing (1) the relation of the villous chorion (fetal part of placenta) to the decidua basalis (maternal part of placenta); (2) the fetal placental circulation; and (3) the maternal placental circulation. Maternal blood flows into the intervillous spaces in funnel-shaped spurts from the spiral arteries, and exchanges occur with the fetal blood as the maternal blood flows around the branch villi. The inflowing arterial blood pushes venous blood out of the intervillous space into the endometrial veins. Note that the umbilical arteries carry poorly oxygenated fetal blood (shown in blue) to the placenta and that the umbilical vein carries oxygenated blood (shown in red) to the fetus. In this drawing, only one stem villus is shown in each cotyledon, but the stumps of those that have been removed are indicated.

numerous **branch villi**, arising from stem chorionic villi, are continuously showered with maternal blood as it circulates through the intervillous space. The blood carries oxygen and nutritional materials that are necessary for fetal growth and development. The maternal blood also contains fetal waste products, such as carbon dioxide, salts, and products of protein metabolism.

Amniochorionic Membrane

The amniotic sac enlarges faster than the chorionic sac. As a result, the amnion and smooth chorion soon fuse to form the amniochorionic membrane (see Figs. 8 - 5 and 8 - 6). This composite membrane fuses with the decidua capsularis and, after disappearance of this capsular part of the decidua, adheres to the decidua parietalis. It is the amniochorionic membrane that ruptures during labor.

Preterm rupture of this membrane is the most common event leading to premature labor. When the amniochorionic membrane ruptures, the amniotic fluid escapes through the cervix and vagina to the exterior of the body.

Placental Circulation

The many *branch chorionic villi* provide a large surface area where materials may be exchanged across the very thin **placental membrane** (barrier) interposed between the fetal and maternal circulations (see Figs. 8 – 6 and 8 – 7). It is through the branch villi that the main exchange of material between the mother and fetus takes place. The circulations of the fetus and the mother are separated by the placental membrane, consisting of extrafetal tissues.

Fetoplacental Circulation

Poorly oxygenated blood leaves the fetus and passes through the **umbilical arteries** to the placenta. At the attachment of the cord to the placenta, these arteries divide into a number of radially disposed **chorionic arteries** that branch freely in the *chorionic plate* before entering the chorionic villi (see Fig. 8 – 6). The blood vessels form an extensive **arteriocapillary venous system** within the chorionic villi (see Fig. 8 – 7*A*), that brings the fetal blood extremely close to the maternal blood. This system provides a very large area for the exchange of metabolic and gaseous products between the maternal and fetal blood. *Normally, no intermingling of fetal and maternal blood occurs*; however, small amounts of fetal blood may enter the maternal circulation through minute defects that sometimes develop in the placental membrane. The well-oxygenated fetal blood in the fetal capillaries passes into thin-walled veins that follow the chorionic arteries to the site of attachment of the umbilical cord, where they converge to form the **umbilical vein**. This large vessel carries oxygen-rich blood to the fetus.

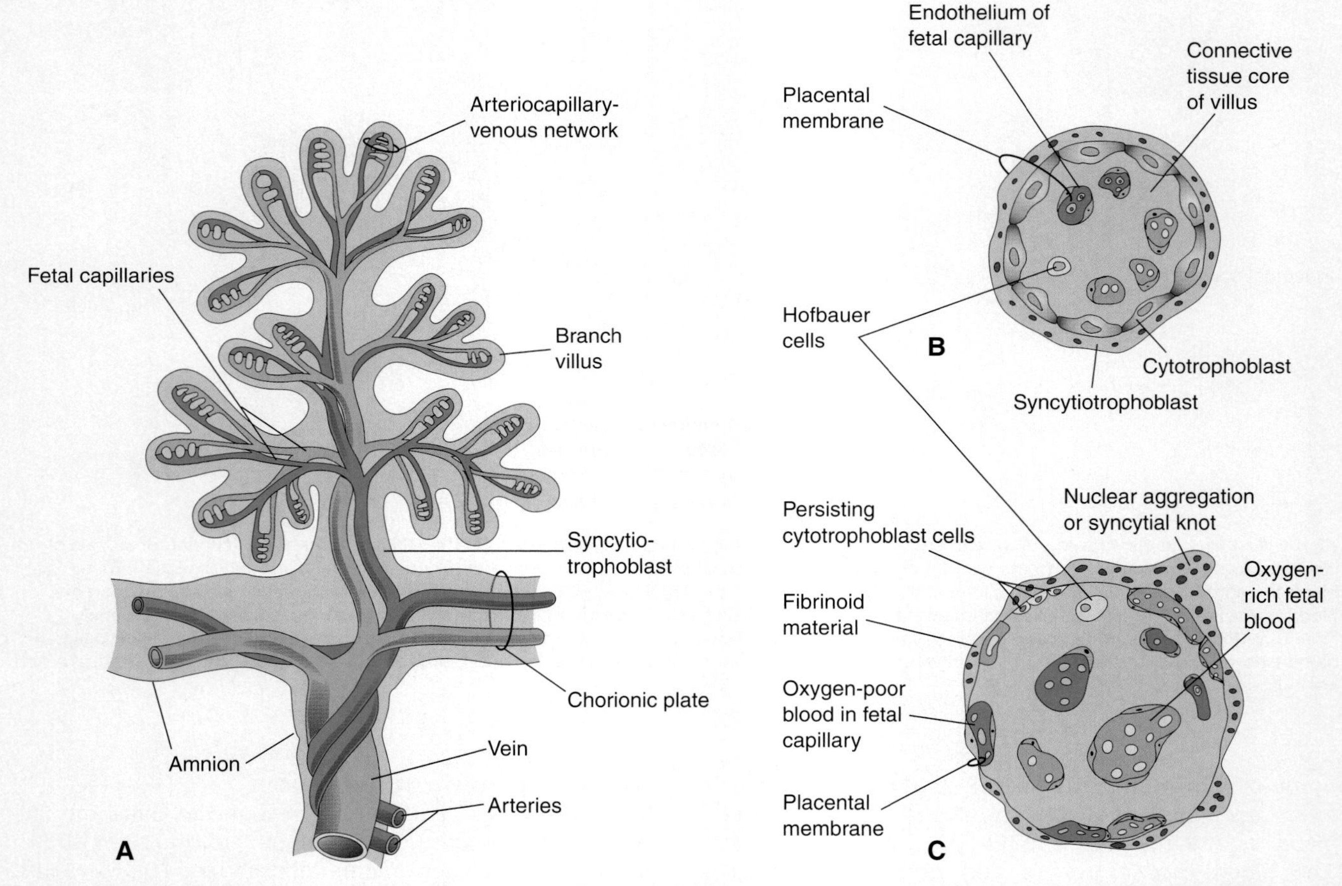

Figure 8 – 7. *A*, Drawing of a stem chorionic villus showing its arteriocapillary-venous system. The arteries carry poorly oxygenated fetal blood and waste products from the fetus, whereas the vein carries oxygenated blood and nutrients to the fetus. *B* and *C*, Drawings of sections through a branch villus at 10 weeks' gestation and at full term, respectively. The placental membrane, composed of extrafetal tissues, separates the maternal blood in the intervillous space from the fetal blood in the capillaries in the villi. Note that the placental membrane becomes very thin at full term. Hofbauer cells are thought to be phagocytic cells.

Maternal-Placental Circulation

The blood enters the intervillous space through 80 to 100 **spiral arteries** in the decidua basalis. The entering blood is at a considerably higher pressure than that in the intervillous space and so it spurts toward the **chorionic plate** (see Fig. 8–6). As the pressure dissipates, the blood flows slowly around the branch villi, allowing an exchange of metabolic and gaseous products with the fetal blood. The blood eventually returns through the endometrial veins to the maternal circulation. Reductions of uteroplacental circulation result in fetal hypoxia (decreased level of oxygen) and intrauterine growth retardation (IUGR). The intervillous space of the mature placenta contains about 150 ml of blood that is replenished three or four times per minute.

Placental Membrane

The placental membrane is a composite membrane that consists of the *extrafetal tissues separating the maternal and fetal blood.* Until about 20 weeks, the placental membrane consists of four layers (see Figs. 8–7 and 8–8):

- syncytiotrophoblast
- cytotrophoblast
- connective tissue of villus
- endothelium of fetal capillaries

After the 20th week, histological changes occur in the branch villi that result in the cytotrophoblast becoming attenuated in many villi. Eventually, cytotrophoblastic cells disappear over large areas of the villi, leaving only thin patches of syncytiotrophoblast. As a result, the placental membrane at full term consists of three layers only in most places (see Fig. 8–7*C*). In some areas, the placental membrane becomes markedly thinned and attenuated. At these sites, the syncytiotrophoblast comes in direct contact with the endothelium of the fetal capillaries to form a *vasculosyncytial placental membrane.* Only a few substances, endogenous or exogenous, are

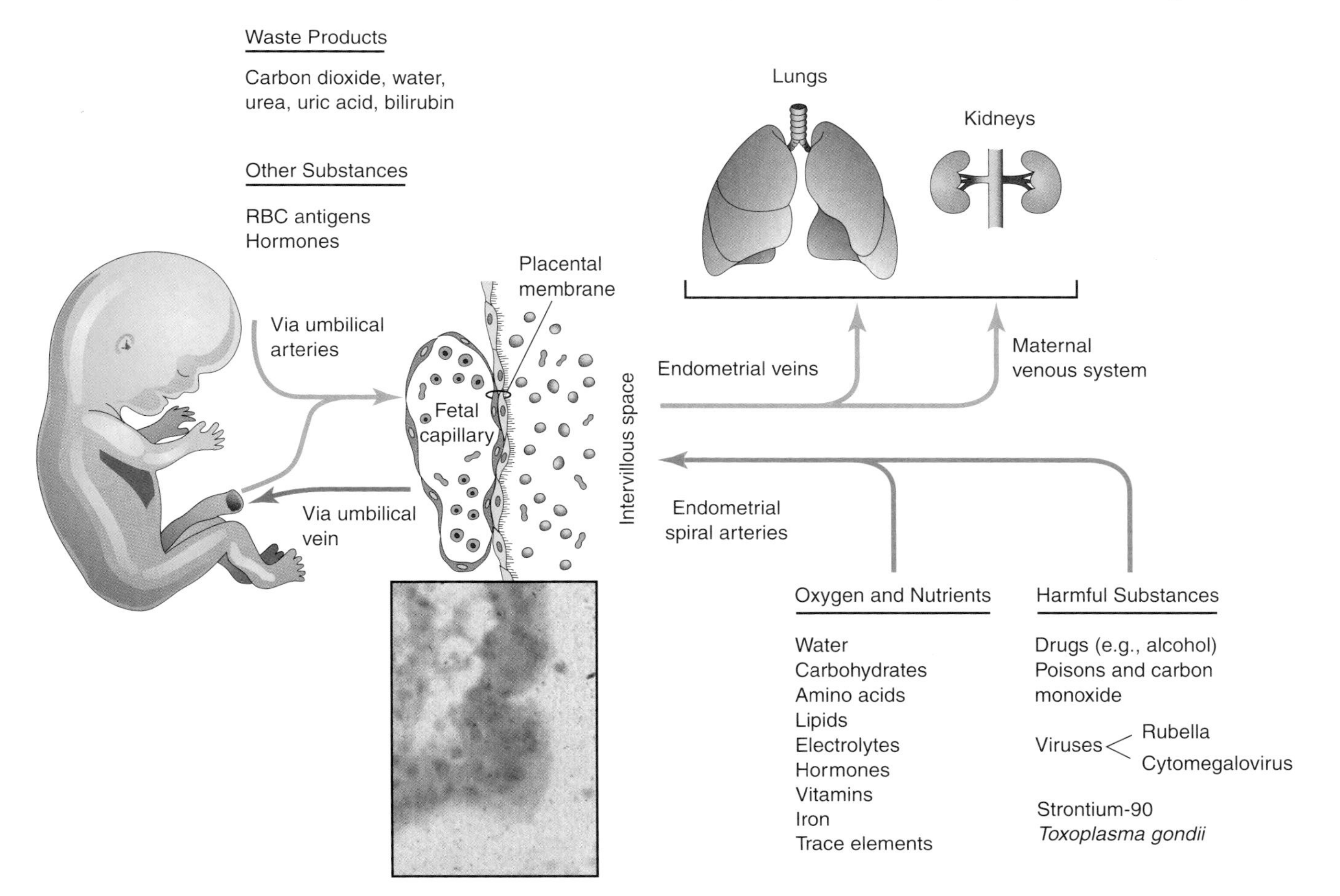

Figure 8–8. Transfer across the placental membrane. The extrafetal tissues, across which transport of substances between the mother and fetus occurs, collectively constitute the placental membrane. IgG, immunoglobulin G; IgS, immunoglobulin S; IgM, immunoglobulin M; RBC, red blood cell.

unable to pass through the placental membrane. The placental membrane acts as a true barrier only when the molecule has a certain size, configuration, and charge (e.g., heparin and bacteria). *Most drugs and other substances in the maternal plasma pass through the placental membrane and are found in the fetal plasma* (see Fig. 8–8).

During the third trimester, numerous nuclei in the syncytiotrophoblast of the villi aggregate to form **syncytial knots**. These nuclear aggregations continually break off and are carried from the intervillous space into the maternal circulation. Some knots lodge in capillaries of the maternal lung, where they are rapidly destroyed by local enzyme action. Toward the end of pregnancy, **fibrinoid material** forms on the surfaces of villi. *The placenta has three main functions:*

- metabolism (e.g., synthesis of glycogen)
- transport of gases and nutrients
- endocrine secretion (e.g., human chorionic gonadotropin [hCG])

Placental Metabolism

The placenta synthesizes glycogen, cholesterol, and fatty acids, which serve as sources of nutrients and energy for the embryo/fetus. Many of the placenta's metabolic activities are critical for its other two major activities: transport and endocrine secretion.

Placental Transport

The transport of substances in both directions between the placenta and maternal blood is facilitated by the great surface area of the placental membrane. Almost all materials are transported across the placental membrane by one of the following *four main transport mechanisms*:

- simple diffusion
- diffusion
- active transport
- pinocytosis

Passive transport by simple diffusion is usually characteristic of substances moving from areas of higher to lower concentration until equilibrium is established. In *facilitated diffusion*, transport occurs through electrical charges. *Active transport* against a concentration gradient requires energy. This mechanism of transport may involve enzymes that temporarily combine with the substances concerned. *Pinocytosis* is a form of endocytosis in which the material being engulfed is a small sample of extracellular fluid. Some proteins are transferred very slowly through the placenta by pinocytosis.

Transfer of Gases. Oxygen, carbon dioxide, and carbon monoxide cross the placental membrane by simple diffusion. The exchange of oxygen and carbon dioxide is limited more by blood flow than by the efficiency of diffusion. *Interruption of oxygen transport for several minutes endangers survival of the embryo or fetus.* The placental membrane approaches the efficiency of the lungs for gas exchange. The quantity of oxygen reaching the fetus is primarily flow-limited, rather than diffusion-limited; hence, fetal hypoxia results primarily from factors that diminish either the uterine blood flow or fetal blood flow through the placenta. Inhaled anesthetics can also cross the placental membrane and affect fetal breathing if used during parturition.

Nutritional Substances. Nutrients constitute the bulk of substances transferred from the mother to the fetus. **Water** is rapidly exchanged by simple diffusion and in increasing amounts as pregnancy advances. **Glucose** produced by the mother and placenta is quickly transferred to the embryo or fetus by diffusion. Little or no maternal cholesterol, triglycerides, or phospholipids are transferred. Although free fatty acids are transported, the amount transferred appears to be relatively small. **Vitamins** cross the placental membrane and are essential for normal development.

Hormones. *Protein hormones* do not reach the embryo or fetus in significant amounts, except for a slow transfer of thyroxine and triiodothyronine. Unconjugated *steroid hormones* cross the placental membrane rather freely. Testosterone and certain synthetic progestins cross the placental membrane and may cause masculinization of female fetuses (see Chapter 9).

Electrolytes. These compounds are freely exchanged across the placental membrane in significant quantities, each at its own rate. When a mother receives *intravenous fluids*, they also pass to the fetus and affect its water and electrolyte status.

Maternal Antibodies. The fetus produces only small amounts of antibodies because of its immature immune system. Some passive immunity is conferred upon the fetus by the placental transfer of maternal antibodies. The alpha and beta globulins reach the fetus in very small quantities, but many gamma globulins, such as the immunoglobulin (IgG) (7S) class, are readily transported to the fetus by pinocytosis. *Maternal antibodies confer fetal immunity* for diseases such as diphtheria, smallpox, and measles; however, no immunity is acquired to pertussis (whooping cough) or chickenpox. A maternal protein, *transferrin*, crosses the placental membrane and carries iron to the embryo or fetus. The placental surface contains special receptors for this protein.

Hemolytic Disease of the Newborn

Small amounts of fetal blood may pass to the maternal blood through microscopic breaks in the placental membrane. If the fetus is Rh-positive and the mother is Rh-negative, the fetal cells may stimulate the formation of anti-Rh antibody by the immune system of the mother. This passes to the fetal blood and causes hemolysis of fetal Rh-positive blood cells and anemia in the fetus. Some fetuses with *hemolytic disease of the newborn* (HDN), or fetal erythroblastosis, fail to make a satisfactory intrauterine adjustment and may die unless delivered early or given intrauterine, intraperitoneal, or intravenous transfusions of packed Rh-negative blood cells in order to maintain the fetus until after birth. HDN is relatively uncommon now because Rh immunoglobulin given to the mother usually prevents development of this disease in the fetus.

Waste Products. Urea and uric acid pass through the placental membrane by simple diffusion, and bilirubin is quickly cleared.

Drugs and Drug Metabolites. Most drugs and drug metabolites cross the placenta by simple diffusion. Some drugs cause major congenital anomalies (see Chapter 9). **Fetal drug addiction** may occur after maternal use of drugs such as heroin, and newborn infants may experience withdrawal symptoms. Except for muscle relaxants such as succinylcholine and curare, most agents used for the management of labor readily cross the placental membrane. Depending on their dose and timing in relation to delivery, these drugs may cause respiratory depression of the newborn infant. All sedatives and analgesics affect the fetus to some degree. Drugs taken by the mother can affect the embryo/fetus directly or indirectly by interfering with maternal or placental metabolism.

Infectious Agents. Cytomegalovirus, rubella, and Coxsackie viruses, as well as viruses associated with variola, varicella, measles, and poliomyelitis, may pass through the placental membrane and cause *fetal infection*. In some cases, as with the **rubella virus**, severe congenital anomalies may result (see Chapter 9). Microorganisms, such as *Treponema pallidum*, which causes syphilis, and *Toxoplasma gondii*, which produces destructive changes in the brain and eyes, also cross the placental membrane. These organisms enter the fetal blood, often causing congenital anomalies and/or death of the embryo or fetus.

Placental Endocrine Synthesis and Secretion

Using precursors derived from the fetus and/or the mother, the *syncytiotrophoblast* of the placenta synthesizes protein and steroid hormones. **Protein hormones** synthesized by the placenta include the following:

- human chorionic gonadotropin (hCG)
- human chorionic somatomammotropin (hCS) or human *placental lactogen* (hPL)
- human chorionic thyrotropin (hCT)
- human chorionic corticotropin (hCACTH)

The glycoprotein hCG, similar to luteinizing hormone (LH), is first secreted by the syncytiotrophoblast during the second week. *hCG maintains the corpus luteum*, preventing the onset of menstrual periods. The concentration of hCG in the maternal blood and urine rises to a maximum by the eighth week and then declines. The placenta also plays a major role in the production of **steroid hormones** — (i.e., *progesterone* and *estrogens*). Progesterone is essential for the maintenance of pregnancy. The ovaries of a pregnant woman can be removed after the first trimester without causing an abortion because the placenta takes over production of progesterone from the corpus luteum of the ovary. Estrogens are also produced in large quantities by the syncytiotrophoblast.

Uterine Growth During Pregnancy

The uterus of a nonpregnant woman lies in the pelvis minor, or true pelvis (Fig. 8 - 9*A*). It increases in size during pregnancy to accommodate the growing fetus. While the uterus is enlarging, it also increases in weight and its walls become thinner (see Fig. 8 - 9*B* and *C*). During the first trimester, the uterus moves out of the pelvic cavity by 20 weeks and usually reaches the level of the umbilicus. By 28 to 30 weeks, it reaches the epigastric region, the area between the xiphoid process of the sternum and the umbilicus.

Parturition

Parturition (childbirth) is the process during which the fetus, placenta, and fetal membranes are expelled from the mother's reproductive tract (Fig. 8 - 10). **Labor** is the sequence of *involuntary uterine contractions* that results in dilation of the cervix and delivery of the fetus and placenta from the uterus. The factors that trigger labor are not completely understood, but several hormones are related to the initiation of contractions. The fetal hypothalamus secretes **corticotropin-releasing hormone**, which stimulates the anterior hypophysis or pituitary gland to produce **adrenocorticotropin hormone** (ACTH), which causes the secretion of cortisol from the suprarenal (adrenal) cortex. **Cortisol** is involved in the synthesis of estrogens. These steroids stimulate uterine contraction.

Peristaltic contractions of uterine smooth muscle are elicited by **oxytocin**, which is released by the posterior cerebral hypophysis. This hormone is administered clinically when it is necessary to induce labor. Oxytocin also stimulates the release of **prostaglandins** from the

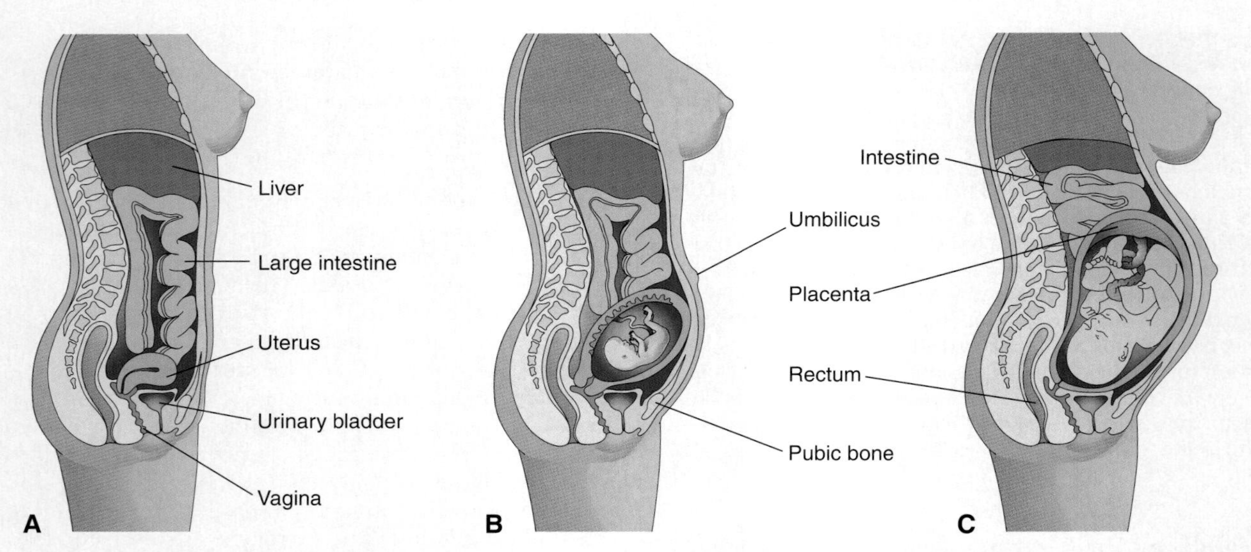

Figure 8 – 9. Median sections of a woman's body. *A,* Not pregnant. *B,* 20 weeks pregnant. *C,* 30 weeks pregnant. Note that as the conceptus enlarges, the uterus increases in size to accommodate the rapidly growing fetus. By 20 weeks, the uterus and fetus reach the level of the umbilicus, and by 30 weeks, they reach the epigastric region.

decidua; that in turn, stimulate myometrial contractility by sensitizing the myometrial cells to oxytocin. **Estrogens** also increase myometrial contractile activity and stimulate the release of oxytocin and prostaglandins.

The Four Stages of Labor

Labor begins with the onset of regular painful contractions of the uterus (spaced less than 10 minutes apart).

- **The first stage of labor** (dilation stage) begins with objective evidence of progressive dilation of the cervix (see Fig. 8 – 10*A* and *B*). The dilation is mediated by changes in the circulating hormones and other regulatory factors, such as prostaglandins. The first stage ends with complete dilation of the cervix. The average duration of the first stage is about 12 hours for first pregnancies (nulliparous women, or *primigravidas*) and about 7 hours for women who have had a child previously (multiparous women, or *multigravidas*).
- **The second stage of labor** (expulsion stage) begins when the cervix is fully dilated and ends with delivery of the baby (see Fig. 8 – 10*C* to *E*). *During this stage, the fetus descends through the cervix and vagina*. As soon as the fetus is outside the mother, it is called a *newborn infant*, or neonate. The average duration of this stage is 50 minutes for primigravidas and 20 minutes for multigravidas.
- **The third stage of labor** (placental stage) begins as soon as the baby is born and ends when the placenta and membranes are expelled (see Fig. 8 – 10*F* to *H*). The duration of this stage is 15 minutes in most pregnancies. *Retraction of the uterus and manual compression of the abdomen reduce the area of placental attachment* (see Fig. 8 – 10*G*). A **hematoma** soon forms deep to the placenta, separating it from the uterine wall. The placenta and fetal membranes are then expelled.
- **The fourth stage of labor** (recovery stage) begins as soon as the placenta and fetal membranes are expelled. This stage lasts about 2 hours. Contractions of the uterus constrict the spiral arteries, preventing excessive uterine bleeding.

The Placenta and Fetal Membranes after Birth

The extruded placenta and fetal membranes are called the *afterbirth*. The placenta is commonly discoid in shape, with a diameter of 15 to 20 cm and a thickness of 2 to 3 cm. The margins of the placenta are continuous with the ruptured amniotic and chorionic sacs.

Variations in Placental Shape

As the placenta develops, chorionic villi usually persist only where the villous chorion is in contact with the decidua basalis. This typically produces a discoid placenta (Fig. 8 – 11). When villi persist elsewhere, several variations in placental shape occur, such as *accessory placenta* (Fig. 8 – 12). Examination of the placenta, prenatally by ultrasonography or postnatally by gross and microscopic study, may provide clinical information about the causes of:

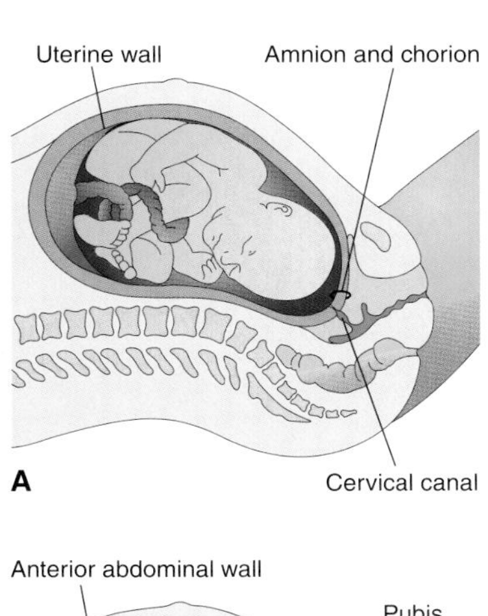

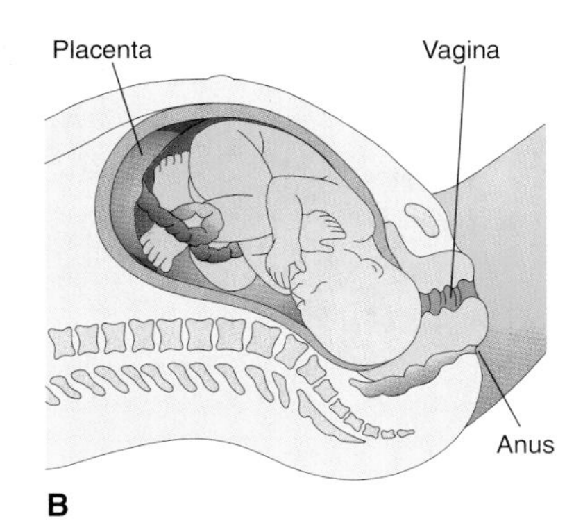

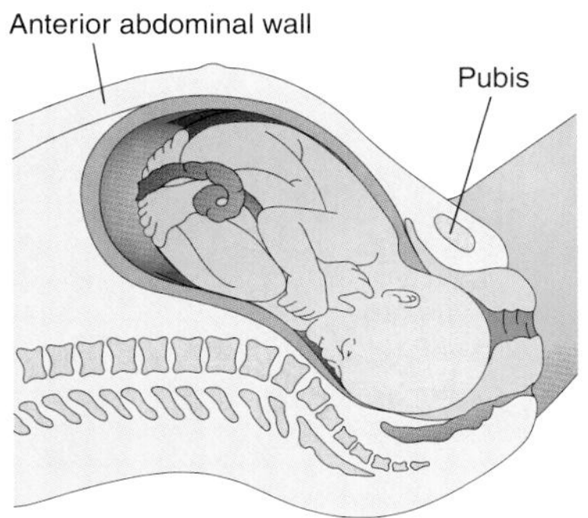

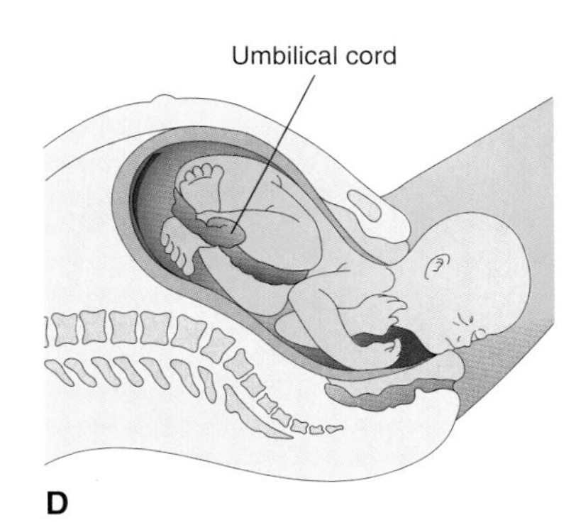

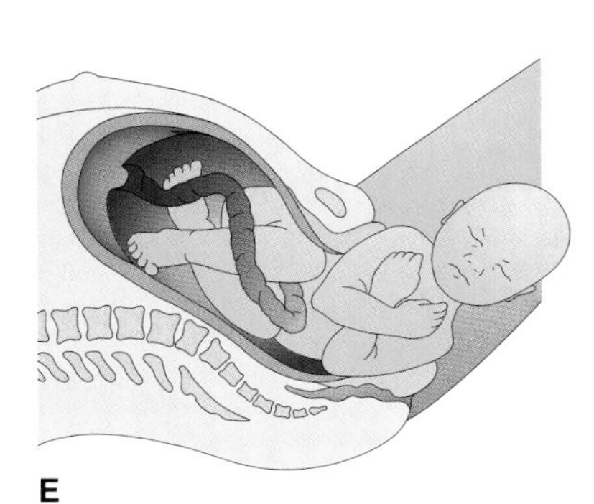

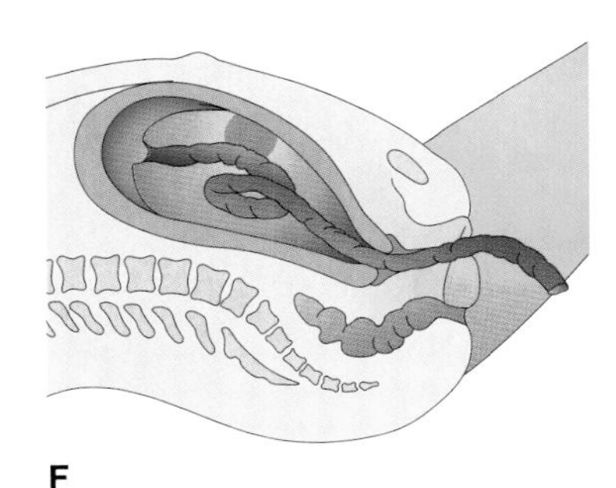

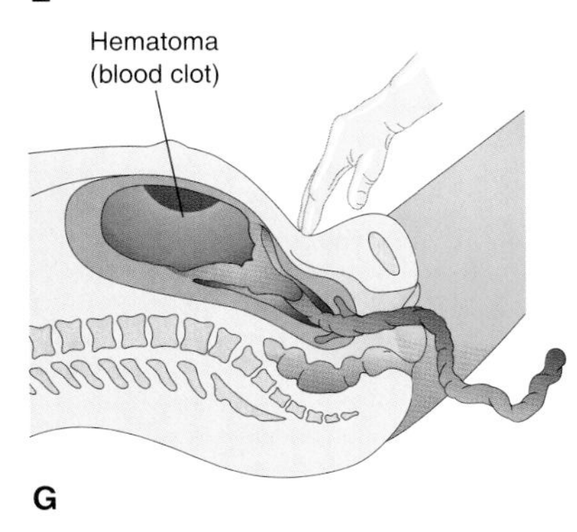

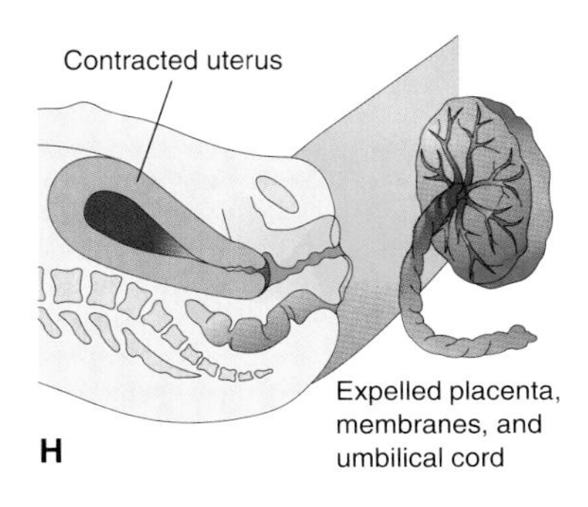

Figure 8 - 10. Drawings illustrating parturition. *A* and *B*, The cervix is dilating during the first stage of labor. *C* to *E*, The fetus is passing through the cervix and vagina during the second stage of labor. *F* and *G*, As the uterus contracts during the third stage of labor, the placenta folds and pulls away from the uterine wall. Separation of the placenta results in bleeding and the formation of a large hematoma (mass of blood). Pressure on the abdomen facilitates placental separation. *H*, The placenta is expelled, and the uterus contracts during the fourth stage of labor.

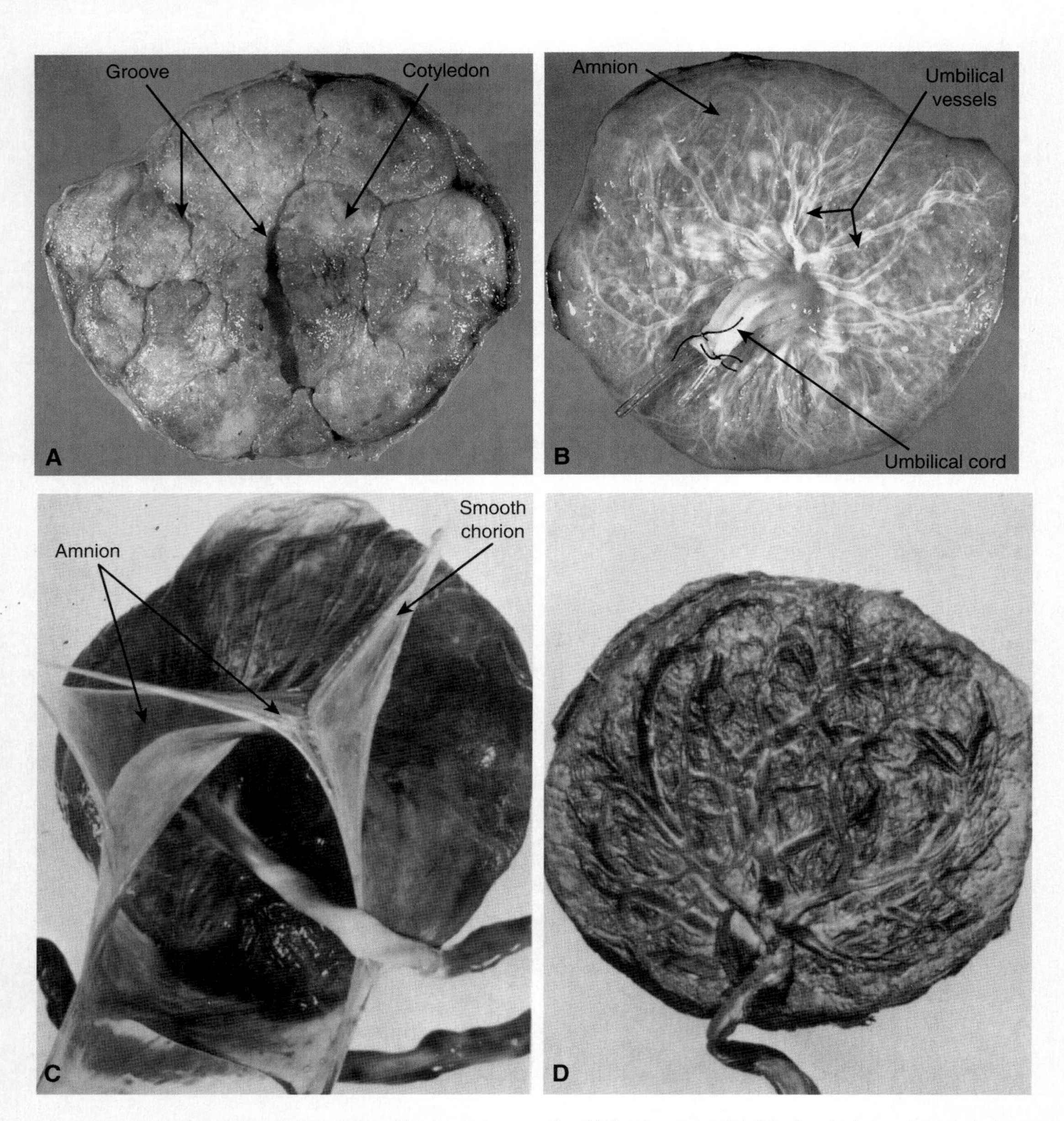

Figure 8-11. Photographs of placentas and fetal membranes after birth, about one-third their actual size. *A,* Maternal surface, showing cotyledons and the grooves around them. Each convex cotyledon consists of a number of main stem villi with their many branch villi. The grooves were occupied by the placental septa when the maternal and fetal parts of the placenta were together (see Fig. 8-6). *B,* Fetal surface, showing blood vessels running in the chorionic plate deep to the amnion and converging to form the umbilical vessels at the attachment of the umbilical cord. *C,* The amnion and smooth chorion are arranged to show that they are fused and continuous with the margins of the placenta. *D,* Placenta with a marginal attachment of the cord.

- placental dysfunction
- IUGR
- fetal distress and death
- neonatal illness

Postnatal placental studies can also determine whether the placenta is complete. Retention of a cotyledon or an accessory placenta in the uterus causes *uterine hemorrhage.*

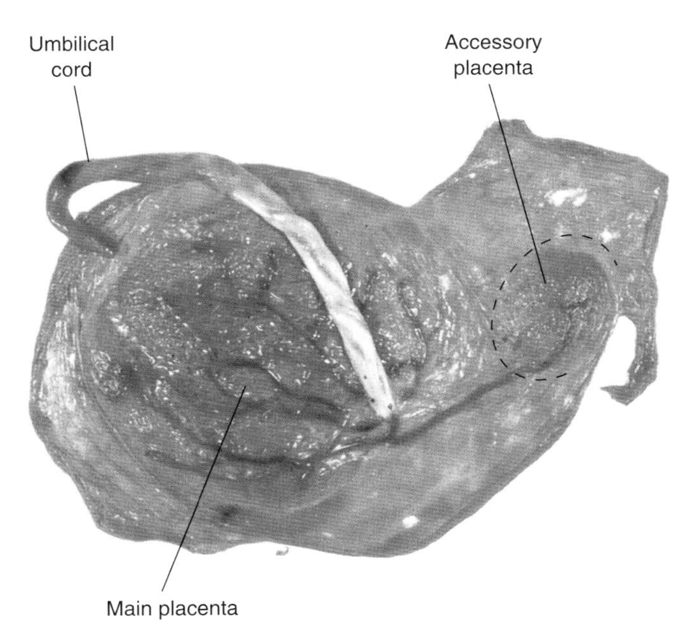

Figure 8 - 12. Photograph of the maternal surface of a full-term placenta and an accessory placenta, about one-quarter its actual size.

Gestational Choriocarcinoma

Abnormal proliferation of the trophoblast results in *gestational trophoblastic disease*, which encompasses a spectrum of lesions, including highly malignant tumors. The cells invade the decidua basalis, penetrate its blood vessels and lymphatics, and metastasize to the maternal lungs, bone marrow, liver, and other organs.

Maternal Surface of the Placenta

The characteristic *cobblestone appearance* of the maternal surface of the placenta is produced by slightly bulging villous areas — the **cotyledons** — which are separated by grooves formerly occupied by **placental septa** (see Fig. 8 – 11*A*). The surface of the cotyledons is covered by thin, grayish shreds of decidua basalis that separate from the uterine wall with the placenta.

Fetal Surface of the Placenta

The **umbilical cord** usually attaches to the fetal surface, and its epithelium is continuous with the amnion adhering to the chorionic plate of the placenta (see Fig. 8 – 11*B* and *C*). The chorionic vessels radiating to and from the umbilical cord are clearly visible through the smooth, transparent amnion. The *umbilical vessels* branch on the fetal surface, forming the *chorionic vessels*, which enter the chorionic villi (see Fig. 8 – 6).

Placental Abnormalities

Abnormal adherence of chorionic villi to the myometrium of the uterine wall is called **placenta accreta** (Fig. 8 – 13). When chorionic villi penetrate the myometrium all the way to the perimetrium (peritoneal covering), the abnormality is called **placenta percreta**. *Third-trimester bleeding is the most common presenting sign of these placental abnormalities.* After birth, the placenta fails to separate from the uterine wall; attempts to remove it may cause hemorrhage that is difficult to control. When the blastocyst implants close to or overlying the internal os of the uterus, the abnormality is called **placenta previa**. Late-pregnancy bleeding can result from this placental abnormality. In such cases, the fetus is delivered by cesarean section because the placenta blocks the cervical canal.

The Placenta as an Allograft*

The placenta, a part of the conceptus that inherits both paternal and maternal genes, can be regarded as an allograft with respect to the mother. What protects the placenta from rejection by the mother's immune system? This question remains a major biological enigma in nature. The syncytiotrophoblastic layer of the chorionic villi, although exposed to maternal immune cells within the blood sinusoids, lacks major histocompatibility complex (MHC) antigens, and thus does not evoke rejection responses. However, extravillous trophoblastic (EVT) cells that invade the uterine decidua and its vasculature (spiral arteries) express class I MHC antigens. These antigens include human leukocyte antigen G (HLA-G), which, because it is nonpolymorphic (class Ib), is poorly recognizable by T lymphocytes as an alloantigen, as well as HLA-C, which, being polymorphic (class Ia), is recognizable by T cells. In addition to averting T cells, trophoblastic cells must also shield themselves from potential attack by natural killer (NK) lymphocytes and from injury inflicted by activation of complement. Multiple mechanisms appear to be in place to guard the placenta:

1. Expression of HLA-G is restricted to a few tissues, including placental EVT cells. Its strategic location in the placenta is believed to provide a dual immunoprotective role: evasion of T-cell recognition owing to its nonpolymorphic nature, and a recognition by the "killer-inhibitory receptors" on NK cells, which turns off their killer function. The inadequacy of this hypothesis is suggested by several observations: (a) healthy individuals showing biallelic loss of HLA-G1 have been identified, indicating that HLA-G is not essential for feto-placental survival; (b) human EVT cells have been found to be vulnerable to NK cell–mediated killing; and (c) it does not explain why HLA-C, a polymorphic antigen, that is also expressed by EVT cells, does not evoke a rejection response in situ. Because both HLA-G and HLA-C have been shown to have the unique ability to resist human cytomegalovirus (HCMV)–mediated MHC

*The authors thank Dr P. K. Lalla for the preparation of this section.

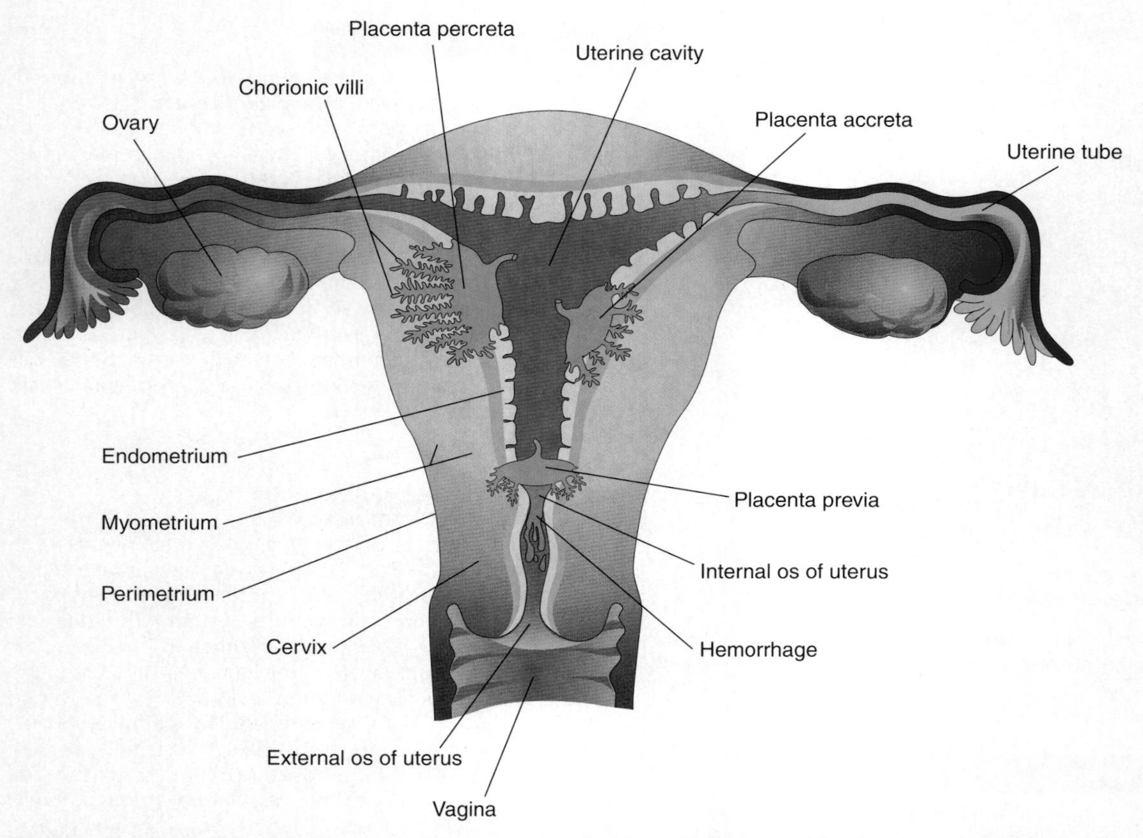

Figure 8 - 13. Placental abnormalities. In *placenta accreta,* there is abnormal adherence of the placenta to the myometrium. In *placenta percreta,* the placenta has penetrated the full thickness of the myometrium (muscle layer). In *placenta previa,* the placenta overlies the internal os of the uterus and blocks the cervical canal.

class 1 degradation, it is speculated that a selective location of these two antigens at the fetomaternal interface may help to withstand viral assault.

2. Immunoprotection is provided locally by certain immunosuppressor molecules (e.g., prostaglandin (PG) E_2 [PGE_2], transforming growth factor ([TGF]), and interleukin-(IL)10 [IL-10], and decidua-derived PGE_2 has been shown to block activation of maternal T cells, as well as NK cells, in situ. Indeed, the immunoregulatory function of decidual cells is consistent with their genealogy. It has been shown that uterine endometrial stromal cells, which differentiate into decidual cells during pregnancy, are derived from progenitor (stem) cells that migrate from hematopoietic organs, such as the fetal liver and the bone marrow, during ontogeny.
3. Transient tolerance of the maternal T-cell repertoire to fetal MHC antigens may serve as a back-up mechanism for placental immunoprotection. A similar B-cell tolerance has also been reported.
4. A trafficking of activated maternal leukocytes into the placenta or the fetus is prevented by deletion of these cells, triggered by apoptosis-inducing ligands present on the trophoblast.
5. Based on genetic manipulation in mice, it has been shown that the presence of complement regulatory proteins (Crry in the mouse; membrane co-factor protein, [MCP] or CD46 in the human), which can block activation of the third component of complement (C3) in the complement cascade, protects the placenta from complement-mediated destruction, which may happen. This destruction might otherwise occur because of residual C3 activation remaining once a defense has been mounted against pathogens. Crry gene knockout mice died in utero because of complement-mediated placental damage, which could be averted by additional knockout of the C3 gene.
6. Experiments in mice have revealed that the presence of the enzyme indoleamine 2,3 deoxygenase (IDO) in trophoblastic cells is critical for immunoprotection of the allogeneic conceptus. The mechanism for this immunoprotection is the suppression of T-cell-driven local inflammatory responses, including complement activation. Treatment

of pregnant mice with an IDO inhibitor, 1-methyltryptophan, is known to cause selective death of allogeneic (but not syngeneic) conceptuses owing to massive deposition of complement and hemorrhagic necrosis at the placental sites.

Umbilical Cord

The attachment of the umbilical cord, which connects the embryo/fetus to the placenta, is usually near the center of the fetal surface (see Fig. 8 - 11*B*), but it may be found at any point. For example, insertion of the umbilical cord at the placental margin produces a *battledore placenta* (see Fig. 8 - 11*D*); its attachment to the membranes is known as a *velamentous insertion of the cord* (Fig. 8 - 14). Color flow *Doppler ultrasonography* may be used for prenatal diagnosis of the position and structural abnormalities of the umbilical cord. The umbilical cord is usually 1 to 2 cm in diameter and 30 to 90 cm in length (average of 55 cm). Long cords have a tendency to prolapse and/or to coil around the fetus (Fig. 8 - 15). Prompt recognition of *prolapse of the cord* is important because the cord may be compressed between the presenting body part of the fetus and the mother's bony pelvis, causing *fetal anoxia*. If the deficiency of oxygen persists for more than 5 minutes, the baby's brain may be damaged.

The umbilical cord usually has two arteries and one vein that are surrounded by mucoid connective tissue (*Wharton jelly*). Because the umbilical vessels are longer than the cord, twisting and bending of the vessels is common. They frequently form loops, producing *false knots* that are of no significance; however, in about 1% of pregnancies, **true knots** form in the cord. These may tighten and cause fetal death secondary to *fetal anoxia*. In most cases, the knots form during labor as a result of the fetus passing through a loop of the cord. Because these knots are usually loose, they have no clinical significance. Simple *looping of the cord around the fetus* occasionally occurs (see Fig. 8 - 15*B*). In about one fifth of all deliveries, the cord is loosely looped around the neck without causing increased fetal risk.

Percutaneous umbilical cord blood sampling (*PUBS*) may be performed to assess fetal acid-base status as a means of monitoring the health of the fetus and neonate.

Absence of an Umbilical Artery

In about 1 in 200 newborn infants, only *one umbilical artery* is present (Fig. 8 - 16), a condition that may be associated with chromosomal and fetal abnormalities, particularly of the cardiovascular system. *Absence of an umbilical artery is accompanied by a 15 to 20% incidence of cardiovascular anomalies in the fetus.* Absence of an artery results from either agenesis or degeneration of this vessel early in development.

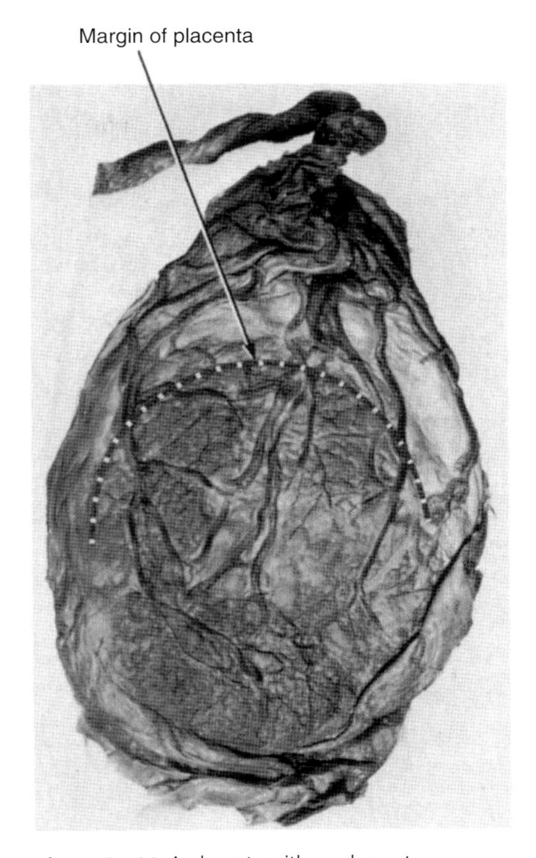

Figure 8 - 14. A placenta with a velamentous insertion of the umbilical cord. The cord is attached to the membranes (amnion and chorion). The umbilical vessels leave the cord and run between the amnion and chorion before spreading over the placenta. The vessels are easily torn in this location.

Amnion and Amniotic Fluid

The *amnion* forms a fluid-filled, membranous *amniotic sac* that surrounds the embryo and fetus (Fig. 8 - 17*A*). Because the amnion is attached to the margins of the embryonic disc, its junction with the embryo (future umbilicus) is located on the ventral surface after embryonic folding (see Fig. 8 - 17*B*). As the amnion enlarges, it gradually obliterates the chorionic cavity and forms the epithelial covering of the umbilical cord (see Fig. 8 - 17*C* and *D*). *Amniotic fluid* plays a major role in fetal growth and development. Initially, some fluid may be secreted by amniotic cells; however, most amniotic fluid is derived from *maternal tissue fluid* by diffusion across the amniochorionic membrane from the decidua parietalis (see Fig. 8 - 6). Later, there is diffusion of fluid through the chorionic plate from blood in the intervillous space of the placenta. Before

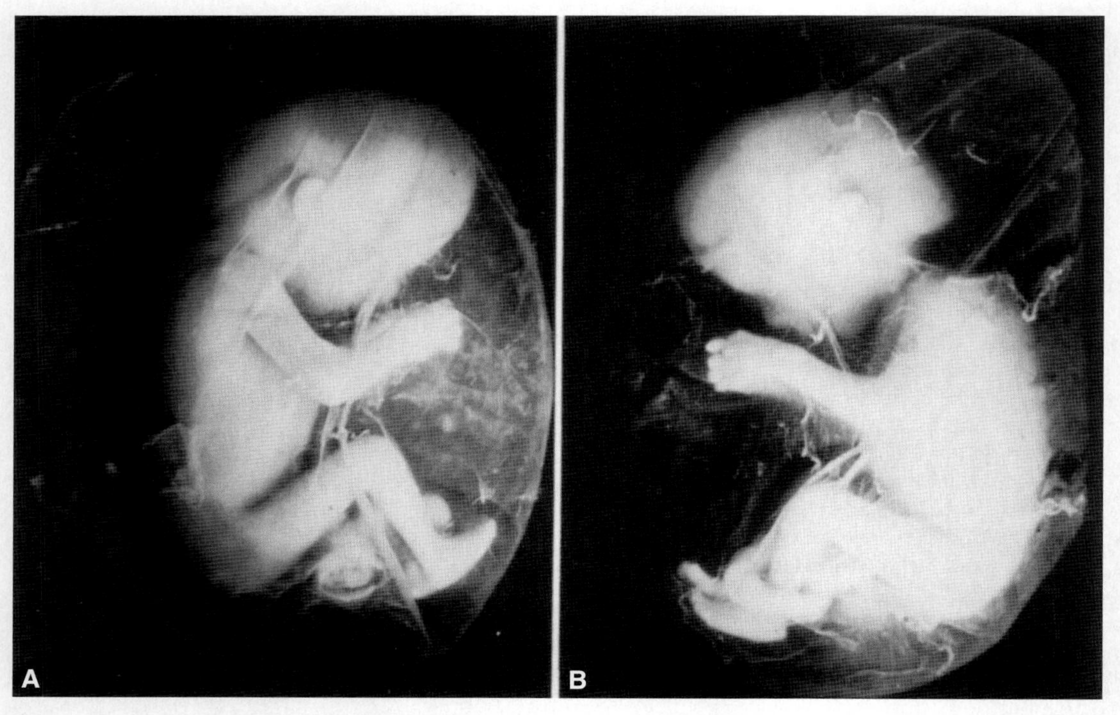

Figure 8-15. *A* and *B*, Photographs of a 12-week fetus within its amniotic sac (actual size). The fetus and its membranes aborted spontaneously. It was removed from its chorionic sac with its amniotic sac intact. In *B*, note that the umbilical cord is looped around the left ankle of the fetus.

keratinization of the skin occurs, a major pathway for passage of water and solutes in tissue fluid from the fetus to the amniotic cavity is through the skin; thus, amniotic fluid is similar to fetal tissue fluid. Fluid is also secreted by the fetal respiratory tract and enters the amniotic cavity. Beginning in the 11th week, the fetus contributes to the amniotic fluid by expelling urine into the amniotic cavity. Normally, the volume of amniotic fluid increases slowly, reaching about 30 mL at 10 weeks, 350 ml at 20 weeks, and 700 to 1000 mL by 37 weeks.

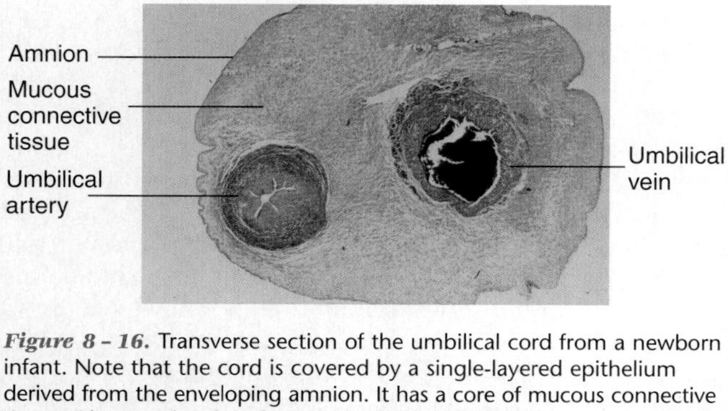

The water content of amniotic fluid changes every 3 hours. Large amounts of water pass through the amniochorionic membrane into the maternal tissue fluid and into the uterine capillaries. An exchange of fluid with fetal blood also occurs through the umbilical cord and at the site where the amnion adheres to the chorionic plate on the fetal surface of the placenta (see Figs. 8-6 and 8-11*B*); thus, amniotic fluid is in balance with the fetal circulation. *Amniotic fluid is swallowed by the fetus* and absorbed by the fetus's respiratory and digestive tracts. It has been estimated that, during the final stages of pregnancy, the fetus swallows up to 400 ml of amniotic fluid per day. The fluid passes into the fetal bloodstream, and the waste products in it cross the placental membrane and enter the maternal blood in the intervillous space. Excess water in the fetal blood is excreted by the fetal kidneys and returned to the amniotic sac through the fetal urinary tract.

Amnion
Mucous connective tissue
Umbilical artery
Umbilical vein

Figure 8-16. Transverse section of the umbilical cord from a newborn infant. Note that the cord is covered by a single-layered epithelium derived from the enveloping amnion. It has a core of mucous connective tissue. Observe also that the cord has one umbilical artery and one vein. Usually, there are two arteries. (Courtesy of Professor V. Becker, Pathologisches Institut der Universität, Erlangen, Germany.)

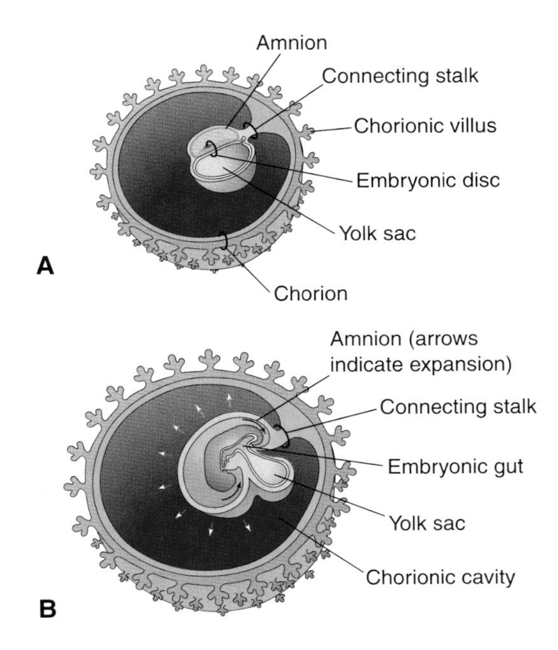

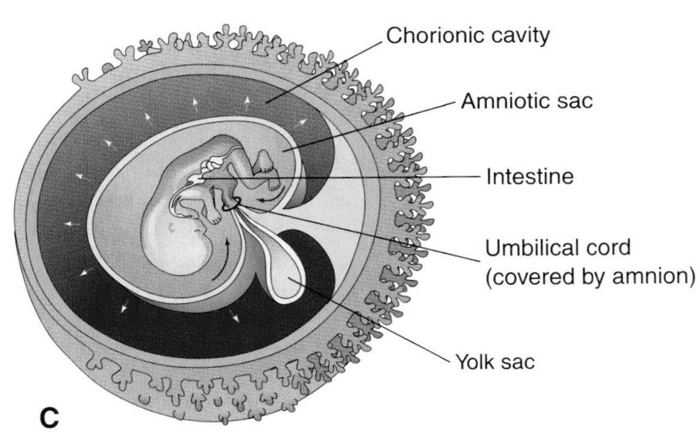

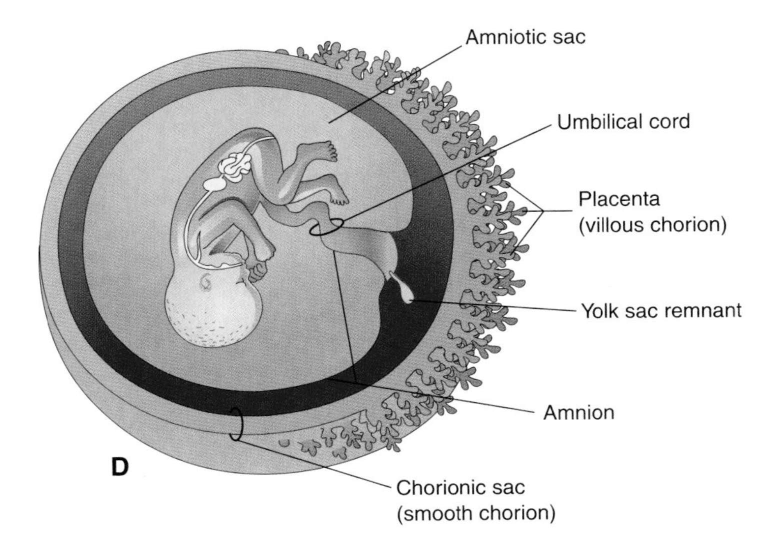

Figure 8 - 17. Drawings illustrating how the amnion enlarges, fills the chorionic sac, and envelops the umbilical cord. Observe that part of the yolk sac is incorporated into the embryo as the primordial gut. Formation of the fetal part of the placenta and degeneration of chorionic villi are also shown. *A*, Three weeks. *B*, Four weeks. *C*, Ten weeks. *D*, Twenty weeks.

Disorders of Amniotic Fluid Volume

A low volume of amniotic fluid for any particular gestational age (e.g., 400 ml in the third trimester) — termed **oligohydramnios** — (e.g., results, in most cases, from placental insufficiency with diminished placental blood flow. Preterm rupture of the amniochorionic membrane occurs in approximately 10% of pregnancies and is the most common cause of oligohydramnios. In the presence of **renal agenesis** (failure of kidney formation), the lack of fetal urine in the amniotic fluid is the main cause of oligohydramnios. A similar decrease in amniotic fluid occurs with **obstructive uropathy** (urinary tract obstruction). Complications of oligohydramnios include fetal abnormalities (pulmonary hypoplasia, facial defects, and limb defects) caused by fetal compression by the uterine wall.

A high volume of amniotic fluid — **polyhydramnios (hydramnios)** — in excess of 2000 ml, for example, results when the fetus does not swallow the usual amount of amniotic fluid. Most cases of polyhydramnios (60%) are idiopathic (of unknown cause); 20% of cases are caused by maternal factors, whereas 20% are fetal in origin. Polyhydramnios may be associated with severe anomalies of the central nervous system, such as meroanencephaly or anencephaly (see Chapter 19). With other anomalies, such as **esophageal atresia**, the fetus is unable to swallow amniotic fluid (see Chapter 13); in such cases, amniotic fluid accumulates because it is unable to pass to the fetal stomach and the intestines for absorption. *Ultrasonography* has become the technique of choice for diagnosing polyhydramnios.

Large volumes of amniotic fluid move in both directions between the fetal and maternal circulations, mainly via the placental membrane. Fetal swallowing of amniotic fluid is also a normal occurrence. Most fluid passes into the gastrointestinal tract, but some passes into the lungs. In either case, the fluid is absorbed and enters the fetal circulation. It then passes into the maternal circulation via the placental membrane. About 99% of the fluid in the amniotic cavity is water. Amniotic fluid is a solution in which undissolved material is suspended, such as desquamated fetal epithelial cells and approximately equal portions of organic and inorganic salts. Half of the organic constituents are protein; the other half is composed of carbohydrates, fats, enzymes, hormones, and pigments. As pregnancy advances, the composition of the amniotic fluid changes as fetal excreta (*meconium* and urine) are added. Because fetal urine enters the amniotic fluid, fetal enzyme systems, amino acids, hormones, and other substances can be studied by examining fluid removed by **amniocentesis**. Studies of cells in the amniotic fluid permit diagnosis of the sex of the fetus and detection of fetuses with chromosomal abnormalities, such as trisomy 21 in Down syndrome. High levels of *alpha-fetoprotein* (AFP) in the amniotic fluid usually indicate the presence of a severe neural tube defect (e.g., meroanencephaly). Low levels of AFP may indicate chromosomal aberrations (e.g., trisomy 21, which results in Down syndrome).

Significance of Amniotic Fluid

The buoyant amniotic fluid:

- permits symmetric external growth of the embryo
- acts as a barrier to infection
- permits normal fetal lung development
- prevents adherence of the amnion to the embryo
- cushions the embryo against injuries by distributing impacts the mother receives
- helps control the embryo's body temperature by maintaining a relatively constant temperature
- enables the fetus to move freely, thereby aiding muscular development (e.g., in the limbs)
- is involved in maintaining homeostasis of fluid and electrolytes

Premature Rupture of Membranes

Premature rupture of the amniochorionic membrane is the most common event leading to premature labor and delivery and the most common complication resulting in oligohydramnios. The absence of amniotic fluid also removes the major protection the fetus has against infection. Rupture of the amnion may cause various fetal anomalies that constitute the *amniotic band syndrome* (ABS), or *amniotic band disruption complex* (ABDC). These anomalies are associated with a variety of abnormalities ranging from digital constriction to major scalp, craniofacial, and visceral defects. The cause of these anomalies is probably related to constriction by encircling amniotic bands (Fig. 8 – 18). The incidence of ABS is about 1 in every 1200 live births. Prenatal ultrasound diagnosis of ABS is now possible.

Yolk Sac

Early development of the yolk sac was described in Chapter 4. At 32 days, the yolk sac is large (see Fig. 8 – 2). By 10 weeks, the yolk sac has shrunk to a pear-shaped remnant measuring about 5 mm in diameter (see Fig. 8 – 17*C*), and it is connected to the midgut by a narrow *yolk stalk*. By 20 weeks, the yolk sac is very small (see Fig. 8 – 17*D*); thereafter, it is usually not visible. The yolk sac can be observed sonographically early in the fifth week. The presence of the amnion and yolk sac enables early recognition and measurement of the embryo. The yolk sac is recognizable in ultrasound examinations until the end of the first trimester.

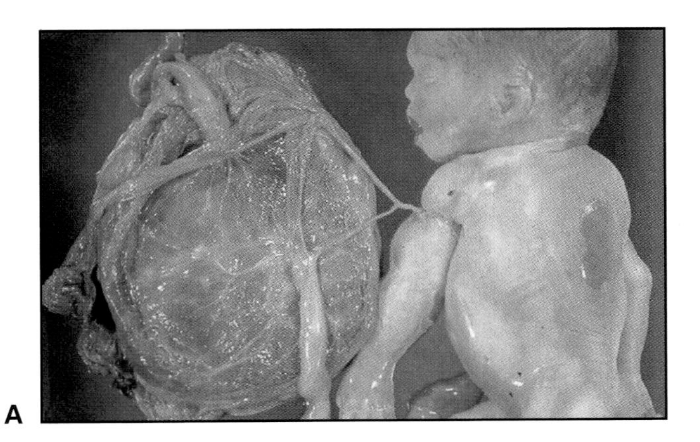

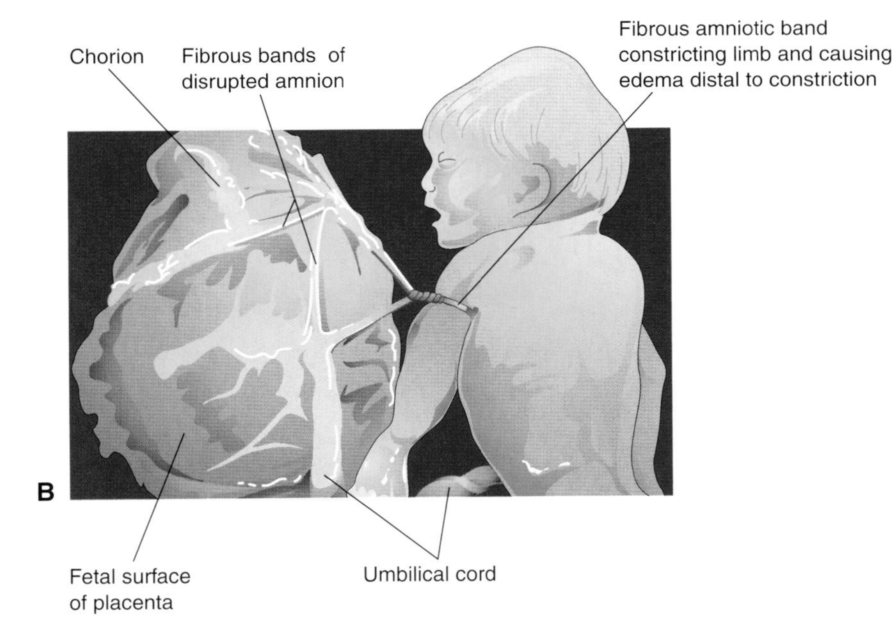

Figure 8 - 18. *A*, Photograph of a fetus with the amniotic band syndrome (ABS), showing amniotic bands constricting the left arm. (Courtesy of Professor V. Becker, Pathologisches Institut der Universität, Erlangen, Germany.) *B*, Drawing indicating the structures shown in *A*.

Significance of the Yolk Sac

Although the yolk sac is nonfunctional as far as yolk storage is concerned, its presence is essential for several reasons:

- It has a role in the *transfer of nutrients* to the embryo during the second and third weeks when the uteroplacental circulation is being established.
- *Blood* first develops in the well-vascularized, extraembryonic mesoderm covering the wall of the yolk sac beginning in the third week (see Chapter 5), and continues to develop there until hematopoietic activity begins in the liver during the sixth week.
- During the fourth week, the dorsal part of the yolk sac is incorporated into the embryo as the *primordial gut* (see Fig. 6 - 1). Its endoderm, derived from epiblast, gives rise to the epithelium of the trachea, bronchi, lungs, and digestive tract.
- *Primordial germ cells* appear in the endodermal lining of the wall of the yolk sac in the third week and subsequently migrate to the developing sex glands (see Chapter 14). They differentiate into the germ cells (spermatogonia in males and oogonia in females).

Fate of Yolk Sac

At 10 weeks, the small yolk sac lies in the chorionic cavity between the amnion and chorionic sac (Fig. 8 - 17*C*). The *yolk stalk* usually detaches from the midgut loop by the end of the sixth week. In about 2% of adults, the proximal intra-abdominal part of the yolk stalk persists

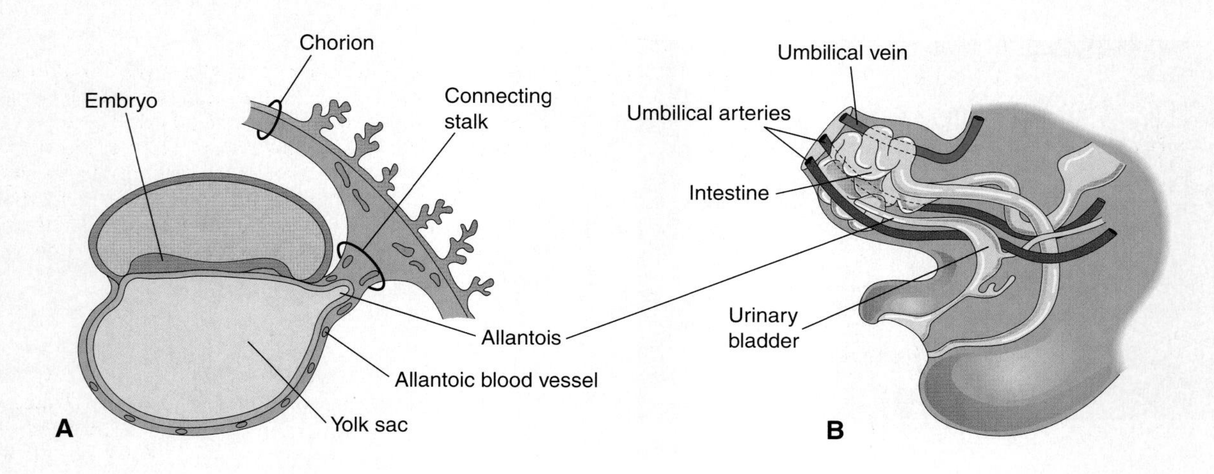

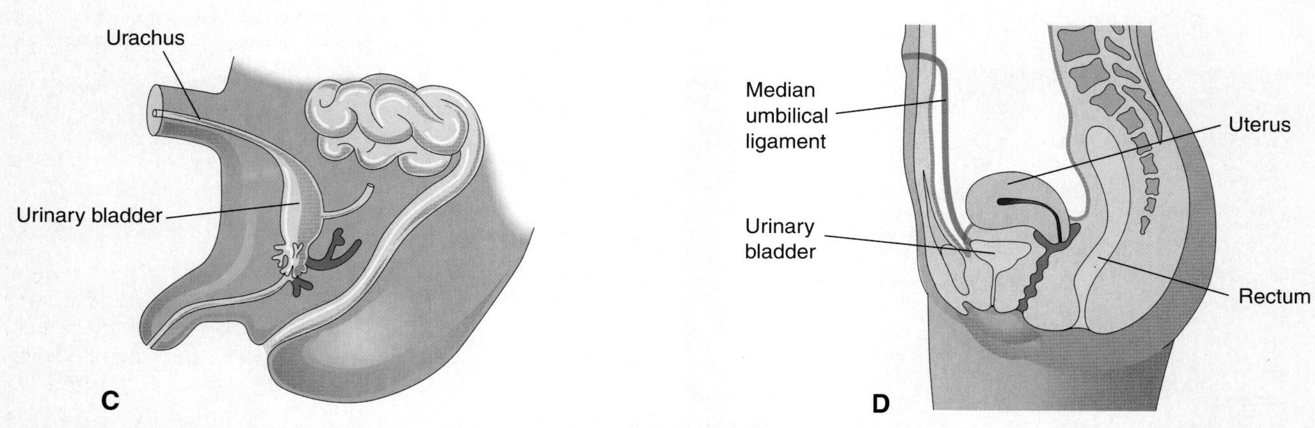

Figure 8 - 19. Development and usual fate of the allantois. *A*, Three-week embryo. *B*, Nine-week fetus. *C*, Three-month male fetus. *D*, Adult female. The nonfunctional allantois forms the urachus in the fetus and the median umbilical ligament in the adult.

as an *ileal diverticulum* known clinically as a *Meckel diverticulum* (see Chapter 13).

Allantois

The early development of the allantois was described in Chapter 5. During the second month of development, the extraembryonic part of the allantois degenerates (see Fig. 8 - 19*B*). Although the allantois is not functional in human embryos, it is important for four reasons:

- Blood formation occurs in its wall during the third to fifth weeks of development.
- Its blood vessels become the umbilical vein and arteries.
- Fluid from the amniotic cavity diffuses into the umbilical vein and enters the fetal circulation for transfer to the maternal blood through the placental membrane.
- The intraembryonic portion of the allantois runs from the umbilicus to the urinary bladder, with which it is continuous (see Fig. 8 - 19*B*). As the bladder enlarges, the allantois involutes to form a thick tube, the *urachus* (see Fig. 8 - 19*C*). After birth, the urachus becomes a fibrous cord, the *median umbilical ligament*, which extends from the apex of the urinary bladder to the umbilicus (see Fig. 8 - 19*D*).

Multiple Pregnancies

Multiple gestations are associated with higher risks of fetal morbidity and mortality than single gestations. The risks are progressively greater as the number of fetuses increases. Multiple births are more common now owing to the stimulation of ovulation that occurs when exogenous gonadotropins are administered to women with ovulatory failure, and to those being treated for infertility by in vitro fertilization and embryo transfer. In North America, **twins** normally occur about once in every 85 pregnancies, **triplets** about once in every 90^2 pregnancies, **quadruplets** about once in every 90^3 pregnancies, and **quintuplets** about once in every 90^4 pregnancies. These estimates increase when ovulations are primed with hormones, a technique that is often used to treat women who are sterile because of tubal occlusion.

Twins and Fetal Membranes

Twins that originate from two zygotes are **dizygotic (DZ) twins**, or fraternal twins (Fig. 8-20), whereas twins that originate from one zygote are **monozygotic (MZ) twins**, or identical twins (Fig. 8-21). The fetal membranes and placenta(s) vary according to the origin of the twins and, in the case of MZ twins, the type of placenta and membranes formed depends on when the twinning process occurs. *About two thirds of twins are dizygotic.* The frequency of DZ twinning shows marked racial differences, but *the incidence of MZ twinning is about the same in all populations.* In addition, the rate of MZ twinning shows little variation with the mother's age, whereas *the rate of DZ twinning increases with maternal age.*

The study of twins is important in human genetics because it is useful for comparing the effects of genes and environment on development. If an abnormal condition does not show a simple genetic pattern, comparison of its incidence in MZ and DZ twins may reveal that heredity is involved. The tendency for DZ but not MZ twins to repeat in families is evidence of hereditary influence. Additionally, if the firstborn are twins, a repetition of twinning or some other form of multiple birth is about five times more likely to occur at the next pregnancy than in the general population.

Anastomosis of Placental Blood Vessels

Anastomoses between blood vessels of fused placentas of DZ twins may occur and result in **erythrocytic mosaicism**. In such cases, the DZ twins will have red cells of two different types because of the exchange of red cells between the two circulations.

Twin-Twin Transfusion Syndrome

The twin-twin transfusion syndrome occurs in 15 to 30% of monochorionic-diamniotic MZ twins. Arterial blood is shunted from one twin through arteriovenous anastomoses into the venous circulation of the other twin. The donor twin is small, pale, and anemic (Fig. 8-22), whereas the recipient twin is large and polycythemic (i.e., having a higher than normal red blood cell count). The placenta shows similar abnormalities; the part of the placenta supplying the anemic twin is pale, whereas the part supplying the polycythemic twin is dark red. In lethal cases, death results from anemia in the donor twin and from congestive heart failure in the recipient twin.

Dizygotic Twins

Because they result from fertilization of two oocytes by two different sperms, DZ twins develop from two zygotes and may be of the same sex or different sexes (see Fig. 8-20). For the same reason, they are no more alike genetically than brothers or sisters born at different times. *DZ twins always have two amnions and two chorions*, but the chorions and placentas may be fused. *DZ twinning shows a hereditary tendency.* The recurrence risk in families with one set of DZ twins is about triple that of the general population. The incidence of DZ twinning shows considerable variation, ranging from 1 in 500 in Asians, to 1 in 125 in Caucasians, to as high as 1 in 20 in some African populations.

Monozygotic Twins

Because they result from the fertilization of one oocyte and develop from one zygote (see Fig. 8-21), *MZ twins are of the same sex, are genetically identical, and are very similar in physical appearance.* Physical differences between MZ twins are environmentally induced, such as anastomosis of placental vessels resulting in differences in blood supply from the placenta (see Fig. 8-22). MZ twinning usually begins in the blastocystic stage, around the end of the first week, and results from division of the inner cell mass or embryoblast into two embryonic primordia. Subsequently, two embryos, each in its own amniotic sac, develop within one chorionic sac and share a **common placenta**, a monochorionic-diamniotic twin placenta. Uncommonly, early separation of embryonic blastomeres (e.g., during the two- to eight-cell stage) results in MZ twins with two amnions, two chorions, and two placentas that may or may not be fused (Fig. 8-23). In such cases, it is impossible to determine, from the membranes alone, whether the twins are monozygotic or dizygotic. To determine the relationship of twins of the same sex with similar blood groups, one must wait until other characteristics, such as eye color and fingerprints, develop.

Establishing the Zygosity of Twins

Establishment of the zygosity of twins has become important, particularly since the introduction of tissue and organ

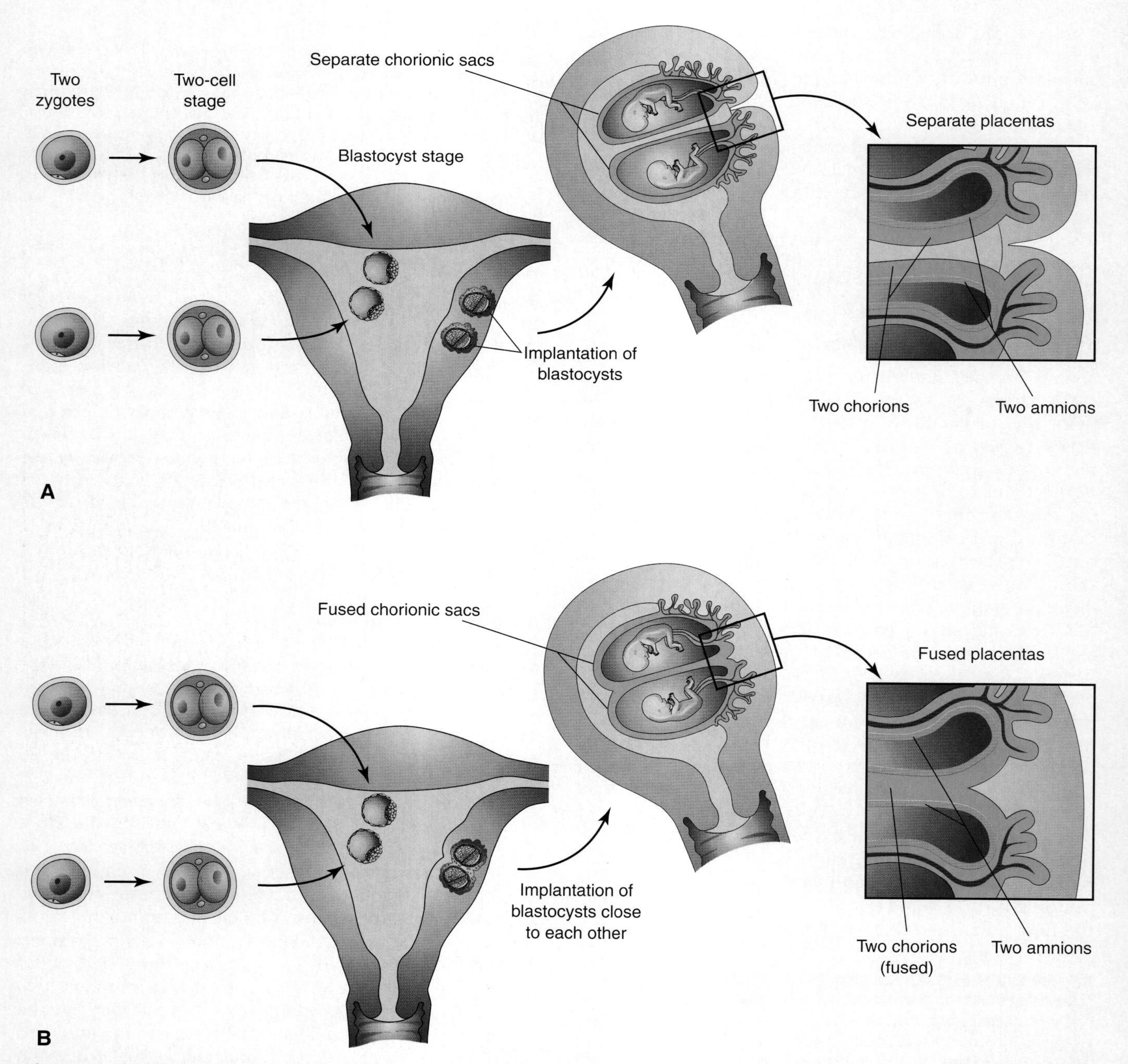

Figure 8 – 20. Dizygotic (DZ) twins developing from two zygotes. The relationship of the fetal membranes and placentas are shown for instances in which the blastocysts implant separately (*A*) and the blastocysts implant close together (*B*). In both cases, there are two amnions and two chorions. The placentas are usually fused when they implant close together.

transplantation (e.g., bone marrow transplants). Twin zygosity is now determined by *molecular diagnosis* because any two people who are not MZ twins are virtually certain to show differences in some of the large number of DNA markers that can be studied.

Late division of early embryonic cells (i.e., division of the embryonic disc during the second week) results in MZ twins that are in one amniotic sac and one chorionic sac (Fig. 8 – 24*A*). A *monochorionic-monoamniotic twin placenta* is associated with a fetal mortality rate approaching 50%. Such twins are

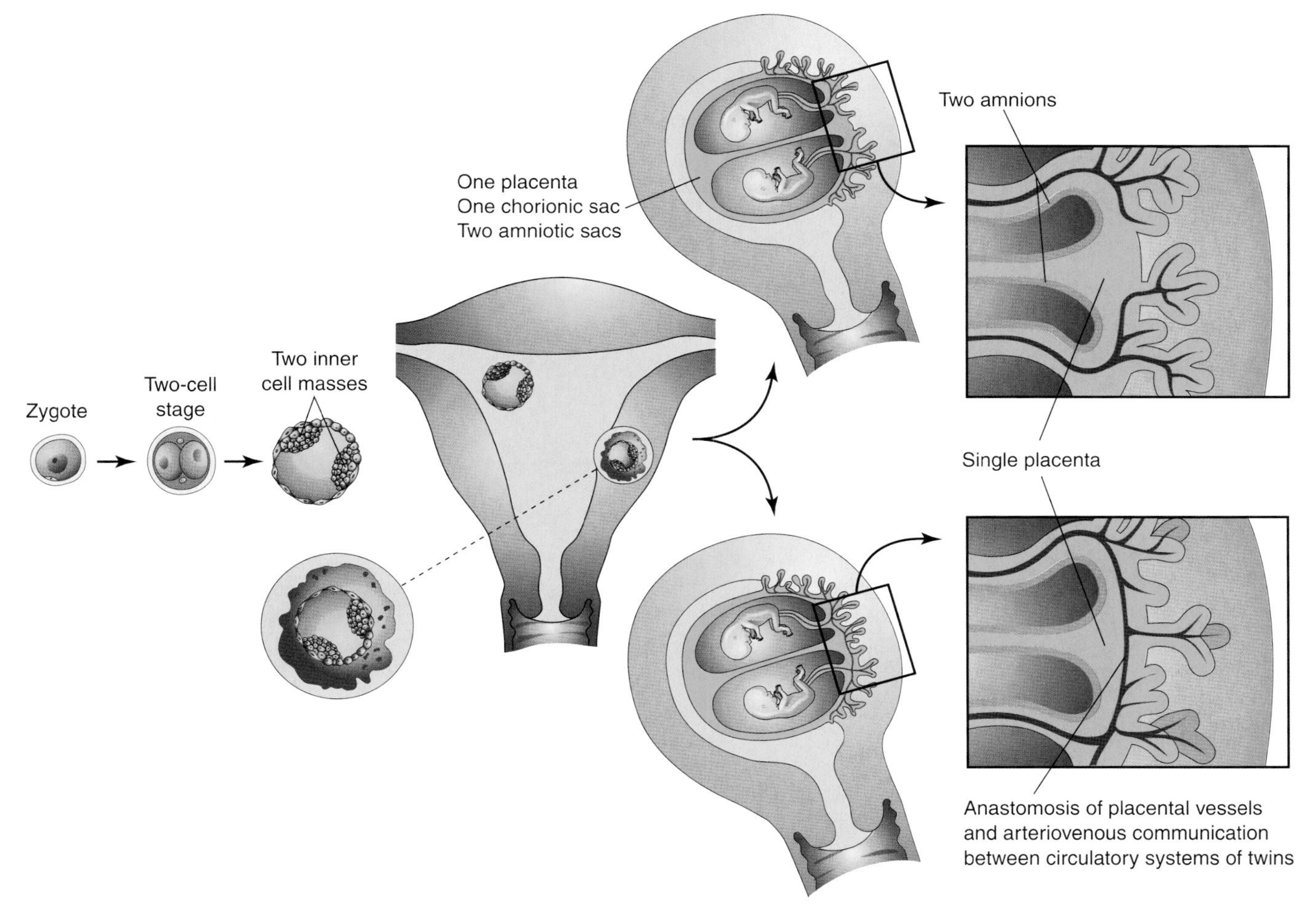

Figure 8 – 21. Diagrams illustrating how about 65% of monozygotic (MZ) twins develop from one zygote by division of the inner cell mass. These twins always have separate amnions, a single chorionic sac, and a common placenta. If there is anastomosis of the placental vessels, one twin may receive most of the nutrition from the placenta (see Fig. 8 – 22).

rarely delivered alive because the umbilical cords are frequently so entangled that circulation of the blood through their vessels ceases, and one or both fetuses die. Ultrasonography plays an important role in the diagnosis and management of twin pregnancies. Ultrasound evaluation is necessary to identify various conditions that may complicate MZ twinning, such as IUGR, intrauterine fetal distress, and premature labor.

Early Death of a Twin

Because ultrasonographic studies are a common part of prenatal care, it is known that early death with resorption of one member of a twin pair is fairly common. This possibility must be considered when discrepancies occur between prenatal cytogenetic findings and the karyotype of an infant. Errors in prenatal diagnosis may arise if extraembryonic tissues (e.g., part of a chorionic villus) from the resorbed twin are examined.

Conjoined Twins

If the embryonic disc does not divide completely, various types of conjoined (MZ) twins may form (see Fig. 8 – 24*B* and *C*). These twins are named according to the regions of the body that are attached; for example, *thoracopagus* indicates anterior union of the thoracic regions. In some cases, the twins are connected to each other by skin only or by cutaneous and other tissues, such as fused livers. Some conjoined twins can be separated successfully by surgery. The incidence of conjoined twins is 1 in 50,000 to 100,000 births.

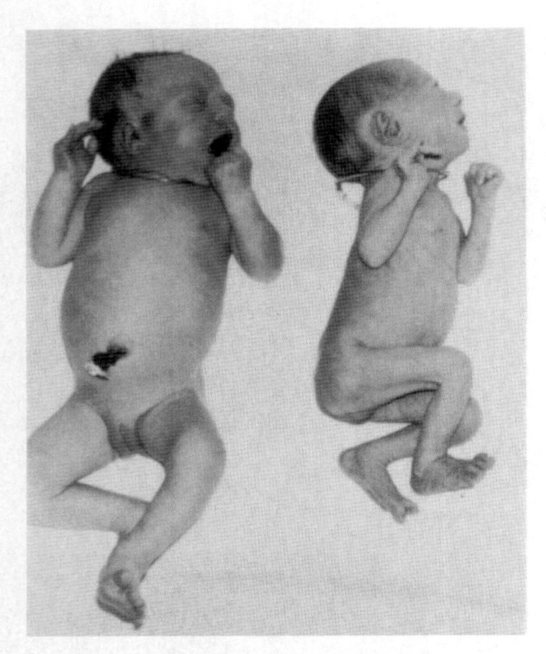

Figure 8 - 22. MZ, monochorionic, diamniotic twins. Note the wide discrepancy in size resulting from an uncompensated arteriovenous anastomosis of placental vessels. Blood was shunted from the smaller twin to the larger one, producing the twin-twin transfusion syndrome.

Other Types of Multiple Births

Triplets may be derived from:

- one zygote and be identical
- two zygotes and consist of identical twins and a singleton
- three zygotes and be of the same sex or of different sexes, in which case the infants are no more similar than infants from three separate pregnancies.

Similar combinations occur in quadruplets, quintuplets, sextuplets, and septuplets.

Summary of the Placenta and Fetal Membranes

In addition to the embryo and fetus, the fetal membranes and the major part of the placenta originate from the zygote. The placenta consists of two parts:

- a comparatively larger fetal part derived from the villous chorion
- a smaller maternal part developed from the decidua basalis

The two parts are held together by stem chorionic villi that attach to the cytotrophoblastic shell surrounding the chorionic sac.

The principal *activities of the placenta* are:

- metabolism, such as synthesis of glycogen, cholesterol, and fatty acids
- respiratory gas exchange (oxygen, carbon dioxide, and carbon monoxide)
- transfer of nutrients, such as vitamins, hormones, and antibodies
- elimination of waste products
- endocrine secretion (e.g., hCG) for maintenance of pregnancy

All of these activities are essential for maintaining pregnancy and promoting normal fetal development.

The fetal circulation is separated from the maternal circulation by a thin layer of extrafetal tissues called the **placental membrane**. It is a permeable membrane that allows water, oxygen, nutritive substances, hormones, and noxious agents to pass from the mother to the embryo or fetus. Excretory products pass through the placental membrane from the fetus to the mother.

The fetal membranes and placenta(s) in *multiple pregnancies* vary considerably, depending on the derivation of the embryos and the time at which division of embryonic cells occurs. The most common type of twin is the *dizygotic (DZ) twins*. This type of twin has two amnions, two chorions, and two placentas that may or may not be fused. *Monozygotic (MZ) twins*, the less common type, represent about a third of all twins; they are derived from one zygote. Monozygotic twins commonly have one chorion, two amnions, and one placenta. Twins with one amnion, one chorion, and one placenta are always monozygotic, and their umbilical cords are often entangled. Other types of multiple birth (e.g., triplets) may be derived from one or more zygotes.

The *yolk sac* and *allantois* are vestigial structures; however, their presence is essential to normal embryonic development. Both are early sites of blood formation and both are partly incorporated into the embryo. Primordial germ cells also originate in the wall of the yolk sac.

The *amnion* forms a sac for amniotic fluid and provides a covering for the umbilical cord. The amniotic fluid has three main functions:

- to provide a protective buffer for the embryo or fetus
- to allow room for fetal movements
- to assist in the regulation of fetal body temperature

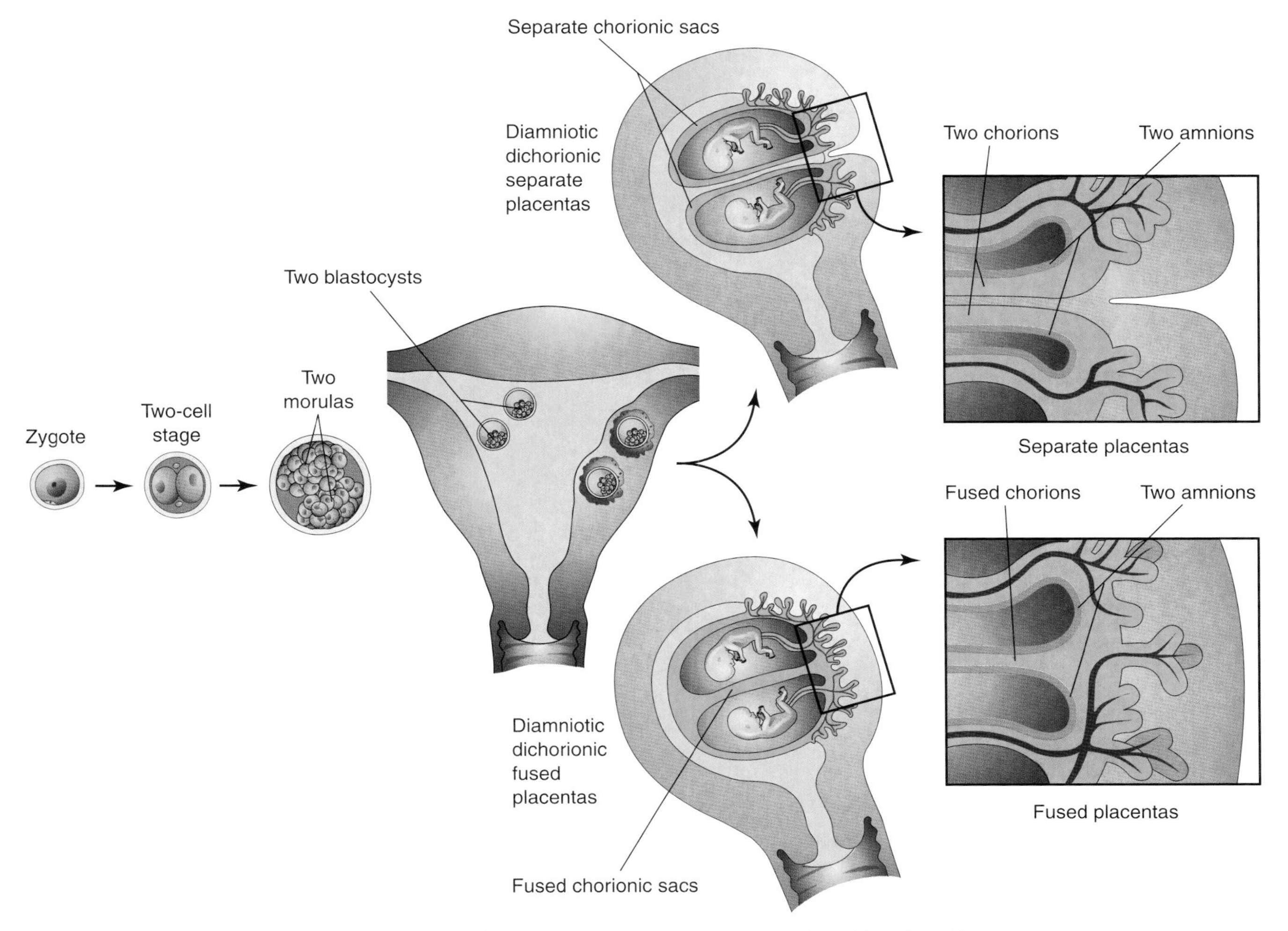

Figure 8 - 23. Diagrams illustrating how about 35% of MZ twins develop from one zygote. Separation of the blastomeres may occur anywhere from the two-cell (blastomere) stage to the morula stage, producing two identical blastocysts. Each embryo subsequently develops its own amniotic and chorionic sacs. The placentas may be separate or fused. In 25% of cases, there is a single placenta resulting from secondary fusion, whereas in 10% of cases, there are two placentas. In the latter cases, examination of the placenta suggests that they are DZ twins. This explains why some MZ twins are wrongly stated to be DZ twins at birth.

Clinically Oriented Questions

1. What is meant by the term *stillbirth*? Do older women have more stillborn infants? Some reports suggest that more male than female infants are born dead. Is this true?
2. A baby was born dead, reportedly because of a "cord accident." What does this mean? Do these accidents always kill the baby? If not, what defects may be present?
3. What is the scientific basis of the home pregnancy tests that are sold in drug stores? Are they accurate?
4. What is the proper name for what lay people refer to as the "bag of waters"? What is meant by a "dry birth"? Does premature rupture of this "bag" induce the birth of the baby?
5. What does the term *fetal distress* mean? How is the condition recognized? What causes fetal stress and distress?
6. Some say that twins are born more commonly to older mothers. Is this true? Others maintain that twinning is hereditary. Is this correct?

The answers to these questions are at the back of the book.

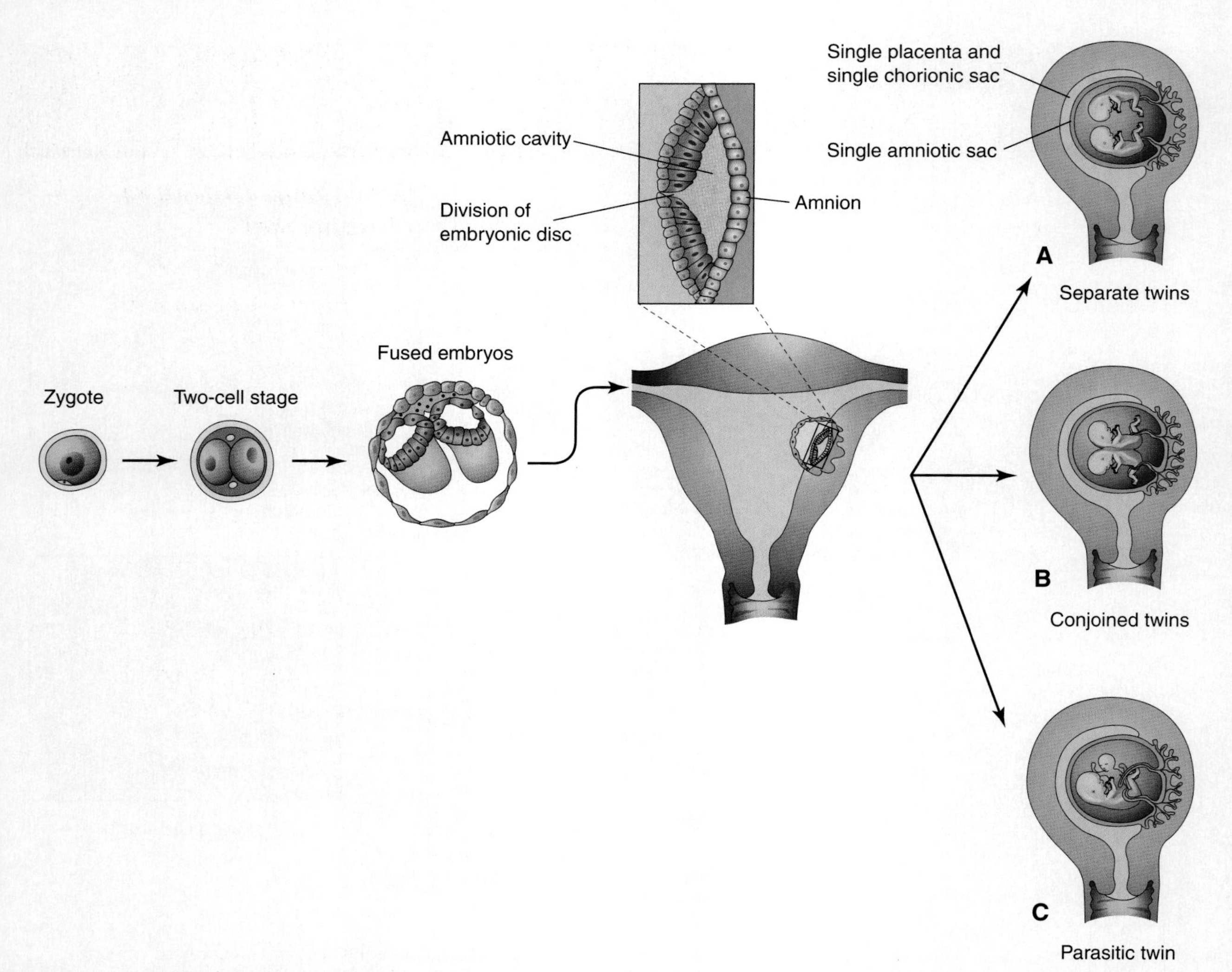

Figure 8 - 24. Diagrams illustrating how some MZ twins develop. This method of development is very uncommon. Division of the embryonic disc results in two embryos within one amniotic sac. *A,* Complete division of the embryonic disc gives rise to twins. Such twins rarely survive because their umbilical cords are often so entangled that interruption of the blood supply to the fetuses occurs. *B* and *C,* Incomplete division of the disc results in various types of conjoined twins.)

9

Human Birth Defects

Congenital anomalies, birth defects, and *malformations* are terms currently used to describe developmental disorders present at birth (L., *congenitus*, born with). Birth defects are the leading cause of infant mortality and may be structural, functional, metabolic, behavioral, or hereditary. A **congenital anomaly** is a structural abnormality of any type; *however, not all variations are anomalies.* Anatomical variations are common. *Congenital anomalies are of four clinically significant types:* malformation, disruption, deformation, and dysplasia. For more information, see Moore and Persaud: *The Developing Human: Clinically Oriented Embryology*, 7th ed. Philadelphia, 2003.

Teratology: The Study of Abnormal Development

Teratology is the branch of science that studies the causes, mechanisms, and patterns of abnormal development. A fundamental concept in teratology is that certain stages of embryonic development are more vulnerable to disruption than others. At one time, it was generally believed that the human embryo was protected from environmental agents, such as drugs and viruses. In 1941, an Australian ophthalmologist reported an unusual number of cases of cataract and other anomalies, including microcephaly, deafness, and heart defects in the infants of mothers who had contracted **rubella** during the first trimester of pregnancy. Two decades later, reports of severe limb anomalies and other developmental defects were observed in the offspring of mothers who had taken **thalidomide**, a mild sedative, during early pregnancy. It is estimated that 7 to 10% of human birth defects result from the disruptive actions of drugs, viruses, and other environmental factors.

More than 20% of infant deaths in North America are attributable to birth defects. Major structural anomalies, such as spina bifida cystica — a severe type of defect in which part of the neural tube fails to fuse — are observed in about 3% of newborn infants. Additional anomalies can be detected after birth; thus, the incidence approaches about 6% in 2-year-old children and 8% in 5-year-old children.

Congenital anomalies may be caused by:

- *genetic factors*, such as chromosomal abnormalities
- *environmental factors*, such as drugs and viruses

However, many common congenital anomalies are the result of **multifactorial inheritance**; that is, they are caused by genetic and environmental factors acting together. For 50 to 60% of congenital anomalies, the causes are unknown (Fig. 9 – 1). Congenital anomalies may be single or multiple and of major or minor clinical significance. Single **minor anomalies** are present in about 14% of neonates. Anomalies of the external ear, for example, are of no serious medical significance, but they indicate the possible presence of associated major anomalies. For example, the presence of a single umbilical artery alerts the clinician to the possible presence of cardiovascular and renal anomalies. Major developmental defects are much more common in early embryos (10 to 15%), but most embryos abort spontaneously during the first 6 weeks. Chromosomal abnormalities are present in more than 50% of spontaneously aborted conceptuses.

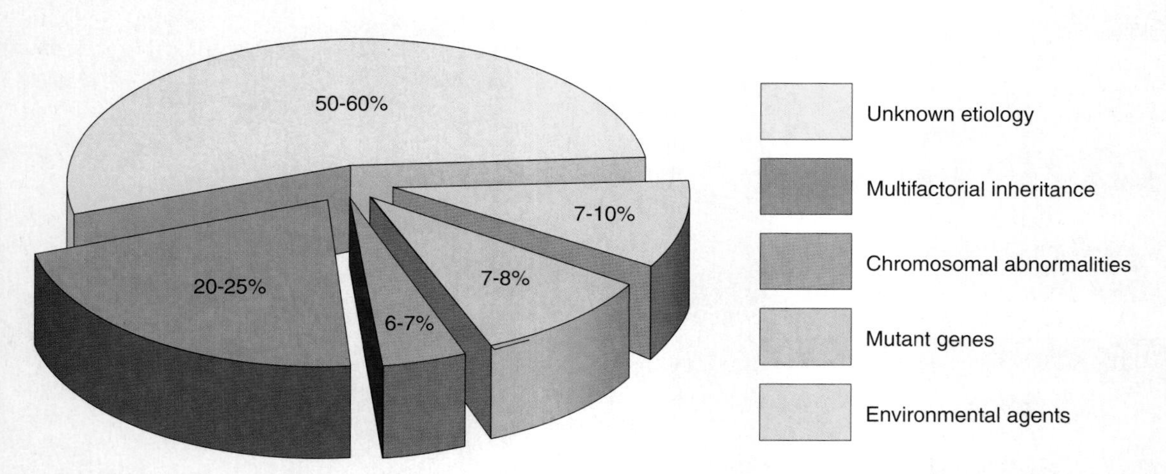

Figure 9 – 1. The causes of human birth defects. Note that the causes of most anomalies are unknown and that 20 to 25% of them are caused by a combination of genetic and environmental factors (multifactorial inheritance).

Anomalies Caused by Genetic Factors*

In terms of sheer number of cases, genetic factors are the most important causes of congenital anomalies. It has been estimated that they cause about one third of all birth defects (see Fig. 9 - 1) and nearly 85% of anomalies with known causes. Any mechanism as complex as mitosis or meiosis may occasionally malfunction; thus, *chromosomal aberrations are common and are present in 6 to 7% of zygotes.* Many of these early embryos never undergo normal cleavage to become blastocysts. The changes may affect the sex chromosomes and/or the autosomes (chromosomes other than sex chromosomes). In some instances, both kinds of chromosome are affected. Persons with chromosomal abnormalities usually have characteristic phenotypes, such as the physical characteristics of infants with Down syndrome. *Numerical and structural changes occur in chromosome complements.*

Numerical Chromosome Abnormalities

Numerical aberrations of chromosomes usually result from **nondisjunction**, an error in cell division in which a chromosome pair or two chromatids of a chromosome fail to disjoin during mitosis or meiosis. As a result, the chromosome pair or chromatids pass to one daughter cell, and the other daughter cell receives neither (Fig. 9 - 2). Nondisjunction may occur during maternal or paternal gametogenesis (see Chapter 2). The chromosomes in somatic (body) cells are normally paired; the homologous chromosomes making up a pair are *homologs*. Normal human females have 22 pairs of autosomes plus two X chromosomes, whereas normal males have 22 pairs of autosomes plus one X and one Y chromosome.

Inactivation of Genes

During embryogenesis, one of the two X chromosomes in female somatic cells is randomly inactivated and appears as a mass of **sex chromatin** (see Chapter 7). Inactivation of genes on one X chromosome in somatic cells of female embryos occurs during implantation. *X-inactivation is important clinically* because it means that each cell from a carrier of an X-linked disease has the mutant gene causing the disease, either on the active X chromosome or on the inactivated X chromosome that is represented by sex chromatin. Uneven X-inactivation in monozygotic twins is one reason given for discordance in a variety of congenital anomalies. The genetic basis for discordance is that one twin preferentially expresses the paternal X, the other the maternal X.

Aneuploidy and Polyploidy

Changes in chromosome number result in either aneuploidy or polyploidy. **Aneuploidy** is any deviation from the human diploid number of 46 chromosomes. *An aneuploid* is an individual or a cell that has a chromosome number that is not an exact multiple of the haploid number of 23 (e.g., 45 or 47). The principal cause of aneuploidy is nondisjunction during cell division (see Fig. 9 – 2), resulting in an unequal distribution of one pair of homologous chromosomes to the daughter cells. One cell has two chromosomes and the other has neither chromosome of the pair. As a result, the embryo's cells may be *hypodiploid* (45,X, as in *Turner syndrome* [Fig. 9 – 3]), or *hyperdiploid* (usually 47, as in trisomy 21, or *Down syndrome* [Fig. 9 – 4]). Embryos with **monosomy** — missing a chromosome — usually die. Monosomy of an autosome is extremely uncommon, and about 99% of embryos lacking a sex chromosome (45,X) abort spontaneously.

Turner Syndrome

About 1% of monosomy X female embryos survive. The incidence of 45,X or Turner syndrome in newborn females is approximately 1 in 8000 live births. Half the affected individuals have 45,X; the other half have a variety of abnormalities of a sex chromosome. *The phenotype of Turner syndrome is female* and is illustrated in Figure 9 - 3. **Phenotype** refers to the morphological characteristics of an individual as determined by the genotype and the environment in which it is expressed. Secondary sexual characteristics do not develop in 90% of girls with Turner syndrome, necessitating hormone replacement therapy. *The monosomy X chromosome abnormality is the most common cytogenetic abnormality observed in liveborn humans and fetuses that abort spontaneously*, and it accounts for about 18% of all abortions caused by chromosomal abnormalities. The error in gametogenesis (nondisjunction) that causes monosomy X, when it can be traced, is in the paternal gamete in about 75% of cases; that is, it is the paternal X chromosome that is usually missing.

If three chromosomes are present instead of the usual pair, the abnormality is known as trisomy. Trisomies are the most common abnormalities of chromosome number. The usual cause of this numerical error is **meiotic nondisjunction of chromosomes** (see Fig. 9 - 2), resulting in a gamete with 24 instead of 23 chromosomes and, subsequently, in a zygote with 47 chromosomes.

*The authors are grateful to Dr. A. E. Chudley, MD, F.R.C.P.C., F.C.C.M.G., Professor of Pediatrics and Child Health, and Head, Section of Genetics and Metabolism, Children's Hospital, Health Sciences Centre, University of Manitoba, Winnipeg, Manitoba, Canada, for assistance with the preparation of this section.

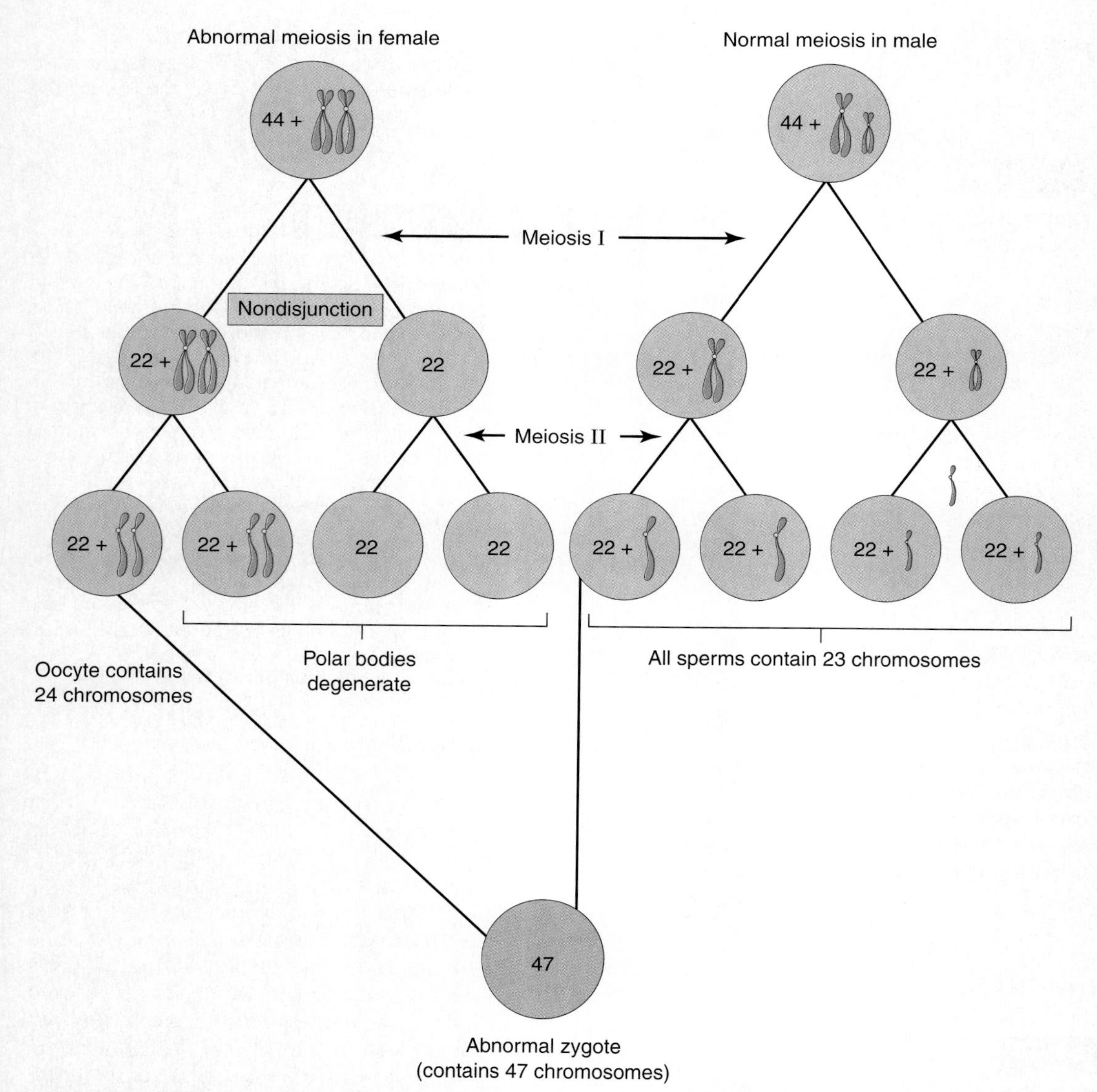

Figure 9 – 2. Diagram showing nondisjunction of chromosomes during the first meiotic division of a primary oocyte, resulting in an abnormal oocyte with 24 chromosomes. Subsequent fertilization by a normal sperm produces a zygote with 47 chromosomes — aneuploidy — a deviation from the human diploid number of 46.

Trisomy of Autosomes

Trisomy of the autosomes is associated mainly with three syndromes (Table 9 – 1):

- trisomy 21, or Down syndrome (see Fig. 9 – 4)
- trisomy 18, or Edwards syndrome (Fig. 9 – 5)
- trisomy 13, or Patau syndrome (Fig. 9 – 6)

Infants with trisomy 13 and trisomy 18 are severely malformed and mentally retarded. They usually die early in infancy. More than 50% of trisomic conceptions abort early in pregnancies. *Trisomy of the autosomes occurs with increasing frequency as maternal age increases* (Table 9 – 2).

Mosaicism — two or more cell types containing different numbers of chromosomes (normal and abnormal) — leads to a less severe phenotype and the affected individuals may have a nearly normal IQ.

Trisomy of Sex Chromosomes

Trisomy of the sex chromosomes is a common condition (Table 9 – 3); however, because no characteristic physical findings are seen in infants or children, this disorder is

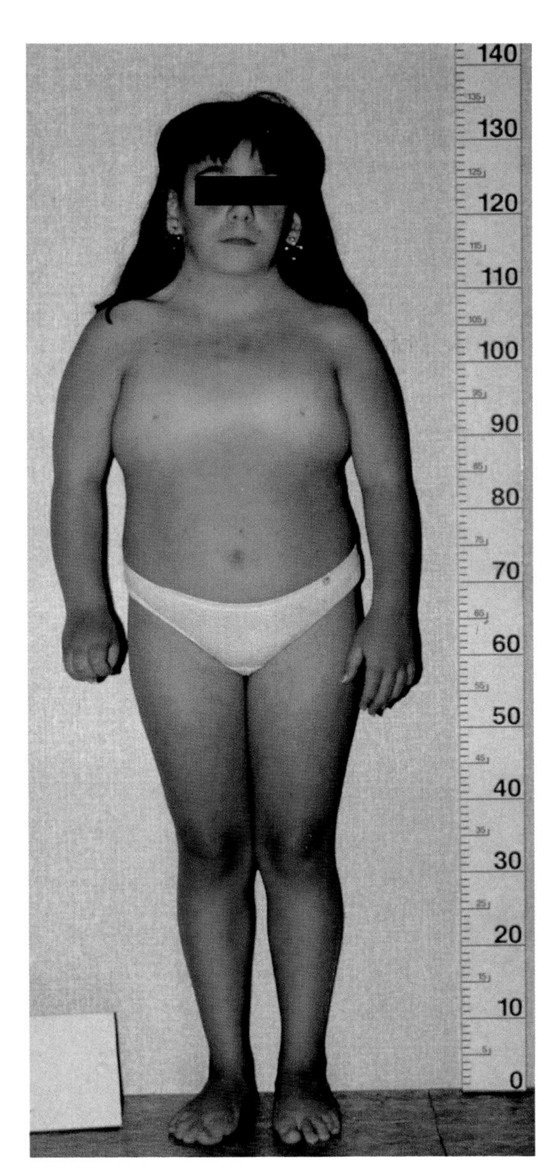

Figure 9 - 3. Turner syndrome in a 14-year-old girl. Note the classic features of the syndrome: short stature; webbed neck; absence of sexual maturation; broad, shieldlike chest with widely spaced nipples; and lymphedema of the hands and feet. (Courtesy of Dr. F. Antoniazzi and Dr. V. Fanos, Department of Pediatrics, University of Verona, Verona, Italy.)

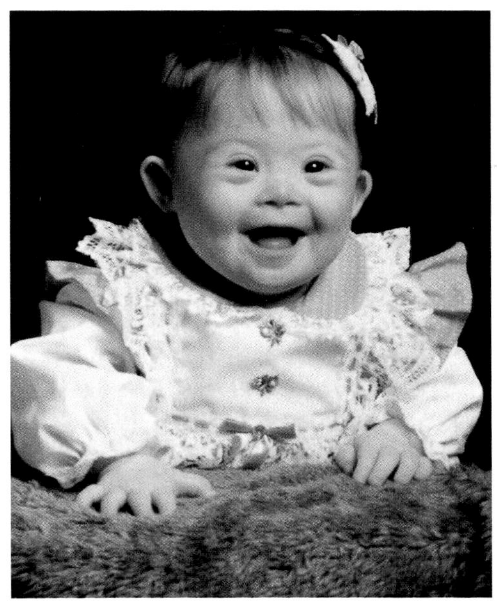

Figure 9 - 4. Photograph of an infant with Down syndrome (trisomy 21). Note the round face, upslanted palpebral fissures, and short digits with incurving of the fifth digit (clinodactyly). (Courtesy of Dr. AE Chudley, Professor of Pediatrics and Child Health, Children's Hospital and University of Manitoba, Winnipeg, Manitoba, Canada.)

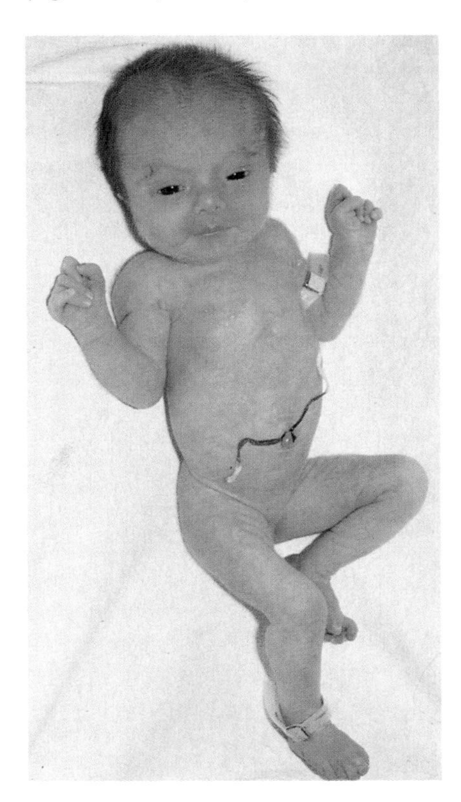

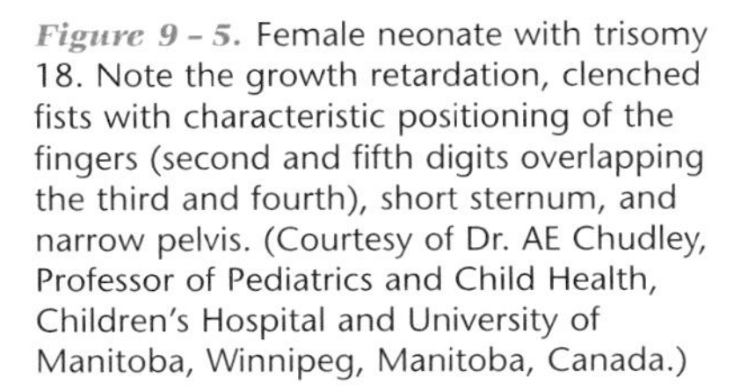

Figure 9 - 5. Female neonate with trisomy 18. Note the growth retardation, clenched fists with characteristic positioning of the fingers (second and fifth digits overlapping the third and fourth), short sternum, and narrow pelvis. (Courtesy of Dr. AE Chudley, Professor of Pediatrics and Child Health, Children's Hospital and University of Manitoba, Winnipeg, Manitoba, Canada.)

Table 9 – 1. **Trisomy of the Autosomes**

Chromosomal Aberration/Syndrome	Incidence	Usual Morphologic Characteristics	Figures
Trisomy 21 or Down syndrome*	1:800	Mental deficiency; brachycephaly, flat nasal bridge; upward slant to palpebral fissures; protruding tongue; simian crease; clinodactyly of 5th digit; congenital heart defects	9–4
Trisomy 18 syndrome†	1:8000	Mental deficiency; growth retardation; prominent occiput; short sternum; ventricular septal defect; micrognathia; low-set, malformed ears; flexed digits, hypoplastic nails; rocker-bottom feet	9–5
Trisomy 13 syndrome†	1:25,000	Mental deficiency; severe central nervous system malformations; sloping forehead; malformed ears, scalp defects; microphthalmia; bilateral cleft lip and/or palate; polydactyly; posterior prominence of the heels	9–6

*The importance of this disorder in the overall problem of mental retardation is indicated by the fact that persons with Down syndrome represent 10 to 15% of institutionalized, mentally defective individuals. *The incidence of trisomy 21 at fertilization is greater than at birth*; however, 75% of affected embryos are spontaneously aborted and at least 20% are stillborn.
†Infants with this syndrome rarely survive beyond 6 months of age.

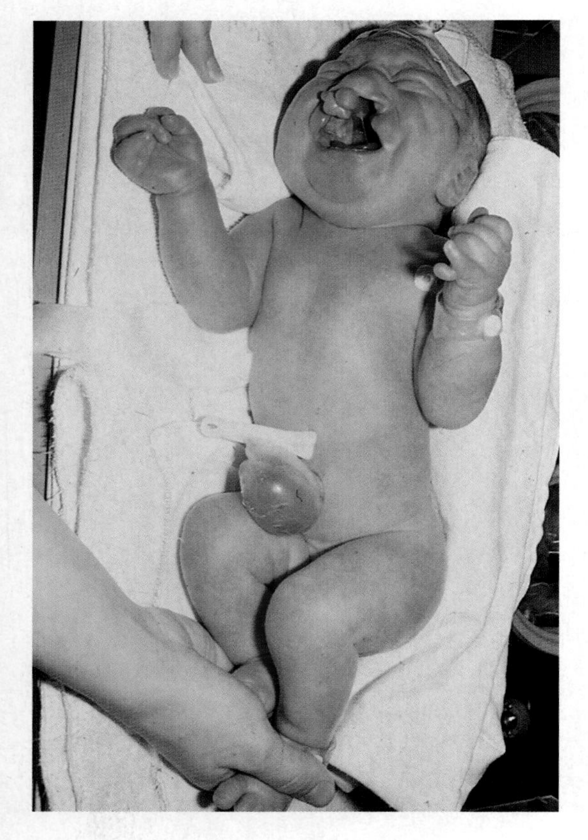

Figure 9 – 6. Female neonate with trisomy 13. Note, in particular, the bilateral cleft lip; low-set, malformed ears; and polydactyly (extra digits). A small omphalocele (herniation of viscera into the umbilical cord) is also present. (Courtesy of Dr. AE Chudley, Professor of Pediatrics and Child Health, Children's Hospital and University of Manitoba, Winnipeg, Manitoba, Canada.)

Table 9-2. Incidence of Down Syndrome in Newborn Infants

Maternal Age (Years)	Incidence
20–24	1:1400
25–29	1:1100
30–34	1:700
35	1:350
37	1:225
39	1:140
41	1:85
43	1:50
45+	1:25

not usually detected before puberty (Fig. 9-7). The diagnosis is best established by chromosome and molecular analysis.

Mosaicism

A person who has at least two cell lines with *two or more different genotypes* (genetic constitutions) is known as a **mosaic**. Either the autosomes or sex chromosomes may be involved. Usually, the anomalies are less serious than in persons with monosomy or trisomy. Mosaicism usually results from nondisjunction during early cleavage of the zygote (see Chapter 3). Mosaicism resulting from loss of a chromosome by *anaphase lagging* also occurs; the chromosomes separate normally, but one of them is delayed in its migration and is eventually lost.

Triploidy

The most common type of polyploidy is **triploidy** (69 chromosomes). Triploid fetuses have severe intrauterine growth retardation (IUGR) with a disproportionately small trunk, as well as other anomalies. Triploidy can result from the second polar body failing to separate from the oocyte during the second meiotic division (see Chapter 2); more likely, however, triploidy results when an oocyte is fertilized by two sperms (dispermy) almost simultaneously. Triploidy occurs in about 2% of embryos, but most of them abort spontaneously. Triploid fetuses account for about 20% of chromosomally abnormal miscarriages. In rare cases, triploid fetuses have been born alive. They all died a few days after birth.

Tetraploidy

Doubling the diploid chromosome number to 92 (*tetraploidy*) probably occurs during the first cleavage division. Division of this abnormal zygote would subsequently result in an embryo with cells containing 92 chromosomes. *Tetraploid embryos* abort very early, and often, all that is recovered is an empty chorionic sac.

Tetrasomy and Pentasomy

Tetrasomy and pentasomy of the sex chromosomes also occur. Persons with these abnormalities have four or five sex chromosomes; the following chromosome complexes have been reported: in *females*, 48,XXXX and 49,XXXXX; in *males*, 48,XXXY, 48,XXYY, 49,XXXYY, and 49,XXXXY. The extra sex chromosomes do not accentuate sexual characteristics; however, usually, the greater the number of sex chromosomes present, the greater the severity of mental retardation and physical impairment.

Structural Chromosome Abnormalities

Most abnormalities of chromosome structure result from chromosome breakage followed by reconstitution in an abnormal combination (Fig. 9-8). **Chromosome breaks** may be induced by various environmental factors, such as irradiation, drugs, chemicals, and viruses. The resulting abnormality in chromosome structure depends upon what happens to the broken pieces. The only two aberrations of chromosome structure that are likely to be transmitted from parent to child are structural rearrangements, such as inversion and translocation.

Translocation

Translocation is the transfer of a piece of one chromosome to a nonhomologous chromosome. If two nonhomologous chromosomes exchange pieces, it is called a *reciprocal translocation* (see Fig. 9-8*A* and *G*). Translocation does not necessarily cause abnormal development. Persons with a translocation between a number 21 chromosome and a number 14, for example (see Fig. 9-8*G*), are phenotypically normal. Such persons are called *balanced translocation carriers*. They have a tendency, independent of age, to produce germ cells with an abnormal translocation chromosome. Three to 4% of persons with Down syndrome have translocation trisomies; that is, the extra 21 chromosome is attached to another chromosome.

Deletion

When a chromosome breaks, a portion of the chromosome may be lost (see Fig. 9-8*B*). A partial terminal deletion from the short arm of chromosome 5 causes the **cri du chat syndrome**. Affected infants have a weak, catlike cry at birth; growth delay with microcephaly (an abnormally small head); hypertelorism; low-set ears; micrognathia; severe mental retardation; and congenital heart disease. A **ring chromosome** is a type of deletion chromosome from which both ends have been lost, and the broken ends have rejoined to form a ring-shaped chromosome

Table 9-3. Trisomy of the Sex Chromosomes

Chromosome Complement*	Sex	Incidence†	Usual Characteristics
47,XXX	Female	1:1000	Normal appearance; usually fertile; 15-25% have mild mental retardation
47,XXY	Male	1:1000	Klinefelter syndrome; small testes; hyalinization of seminiferous tubules; aspermato genesis; often tall with disproportionately long lower limbs; intelligence is less than in normal siblings; gynecomastia in about 40%
47,XYY	Male	1:1000	Normal appearance; usually tall; often exhibiting aggressive behavior

*The numbers designate the total number of chromosomes, including the sex chromosomes shown after the comma.
†See Nussbaum RL, McInnes RR, Willard HF: *Thompson & Thompson Genetics in Medicine*, 6th ed. Philadelphia, WB Saunders Co, 2001.

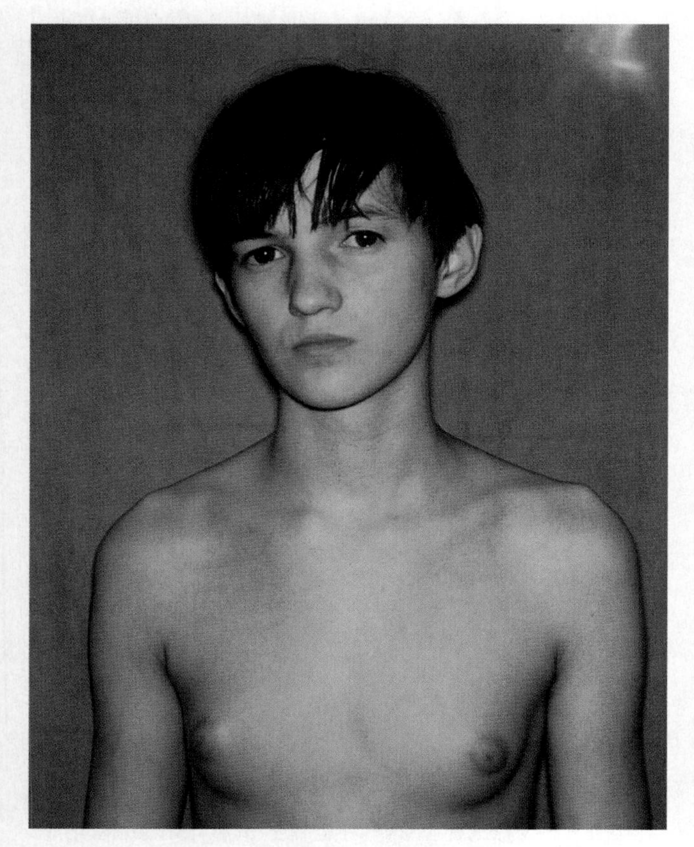

Figure 9-7. Young male with Klinefelter syndrome (XXY trisomy). Note the presence of developed breasts; about 40% of males with this syndrome have gynecomastia (excessive development of male mammary glands) and small testes. (Courtesy of Dr. AE Chudley, Professor of Pediatrics and Child Health, Children's Hospital and University of Manitoba, Winnipeg, Manitoba, Canada.)

(see Fig. 9-8*C*). Ring chromosomes are very rare, but they have been found for all chromosomes. These abnormal chromosomes have been described in persons with Turner syndrome, trisomy 18, and other abnormalities.

Microdeletions and Microduplications

High-resolution banding techniques have allowed detection of very small interstitial and terminal deletions in a number of disorders. Normal-resolution chromosome banding reveals 350 bands per haploid set, whereas *high-resolution chromosome banding* reveals up to 1300 bands per haploid set. Because the deletions span several contiguous genes, these disorders, as well as those with microduplications, are referred to as **contiguous gene syndromes**. Two examples are:

- *Prader-Willi syndrome* (PWS), a sporadically occurring disorder associated with short stature, mild mental retardation, obesity, hyperphagia (overeating), and hypogonadism (inadequate gonadal function)
- *Angelman syndrome* (AS), characterized by severe mental retardation, microcephaly, brachycephaly (shortness of the head), seizures, and ataxic (jerky) movements of the limbs and trunk

The clinical phenotype is determined by the parental origin of the deleted chromosome 15. If the deletion arises in the mother, AS occurs; if passed on by the father, the child exhibits the PWS phenotype. This suggests the phenomenon of **genetic imprinting**, whereby differential expression of genetic material depends on the sex of the transmitting parent.

Molecular Cytogenetics

Several new methods for merging classic cytogenetics with DNA technology have facilitated a more precise definition of

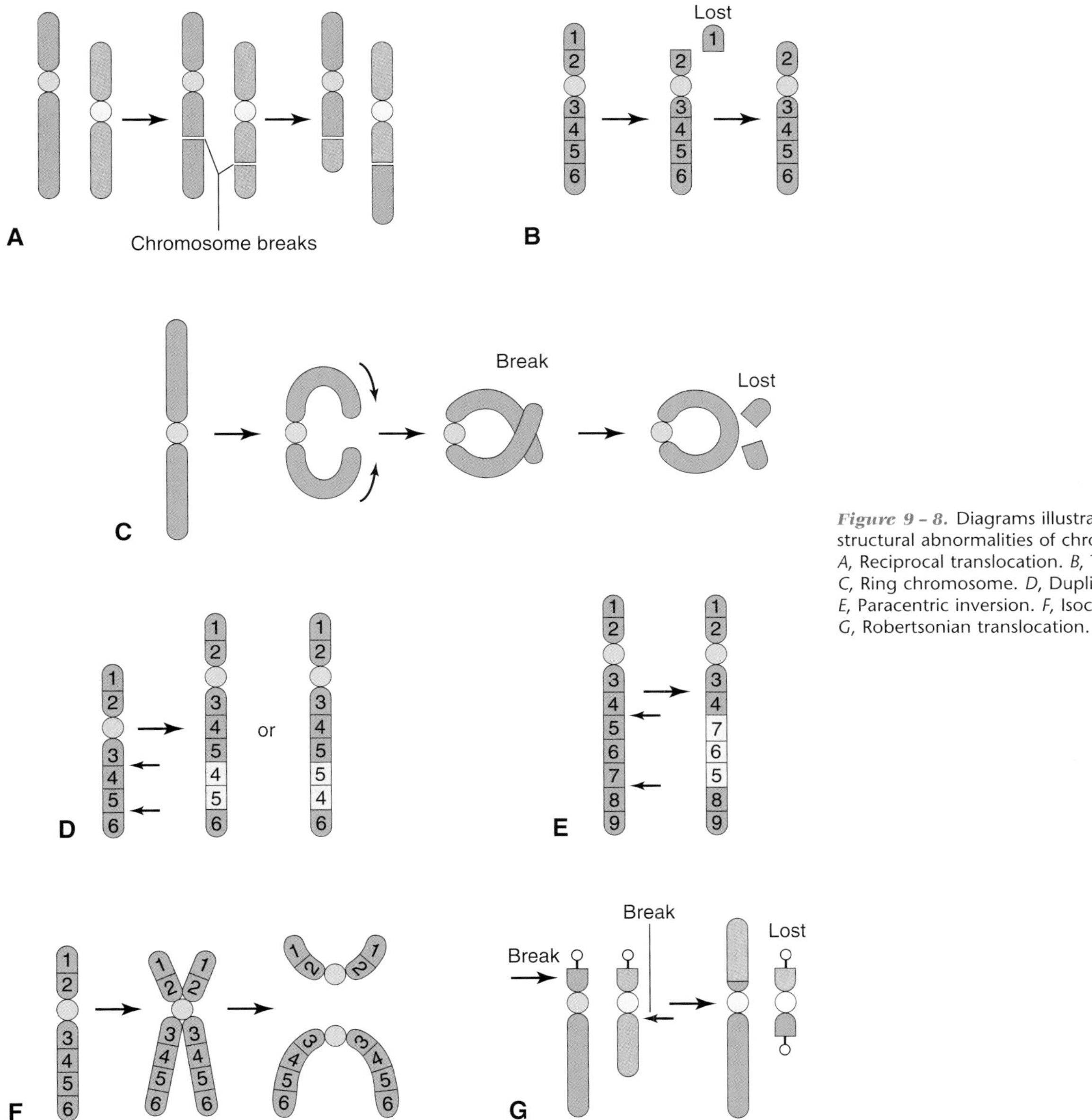

Figure 9 - 8. Diagrams illustrating various structural abnormalities of chromosomes. *A*, Reciprocal translocation. *B*, Terminal deletion. *C*, Ring chromosome. *D*, Duplication. *E*, Paracentric inversion. *F*, Isochromosome. *G*, Robertsonian translocation.

chromosome abnormalities, location, or origins, including unbalanced translocations, accessory or marker chromosomes, and *gene mapping*. One new approach to chromosome identification is based on *fluorescent in situ hybridization* (FISH), whereby chromosome-specific DNA probes can adhere to complementary regions located on specific chromosomes. This allows improved identification of chromosome location and number in metaphase spreads or even in interphase cells. The FISH techniques using interphase cells may soon obviate the need to culture cells for specific chromosome analysis, such as in the case of prenatal diagnosis of fetal trisomies.

Duplications

Duplications may be manifested as a duplicated part of a chromosome, within a chromosome (see Fig. 9–8*D*), attached to a chromosome, or as a separate fragment. *Duplications are more common than deletions, and they are less harmful* because no loss of genetic material occurs. Duplication may involve part of a gene, whole genes, or a series of genes.

Inversion

Inversion is a chromosomal aberration in which a segment of a chromosome is reversed. *Paracentric inversion* is confined to a single arm of the chromosome (Fig. 9–8*E*), whereas *pericentric inversion* involves both arms and includes the centromere. Carriers of pericentric inversions are at risk of having offspring with abnormalities because of unequal crossing-over and malsegregation at meiosis.

Isochromosomes

The abnormality resulting in isochromosomes occurs when the centromere divides transversely instead of longitudinally (see Fig. 9–8*F*). An isochromosome is a chromosome in which one arm is missing and the other is duplicated. It appears to be the *most common structural abnormality of the X chromosome.* Persons with this chromosomal abnormality are often short in stature and have other stigmata of Turner syndrome. These characteristics are related to the loss of an arm of an X chromosome.

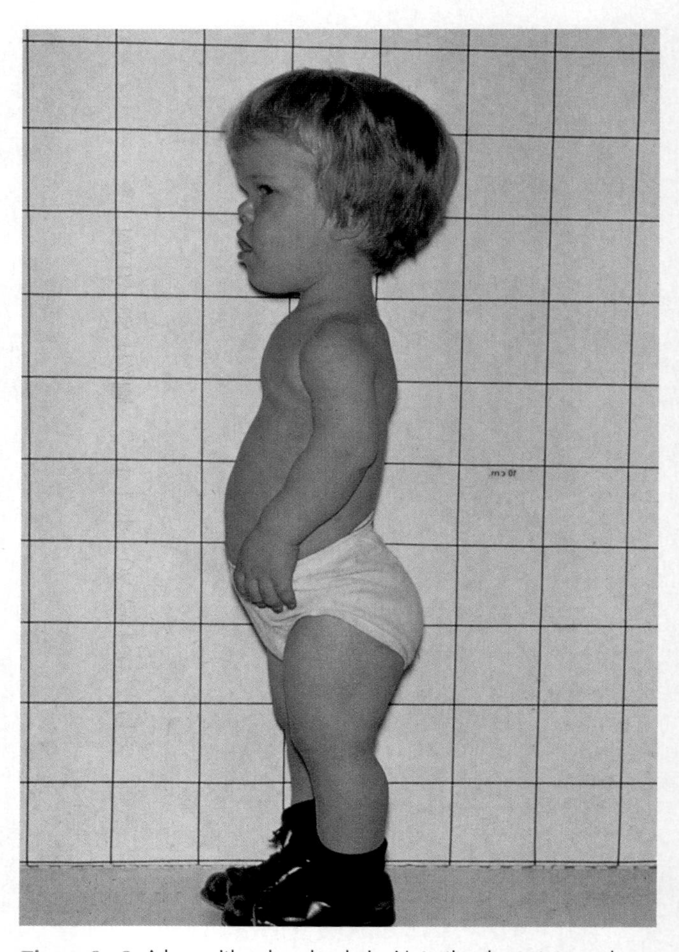

Figure 9–9. A boy with achondroplasia. Note the short stature, short limbs and fingers, normal length of the trunk, relatively large head, prominent forehead, and depressed nasal bridge. (Courtesy of Dr. AE Chudley, Professor of Pediatrics and Child Health, Children's Hospital and University of Manitoba, Winnipeg, Manitoba, Canada.)

Anomalies Caused by Mutant Genes

Seven to 8% of congenital anomalies are caused by gene defects (see Fig. 9–1). A mutation usually involves a loss or change in the function of a gene and is any permanent, heritable change in the sequence of genomic DNA. Because a random change is unlikely to lead to an improvement in development, *most mutations are deleterious and some are lethal.* The mutation rate can be increased by a number of environmental agents, such as large doses of radiation. Anomalies resulting from gene mutations are inherited according to mendelian laws; consequently, predictions can be made about the probability of their occurrence in the affected person's children and other relatives. An example of a *dominantly inherited congenital anomaly* is **achondroplasia** (Fig. 9–9), which results from a mutation of the complementary DNA in the *fibroblast growth factor receptor 3 (FGFR-3) gene* on chromosome 4p. Other congenital anomalies are attributable to *autosomal recessive inheritance.* Autosomal recessive genes manifest themselves only when homozygous; as a consequence, many carriers of these genes (heterozygous persons) remain undetected.

The **fragile X syndrome** is the most common inherited cause of moderate mental retardation (Fig. 9–10), and is second only to *Down syndrome among all causes of moderate mental retardation in males.* The fragile X syndrome has a frequency of 1 in 1500 male births and may account for much of the predominance of males in the mentally retarded population. The diagnosis can be confirmed by chromosomal analysis, which will demonstrate the fragile X chromosome, or by DNA studies, which will indicate an expansion of CGG nucleotides in a specific region of the fragile mental retardation I (FMRI) gene.

Several genetic disorders have been linked to the expansion of trinucleotides in specific genes. Some examples include myotonic dystrophy, Huntington chorea, spinobulbar atrophy (Kennedy disease), Friedreich ataxia, among others. X-linked recessive genes are usually

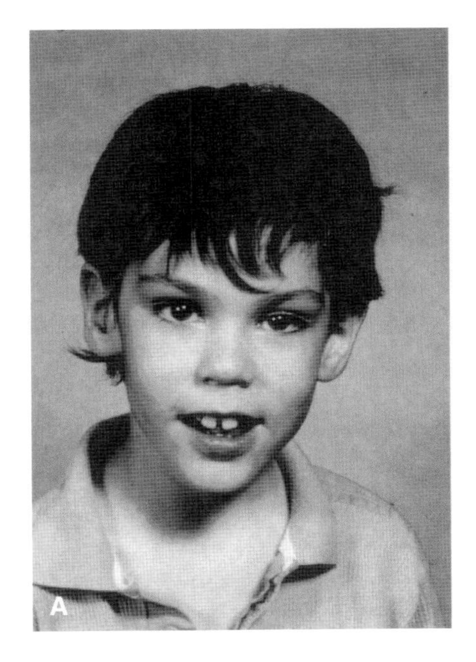

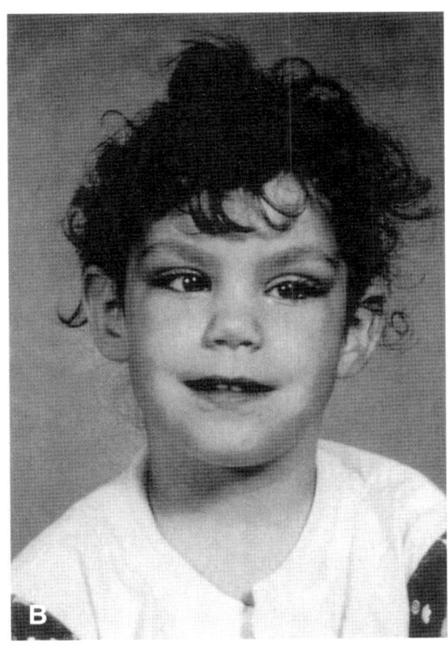

Figure 9 - 10. Fragile X syndrome. *A*, An 8-year-old, mentally retarded boy exhibiting a relatively normal appearance with a long face and prominent ears. *B*, His 6-year-old sister also has this syndrome. She has a mild learning disability and similar features of long face and prominent ears. Note the strabismus (crossed right eye). Although this is an X-linked disorder, sometimes, female carriers express the disease. (Courtesy of Dr. AE Chudley, Professor of Pediatrics and Child Health, Children's Hospital and University of Manitoba, Winnipeg, Manitoba, Canada.)

manifested in affected (homozygous) males and occasionally in carrier (heterozygous) females (for example, fragile X syndrome).

The **human genome** comprises an estimated 30,000 to 40,000 genes per haploid set, or 3 billion base pairs. Because of the *Human Genome Project* and international research collaboration, many disease- and birth defect-causing mutations in genes have been and will continue to be identified. Most genes will be sequenced and their specific function determined. Understanding the cause of birth defects will require an improvement in our understanding of gene expression during early development. Most genes that are expressed in a cell are expressed in a wide variety of cells and are involved in basic cellular metabolic functions, such as nucleic acid and protein synthesis, cytoskeleton and organelle biogenesis, and nutrient transport and mechanisms. These genes are referred to as *house keeping genes*. The *specialty genes* are expressed at specific times in specific cells and define the hundreds of different cell types that make up the human organism. An essential aspect of developmental biology is regulation of gene expression. Regulation is often achieved by *transcription factors*, which bind to regulatory or promoter elements of specific genes.

Genomic imprinting is an epigenetic process whereby the female and male germlines confer a sex-specific mark on a chromosome subregion, so that only the paternal or maternal allele of a gene is active in the offspring. In other words, the sex of the transmitting parent influences expression or nonexpression of certain genes in the offspring. This is the reason for PWS and AS, in which the phenotype is determined by whether the microdeletion is transmitted by the father (PWS) or the mother (AS).

Homeobox genes are a group of genes found in all vertebrates. They have highly conserved sequences and order. They are involved in early embryonic development and specify identity and spatial arrangements of body segments. Protein products of these genes bind to DNA and form transcriptional factors that regulate gene expression.

Anomalies Caused by Environmental Factors

Although the human embryo is well protected in the uterus, certain environmental agents, called **teratogens**, may cause developmental disruptions following maternal exposure to them (Table 9 – 4). A teratogen is any agent that can produce a congenital anomaly or raise the incidence of an anomaly in a population. Environmental factors, such as infection and drugs, may simulate genetic conditions, such as when two or more children of normal parents are affected. The *important principle* to remember is that "not everything that is familial is genetic." The organs and parts of an embryo are most sensitive to teratogenic agents during periods of rapid differentiation (Fig. 9 – 11). *Environmental factors cause 7 to 10% of congenital anomalies* (see Fig. 9 – 1). Because molecular signaling and embryonic induction

Table 9 - 4. Teratogens Known to Cause Human Birth Defects

Agents	Most Common Congenital Anomalies
DRUGS	
Alcohol	*Fetal alcohol syndrome (FAS)*: intrauterine growth retardation (IUGR); mental retardation, microcephaly; ocular anomalies; joint abnormalities; short palpebral fissures
Androgens and high doses of progestogens	Varying degrees of masculinization of female fetuses; ambiguous external genitalia manifested by labial fusion and clitoral hypertrophy
Aminopterin	IUGR; skeletal defects; malformations of the central nervous system, notably meroanencephaly (absence of most of the brain)
Busulfan	Stunted growth; skeletal abnormalities; corneal opacities; cleft palate; hypoplasia of various organs
Cocaine	IUGR; prematurity; microcephaly; cerebral infarction; urogenital anomalies, neurobehavioral disturbances
Diethylstilbestrol (DES)	Abnormalities of the uterus and vagina; cervical erosion and ridges
Isotretinoin (13-*cis*-retinoic acid)	Craniofacial abnormalities; neural tube defects (NTDs), such as spina bifida cystica; cardiovascular defects; cleft palate; thymic aplasia
Lithium carbonate	Various anomalies usually involving the heart and great vessels
Methotrexate	Multiple anomalies, especially skeletal, involving the face, cranium, limbs, and vertebral column
Phenytoin (Dilantin)	*Fetal hydantoin syndrome (FHS)*: IUGR; microcephaly; mental retardation; ridged metopic suture; inner epicanthal folds; eyelid ptosis; broad, depressed nasal bridge; phalangeal hypoplasia
Tetracycline	Stained teeth; hypoplasia of enamel
Thalidomide	Abnormal development of the limbs; meromelia (partial absence of limb) and amelia (complete absence of limb); facial anomalies; systemic anomalies (e.g., cardiac and kidney defects)
Trimethadione	Developmental delay; V-shaped eyebrows; low-set ears; cleft lip and/or palate
Valproic acid	Craniofacial anomalies; neural tube defects (NTDs); often hydrocephalus; heart and skeletal defects
Warfarin	Nasal hypoplasia; stippled epiphyses; hypoplastic phalanges; eye anomalies; mental retardation
CHEMICALS	
Methylmercury	Cerebral atrophy; spasticity; seizures; mental retardation
Polychlorinated biphenyls (PCBs)	IUGR; skin discoloration
INFECTIONS	
Cytomegalovirus (CMV)	Microcephaly; chorioretinitis; sensorineural loss; delayed psychomotor/ mental development; hepatosplenomegaly; hydrocephaly; cerebral palsy; brain (periventricular) calcification
Herpes simplex virus (HSV)	Skin vesicles and scarring; chorioretinitis; hepatomegaly; thrombocytopenia; petechiae; hemolytic anemia; hydranencephaly
Human immunodeficiency virus (HIV)	Growth failure; microcephaly; prominent, boxlike forehead; flattened nasal bridge; hypertelorism; triangular philtrum and patulous lips
Human parvovirus B19	Eye defects; degenerative changes in fetal tissues
Rubella virus	IUGR; postnatal growth retardation; cardiac and great vessel abnormalities; microcephaly; sensorineural deafness; cataract; microphthalmos; glaucoma; pigmented retinopathy; mental retardation; neonatal bleeding; hepatosplenomegaly; osteopathy; tooth defect
Toxoplasma gondii	Microcephaly; mental retardation; microphthalmia; hydrocephaly; chorioretinitis; cerebral calcifications; hearing loss; neurologic disturbances
Treponema pallidum	Hydrocephalus; congenital deafness; mental retardation; abnormal teeth and bones
Varicella virus	Cutaneous scars (dermatome distribution); neurologic anomalies (e.g., limb paresis, hydrocephaly, seizures); cataracts; microphthalmia; Horner syndrome; optic atrophy; nystagmus; chorioretinitis; microcephaly; mental retardation; skeletal anomalies (e.g., hypoplasia of limbs, fingers, and toes); urogenital anomalies
HIGH LEVELS OF IONIZING RADIATION	Microcephaly; mental retardation; skeletal anomalies; growth retardation; cataracts

precede morphologic differentiation, the period during which structures are sensitive to interference by teratogens often precedes the stage of their visible development. Teratogens do not appear to be effective in causing anomalies until cellular differentiation has begun; however, their early actions may cause the death of an embryo, for example, during the first 2 weeks of development. The exact *mechanisms* by which drugs, chemicals, and other environmental factors disrupt embryonic development and induce abnormalities are unclear.

Rapid progress in molecular biology is providing additional information on the genetic control of differentiation, as well as the cascade of molecular signals and factors controlling gene expression and pattern formation. Researchers are now directing increasing attention to the molecular mechanisms of abnormal development in an attempt to understand better the pathogenesis of congenital anomalies.

Basic Principles in Teratogenesis

When considering the possible teratogenicity of an agent, such as a drug or chemical, *three factors are important* to consider:

- critical periods of development
- dosage of the drug or chemical
- genotype (genetic constitution) of the embryo

Critical Periods of Human Development

An embryo's susceptibility to a teratogen depends on its stage of development when an agent, such as a drug, is present (see Fig. 9 - 11). The most critical period in development is when cell differentiation and morphogenesis are at their peak. Table 9 - 5 indicates the relative frequencies of anomalies for certain organs. *The most critical period for brain development is from 3 to 16 weeks*, but its development may be disrupted after this time because the brain is differentiating and growing rapidly at birth and continues to do so throughout the first 2 years after birth. Teratogens (e.g., alcohol) may produce mental retardation during the embryonic and fetal periods. *Tooth development continues long after birth* (see Chapter 21); hence, development of permanent teeth may be disrupted by *tetracyclines* from 18 weeks (prenatal) to 16 years. *The skeletal system has a prolonged critical period of development* extending into childhood; hence, the growth of skeletal tissues provides a good gauge of general growth. Environmental disturbances during the first 2 weeks after fertilization may interfere with cleavage of the zygote and implantation of the blastocyst and/or may cause early death and spontaneous abortion of the embryo (see Fig. 9 - 11). In some cases, teratogens acting during the first 2 weeks of development kill the embryo; in other cases, their disruptive effects are mitigated by powerful regulatory properties of the early embryo.

Development of the embryo is most easily disrupted when the tissues and organs are forming (see Fig. 9 - 11). During this **organogenetic period**, teratogenic agents may induce major congenital anomalies. Physiological defects — minor morphological anomalies of the external ear, for example — and functional disturbances, such as mental retardation, are likely to result from disruption of development during the fetal period. *Each part, tissue, and organ of an embryo has a critical period during which its development may be disrupted* (see Fig. 9 - 11). The type of congenital anomalies produced depends on which parts, tissues, and organs are most susceptible at the time the teratogen is active.

Embryologic timetables such as the one in Figure 9 - 11 are helpful when considering the cause of human birth defects. However, it is wrong to assume that anomalies always result from a single event occurring during the critical period of development, or that one can determine from these tables the day on which an anomaly was produced. What is known is that the teratogen would have to disrupt development of the tissue, part, or organ before the end of the critical period. The *critical period for limb development, for example, is 24 to 36 days after fertilization.*

Dosage of Drug or Chemical

Animal research has shown a dose-response relationship for teratogens; however, *the dose used to produce anomalies in animals is often much higher than typical human dose exposure. Consequently, animal studies are not readily applicable to human pregnancies.* For a drug to be considered a human teratogen, a dose-response relationship has to be observed; that is, it must be shown that the greater the exposure during pregnancy, the more severe the phenotypic effect.

Genotype of Embryo

Numerous studies in experimental animals and several suspected cases in humans have revealed genetic differences in the way individual subjects respond to a teratogen. *Phenytoin*, for example, is a well-known human teratogen (see Table 9 - 4). Five to 10% of embryos exposed to this anticonvulsant medication develop the *fetal hydantoin syndrome*. However, about one third of exposed embryos develop only some of the associated congenital anomalies, and more than half of the embryos are unaffected. It appears, therefore, that the genotype of the embryo determines whether a teratogenic agent disrupts its development.

Known Human Teratogens

Awareness that certain agents can disrupt human prenatal development offers the opportunity to prevent

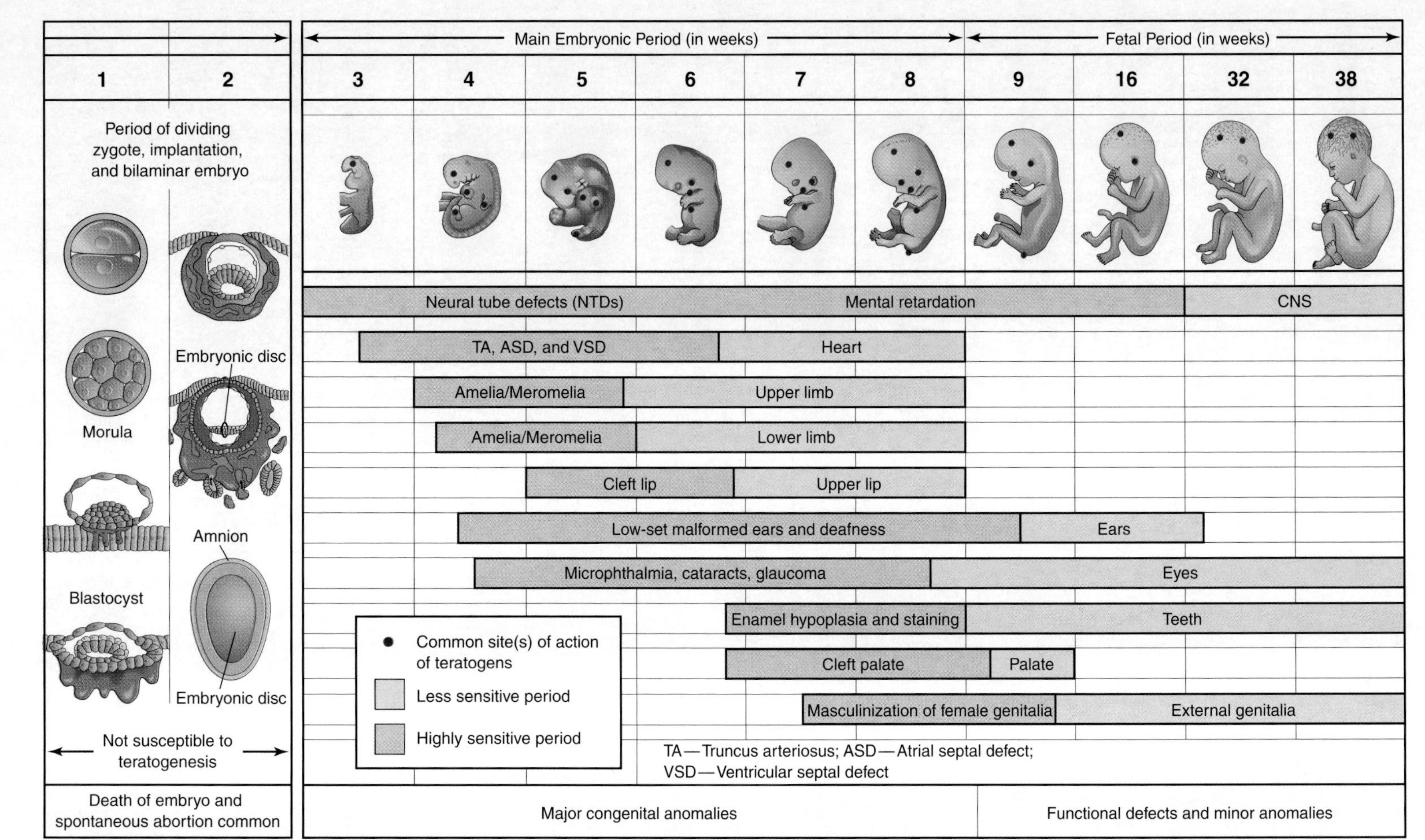

Figure 9 – 11. Schematic illustration of critical periods in human prenatal development. During the first 2 weeks of development, the embryo is usually not susceptible to teratogens; at this point, a teratogen either damages all or most of the cells, resulting in death of the embryo, or damages only a few cells, allowing the conceptus to recover and the embryo to develop without birth defects. The mauve areas denote highly sensitive periods when major defects may be produced (e.g., amelia, absence of limbs). The green sections indicate stages that are less sensitive to teratogens when minor defects may be induced (e.g., hypoplastic thumbs).

Table 9 – 5. Incidence of Major Malformations in Human Organs at Birth

Organ	Incidence of Malformations
Brain	10:1000
Heart	8:1000
Kidneys	4:1000
Limbs	2:1000
All other	6:1000
Total	30:1000

Data from Connor JM, Ferguson-Smith MA: *Essential Medical Genetics*, 2nd ed. Oxford, Blackwell Scientific Publications, 1987.

some congenital anomalies. For example, if women are made aware of the harmful effects of drugs (e.g., alcohol), environmental chemicals (e.g., polychlorinated biphenyls), and viruses, most pregnant women will avoid exposure to these teratogenic agents.

Proof of Teratogenicity

To prove that an agent is a teratogen, one must show either that the frequency of anomalies is increased above the spontaneous rate in pregnancies in which the mother is exposed to the agent (*the prospective approach*) or that malformed infants have a history of maternal exposure to the agent more often than normal children (*the retrospective approach*). Both types of data are difficult to obtain in an unbiased form. *Case reports are not convincing* unless both the causative agent and the type of anomaly are so uncommon that their association in several cases can be judged not to be coincidental.

Drugs As Teratogens

Drugs vary considerably in their teratogenicity. Some teratogens, such as thalidomide, cause severe disruption of development if administered during the organogenetic period of certain parts of the embryo. Other teratogens cause mental and growth retardation and other anomalies if used excessively throughout development.

The use of prescription and nonprescription drugs during pregnancy is surprisingly high. Most pregnant women consume at least one drug during pregnancy. Drug consumption also tends to be higher during the critical period of development among heavy smokers and drinkers. Despite this, *less than 2% of congenital anomalies are caused by drugs and chemicals.* Only a few drugs have been positively implicated as human teratogenic agents (see Table 9 – 4). Although only 7 to 10% of anomalies are caused by recognizable teratogens (see Fig. 9 – 1), new agents continue to be identified. It is best for women to avoid using all medication during the first trimester unless a strong medical reason exists for its use, and then only if it is recognized as reasonably safe for the human embryo.

Cigarette Smoking. *Maternal smoking during pregnancy is a well-established cause of IUGR.* Despite warnings that cigarette smoking is harmful to the fetus, more than 25% of women continue to smoke during their pregnancies. In heavy cigarette smokers (20 or more cigarettes per day), premature delivery is twice as frequent as in mothers who do not smoke, and their infants weigh less than normal. *Low birth weight (less than 2000 gm) is the chief predictor of infant death.*

Nicotine constricts uterine blood vessels, thereby causing a decrease in uterine blood flow and lowering the supply of oxygen and nutrients available to the embryo/fetus from the maternal blood in the intervillous space of the placenta. High levels of *carboxyhemoglobin*, resulting from cigarette smoking, appear in the maternal and fetal blood and may alter the capacity of the blood to transport oxygen. As a result, chronic *fetal hypoxia* (a decrease in oxygen level to below normal) may occur, affecting fetal growth and development.

Caffeine. Caffeine is the most popular drug in North America because it is present in several widely consumed beverages (e.g., coffee, tea, and cola drinks), chocolate products, and some drugs. *Caffeine is not known to be a human teratogen*; however, there is no assurance that heavy maternal consumption of it is safe for the embryo.

Alcohol. Alcoholism is a drug abuse problem that affects 1 to 2% of women of childbearing age. Both moderate and high levels of alcohol intake during early pregnancy may result in alterations in growth and morphogenesis of the fetus; the greater the intake, the more severe the signs. Infants born to chronic alcoholic mothers exhibit a specific pattern of defects, including prenatal and postnatal growth deficiency, mental retardation, and other anomalies (Fig. 9 – 12; see Table 9 – 4). This pattern of anomalies, called **fetal alcohol syndrome (FAS)**, is detected in 1 to 2 infants per 1000 live births. *Maternal alcohol abuse is now thought to be the most common cause of mental retardation.* Even moderate maternal alcohol consumption (e.g., 1 to 2 ounces per day) may produce **fetal alcohol effects (FAE)** — children with behavioral and learning difficulties, for example — especially if the drinking is associated with malnutrition. **Binge drinking** (heavy consumption of alcohol for 1 to 3 days during pregnancy) is very likely to produce FAE. The susceptible period of brain development spans the major part of gestation;

therefore, the safest advice is total abstinence from alcohol during pregnancy.

Androgens and Progestogens. Androgens and progestogens may affect the female fetus, producing masculinization of the external genitalia (Fig. 9 – 13; and see Table 9 – 4). The preparations that should be avoided are the *progestins ethisterone* and *norethisterone*. From a practical standpoint, the teratogenic risk of these hormones is low. However, progestin exposure during the critical period of development is also associated with an increased prevalence of cardiovascular abnormalities, and exposure of male fetuses during this period may double the incidence of *hypospadias* in the offspring (see Chapter 14). Obviously, the administration of *testosterone* also produces masculinizing effects in female fetuses.

Many women use contraceptive hormones (birth control pills). *Oral contraceptives* containing progestogens and estrogens, when taken during the early stages of an unrecognized pregnancy, are thought to be teratogenic agents. Many infants of mothers who took *progestogen-estrogen birth control pills* during the critical period of development have been found to exhibit the VACTERL syndrome. The acronym VACTERL stands for **v**ertebral, **a**nal, **c**ardiac, **t**racheal, **e**sophageal, **r**enal, and **l**imb anomalies. As a precaution, use of oral contraceptives should be discontinued as soon as pregnancy is detected because of the risk of possible teratogenic effects.

Diethylstilbestrol (DES, stilbestrol) is recognized as a human teratogen. Vaginal adenosis, cervical erosions, and transverse vaginal ridges have been detected in women exposed to DES in utero. Moreover, a number of young women with a common history of exposure to this synthetic estrogen in utero have been found to develop *adenocarcinoma of the vagina*. However, the probability of cancers developing at this early age in females exposed to DES in utero now appears to be low. In males exposed to DES in utero before the 11th week of gestation, an increased incidence of genital tract anomalies has been reported.

Antibiotics. Tetracyclines cross the placental membrane and are deposited in the embryo's bones and teeth at sites of active calcification. As little as 1 gm per day of **tetracycline** during the third trimester of pregnancy can produce yellow staining of the primary and/or deciduous teeth. Tetracycline therapy during the 4th to

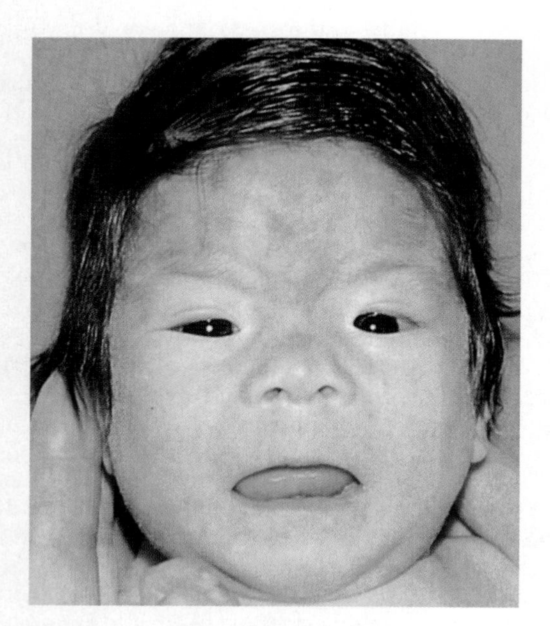

Figure 9 - 12. Infant with fetal alcohol syndrome. Note the thin upper lip, short palpebral fissures, flat nasal bridge, short nose, and elongated and poorly formed philtrum (vertical groove in median part of upper lip). Maternal alcohol abuse is thought to be the most common environmental cause of mental retardation. (Courtesy of Dr. AE Chudley, Professor of Pediatrics and Child Health, Children's Hospital and University of Manitoba, Winnipeg, Manitoba, Canada.)

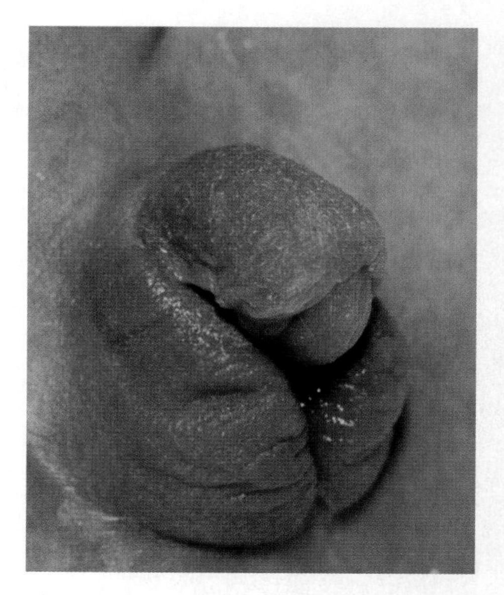

Figure 9 - 13. Masculinized external genitalia of a 46,XX female infant. Observe the enlarged clitoris and fused labia majora. Just below the clitoris, there is a single orifice of a urogenital sinus. The virilization was caused by excessive androgens produced by the suprarenal (adrenal) glands during the fetal period (congenital adrenal hyperplasia). (Courtesy of Dr. Heather Dean, Department of Pediatrics and Child Health and University of Manitoba, Winnipeg, Canada.)

10th months of pregnancy may also cause tooth defects (e.g., enamel hypoplasia), yellow to brown discoloration of the teeth, and diminished growth of long bones. Moreover, more than 30 cases of hearing deficit and eighth cranial nerve damage have been reported in infants exposed to **streptomycin derivatives** in utero. By contrast, *penicillin* has been used extensively during pregnancy and appears to be harmless to the human embryo and fetus.

Anticoagulants. All anticoagulants except heparin cross the placental membrane and may cause hemorrhage in the embryo or fetus. **Warfarin**, an anticoagulant, is definitely a teratogen (see Table 9 - 4). The period of greatest sensitivity is between 6 and 12 weeks after fertilization, or 8 to 14 weeks after the last normal menstrual period (LNMP). Second- and third-trimester exposure may result in mental retardation, optic atrophy, and microcephaly. **Heparin** does not cross the placental membrane and so is the drug of choice for pregnant women requiring anticoagulant therapy.

Anticonvulsants. Approximately 1 of 200 pregnant women is epileptic and requires treatment with an anticonvulsant. Of the anticonvulsant drugs available, **phenytoin (Dilantin, Novophenytoin) has been definitively identified as a teratogen** (Fig. 9 - 14). The *fetal hydantoin syndrome* occurs in 5 to 10% of children born to mothers treated with phenytoins or hydantoin anticonvulsants (see Table 9 - 4).

Valproic acid has been the drug of choice for the management of different types of epilepsy; however, its use by pregnant women has led to a *pattern of anomalies* consisting of craniofacial, heart, and limb defects. There is also an increased risk of neural tube defects. *Phenobarbital is considered to be a safe antiepileptic drug* for use during pregnancy.

Antinauseants. There has been extensive debate in the lay press and in the courts as to whether *Bendectin* (Debendox, Lenotan, Diclectin) is a teratogenic drug in humans. Teratologists consider Bendectin to be nonteratogenic in humans because large-scale epidemiologic studies of infants have failed to show an increased risk of birth defects after its administration to pregnant women.

Antineoplastic Agents. Tumor-inhibiting chemicals are highly teratogenic. This is not surprising because these agents inhibit mitosis in rapidly dividing cells. It is recommended that they be avoided, especially during the first trimester of pregnancy.

Aminopterin is a known potent teratogen that produces major congenital anomalies. Aminopterin, an antimetabolite, is a *folic acid antagonist*. Multiple skeletal and other congenital anomalies have also been reported in an infant born to a mother who attempted to terminate her pregnancy by taking *methotrexate*, a derivative of aminopterin that is also a folic acid antagonist (see Table 9 - 4).

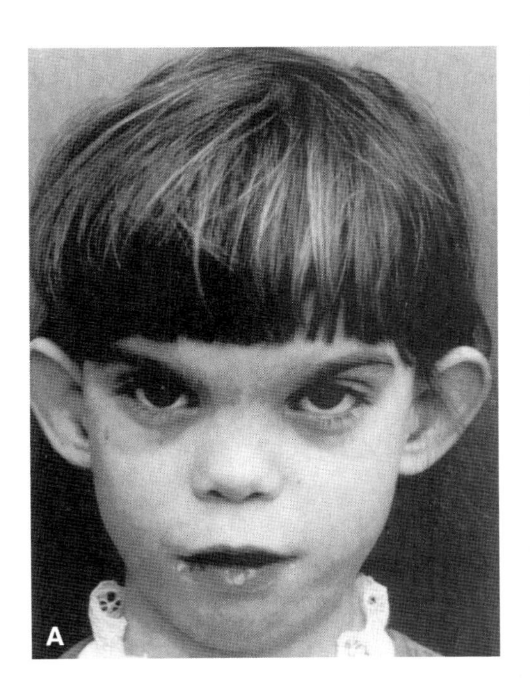

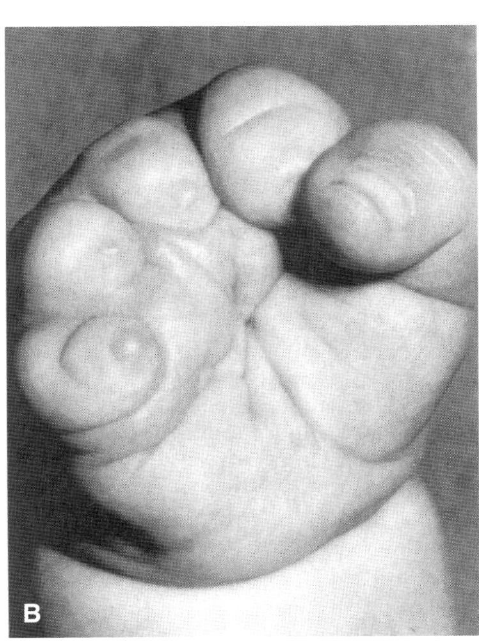

Figure 9 - 14. Fetal hydantoin syndrome. *A*, This young girl has a learning disability. Note the unusual ears, wide spacing of the eyes, epicanthal folds, short nose, and long philtrum. Her mother has epilepsy and took Dilantin throughout her pregnancy. (Courtesy of Dr. AE Chudley, Professor of Pediatrics and Child Health, Children's Hospital and University of Manitoba, Winnipeg, Manitoba, Canada.) *B*, Right hand of an infant with severe digital hypoplasia (short fingers), born to a mother who took Dilantin throughout her pregnancy. (From Chodirker, BN, Chudley AE, Persaud TVN: Possible prenatal hydantoin effect in child born to a nonepileptic mother. *Am J Med Genet* 27:373, copyright (c) 1987. Reprinted by permission of Wiley-Liss, a division of John Wiley and Sons, Inc.)

Corticosteroids. The teratogenic risk of corticosteroids is minimal to nonexistent.

Angiotensin-Converting Enzyme (ACE) Inhibitors. Exposure of the fetus to ACE inhibitors, used as antihypertensive agents, causes oligohydramnios, fetal death, long-lasting hypoplasia of the bones of the calvaria, IUGR, and renal dysfunction. During early pregnancy, the risk to the embryo is apparently diminished, so there is no indication in such a case to terminate a wanted pregnancy.

Insulin and Hypoglycemic Drugs. Insulin is not teratogenic in human embryos except possibly in maternal insulin coma therapy. Hypoglycemic drugs (e.g., tolbutamide) have been implicated in some cases, but evidence for their teratogenicity is very weak. No convincing evidence exists that oral hypoglycemic agents are teratogenic in human embryos. The incidence of congenital anomalies is increased two to three times in the offspring of diabetic mothers. Women with insulin-dependent diabetes mellitus may significantly decrease their risk of having infants with birth defects by achieving good control of their disease *before conception*.

Retinoic Acid (Vitamin A). Isotretinoin (13-*cis*-retinoic acid), used for the oral treatment of severe cystic acne, is teratogenic in humans, even at very low doses (see Table 9 – 4). The critical period for exposure appears to be from the third week to the fifth week (5 to 7 weeks after the LNMP). The risk of spontaneous abortion and birth defects after exposure to retinoic acid is high. Postnatal follow-up studies of children exposed in utero to isotretinoin have revealed significant neuropsychological impairment. Vitamin A is a valuable and necessary nutrient during pregnancy, but long-term exposure to large doses of vitamin A is unwise because of insufficient evidence to rule out a teratogenic risk.

Salicylates. Some evidence indicates that large doses of *acetylsalicylic acid* (ASA) or *aspirin*, the most commonly ingested drug during pregnancy, is potentially harmful to the embryo or fetus. However, epidemiological studies indicate that aspirin is not a teratogenic agent but large doses should be avoided.

Thyroid Drugs. Iodides readily cross the placental membrane and interfere with thyroxin production. They may also cause thyroid enlargement and **cretinism** (arrested physical and mental development and dystrophy of bones and soft parts). *Maternal iodine deficiency* may cause congenital cretinism. The administration of *antithyroid drugs* for the treatment of maternal thyroid disorders may cause congenital goiter if the dose administered exceeds that required to control the disease.

Tranquilizers. *Thalidomide is a potent teratogen.* Nearly 12,000 infants have been born with defects caused by this drug. The characteristic feature of the **thalidomide syndrome** is *meromelia*—phocomelia, or "seal limbs" (Fig. 9 – 15; and see Table 9 – 4). It has been well established clinically that the period when thalidomide causes congenital anomalies is from 24 to 36 days after fertilization (38 to 50 days after the LNMP). This sensitive period coincides with the critical periods for the development of the affected parts and organs (see Fig. 9 – 11). *Thalidomide is absolutely contraindicated in women of childbearing age.* With appropriate precautions, thalidomide is now used for the treatment of leprosy and other disorders because of its immunosuppressive properties.

Lithium is the drug of choice for long-term maintenance therapy in patients with manic-depressive psychosis; however, it has been known to cause congenital anomalies, mainly of the heart and great

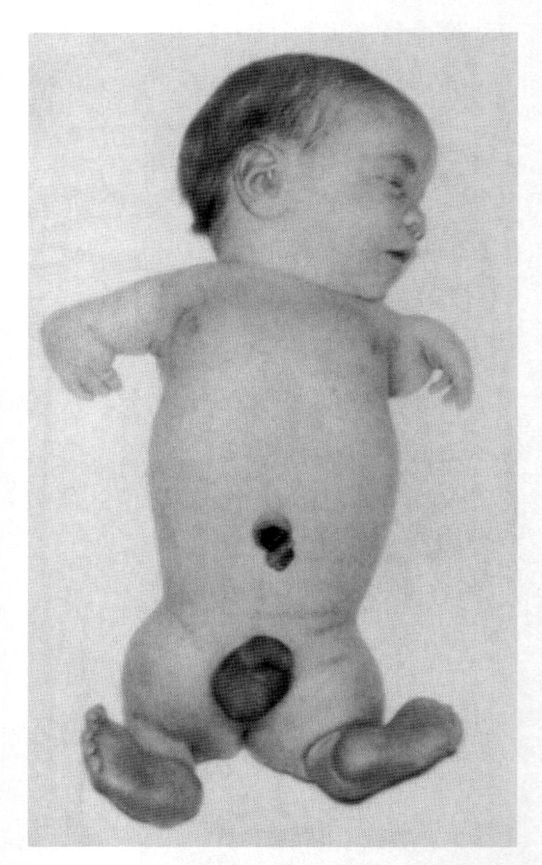

Figure 9 – 15. Newborn male infant with malformed limbs (meromelia—limb reduction) caused by maternal ingestion of thalidomide during the critical period of limb development. (From Moore KL: The vulnerable embryo. Causes of malformation in man. *Manitoba Med Rev* 43:306, 1963.)

vessels, in infants born to mothers given the drug early in pregnancy. Although **lithium carbonate is a human teratogen**, the Food and Drug Administration (FDA) has stated that the agent may be used during pregnancy if "in the opinion of the physician the potential benefits outweigh the possible hazards." **Benzodiazepine derivatives** are psychoactive drugs frequently used by pregnant women. These include *diazepam* and *oxazepam*, which readily cross the placental membrane. The use of these drugs during the first trimester of pregnancy is associated with transient withdrawal symptoms and **craniofacial anomalies** in the neonate.

Illicit Drugs. Limb defects and an increased incidence of nervous system defects have been reported in infants born to mothers who use *lysergic acid diethylamide* (*LSD*) during pregnancy. There is little evidence that **marijuana** is a human teratogen, although there is some indication that marijuana use during the first 2 months of pregnancy affects fetal length and birth weight. In addition, alterations in sleep and electroencephalogram (EEG) patterns have been noted in newborn infants exposed prenatally to marijuana.

Cocaine is one of the most commonly abused illicit drugs in North America, and its increasing use by women of childbearing age is of major concern. Many reports deal with the prenatal effects of cocaine. These include spontaneous abortion, prematurity, and diverse anomalies in the offspring (see Table 9-4).

Methadone, used for the treatment of heroin addiction, is considered to be a "behavioral teratogen," as is heroin. Infants born to narcotic-dependent women receiving maintenance methadone therapy have been found to have central nervous system (CNS) dysfunction and smaller birth weights and head circumferences than nonexposed infants. There is also concern about the long-term postnatal developmental effects of methadone.

Environmental Chemicals as Teratogens

In recent years, there has been increasing concern about the possible teratogenicity of environmental, industrial, and agricultural chemicals, pollutants, and food additives.

Organic Mercury. Infants of mothers whose main diet during pregnancy consists of fish containing abnormally high levels of organic mercury acquire **fetal Minamata disease** and exhibit neurologic and behavioral disturbances resembling cerebral palsy. **Methylmercury is a teratogen** that causes cerebral atrophy, spasticity, seizures, and mental retardation.

Lead. Lead crosses the placental membrane and accumulates in fetal tissues. Prenatal exposure to lead is associated with an increased incidence of abortions, fetal anomalies, IUGR, and functional deficits. Children born to mothers exposed to subclinical levels of lead exhibit neurobehavioral and psychomotor disturbances.

Polychlorinated Biphenyls (PCBs). PCBs are teratogenic chemicals that produce IUGR and skin discoloration in infants exposed to these agents in utero. The main dietary source of PCBs in North America is probably sport fish caught in contaminated waters. In Japan and Taiwan, the teratogenic chemical has been detected in contaminated cooking oil.

Infectious Agents as Teratogens

Throughout prenatal life, the embryo and fetus are endangered by a variety of microorganisms. In most cases, the microbial assault is resisted; in some cases, an abortion or stillbirth occurs, and in other cases, the infants are born with IUGR, congenital anomalies, or neonatal diseases (see Table 9-4).

Rubella (German or Three-Day Measles). The **rubella virus** crosses the placental membrane and infects the embryo/fetus. In cases of primary maternal infection during the first trimester of pregnancy, the overall risk of embryonic/fetal infection is about 20%. The clinical features of **congenital rubella syndrome** are *cataract*, *cardiac defects*, and *deafness*; however, other abnormalities are occasionally observed (Fig. 9-16; and see Table 9-4). The earlier in pregnancy the maternal rubella infection occurs, the greater the danger that the embryo will be malformed.

Cytomegalovirus. Infection with the cytomegalovirus (CMV) is the most common viral infection of the human fetus. Because the disease seems to be fatal when it affects the embryo, most pregnancies likely end in spontaneous abortion when the infection occurs during the first trimester. Newborn infants infected during the early fetal period usually show no clinical signs and are identified through screening programs. Later in pregnancy, *CMV infection may result in IUGR and severe fetal anomalies* (see Table 9-4). Of particular concern are cases of *asymptomatic CMV infection*, which are often associated with audiological, neurological, and neurobehavioral disturbances in infancy.

Herpes Simplex Virus. Maternal infection with herpes simplex virus (HSV) in early pregnancy increases the abortion rate threefold, and infection after the 20th week is associated with an increased rate of prematurity and congenital anomalies (see Table 9-4). Infection of the fetus with HSV usually occurs very late in pregnancy, probably most often during delivery.

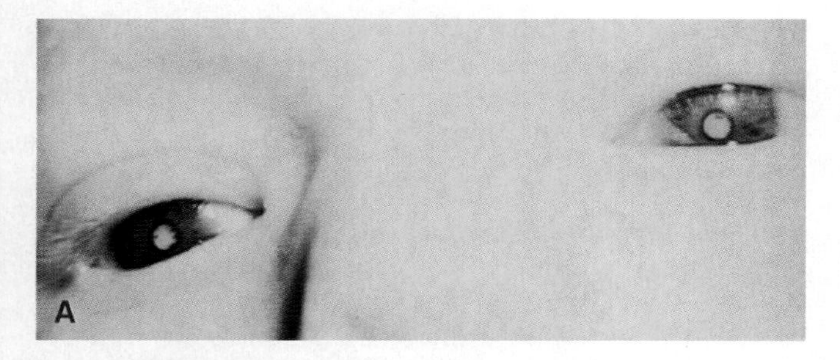

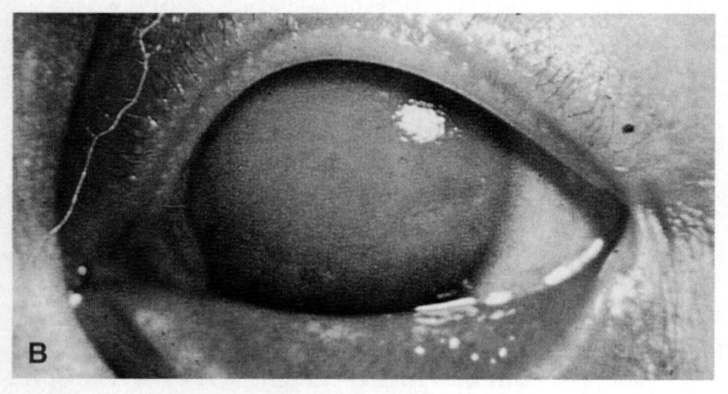

Figure 9 - 16. *A*, Typical bilateral congenital cataracts caused by the rubella virus. Cardiac defects and deafness are other congenital defects common to this syndrome. *B*, Severe congenital glaucoma caused by the rubella virus. Observe the dense corneal haze, enlarged corneal diameter, and deep anterior chamber. (*A*, Courtesy of Dr. Richard Bargy, Department of Ophthalmology, Cornell-New York Hospital, New York, New York. *B*, Courtesy of Dr. Daniel I. Weiss, Department of Ophthalmology, New York University College of Medicine, New York, New York. From Cooper LA, et al: Neonatal thrombocytopenic purpura and other manifestations of rubella contracted in utero. *Am J Dis Child* 110:416, 1965. Copyright 1965, American Medical Association.)

Varicella (Chickenpox). Varicella and herpes zoster (shingles) are caused by the same virus, *varicella-zoster virus*. There is convincing evidence that maternal *varicella infection during the first 4 months of pregnancy causes congenital anomalies* (see Table 9 - 4). There is a 20% incidence of these or other anomalies when the infection occurs during the critical period of development (see Fig. 9 - 11). After 20 weeks of gestation, there is no proven teratogenic risk.

Human Immunodeficiency Virus. Human immunodeficiency virus (HIV) is the retrovirus that causes acquired immunodeficiency syndrome (AIDS). Infection of pregnant women with HIV is now a prevalent and serious health problem. Information is conflicting on the fetal effects of in utero infection with HIV. Preventing the transmission of the virus to women and their infants is important because of potential embryopathic effects.

Toxoplasma gondii. *Maternal infection* with the intracellular parasite *Toxoplasma gondii* is usually acquired by:

- eating raw or poorly cooked meat (usually pork or lamb containing *Toxoplasma* cysts)
- close contact with infected domestic animals (usually *cats*) or soil

The *Toxoplasma gondii organism crosses the placental membrane and infects the fetus*, causing destructive changes in the brain and eyes that result in **mental deficiency** and other anomalies (see Table 9 - 4). Mothers of congenitally defective infants are often unaware of having had **toxoplasmosis**, the disease caused by the parasitic organism. Because animals (cats, dogs, rabbits, and other domestic and wild animals) may be infected with this parasite, pregnant women should avoid them. In addition, eggs from domestic fowl should be well cooked, and unpasteurized milk should be avoided.

Congenital Syphilis. About 3 in 10,000 liveborn infants in the United States are infected with syphilis. *Treponema pallidum*, the small, spiral microorganism that causes syphilis, rapidly crosses the placental membrane as early as 9 to 10 weeks of gestation. The fetus can become infected at any stage of the disease or at any stage of pregnancy. *Primary maternal infections* (acquired during pregnancy and left untreated) nearly always cause serious fetal infection and congenital anomalies (see Table 9 - 4). However, adequate treatment of the mother kills the organism. *Secondary maternal infections* (acquired before pregnancy) seldom result in fetal disease and anomalies. If the mother remains untreated, stillbirths occur in about 25% of cases.

Radiation as a Teratogen

Exposure to **high levels of ionizing radiation** may injure embryonic cells, resulting in cell death, chromosomal injury, and retardation of mental development and physical growth. The severity of the embryonic damage is related to the absorbed dose, the dose rate, and the stage of embryonic or fetal development when the exposure occurs. Accidental exposure of pregnant women to radiation is a common cause for anxiety.

No conclusive proof exists that human congenital anomalies have been caused by diagnostic levels of radiation. Scattered radiation from a radiographic (x-ray) examination of a part of the body that is not near the uterus (e.g., the thorax, sinuses, teeth) produces a dose of only a few millirads, which is not teratogenic to the embryo. If the embryonic radiation exposure is 5 rad or less, the radiation risks to the embryo are minuscule. The recommended limit of maternal exposure of the whole body to radiation from all sources is 500 mrad for the entire gestational period.

Electromagnetic Fields. No evidence exists that the risk of IUGR or other developmental defects is increased by maternal exposure to low-frequency electromagnetic fields (e.g., electric blankets and video display terminals).

Ultrasonic Waves. Ultrasonography is widely used during pregnancy for fetal diagnosis and prenatal care. There are no confirmed biologic effects in patients and their fetuses from the use of diagnostic ultrasound evaluation. Moreover, the benefits to patients exposed to prudent use of this modality outweigh the risks, if any.

Maternal Factors as Teratogens

Maternal diseases can sometimes lead to an increased risk of abnormalities in offspring. Poorly controlled *diabetes mellitus* in the mother with persisting hyperglycemia and ketosis, particularly during embryogenesis, is associated with a two- to three-fold higher incidence of birth defects. No specific diabetic embryopathic syndrome exists, but the infant of the diabetic mother is usually large (*macrosomia*). The common anomalies include *holoprosencephaly* (failure of the forebrain to divide into hemispheres), meroencephaly (partial absence of the brain), sacral agenesis, vertebral anomalies, congenital heart defects, and limb anomalies. If left untreated, women who are homozygous for phenylalanine hydroxylase deficiency — **phenylketonuria** (PKU) — and those with hyperphenylalaninemia are at increased risk of having offspring with microcephaly, cardiac defects, mental retardation, and IUGR. The congenital anomalies can be prevented if the mother with PKU is placed on a phenylalanine-restricted diet prior to and during the pregnancy.

Mechanical Factors as Teratogens

The amniotic fluid absorbs mechanical pressures, thereby protecting the embryo from most external trauma. It is generally accepted that congenital abnormalities caused by external injury to the mother are extremely rare, but possible. *Congenital dislocation of the hip and clubfoot* may be caused by mechanical forces, particularly in a malformed uterus. Such deformations may be caused by any factor that restricts the mobility of the fetus, thereby causing prolonged compression in an abnormal posture. A significantly reduced quantity of amniotic fluid (*oligohydramnios*) may result in mechanically induced deformation of the limbs (see Chapter 8), such as hyperextension of the knee. Intrauterine amputations or other anomalies caused by local constriction during fetal growth may result from *amniotic bands*, rings formed as a result of rupture of the amnion during early pregnancy (see Fig. 8 – 18).

Anomalies Caused by Multifactorial Inheritance

Many common congenital anomalies (e.g., cleft lip with or without cleft palate) have familial distributions consistent with multifactorial inheritance (MFI) (see Fig. 9 – 1). MFI may be represented by a model in which one's "liability" for a disorder is a continuous variable determined by a combination of genetic and environmental factors, with a developmental threshold dividing individuals with the anomaly from those without it (Fig. 9 – 17). *Multifactorial traits are often single major anomalies*, such as cleft lip, isolated cleft palate, and neural tube defects. Some of these anomalies may also occur as part of the phenotype in syndromes determined by single-gene inheritance, chromosomal abnormality, or an environmental teratogen. The *recurrence risks* used for genetic counseling of families having congenital anomalies that have been determined by MFI are *empirical risks* based on the frequency of the anomaly in the general population and in different categories of relatives. In individual families, such estimates may be inaccurate because they are usually averages for the population rather than precise probabilities for the individual family.

Summary of Human Birth Defects

A congenital anomaly is a structural abnormality of any type that is present at birth. It may be macroscopic or microscopic, on the surface or within the body. Four clinically significant types of anomalies occur: malformation, disruption, deformation, and dysplasia. Congenital anomalies may be induced by genetic factors or by environmental factors. Most common congenital anomalies, however, show the familial patterns expected of MFI with a threshold and are determined by a combination of genetic and environmental factors. *About 3% of all liveborn infants have an obvious major anomaly*. Additional anomalies are detected after birth; thus, the incidence of congenital anomalies is about 6% in 2-year-olds and 8% in 5-year-olds. Other anomalies (about 2%) are detected later in life (e.g., during surgery or autopsy). *Single minor anomalies are present in about 14% of newborn infants*. Of the 3% of infants born with a major congenital anomaly, 0.7% have multiple major anomalies.

Major anomalies are more common in early embryos (up to 15%) than they are in newborn infants (up to 3%). Most severely malformed embryos are usually spontaneously aborted during the first 6 to 8 weeks of development. Some congenital anomalies are caused by *genetic factors* (chromosomal abnormalities and mutant genes). A few congenital abnormalities are caused by

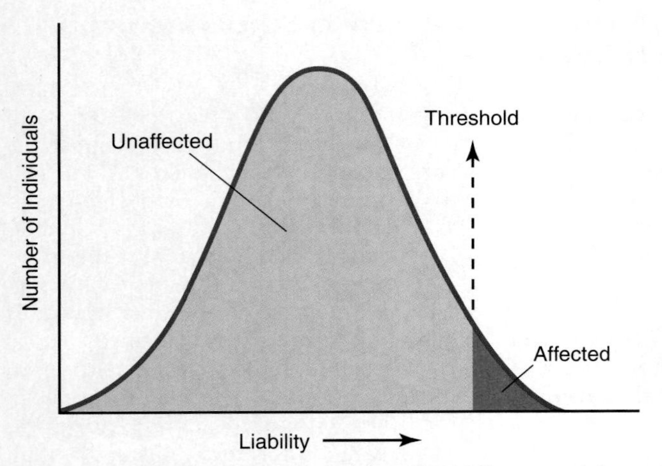

Figure 9 - 17. Multifactorial threshold model. Liability to a trait is distributed normally, with a threshold dividing the population into unaffected and affected classes. (From Thompson MW, McInnes RR, Willard FH: *Thompson and Thompson Genetics in Medicine*, 5th ed. Philadelphia, WB Saunders, 1991.)

environmental factors (infectious agents, environmental chemicals, and drugs); however, most common anomalies result from a complex *interaction between genetic and environmental factors*. The cause of most congenital anomalies is unknown. During the first 2 weeks of development, teratogens usually kill the embryo. During the *organogenetic period*, teratogenic agents disrupt development and may cause *major congenital anomalies*. During the fetal period, teratogens may produce morphologic and functional abnormalities, particularly of the brain and eyes.

Clinically Oriented Questions

1. If a pregnant woman takes aspirin in normal doses, will it cause congenital anomalies?
2. If a woman is a drug addict, will her child show signs of drug addiction?
3. Are all drugs tested for teratogenicity before they are marketed? If the answer is "yes," why are these teratogens still sold?
4. Is cigarette smoking during pregnancy harmful to the embryo or fetus? If the answer is "yes," would refraining from inhaling cigarette smoke be safer?
5. Are any drugs safe to take during pregnancy? If so, what are they?

The answers to these questions are at the back of the book.

10

Body Cavities, Mesenteries, and Diaphragm

Early in the fourth week, the **intraembryonic coelom** — the primordium of the embryonic body cavity — appears as a horseshoe-shaped cavity in the cardiogenic and lateral mesoderm (Fig. 10 - 1*A*). The curve or bend in this cavity at the cranial end of the embryo represents the future *pericardial cavity*, and its limbs (lateral extensions) indicate the future *pleural and peritoneal cavities*. The distal part of each limb of the intraembryonic coelom is continuous with the **extraembryonic coelom** at the lateral edges of the embryonic disc (see Fig. 10 - 1*B*). This communication is important because most of the midgut normally herniates through this communication into the umbilical cord, where it develops into most of the small intestine and part of the large intestine. The coelom provides room for the organs to develop and move. During embryonic folding in the horizontal plane, the limbs of the intraembryonic coelom are brought together on the ventral aspect of the embryo (Fig. 10 - 2*A* to *F*).

The ventral mesentery degenerates, resulting in a large embryonic peritoneal cavity extending from the heart to the pelvic region (see Figs. 10 - 2*F* and 10 - 3*A* to *E*).

The Embryonic Body Cavity

The intraembryonic coelom, or primordial embryonic body cavity, gives rise to three well-defined coelomic or body cavities during the fourth week (see Figs. 10 - 2 and 10 - 4):

- a *pericardial cavity*
- two *pericardioperitoneal canals* connecting the pericardial and peritoneal cavities
- a large *peritoneal cavity*

These body cavities have a parietal wall lined by mesothelium (a major part of the future parietal layer of peritoneum) derived from somatic mesoderm, as well as a visceral wall covered by mesothelium (the future visceral layer of peritoneum) derived from splanchnic mesoderm (see Fig. 10 - 3*E*). The peritoneal cavity (a major part of the intraembryonic coelom) is connected to the extraembryonic coelom at the umbilicus (see Fig. 10 - 4*C* and *D*). The **peritoneal cavity** loses its connection with the extraembryonic coelom during the 10th week as the intestines return to the abdomen from the umbilical cord (see Chapter 13). During formation of the head fold, the heart and **pericardial cavity** move ventrocaudally, anterior to the foregut (see Fig. 10 - 2*B*). As a result, the pericardial cavity opens into the **pericardioperitoneal canals**, which pass dorsal to the foregut (see Fig. 10 - 4*B* and *D*). After embryonic folding, the caudal parts of the foregut, midgut, and hindgut are suspended in the peritoneal cavity from the posterior abdominal wall by the **dorsal mesentery** (see Figs. 10 - 2*F* and 10 - 3*C* to *E*).

Mesenteries

A mesentery is a double layer of peritoneum that begins as an extension of the visceral peritoneum covering an

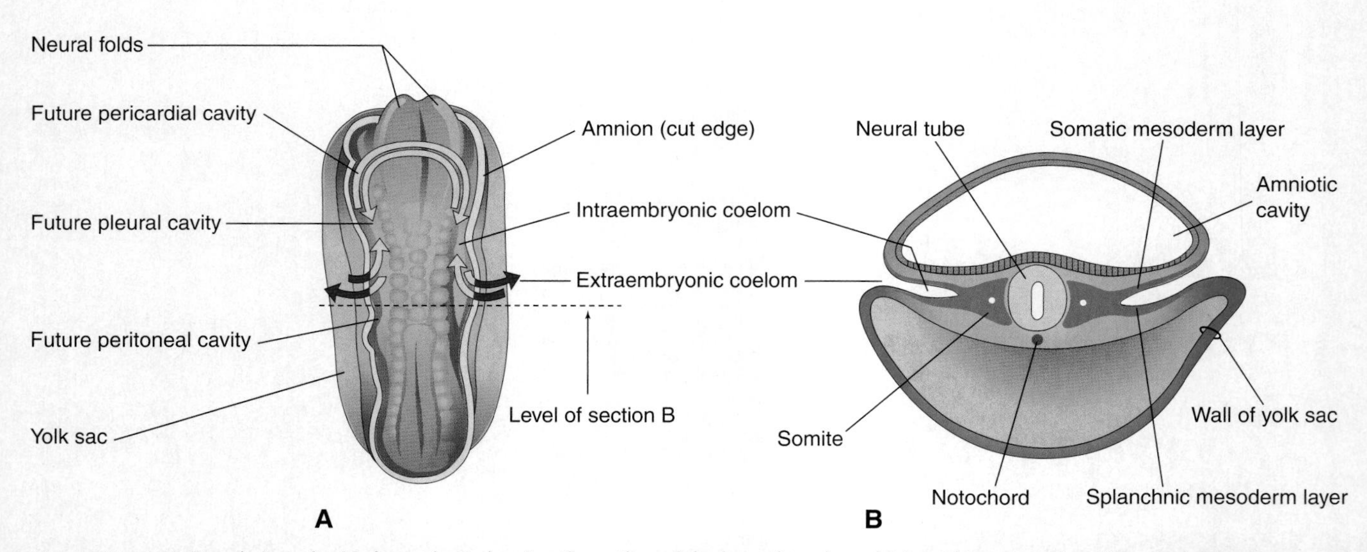

Figure 10 - 1. *A*, Dorsal view of a 22-day embryo showing the outline of the horseshoe-shaped intraembryonic coelom. The amnion has been removed and the coelom is shown as if the embryo were translucent. The continuity of the intraembryonic coelom, as well as the communication of its right and left limbs with the extraembryonic coelom, is indicated by arrows. *B*, Transverse section through the embryo at the level shown in *A*.

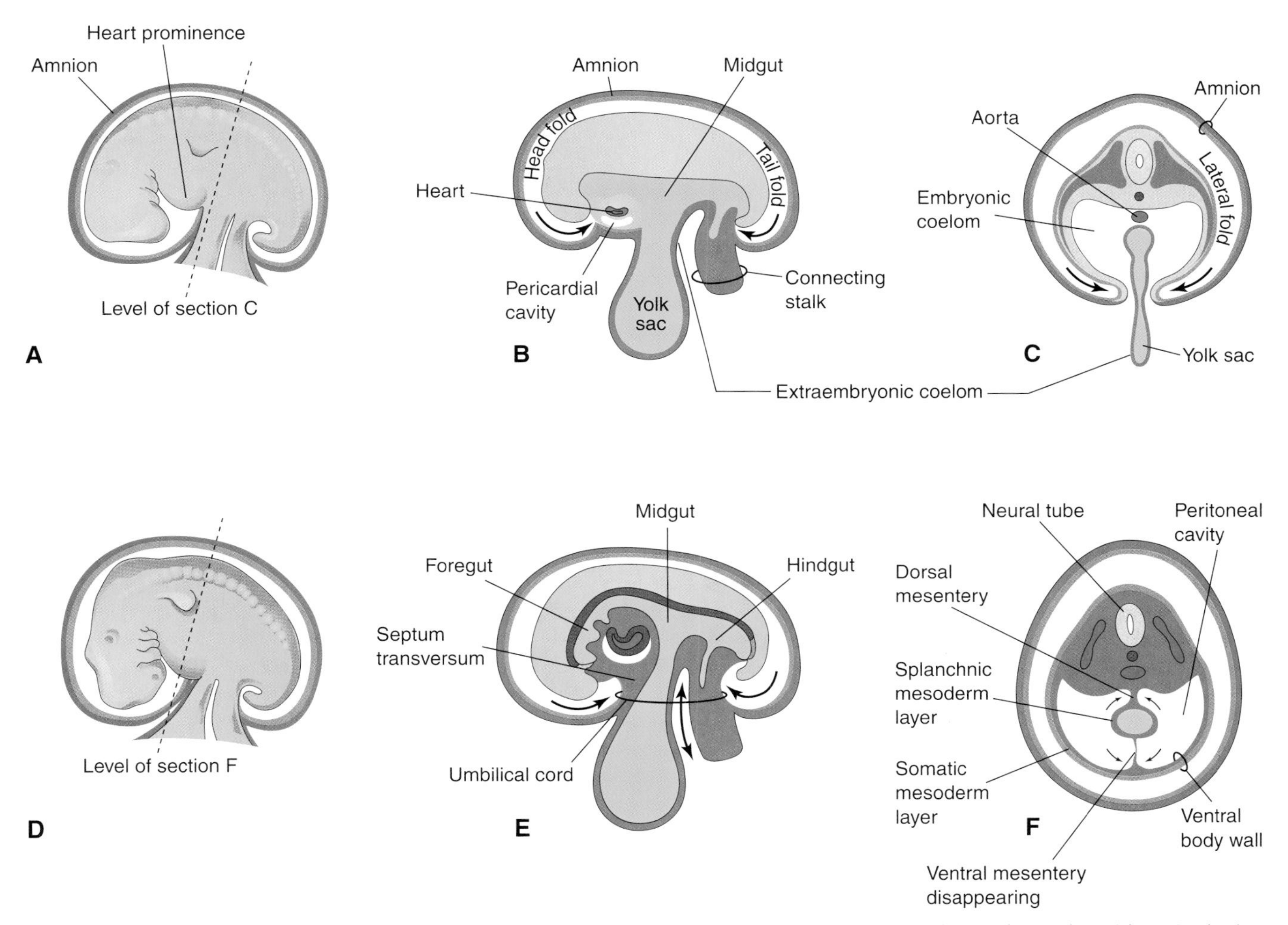

Figure 10 - 2. Embryonic folding and its effects on the intraembryonic coelom and other structures. *A,* Lateral view of an embryo (about 26 days). *B,* Schematic sagittal section of this embryo showing the head and tail folds. *C,* Transverse section at the level shown in *A,* indicating how fusion of the lateral folds gives the embryo a cylindrical form. *D,* Lateral view of an embryo (about 28 days). *E,* Schematic sagittal section of this embryo showing the reduced communication between the intraembryonic and extraembryonic coeloms (*double-headed arrow*). *F,* Transverse section, as indicated in *D,* illustrating formation of the ventral body wall and disappearance of the ventral mesentery. The arrows indicate the junction of the somatic and splanchnic layers of mesoderm. The somatic mesoderm will become the parietal peritoneum lining the abdominal wall, and the splanchnic mesoderm will become the visceral peritoneum covering the organs (e.g., the stomach).

organ; *it connects the organ to the body wall and conveys its vessels and nerves.* Transiently, the dorsal and ventral mesenteries divide the peritoneal cavity into right and left halves (see Fig. 10 - 3*C*). However, the ventral mesentery soon disappears (see Fig. 10 - 3*E*), except where it is attached to the caudal part of the foregut (primordium of the stomach and proximal part of the duodenum). The peritoneal cavity then becomes a continuous space (see Figs. 10 - 3 and 10 - 4). The arteries supplying the primordial gut — the *celiac trunk* (foregut), *superior mesenteric artery* (midgut), and *inferior mesenteric artery* (hindgut) — pass between the layers of the dorsal mesentery (see Fig. 10 - 3*C*).

Division of the Embryonic Body Cavity

Each pericardioperitoneal canal lies lateral to the foregut (future esophagus) and dorsal to the **septum transversum**, a thick plate of mesoderm that occupies the space between the thoracic cavity and yolk stalk (see Fig. 10 - 4*A* and *B*). The septum transversum is the primordium of the **central tendon of the diaphragm**. Partitions form in each pericardioperitoneal canal, separating the pericardial cavity from the pleural cavities and the pleural cavities from the peritoneal cavity. Because of the growth of the **bronchial buds** (primordia of the bronchi and lungs) into the pericardioperitoneal

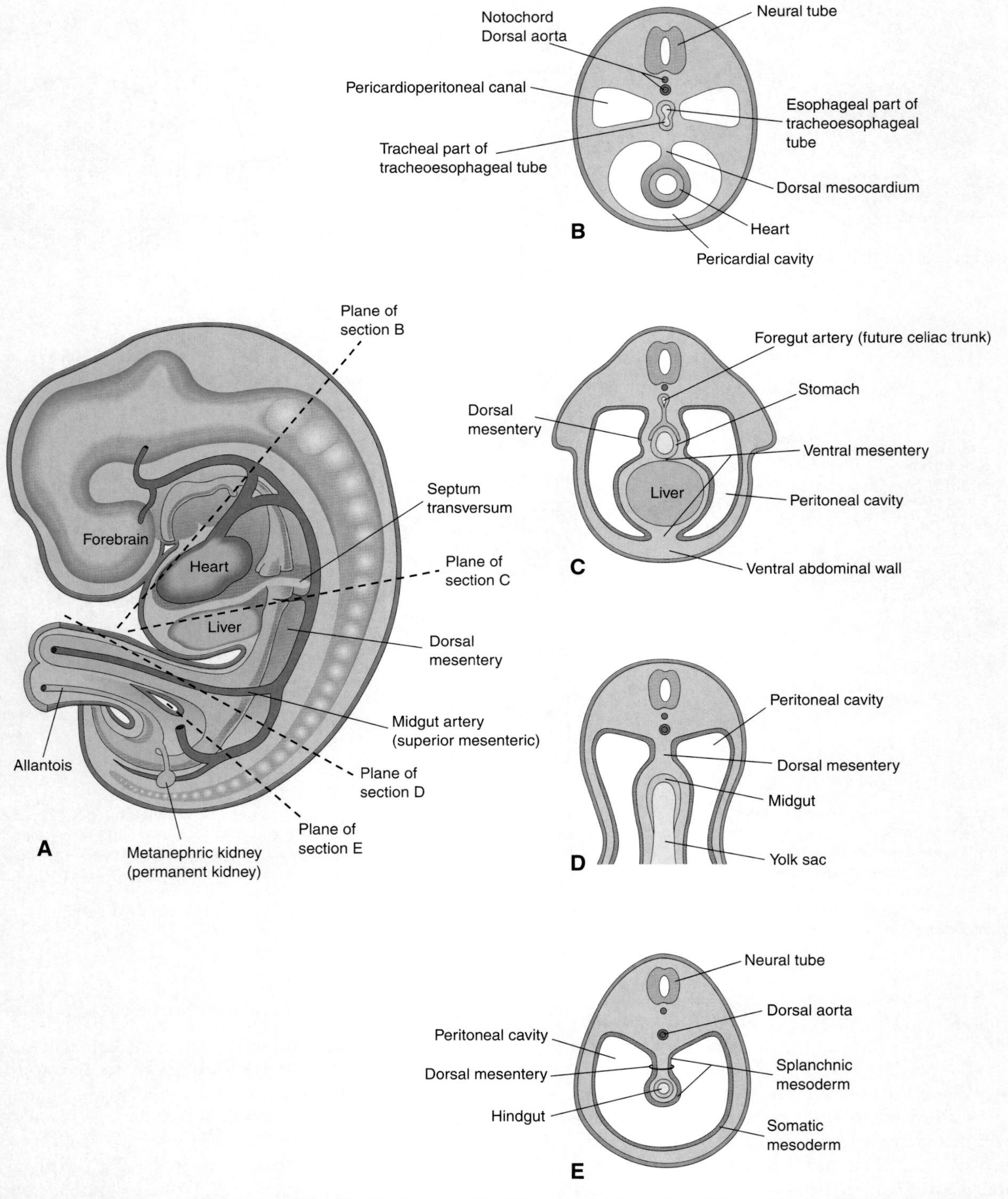

Figure 10 – 3. Mesenteries and body cavities at the beginning of the fifth week. *A,* Schematic sagittal section. Note that the dorsal mesentery serves as a pathway for the arteries supplying the developing gut. Nerves and lymphatics also pass between the layers of this mesentery. *B* to *E,* Transverse sections through the embryo at the levels indicated in *A.* The ventral mesentery disappears, except in the region of the terminal esophagus, stomach, and first part of the duodenum. Note that the right and left parts of the peritoneal cavity, which are separate in *C,* are continuous in *E.*

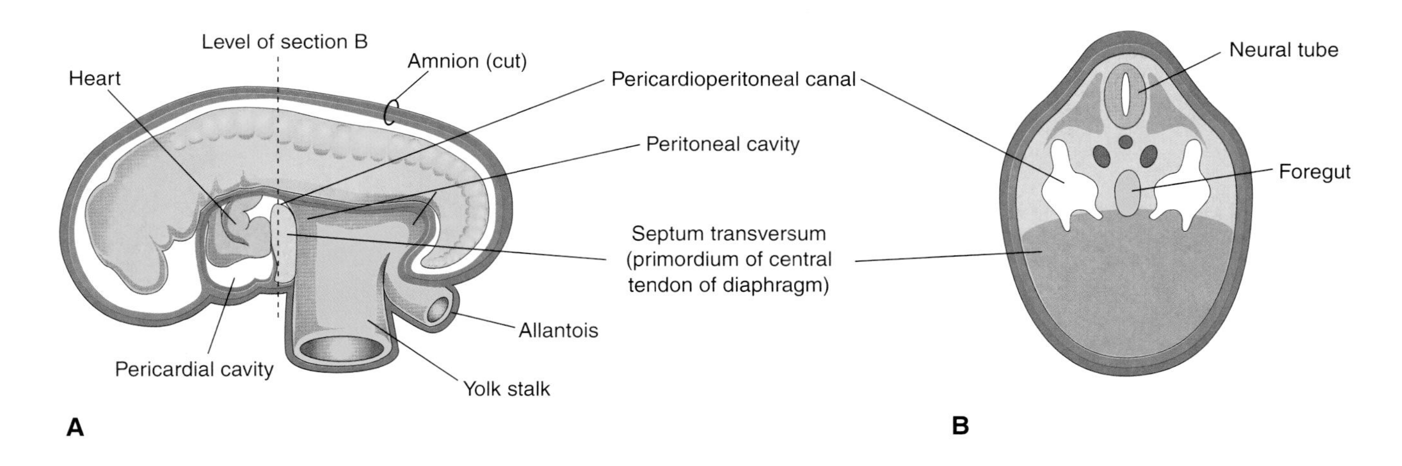

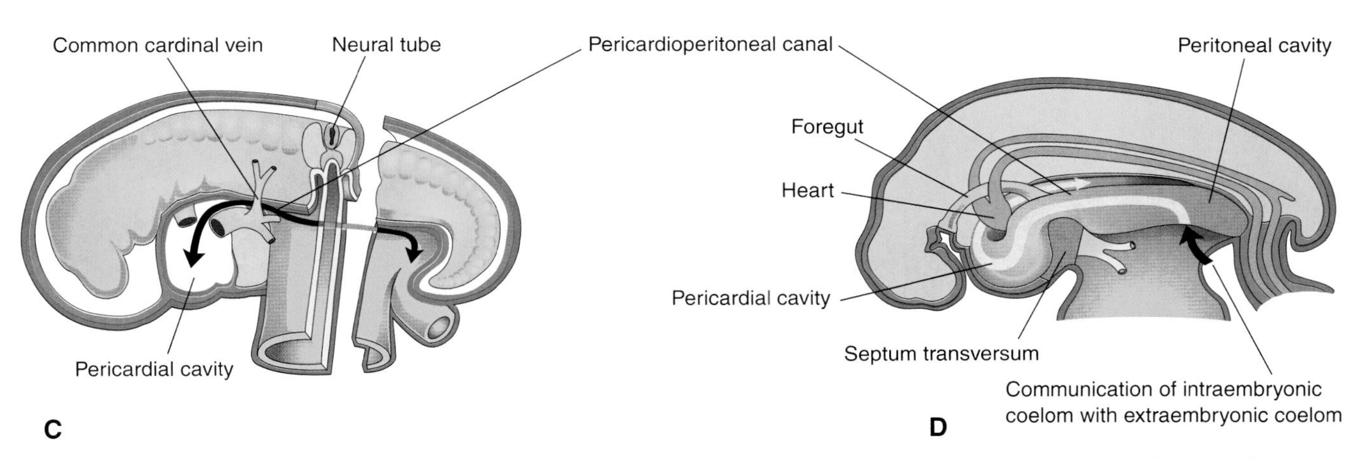

Figure 10 – 4. Schematic drawings of an embryo (about 24 days). *A,* The lateral wall of the pericardial cavity has been removed to show the primordial heart. *B,* Transverse section of the embryo illustrating the relationship of the pericardioperitoneal canals to the septum transversum (primordium of the central tendon of the diaphragm) and the foregut. *C,* Lateral view of the embryo with the heart removed. The embryo has also been sectioned transversely to show the continuity of the intraembryonic and extraembryonic coeloms. *D,* Sketch showing the pericardioperitoneal canals arising from the dorsal wall of the pericardial cavity and passing on each side of the foregut to join the peritoneal cavity. The arrows show the communication of the extraembryonic coelom with the intraembryonic coelom and the continuity of the intraembryonic coelom at this stage.

canals (Fig. 10 – 5*A*), a pair of membranous ridges is produced in the lateral wall of each canal.

- The cranial ridges — the *pleuropericardial folds* — are located superior to the developing lungs.
- The caudal ridges — the *pleuroperitoneal folds* — are located inferior to the lungs.

As the **pleuropericardial folds** enlarge, they form partitions that separate the pericardial cavity from the pleural cavities. These partitions, known as the **pleuropericardial membranes**, contain the **common cardinal veins** (see Fig. 10 – 5*A* and *B*). These large veins drain the primordial venous system into the sinus venosus of the primordial heart (see Chapter 15). Initially, the **bronchial buds** are small relative to the heart and pericardial cavity (see Fig. 10 – 5*A*). Subsequently, they grow laterally from the caudal end of the trachea into the pericardioperitoneal canals (future pleural canals). As the **primordial pleural cavities** expand ventrally around the heart, they extend into the body wall, splitting the mesenchyme into two layers:

- an outer layer that becomes the thoracic wall
- an inner layer (the pleuropericardial membrane) that becomes the **fibrous pericardium**, the outer layer of the pericardial sac enclosing the heart (see Fig. 10 – 5*C* and *D*).

The **pleuropericardial membranes** project into the cranial ends of the **pericardioperitoneal canals** (see Fig. 10 – 5B). With subsequent growth of the common

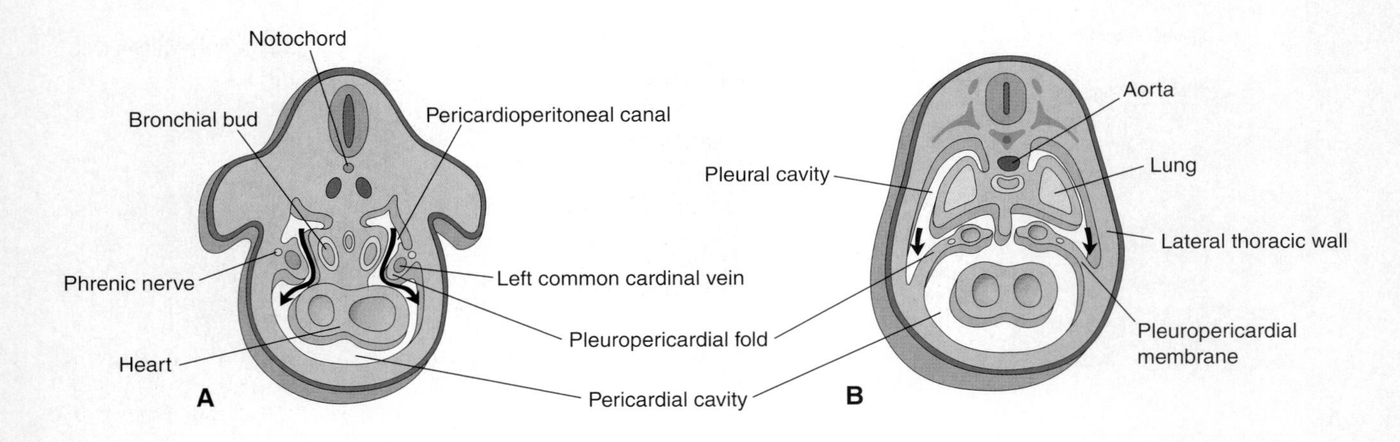

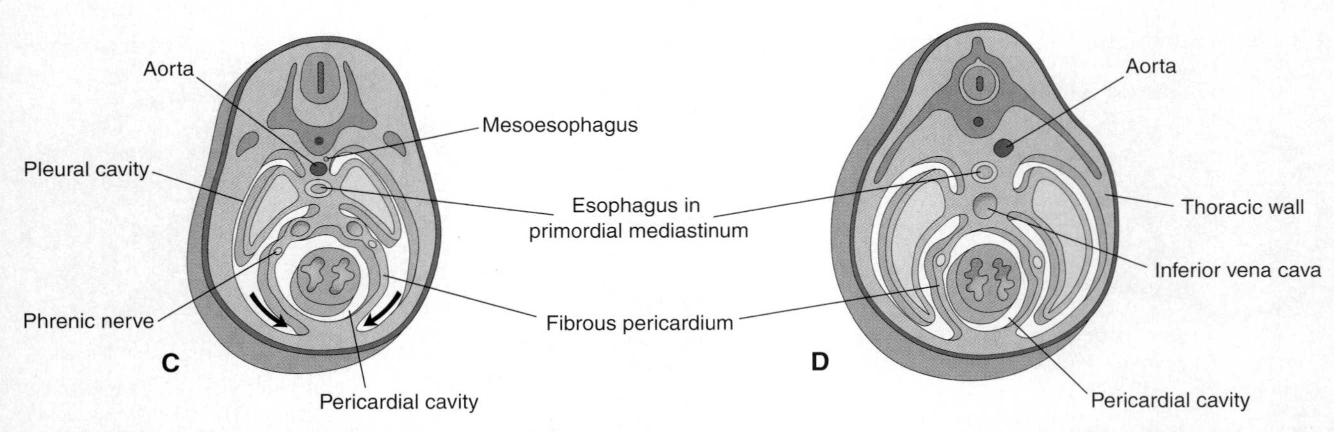

Figure 10-5. Transverse sections through embryos cranial to the septum transversum, illustrating successive stages in the separation of the pleural cavities from the pericardial cavity. Growth and development of the lungs, expansion of the pleural cavities, and formation of the fibrous pericardium are also shown. *A,* Five weeks. The arrows indicate the communications between the pericardioperitoneal canals and the pericardial cavity. *B,* Six weeks. The arrows indicate development of the pleural cavities as they expand into the body wall. *C,* Seven weeks. Expansion of the pleural cavities ventrally around the heart is evident. The pleuropericardial membranes are now fused in the median plane with each other and with the mesoderm ventral to the esophagus. *D,* Eight weeks. Continued expansion of the lungs and pleural cavities and formation of the fibrous pericardium and thoracic wall are illustrated.

cardinal veins, descent of the heart, and expansion of the pleural cavities, the pleuropericardial membranes become mesenterylike folds extending from the lateral thoracic wall. By the seventh week, the pleuropericardial membranes fuse with the mesenchyme ventral to the esophagus, forming the **primordial mediastinum** and separating the pericardial cavity from the pleural cavities (see Fig. 10-5*C*). The right pleuropericardial opening closes slightly earlier than the left one, probably because the right common cardinal vein is larger than the left one and produces a larger pleuropericardial membrane.

As the **pleuroperitoneal folds** enlarge, they project into the pericardioperitoneal canals. Gradually, the folds become membranous, forming the **pleuroperitoneal membranes** (Fig. 10-6*A* and *B*). Eventually, these membranes separate the pleural cavities from the peritoneal cavity. The pleuroperitoneal membranes are produced as the developing lungs and pleural cavities expand and invade the body wall. They are attached dorsolaterally to the abdominal wall, and initially, their crescentic free edges project into the caudal ends of the **pericardioperitoneal canals**. They become relatively more prominent as the lungs enlarge cranially and the liver expands caudally. During the sixth week, the pleuroperitoneal membranes extend ventromedially until their free edges fuse with the dorsal mesentery of the esophagus and septum transversum (see Fig. 10-6*C*). This separates the pleural cavities from the peritoneal cavity. *Closure of the pleuroperitoneal openings* is assisted

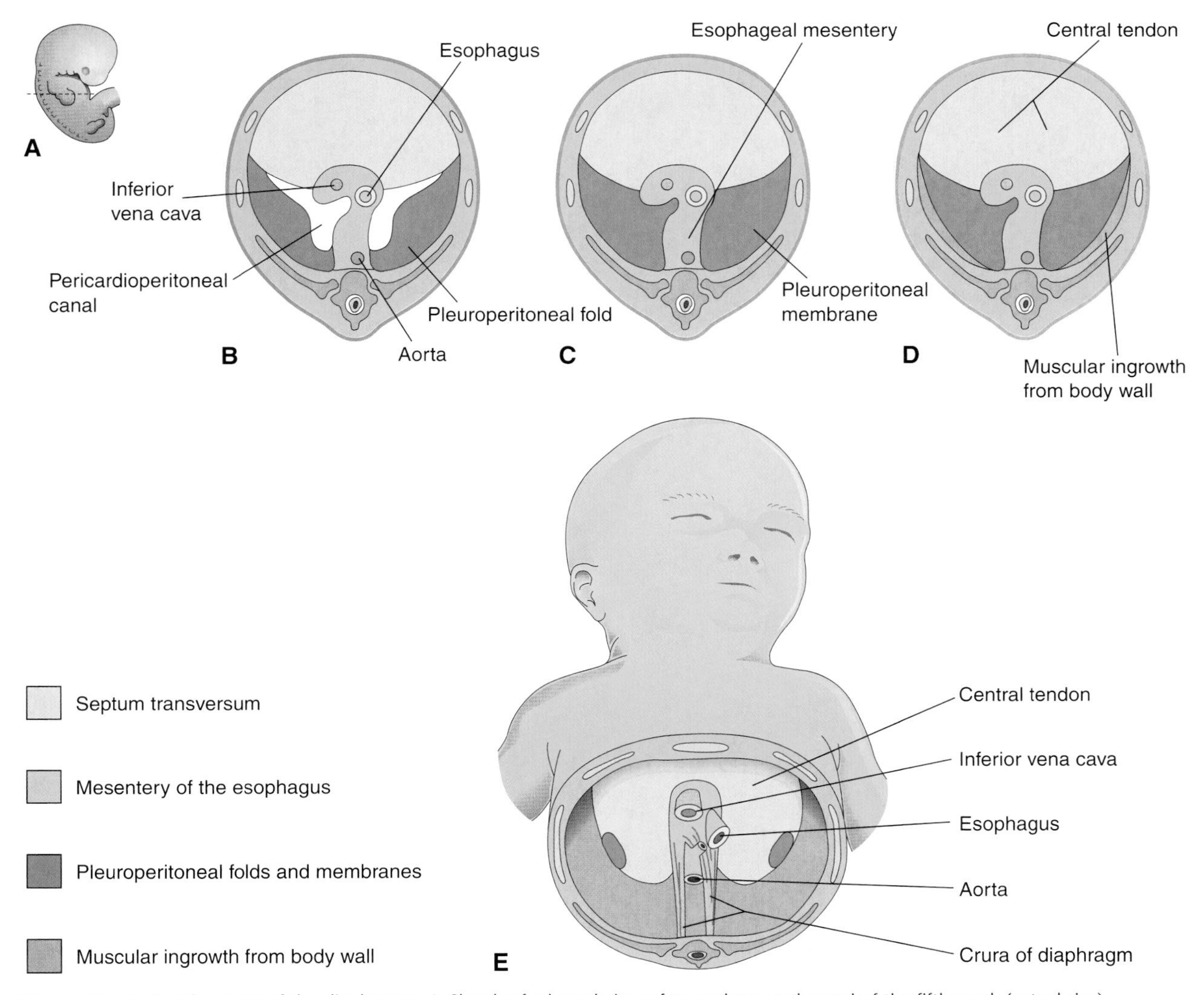

Figure 10 - 6. Development of the diaphragm. *A*, Sketch of a lateral view of an embryo at the end of the fifth week (actual size), indicating the level of sections in *B* to *D*. *B* to *E* show the developing diaphragm as viewed inferiorly. *B*, Transverse section showing the unfused pleuroperitoneal membranes. *C*, Similar section at the end of the sixth week after fusion of the pleuroperitoneal membranes with the other two diaphragmatic components. *D*, Transverse section of a 12-week embryo after ingrowth of the fourth diaphragmatic component from the body wall. *E*, View of the diaphragm of a newborn infant, indicating the embryologic origin of its components.

by the migration of myoblasts (primordial muscle cells) into the pleuroperitoneal membranes (see Fig. 10 - 6*E*). The pleuroperitoneal opening on the right side closes slightly before the left one. The reason for this is uncertain, but it may be related to the relatively large size of the right lobe of the liver at this stage of development.

Development of the Diaphragm

The diaphragm is a dome-shaped, musculotendinous partition that separates the thoracic and abdominal cavities. It is a composite structure that develops from four embryonic components (see Fig. 10 - 6):

- septum transversum
- pleuroperitoneal membranes
- dorsal mesentery of esophagus
- lateral body walls

Septum Transversum

The septum transversum (transverse septum), composed of mesodermal tissue, is the **primordium of the central**

tendon of the diaphragm (see Fig. 10 - 6*D* and *E*). The septum transversum grows dorsally from the ventrolateral body wall and forms a semicircular shelf that separates the heart from the liver. During its early development, a large part of the liver is embedded in the septum transversum. The septum transversum is located caudal to the pericardial cavity and partially separates it from the developing peritoneal cavity. The septum transversum is first identifiable at the end of the third week as a mass of mesodermal tissue cranial to the pericardial cavity (see Chapter 6). After the head folds ventrally during the fourth week, the septum transversum forms a thick, incomplete partition between the pericardial and abdominal cavities (see Fig. 10 - 4). A large opening, the **pericardioperitoneal canal**, is found on each side of the esophagus (see Fig. 10 - 6*B*). The septum transversum expands and fuses with the mesenchyme ventral to the esophagus and the pleuroperitoneal membranes (see Fig. 10 - 6*C*).

Pleuroperitoneal Membranes

The pleuroperitoneal membranes fuse with the dorsal mesentery of the esophagus and septum transversum (see Fig. 10 - 6*C*). This completes the partition between the thoracic and abdominal cavities and forms the **primordial diaphragm**. Although the pleuroperitoneal membranes form large portions of the fetal diaphragm, they represent relatively small parts of the newborn infant's diaphragm (see Fig. 10 - 6*E*).

Mesentery of the Esophagus

The septum transversum and pleuroperitoneal membranes fuse with the dorsal mesentery of the esophagus. This mesentery becomes the median portion of the diaphragm. The **crura of the diaphragm** — a leglike pair of diverging muscle bundles that cross in the median plane anterior to the aorta (see Fig. 10 - 6*E*) — develop from myoblasts that grow into the dorsal mesentery of the esophagus.

Muscular Ingrowth from the Body Wall

During the 9th to 12th weeks, the lungs and pleural cavities enlarge, "burrowing" into the lateral body walls (see Fig. 10 - 5). During this process, the body-wall tissue is split into two layers:

- an external layer that becomes part of the definitive abdominal wall
- an internal layer that contributes muscle to peripheral portions of the diaphragm, external to the parts derived from the pleuroperitoneal membranes (see Fig. 10 - 6*D* and *E*).

Further extension of the developing pleural cavities into the lateral body walls forms the right and left **costodiaphragmatic recesses** (Fig. 10 - 7), establishing the characteristic dome-shaped configuration of the diaphragm. After birth, the costodiaphragmatic recesses become alternately smaller and larger as the lungs move in and out of them during inspiration and expiration.

Positional Changes and Innervation of the Diaphragm

During the fourth week of development, the septum transversum, prior to its descent with the heart, lies opposite the third to fifth *cervical somites* (Fig. 10 - 8*A*). During the fifth week, myoblasts (primordial muscle cells) from these somites migrate into the developing diaphragm, bringing their nerve fibers with them. Consequently, the **phrenic nerves** that supply motor innervation to the diaphragm arise from the ventral primary rami of the third, fourth, and fifth cervical spinal nerves. The three twigs on each side join together to form a phrenic nerve. The phrenic nerves also supply sensory fibers to the superior and inferior surfaces of the right and left domes of the diaphragm.

Rapid growth of the dorsal part of the embryo's body results in an *apparent descent of the diaphragm*. By the sixth week, the developing diaphragm is at the level of the thoracic somites (see Fig. 10 - 8*B*). The phrenic nerves now have a descending course. As the diaphragm "moves" relatively farther caudally in the body, the nerves are correspondingly lengthened. By the beginning of the eighth week, the dorsal part of the diaphragm lies at the level of the first lumbar vertebra (see Fig. 10 - 8*C*). Because of the embryonic origin of the phrenic nerves in the neck, they are about 30 cm long in adults. The

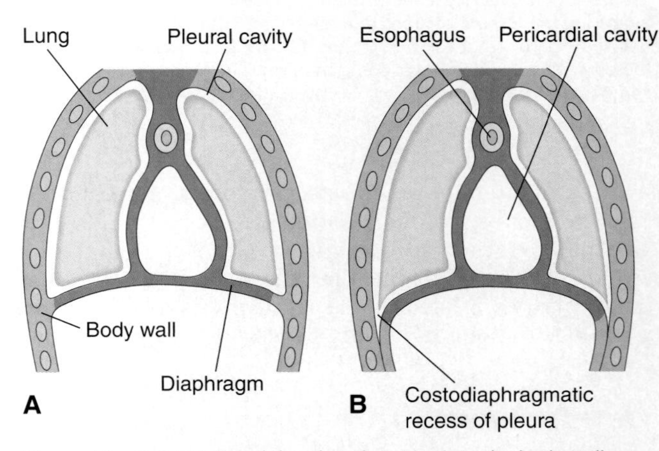

Figure 10 - 7. Extension of the pleural cavities into the body walls to form peripheral parts of the diaphragm, the costodiaphragmatic recesses, and the characteristic dome-shaped configuration of the diaphragm.

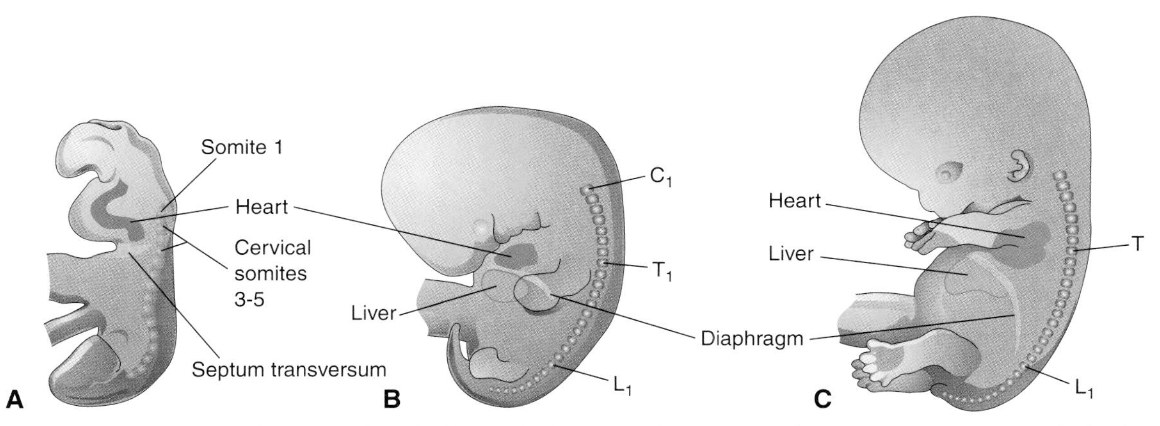

Figure 10 - 8. The positional changes of the developing diaphragm. *A*, About 24 days. The septum transversum is at the level of the third, fourth, and fifth cervical segments. *B*, About 41 days. *C*, About 52 days. C, cervical; T, thoracic; L, lumbar.

embryonic phrenic nerves enter the diaphragm by passing through the pleuropericardial membranes. This explains why the phrenic nerves subsequently lie on the **fibrous pericardium**, the derivation of the pleuropericardial membranes (see Fig. 10 - 5*C* and *D*).

As the four parts of the diaphragm fuse (see Fig. 10 - 6), mesenchyme in the septum transversum extends into the other three parts. It forms myoblasts that differentiate into the skeletal muscle of the diaphragm; hence, the motor nerve supply to the diaphragm is from the phrenic nerves. The sensory innervation of the diaphragm is also from the phrenic nerves, but its costal rim receives sensory fibers from the lower intercostal nerves because of the origin of the peripheral part of the diaphragm from the lateral body walls (see Fig. 10 - 6*D* and *E*).

Congenital Diaphragmatic Hernia

A posterolateral defect of the diaphragm through which hernias occur is the most common anomaly. A congenital diaphragmatic hernia (**CDH**) is characterized by the presence of abdominal viscera in the thoracic cavity.

Posterolateral Defect of Diaphragm

A posterolateral defect of the diaphragm is the only relatively common congenital anomaly involving the diaphragm (Fig. 10 – 9*A* and *B*). This diaphragmatic defect occurs in about 1 in 2200 newborn infants and is associated with **CDH**, or herniation of abdominal contents into the thoracic cavity. Life-threatening breathing difficulties may be associated with CDH because of inhibition of development and inflation of the lungs (Fig. 10 – 10).

Moreover, fetal lung maturation may be delayed. *CDH is the most common cause of pulmonary hypoplasia.* **Polyhydramnios** (excess amniotic fluid) may also be present. Usually unilateral, CDH results from defective formation or fusion of the pleuroperitoneal membrane with the other three parts of the diaphragm (see Fig. 10 – 6). This produces a large opening in the posterolateral region of the diaphragm. As a result, the peritoneal and pleural cavities are continuous with one another along the posterior body wall. The defect, — which is sometimes referred to clinically as the foramen of Bochdalek — usually (in 85 to 90% of cases) occurs on the left side. The preponderance of left-sided defects is likely related to the earlier closure of the right pleuroperitoneal opening. **Prenatal diagnosis of CDH** (Fig. 10 – 11) depends on the ultrasonographic and magnetic resonance imaging (MRI) of the abdominal organs in the thorax.

The pleuroperitoneal membranes normally fuse with the other three diaphragmatic components by the end of the sixth week of development (see Fig. 10 – 6*C*). If a pleuroperitoneal canal is still open when the intestines return to the abdomen from the umbilical cord in the 10th week (see Chapter 13), some intestine and other viscera may pass into the thorax. The presence of abdominal viscera in the thorax pushes the lungs and heart anteriorly, and compression of the lungs occurs. Often the stomach, spleen, and most of the intestines herniate (see Figs. 10 – 10 and 10 – 11). Most babies born with CDH die not because there is a defect in the diaphragm or viscera in the chest, but because the lungs are hypoplastic secondary to compression during development. If severe **lung hypoplasia** is present, some primordial alveoli may rupture, causing air to enter the pleural cavity, a condition known as *pneumothorax*.

Eventration of the Diaphragm

In the uncommon condition of diaphragmatic eventration, half the diaphragm has defective musculature, causing it

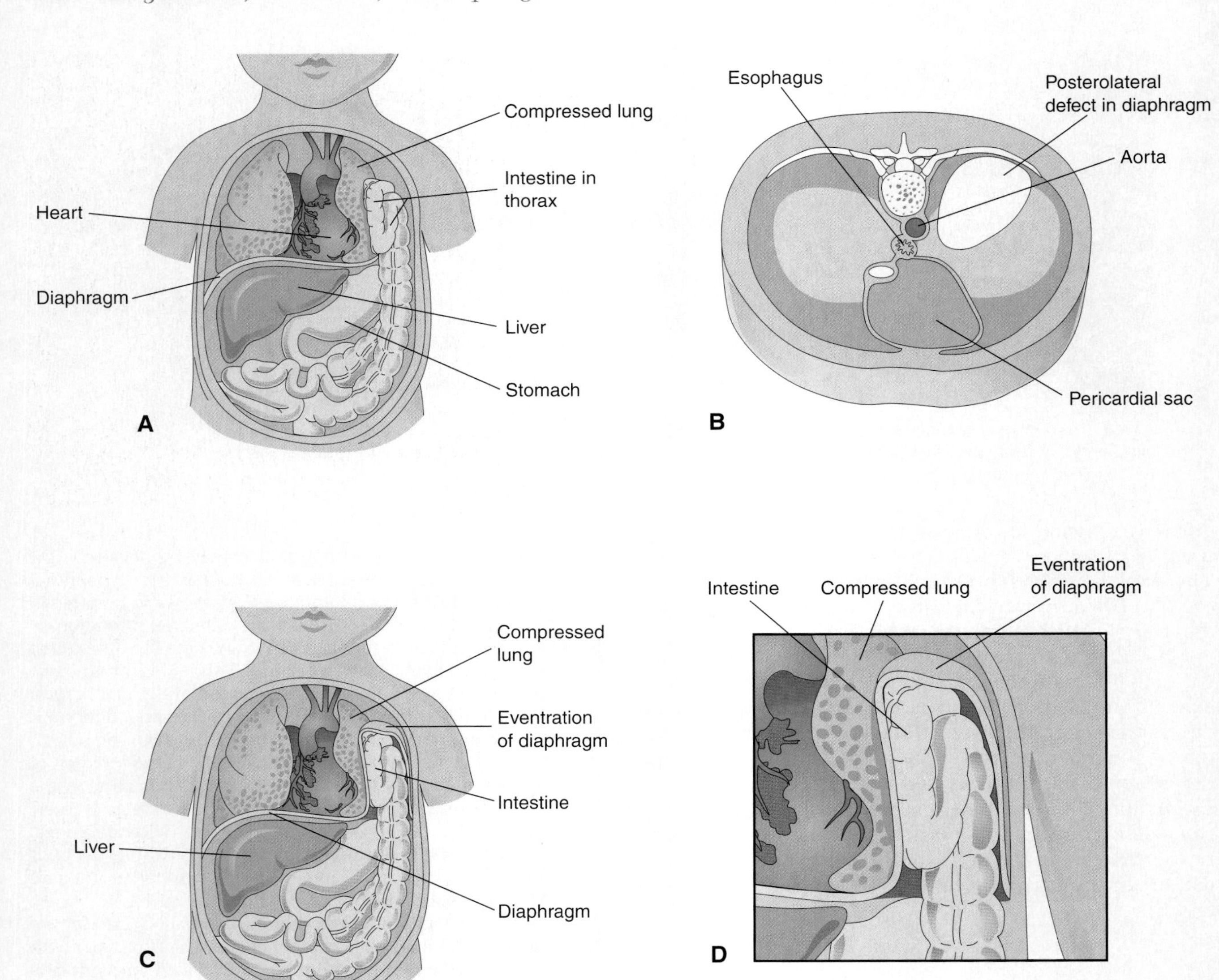

Figure 10–9. *A,* A "window" view overlooking the thorax and abdomen shows the herniation of the intestine into the thorax through a posterolateral defect in the left side of the diaphragm. Note that the left lung is compressed and hypoplastic. *B,* Drawing of a diaphragm with a large posterolateral defect on the left side due to abnormal formation and/or fusion of the pleuroperitoneal membrane on the left side with the mesoesophagus and septum transversum. *C* and *D,* Eventration of the diaphragm resulting from defective muscular development of the diaphragm. The abdominal viscera are displaced into the thorax along with a pouch of diaphragmatic tissue.

to balloon into the thoracic cavity as an aponeurotic (membranous) sheet, forming a diaphragmatic pouch (see Fig. 10–9*C* and *D*). Consequently, the abdominal viscera are displaced superiorly into the pocketlike outpouching of the diaphragm. This congenital anomaly results mainly from failure of muscular tissue from the body wall to extend into the pleuroperitoneal membrane on the affected side.

Gastroschisis and Congenital Epigastric Hernia

Gastroschisis and congenital epigastric hernia represent an uncommon type of hernia that occurs in the median plane between the xiphoid process and the umbilicus. These defects are similar to umbilical hernias (see Chapter 13) except for their location. Gastroschisis and epigastric hernias result from failure of the lateral body folds to fuse completely when forming

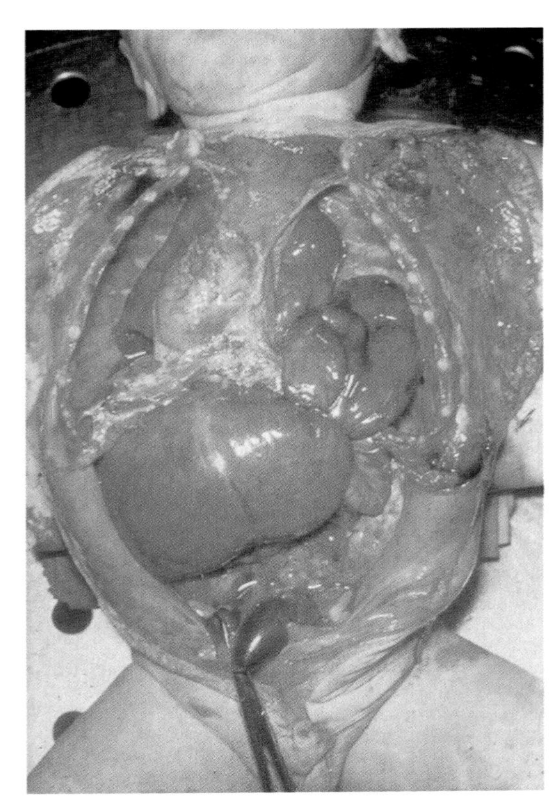

Figure 10–10. Diaphragmatic hernia. Note the herniation of the stomach and small intestine into the thorax through a posterolateral defect in the left side of the diaphragm, similar to that shown in Figure 10–9*A* and *B*. (Courtesy of Dr. Nathan E. Wiseman, Professor of Surgery, Children's Hospital, University of Manitoba, Winnipeg, Manitoba, Canada.)

the anterior abdominal wall during folding in the fourth week (see Fig. 10–2*C* and *F*). The small intestine herniates into the amniotic cavity and amniotic fluid, a condition that can be detected prenatally by ultrasonography.

Retrosternal (Parasternal) Hernia

Herniations may occur through the *sternocostal hiatus* (foramen of Morgagni), the opening for the superior epigastric vessels in the retrosternal area. This hiatus is located between the sternal and costal parts of the diaphragm. Herniation of intestine into the pericardial sac may occur or, conversely, part of the heart may descend into the peritoneal cavity in the epigastric region. Large defects are commonly associated with body wall defects in the umbilical region (e.g., omphalocele; see Chapter 13).

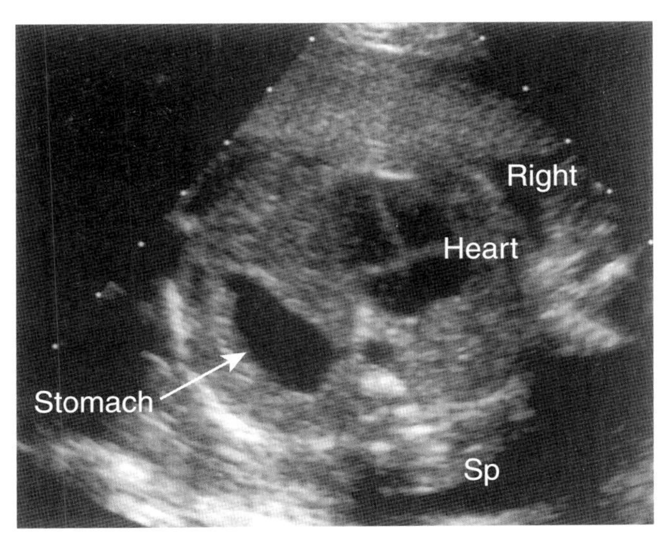

Figure 10–11. An ultrasound scan of the thorax showing the heart shifted to the right and the stomach on the left. The diaphragmatic hernia was detected at 23.4 weeks' gestation. The stomach herniated through a posterolateral defect in the diaphragm (congenital diaphragmatic hernia). Sp, vertebral column or spine. (Courtesy of Dr. Wesley Lee, William Beaumont Hospital, Royal Oak, Michigan.)

Summary of Development of the Body Cavities

The intraembryonic coelom, the primordium of the body cavities, begins to develop near the end of the third week. By the fourth week, it appears as a horseshoe-shaped cavity in the cardiogenic and lateral mesoderm. The curve of the "horseshoe" represents the future pericardial cavity, and its lateral extensions represent the future pleural and peritoneal cavities. During folding of the embryonic disc in the fourth week, lateral parts of the intraembryonic coelom move together on the ventral aspect of the embryo. When the caudal part of the ventral mesentery disappears, the right and left parts of the intraembryonic coelom merge to form the peritoneal cavity. As peritoneal parts of the intraembryonic coelom come together, the splanchnic layer of mesoderm encloses the primordial gut and suspends it from the dorsal body wall by a double-layered peritoneal membrane known as the dorsal mesentery. The parietal layer of mesoderm lining the peritoneal, pleural, and pericardial cavities becomes the parietal peritoneum, parietal pleura, and serous pericardium, respectively.

Until the seventh week of development, the embryonic pericardial cavity communicates with the peritoneal cavity through paired *pericardioperitoneal canals*. During the fifth and sixth weeks, folds form near the cranial and caudal ends of these canals. Fusion of the cranial *pleuropericardial membranes* with mesoderm ventral to the esophagus separates the pericardial cavity from the pleural cavities. Fusion of the caudal *pleuroperitoneal membranes* during formation of the diaphragm separates the pleural cavities from the peritoneal cavity. The diaphragm develops from four structures:

- septum transversum
- pleuroperitoneal membranes
- dorsal mesentery of the esophagus
- muscular ingrowth from the lateral body walls

Clinically Oriented Questions

1. There have been reports of a baby who was born with its stomach and liver in its chest. Is this possible?
2. Can a baby with most of its abdominal viscera in its chest survive? Some say that diaphragmatic defects can be operated on before birth. Is this true?
3. Do the lungs develop normally in babies who are born with CDH?
4. An individual underwent routine chest radiography about a year ago and was told that a small part of his small intestine was in his chest. Is it possible for him to have a CDH without being aware of it? Would his lung on the affected side be normal?

The answers to these questions are at the back of the book.

11

The Pharyngeal Apparatus

At about four weeks of development, the head and neck regions of the human embryo somewhat resemble those regions of a fish embryo at a comparable stage of development. By the end of the embryonic period, the pharyngeal apparatus has become rearranged and adapted to new functions or have disappeared.

The **pharyngeal (branchial) apparatus** (Fig. 11 – 1) consists of the following:

- pharyngeal arches
- pharyngeal pouches
- pharyngeal grooves
- pharyngeal membranes

These embryonic structures contribute to the formation of the head and neck. Most congenital anomalies in these regions originate during transformation of the pharyngeal apparatus into its adult derivatives. Because gills do not form in human embryos, the term ***pharyngeal arch*** is now used instead of *branchial arch*.

Pharyngeal Arches

The pharyngeal arches begin to develop early in the fourth week of development as **neural crest cells** migrate from specific hindbrain segments (rhombomeres 2 to 7) into the future head and neck regions (see Chapter 6). The first pair of pharyngeal arches, the primordium of the jaws, appears as surface elevations lateral to the developing pharynx (see Fig. 11 – 1*A* and *B*). Rhombomere 2 contributes neural crest cells to the first pharyngeal arches. Soon thereafter, other arches appear as obliquely disposed, rounded ridges on each side of the future head and neck regions (see Fig. 11 – 1*C* and *D*). By the end of the fourth week, four well-defined pairs of pharyngeal arches are visible (Fig. 11 – 2). The fifth and sixth arches are rudimentary and are not visible on the surface of the embryo.

The arches are separated from each other by prominent clefts called **pharyngeal grooves**. Like the pharyngeal arches, the grooves are numbered in a craniocaudal sequence.

The **first pharyngeal arch** (mandibular arch) develops two prominences (see Figs. 11 – 1*E* and *F* and 11 – 2):

- The smaller **maxillary prominence** gives rise to the maxilla (upper jaw), zygomatic bone, and squamous part of the temporal bone.
- The larger **mandibular prominence** forms the mandible (lower jaw).

Consequently, the first pair of pharyngeal arches plays a major role in the development of the face.

The **second pharyngeal arch** (hyoid arch) makes a major contribution to the formation of the hyoid bone. The pharyngeal arches support the lateral walls of the primordial pharynx, which is derived from the cranial part of the foregut. The primordial mouth or **stomodeum** initially appears as a slight depression of the surface ectoderm (see Fig. 11 – 1*D* and *E*). It is separated from the cavity of the primordial pharynx by a bilaminar membrane — the **oropharyngeal membrane**. It is composed of ectoderm externally and endoderm internally. The oropharyngeal membrane ruptures at about 26 days, bringing the primordial pharynx and foregut into communication with the amniotic cavity (see Fig. 11 – 1*F* and *G*).

Pharyngeal Arch Components

Initially, each pharyngeal arch consists of a core of mesenchyme (embryonic connective tissue) and is covered externally by ectoderm and internally by endoderm (see Fig. 11 – 1*H* and *I*). The original mesenchyme is derived from mesoderm in the third week of development. During the fourth week, most of the mesenchyme is derived from **neural crest cells** that migrate from the rhombomeres of the hindbrain into the pharyngeal arches. The migration of the neural crest cells into the arches and their differentiation into mesenchyme (epithelial-mesenchymal transformation) produces the maxillary and mandibular prominences of the first arch (see Fig. 11 – 2). **Homeobox genes** regulate the migration of neural crest cells and patterning of the pharyngeal arches. Other genes and signaling factors, such as *Sonic hedgehog* (*Shh*), *distal-less* (*Dlx*), *ephrins*, *fibroblast growth factors*, and retinoic acid receptor (*RAR*), are also involved. Neural crest cells are unique in that, despite their neuroectodermal origin, they make a major contribution to mesenchyme in the head as well as to structures in many other regions (see Chapter 6). Skeletal musculature and vascular endothelia, however, are derived from the original mesenchyme in the pharyngeal arches.

Fate of the Pharyngeal Arches

The pharyngeal arches contribute extensively to the formation of the face, nasal cavities, mouth, larynx, pharynx, and neck (Figs. 11 – 3 and 11 – 4). During the fifth week of development, the second pharyngeal arch enlarges and overgrows the third and fourth arches, forming an ectodermal depression known as the **cervical sinus** (see Fig. 11 – 4*A* to *D*). By the end of the seventh week, the second to fourth pharyngeal grooves and the cervical sinus have disappeared, giving the neck a smooth contour. A **typical pharyngeal arch** is composed of the following components:

- an *aortic arch*, an artery that arises from the **truncus arteriosus** of the primordial heart (see

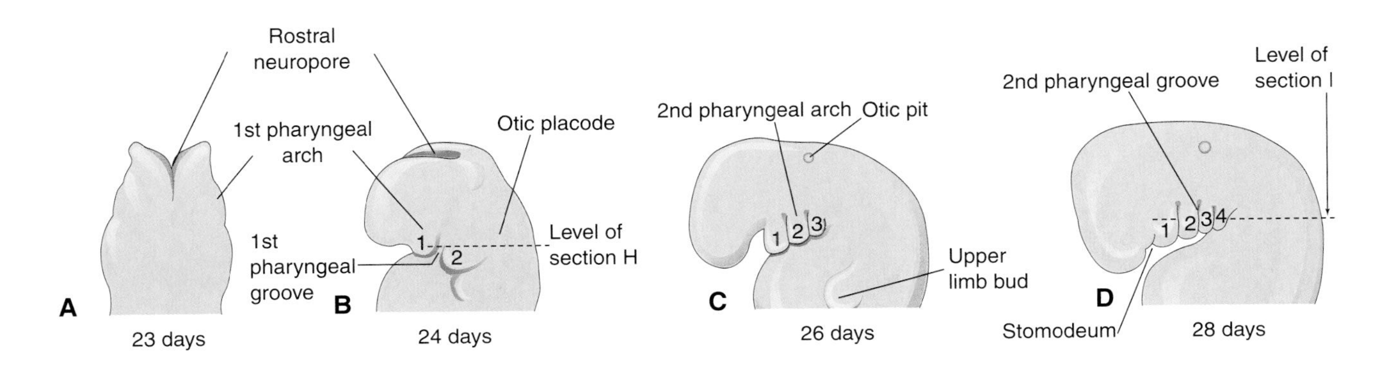

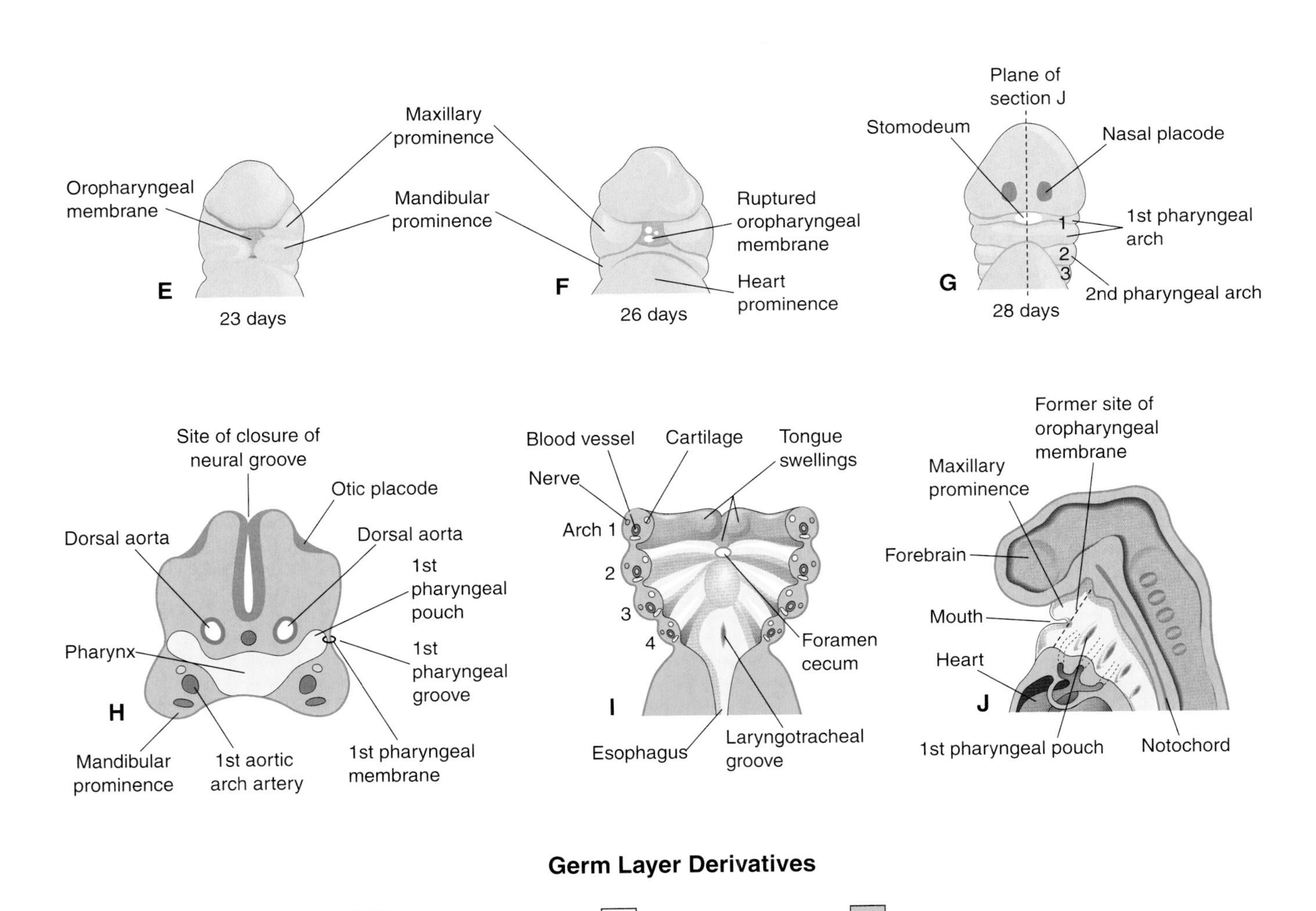

Figure 11 - 1. Drawings illustrating the human pharyngeal apparatus. *A*, Dorsal view of the cranial part of an early embryo. *B* to *D*, Lateral views showing later development of the pharyngeal arches. *E* to *G*, Ventral or facial views illustrating the relationship of the first pharyngeal arch to the stomodeum. *H*, Horizontal section through the cranial region of an embryo. *I*, Similar section illustrating the arch components and floor of the primordial pharynx. *J*, Sagittal section of the cranial region of an embryo, illustrating the openings of the pharyngeal pouches in the lateral wall of the primordial pharynx.

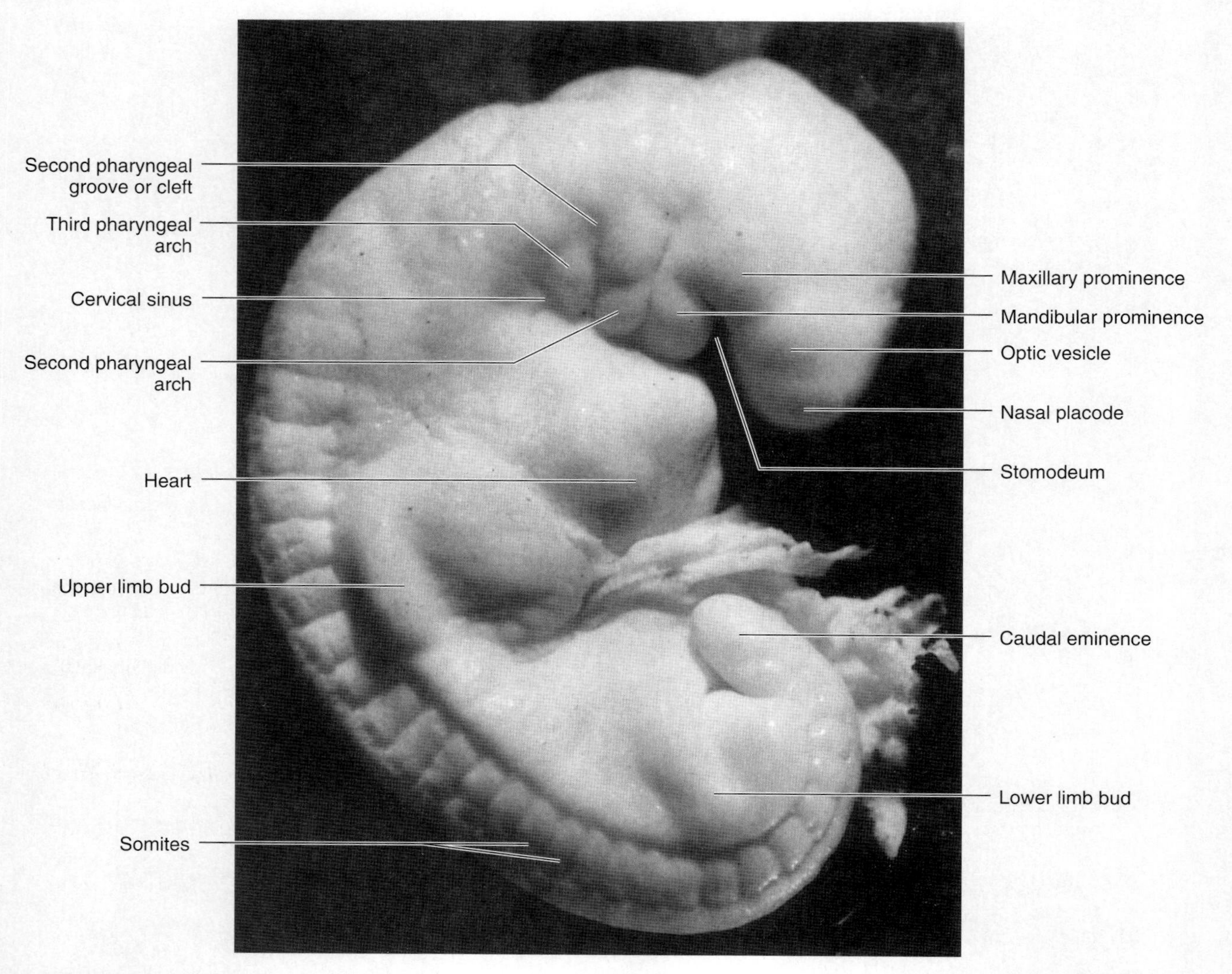

Figure 11 - 2. Macrophotograph of a Carnegie stage 13, $4^1/_2$-week human embryo. (Courtesy of Professor Emeritus Dr. KV Hinrichsen, Medizinische Fakultät, Institut für Anatomie, Ruhr-Universität Bochum, Germany.)

Fig. 11 - 3*B*) and runs around the primordial pharynx to enter the **dorsal aorta**

- a *cartilaginous rod* that forms the skeleton of the arch
- a *muscular component* that differentiates into muscles in the head and neck
- a *nerve* that supplies the mucosa and muscles derived from the arch

The nerves that grow into the arches are derived from neuroectoderm of the primordial brain.

Derivatives of the Aortic Arches

The transformation of the aortic arches (pharyngeal arch arteries) into the adult arterial pattern of the head and neck is described in the section on the aortic arch derivatives presented in Chapter 15. Blood in the aortic arches supplies the pharyngeal arches and then enters the dorsal aorta.

Derivatives of the Pharyngeal Arch Cartilages

The dorsal end of the **first arch cartilage** (Meckel cartilage) becomes ossified to form two middle ear bones, the **malleus** and **incus** (Fig. 11 - 5*A* and *B* and Table 11 - 1). The middle part of the cartilage regresses, but its perichondrium forms the **anterior ligament of malleus** and the **sphenomandibular ligament**. Ventral parts of the first arch cartilages form the horseshoe-shaped primordium of the mandible. Each half of the

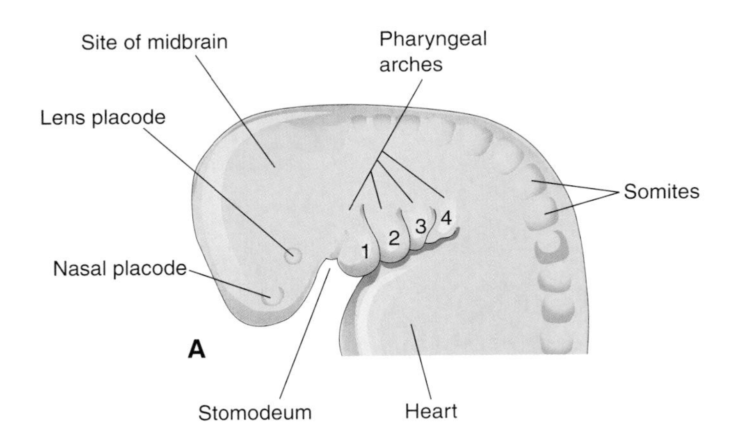

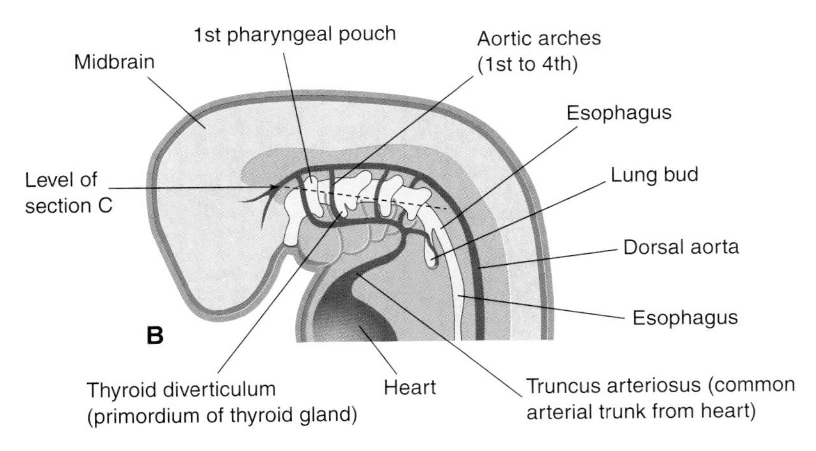

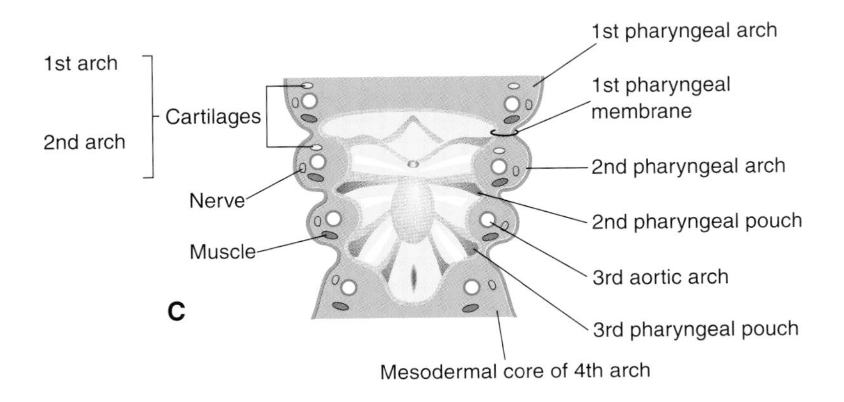

Germ Layer Derivatives

Ectoderm　Endoderm　Mesoderm

Figure 11 – 3. *A*, Drawing of the head, neck, and thoracic regions of a human embryo (about 28 days), illustrating the pharyngeal apparatus. *B*, Schematic drawing showing the pharyngeal pouches and aortic arches. *C*, Horizontal section through the embryo showing the floor of the primordial pharynx and illustrating the germ layer of origin of the pharyngeal arch components.

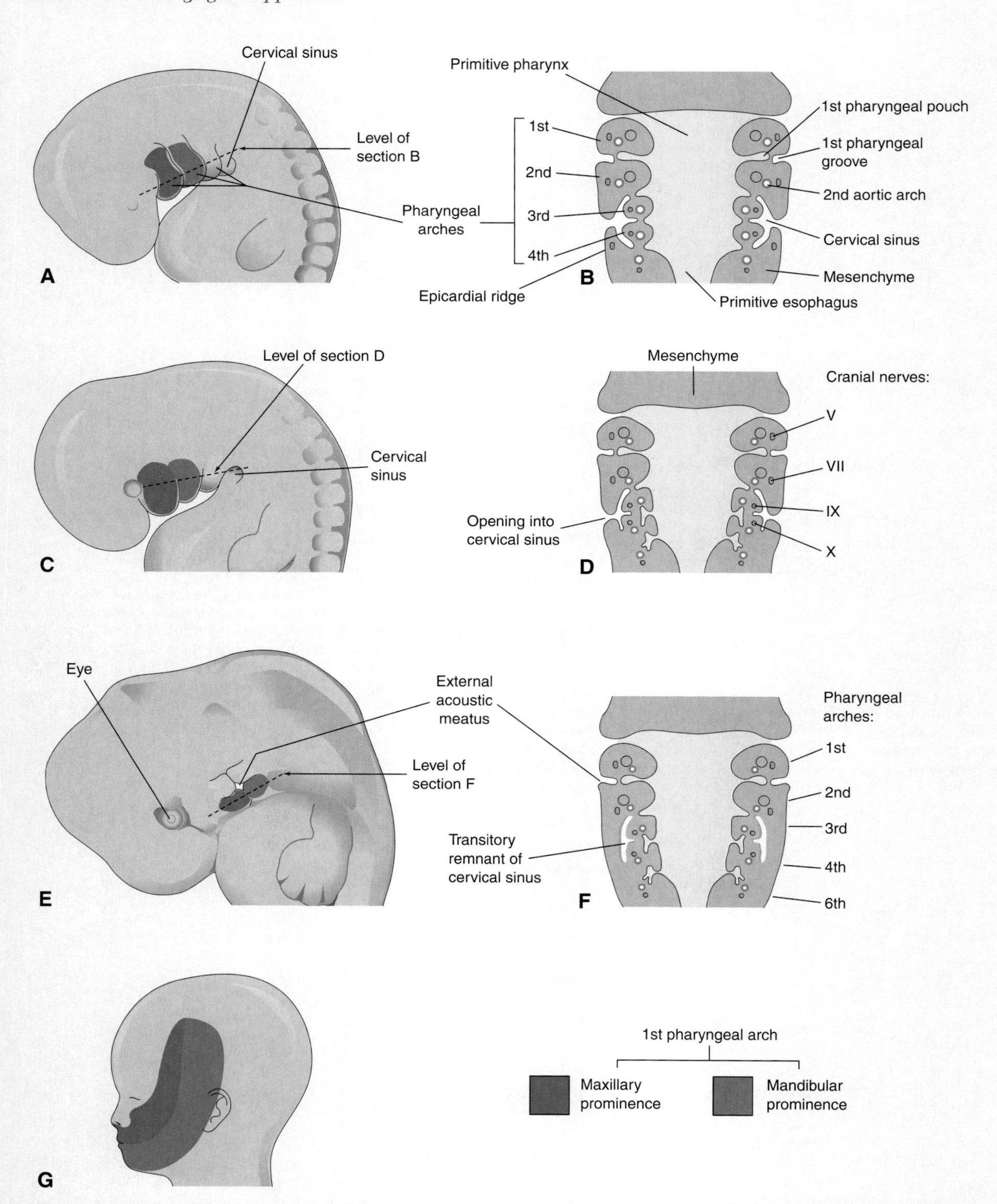

Figure 11 - 4. *A*, Lateral view of the head, neck, and thoracic regions of an embryo (about 32 days), showing the pharyngeal arches and cervical sinus. *B*, Diagrammatic section through the embryo at the level shown in *A*, illustrating growth of the second arch over the third and fourth arches. *C*, An embryo of about 33 days. *D*, Section of the embryo at the level shown in *C*, illustrating early closure of the cervical sinus. *E*, An embryo of about 41 days. *F*, Section of the embryo at the level shown in *E*, showing the transitory cystic remnant of the cervical sinus. *G*, Drawing of a 20-week fetus illustrating the area of the face derived from the first pair of pharyngeal arches.

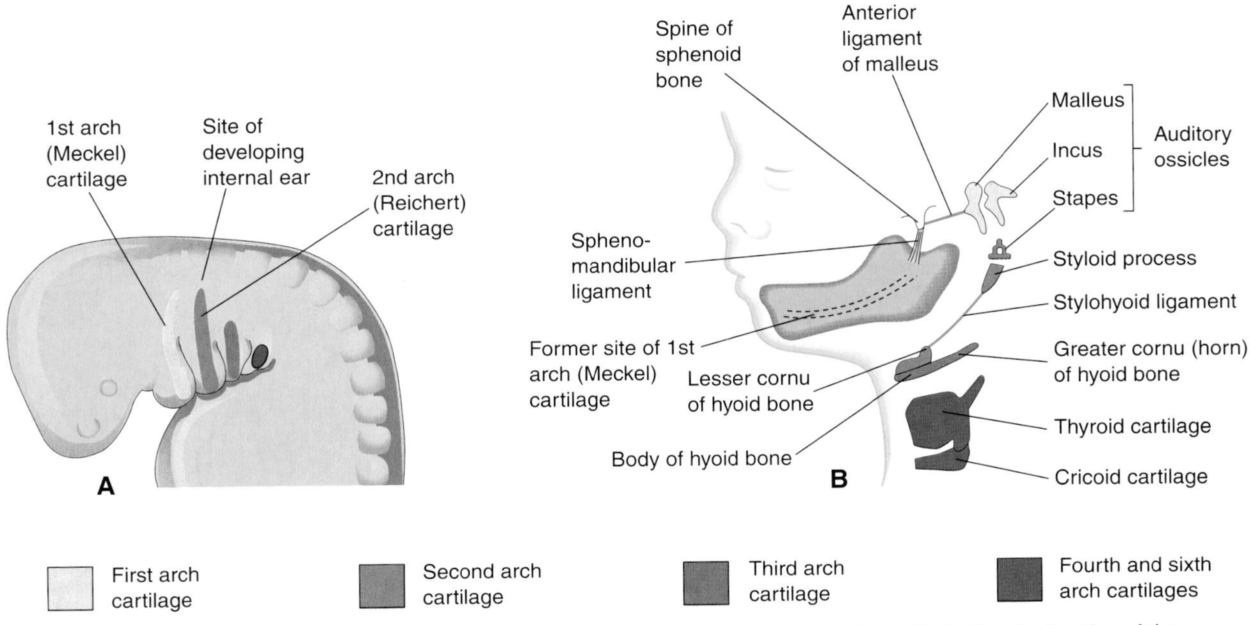

Figure 11 - 5. *A*, Schematic lateral view of the head, neck, and thoracic regions of a 4-week embryo, illustrating the location of the cartilages in the pharyngeal arches. *B*, Similar view of a 24-week fetus illustrating the adult derivatives of the arch cartilages. Note that the mandible is formed by intramembranous ossification of mesenchymal tissue surrounding the first arch cartilage.

Table 11 - 1. Structures Derived from Pharyngeal Arch Components*

Arch	Nerve	Muscles	Skeletal Structures	Ligaments
First (mandibular)	Trigeminal† (CN V)	Muscles of mastication‡ Mylohyoid and anterior belly of digastric Tensor tympani Tensor veli palatini	Malleus Incus	Anterior ligament of malleus Sphenomandibular ligament
Second (hyoid)	Facial (CN VII)	Muscles of facial expression§ Stapedius Stylohyoid Posterior belly of digastric	Stapes Styloid process Lesser cornu of hyoid Upper part of body of hyoid bone	Stylohyoid ligament
Third	Glossopharyngeal (CN IX)	Stylopharyngeus	Greater cornu of hyoid Lower part of body of hyoid bone	
Fourth and Sixth‖	Superior laryngeal branch of vagus (CN X) Recurrent laryngeal branch of vagus (CN X)	Cricothyroid Levator veli palatini Constrictors of pharynx Intrinsic muscles of larynx Striated muscles of esophagus	Thyroid cartilage Cricoid cartilage Arytenoid cartilage Corniculate cartilage Cuneiform cartilage	

*The derivatives of the aortic arch arteries are described in Chapter 15.
†The ophthalmic division does not supply any pharyngeal arch components.
‡Temporalis, masseter, medial, and lateral pterygoids.
§Buccinator, auricularis, frontalis, platysma, orbicularis oris and oculi.
‖The fifth pharyngeal arch regresses. The cartilaginous components of the fourth and sixth arches fuse to form the cartilages of the larynx.

mandible forms lateral to and in close association with its cartilage. The cartilage disappears as the mandible develops around it by intramembranous ossification (see Fig. 11 – 5*B*). The dorsal end of the **second arch cartilage** (Reichert cartilage) ossifies to form the **stapes** of the middle ear and the **styloid process** of the temporal bone (see Fig. 11 – 5*B*). The part of cartilage between the styloid process and hyoid bone regresses; its perichondrium forms the **stylohyoid ligament**. The ventral end of the second arch cartilage ossifies to form the lesser cornu (L., horn) and the superior part of the body of the **hyoid bone** (see Fig. 11 – 5*B*). The **third arch cartilage** ossifies to form the greater cornu and the inferior part of the body of the hyoid bone. The **fourth and sixth arch cartilages** fuse to form the **laryngeal cartilages** (see Fig. 11 – 5*B* and Table 11 – 1), except for the epiglottis. The cartilage of the epiglottis develops from mesenchyme in the *hypopharyngeal eminence* (*hypobranchial eminence*), a prominence in the floor of the embryonic pharynx that is derived from the third and fourth pharyngeal arches (see Fig. 11 – 22*A*).

Derivatives of the Pharyngeal Arch Muscles

The muscular components of the arches form various muscles in the head and neck; for example, the musculature of the first pharyngeal arch forms the **muscles of mastication** and other muscles (Fig. 11 – 6A and B; see Table 11 – 1).

Derivatives of the Pharyngeal Arch Nerves

Each arch is supplied by its own cranial nerve (CN). Rhombomeres contribute motor fibers to each cranial nerve. The *special visceral efferent (branchial) components* of the cranial nerves supply muscles derived from the pharyngeal arches (Fig. 11 – 7*A*; see Table 11 – 1). Because mesenchyme from the pharyngeal arches contributes to the dermis and mucous membranes of the head and neck, these areas are supplied with *special visceral afferent nerves*. The facial skin is supplied by the fifth cranial nerve (CN V), or the **trigeminal nerve**; however, only its caudal two branches (*maxillary and mandibular*) supply derivatives of the first pharyngeal arch (see Fig. 11 – 7*B*). CN V is the principal sensory nerve of the head and neck and is the motor nerve for the muscles of mastication (see Table 11 – 1). Its sensory branches innervate the face, teeth, and mucous membranes of the nasal cavities, palate, mouth, and tongue (see Fig. 11 – 7*C*). The seventh cranial nerve (CN VII) — or **facial nerve**, the ninth cranial nerve (CN IX) — or **glossopharyngeal nerve**, and the tenth cranial nerve (CN X) — the **vagus nerve** supply the second, third, and caudal (fourth to sixth) arches, respectively. The fourth arch is supplied by the

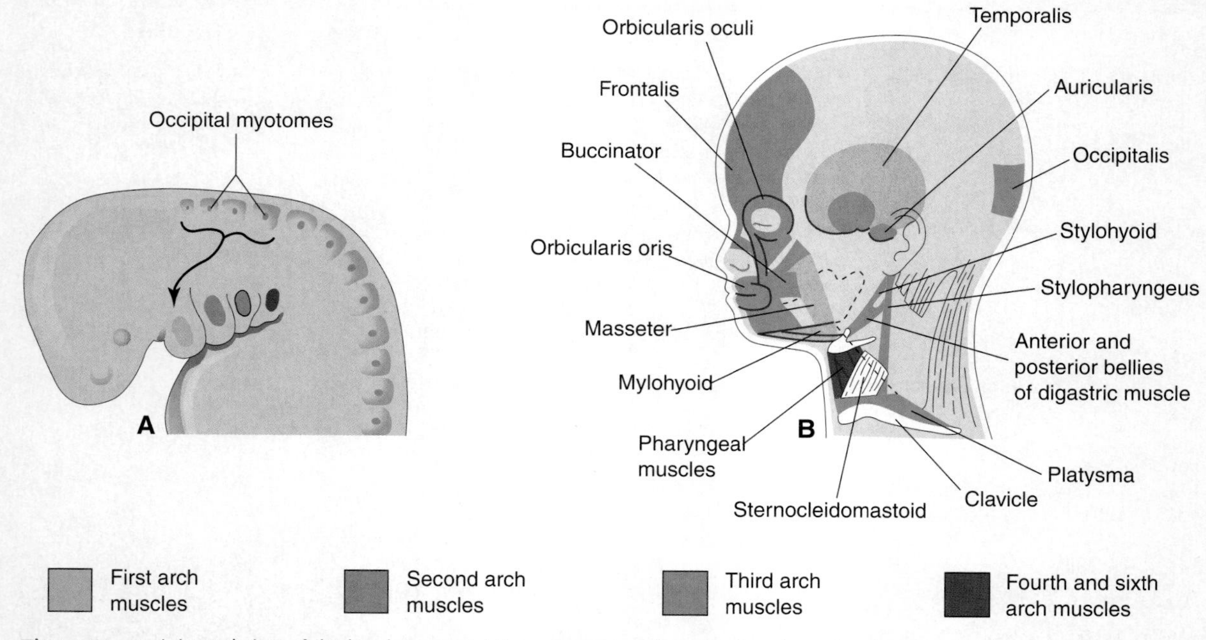

Figure 11 – 6. *A*, Lateral view of the head, neck, and thoracic regions of a 4-week embryo showing the muscles derived from the pharyngeal arches. The arrow shows the pathway taken by myoblasts from the occipital myotomes to form the tongue musculature. *B*, Sketch of the head and neck regions of a 20-week fetus, dissected to show the muscles derived from the pharyngeal arches. Parts of the platysma and sternocleidomastoid muscles have been removed to show the deeper muscles. Note that myoblasts from the second arch migrate from the neck to the head, where they give rise to the muscles of facial expression. These muscles are supplied by the facial nerve (CN VII), the nerve of the second pharyngeal arch.

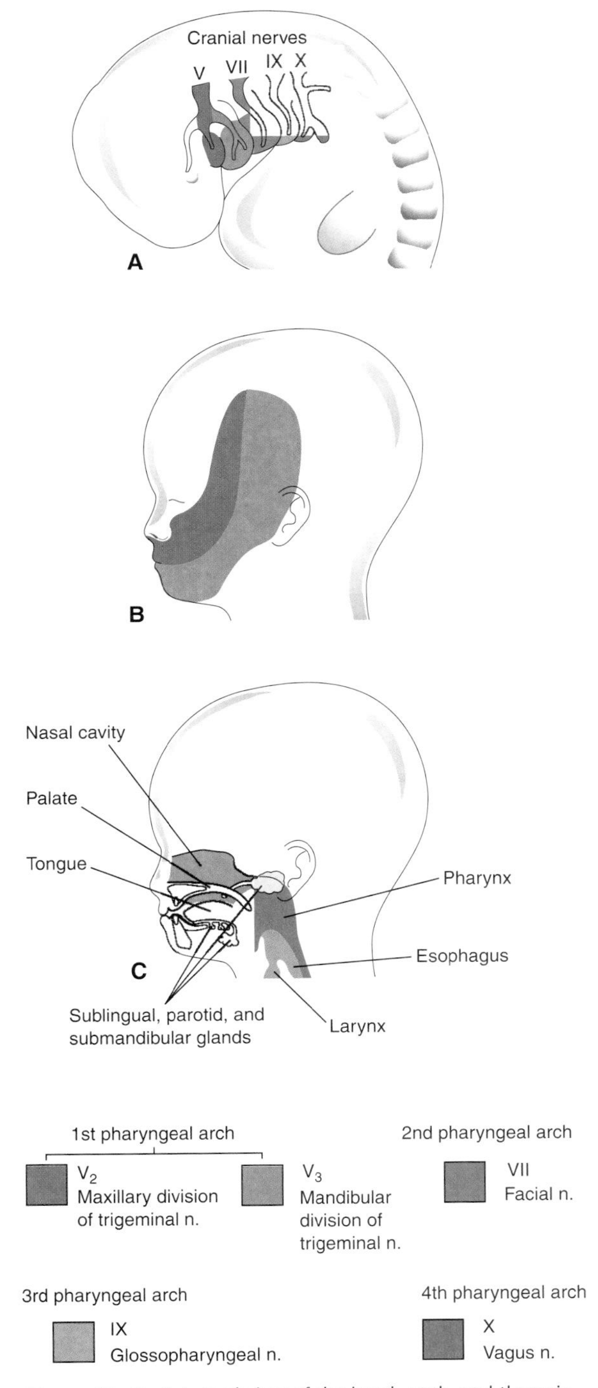

Figure 11 - 7. *A*, Lateral view of the head, neck, and thoracic regions of a 4-week embryo showing the cranial nerves supplying the pharyngeal arches. *B*, Sketch of the head and neck regions of a 20-week fetus showing the superficial distribution of the two caudal branches of the first arch nerve (CN V). *C*, Sagittal section of the fetal head and neck, showing the deep distribution of sensory fibers of the nerves to the teeth and mucosa of the tongue, pharynx, nasal cavity, palate, and larynx.

superior laryngeal branch of the vagus nerve, whereas the sixth arch is supplied by its recurrent laryngeal branch. The nerves of the second to sixth pharyngeal arches have little cutaneous distribution (see Fig. 11 - 7*C*); however, they innervate the mucous membranes of the tongue, pharynx, and larynx.

Pharyngeal Pouches

The *primordial pharynx*, derived from the foregut, widens cranially where it joins the primordial mouth or *stomodeum* and narrows caudally where it joins the *esophagus* (see Figs. 11 - 3*A* and *B* and 11 - 4*B*). The endoderm of the pharynx lines the internal aspects of the pharyngeal arches and passes into balloonlike diverticula known as the **pharyngeal pouches** (see Figs. 11 - 1*H* to *J* and 11 - 3*B* and *C*). These pairs of pouches develop in a craniocaudal sequence between the arches. The first pair of pouches, for example, lies between the first and second pharyngeal arches. Four pairs of pharyngeal pouches are well defined; the fifth pair is absent or rudimentary. The endoderm of the pouches contacts the ectoderm of the pharyngeal grooves, and together they form the double-layered **pharyngeal membranes** that separate the pharyngeal pouches from the pharyngeal grooves (see Figs. 11 - 1*H* and 11 - 3*C*).

Derivatives of the Pharyngeal Pouches

The endodermal epithelial lining of the pharyngeal pouches (Fig. 11 - 8*A*) gives rise to important organs in the head and neck (e.g., the thymus and parathyroid glands).

First Pharyngeal Pouch

The first pharyngeal pouch gives rise to the **tubotympanic recess** (see Fig. 11 - 8*B*). The expanded distal part of this recess contacts the first pharyngeal groove, where it later contributes to the formation of the **tympanic membrane** (eardrum). The cavity of the tubotympanic recess gives rise to the **tympanic cavity** and **mastoid antrum**. The connection of the tubotympanic recess with the pharynx forms the **pharyngotympanic tube** (auditory tube).

Second Pharyngeal Pouch

The second pharyngeal pouch is largely obliterated as the **palatine tonsil** develops (see Figs. 11 - 8*C* and 11 - 9). Part of this pouch remains as the **tonsillar sinus** or **fossa**. The endoderm of the second pouch proliferates and grows into the underlying mesenchyme. The central parts of these buds break down, forming crypts (pitlike depressions). The pouch endoderm forms the surface epithelium and lining of the **tonsillar crypts**.

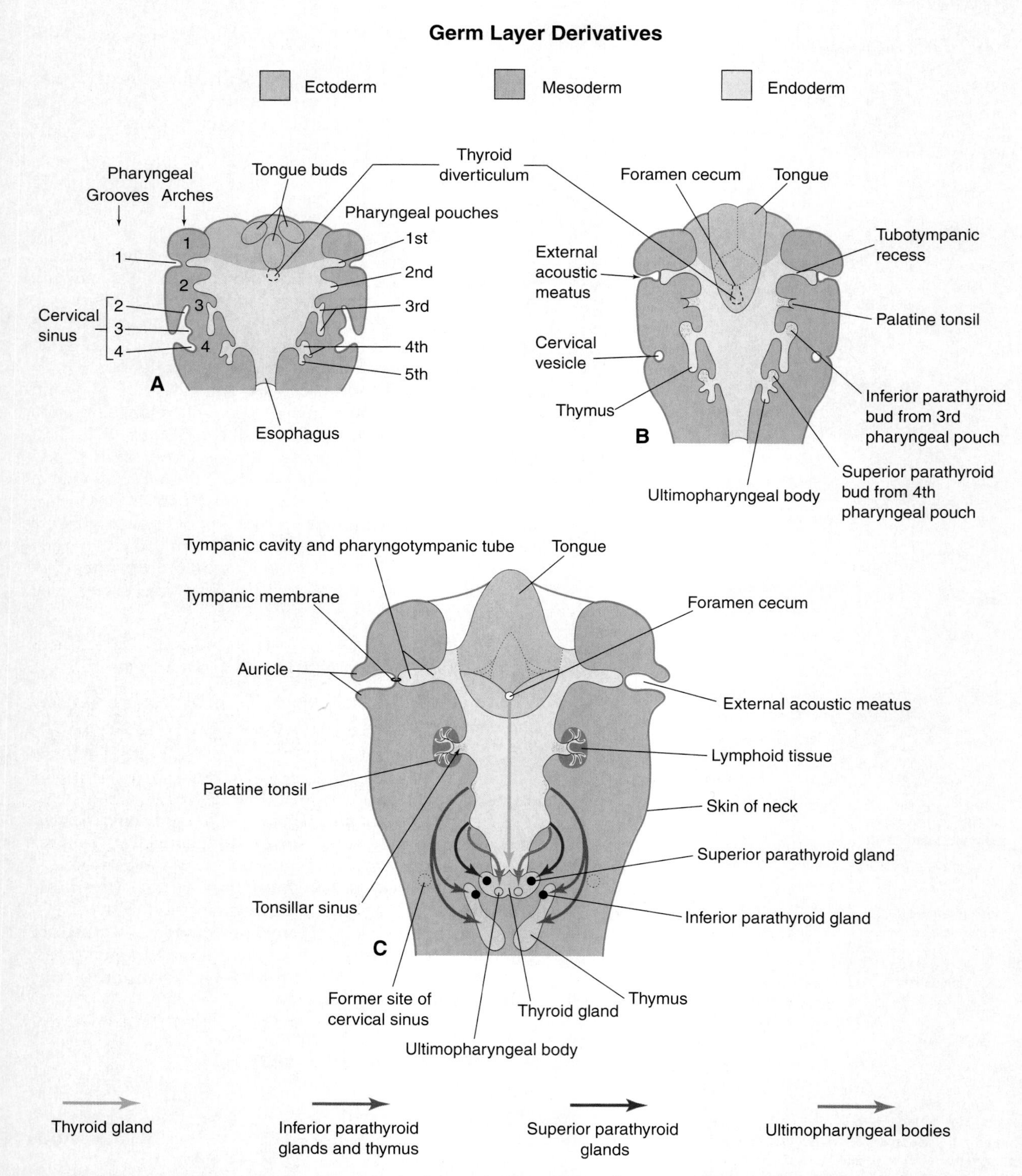

Figure 11 - 8. Schematic horizontal sections at the level shown in Figure 11 – 4*A*, illustrating the adult derivatives of the pharyngeal pouches. *A*, Five weeks. Note that the second pharyngeal arch grows over the third and fourth arches, burying the second to fourth pharyngeal grooves in the cervical sinus. *B*, Six weeks. *C*, Seven weeks. Note the migration of the developing thymus, parathyroid, and thyroid glands into the neck.

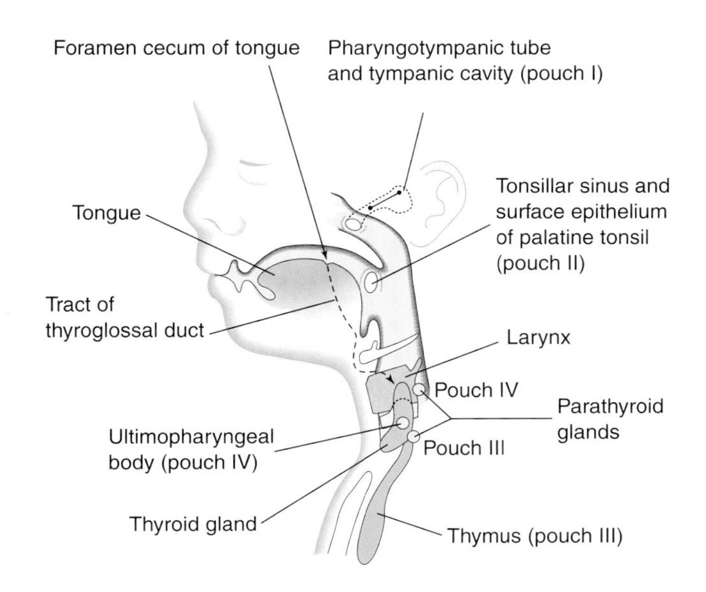

Figure 11 - 9. Schematic drawing of a sagittal section of the head, neck, and upper thoracic regions of a 20-week fetus, showing the adult derivatives of the pharyngeal pouches and the descent of the thyroid gland into the neck.

The mesenchyme around the crypts differentiates into lymphoid tissue, which soon organizes into the *lymphatic nodules* of the palatine tonsil.

Third Pharyngeal Pouch

The third pharyngeal pouch expands and develops a solid, dorsal, bulbar part and a hollow, elongate, ventral part (see Fig. 11 - 8*B*). Its connection with the pharynx is reduced to a narrow duct that soon degenerates. By the sixth week of development, the epithelium of each dorsal bulbar part begins to differentiate into an **inferior parathyroid gland**. The epithelium of the elongate ventral parts of the third pair of pouches proliferates, obliterating their cavities. These bilateral primordia of the thymus come together in the median plane to form the **thymus**, which descends into the superior mediastinum (superior part of the median septum of the thorax). The primordia of the thymus and parathyroid glands lose their connections with the pharynx and migrate into the neck. Later, the parathyroid glands separate from the thymus and lie on the dorsal surface of the thyroid gland (see Figs. 11 - 8*C* and 11 - 9). The mesenchyme surrounding the thymic primordium and certain epithelial cells in the thymus are derived from *neural crest cells*. Growth and development of the thymus are not complete at birth. It is a relatively large organ during the perinatal period and may extend superiorly through the superior aperture of the thorax into the root of the neck. During late childhood, as puberty is reached, the thymus begins to diminish in relative size. By adulthood, it is often scarcely recognizable because of fat infiltrating the cortex of the gland.

Fourth Pharyngeal Pouch

The dorsal part of the fourth pharyngeal pouch (see Fig. 11 - 8*B*) develops into a **superior parathyroid gland**, which lies on the dorsal surface of the thyroid gland. As described, the parathyroid glands derived from the third pouches descend with the thymus and are carried to a more inferior position than the parathyroid glands derived from the fourth pouches. This explains why the parathyroid glands derived from the third pair of pouches are located inferior to those from the fourth pouches (see Fig. 11 - 9). The elongated ventral part of each fourth pouch develops into an **ultimopharyngeal (ultimobranchial) body**. This body fuses with the thyroid gland, and its cells disseminate within it, giving rise to the **parafollicular cells** of the thyroid gland; they are also called **C cells** to indicate that they produce *calcitonin*, a hormone involved in the regulation of the normal calcium level in body fluids. *C cells differentiate from neural crest cells* that migrate from the pharyngeal arches into the fourth pair of pharyngeal pouches.

Fifth Pharyngeal Pouch

When the fifth pharyngeal pouch develops, this rudimentary structure becomes part of the fourth pharyngeal pouch and helps to form the ultimopharyngeal body.

Pharyngeal Grooves

The head and neck regions of the human embryo exhibit four pharyngeal grooves (clefts) on each side during the fourth and fifth weeks (see Fig. 11 - 1*B* to *D*). These grooves separate the pharyngeal arches externally. Only one pair of grooves contributes to adult structures; the first pair persists as the **external acoustic meatus** (see Fig. 11 - 8*C*). The other grooves lie in a slitlike depression — the **cervical sinus** — and are normally obliterated with it as the neck develops (see Fig. 11 - 4*B*, *D*, and *F*).

Pharyngeal Membranes

The pharyngeal membranes appear in the floors of the pharyngeal grooves on each side of the head and neck regions of the human embryo during the fourth week (see Figs. 11 - 1*H* and 11 - 3*C*). These membranes form where the epithelia of a groove and a pouch approach each other. The endoderm of the pouches and the ectoderm of the grooves are separated by mesenchyme. Only one pair of membranes contributes to the formation of adult structures; the *first pharyngeal membrane*, along with the intervening layer of mesenchyme, becomes the **tympanic membrane** (see Fig. 11 - 8C).

Anomalies of the Head and Neck

Many congenital anomalies of the head and neck (e.g., branchial cysts) originate during transformation of the pharyngeal apparatus into adult structures (Figs. 11 - 10 and 11 - 12). Most defects represent remnants of the pharyngeal apparatus that normally disappear as the adult structures develop.

Congenital Auricular Sinuses and Cysts

Small auricular sinuses (pits) and cysts are usually located in a triangular area of skin anterior to the auricle of the external ear (see Fig. 11 - 10*F*); however, they may occur in other sites around the auricle or in its lobule (earlobe). Although some sinuses and cysts are remnants of the first pharyngeal groove, others represent ectodermal folds sequestered during formation of the auricle from the auricular hillocks (swellings that form the auricle). These small sinuses and cysts are classified as minor anomalies that are of no serious medical consequence.

Branchial Sinuses

Branchial sinuses are uncommon, and almost all that open externally on the side of the neck result from failure of the second pharyngeal groove and cervical sinus to obliterate (see Figs. 11 - 10*D* and 11 - 11). The blind pit or sinus typically opens along the anterior border of the sternocleidomastoid muscle in the inferior third of the neck. Anomalies of the other pharyngeal grooves (first, third, or fourth) occur in about 5% of cases. **External branchial sinuses** are commonly detected during infancy because of the discharge of mucous material from their orifices in the neck (see Fig. 11 - 11). These *lateral cervical sinuses* are bilateral in about 10% of cases and are commonly associated with auricular sinuses. **Internal branchial sinuses** open into the pharynx and are very rare. Because they usually open into the tonsillar sinus or near the palatopharyngeal arch (see Fig. 11 - 10*D* and *F*), almost all of these sinuses result from persistence of the proximal part of the second pharyngeal pouch. Normally, this pouch disappears as the palatine tonsil develops; its normal remnant is the tonsillar sinus or fossa.

Branchial Fistula

An abnormal canal that opens internally into the tonsillar sinus and externally on the side of the neck is a *branchial fistula*. This rare anomaly results from persistence of parts of the second pharyngeal groove and second pharyngeal pouch (see Figs. 11 - 10*E* and *F* and 11 - 11). The fistula ascends from its opening in the neck through the subcutaneous tissue and platysma muscle to reach the tonsillar sinus.

Branchial Cysts

The third and fourth pharyngeal arches are buried in the *cervical sinus* (see Fig. 11 - 10*B*). Remnants of parts of the cervical sinus and/or the second pharyngeal groove may persist and form a spherical or elongate cyst (see Fig. 11 - 10*F*). Branchial cysts often do not become apparent until late childhood or early adulthood, when they produce a slowly enlarging, painless swelling in the neck (Fig. 11 - 12). The cysts enlarge because of the accumulation of fluid and cellular debris derived from desquamation of their epithelial linings (Fig. 11 - 13).

Branchial Vestiges

Normally, the pharyngeal cartilages disappear, except for parts that form ligaments or bones; however, in unusual cases cartilaginous or bony remnants of pharyngeal arch cartilages appear under the skin in the side of the neck. These are usually found anterior to the inferior third of the sternocleidomastoid muscle (see Fig. 11 - 10*F*).

First Arch Syndrome

Abnormal development of the first pharyngeal arch results in various congenital anomalies of the eyes, ears, mandible, and palate that together constitute the first arch syndrome (Fig. 11 - 14). This syndrome is believed to result from insufficient migration of neural crest cells into the first arch during the fourth week. *There are two main manifestations of the first arch syndrome*:

(continues on page 165)

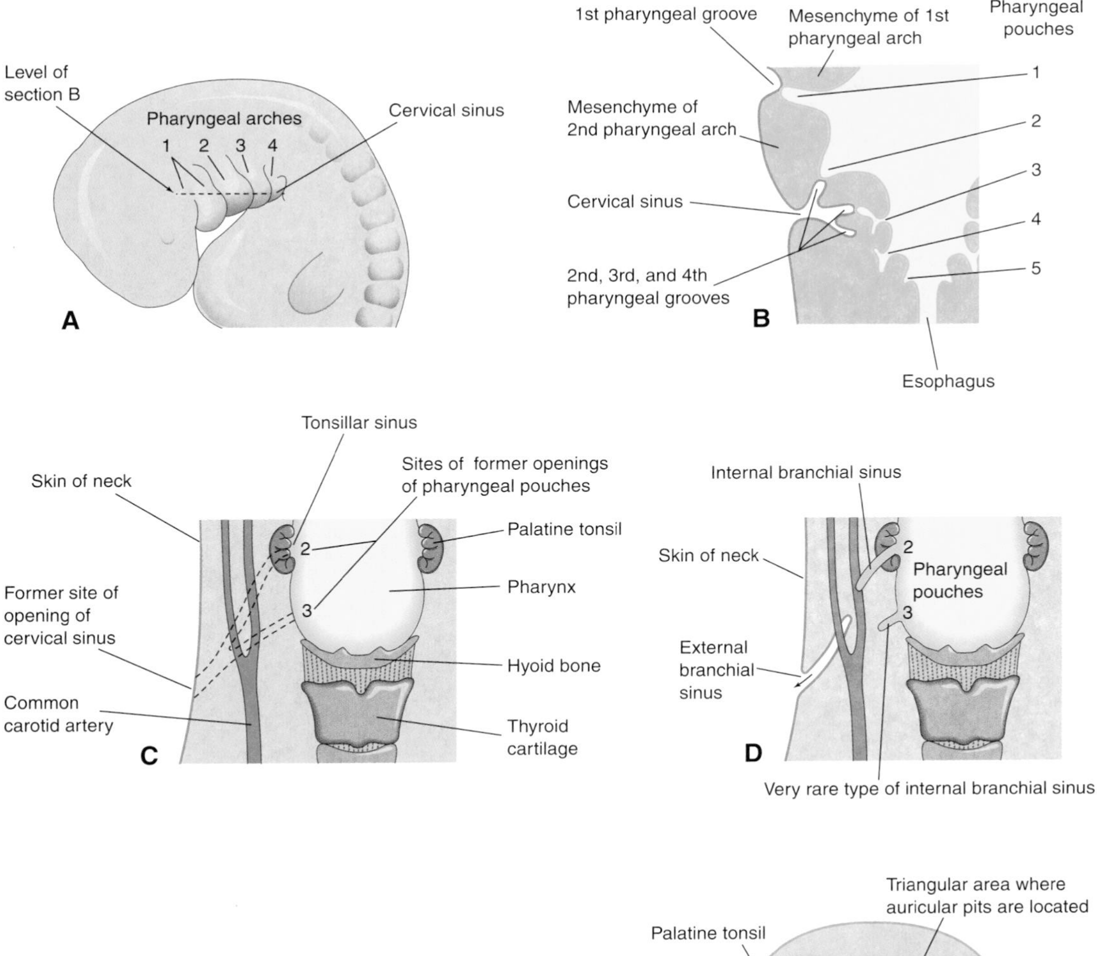

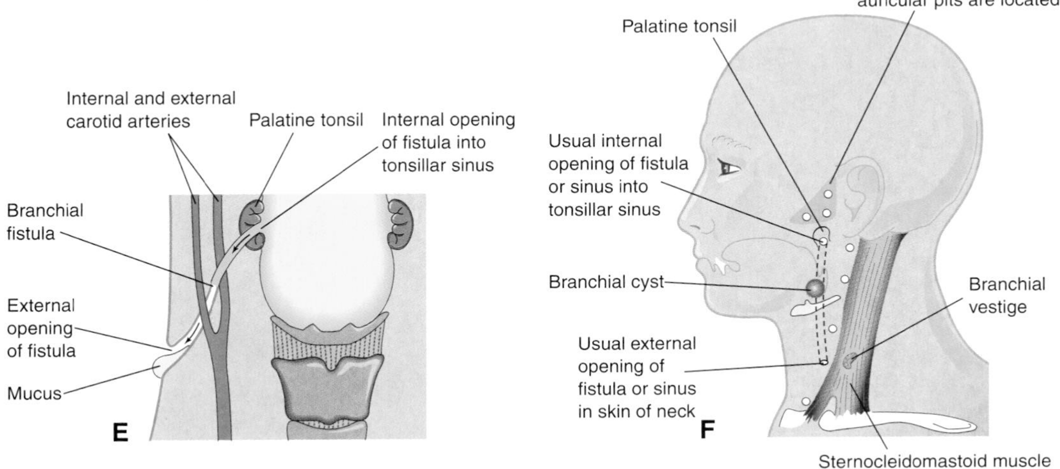

Figure 11 – 10. *A*, Lateral view of the head, neck, and thoracic regions of a 5-week embryo, showing the cervical sinus that is normally present at this stage. *B*, Horizontal section of the embryo, at the level shown in *A*, illustrating the relationship of the cervical sinus to the pharyngeal arches and pouches. *C*, Diagrammatic sketch of the adult pharyngeal and neck regions, indicating the former sites of openings of the cervical sinus and pharyngeal pouches. The *broken lines* indicate possible courses of branchial fistulas. *D*, Similar sketch showing the embryological basis for various types of branchial sinuses. *E*, Drawing of a branchial fistula resulting from persistence of parts of the second pharyngeal groove and second pharyngeal pouch. *F*, Sketch showing possible sites of branchial cysts and openings of branchial sinuses and fistulas. A branchial vestige is also illustrated.

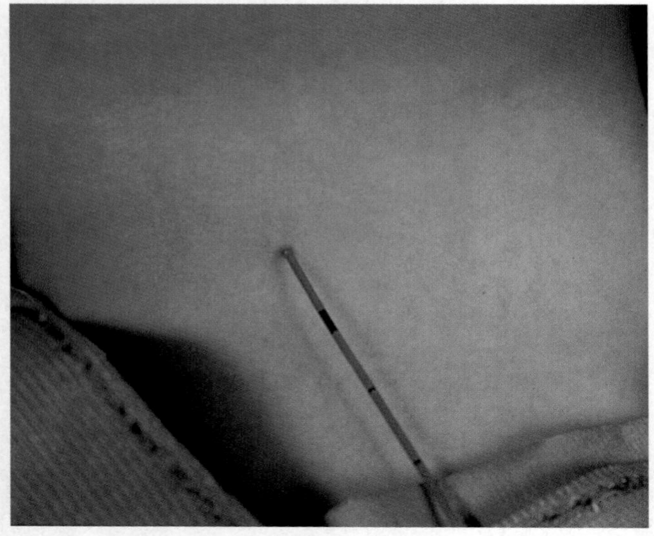

A

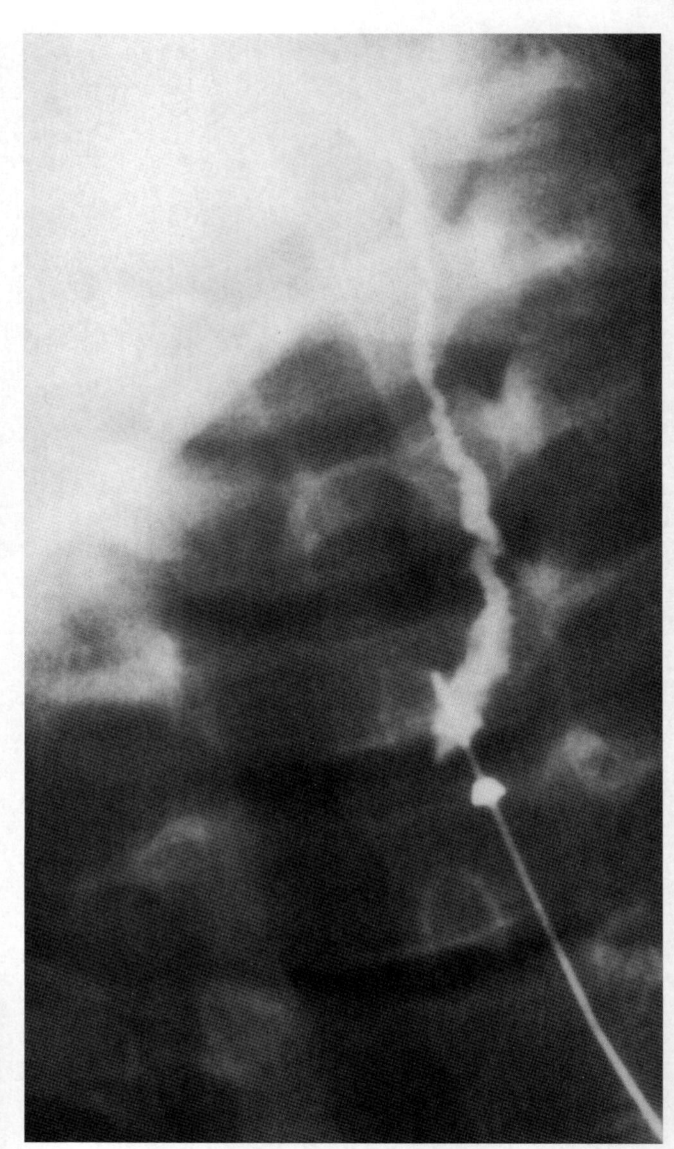

B

Figure 11 - 11. *A*, Photograph of a child's neck showing a catheter inserted into the external opening of a branchial sinus. The catheter allows definition of the length of the tract, which facilitates surgical excision. *B*, A fistulogram of a complete branchial fistula. The radiograph was taken after injection of a contrast medium to show the course of the fistula through the neck. (Courtesy of Dr. Pierre Soucy, Division of Paediatric Surgery, Children's Hospital of Eastern Ontario, Ottawa, Canada.)

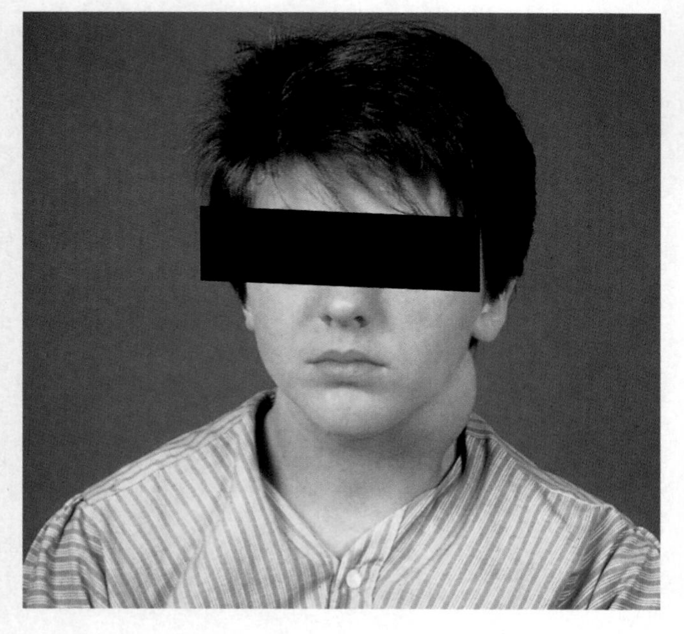

Figure 11 - 12. Photograph of a boy showing the swelling in the neck produced by a branchial cyst. Branchial cysts often lie free in the neck just inferior to the angle of the mandible, or they may develop anywhere along the anterior border of the sternocleidomastoid muscle. (Photo courtesy of Dr. Pierre Soucy, Division of Paediatric Surgery, Children's Hospital of Eastern Ontario, Ottawa, Canada.)

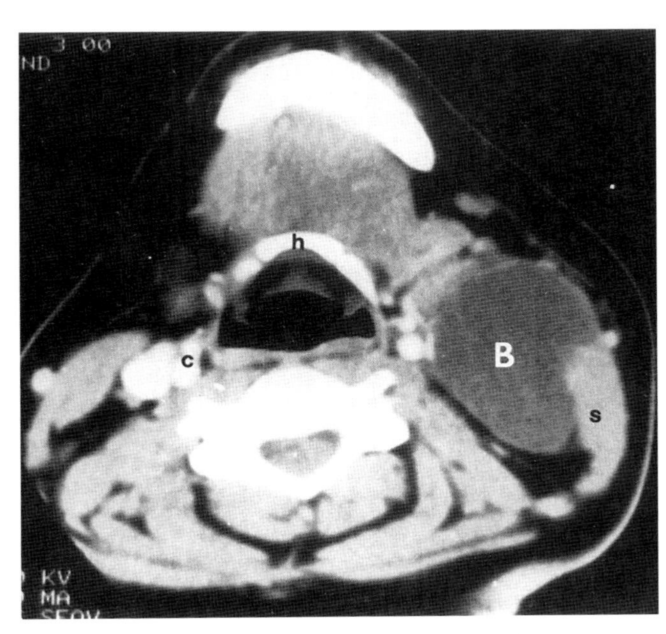

Figure 11 - 13. Branchial (cleft) cyst (B) demonstrated by computed tomography (CT) of the neck region of a woman who presented with a "lump" in the neck, similar to that shown in Figure 11 – 12. The low-density cyst is anterior to the right sternocleidomastoid muscle(*s*) at the level of the hyoid bone (*h*). The normal appearance of the carotid sheath (*c*) is shown for comparison with the compressed sheath on the right side. (From McNab T, McLennan MK, Margolis M: Radiology Rounds. *Can Fam Physician* 41:1673, 1995.)

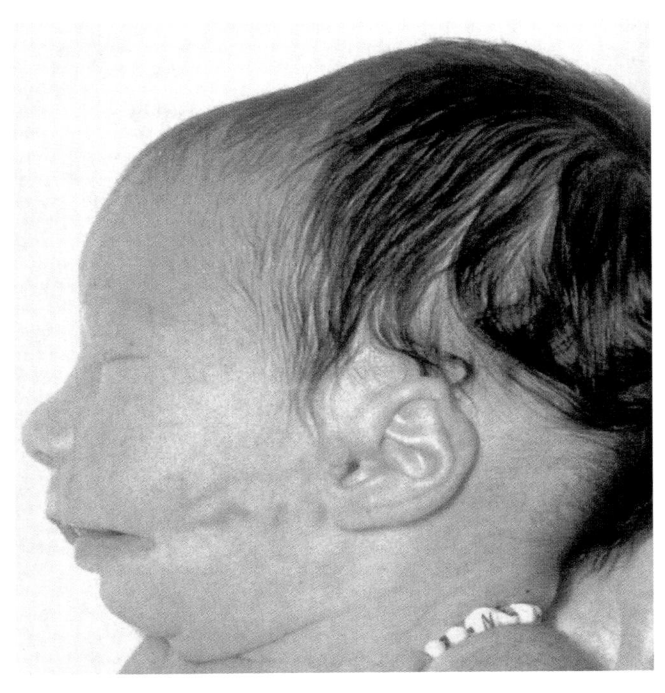

Figure 11 - 14. Photograph of an infant with the first arch syndrome, a pattern of anomalies resulting from insufficient migration of the neural crest cells into the first pharyngeal arch. Note the following characteristics: deformed auricle of the external ear, preauricular appendage, defect in the cheek between the auricle and the mouth, hypoplasia of the mandible, and macrostomia (large mouth).

- The **Treacher Collins syndrome** (mandibulofacial dysostosis), caused by an autosomal dominant gene, causes malar hypoplasia (underdevelopment of the zygomatic bones of the face). Characteristic features of the syndrome include downslanting palpebral fissures, defects of the lower eyelids, deformed external ears, and sometimes, abnormalities of the middle and internal ears.
- The **Pierre Robin syndrome** consists of hypoplasia of the mandible, cleft palate, and defects of the eye and ear. Many cases of this syndrome are sporadic; however, some appear to have a genetic basis. In the *Robin morphogenetic complex*, the initiating defect is a small mandible (micrognathia), which results in posterior displacement of the tongue and obstruction to full closure of the palatine processes, resulting in a bilateral cleft palate.

DiGeorge Syndrome: Congenital Thymic Aplasia and Absence of the Parathyroid Glands

Infants with the DiGeorge syndrome are born without a thymus and parathyroid glands although in some cases, ectopic glandular tissue has been found. The disease is characterized by *congenital hypoparathyroidism*, increased susceptibility to infections, anomalies of the mouth (shortened philtrum of the lip [fish-mouth deformity]), low-set notched ears, nasal clefts, *thyroid hypoplasia*, and cardiac abnormalities (defects of the arch of the aorta and heart). The *DiGeorge syndrome* occurs because the third and fourth pharyngeal pouches fail to differentiate into the thymus and parathyroid glands. The facial abnormalities result primarily from abnormal development of the first arch components during formation of the face and ears. The DiGeorge syndrome involves a microdeletion (22q11.2 region), *Hox* genes mutation, and neural crest cell defects.

Ectopic Parathyroid Glands

The parathyroids are highly variable in number and location. They may be found anywhere near or within the thyroid gland or thymus (Fig. 11 – 15). The superior glands are more constant in position than the inferior ones. Occasionally, an inferior parathyroid gland fails to descend and remains near the bifurcation of the common carotid artery. In other cases, it may accompany the thymus into the thorax.

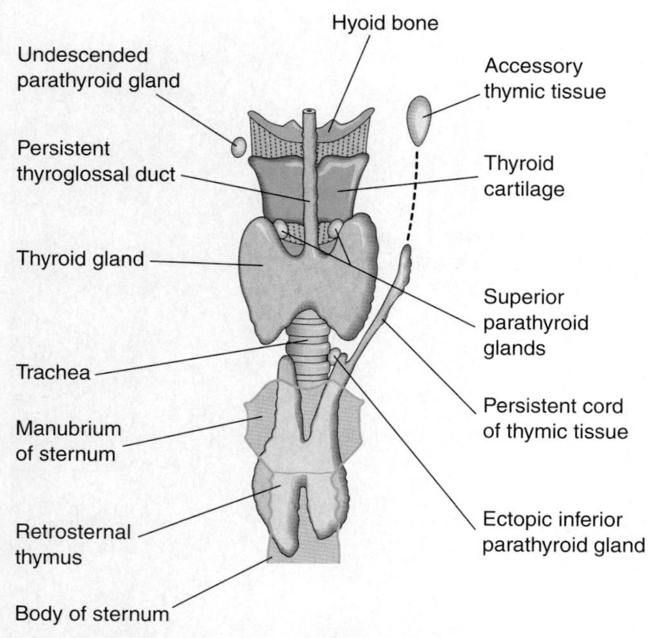

Figure 11 - 15. Anterior view of the thyroid gland, thymus, and parathyroid glands illustrating various possible congenital anomalies.

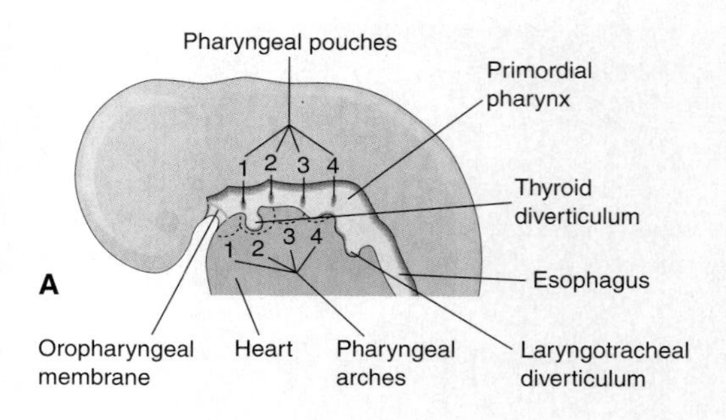

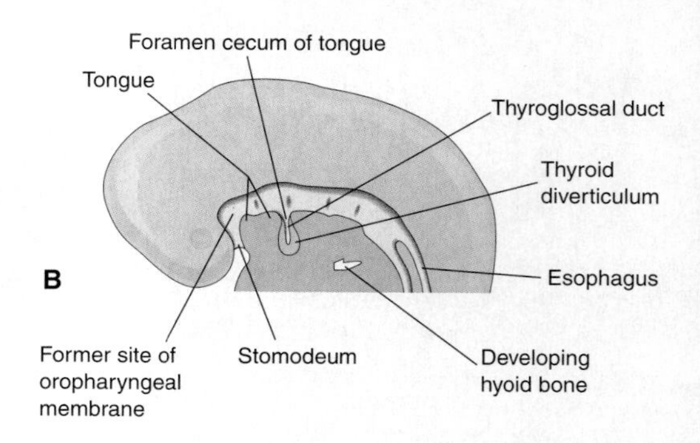

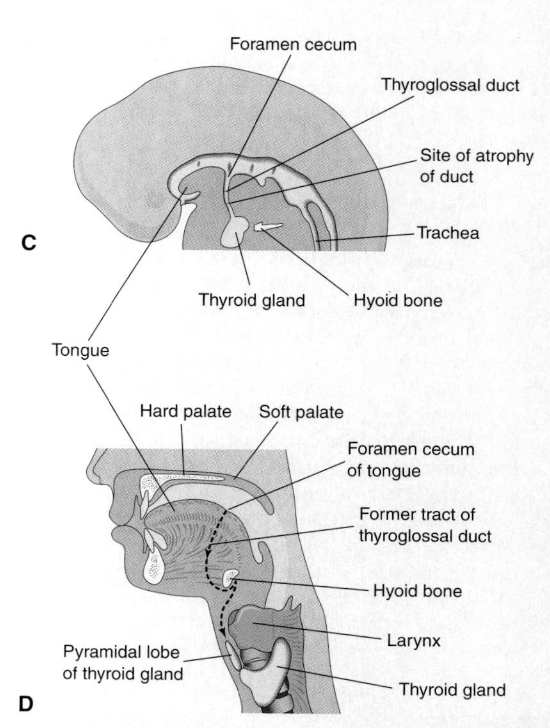

Figure 11 - 16. Development of the thyroid gland. *A-C*, Schematic sagittal sections of the head and neck regions of 4-week, 5-week, and 6-week embryos, illustrating successive stages in the development of the thyroid gland. *D*, Similar section of an adult head and neck, showing the path taken by the thyroid gland during its embryonic descent (indicated by the former tract of the thyroglossal duct).

Abnormal Number of Parathyroid Glands

In unusual cases, there may be more than four parathyroid glands. Supernumerary parathyroid glands probably result from division of the primordia of the original glands. Absence of a parathyroid gland results from failure of one of the primordia to differentiate or atrophy of a gland early in development.

Development of the Thyroid Gland

The thyroid gland is the first endocrine gland to develop in the embryo. It begins to form about 24 days after fertilization from a median endodermal thickening in the floor of the primordial pharynx (Fig. 11 - 16*A*). This thickening soon forms a small outpouching known as the **thyroid diverticulum** (primordium). As the embryo and tongue grow, the developing thyroid gland descends in the neck, passing ventral to the developing hyoid bone and laryngeal cartilages. For a short time, the developing thyroid gland is connected to the tongue by a narrow tube, the **thyroglossal duct** (see Fig. 11 - 16*B* and *C*). As a result of rapid cell proliferation, the lumen of the thyroid primordium soon obliterates. The solid thyroid primordium then divides into right and left lobes that are connected by the *isthmus of the thyroid gland.* By 7 weeks, the thyroid gland has assumed its definitive shape and has usually reached its final site in the neck (see Fig. 11 – 16*D*). By this time, the thyroglossal duct has usually degenerated and disappeared. The proximal opening of the thyroglossal duct persists as a small blind pit, the **foramen cecum of the tongue**. A pyramidal lobe extends superiorly from the isthmus in about 50% of individuals. The **pyramidal lobe** may be attached to the hyoid bone by fibrous and/or smooth muscle. A pyramidal lobe and the associated smooth muscle represent a persistent part of the distal end of the thyroglossal duct.

At about 11 weeks of development, colloid begins to appear in the **thyroid follicles**; thereafter, iodine concentration and the synthesis of *thyroid hormones* can be demonstrated. *Molecular studies* have shown that expression of thyroid transcription factors *TTF-1* and *TTF-2*, as well as *Pax-8* and *Hox3*, are essential for thyroid morphogenesis.

Thyroglossal Duct Cysts and Sinuses

Cysts may form anywhere along the course followed by the thyroglossal duct during descent of the primordial thyroid gland from the tongue (Fig. 11 – 17*A* and *B*). Normally, the thyroglossal duct atrophies and disappears, but a remnant of it may persist and form a cyst in the tongue or in the anterior part of the neck, usually just inferior to the hyoid bone (Fig. 11 – 18). The swelling produced by a *thyroglossal duct cyst* usually develops as a painless, progressively enlarging, movable median mass (Fig. 11 – 19).

The cyst may contain some thyroid tissue. Following infection of a cyst, a perforation of the skin occurs in some cases, forming a **thyroglossal duct sinus** that usually opens in the median plane of the neck, anterior to the laryngeal cartilages (see Fig. 11 – 17*A*).

Ectopic Thyroid Gland

An ectopic thyroid gland is an infrequent congenital anomaly that is usually located along the normal route of its descent from the tongue (see Fig. 11 – 16*C*). **Lingual thyroid glandular tissue** is the most common type of ectopic thyroid tissue. Incomplete descent of the thyroid gland results in a **sublingual thyroid gland** appearing high in the neck, at or just inferior to the hyoid bone (Figs. 11 – 20 and 11 – 21). As a rule, an ectopic sublingual thyroid gland is the only thyroid tissue present. It is clinically important to differentiate an ectopic thyroid gland from a thyroglossal duct cyst or accessory thyroid tissue in order to prevent *inadvertent surgical removal of the thyroid gland* because this may be the only thyroid tissue present. Failure to recognize the thyroid gland may leave the person permanently dependent on thyroid medication.

Development of the Tongue

Near the end of the fourth week, a median triangular elevation appears in the floor of the primordial pharynx,

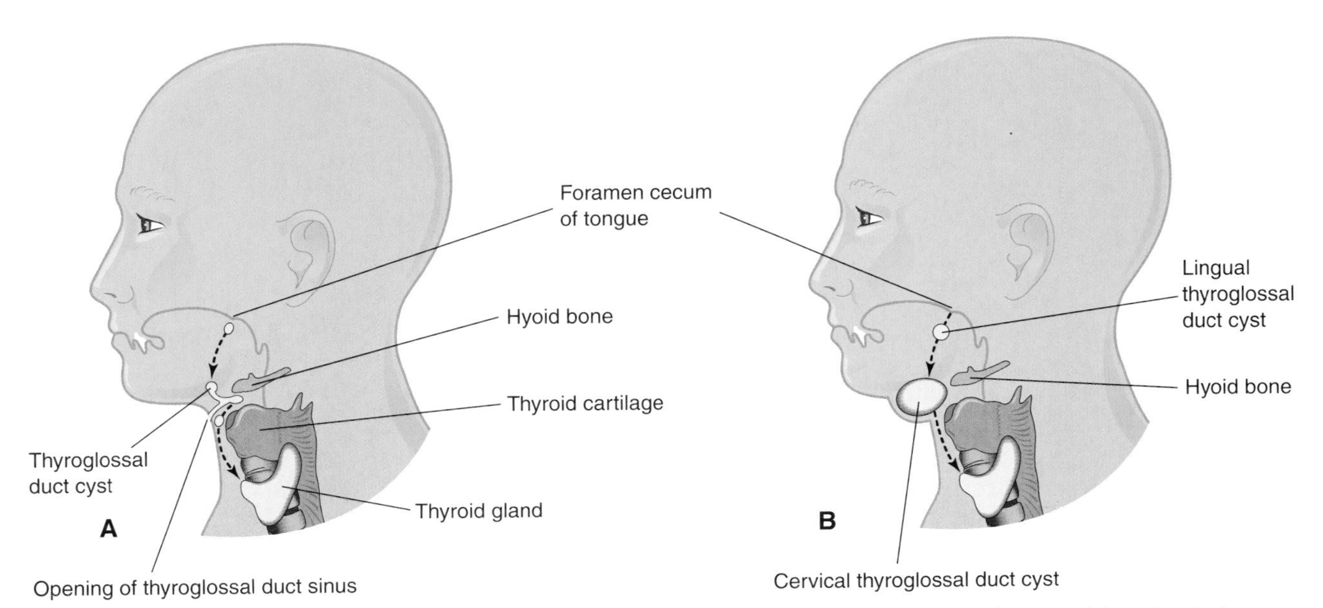

Figure 11 - 17. *A*, Sketch of the head and neck showing the possible locations of thyroglossal duct cysts. A thyroglossal duct sinus is also illustrated. The *broken line* indicates the course taken by the thyroglossal duct during descent of the developing thyroid gland from the foramen cecum to its final position in the anterior part of the neck. *B*, Similar sketch illustrating lingual and cervical thyroglossal duct cysts. Most thyroglossal duct cysts are located just inferior to the hyoid bone.

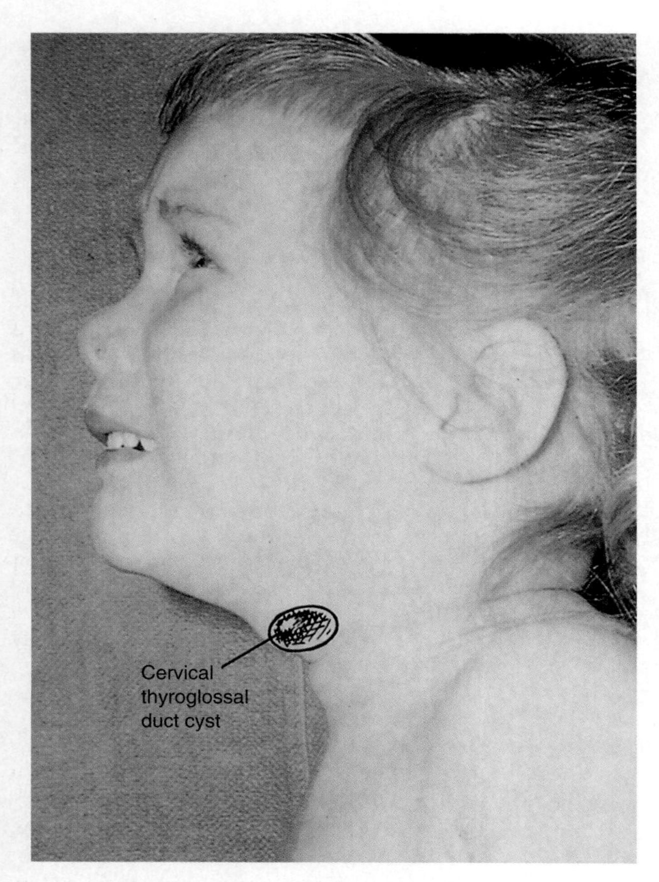

Figure 11 - 18. Typical thyroglossal duct cyst in a female child. The round, firm mass (indicated by the sketch) produced a swelling in the median plane of the neck just inferior to the hyoid bone.

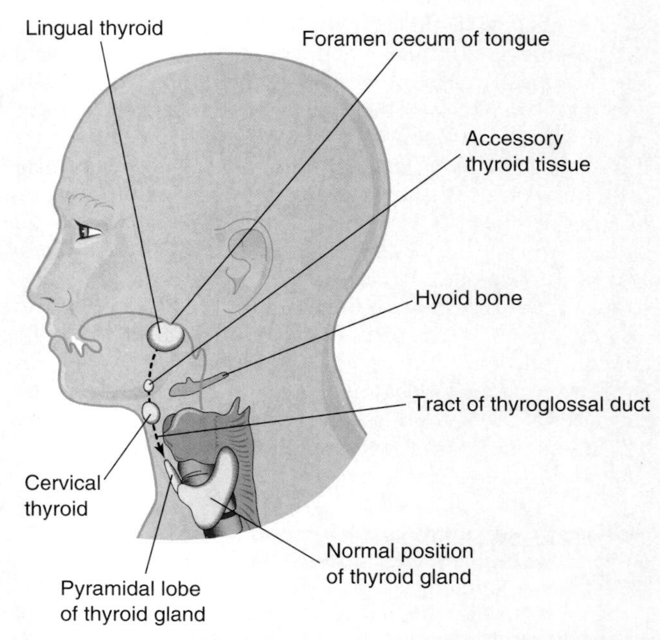

Figure 11 - 20. Sketch of the head and neck showing the usual sites of ectopic thyroid tissue. The *broken line* indicates the path followed by the thyroid gland during its descent, as well as the former tract of the thyroglossal duct.

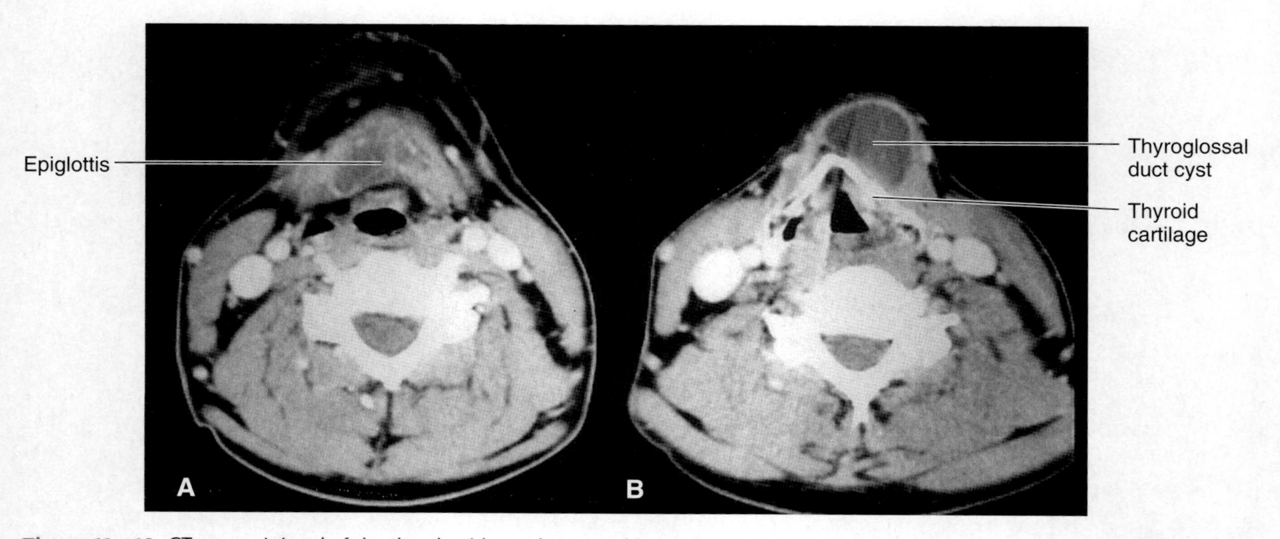

Figure 11 - 19. CT scans. *A*, Level of the thyrohyoid membrane and base of the epiglottis. *B*, Level of the thyroid cartilage, which is calcified. The thyroglossal duct cyst extends cranially to the margin of the hyoid bone. (Courtesy of Dr. Gerald S. Smyser, Altru Health System, Grand Forks, ND.)

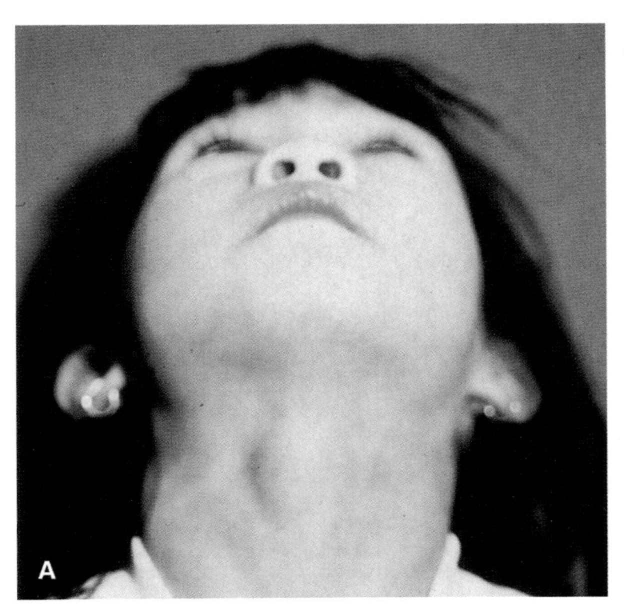

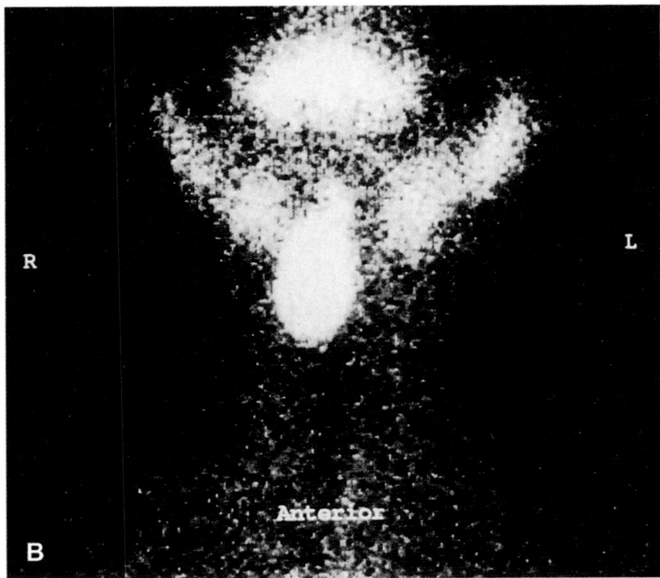

Figure 11 - 21. *A*, Photograph of a sublingual thyroid mass in a 5-year-old girl. *B*, Technetium-99m pertechnetate scan showing a sublingual thyroid gland without evidence of functioning thyroid tissue in the lower neck. (From Leung AKC, Wong AL, Robson WLLM: Ectopic thyroid gland simulating a thyroglossal duct cyst: A case report. Can J Surg 38:87, 1995.)

just rostral to the foramen cecum (Fig. 11 - 22*A*). This swelling — the **median tongue bud** (tuberculum impar) — is the first indication of tongue development. Soon, two oval **distal tongue buds** (lateral lingual swellings) develop on each side of the median tongue bud. The three lingual buds result from the proliferation of mesenchyme in ventromedial parts of the first pair of pharyngeal arches. The distal tongue buds rapidly increase in size, merge with each other, and overgrow the median tongue bud. *The merged distal tongue buds form the anterior two thirds (oral part) of the tongue* (see Fig. 11 - 22*C*). The plane of fusion of the distal tongue buds is indicated superficially by the midline groove of the tongue and internally by the fibrous *lingual septum*. The median tongue bud forms no recognizable part of the adult tongue.

Formation of the posterior third (pharyngeal part) of the tongue is indicated by two elevations that develop caudal to the foramen cecum (see Fig. 11 - 22*A*):

- The *copula* (L., bond, tie) forms by fusion of the ventromedial parts of the second pair of pharyngeal arches.
- The *hypopharyngeal eminence* develops caudal to the copula from mesenchyme in the ventromedial parts of the third and fourth pairs of arches.

As the tongue develops, the copula is gradually overgrown by the hypopharyngeal eminence and disappears (see Fig. 11 - 22*B* and *C*). As a result, the pharyngeal part of the tongue develops from the rostral part of the hypopharyngeal eminence, a derivative of the third pair of pharyngeal arches. The line of fusion of the anterior and posterior parts of the tongue is roughly indicated by a V-shaped groove called the **terminal groove**, or sulcus (see Fig. 11 - 22*C*). Pharyngeal arch mesenchyme forms the connective tissue and vasculature of the tongue. The *tongue muscles* are derived from myoblasts that migrate from the paraxial mesoderm of the occipital somites (see Fig. 11 - 6*A*). The *hypoglossal nerve* (CN XII) accompanies the myoblasts during their migration and innervates the tongue muscles as they develop.

Papillae and Taste Buds of the Tongue

The **lingual papillae** appear by the end of the eighth week of development. The *vallate* and *foliate papillae* appear first, close to terminal branches of the glossopharyngeal nerve. The *fungiform papillae* appear later near terminations of the chorda tympani branch of the facial nerve. *Filiform papillae*, the most common lingual papillae, develop during the early fetal period (10 to 11 weeks). They contain afferent nerve endings that are *sensitive to touch*. **Taste buds** develop during weeks 11 to 13 by inductive interaction between the epithelial cells of the tongue and invading gustatory nerve cells from the chorda tympani, glossopharyngeal, and vagus nerves. Fetal responses in the face can be induced by bitter-tasting substances at 26 to 28 weeks,

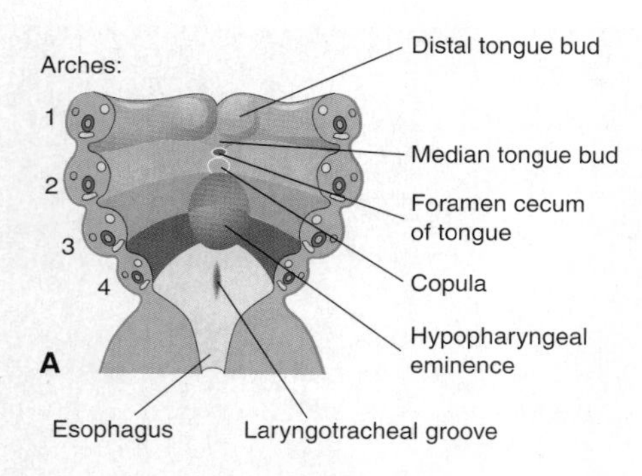

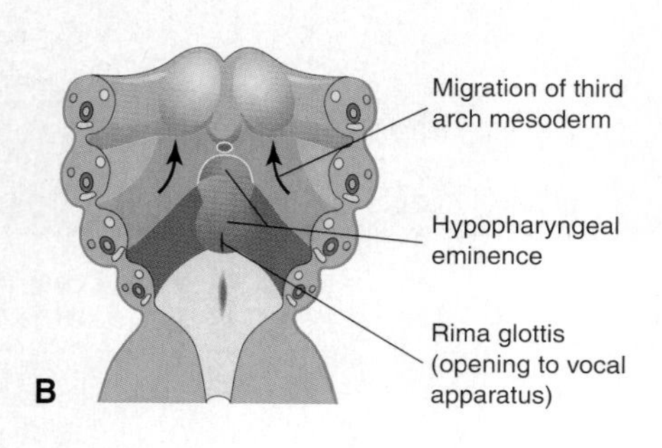

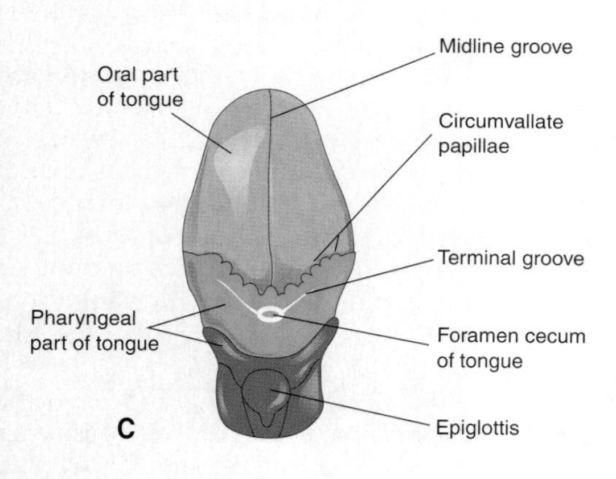

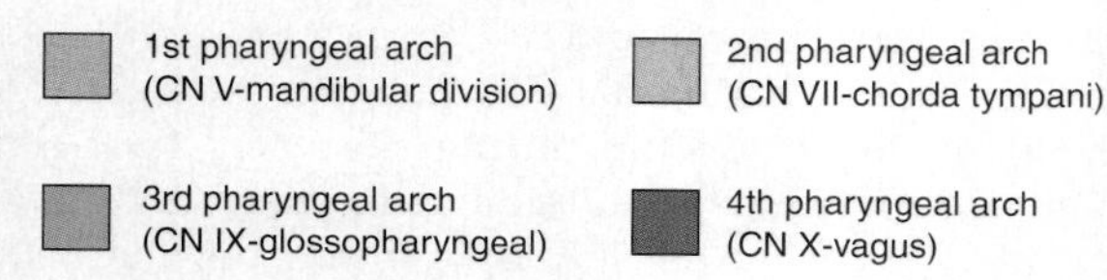

Figure 11 – 22. *A* and *B*, Schematic horizontal sections through the pharynx at the level shown in Figure 11 – 4*A*, showing successive stages in the development of the tongue during the fourth and fifth weeks. *C*, Drawing of the adult tongue showing the pharyngeal arch derivation of the nerve supply of its mucosa.

indicating that reflex pathways between taste buds and facial muscles are established by this stage.

Nerve Supply of the Tongue

The development of the tongue explains its nerve supply. The sensory supply to the mucosa of almost the entire *anterior two thirds of the tongue* (oral part) is from the lingual branch of the mandibular division of the **trigeminal nerve** (CN V), the nerve of the first pharyngeal arch (see Fig. 11 – 22). Although the facial nerve is the nerve of the second pharyngeal arch, its chorda tympani branch supplies the taste buds in the anterior two thirds of the tongue, except for the vallate papillae. Because the second arch component, the copula, is overgrown by the third arch, the facial nerve does not supply any of the tongue mucosa, except for the taste buds in the oral part of the tongue. The *vallate papillae* in the oral part of the tongue are innervated by the *glossopharyngeal nerve* of the third pharyngeal arch (see Fig. 11 – 22*C*). The reason usually given for this is that the mucosa of the posterior third of the tongue is pulled slightly anteriorly as the tongue develops. The *posterior third of the tongue* (pharyngeal part) is innervated mainly by the **glossopharyngeal nerve** (CN IX). The superior laryngeal branch of the vagus nerve of the fourth arch supplies a small area of the tongue anterior to the epiglottis (see Fig. 11 – 22*C*). All **muscles of the tongue** are supplied by the **hypoglossal nerve** (CN XII), except for the palatoglossus, which is supplied from the pharyngeal plexus by fibers arising from the *vagus nerve*.

Congenital Anomalies of Tongue

Abnormalities of the tongue are uncommon, except for fissuring of the tongue and hypertrophy of the lingual papillae, which are characteristics of infants with Down syndrome (see Chapter 9).

Congenital Lingual Cysts and Fistulas

Cysts in the tongue may be derived from remnants of the thyroglossal duct (see Fig. 11 – 17). They may enlarge and produce symptoms of pharyngeal discomfort and/or *dysphagia* (difficulty in swallowing). Fistulas may also arise as a result of persistence of lingual parts of the thyroglossal duct; such fistulas open through the *foramen cecum* into the oral cavity.

Ankyloglossia (Tongue-Tie)

The lingual frenulum normally connects the inferior surface of the tongue to the floor of the mouth (Fig. 11 – 23). Tongue-tie (ankyloglossia) occurs in about 1 in 300 North American infants, but is usually of no functional significance. A short frenulum usually stretches with time, making surgical correction of the anomaly unnecessary.

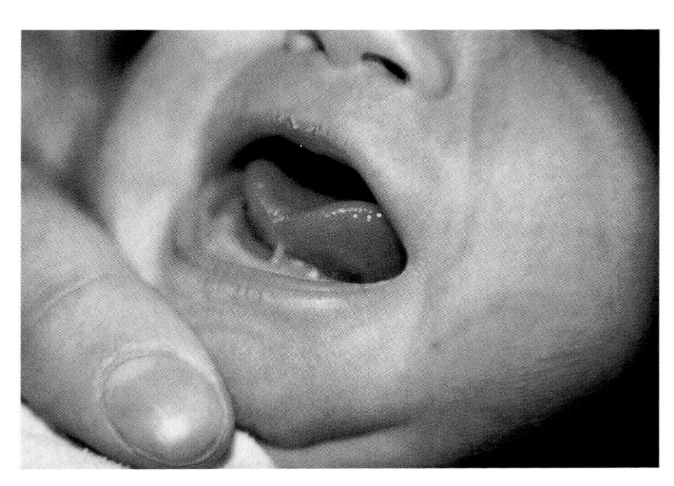

Figure 11 - 23. Photograph of an infant with ankyloglossia or tongue-tie. (Courtesy of Dr. Evelyn Jain, Lakeview Breastfeeding Clinic, Calgary, Alberta, Canada.) Note the short frenulum, which extends to the tip of the tongue. Tongue-tie interferes with protrusion of the tongue and may make breastfeeding difficult.

Development of the Salivary Glands

During the sixth and seventh weeks of development, the salivary glands begin as solid epithelial buds from the primordial oral cavity (see Fig. 11 - 7*C*). The club-shaped ends of these epithelial buds grow into the underlying mesenchyme. The connective tissue in the glands is derived from neural crest cells. All parenchymal (secretory) tissue arises by proliferation of the oral epithelium.

The **parotid glands** are the first to appear (early in the sixth week). They develop from buds that arise from the oral ectodermal lining near the angles of the stomodeum. The buds grow toward the ears and branch to form solid cords with rounded ends. Later, the cords canalize — (i.e., develop lumina) — and become ducts by about 10 weeks. The rounded ends of the cords differentiate into acini. Secretions commence at 18 weeks. The capsule and connective tissue develop from the surrounding mesenchyme.

The **submandibular glands** appear late in the sixth week of development. They develop from endodermal buds in the floor of the stomodeum. Solid cellular processes grow posteriorly, lateral to the developing tongue. Later, they branch and differentiate. Acini begin to form at 12 weeks, and secretory activity begins at 16 weeks. Growth of the submandibular glands continues after birth with the formation of mucous acini. Lateral to the tongue, a linear groove forms that soon closes over to form the *submandibular duct*.

The **sublingual glands** appear in the eighth week, about 2 weeks later than the other salivary glands. They develop from multiple endodermal epithelial buds in the paralingual sulcus (see Fig. 11 - 7*C*). These buds branch and canalize to form 10 to 12 ducts that open independently into the floor of the mouth.

Development of the Face

The facial primordia begin to appear early in the fourth week of development around the large **primordial stomodeum** (Fig. 11 - 24*A* and *B*). Facial development depends upon the inductive influence of the prosencephalic and rhombencephalic organizing centers. The **prosencephalic organizing center**, derived from prechordal mesoderm that migrates from the primitive streak, is located rostral to the notochord and ventral to the prosencephalon or forebrain (see Chapter 19). The **rhombencephalic organizing center** is ventral to the rhombencephalon (hindbrain). The **five facial primordia**, which appear as prominences around the stomodeum, are:

- the single frontonasal prominence
- the paired maxillary prominences
- the paired mandibular prominences

The paired prominences are derivatives of the first pair of pharyngeal arches. The prominences are produced by mesenchyme derived from **neural crest cells** that migrate from the first two rhombomeres (r1 and r2) of the hindbrain into the arches during the fourth week of development. These cells are the major source of connective tissue components, including cartilage, bone, and ligaments in the facial and oral regions. The **frontonasal prominence** (FNP) surrounds the ventrolateral part of the forebrain, which gives rise to the *optic vesicles* that form the eyes (see Figs. 11 - 24*A* to *C* and 11 - 25). The frontal part of the FNP forms the forehead; the nasal part of the FNP forms the rostral boundary of the stomodeum and nose. The **maxillary prominences** form the lateral boundaries of the stomodeum, whereas the **mandibular prominences** constitute the caudal boundary of the primordial mouth (see Figs. 11 - 24 and 11 - 25). The five facial prominences are active **centers of growth** in the underlying mesenchyme. This embryonic connective tissue is continuous from one prominence to the other. Facial development occurs mainly between the fourth and eighth weeks (see Fig. 11 - 24*A* to *G*). By the end of the embryonic period, the face has an unquestionably human appearance. Facial proportions also are established during the fetal period (see Fig. 11 - 24*H* and *I*). The lower jaw and lower lip are the first parts of the face to form. They result from merging of the medial ends of the mandibular prominences in the median plane.

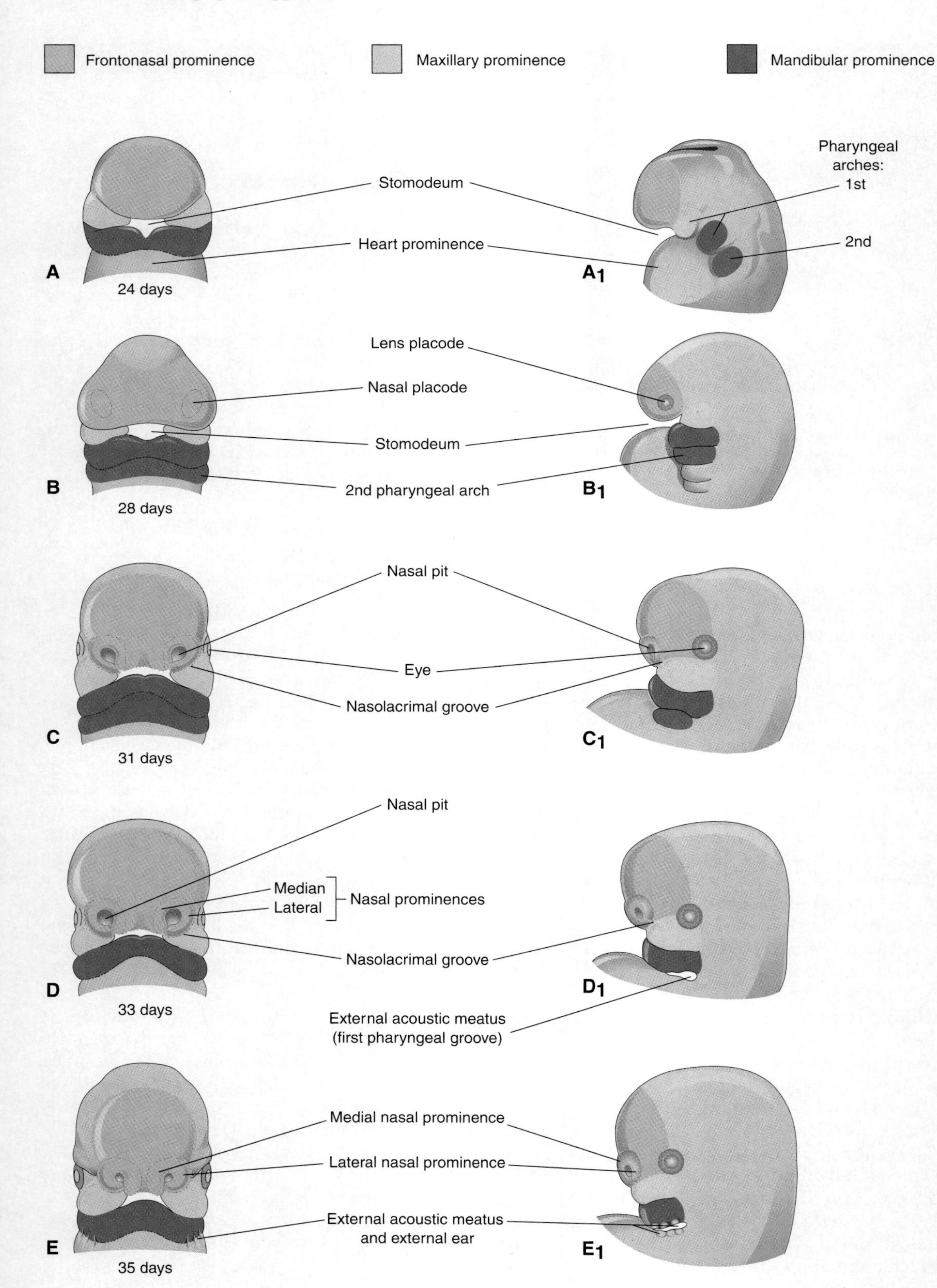

Figure 11 - 24. Diagrams illustrating progressive stages in the development of the human face.

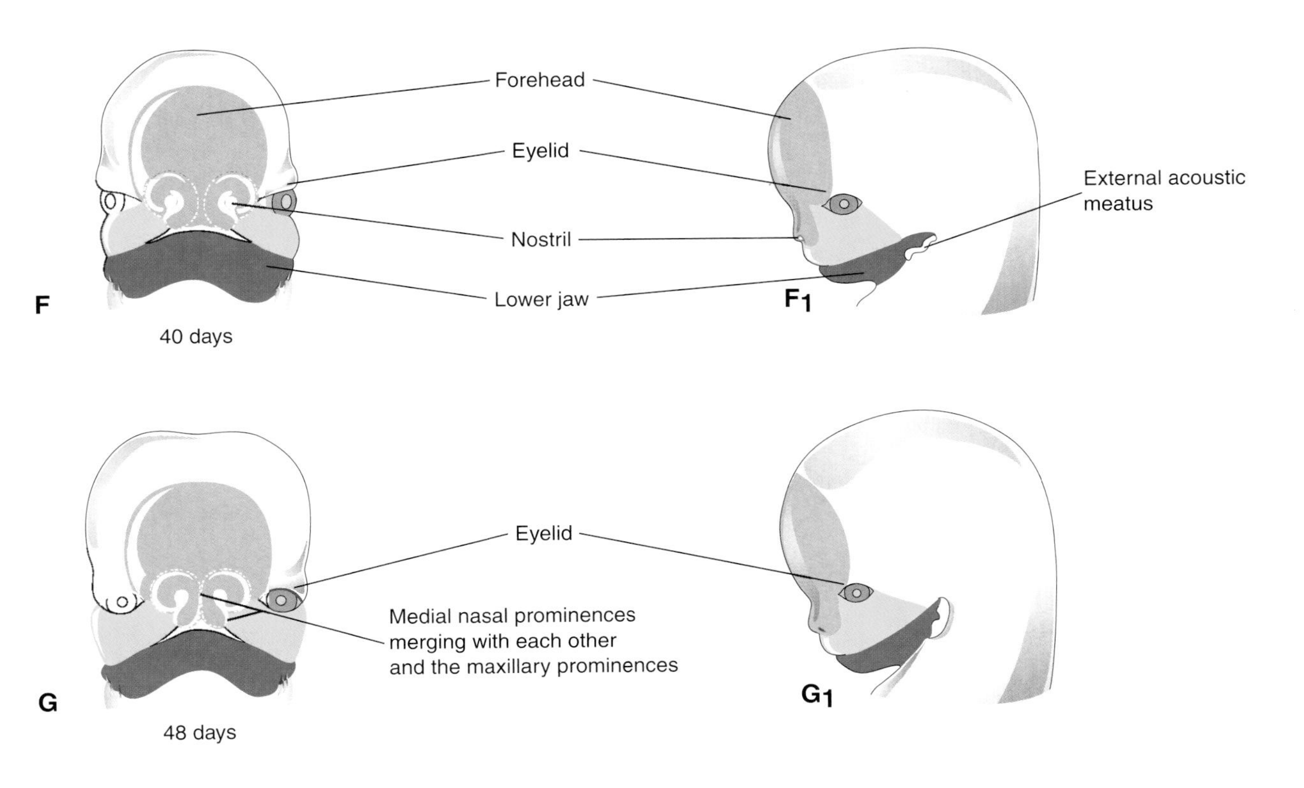

Figure 11 – 24. Continued

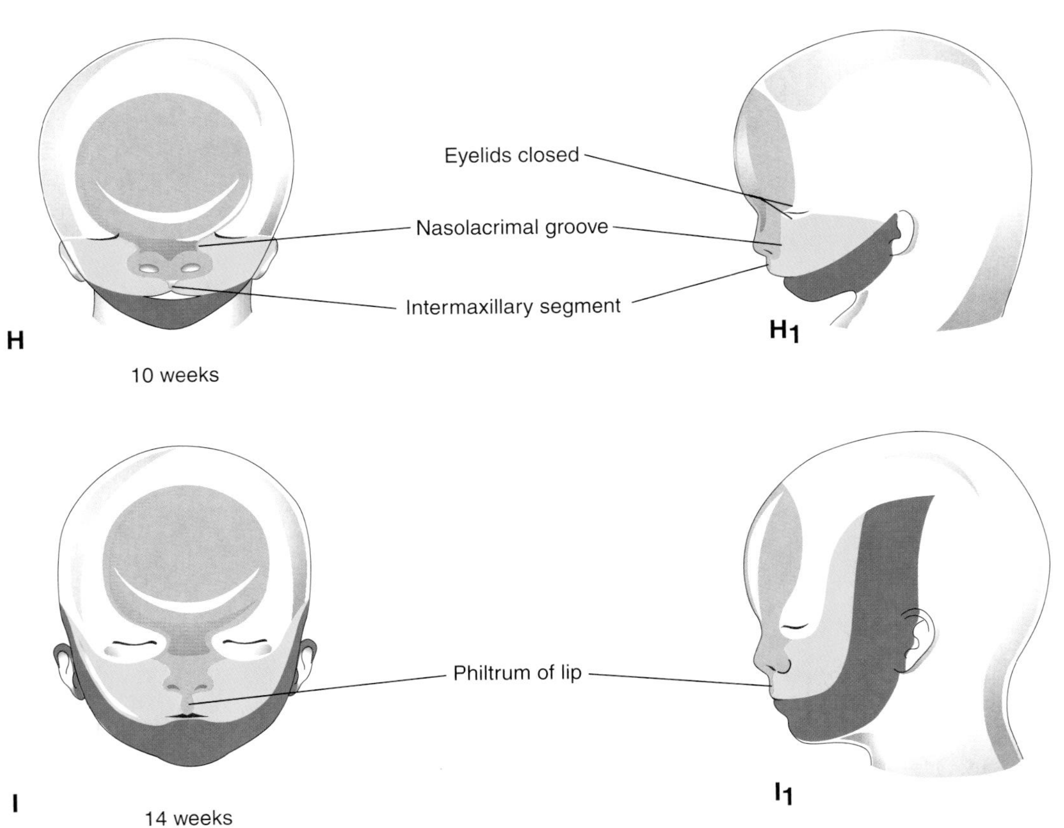

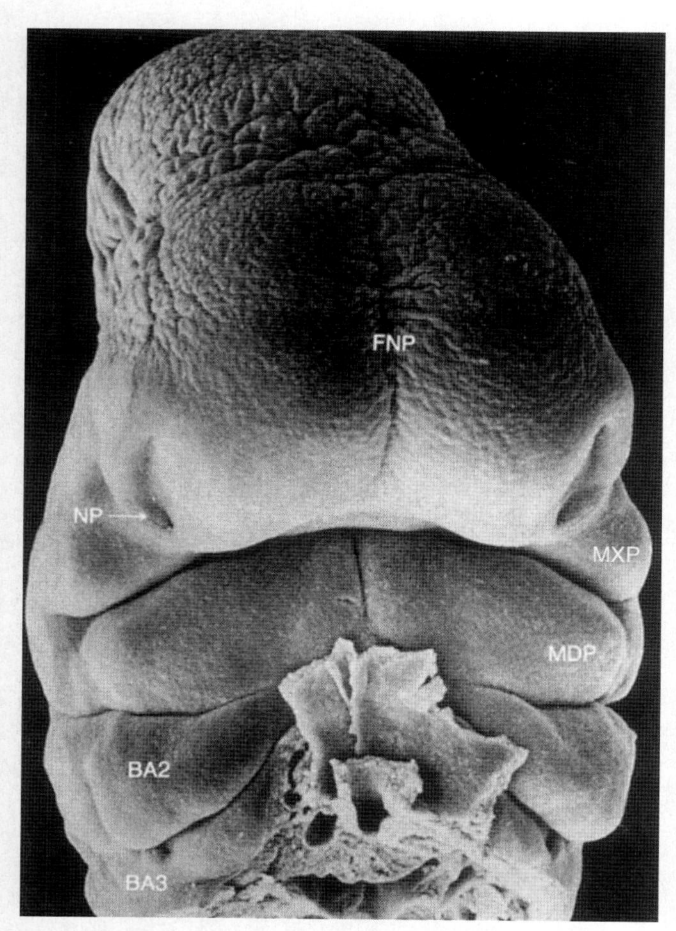

Figure 11 - 25. Scanning electron micrograph; ventral view of a human embryo at about 33 days (stage 15, crown-rump length [CRL] of 8 mm). Observe the prominent frontonasal prominence (FNP) surrounding the telencephalon (forebrain). Also observe the nasal pits (NP) located in the ventrolateral regions of the frontonasal prominence. Medial and lateral nasal prominences surround these pits. The cuneiform, wedge-shaped maxillary prominences (MXP) form the lateral boundaries of the stomodeum. The fusing mandibular prominences (MDP) are located just caudal to the stomodeum. The second pharyngeal (branchial) arch (BA2) is clearly visible and shows overhanging margins (opercula). The third pharyngeal (branchial) arch (BA3) is also clearly visible. (From Hinrichsen K: The early development of morphology and patterns of the face in the human embryo. *Adv Anat Embryol Cell Biol* 98:1, 1985.)

By the end of the fourth week of development, bilateral oval thickenings of the surface ectoderm — **nasal placodes** (primordia of the nose and nasal cavities) — have developed on the inferolateral parts of the frontonasal prominence (see Figs. 11 - 25 and 11 - 26*A* and *B*). Initially, these placodes are convex, but later, they are stretched to produce a flat depression in each placode. Mesenchyme in the margins of the placodes proliferates, producing horseshoe-shaped elevations, the **medial and lateral nasal prominences** (see Fig. 11 - 26*D* and *E*). As a result, the nasal placodes lie in depressions, called **nasal pits** (see Fig. 11 - 26*C* and *D*). These pits are the primordia of the **anterior nares** (nostrils) and nasal cavities (see Fig. 11 - 26*E*). Proliferation of mesenchyme in the maxillary prominences causes them to enlarge and grow medially toward each other and the nasal prominences (see Figs. 11 - 24*D* to *G* and 11 - 25). The medial migration of the maxillary prominences moves the medial nasal prominences toward the median plane and each other. Each lateral nasal prominence is separated from the maxillary prominence by a cleft called the **nasolacrimal groove** (see Fig. 11 - 24*C* and *D*).

By the end of the fifth week of development, the *primordia of the auricles* of the external ears have begun to develop (Fig. 11 - 27; see also Fig. 11 - 24*E*). Six **auricular hillocks** (mesenchymal swellings) form around the first pharyngeal groove (three on each side), the primordia of the auricle, and the external acoustic meatus (canal). Initially, the external ears are located in the neck region; however, as the mandible develops, they ascend to the side of the head at the level of the eyes (see Fig. 11 - 24*H*). By the end of the sixth week, each maxillary prominence has begun to merge with the lateral nasal prominence along the line of the **nasolacrimal groove** (Figs. 11 - 28 and 11 - 29*A* and *B*). This establishes continuity between the side of the nose, formed by the lateral nasal prominence, and the cheek region formed by the maxillary prominence.

The **nasolacrimal duct** develops from a rodlike thickening of ectoderm in the floor of the nasolacrimal groove. This thickening gives rise to a solid epithelial cord that separates from the ectoderm and sinks into the mesenchyme. Later, as a result of cell degeneration, this epithelial cord canalizes to form the nasolacrimal duct. The cranial end of this duct expands to form the **lacrimal sac**. By the late fetal period, the nasolacrimal duct drains into the inferior meatus in the lateral wall of the nasal cavity. The duct usually becomes completely patent only after birth. Part of the nasolacrimal duct occasionally fails to canalize, resulting in a congenital anomaly known as *atresia of the nasolacrimal duct*.

During the seventh week of development, the blood supply of the face shifts from the internal to the external carotid artery. This change is related to transformation of the primordial aortic arch pattern into the postnatal arterial arrangement. Between the 7th and 10th weeks, the medial nasal prominences merge with each other and with the maxillary and lateral nasal prominences (see Fig. 11 - 24*G* and *H*). Merging of these prominences requires disintegration of their contacting surface epithelia. This results in intermingling of the underlying mesenchymal cells. Merging of the medial nasal and maxillary prominences results in continuity of the upper jaw and lip and separation of the nasal pits from the

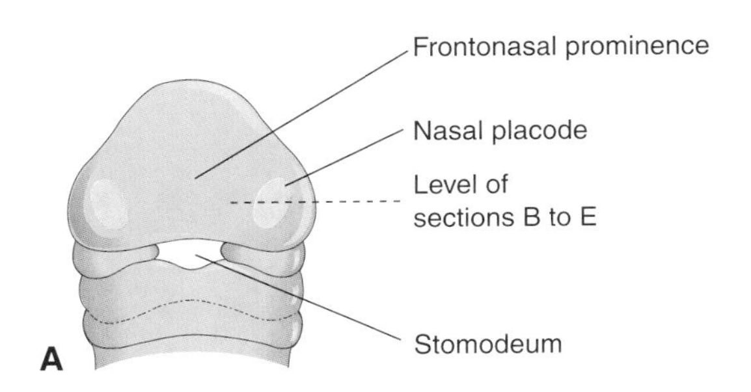

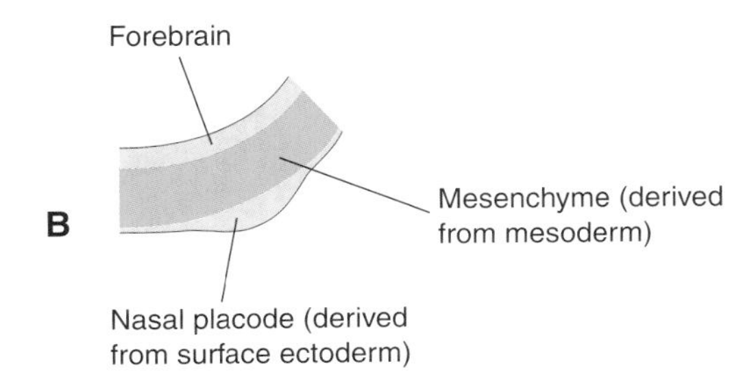

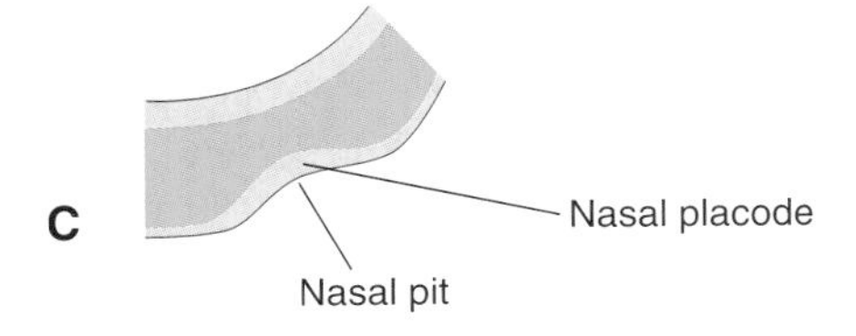

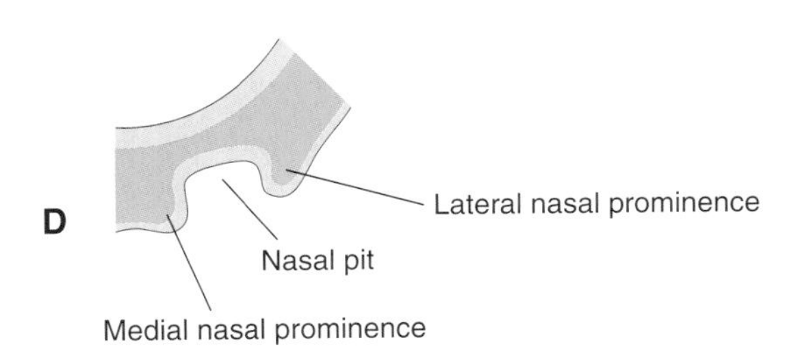

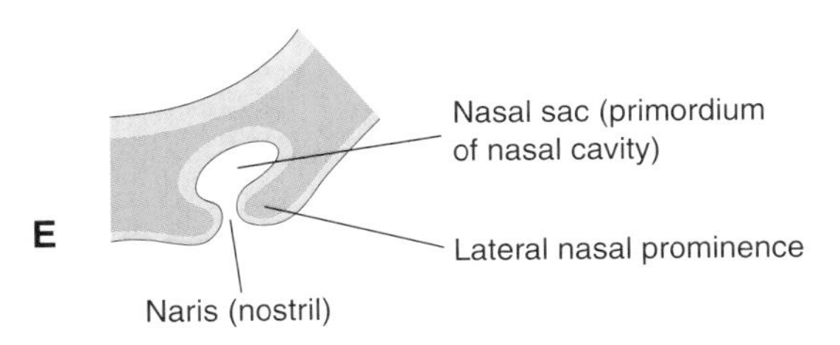

Figure 11 - 26. Progressive stages in the development of a human nasal sac (primordial nasal cavity). *A*, Ventral view of an embryo at about 28 days. *B* to *E*, Transverse sections through the left side of the developing nasal sac.

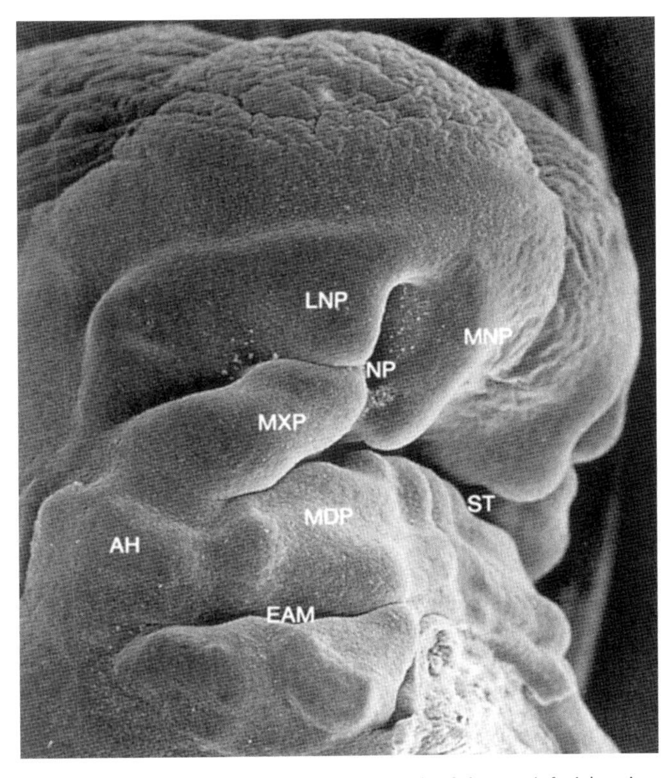

Figure 11 - 27. Scanning electron micrograph of the craniofacial region of a human embryo at about 41 days (stage 16, CRL = 10.8 mm), viewed obliquely. The maxillary prominence (MXP) appears puffed up laterally and is wedged between the lateral (LNP) and medial (MNP) nasal prominences surrounding the nasal pit (NP). Observe the mandibular prominence (MDP) and the stomodeum (ST) just above it. The auricular hillocks (AH) can be seen on both sides of the groove between the mandibular and hyoid arches, which will form the external acoustic meatus (EAM). (From Hinrichsen K: The early development of morphology and patterns of the face in the human embryo. *Adv Anat Embryol Cell Biol* 98:1, 1985.)

stomodeum. As the medial nasal prominences merge, they form an intermaxillary segment (see Figs. 11 - 24*H* and 11 - 29*C* to *F*). The **intermaxillary segment** gives rise to:

- the middle part or philtrum of the upper lip
- the premaxillary part of the maxilla and its associated gingiva (gum)
- the primary palate

Lateral parts of the upper lip, most of the maxilla, and the secondary palate form from the maxillary prominences (see Fig. 11 - 24*H*). These prominences merge laterally with the mandibular prominences. The primitive lips and cheeks are invaded by mesenchyme from the second pair of pharyngeal arches, which differentiates into the facial muscles (see Fig. 11 - 6 and Table 11 - 1). These *muscles of facial expression* are supplied

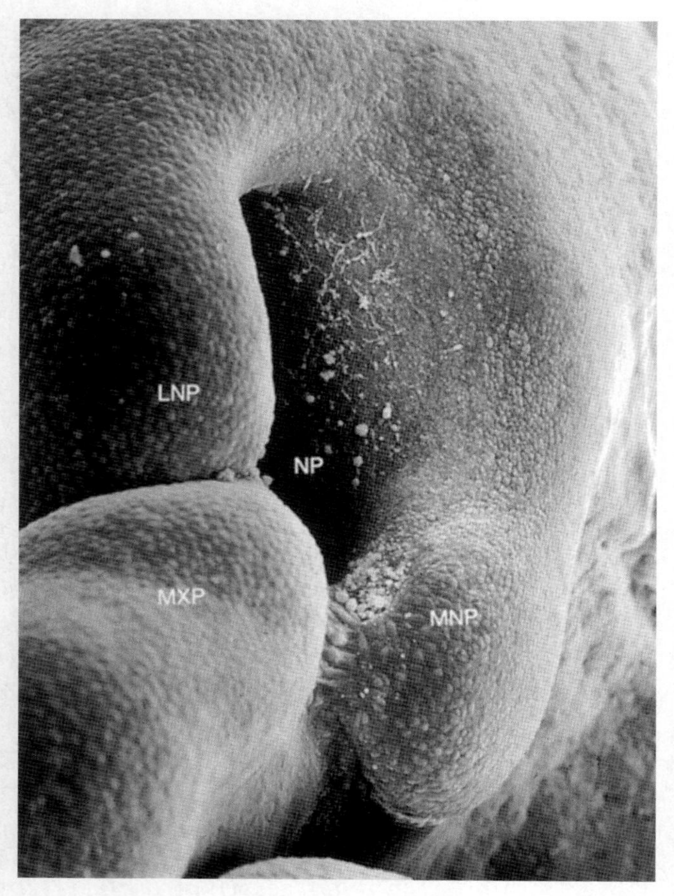

Figure 11 - 28. Scanning electron micrograph of the right nasal region of a human embryo at about 41 days (stage 17, CRL = 10.8 mm) showing the maxillary prominence (MXP) fusing with the medial nasal prominence (MNP). Observe the large nasal pit (NP). Epithelial bridges can be seen between these prominences. Observe the furrow representing the nasolacrimal groove between the MXP and the lateral nasal prominence (LNP). (From Hinrichsen K: The early development of morphology and patterns of the face in the human embryo. *Adv Anat Embryol Cell Biol* 98:1, 1985.)

by the facial nerve, the nerve of the second arch. The mesenchyme in the first pair of arches differentiates into the *muscles of mastication* and a few others, all of which are innervated by the trigeminal nerves, which supply the first pair of arches.

Summary of Facial Development

- The frontonasal prominence forms the forehead and the dorsum and apex of the nose.
- The lateral nasal prominences form the sides of the nose.
- The medial nasal prominences form the nasal septum.
- The maxillary prominences form the upper cheek regions and most of the upper lip.
- The mandibular prominences give rise to the chin, lower lip, and lower cheek regions (see Fig. 11 - 24).

In addition to these fleshy derivatives, various bones are derived from the mesenchyme in the facial prominences (see Fig. 11 - 29). Until the end of the sixth week, the primitive jaws are composed of masses of mesenchymal tissue. The lips and *gingivae* begin to develop when a linear thickening of the ectoderm, the *labiogingival lamina*, grows into the underlying mesenchyme (see Fig. 11 - 32*B*). Gradually, most of the lamina degenerates, leaving a *labiogingival groove* between the lips and the gingivae (see Fig. 11 - 32*H*). A small area of the labiogingival lamina persists in the median plane to form the *frenulum of the upper lip*, which attaches the lip to the gingiva.

Final development of the face occurs mainly from changes in the proportion and relative positions of the facial components. During the early fetal period, the nose is flat and the mandible is underdeveloped (see Fig. 11 - 24*H*); these structures obtain their characteristic form as facial development is completed (see Fig. 11 - 24*I*). As the brain enlarges, a prominent forehead is created and the eyes move medially. As the mandible and head enlarge, the auricles of the external ears rise to the level of the eyes. *The small size of the face prenatally results from*:

- the rudimentary upper and lower jaws
- the unerupted primary teeth
- the small size of the nasal cavities and maxillary sinuses

Development of the Nasal Cavities

As the face develops, the **nasal placodes** become depressed, forming **nasal pits** (see Figs. 11 - 25, 11 - 26, and 11 - 28). Proliferation of the surrounding mesenchyme forms the medial and lateral **nasal prominences** and results in deepening of the nasal pits and formation of primordial **nasal sacs**. Each nasal sac grows dorsally, ventral to the developing forebrain (Fig. 11 - 30*A*). At first, the nasal sacs are separated from the oral cavity by the **oronasal membrane**. This membrane ruptures by the end of the sixth week of development, bringing the nasal and oral cavities into communication (see Fig. 11 - 30*B* and *C*). The regions of continuity between the nasal and oral cavities are the **primordial choanae**, which lie posterior to the primary palate. After the *secondary palate* develops, the choanae are located at the junction of the nasal cavity and pharynx

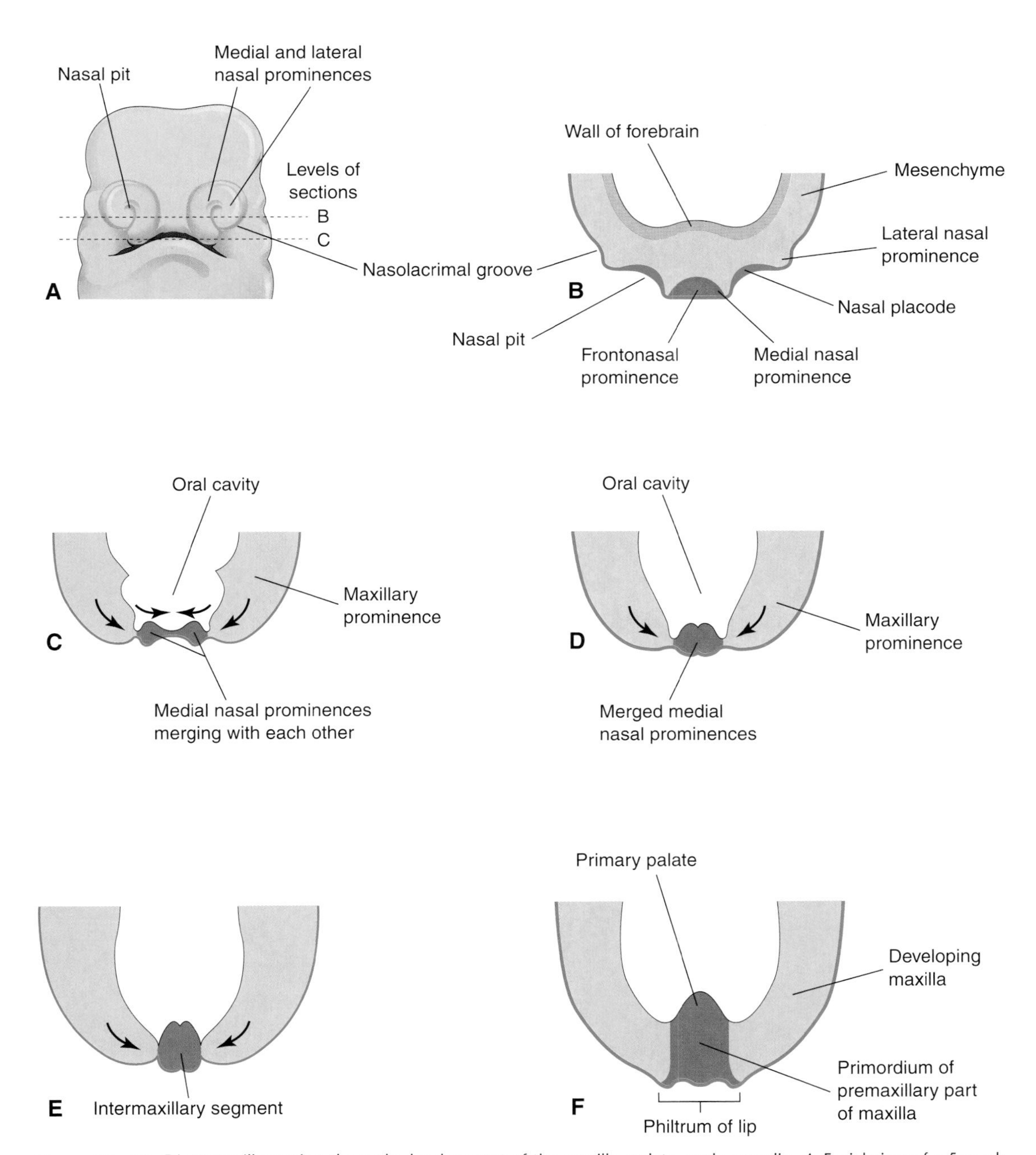

Figure 11 - 29. Diagrams illustrating the early development of the maxilla, palate, and upper lip. *A*, Facial view of a 5-week embryo. *B* and *C*, Sketches of horizontal sections at the levels shown in *A*. The arrows in *C* indicate subsequent growth of the maxillary and medial nasal prominences toward the median plane and merging of the prominences with each other. *D* to *F*, Similar sections of older embryos illustrating merging of the medial nasal prominences with each other and the maxillary prominences to form the upper lip.

(see Fig. 11 - 30*D*). While these changes are occurring, the *superior*, *middle*, and *inferior* **conchae** develop as elevations of the lateral walls of the nasal cavities (see Fig. 11 - 30*D*). Concurrently, the ectodermal epithelium in the roof of each nasal cavity becomes specialized to form the **olfactory epithelium**. Some epithelial cells differentiate into *olfactory receptor cells* (neurons). The axons of these cells constitute the **olfactory nerves**,

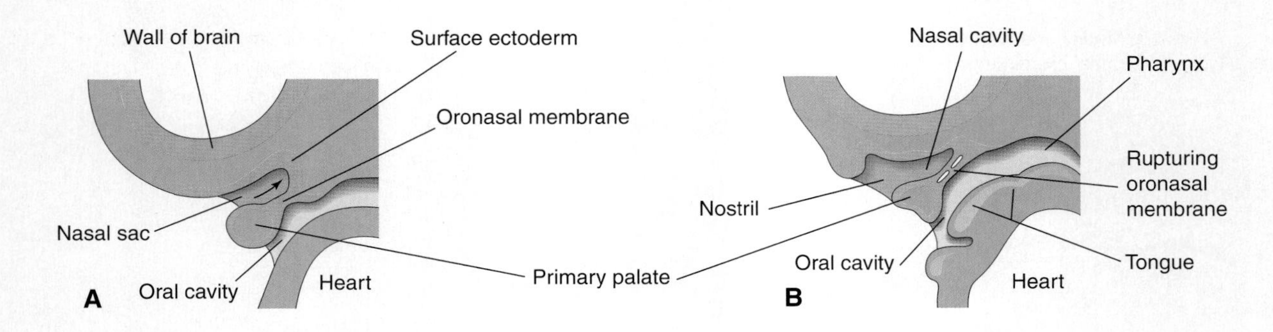

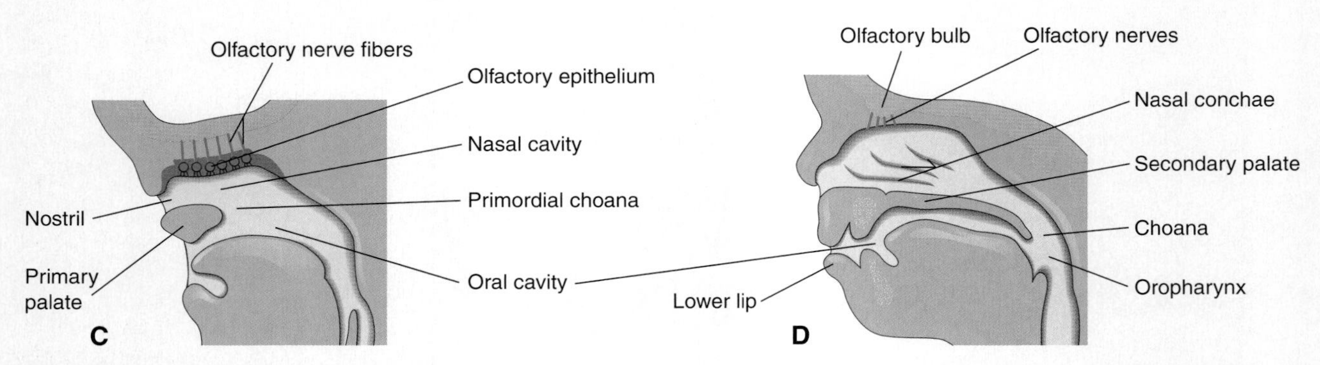

Figure 11 – 30. Drawings of sagittal sections of the head showing development of the nasal cavities. The nasal septum has been removed. *A,* Five weeks. *B,* Six weeks, showing breakdown of the oronasal membrane. *C,* Seven weeks, showing the nasal cavity communicating with the oral cavity and the development of the olfactory epithelium. *D,* Twelve weeks. The palate and the lateral wall of the nasal cavity are evident.

which grow into the **olfactory bulbs** of the brain (see Fig. 11 – 30*C* and *D*).

Paranasal Sinuses

Some paranasal sinuses, in particular, the maxillary sinuses, begin to develop during late fetal life; the remainder of them develop after birth. They form from outgrowths or diverticula of the walls of the nasal cavities, becoming pneumatic (air-filled) extensions of the nasal cavities in the adjacent bones, such as the maxillary sinuses in the maxillae. The original openings of the diverticula persist as the orifices of the adult sinuses.

Paranasal Sinuses in Neonatal and Postnatal Development

Most of the paranasal sinuses are rudimentary or absent in newborn infants. The *maxillary sinuses* are small at birth. They grow slowly until puberty and are not fully developed until all the permanent teeth have erupted in early adulthood. No frontal or sphenoidal sinuses are present at birth. The ethmoidal cells (sinuses) are small before the age of 2 years, and they do not begin to grow rapidly until 6 to 8 years of age. Around the age of 2 years, the two most anterior ethmoidal cells grow into the frontal bone, forming a frontal sinus on each side. Usually, the *frontal sinuses* are visible in radiographs by the seventh year. The two most posterior ethmoidal cells grow into the sphenoid bone at about the age of 2 years, forming two *sphenoid sinuses.* Growth of the paranasal sinuses is important in altering the size and shape of the face during infancy and childhood and in adding resonance to the voice during adolescence.

Development of the Palate

The palate develops from two primordia: the primary palate and the secondary palate.

Palatogenesis begins at the end of the fifth week of development; however, development of the palate is not completed until the 12th week. The *critical period of development of the palate* is from the end of the sixth week until the beginning of the ninth week.

Primary Palate

Early in the sixth week, the primary palate (**median palatine process**) begins to develop from the deep part of the intermaxillary segment of the maxilla (see Figs. 11 - 29*F* and 11 - 30). Initially, this segment, which is formed by the merging of the medial nasal prominences, is a wedge-shaped mass of mesenchyme between the internal surfaces of the maxillary prominences of the developing maxillae. The primary palate forms the **premaxillary part of the maxilla** (Fig. 11 - 31*A* and *B*). It represents only a small part of the adult hard palate (i.e., the part anterior to the incisive fossa).

Secondary Palate

The secondary palate is the primordium of the hard and soft parts of the palate that extend posteriorly from the **incisive fossa** (see Fig. 11 - 31*A* and *B*). The secondary palate begins to develop early in the sixth week from two mesenchymal projections that extend from the internal aspects of the maxillary prominences. Initially, these structures — called the **lateral palatine processes** or palatal shelves — project inferomedially on each side of the tongue (Figs. 11 - 32*A* to *C* and 11 - 33*A* and *B*). As the jaws develop, the tongue becomes relatively smaller and moves inferiorly. During the seventh and eighth weeks, the lateral palatine processes elongate and ascend to a horizontal position superior to the tongue. Gradually, the processes (shelves) approach each other and fuse in the median plane (see Figs. 11 - 32*E* to *H* and 11 - 33*C*). They also fuse with the nasal septum and the posterior part of the primary palate. Elevation of the palatal processes or shelves to the horizontal position is believed to be caused by an intrinsic *shelf-elevating force* that is generated by the hydration of hyaluronic acid in the mesenchymal cells within the palatal processes. The medial epithelial seam at the edges of the palatal shelves breaks down, which is a prerequisite for fusion of the palatal shelves. Experimental results suggest that transforming growth factor (TGF) β^3 is involved in this process.

The **nasal septum** develops in a downward growth pattern from internal parts of the merged medial nasal prominences (see Figs. 11 - 32 and 11 - 33). The fusion between the nasal septum and the palatine processes begins anteriorly during the 9th week and is completed posteriorly by the 12th week, superior to the primordium of the hard palate. Bone gradually develops in the primary palate, forming the premaxillary part of the maxilla, which lodges the incisor teeth (see Fig. 11 - 31*B*). Concurrently, bone extends from the maxillae and palatine bones into the lateral palatine processes (palatal shelves) to form the **hard palate** (see Fig. 11 - 32*E* and *G*). The posterior parts of these processes do not become ossified. They extend posteriorly beyond the nasal septum and fuse to form the **soft palate**, including its soft conical projection, the **uvula** (see Fig. 11 - 32*D*, *F*, and *H*). The *median palatine raphe* indicates the line of fusion of the lateral palatine processes. A small **nasopalatine canal** persists in the median plane of the palate between the premaxillary part of the maxilla and the palatine processes of the maxillae. This canal is represented in the adult hard palate by the **incisive fossa** (see Fig. 11 - 31*B*), which is the common opening for the right and left *incisive*

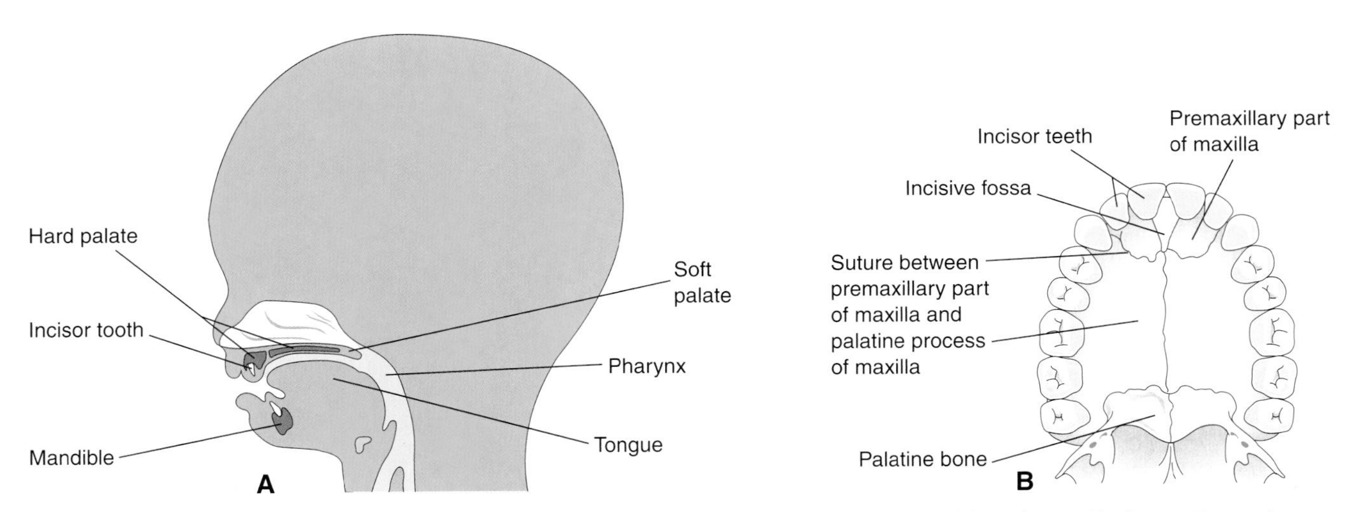

Figure 11 - 31. *A,* Drawing of a sagittal section of the head of a 20-week fetus illustrating the location of the palate. *B,* The bony palate and alveolar arch of a young adult. The suture between the premaxillary part of the maxilla and the fused palatine processes of the maxillae is usually visible in crania of young persons. It is not visible in the hard palates of most dried crania because they are usually from old adults.

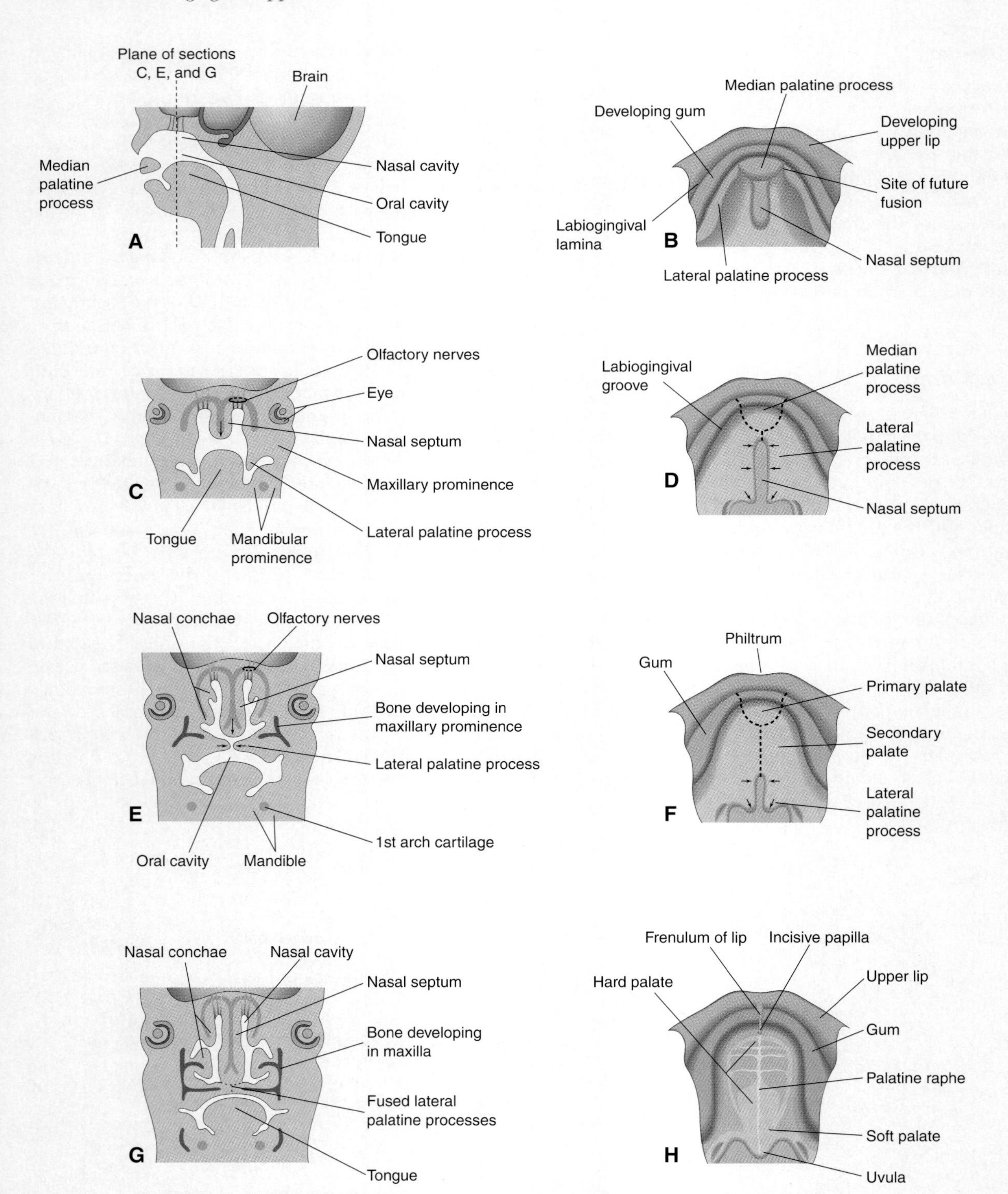

Figure 11 – 32. *A*, Sagittal section of the embryonic head at the end of the sixth week showing the median palatine process or primary palate. *B*, *D*, *F*, and *H*, Drawings of the roof of the mouth from the 6th to 12th weeks illustrating development of the palate. The *broken lines* in *D* and *F* indicate sites of fusion of the palatine processes. The arrows indicate medial and posterior growth of the lateral palatine processes. *C*, *E*, and *G*, Drawings of frontal sections of the head illustrating fusion of the lateral palatine processes with each other and the nasal septum, and separation of the nasal and oral cavities.

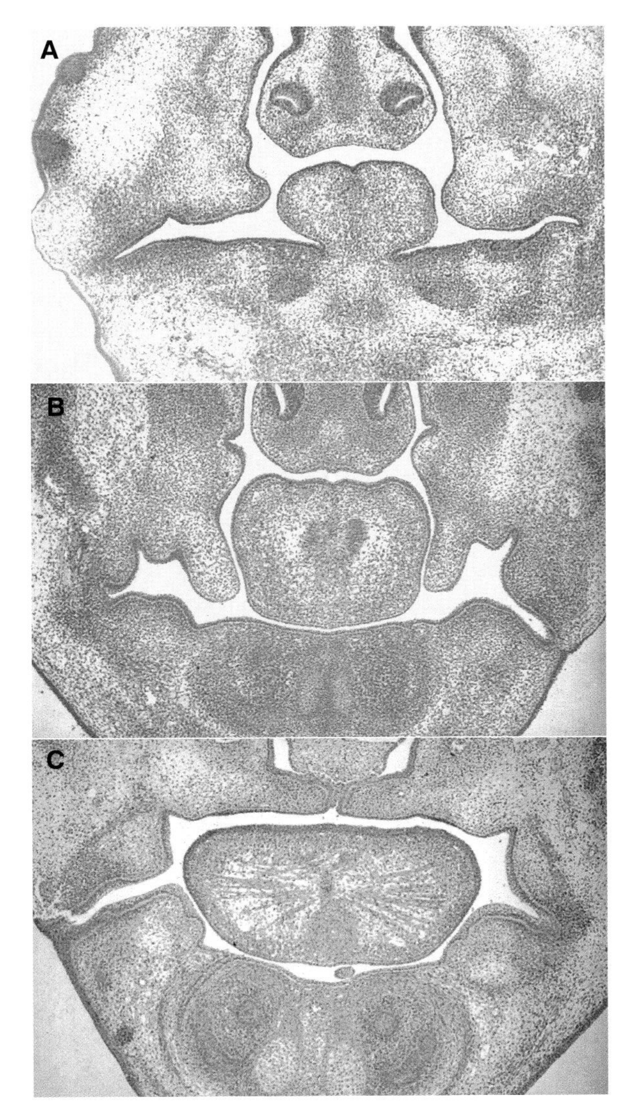

Figure 11 - 33. Coronal sections of human embryonic heads showing development of the palatal processes (shelves) during the eighth week. *A*, Embryo with a CRL of 24 mm. This section shows early development of the palatine processes. *B*, Embryo with a CRL of 27 mm. This section shows the palate just prior to palatal process elevation. *C*, Embryo with a CRL of 29 mm (near the end of the eighth week). The palatine processes are elevated and fused. (From Sandham A: Embryonic facial vertical dimension and its relationship to palatal shelf elevation. *Early Hum Devel* 12:241, 1985.)

canals. An irregular suture runs from the incisive fossa to the alveolar process of the maxilla, between the lateral incisor and canine teeth on each side. It is visible in the anterior region of the palates of young persons. This suture indicates where the embryonic primary and secondary palates fused.

Cleft Lip and Palate

Clefts of the upper lip and palate are common. The defects are usually classified according to developmental criteria, with the incisive fossa and papilla as reference landmarks (see Figs. 11 – 31*B* and 11 – 35*A*). Cleft lip ("harelip") and palate are especially conspicuous because they result in an abnormal facial appearance and defective speech. There are *two major groups of cleft lip and palate* (Figs. 11 – 34 to 11 – 36):

- clefts involving the upper lip and anterior part of the maxilla, with or without involvement of parts of the remaining hard and soft regions of the palate
- clefts involving the hard and soft regions of the palate

Anterior cleft anomalies include cleft lip, with or without cleft of the alveolar part of the maxilla. A complete anterior cleft anomaly is one in which the cleft extends through the lip and the alveolar part of the maxilla to the incisive fossa, separating the anterior and posterior parts of the palate (see Fig. 11 – 35*E* and *F*). Anterior cleft anomalies result from a deficiency of mesenchyme in the maxillary prominence(s) and the intermaxillary segment (see Fig. 11 – 29*E*).

Posterior cleft anomalies include clefts of the secondary or posterior palate that extend through the soft and hard

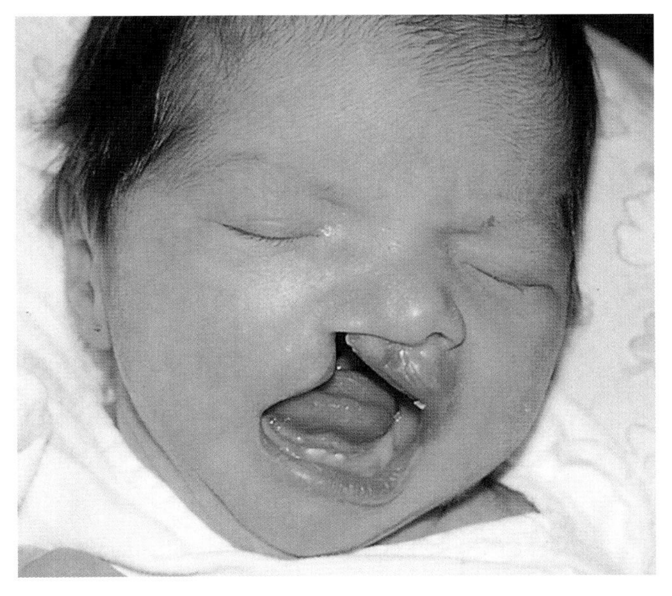

Figure 11 - 34. Infant with unilateral cleft lip and palate. Clefts of the lip, with or without cleft palate, occur in about 1 in every 1000 births; 60 to 80% of affected individuals are male infants. (Courtesy of Dr. AE Chudley, Professor of Pediatrics and Child Health, Children's Hospital and University of Manitoba, Winnipeg, Manitoba, Canada.)

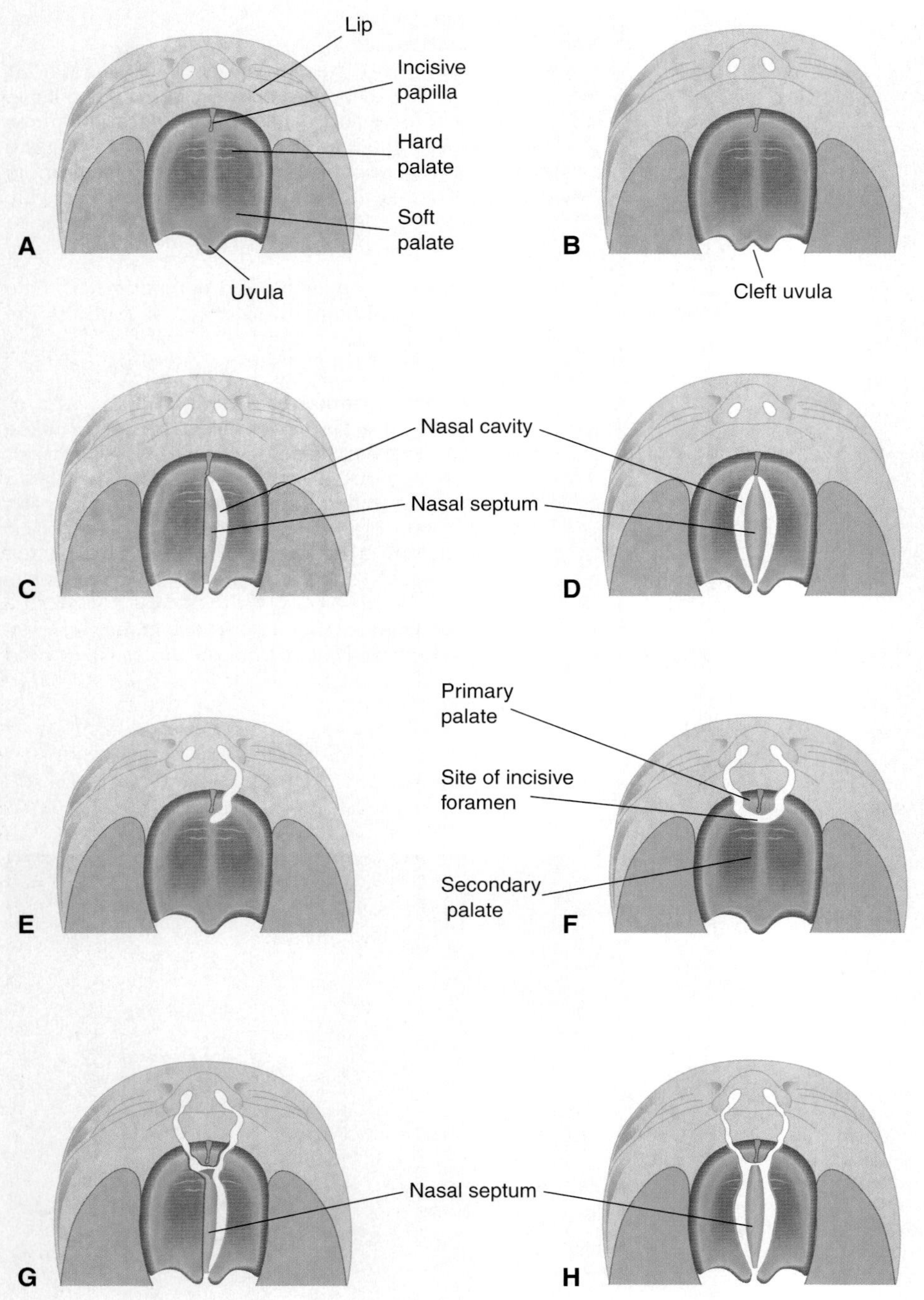

Figure 11 - 35. Various types of cleft lip and palate. *A*, Normal lip and palate. *B*, Cleft uvula. *C*, Unilateral cleft of the posterior or secondary palate. *D*, Bilateral cleft of the posterior palate. *E*, Complete unilateral cleft of the lip and alveolar process of the maxilla with a unilateral cleft of the anterior or primary palate. *F*, Complete bilateral cleft of the lip and alveolar processes of the maxillae with bilateral cleft of the anterior palate. *G*, Complete bilateral cleft of the lip and alveolar processes of the maxillae with bilateral cleft of the anterior palate and unilateral cleft of the posterior palate. *H*, Complete bilateral cleft of the lip and alveolar processes of the maxillae with complete bilateral cleft of the anterior and posterior palate.

regions of the palate to the incisive fossa, separating the anterior and posterior parts of the palate (see Fig. 11 – 35*G* and *H*). Posterior cleft anomalies are caused by defective development of the secondary palate and result from growth distortions in the lateral palatine processes that, in turn, prevent the medial migration and fusion of these processes.

Clefts involving the upper lip, with or without cleft palate, occur in about once in 1000 births; however, their frequency varies widely and 60 to 80% of those affected are male infants. The clefts vary in severity from small notches in the vermilion border of the lip (see Fig. 7 – 37*G*) to larger clefts that extend into the floor of the nostril and through

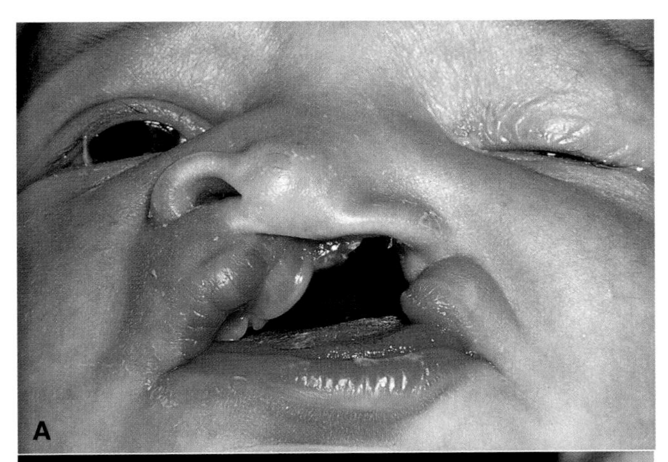

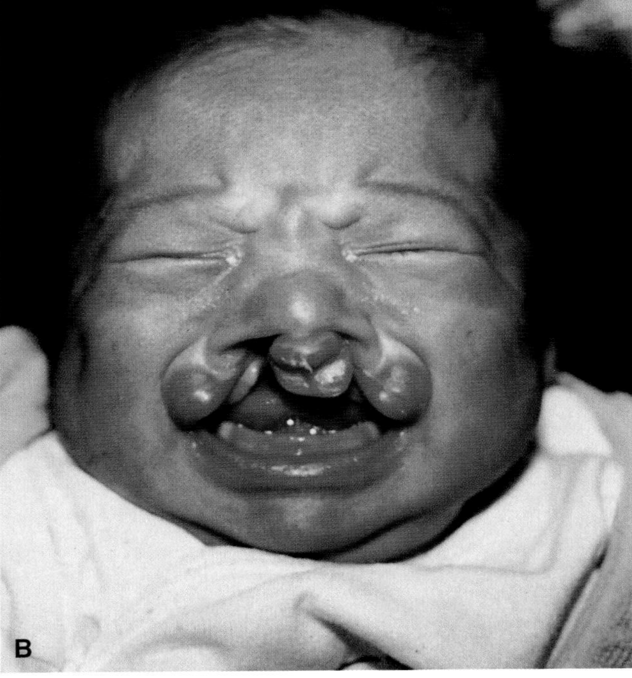

Figure 11 - 36. Photographs illustrating congenital anomalies of the lip and palate. *A*, Infant with a left unilateral cleft lip and cleft palate. *B*, Infant with a bilateral cleft lip and cleft palate. (*A* and *B*, Courtesy of Dr. Barry H. Grayson and Dr. Bruno L. Vendittelli, New York University Medical Center, Institute of Reconstructive Plastic Surgery, New York, New York.)

the alveolar part of the maxilla (see Figs. 11 – 35 and 11 – 36*A*). Cleft lip can be unilateral or bilateral.

Unilateral cleft lip (Figs. 11 – 34 and 11 – 36*A*) results from failure of the maxillary prominence on the affected side to unite with the merged medial nasal prominences (Fig. 11 – 37*A* to *H*). This is the consequence of failure of the mesenchymal masses to merge and the mesenchyme to proliferate and smooth out the overlying epithelium. This results in a *persistent labial groove.* In addition, the epithelium in the labial groove becomes stretched, and the tissues in the floor of the persistent groove break down. As a result, the lip is divided into medial and lateral parts. Sometimes, a bridge of tissue, called a **Simonart band**, joins the parts of the incomplete cleft lip.

Bilateral cleft lip (see Figs. 11 – 35*F* and 11 – 36*B*) results from failure of the mesenchymal masses in the maxillary prominences to meet and unite with the merged medial nasal prominences. The epithelium in both labial grooves becomes stretched and breaks down. In bilateral cases, the defects may be dissimilar, with varying degrees of defect on each side. When there is a complete bilateral cleft of the lip and alveolar part of the maxilla, the intermaxillary segment hangs free and projects anteriorly. These defects are especially deforming because of the loss of continuity of the *orbicularis oris muscle*, which closes the mouth and purses the lips (as when whistling).

Median cleft lip is an extremely rare defect (Fig. 11 – 38*A*). It results from a mesenchymal deficiency, which causes partial or complete failure of the medial nasal prominences to merge and form the intermaxillary segment. A median cleft of the upper lip is a characteristic feature of the *Mohr syndrome*, which is transmitted as an autosomal recessive trait. **Median cleft of the lower lip** (see Fig. 11 – 38*B*) is also very rare and is caused by failure of the mesenchymal masses in the mandibular prominences to merge completely and smooth out the embryonic cleft between them.

A **complete cleft palate** indicates the maximal degree of clefting of any particular type; for example, a *complete cleft of the posterior palate* is an anomaly in which the cleft extends through the soft palate and anteriorly to the incisive fossa. The landmark for distinguishing anterior from posterior cleft anomalies is the *incisive fossa.* Anterior and posterior cleft anomalies are embryologically distinct. Cleft palate, with or without cleft lip, occurs in about 1 in 2500 births and is more common in female than in male infants. The cleft may involve only the uvula, giving it a fishtail appearance (see Fig. 11 – 35*B*), or it may extend through the soft and hard regions of the palate (see Fig. 11 – 35*C* and *D*). In severe cases associated with cleft lip, the cleft in the palate extends through the alveolar part of the maxilla and the lips on both sides (see Fig. 11 – 35*G* and *H*).

The embryologic basis of cleft palate is failure of the mesenchymal masses in the lateral palatine processes to meet and fuse with each other, with the nasal septum, and/or with the posterior margin of the median palatine process (see Figs. 11 – 29*D* and 11 – 35). Unilateral and bilateral clefts in the palate are classified into three groups:

- *Clefts of the anterior palate* (i.e., clefts anterior to the incisive fossa) result from failure of mesenchymal masses in the lateral palatine processes (palatine shelves) to meet and fuse with the mesenchyme in the primary palate (see Fig. 11 – 35*E* and *F*).
- *Clefts of the posterior palate* (i.e., clefts posterior to the incisive fossa) result from failure of mesenchymal masses

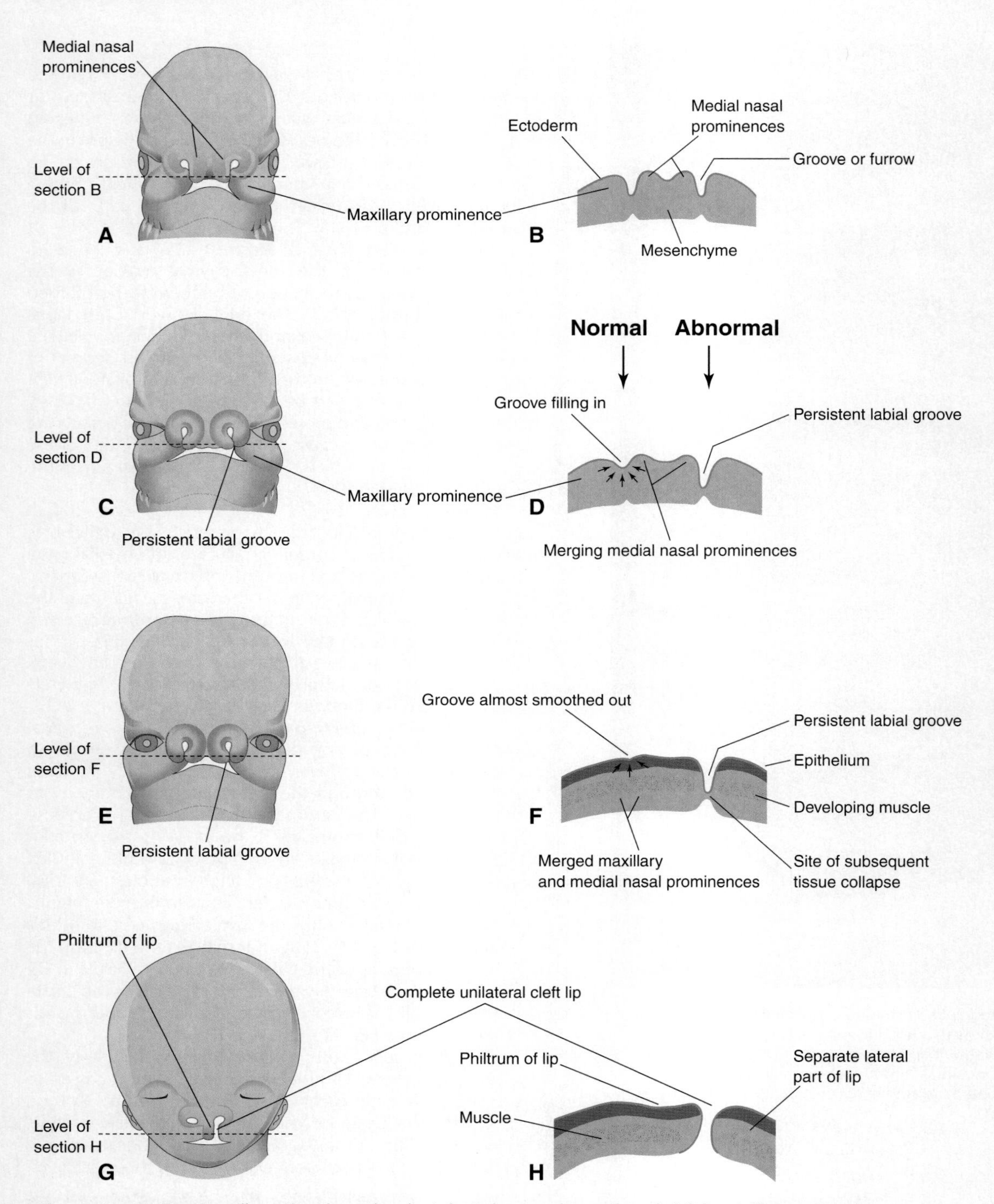

Figure 11 – 37. Drawings illustrating the embryologic basis for complete unilateral cleft lip. *A*, Five-week embryo. *B*, Horizontal section through the head illustrating the grooves between the maxillary prominences and the merging medial nasal prominences. *C*, Six-week embryo showing a persistent labial groove on the left side. *D*, Horizontal section through the head showing the groove gradually filling in on the right side following proliferation of mesenchyme (*arrows*). *E*, Seven-week embryo. *F*, Horizontal section through the head showing that the epithelium on the right has almost been pushed out of the groove between the maxillary and medial nasal prominences. *G*, Ten-week fetus with a complete unilateral cleft lip. *H*, Horizontal section through the head after stretching of the epithelium and breakdown of the tissues in the floor of the persistent labial groove on the left side, resulting in the formation of a complete unilateral cleft lip.

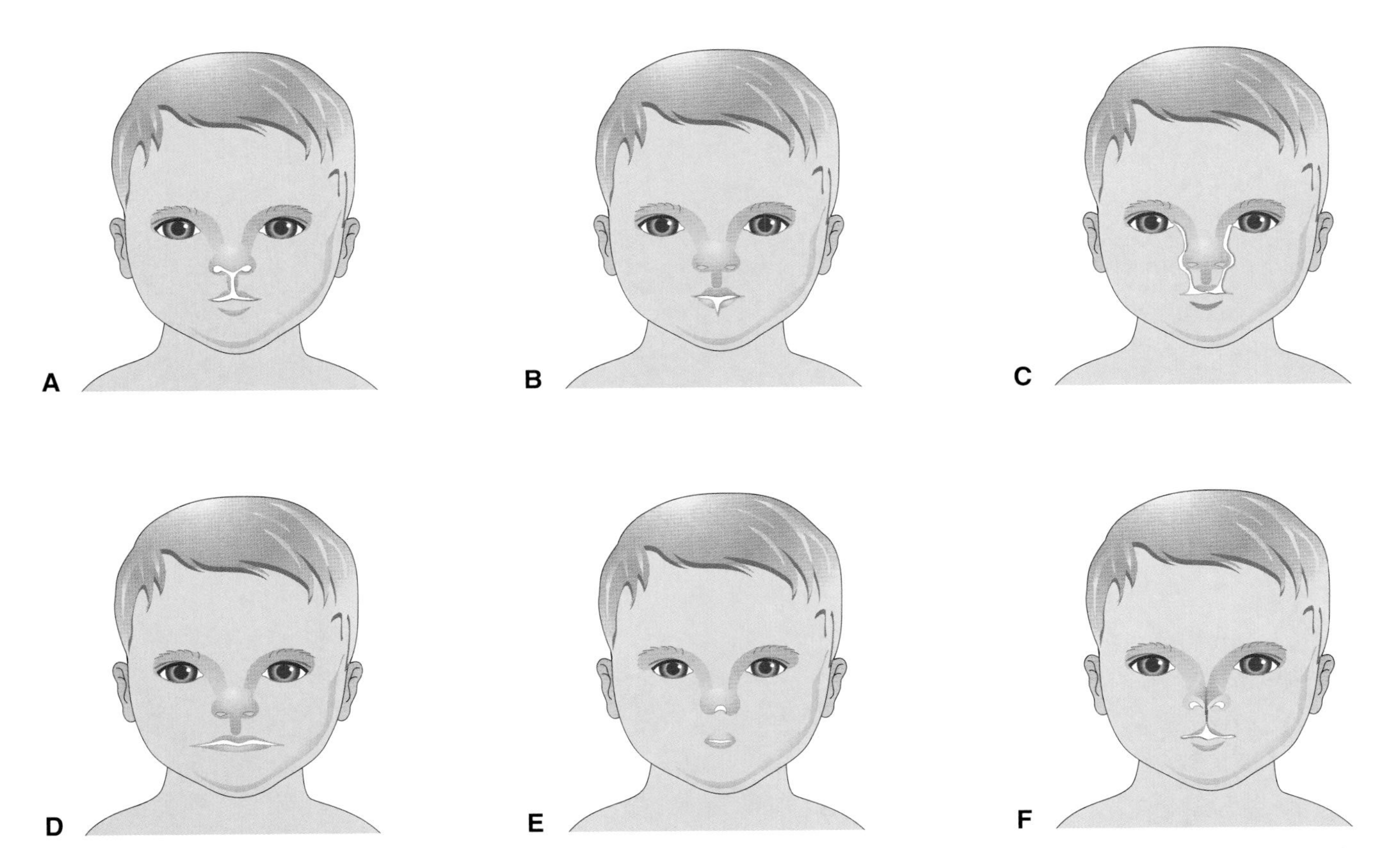

Figure 11 - 38. Drawings of unusual congenital anomalies of the face. *A*, Median cleft of the upper lip. *B*, Median cleft of the lower lip. *C*, Bilateral oblique facial clefts with complete bilateral cleft lip. *D*, Macrostomia. *E*, Single nostril and microstomia; these anomalies are not usually associated with each other. *F*, Bifid nose and incomplete median cleft lip.

in the lateral palatine processes to meet and fuse with each other and the nasal septum (see Fig. 11 – 35*B* to *D*).

- *Clefts of the anterior and posterior parts of the palate* (i.e., clefts of the primary and secondary palates) result from failure of the mesenchymal masses in the lateral palatine processes to meet and fuse with mesenchyme in the primary palate, with each other, and with the nasal septum (see Fig. 11 – 35*G* and *H*).

Most clefts of the lip and palate result from multiple factors (*multifactorial inheritance*; see Chapter 9), both genetic and nongenetic, each causing a minor developmental disturbance. How teratogenic factors induce cleft lip and palate is still unknown. Experimental studies have given us some insight into the cellular and molecular basis of these defects. Some clefts of the lip and/or palate appear as part of syndromes determined by single mutant genes. Other clefts are features of chromosomal syndromes, especially *trisomy 13* (see Chapter 9). A few cases of cleft lip and/or palate appear to be caused by teratogenic agents (e.g., anticonvulsant drugs). Studies of twins indicate that genetic factors are of greater importance in cleft lip, with or without cleft palate, than in cleft palate alone. A sibling of a child with a cleft palate has an elevated risk of developing a cleft palate, but no increased risk of developing a cleft lip. A cleft of the lip and alveolar process of the maxilla that continues through the palate is usually transmitted through a male sex-linked gene. When neither parent is affected, the *recurrence risk* in subsequent siblings (brother or sister) is about 4%.

Facial Clefts

Various types of facial cleft may occur, but they are all extremely rare. Severe clefts are usually associated with gross anomalies of the head. *Oblique facial clefts* (orbitofacial fissures) are often bilateral and extend from the upper lip to the medial margin of the orbit (see Fig. 11 – 38*C*). When this occurs, the nasolacrimal ducts are open grooves (persistent nasolacrimal grooves). Oblique facial clefts associated with cleft lip result from failure of the mesenchymal masses in the maxillary prominences to merge with the lateral and medial nasal prominences. Lateral or transverse facial clefts run from the mouth toward the ear. Bilateral clefts result in a very large mouth, a condition called *macrostomia* (see Fig. 11 – 38*D*). In severe cases, the clefts in the cheeks extend almost to the ears.

Other Facial Anomalies

Congenital microstomia (small mouth) results from excessive merging of the mesenchymal masses in the maxillary and mandibular prominences of the first arch (see Fig. 11 – 38*E*). In severe cases, the abnormality may be associated with underdevelopment (hypoplasia) of the mandible. *Absence of the nose* occurs when no nasal placodes form. *A single nostril* results when only one nasal placode forms. *Bifid nose* results when the medial nasal prominences do not merge completely; in such cases, the nostrils are widely separated and the nasal bridge is bifid (see Fig. 11 – 38*F*). In mild forms of bifid nose, a groove is apparent in the tip of the nose.

Summary of the Pharyngeal Apparatus

During the fourth and fifth weeks of development, the primordial pharynx is bounded laterally by **pharyngeal arches**. Each arch consists of a core of mesenchyme covered externally by ectoderm and internally by endoderm. The original mesenchyme of each arch is derived from mesoderm; later, **neural crest cells** migrate into the arches and are the major source of their connective tissue components — including cartilage, bone, and ligaments — in the oral and facial regions. Each pharyngeal arch contains an artery, a cartilage rod, a nerve, and a muscular component. Externally, the pharyngeal arches are separated by **pharyngeal grooves**. Internally, the arches are separated by evaginations of the pharynx called *pharyngeal pouches*. Where the ectoderm of a groove contacts the endoderm of a pouch, **pharyngeal membranes** are formed. The arches, pouches, grooves, and membranes make up the pharyngeal apparatus. Development of the tongue, face, lips, jaws, palate, pharynx, and neck largely involves transformation of the pharyngeal apparatus into adult structures.

The *pharyngeal grooves* disappear except for the first pair, which persists as the *external acoustic meatus*. The pharyngeal membranes also disappear except for the first pair, which becomes the *tympanic membranes*. The first pharyngeal pouch gives rise to the *tympanic cavity*, mastoid antrum, and *pharyngotympanic tube*. The second pharyngeal pouch is associated with development of the *palatine tonsil*. The *thymus* is derived from the third pair of pharyngeal pouches, and the *parathyroid glands* are formed from the third and fourth pairs of pharyngeal pouches.

The **thyroid gland** develops in a downward pattern of growth from the floor of the primordial pharynx in the region where the tongue develops. The parafollicular (C) cells in the thyroid gland are derived from the *ultimopharyngeal bodies*, which are derived mainly from the fourth pair of pharyngeal pouches. An *ectopic thyroid gland* results when the thyroid gland fails to descend completely from its site of origin in the tongue. The thyroglossal duct may persist, or remnants of it may give rise to *thyroglossal duct cysts* and *ectopic thyroid tissue masses*. Infected cysts may perforate the skin and form *thyroglossal duct sinuses* that open anteriorly in the median plane of the neck.

Most congenital anomalies of the head and the neck originate during transformation of the pharyngeal apparatus into adult structures. Branchial cysts, sinuses, and fistulas may develop from parts of the second pharyngeal groove, the cervical sinus, or the second pharyngeal pouch that fail to obliterate.

Because of the complicated development of the face and palate, congenital anomalies of the face and palate are common. *Anomalies result mainly from maldevelopment of neural crest tissue*, which gives rise to the skeletal and connective tissue primordia of the face. Neural crest cells may be deficient in number, they may not complete their migration to the face, or they may fail in their inductive capacity. Anomalies of the face and palate result from arrested development or a failure of fusion of the facial prominences and palatal processes, or both.

Cleft lip is a common congenital anomaly. Although frequently associated with cleft palate, *cleft lip and palate are etiologically distinct anomalies* that involve different developmental processes occurring at different times. Cleft lip results from failure of mesenchymal masses in the medial nasal and maxillary prominences to merge. By contrast, **cleft palate** results from failure of mesenchymal masses in the palatine processes to meet and fuse. Most cases of cleft lip, with or without cleft palate, are caused by a combination of genetic and environmental factors (*multifactorial inheritance*). These factors interfere with the migration of *neural crest cells* into the maxillary prominences of the first pharyngeal arch. If the number of cells is insufficient, clefting of the lip and/or palate may occur. Other cellular and molecular mechanisms may also be involved.

Clinically Oriented Questions

1. What kind of lip defect is a "harelip"? What is the clinical name for this birth defect?
2. Some say that embryos have cleft lips and that this common facial anomaly represents a persistence of this embryonic condition. Are these statements accurate?

3. Neither Clare's husband nor Clare has a cleft lip or palate, and no one in their families is known to have or to have had these anomalies. What are their chances of having a child with a cleft lip, with or without a cleft palate?
4. Mary's son has a cleft lip and cleft palate. Her brother has a similar defect involving his lip and palate. Although Mary does not plan to have any more children, her husband says that Mary is entirely to blame for their son's birth defects. Was the defect likely inherited only from Mary's side of the family?
5. A patient's son has minor anomalies involving his external ears, but he does not have hearing problems or a facial malformation. Would his ear abnormalities be considered to be branchial defects?

The answers to these questions are at the back of the book.

12

The Respiratory System

Development of the upper respiratory organs, the nasal cavities for example, is described in Chapter 11. The **lower respiratory organs** (larynx, trachea, bronchi, and lungs) begin to form during the fourth week of development. The **respiratory primordium** is present at about 28 days, as evidenced by a median outgrowth from the caudal end of the ventral wall of the primordial pharynx, a structure known as the **laryngotracheal groove** (Figs. 12 – 1*A* to *C* and 12 – 3*A*). This primordium of the tracheobronchial tree develops caudal to the fourth pair of pharyngeal pouches. The endoderm lining the laryngotracheal groove gives rise to the epithelium and glands of the larynx, trachea, bronchi, and pulmonary epithelium. The connective tissue, cartilage, and smooth muscle in these structures develop from the splanchnic mesoderm surrounding the foregut (see Fig. 12 – 4*A*). By the end of the fourth week, the laryngotracheal groove has evaginated to form a pouchlike **respiratory diverticulum** (the lung bud), which is located ventral to the caudal part of the foregut (see Figs. 12 – 1*B* and 12 – 2*A*). As this diverticulum elongates, it is invested with splanchnic mesoderm, and its distal end enlarges to form a globular **tracheal bud** (see Fig. 12 – 2*B*). The respiratory diverticulum soon separates from the **primordial pharynx**; however, it maintains communication with it through the *primordial laryngeal inlet* (see Fig. 12 – 2*C*). Longitudinal **tracheoesophageal folds** develop in the respiratory diverticulum; these folds approach each other and fuse to form a partition known as the **tracheoesophageal septum** (see Fig. 12 – 2*D* and *E*). This septum divides the cranial part of the foregut into a ventral part, the **laryngotracheal tube** (primordium of the larynx, trachea, bronchi, and lungs), and a dorsal part (primordium of the oropharynx and esophagus (see Fig. 12 – 2*F*). The opening of the laryngotracheal tube into the pharynx is the **primordial laryngeal inlet** or aperture (see Figs. 12 – 2*C* and 12 – 3*A* to *C*).

Development of the Larynx

The epithelial lining of the larynx develops from the endoderm of the cranial end of the laryngotracheal tube. The cartilages of the larynx develop from the cartilages in the fourth and sixth pairs of pharyngeal arches (see Chapter 11). The **laryngeal cartilages** develop from mesenchyme that is derived from *neural crest cells*. The mesenchyme at the cranial end of the laryngotracheal tube proliferates rapidly, producing paired **arytenoid swellings** (see Fig. 12 – 3*B*). These swellings grow toward the tongue, converting the slitlike aperture — the *primordial glottis* — into a T-shaped **laryngeal inlet** and reducing the developing laryngeal lumen to a narrow slit. The laryngeal epithelium proliferates rapidly, resulting in *temporary occlusion of the laryngeal lumen*. Recanalization of the larynx occurs by the 10th week. The *laryngeal ventricles* form during this recanalization process. These recesses are bounded by folds of mucous membrane that become the *vocal folds* (cords) and *vestibular folds*. The **epiglottis** develops from the caudal part of the *hypopharyngeal eminence*, a prominence produced by proliferation of mesenchyme in the ventral ends of the third and fourth pharyngeal arches (see Fig. 12 – 3*B* to *D*). The rostral part of this eminence forms the posterior third or pharyngeal part of the tongue (see Chapter 11). Because the **laryngeal muscles** develop from myoblasts in the fourth and sixth pairs of pharyngeal arches, they are innervated by the laryngeal branches of the vagus nerves (cranial nerve [CN] X) that supply these arches (see Table 11 – 1). Growth of the larynx and epiglottis is rapid during the first 3 years after birth. By 3 years of age, the epiglottis has reached its adult form.

Laryngeal Atresia

Laryngeal atresia (obstruction) is a rare anomaly that results in obstruction of the upper fetal airway; it is also known as **congenital high airway obstruction syndrome** (CHAOS). Distal to the atresia or stenosis (narrowing), the airways become dilated, the lungs are enlarged and echogenic (capable of producing echoes), the diaphragm is either flattened or inverted, and fetal ascites and/or hydrops (accumulation of serous fluid) is present. Prenatal ultrasonography permits diagnosis of these anomalies.

Development of the Trachea

The endodermal lining of the laryngotracheal tube distal to the larynx differentiates into the epithelium and glands of the trachea and the pulmonary epithelium. The cartilage, connective tissue, and muscles of the trachea are derived from the splanchnic mesoderm surrounding the laryngotracheal tube (Fig. 12 – 4).

Tracheoesophageal Fistula

A tracheoesophageal fistula (TEF) (abnormal passage) between the trachea and esophagus occurs at a rate of about 1:3000 to 1:4500 live births (Fig. 12 – 5); and predominantly affects male infants. In most cases, the fistula is associated with **esophageal atresia**. TEF results from incomplete division of the cranial part of the foregut into respiratory and esophageal parts during the fourth week. Incomplete fusion of the tracheoesophageal folds results in a **defective tracheoesophageal septum** and communication (TEF) between the trachea and esophagus.

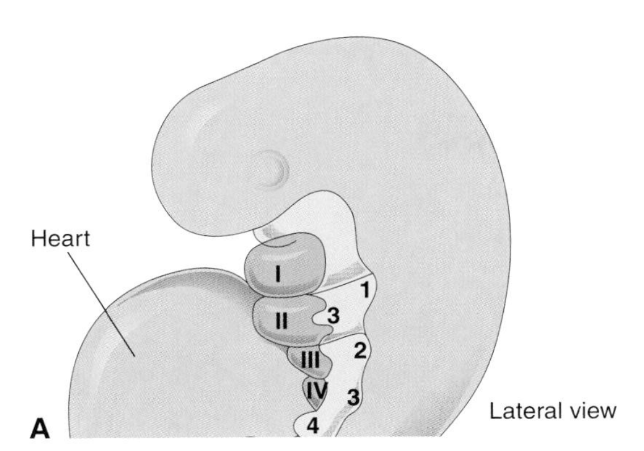

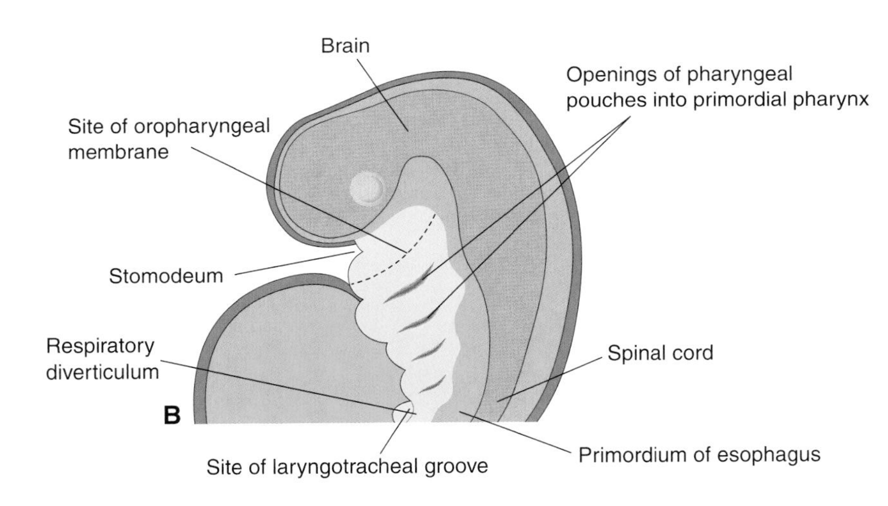

Figure 12 - 1. *A*, Lateral view of a 4-week-old embryo, illustrating the relationship of the pharyngeal apparatus to the developing respiratory system. *B*, Sagittal section of the cranial half of the embryo. *C*, Horizontal section of the embryo, illustrating the floor of the primordial pharynx and the location of the laryngotracheal groove.

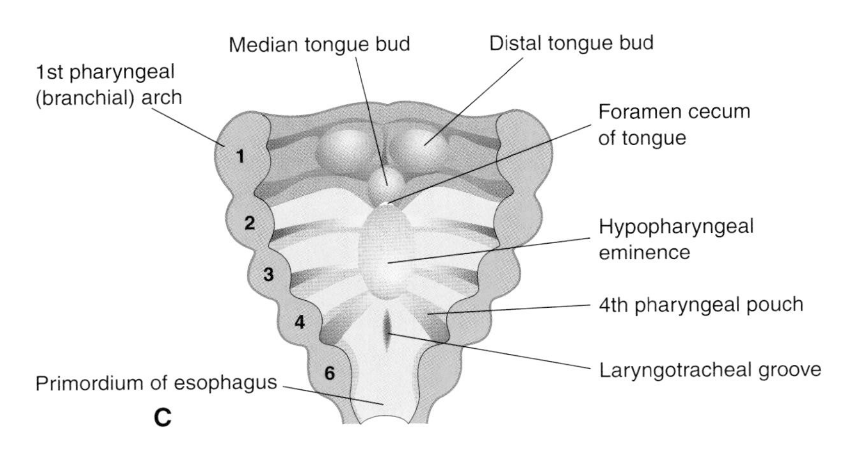

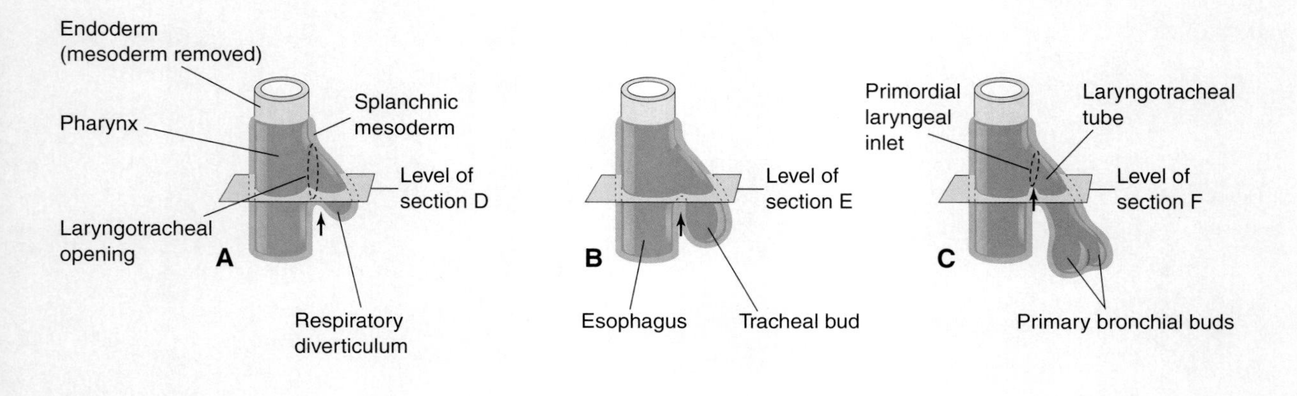

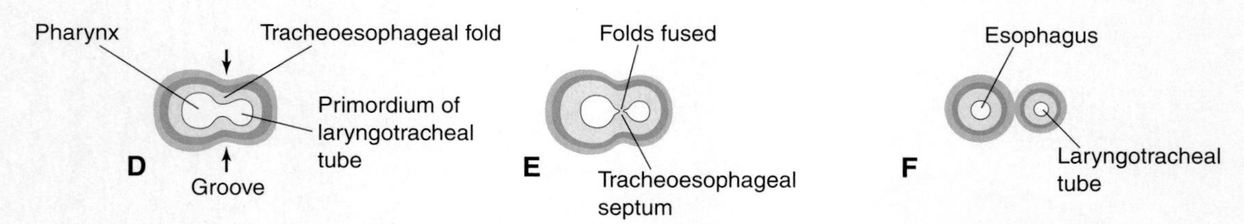

Figure 12 - 2. Successive stages in the development of the tracheoesophageal septum during the fourth and fifth weeks of development. *A* to *C*, Lateral views of the caudal part of the primordial pharynx showing the respiratory diverticulum and partitioning of the foregut into the esophagus and laryngotracheal tube. *D* to *F*, Transverse sections illustrating formation of the tracheoesophageal septum and how it separates the foregut into the laryngotracheal tube and esophagus.

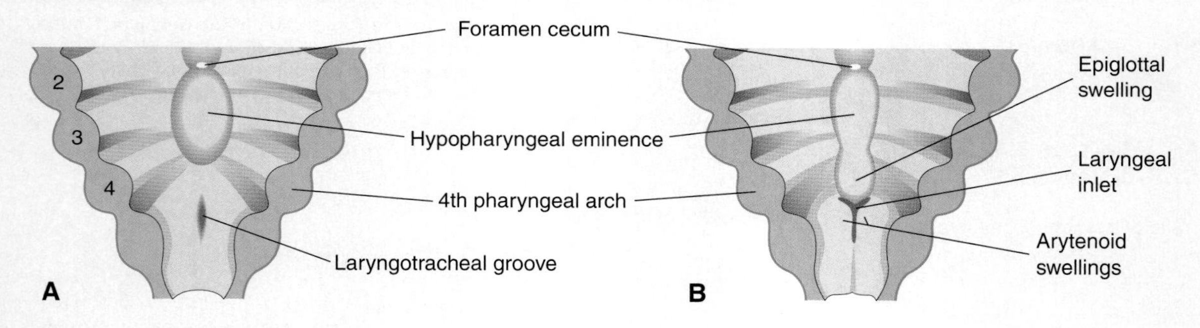

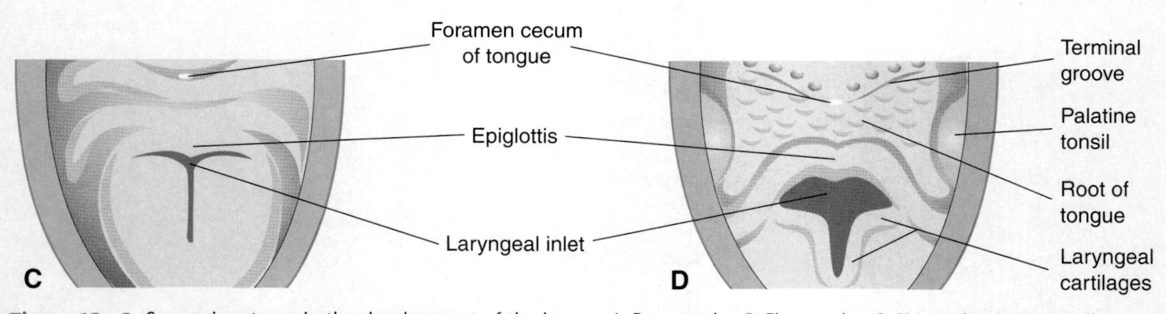

Figure 12 - 3. Successive stages in the development of the larynx. *A*, Four weeks. *B*, Five weeks. *C*, Six weeks. *D*, Ten weeks. The epithelium lining the larynx is of endodermal origin. The cartilages and muscles of the larynx arise from mesenchyme in the fourth and sixth pairs of pharyngeal arches. Note that the laryngeal inlet changes in shape from a slitlike opening to a T-shaped inlet as the mesenchyme surrounding the developing larynx proliferates.

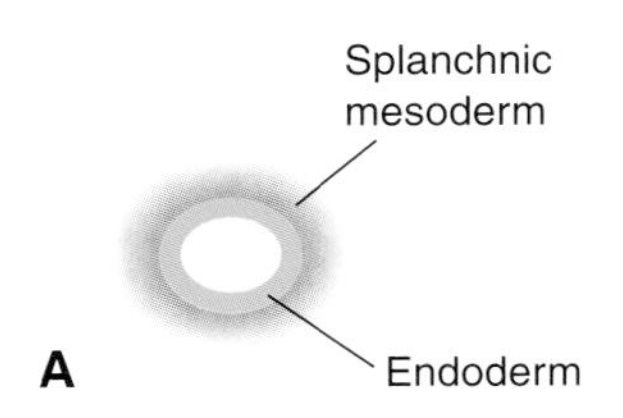

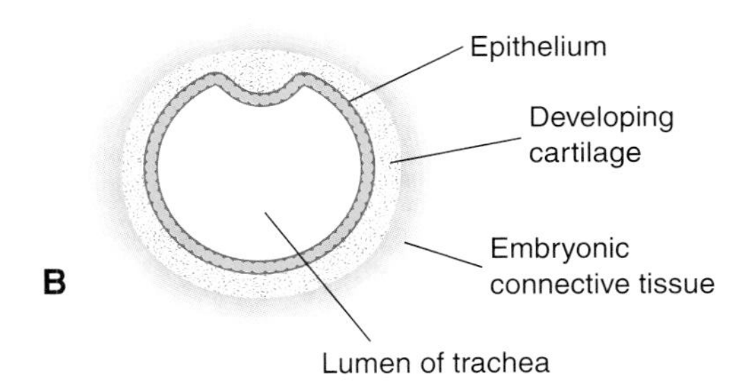

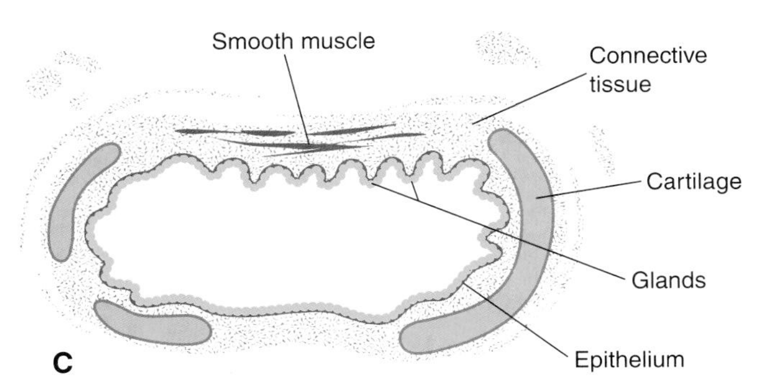

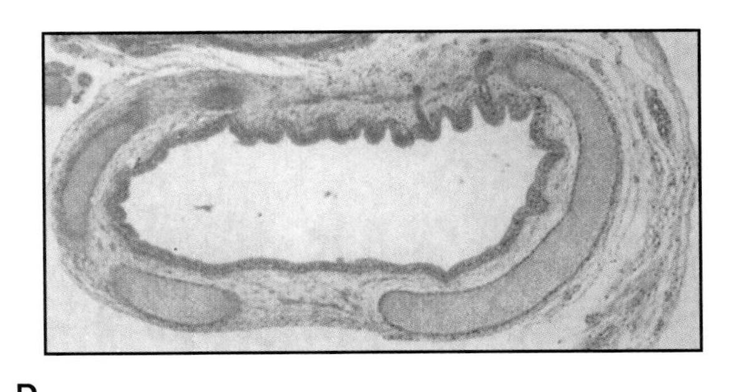

Figure 12-4. Transverse sections through the laryngotracheal tube illustrating progressive stages in the development of the trachea. *A*, Four weeks. *B*, Ten weeks. *C*, Eleven weeks. Note that the endoderm of the tube gives rise to the epithelium and glands of the trachea and that mesenchyme surrounding the tube forms the connective tissue, muscle, and cartilage. *D*, Photomicrograph of a transverse section of the developing trachea at 14 weeks. (From Moore KL, Persaud TVN, Shiota K: *Color Atlas of Clinical Embryology*, 2nd ed. Philadelphia, WB Saunders, 2000.)

TEF is the most common anomaly of the lower respiratory tract. Four main varieties of TEF may develop (see Fig. 12-5). The usual anomaly is a blind ending of the superior part of the esophagus (*esophageal atresia*) and a joining of the inferior part to the trachea near its bifurcation (see Figs 12-5*A* and 12-6). Other varieties of TEF are illustrated in Figure 12-5*B* to *D*. Infants with the common type of TEF and esophageal atresia cough and choke when swallowing because of the accumulation of excessive amounts of saliva in the mouth and upper respiratory tract. When the infant attempts to swallow milk, it rapidly fills the esophageal pouch and is regurgitated. Gastric contents may also reflux from the stomach through the fistula into the trachea and lungs. This causes choking and may result in pneumonia or *pneumonitis* (inflammation of the lungs). **Polyhydramnios** (see Chapter 8) is often associated with esophageal atresia and TEF. The excess amniotic fluid develops because fluid cannot pass to the stomach and intestines for absorption and subsequent transfer through the placenta to the mother's blood for disposal.

Tracheal Stenosis and Atresia

Narrowing (stenosis) and obstruction (atresia) of the trachea are uncommon anomalies that are usually associated with one of the varieties of tracheoesophageal fistula. Stenoses and atresias probably result from unequal partitioning of the foregut into the esophagus and trachea. In some cases, a web of tissue obstructs airflow (*incomplete tracheal atresia*).

Development of the Bronchi and Lungs

The tracheal bud that develops at the caudal end of the respiratory diverticulum during the fourth week (see Fig. 12-2*B*) soon divides into two outpouchings called **primary bronchial buds** (Fig. 12-7*A*). These buds grow laterally into the pericardioperitoneal canals, the primordia of the pleural cavities (see Fig. 12-7*B*). Together with the surrounding splanchnic mesoderm,

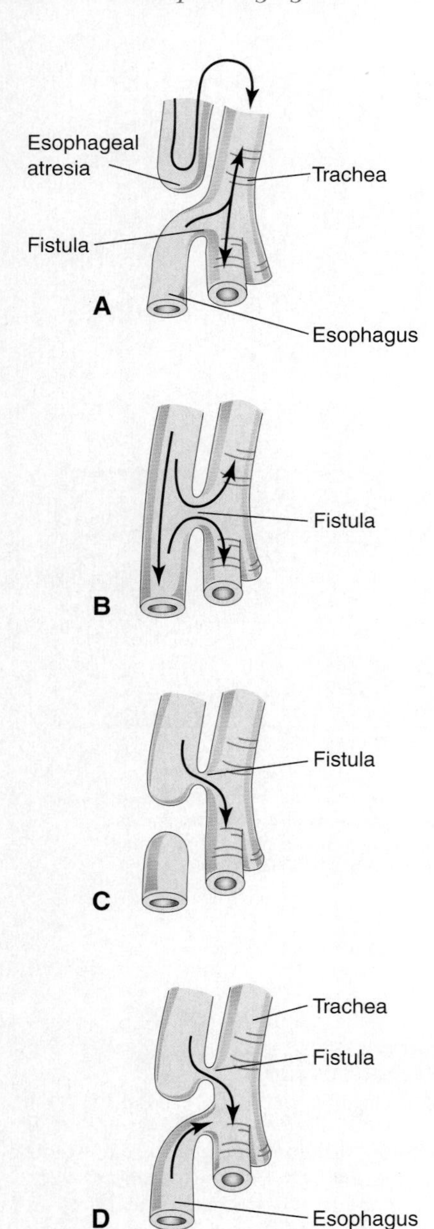

Figure 12–5. Main varieties of tracheoesophageal fistula, shown in order of frequency. Possible directions of the flow of the contents are indicated by arrows. *A,* Esophageal atresia (shown), is associated with tracheoesophageal fistula in more than 85% of cases. *B,* Fistula between the trachea and esophagus; this type of anomaly accounts for about 4% of cases. *C,* Air cannot enter the distal esophagus and stomach. *D,* Atresia of the proximal segment of the esophagus, with fistulas between the trachea and both proximal and distal segments of the esophagus. All infants born with tracheoesophageal fistula have esophageal dysmotility, and most have reflux.

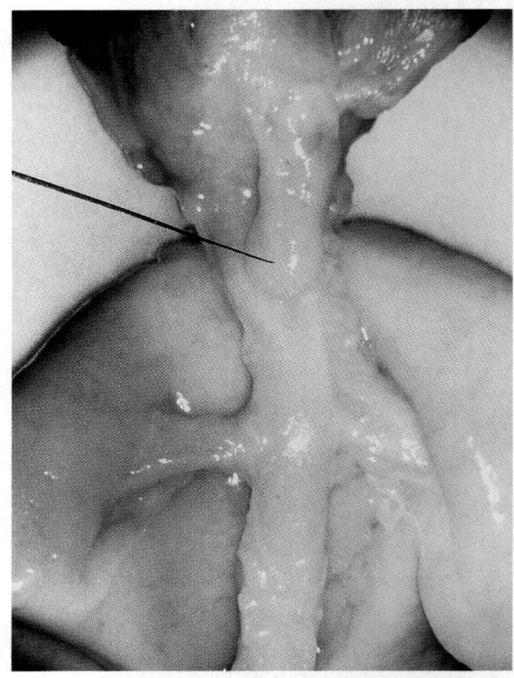

Figure 12–6. Tracheoesophageal fistula in a 17-week male fetus. The upper esophageal segment ends blindly (*pointer*). (From Kalousek DK, Fitch N, Paradice BA: *Pathology of the Human Embryo and Previable Fetus.* New York, Springer Verlag, 1990.)

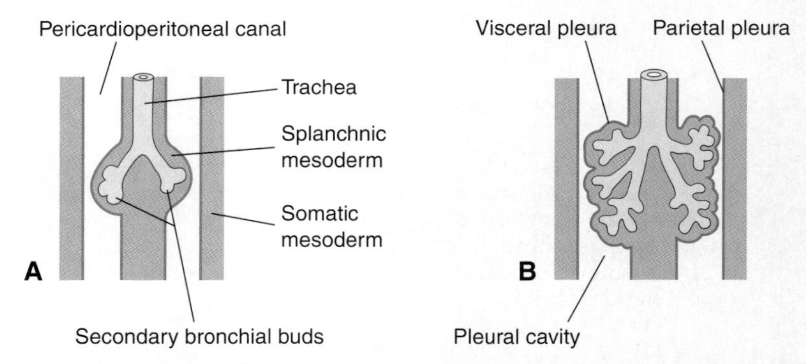

Figure 12–7. Diagrams illustrating growth of the developing lungs into the splanchnic mesoderm adjacent to the medial walls of the pericardioperitoneal canals (primordial pleural cavities). Development of the layers of the pleura is also shown. *A,* Five weeks. *B,* Six weeks.

the bronchial buds differentiate into the bronchi and their ramifications in the lungs. Early in the fifth week, the connection of each bronchial bud with the trachea enlarges to form the primordium of a primary or **main bronchus** (Fig. 12–8). The embryonic right main bronchus is slightly larger than the left one and is oriented more vertically. This embryonic relationship persists in the adult; consequently, a foreign body is

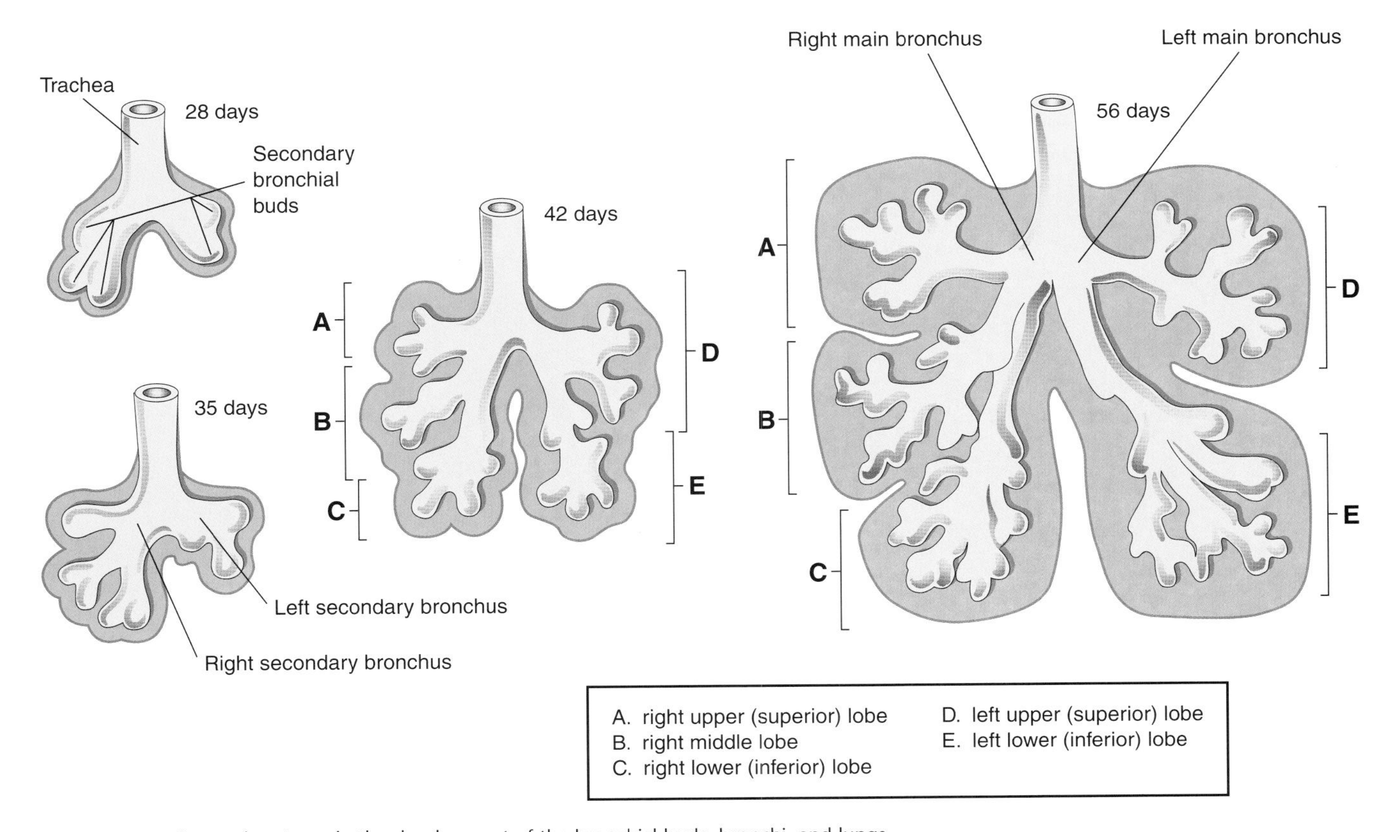

Figure 12 - 8. Successive stages in the development of the bronchial buds, bronchi, and lungs.

more likely to enter the right main bronchus than the left one. The main bronchi subdivide into **secondary bronchi** (see Fig. 12 - 7). On the right, the superior secondary bronchus supplies the upper (superior) lobe of the lung, whereas the inferior secondary bronchus subdivides into two bronchi, one connecting to the middle lobe of the right lung and the other connecting to the lower (inferior) lobe. On the left the two secondary bronchi supply the upper and lower lobes of the lung. Each secondary bronchus undergoes progressive branching. The **segmental bronchi** — 10 in the right lung and 8 or 9 in the left lung begin to form by the seventh week. As this occurs, the surrounding mesenchyme also divides. Each segmental bronchus with its surrounding mass of mesenchyme is the primordium of a **bronchopulmonary segment**. By 24 weeks, about 17 orders of branches have formed, and **respiratory bronchioles** have developed (Fig. 12 - 9*B*). An additional seven orders of airways develop after birth.

As the bronchi develop, cartilaginous plates develop from the surrounding splanchnic mesoderm. The bronchial smooth muscle and connective tissue and the pulmonary connective tissue and capillaries are also derived from this mesoderm. As the lungs develop, they acquire a layer of **visceral pleura** from the splanchnic mesoderm (see Fig. 12 - 7*B*). With expansion, the lungs and pleural cavities grow caudally into the mesenchyme of the body wall and soon lie close to the heart. The thoracic body wall becomes lined by a layer of **parietal pleura**, derived from the somatic mesoderm.

Maturation of the Lungs

Maturation of the lungs is divided into four periods:

- pseudoglandular period
- canalicular period
- terminal saccular period
- alveolar period

Pseudoglandular Period (6 to 16 Weeks)

The developing lung somewhat resembles an exocrine gland during the pseudoglandular period (see Fig. 12 - 9*A*). By 16 weeks, all major elements of the lung have formed except those involved with gas exchange. Respiration is not possible; hence, *fetuses born during this period are unable to survive.*

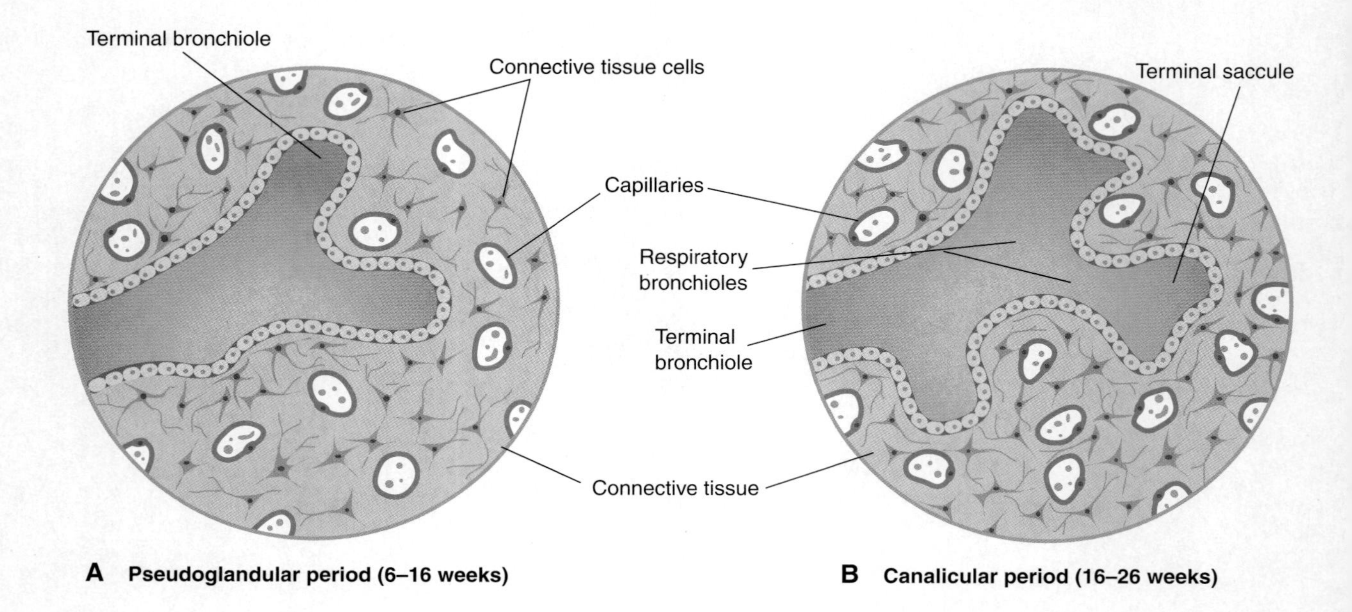

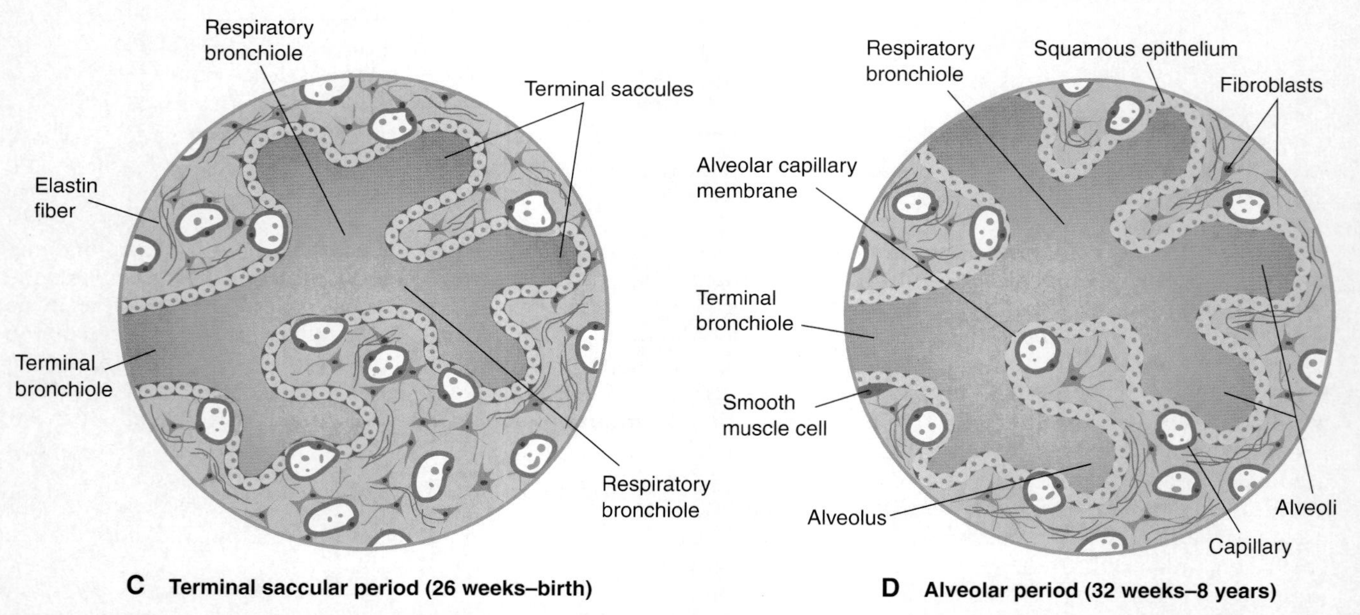

Figure 12 - 9. Diagrammatic of histologic sections, illustrating progressive stages of lung development. In *C* and *D*, note that the alveolocapillary membrane is thin and that some capillaries bulge into the terminal saccules (primordial alveoli).

Canalicular Period (16 to 26 Weeks)

The canalicular period overlaps the pseudoglanduiar period because cranial segments of the lungs mature faster than the caudal segments. During the canalicular period, the lumina of the bronchi and terminal bronchioles become larger, and the lung tissue becomes highly vascular (see Fig. 12 - 9*B*). By 24 weeks, each terminal bronchiole has given rise to two or more **respiratory bronchioles**, each of which then divides into three to six tubular passages called the **alveolar ducts**. Respiration is possible toward the end of the canalicular period because some thin-walled **terminal saccules** (primordial alveoli) have developed at the ends of the respiratory bronchioles, and *the lung tissue is well* *vascularized.* Although a fetus born toward the end of this period may survive if given intensive care, it often dies because its respiratory and other systems are still relatively immature.

Terminal Saccular Period (26 Weeks to Birth)

During this period, many more terminal saccules develop (see Fig. 12 – 9*C*), and *their epithelium becomes very thin.* Capillaries begin to bulge into these developing alveoli. The intimate contact between epithelial and endothelial cells establishes the **blood-air barrier**, which permits adequate gas exchange for survival of the fetus if it is born prematurely. By 26 weeks, the terminal saccules are lined mainly by squamous epithelial cells of endodermal origin — **type I alveolar cells** or pneumocytes — across which gas exchange occurs. The capillary network proliferates rapidly in the mesenchyme around the developing alveoli and there is concurrent active development of lymphatic capillaries. Scattered among the squamous epithelial cells are rounded secretory epithelial cells — **type II alveolar cells** or pneumocytes — *which secrete pulmonary surfactant*, a complex mixture of phospholipids.

Surfactant forms as a monomolecular film over the internal walls of the terminal saccules, lowering surface tension at the air-alveolar interface. The maturation of alveolar type II cells and surfactant production vary widely in fetuses of different ages. The production of surfactant increases during the terminal stages of pregnancy, particularly during the last 2 weeks before birth. *Surfactant counteracts surface tension forces and facilitates expansion of the terminal saccules* (primordial alveoli). Fetuses born prematurely at 24 to 26 weeks after fertilization may survive if given intensive care; however, they may suffer from respiratory distress because of *surfactant deficiency*. Surfactant production begins by 20 weeks, but it is present in only small amounts in premature infants; it does not reach adequate levels until the late fetal period. By 26 to 28 weeks after fertilization, the fetus usually weighs about 1000 gm, and sufficient terminal saccules and surfactant are present to permit survival of a prematurely born infant. Before this, the lungs are usually incapable of providing adequate gas exchange, partly because the alveolar surface area is insufficient and the vascularity underdeveloped. It is not the presence of thin terminal saccules or a primordial alveolar epithelium so much as the development of an adequate pulmonary vasculature and sufficient surfactant that are critical to the survival and neurodevelopmental outcome of premature infants.

Alveolar Period (32 Weeks to 8 Years)

Exactly when the terminal saccular period ends and the alveolar period begins depends on the definition of the term **alveolus**. Structures analogous to alveoli are present at 32 weeks. The epithelial lining of the terminal saccules attenuates to an extremely thin squamous epithelial layer. The type I alveolar cells become so thin that the adjacent capillaries bulge into the terminal saccules (see Fig. 12 – 9*D*). By the late fetal period, the lungs are capable of respiration because the **alveolocapillary membrane** (pulmonary diffusion barrier or respiratory membrane) is sufficiently thin to allow gas exchange. Although the lungs do not begin to perform this vital function until birth, they must be well developed so that they are capable of functioning as soon as the baby is born. At the beginning of the alveolar period, each respiratory bronchiole terminates in a cluster of thin-walled terminal saccules, separated from one another by loose connective tissue. These terminal saccules represent future alveolar ducts. The transition from dependence on the placenta for gas exchange to autonomous gas exchange requires the following adaptive changes in the lungs:

- production of adequate surfactant in the alveoli
- transformation of the lungs from secretory into gas-exchanging organs
- establishment of parallel pulmonary and systemic circulations

Characteristic mature alveoli do not form until after birth; about 95% of alveoli develop postnatally. Before birth, the primordial alveoli appear as small bulges on the walls of respiratory bronchioles and terminal saccules (future alveolar ducts). After birth, the primordial alveoli enlarge as the lungs expand; however, most of the increase in the size of the lungs results from an increase in the number of respiratory bronchioles and primordial alveoli rather than from an increase in the size of the alveoli. From the third to the eighth year or so, the number of immature alveoli continues to increase. Unlike mature alveoli, immature alveoli have the potential for forming additional primordial alveoli. As these alveoli increase in size, they become mature alveoli.

Lung development during the first few months after birth is characterized by an exponential increase in the surface of the air-blood barrier. This increase is accomplished by the multiplication of alveoli and capillaries. About 50 million alveoli, one sixth the number in adults, are present in the lungs of a full-term newborn infant. On chest radiographs, therefore, the lungs of newborn infants appear denser than adult lungs. By about the eighth year, the adult complement of 300 million alveoli is present.

Molecular studies have revealed that several regulatory substances participate in mesenchymal-epithelial interactions and in lung development. For example, *keratinocyte growth factor*, a member of the family of fibroblast growth factors, has been shown to be involved in lung morphogenesis by influencing branching, epithelial growth differentiation, and patterning of lung explants in culture.

Breathing movements occur before birth, exerting sufficient force to cause aspiration of some amniotic fluid into the lungs. These fetal breathing movements, which can be detected by real-time ultrasonography, are not continuous; however, they are essential for normal lung development. The pattern of fetal breathing movements is widely used in the diagnosis of labor and as a predictor of fetal outcome in preterm delivery. By birth the fetus has had the advantage of several months of breathing exercise. *Fetal breathing movements*, which increase as the time of delivery approaches, probably condition the respiratory muscles. In addition, these movements stimulate lung development, possibly by creating a pressure gradient between the lungs and the amniotic fluid.

At birth, the lungs are about half-filled with fluid derived from the amniotic cavity, lungs, and tracheal glands. Aeration of the lungs at birth is not so much the inflation of empty collapsed organs as the rapid replacement of intra-alveolar fluid by air. The fluid in the lungs is cleared at birth by three routes:

- through the mouth and nose by pressure on the thorax during delivery
- into the pulmonary capillaries
- into the lymphatics and pulmonary arteries and veins

In the near-term fetus, the pulmonary lymphatic vessels are relatively larger and more numerous than in the adult. Lymph flow is rapid during the first few hours after birth and then diminishes. *Three factors are important for normal lung development*:

- adequate thoracic space for lung growth
- fetal breathing movements
- adequate amniotic fluid volume

Oligohydramnios and Lung Development

The fluid in the lungs is an important stimulus for lung development. When oligohydramnios (an insufficient amount of amniotic fluid) is severe and chronic (e.g., because of amniotic fluid leakage), lung development is retarded and severe pulmonary hypoplasia results.

Lungs of Newborn Infants

Fresh, healthy lungs always contain some air; consequently, pulmonary tissue samples float in water. By contrast, a diseased lung that is partially filled with fluid may not float. Of medicolegal significance is the fact that the lungs of a stillborn infant are firm and sink when placed in water because they contain fluid, not air.

Respiratory Distress Syndrome

Respiratory distress syndrome (RDS) affects about 2% of live newborn infants, and those born prematurely are most susceptible to the disease. These affected infants develop rapid, labored breathing shortly after birth. RDS is also known as *hyaline membrane disease* (HMD). An estimated 30% of all neonatal disease results from HMD or its complications. *Surfactant deficiency is a major cause of RDS or HMD*. The lungs are underinflated, and the alveoli contain a fluid of high protein content that resembles a glassy or *hyaline membrane*. This membrane is believed to be derived from a combination of substances in the circulation and from the injured pulmonary epithelium. Prolonged *intrauterine asphyxia* may produce irreversible changes in the type II alveolar cells, making them incapable of producing surfactant. Not all the growth factors and hormones controlling surfactant production have been identified, but *thyroxine* is a potent stimulator of surfactant production.

Glucocorticoid treatment during pregnancy accelerates fetal lung development and surfactant production. This finding has led to the routine clinical use of corticosteroids (betamethasone) for the *prevention of RDS*. In addition, administration of exogenous surfactant (**surfactant replacement therapy**) reduces the severity of RDS and neonatal mortality.

Lung Hypoplasia

In infants with congenital diaphragmatic hernia (CDH; see Chapter 10), the lung is unable to develop normally because it is compressed by the abnormally positioned abdominal viscera. Lung hypoplasia is characterized by a markedly reduced lung volume. Most infants with CDH die of pulmonary insufficiency, despite optimal postnatal care, because their lungs are too hypoplastic to support extrauterine life.

Summary of the Respiratory System

Around the middle of the fourth week of development, the lower respiratory system begins to develop from a median *laryngotracheal groove* in the floor of the primordial pharynx. The groove deepens to produce a *laryngotracheal diverticulum* (respiratory primordium), which soon becomes separated from the foregut by tracheoesophageal folds that fuse to form a *tracheoesophageal septum*. This septum results in the formation of the esophagus and laryngotracheal tube. The endoderm of the *laryngotracheal tube* gives rise to the epithelium of the lower respiratory organs and the tracheobronchial glands. The splanchnic mesoderm surrounding the laryngotracheal tube forms the connective tissue, cartilage, muscle, and blood and lymphatic vessels of these organs. Pharyngeal arch mesenchyme contributes to formation of the epiglottis and connective tissue of the larynx. The *laryngeal muscles* are derived from mesenchyme in the caudal pharyngeal arches. The *laryngeal cartilages* are derived from the cartilaginous bars in the fourth and sixth pairs of pharyngeal arches, which are derived from *neural crest cells* (see Table 11 – 1).

During the fourth week, the respiratory diverticulum develops a *tracheal bud* at its distal end; this bud subsequently divides into two *bronchial buds* during the early part of the fifth week. Each bronchial bud soon enlarges to form a *main bronchus*, and then each bronchus gives rise to two new bronchial buds, which develop into *secondary bronchi*. The right inferior secondary bronchus soon divides into two bronchi. The secondary bronchi supply the lobes of the developing lungs. Each bronchus undergoes progressive branching to form segmental bronchi. Each segmental bronchus, along with its surrounding mesenchyme, is the primordium of a *bronchopulmonary segment*. Branching continues until about 17 orders of branches have formed. Additional airways are formed after birth until about 24 orders of branches are present.

Lung development is divided into four periods. During the *pseudoglandular period* (6 to 16 weeks), the bronchi and terminal bronchioles form. During the *canalicular period* (16 to 26 weeks), the lumina of the bronchi and terminal bronchioles enlarge, the respiratory bronchioles and alveolar ducts develop, and the lung tissue becomes highly vascular. During the *terminal saccular period* (26 weeks to birth), the alveolar ducts give rise to terminal saccules (primordial alveoli). The terminal saccules are initially lined with cuboidal epithelium that begins to attenuate to squamous epithelium at about 26 weeks. By this time, capillary networks have proliferated close to the alveolar epithelium and the lungs are usually sufficiently well developed to permit survival of the fetus if it is born prematurely. The *alveolar period*, the final stage of lung development, extends from 32 weeks to about 8 years of age. As the lungs mature, the number of respiratory bronchioles and primordial alveoli increases.

The respiratory system develops in such a way that it is capable of immediate function at birth. To be capable of respiration, the lungs must acquire an *alveolocapillary membrane* that is sufficiently thin, as well as an adequate amount of *surfactant*. A deficiency of surfactant appears to be responsible for the failure of primordial alveoli to remain open, resulting in **RDS**. Growth of the lungs after birth results mainly from an increase in the number of respiratory bronchioles and alveoli. New alveoli form for at least 8 years after birth. Major congenital anomalies of the lower respiratory system are uncommon except for *tracheoesophageal fistula*, which is usually associated with *esophageal atresia*. These anomalies result from faulty partitioning of the foregut into the esophagus and trachea during the fourth and fifth weeks of development.

Clinically Oriented Questions

1. Does the fetus breathe before birth?
2. What stimulates the baby to start breathing when it is born? Is "slapping the buttocks" necessary?
3. A baby reportedly died about 72 hours after birth from the effects of *respiratory distress syndrome* (*RDS*). What is RDS? By what other name is this condition known? Is its cause genetic or environmental?
4. Can an infant born 22 weeks after fertilization survive?

The answers to these questions are at the back of the book.

13

The Digestive System

The **primordial gut** at the beginning of the fourth week of development is closed at its cranial end by the **oropharyngeal membrane** (see Fig. 11 – 1) and at its caudal end by the **cloacal membrane** (Fig. 13 – 1). The primordial gut forms during the fourth week as the head, tail, and lateral folds incorporate the dorsal part of the yolk sac into the embryo (see Chapter 6). The endoderm of the primordial gut gives rise to most of the epithelium and glands of the digestive tract. The epithelium at the cranial and caudal extremities of the tract is derived from ectoderm of the **stomodeum** (primordial mouth) and **proctodeum** (anal pit), respectively (see Fig. 13 – 1). The muscular tissue, connective tissue, and other layers of the wall of the digestive tract are derived from the splanchnic mesenchyme surrounding the primordial gut. For descriptive purposes, the primordial gut is divided into three parts: foregut, midgut, and hindgut. The regional differentiation of the primordial gut is established by *homeobox genes* that are expressed in the endoderm and surrounding mesoderm. The endoderm provides temporal and positional information for the development of the gut.

The Foregut

The derivatives of the foregut are as follows:

- the *primordial pharynx* and its derivatives (oral cavity, pharynx, tongue, tonsils, salivary glands, and upper respiratory system), which are discussed in Chapter 11
- the *lower respiratory system* (described in Chapter 12)
- the *esophagus and stomach*
- the *duodenum,* proximal to the opening of the bile duct
- the *liver, biliary apparatus* (hepatic ducts, gallbladder, and bile duct), and *pancreas*

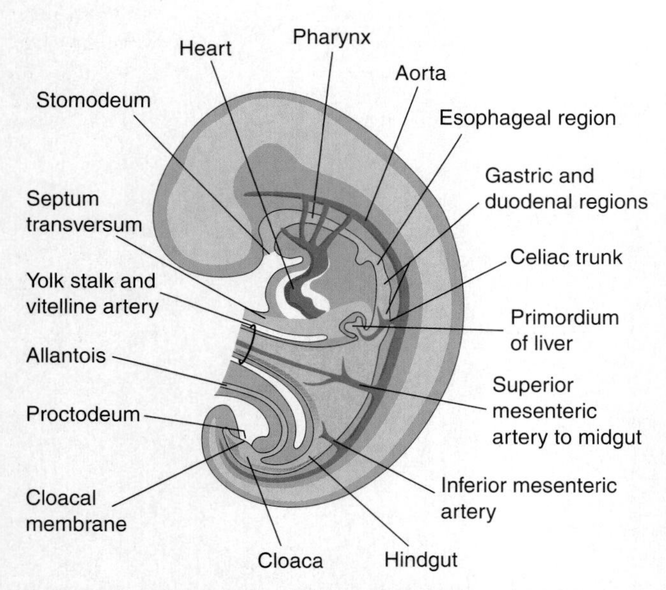

Figure 13 – 1. Drawing of a median section of a 4-week embryo showing the early digestive system and its blood supply. The primordial gut is a long tube extending the length of the embryo. Its blood vessels are derived from the vessels that supplied the yolk sac.

All of these foregut derivatives *except* the pharynx, respiratory tract, and most of the esophagus are supplied by the *celiac trunk,* the artery of the foregut (see Fig. 13 – 1).

Development of the Esophagus

The esophagus develops from the foregut immediately caudal to the pharynx (see Fig. 13 – 1*B*). The partitioning of the trachea from the esophagus by the **tracheoesophageal septum** is described in Chapter 12. Initially, the esophagus is short but it elongates rapidly and reaches its final relative length by the seventh week of development. Its epithelium and glands are derived from endoderm. The epithelium proliferates and partly or completely obliterates the lumen; however, recanalization of the esophagus normally occurs by the end of the embryonic period. The striated muscle forming the muscularis externa of the superior third of the esophagus is derived from mesenchyme in the caudal pharyngeal arches. The smooth muscle, mainly in the inferior third of the esophagus, develops from the surrounding splanchnic mesenchyme. Both types of muscle are innervated by branches of the vagus nerves (cranial nerve [CN] X), which supply the caudal pharyngeal arches (see Table 11 – 1).

Esophageal Atresia

Blockage of the esophagus occurs in approximately 1 in 3000 to 4500 live births. About one third of affected infants are born prematurely. Esophageal atresia is frequently associated with **tracheoesophageal fistula** (see Fig. 12 – 5). It may occur as a separate anomaly, but this is less common. Esophageal atresia results from deviation of the *tracheoesophageal septum* in a posterior direction (see Figs. 12 – 2 and 12 – 6); as a result, separation of the esophagus from the laryngotracheal tube is incomplete. In some cases, the atresia results from *failure of esophageal recanalization* during the eighth week of development.

A fetus with esophageal atresia is unable to swallow amniotic fluid; consequently, this fluid cannot pass to the intestine for absorption and transfer through the placenta to the maternal blood for disposal. This results in **polyhydramnios,** the accumulation of an excessive amount of amniotic fluid.

Esophageal Stenosis

Narrowing of the lumen of the esophagus can occur anywhere along the esophagus, but it usually occurs in the distal third, either as a web or as a long segment of esophagus with a threadlike lumen. The stenosis usually results from incomplete recanalization of the esophagus during the eighth week of development, but it may result from a failure of esophageal blood vessels to develop in the affected area. As a result, atrophy of a segment of the esophageal wall occurs.

Development of the Stomach

The distal part of the foregut is initially a simple tubular structure (see Fig. 13 - 1*B*). Around the middle of the fourth week of development, a slight dilation of the foregut indicates the site of the primordial stomach. It first appears as a fusiform enlargement of the caudal part of the foregut and is initially oriented in the median plane (see Figs. 13 - 1 and 13 - 2*B*). This primordium soon enlarges and broadens ventrodorsally. During the next 2 weeks, the dorsal border of the primordial stomach grows faster than its ventral border; this demarcates the **greater curvature of the stomach** (see Fig. 13 - 2*D, F,* and *G*).

Rotation of the Stomach

As the stomach enlarges and acquires its adult shape, it slowly rotates 90 degrees in a clockwise direction around its longitudinal axis. The effects of rotation on the stomach are as follows (see Figs. 13 - 2 and 13 - 3):

- The ventral border (lesser curvature) moves to the right, and the dorsal border (greater curvature) moves to the left.
- The original left side becomes the ventral surface, and the original right side becomes the dorsal surface.
- Before rotation, the cranial and caudal ends of the stomach are in the median plane (see Fig. 13 - 2*B*). During rotation and growth of the stomach, its cranial region moves to the left and slightly inferiorly, and its caudal region moves to the right and superiorly.
- After rotation, the stomach assumes its final position with its long axis almost transverse to the long axis of the body (see Fig. 13 - 2*E*). The rotation and growth of the stomach explain why the left vagus nerve supplies the anterior wall of the adult stomach and the right vagus nerve innervates its posterior wall.

Mesenteries of the Stomach

The stomach is suspended from the dorsal wall of the abdominal cavity by a dorsal mesentery — the **dorsal mesogastrium** (see Figs. 13 - 2*B* and C and 13 - 3*A*). This mesentery, originally located in the median plane, is carried to the left during rotation of the stomach and formation of the *omental bursa* or lesser sac of peritoneum (see Fig. 13 - 3*A* to *E*). A ventral mesentery or **ventral mesogastrium** attaches the stomach and duodenum to the liver and the ventral abdominal wall (see Fig. 13 - 2*C*).

Omental Bursa

Isolated clefts (cavities) develop in the mesenchyme, forming the thick dorsal mesogastrium (see Fig. 13 - 3*A* and *B*). The clefts soon coalesce to form a single cavity — the **omental bursa** (see Figs. 13 - 2*F* and *G* and 13 - 3*C* and *D*). Rotation of the stomach pulls the dorsal mesogastrium to the left, thereby enlarging the bursa, a large recess of the peritoneal cavity. The omental bursa lies between the stomach and the posterior abdominal wall. The superior part of the omental bursa is cut off as the diaphragm develops, forming a closed space known as the *infracardiac bursa*. Part of the omental bursa persists as the **superior recess of the omental bursa.** As the stomach enlarges, the omental bursa expands and acquires an **inferior recess of the omental bursa** between the layers of the elongated dorsal mesogastrium — called the **greater omentum.** This membrane overhangs the developing intestines (see Fig. 13 - 3*G* to *J*). The inferior recess of the omental bursa disappears as the layers of the greater omentum fuse (see Fig. 13 - 14*F*). The omental bursa communicates with the main part of the peritoneal cavity through a small opening known as the **omental foramen** (see Figs. 13 - 2*D* and *F* and 13 - 3*C* and *F*). In the adult, this foramen is located posterior to the free edge of the lesser omentum.

Congenital Hypertrophic Pyloric Stenosis

Anomalies of the stomach are uncommon except for hypertrophic pyloric stenosis. This anomaly affects 1 in every 150 male infants and 1 in every 750 female infants. Infants with this anomaly have a marked **thickening of the pylorus,** the distal sphincteric region of the stomach. The circular and, to a lesser degree, the longitudinal muscles in the pyloric region are hypertrophied. This results in severe *stenosis (narrowing) of the pyloric canal* and obstruction to the passage of food. As a result, the stomach becomes markedly distended and the infant expels the stomach's contents with considerable force **(projectile vomiting).** Surgical relief of the pyloric obstruction is the usual treatment. The cause of congenital pyloric stenosis is unknown, but the high incidence of the condition in both infants of monozygotic twins suggests the involvement of genetic factors.

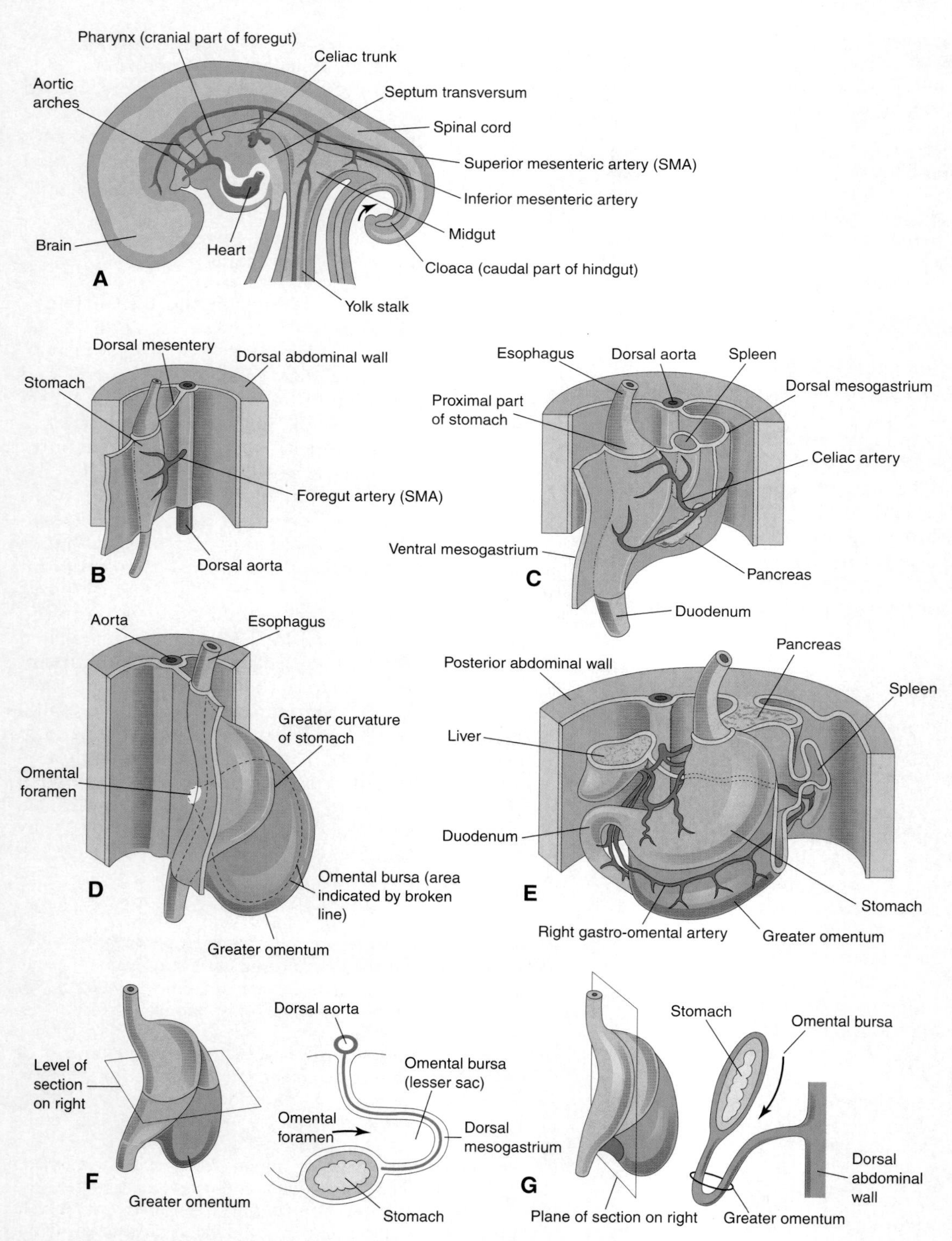

Figure 13-2. Drawings illustrating the development and rotation of the stomach and formation of the omental bursa (lesser sac) and greater omentum. *A,* Drawing of a median section of a 28-day embryo. *B,* Anterolateral view of a 28-day embryo. *C,* Embryo at about 35 days. *D,* Embryo at about 40 days. *E,* Embryo at about 48 days. *F,* Lateral view of the stomach and greater omentum of an embryo at about 52 days. The transverse section shows the omental foramen and omental bursa. *G,* Sagittal section showing the omental bursa and greater omentum.

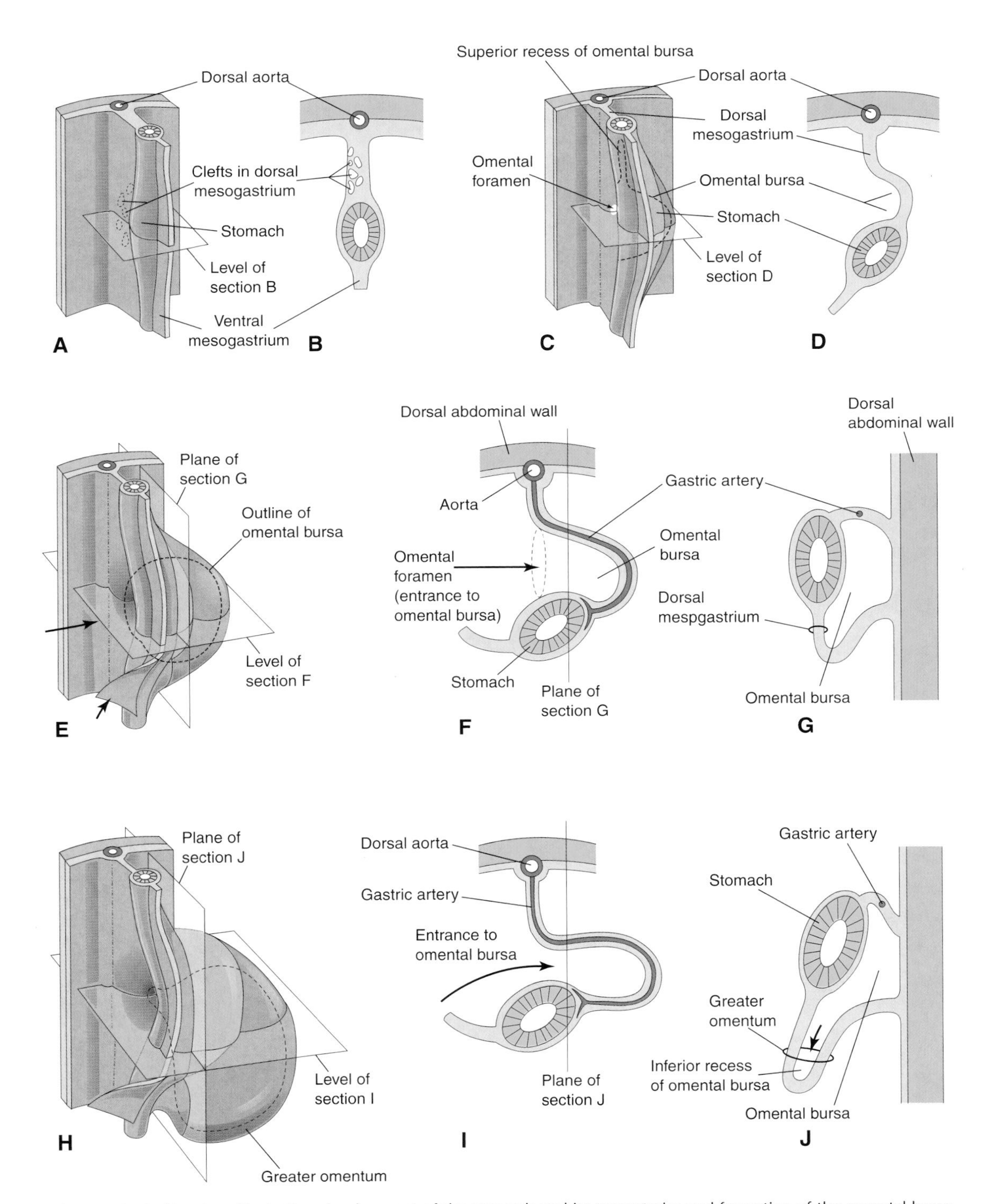

Figure 13 – 3. Drawings illustrating development of the stomach and its mesenteries and formation of the omental bursa (lesser sac). *A,* Five weeks. *B,* Transverse section showing clefts in the dorsal mesogastrium. *C,* Later stage, after coalescence of the clefts to form the omental bursa. *D,* Transverse section showing the initial appearance of the omental bursa. *E,* The dorsal mesentery has elongated and the omental bursa has enlarged. *F* and *G,* Transverse and sagittal sections, respectively, showing elongation of the dorsal mesogastrium and expansion of the omental bursa. *H,* Six weeks, showing the greater omentum and expansion of the omental bursa. *I* and *J,* Transverse and sagittal sections, respectively, showing the inferior recess of the omental bursa and the omental foramen.

Development of the Duodenum

Early in the fourth week of development, the duodenum begins to develop from the caudal part of the foregut, the cranial part of the midgut, and the splanchnic mesenchyme associated with these endodermal parts of the primordial gut (Fig. 13 – 4*A*). The two parts of the duodenum are joined just distal to the origin of the bile duct. The developing duodenum grows rapidly, forming a C-shaped loop that projects ventrally (see Fig. 13 – 4*B* to *D*). As the stomach rotates, the duodenal loop rotates to the right and comes to lie retroperitoneally (external to the peritoneum). Because of its derivation from the foregut and midgut, the duodenum is supplied by branches of the celiac and superior mesenteric arteries that supply these parts of the primordial gut (see Fig. 13 – 1). During the fifth and sixth weeks of development, the lumen of the duodenum becomes progressively smaller

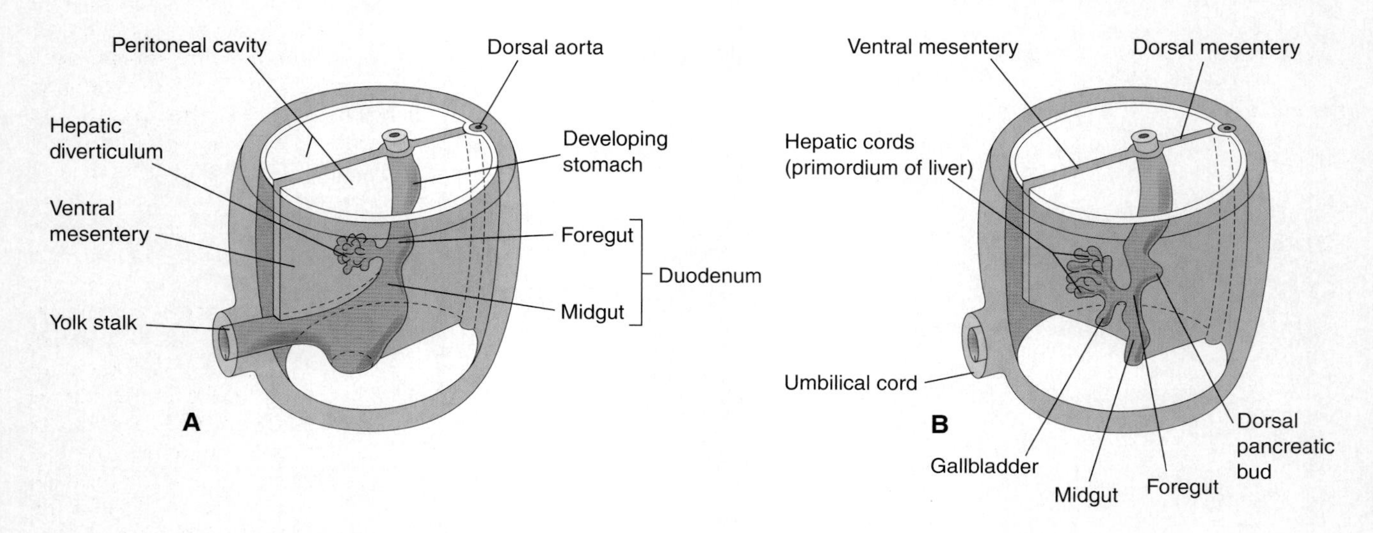

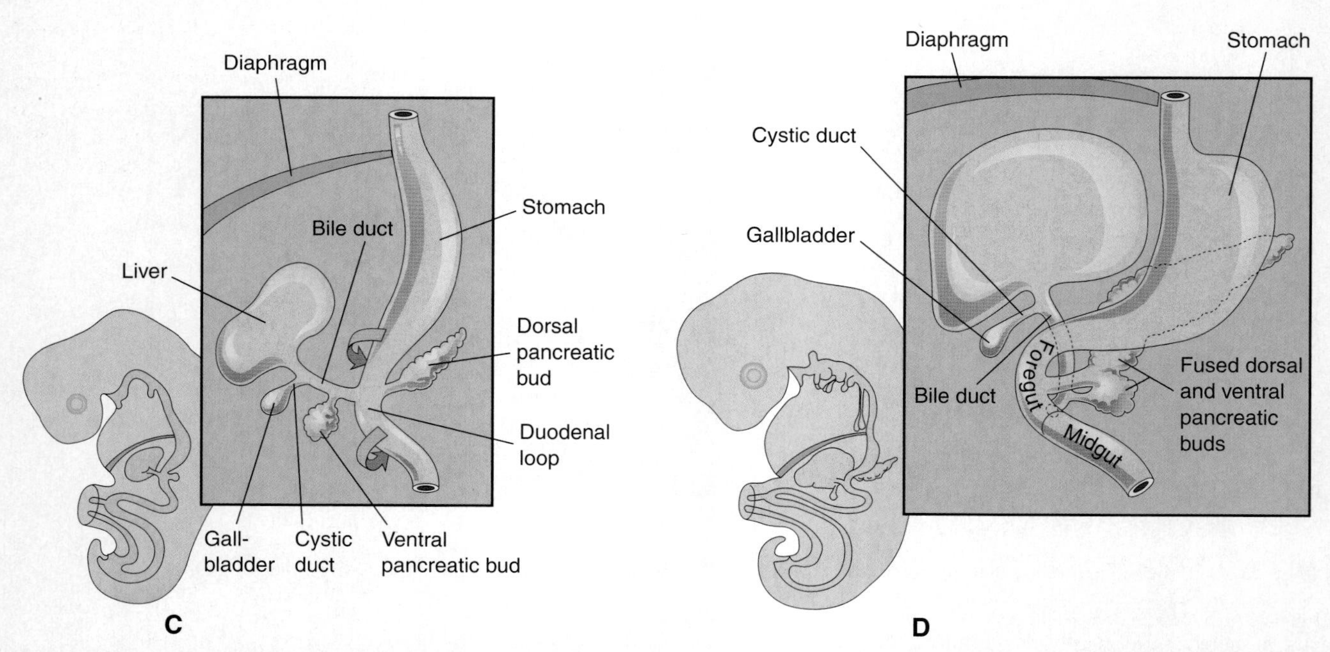

Figure 13 – 4. Drawings illustrating progressive stages in the development of the duodenum, liver, pancreas, gallbladder, and extrahepatic biliary apparatus. *A,* Four weeks. *B* and *C,* Five weeks. *D,* Six weeks. Note that the entrance of the bile duct into the duodenum gradually shifts from its initial position to a posterior one. This explains why the bile duct in the adult passes posterior to the duodenum and the head of the pancreas.

and is temporarily obliterated because of the proliferation of its epithelial cells. Normally, vacuolation occurs as degeneration of the epithelial cells occurs; as a result, the duodenum generally becomes recanalized by the end of the embryonic period. By this time, most of the ventral mesentery of the duodenum has disappeared.

Duodenal Stenosis

Partial occlusion of the duodenal lumen — termed duodenal stenosis (Fig. 13–5*A*) — is usually caused by incomplete recanalization of the duodenum resulting from defective vacuolization. Most stenoses involve the horizontal (third) or ascending (fourth) parts of the duodenum, or both. Because of the occlusion, the stomach's contents (usually containing bile) are often vomited.

Duodenal Atresia

Complete occlusion of the lumen of the duodenum — or duodenal atresia (see Fig. 13–5*B*) — is not uncommon. Twenty to 30% of affected infants have Down syndrome, and an additional 20% are premature. In about 20% of cases, the bile duct enters the duodenum just distal to the opening of the hepatopancreatic ampulla. During duodenal development, the lumen is completely occluded by epithelial cells. If reformation of the lumen fails to occur, a short segment of the duodenum is occluded. Most atresias involve the descending (second) and horizontal (third) parts of the duodenum and are located distal to the opening of the bile duct. In infants with duodenal atresia, vomiting begins within a few hours of birth. The vomitus almost always contains bile. **Polyhydramnios** also occurs because duodenal atresia prevents normal absorption of amniotic fluid by the intestines. The diagnosis of duodenal atresia is suggested by the presence of a "double-bubble sign" on plain radiographs or ultrasound scans (Fig. 13–6). This double-bubble sign is caused by a distended, gas-filled stomach and the proximal duodenum.

Development of the Liver, Gallbladder, and Biliary Apparatus

The liver, gallbladder, and biliary duct system arise as a ventral outgrowth from the caudal part of the foregut early in the fourth week of development (see Figs. 13–4*A* and 13–7*A*). The **hepatic diverticulum** extends into the septum transversum (see Fig. 13–7*B*), a mass of splanchnic mesoderm between the developing heart and midgut. The septum transversum forms the ventral mesentery in this region. The hepatic diverticulum enlarges rapidly and divides into two parts as it grows between the layers of the **ventral mesentery** (see Fig. 13–4*A*). The larger cranial part of the hepatic diverticulum is the **primordium of the liver.** The proliferating endodermal cells give rise to interlacing cords of hepatic cells and to the epithelial lining of the intrahepatic part of the biliary apparatus. The **hepatic cords** anastomose around endothelium-lined spaces, the primordia of the *hepatic sinusoids.* The fibrous and *hematopoietic tissue* and *Kupffer cells* of the liver are derived from mesenchyme in the septum transversum. The liver grows rapidly and, from the 5th to 10th weeks of development, fills a large part of the abdominal cavity (see Fig. 13–7*C* and *D*). Initially, the right and left lobes of the liver are about the same size, but the right lobe soon becomes larger. *Hematopoiesis* (formation of various types of blood cells and other formed elements), *which begins during the sixth week,* is mainly responsible for the relatively large size of the liver. By the ninth week, the liver accounts for about 10% of the total weight of the fetus. *Bile formation* by the hepatic cells begins during the 12th week of development.

The small caudal part of the hepatic diverticulum becomes the **gallbladder,** and the stalk of the diverticulum forms the **cystic duct** (see Fig. 13–4*C*). Initially, the extrahepatic biliary apparatus is occluded with epithelial

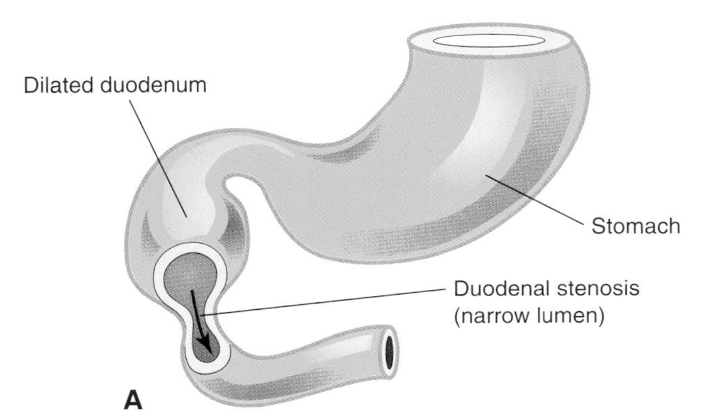

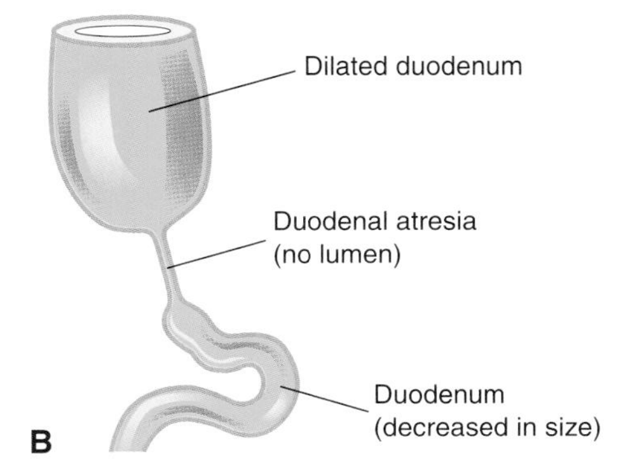

Figure 13–5. Diagrams illustrating the embryological basis of the two common types of congenital intestinal obstruction. *A,* Duodenal stenosis. *B,* Duodenal atresia. Most duodenal atresias occur in the descending (second) and horizontal (third) parts of the duodenum.

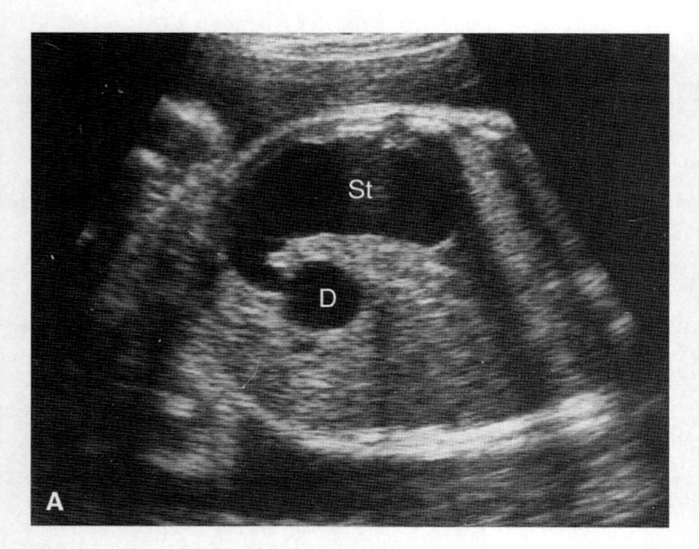

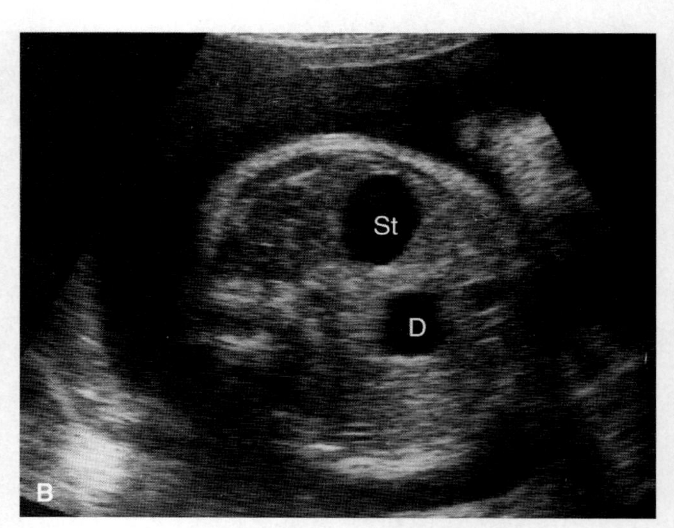

Figure 13 – 6. Ultrasound scans of a fetus at 33 weeks' gestation (31 weeks after fertilization) showing duodenal atresia. *A,* An oblique scan demonstrates the dilated, fluid-filled stomach (**St**) entering the proximal duodenum (**D**), which is also enlarged because of the atresia (blockage) distal to it. *B,* Transverse scan illustrating the characteristic "double-bubble" appearance of the stomach and duodenum when there is duodenal atresia. (Courtesy of Dr. Lyndon M. Hill, Magee-Women's Hospital, Pittsburgh, PA.)

cells, but it is later canalized because of vacuolation resulting from degeneration of these cells. The stalk connecting the hepatic and cystic ducts to the duodenum becomes the **bile duct** (common bile duct). Initially, this duct attaches to the ventral aspect of the duodenal loop; however, as the duodenum grows and rotates, the entrance of the bile duct is carried to the dorsal aspect of the duodenum (see Fig. 13 – 4*C* and *D*). The bile entering the duodenum through the bile duct after the 13th week of development gives the **meconium** (intestinal contents) a dark green color.

Ventral Mesentery

The thin, double-layered ventral mesentery (Fig. 13 – 7) gives rise to two structures:

- the **lesser omentum**, which passes from the liver to the lesser curvature of the stomach (*hepatogastric ligament*) and from the liver to the duodenum (*hepatoduodenal ligament*)
- the **falciform ligament**, which extends from the liver to the ventral abdominal wall

The **umbilical vein** passes in the free border of the falciform ligament on its way from the umbilical cord to the liver. The ventral mesentery also forms the *visceral peritoneum of the liver.* The liver is covered by peritoneum except for the *bare area* that is in direct contact with the diaphragm (Fig. 13 – 8).

Anomalies of Liver

Minor variations of liver lobulation are common, but congenital anomalies of the liver are rare. Variations of the hepatic ducts, bile duct, and cystic duct are common and clinically significant. *Accessory hepatic ducts* may be present, and an awareness of their possible presence is important from a surgical perspective.

Extrahepatic Biliary Atresia

Extrahepatic biliary atresia, the most serious anomaly involving the extrahepatic biliary system, occurs in 1 in 10,000 to 15,000 live births. The most common form of extrahepatic biliary atresia is obstruction of the ducts at or superior to the *porta hepatis* — a deep, transverse fissure on the visceral surface of the liver. Failure of the bile ducts to canalize often results from persistence of the solid stage of duct development. *Jaundice* occurs soon after birth. When biliary atresia cannot be corrected surgically, the child may die unless a liver transplant is performed.

Development of the Pancreas

The pancreas develops between the layers of the mesentery from the dorsal and ventral **pancreatic buds** of endodermal cells, which arise from the caudal part of the foregut (Fig. 13 – 9). Most of the pancreas is derived

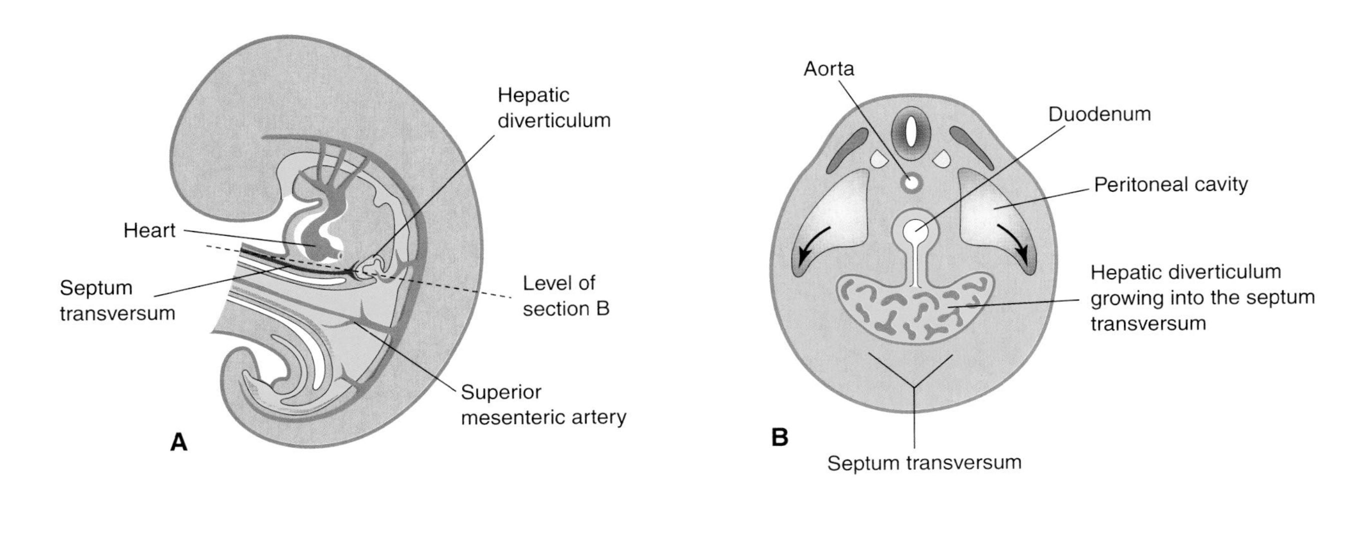

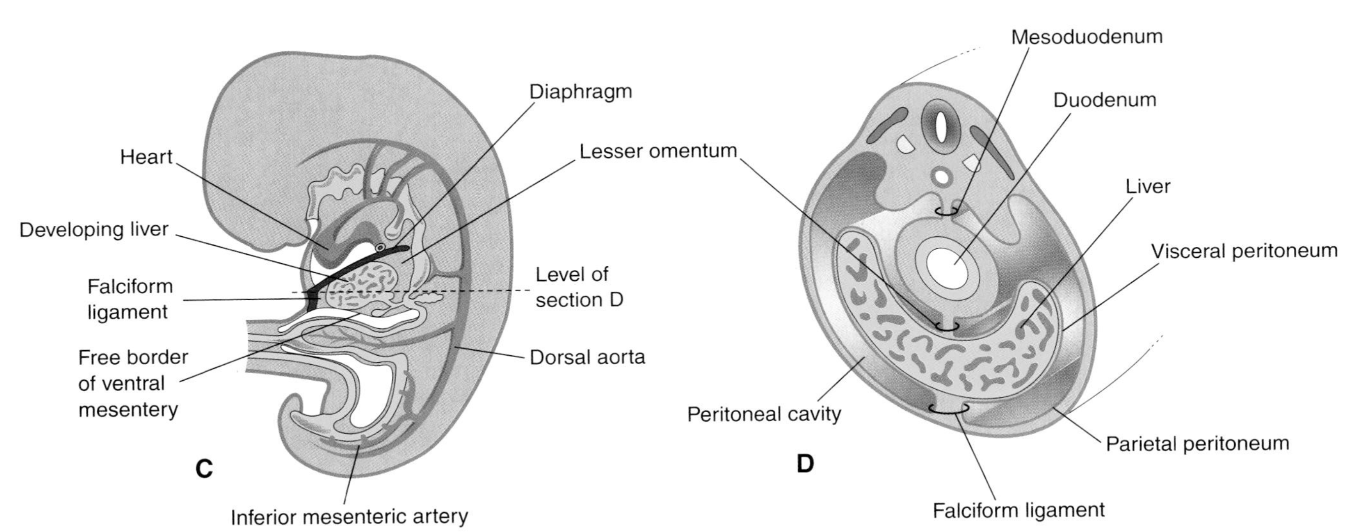

Figure 13-7. Drawings illustrating how the caudal part of the septum transversum becomes stretched and membranous as it forms the ventral mesentery. *A,* Median section of a 4-week embryo. *B,* Transverse section of the embryo, showing expansion of the peritoneal cavity (*arrows*). *C,* Sagittal section of a 5-week embryo. *D,* Transverse section of the embryo after formation of the dorsal and ventral mesenteries. Note that the liver is joined to the ventral abdominal wall and to the stomach and the duodenum by the falciform ligament and lesser omentum, respectively.

from the dorsal pancreatic bud. The larger **dorsal pancreatic bud** appears first and develops a slight distance cranial to the ventral bud. It grows rapidly between the layers of the dorsal mesentery. The **ventral pancreatic bud** develops near the entry of the bile duct into the duodenum and grows between the layers of the ventral mesentery (Fig. 13-9*A* and *B*). As the duodenum rotates to the right and becomes C-shaped, the ventral pancreatic bud is carried dorsally with the bile duct (Fig. 13-9*C* to *F*). It soon lies posterior to the dorsal pancreatic bud and later fuses with it. *Molecular studies* show that the ventral pancreas develops from a bipotential cell population in the ventral part of the endoderm. A default mechanism involving *fibroblast growth factor* (FGF), which is secreted by the developing heart, appears to play an important role. Formation of the dorsal pancreatic bud depends on signals from the notochord (*activin and FGF-2*) that block the expression of *Sonic hedgehog* (*Shh*) in the endoderm.

The ventral pancreatic bud forms the *uncinate process* and part of the *head of the pancreas*. As the stomach, duodenum, and ventral mesentery rotate, the pancreas comes to lie along the dorsal abdominal wall. As the pancreatic buds fuse, their ducts anastomose. The **main**

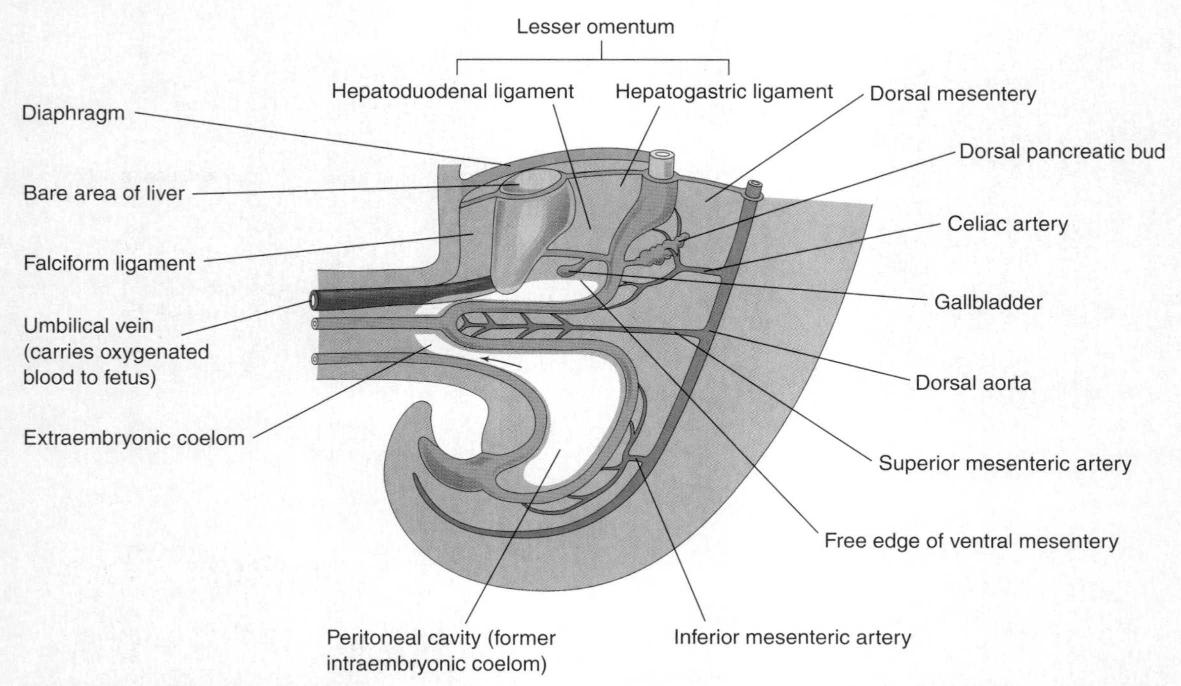

Figure 13-8. Diagram of a median section of the caudal half of an embryo at the end of the fifth week of development showing the liver and its associated ligaments. The arrow indicates the communication of the peritoneal cavity with the extraembryonic coelom. Because of the rapid growth of the liver and the midgut loop, the abdominal cavity temporarily becomes too small to contain the developing intestines; consequently, they enter the extraembryonic coelom in the proximal part of the umbilical cord.

pancreatic duct forms from the duct of the ventral bud and the distal part of the duct of the dorsal bud (see Fig. 13-9*G*). The proximal part of the duct of the dorsal bud often persists as an **accessory pancreatic duct** that opens into the *minor duodenal papilla.* In about 9% of people, the pancreatic ducts fail to fuse and the original two ducts persist.

Insulin secretion begins during the early fetal period (at about 10 weeks). The *glucagon-* and *somatostatin-containing cells* develop before differentiation of the insulin-secreting cells. Glucagon has been detected in fetal plasma at 15 weeks. With increasing fetal age, the total pancreatic insulin and glucagon content also increases. The connective tissue sheath and interlobular septa of the pancreas develop from the surrounding splanchnic mesenchyme.

Accessory Pancreatic Tissue

Accessory pancreatic tissue may be located in the wall of the stomach or duodenum or in an ileal diverticulum.

Anular Pancreas

Anular pancreas is a rare anomaly, but it warrants description because it may cause duodenal obstruction (Fig. 13-10*C*).

The ringlike or anular part of the pancreas consists of a thin, flat band of pancreatic tissue surrounding the descending or second part of the duodenum. An anular pancreas may cause obstruction of the duodenum shortly after birth or much later. Male infants are affected much more frequently than female infants. Anular pancreas probably results from the growth of a bifid ventral pancreatic bud around the duodenum (see Fig. 13-10*A* to *C*). The parts of the bifid ventral bud then fuse with the dorsal bud, forming a pancreatic ring (L. *anulus*).

Development of the Spleen

The spleen is derived from a mass of mesenchymal cells located between the layers of the dorsal mesogastrium (Fig. 13-11*A* and *B*). The spleen begins to develop during the fifth week of development but does not acquire its characteristic shape until early in the fetal period. The spleen is lobulated in the fetus but these lobules normally disappear before birth. The notches in the superior border of the adult spleen are remnants of the grooves that separated the fetal lobules. *Homeobox genes* (*NKx2-5, Hox11,* and *Bapx1*) are essential regulators in spleen development.

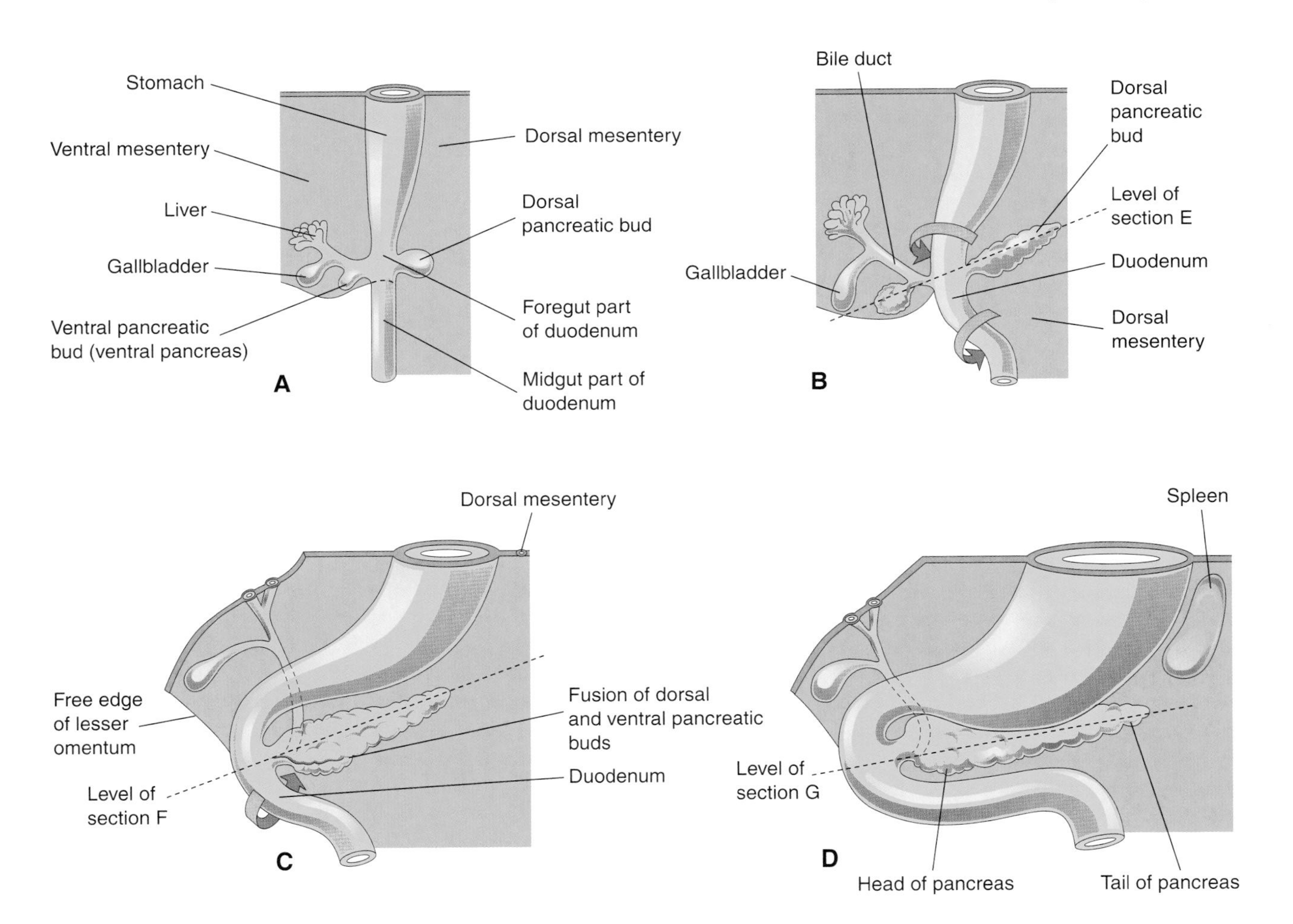

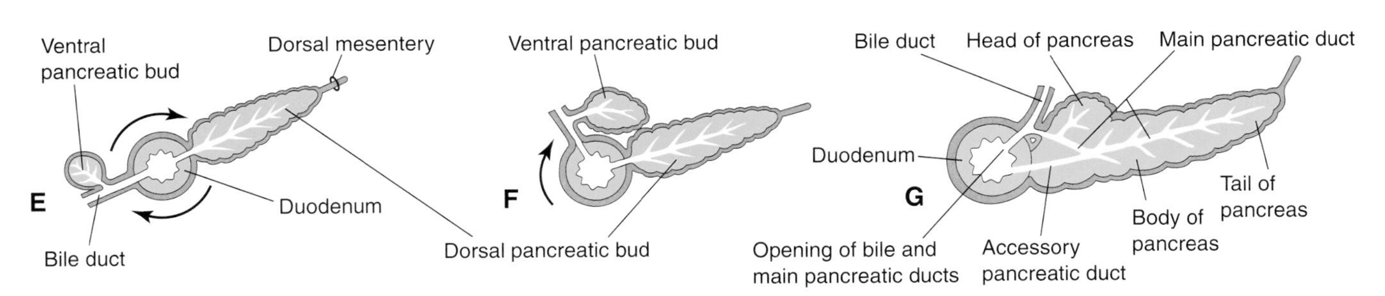

Figure 13-9. *A* to *D*, Schematic drawings showing successive stages in the development of the pancreas from the fifth to the eighth weeks. *E* to *G*, Diagrammatic transverse sections through the duodenum and developing pancreas. Growth and rotation (*arrows*) of the duodenum bring the ventral pancreatic bud toward the dorsal bud; these two structures subsequently fuse. Note that the bile duct initially attaches to the ventral aspect of the duodenum and is carried around to the dorsal aspect as the duodenum rotates. The main pancreatic duct is formed by the union of the distal part of the dorsal pancreatic duct and the entire ventral pancreatic duct. The proximal part of the dorsal pancreatic duct usually obliterates, but it may persist as an accessory pancreatic duct.

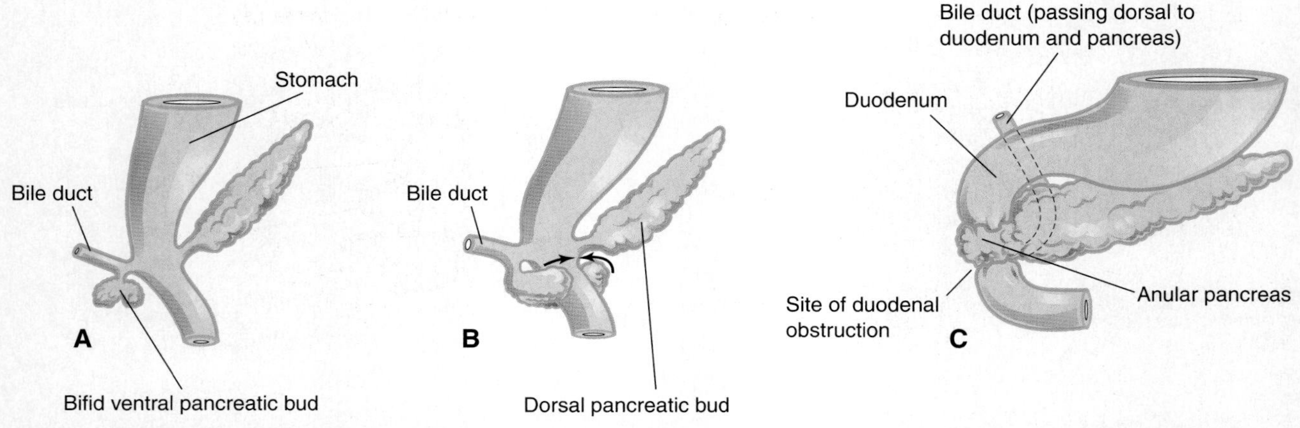

Figure 13 - 10. *A* and *B*, Drawings illustrating the probable embryologic basis of an anular pancreas. *C*, An anular pancreas encircling the duodenum. This anomaly sometimes produces complete obstruction (atresia) or partial obstruction (stenosis) of the duodenum. In most cases, the anular pancreas encircles the second part of the duodenum, distal to the hepatopancreatic ampulla.

Accessory Spleen

One or more small splenic masses may develop in one of the peritoneal folds, commonly near the hilum of the spleen or adjacent to the tail of the pancreas. An accessory spleen (about 1 cm in diameter) occurs in about 10% of people.

The Midgut

The derivatives of the midgut are as follows:

- the small intestine, including most of the duodenum
- the cecum, appendix, ascending colon, and right half to two thirds of the transverse colon

All of these midgut derivatives are supplied by the **superior mesenteric artery,** the artery of the midgut (see Fig. 13 - 1). The midgut loop is suspended from the dorsal abdominal wall by an elongated mesentery (Fig. 13 - 12*A*). As the midgut elongates, it forms a ventral, U-shaped loop of gut — the **midgut loop** — which projects into the remains of the extraembryonic coelom in the proximal part of the umbilical cord. At this stage, the intraembryonic coelom communicates with the extraembryonic coelom at the umbilicus (see Fig. 13 - 8). This movement of the intestine is termed a **physiological umbilical herniation.** It occurs at the beginning of the sixth week of development and is the normal migration of the midgut into the umbilical cord (see Figs. 13 - 12 and 13 - 13). The midgut loop communicates with the yolk sac through the narrow *yolk stalk* or *vitelline duct* until the 10th week. Umbilical herniation occurs because there is not enough room in the abdomen for the rapidly growing midgut. The midgut loop has both cranial and caudal limbs. The **yolk stalk** is attached to the apex of the midgut loop where the two limbs join (see Fig. 13 - 12*A*). The cranial limb grows rapidly and forms small intestinal loops, but the caudal limb undergoes very little change except for development of the **cecal diverticulum,** the primordium of the cecum and appendix (see Fig. 13 - 12*C* and *E*).

Rotation of the Midgut Loop

While it is in the umbilical cord, the midgut loop rotates 90 degrees counterclockwise around the axis of the **superior mesenteric artery** (see Fig. 13 - 12*B*). This brings the cranial limb of the midgut loop to the right and the caudal limb to the left. During rotation, the midgut elongates and forms loops of small intestine (e.g., the jejunum and ileum).

Return of the Midgut to the Abdomen

During the 10th week of development, the intestines return to the abdomen (see Fig. 13 - 12*C* and *D*). It is not known what causes the intestine to return, but the decrease in the size of the liver and kidneys and the enlargement of the abdominal cavity are thought to be important factors. This process has been called *reduction of the physiological midgut hernia.* The small intestine (formed from the cranial limb) returns first, passing posterior to the superior mesenteric artery and occupies the central part of the abdomen. As the large intestine returns, it undergoes a further 180-degree counterclockwise rotation (see Fig. 13 - 12C_1 and D_1). Later, it comes to occupy the right side of the abdomen. The ascending colon becomes recognizable as the posterior abdominal wall progressively elongates (see Figs. 13 - 12*E* and 13 - 14*A*).

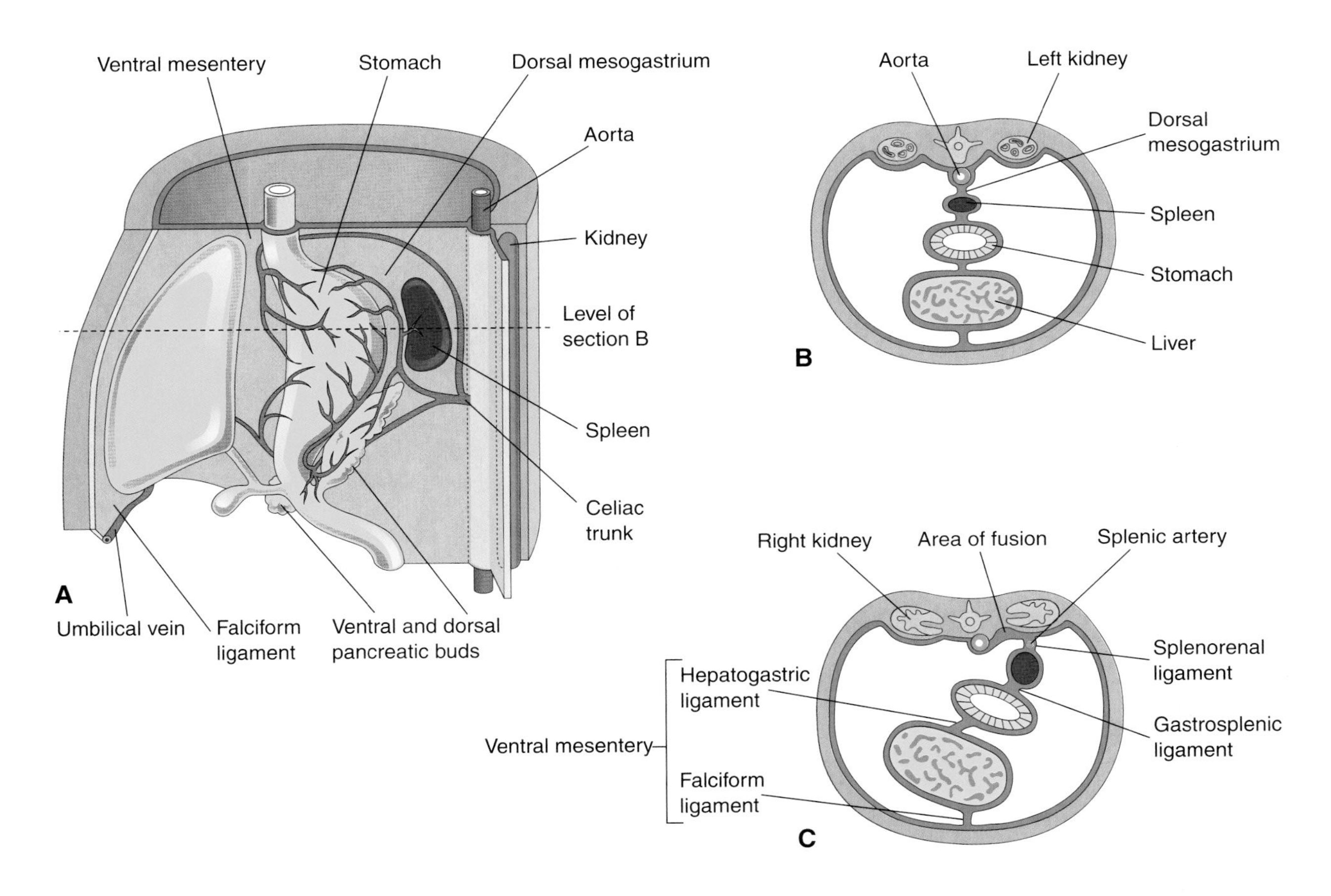

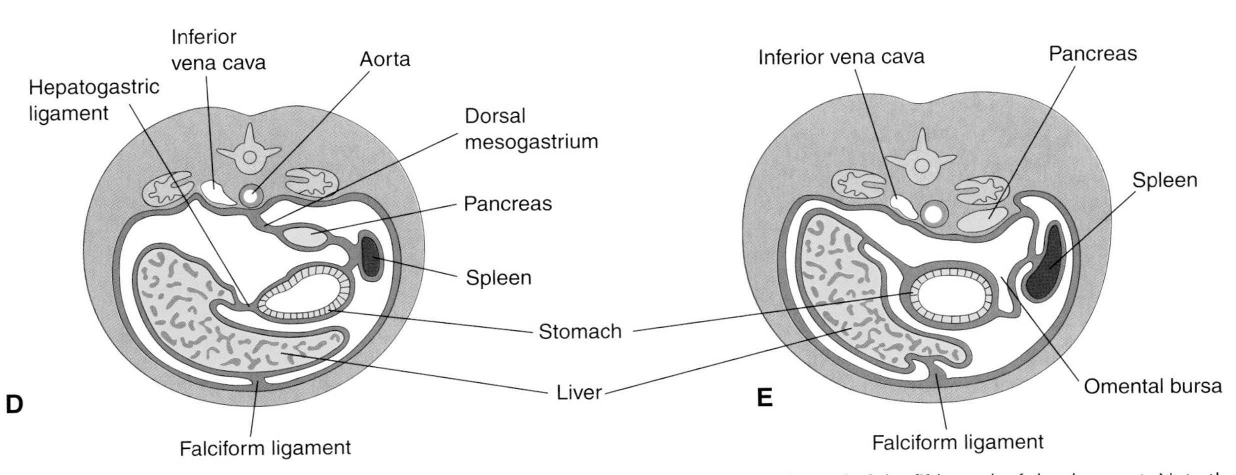

Figure 13 - 11. *A,* Drawing of the left side of the stomach and associated structures at the end of the fifth week of development. Note that the pancreas, spleen, and celiac trunk are located between the layers of the dorsal mesogastrium. *B,* Transverse section of the liver, stomach, and spleen at the level shown in *A,* illustrating their relationship to the dorsal and ventral mesenteries. *C,* Transverse section of a fetus showing fusion of the dorsal mesogastrium with the peritoneum on the posterior abdominal wall. *D* and *E,* Similar sections showing movement of the liver to the right and rotation of the stomach. Observe the fusion of the dorsal mesogastrium to the dorsal abdominal wall, which results in the pancreas becoming retroperitoneal.

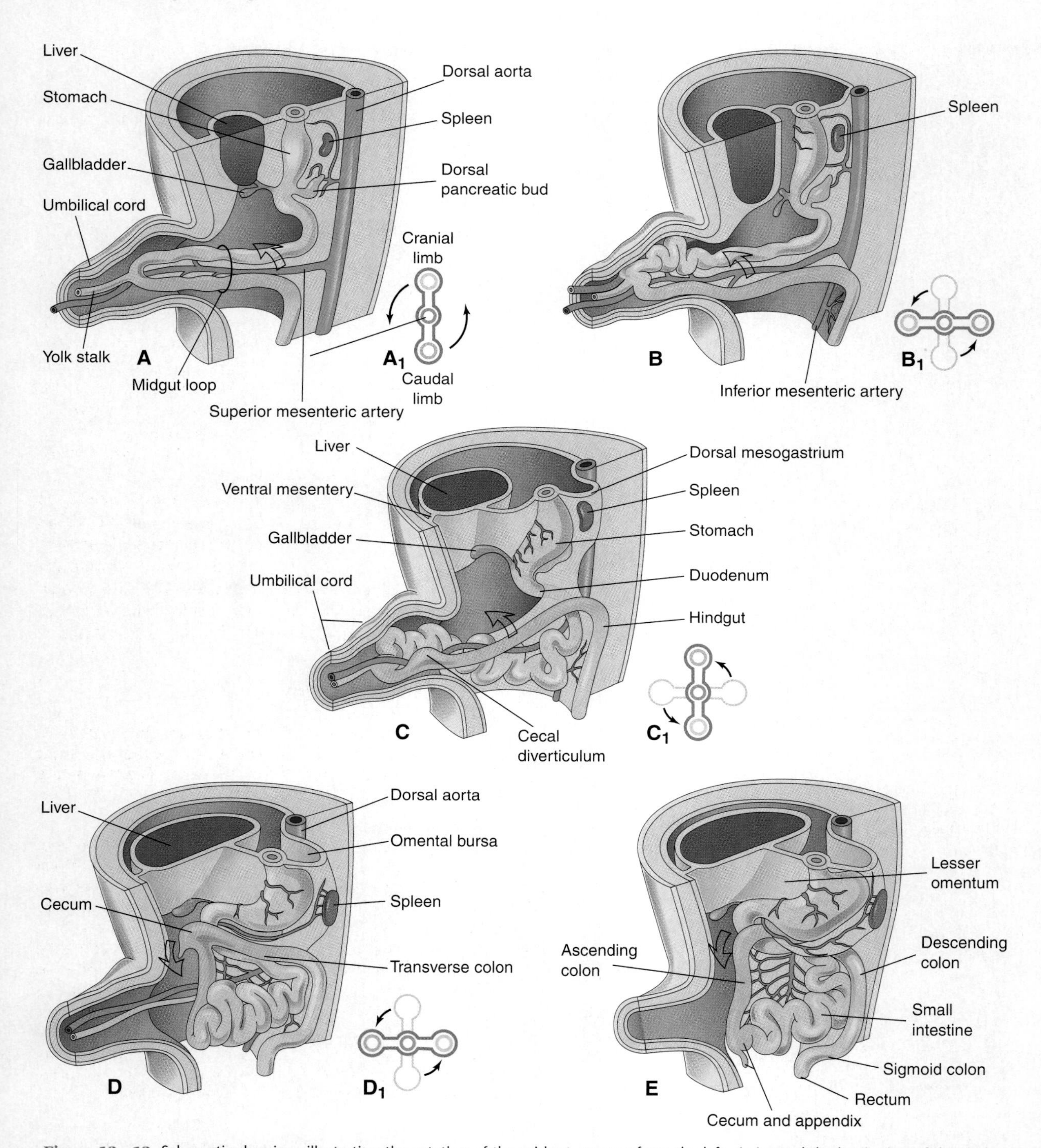

Figure 13 - 12. Schematic drawings illustrating the rotation of the midgut, as seen from the left. *A,* Around the beginning of the sixth week, the midgut loop is situated in the proximal part of the umbilical cord. *A_1,* Transverse section through the midgut loop, illustrating the initial relationship of the limbs of the midgut loop to the superior mesenteric artery. *B,* Later stage showing the beginning of midgut rotation. *B_1,* Drawing illustrating the 90-degree counterclockwise rotation that carries the cranial limb of the midgut to the right. *C,* At about 10 weeks, the intestines return to the abdomen. *C_1,* Drawing illustrating a further rotation of 90 degrees. *D,* At about 11 weeks, the intestines return to the abdomen. *D_1,* Drawing showing a further 90-degree rotation of the gut, for a total of 270 degrees. *E,* Later fetal period, showing the cecum rotating to its normal position in the lower right quadrant of the abdomen.

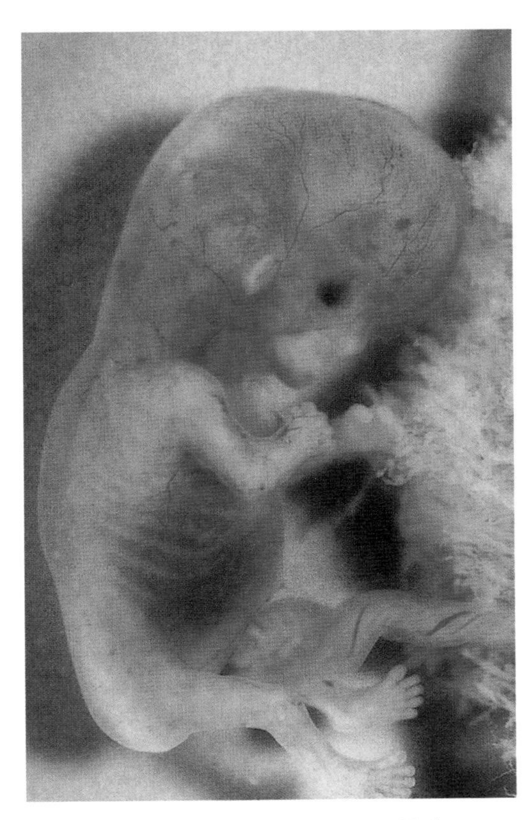

Figure 13 - 13. Physiological hernia in a 58-day embryo attached to its chorionic sac. Note the herniated intestine derived from the midgut loop in the proximal part of the umbilical cord. Also note the umbilical blood vessels. Observe also the cartilaginous ribs, the prominent eye, the large liver, and the relatively well-developed brain. (Courtesy of Dr. DK Kalousek, Department of Pathology, University of British Columbia, Children's Hospital, Vancouver, British Columbia, Canada.)

Fixation of the Intestines

Rotation of the stomach and duodenum causes the duodenum and pancreas to fall to the right, where they are pressed against the posterior abdominal wall by the colon. The adjacent layers of peritoneum fuse and subsequently disappear (see Fig. 13 - 14*C* and *F*); consequently, most of the duodenum and the head of the pancreas become retroperitoneal (posterior to the peritoneum). The attachment of the dorsal mesentery to the posterior abdominal wall is greatly modified after the intestines return to the abdominal cavity. The mesentery of the ascending colon fuses with the parietal peritoneum on this wall and disappears; consequently, the ascending colon also becomes retroperitoneal (see Fig. 13 - 14*B* and *E*). The enlarged colon presses the duodenum against the posterior abdominal wall; as a result, most of the duodenal mesentery is absorbed (see Fig. 13 - 14*C*, *D*, and *F*). Consequently, the duodenum, except for about the first 2.5 cm (derived from the foregut), has no mesentery and lies retroperitoneally. Other derivatives of the midgut loop (e.g., the jejunum and ileum) retain their mesenteries.

The Cecum and Appendix

The primordium of the cecum and the wormlike (L., vermiform) appendix — collectively known as the **cecal diverticulum** — appears in the sixth week of development as a swelling on the antimesenteric border of the caudal limb of the midgut loop (see Figs. 13 - 12*C* and *D* and 13 - 15*A*). The apex of the cecal diverticulum does not grow as rapidly as the rest of it; thus, the appendix is initially a small diverticulum of the cecum. The appendix subsequently increases rapidly in length so that, at birth, it is a relatively long tube arising from the distal end of the cecum (see Fig. 13 - 15*D*). After birth the wall of the cecum grows unequally, with the result that the appendix comes to enter its medial side (see Fig. 13 - 15*E*). The appendix is subject to considerable variation in position. As the ascending colon elongates, the appendix may pass posterior to the cecum (*retrocecal appendix*) or colon (*retrocolic appendix*). It may also descend over the brim of the pelvis (*pelvic appendix*). *In about 64% of people, the appendix is located retrocecally.*

Anomalies of the Midgut

Congenital abnormalities of the intestine are common; most of them are anomalies of gut rotation — malrotation of gut — that result from incomplete rotation or fixation of the intestines, or both.

Congenital Omphalocele

Congenital omphalocele is a persistence of the herniation of abdominal contents into the proximal part of the umbilical cord (Figs. 13 - 16 and 13 - 17). Herniation of the intestines into the cord occurs in about 1 in 5000 births, whereas herniation of the liver and intestines occurs in about 1 in 10,000 births. The size of the hernia depends on its contents. The abdominal cavity is proportionately small when an omphalocele is present because the impetus for it to grow is absent. Immediate surgical repair is required. Omphalocele results from failure of the intestines to return to the abdominal cavity during the 10th week of development. The covering of the hernial sac is the epithelium of the umbilical cord, a derivative of the amnion.

Umbilical Hernia

When the intestines return to the abdominal cavity during the 10th week of development and then herniate through an imperfectly closed umbilicus, an umbilical hernia forms. This common type of hernia differs from an omphalocele. In umbilical hernias, the protruding mass (usually consisting of the greater omentum and some of the small intestine) is

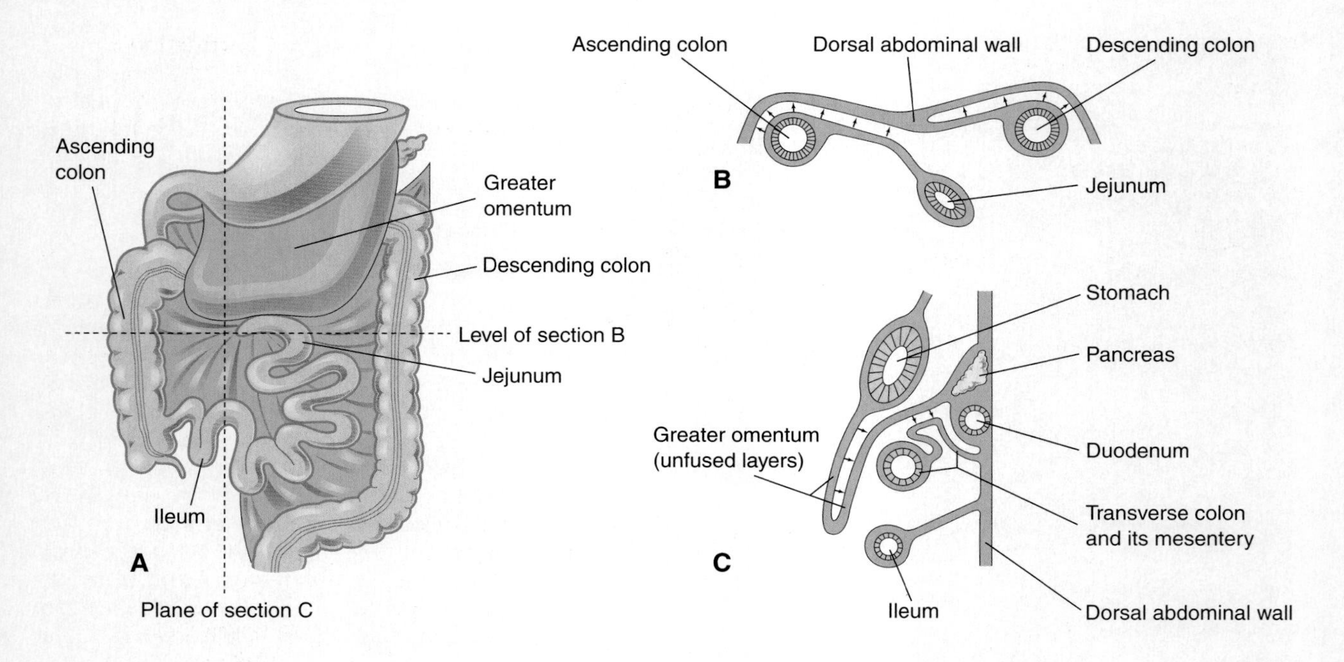

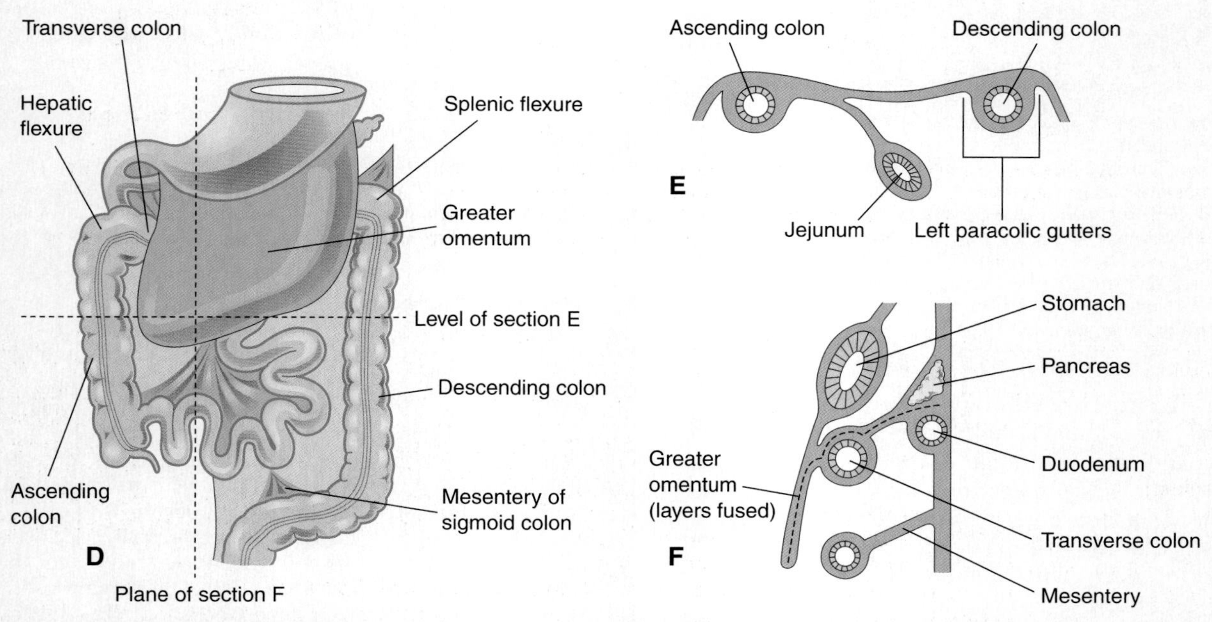

Figure 13-14. Fixation of the intestines. *A,* Ventral view of the intestines prior to their fixation. *B,* Transverse section at the level shown in *A.* The arrows indicate areas of subsequent fusion. *C,* Sagittal section at the plane shown in *A,* illustrating the greater omentum overhanging the transverse colon. The arrows indicate areas of subsequent fusion. *D,* Ventral view of the intestines after their fixation. *E,* Transverse section at the level shown in *D* after disappearance of the mesentery of the ascending and descending colon. *F,* Sagittal section at the plane shown in *D,* illustrating fusion of the greater omentum with the mesentery of the transverse colon and fusion of the layers of the greater omentum.

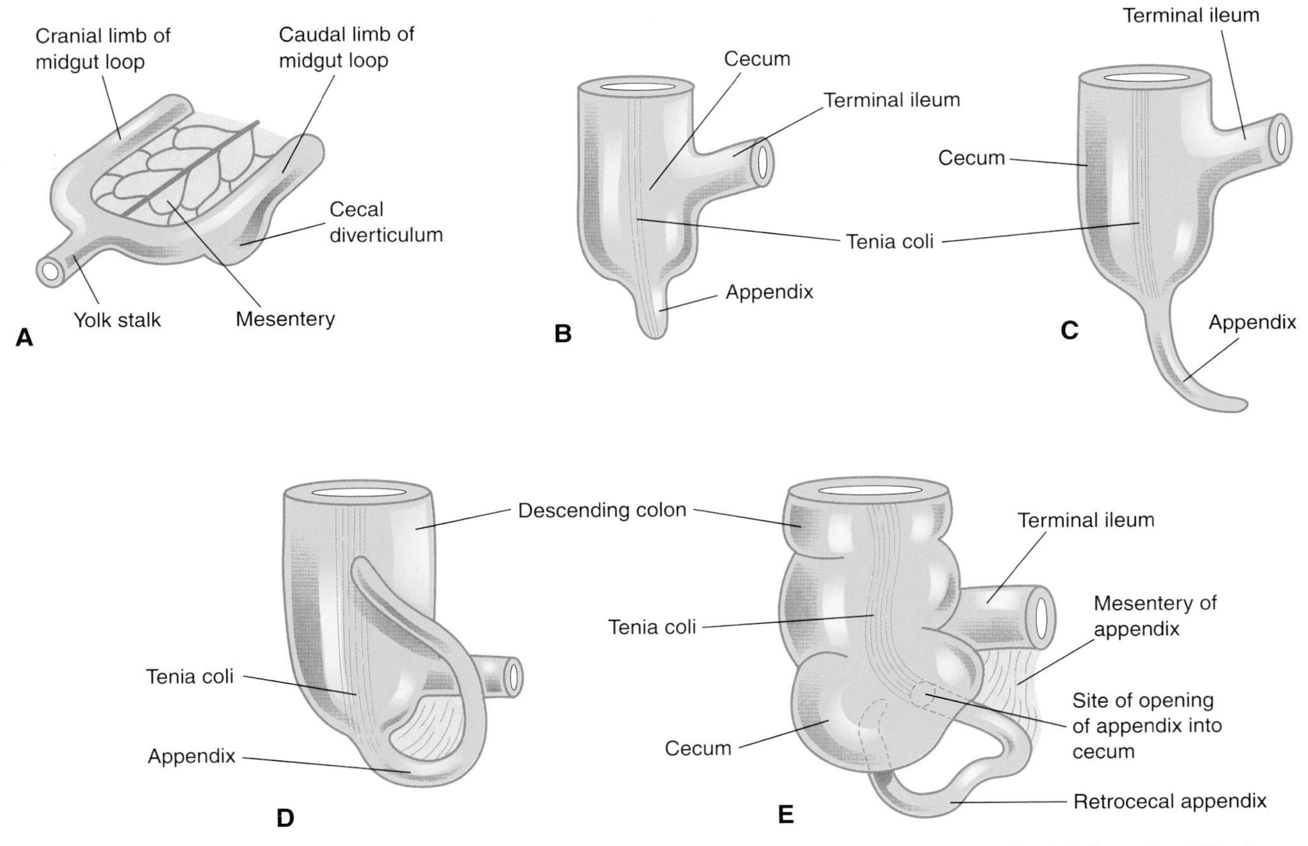

Figure 13–15. Drawings showing successive stages in the development of the cecum and appendix. *A,* Six weeks. *B,* Eight weeks. *C,* Twelve weeks. *D,* At birth. Note that the appendix is relatively long and is continuous with the apex of the cecum. *E,* Adult. Note that the appendix is now relatively short and lies on the medial side of the cecum. In about 64% of people, the appendix is located posterior to the cecum (retrocecal) or posterior to the ascending colon (retrocolic). The tenia coli, a thickened band of longitudinal muscle in the wall of the colon, ends at the base of the appendix.

covered by subcutaneous tissue and skin. The hernia protrudes during crying, straining, or coughing and can easily be reduced through the fibrous ring at the umbilicus. Surgery is not usually performed unless the hernia persists to the age of 3 to 5 years.

Gastroschisis

Gastroschisis results from a defect near the median plane of the abdominal wall (Fig. 13–18). The viscera protrude into the amniotic cavity and are bathed by amniotic fluid. The term *gastroschisis,* which literally means a "split or open stomach," is a misnomer because the anterior abdominal wall, not the stomach, is split. The defect usually occurs on the right side lateral to the median plane and is more common in male infants than in females infants. The anomaly results from incomplete closure of the lateral folds during the fourth week of development (see Chapter 6).

Nonrotation of the Midgut

Nonrotation of the midgut, a relatively common condition that is sometimes called *left-sided colon,* is generally asymptomatic, but twisting of the intestines (*volvulus*) may occur (Fig. 13–19*A* and *B*). Nonrotation occurs when the midgut loop does not rotate as it reenters the abdomen. As a result, the caudal limb of the loop returns to the abdomen first, the small intestine lies on the right side of the abdomen, and the entire large intestine lies on the left. When volvulus occurs, the superior mesenteric artery may be obstructed, resulting in infarction and gangrene of the intestine supplied by it.

Mixed Rotation and Volvulus

With mixed rotation and volvulus, the cecum lies just inferior to the pylorus of the stomach and is fixed to the posterior abdominal wall by peritoneal bands that pass over the duodenum (see Fig. 13–19*B*). These bands and the volvulus of the intestines usually cause **duodenal obstruction.** This type of malrotation results from failure of the midgut loop to complete the final 90 degrees of rotation (see Fig. 13–12*D*); consequently, the terminal part of the ileum returns to the abdomen first.

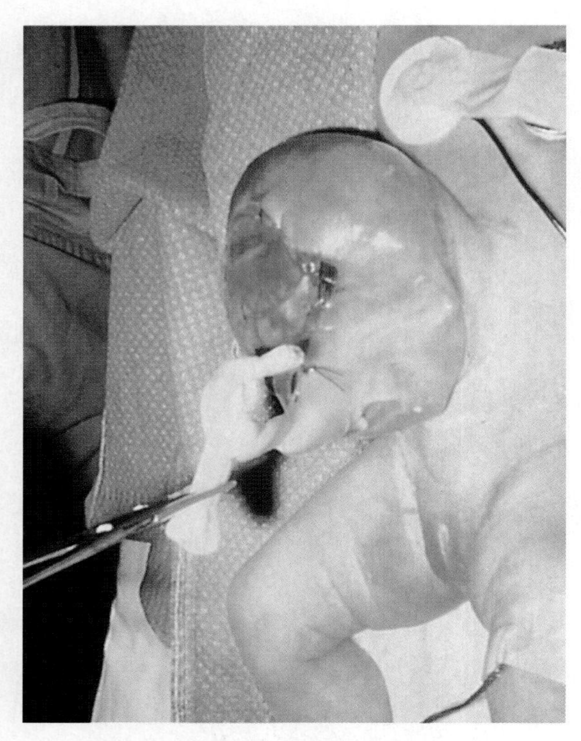

Figure 13 - 16. Photograph of an infant with an omphalocele. (Courtesy of Dr. NE Wiseman, Pediatric Surgeon, Children's Hospital, Winnipeg, Manitoba, Canada.) The defect resulted in herniation of intra-abdominal structures (liver and intestine) into the proximal end of the umbilical cord. The omphalocele is covered by a membrane composed of peritoneum and amnion.

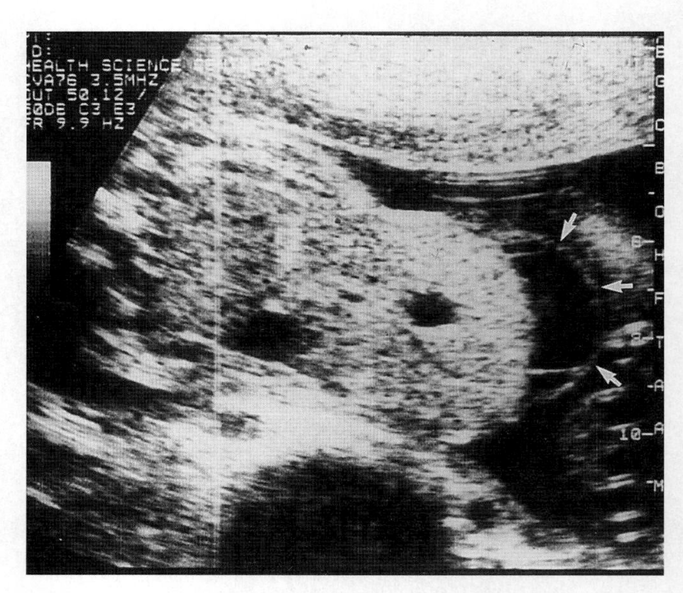

Figure 13 - 17. Sonogram of the abdomen of a fetus (28 weeks' gestation) showing a large omphalocele, with much of the liver protruding from the abdomen. The mass also contains a small, membrane-covered sac (*small arrows*). The umbilical cord was integrally involved in the anomaly. (Courtesy of Dr. CR Harman, Department of Obstetrics, Gynecology and Reproductive Sciences, Women's Hospital and University of Maryland, Baltimore, Maryland.)

Reversed Rotation

In very unusual cases, the midgut loop rotates in a clockwise rather than a counterclockwise direction (see Fig. 13 – 19*C*). As a result, the duodenum lies anterior to the superior mesenteric artery (SMA) rather than posterior to it, and the transverse colon lies posterior to the SMA instead of anterior to it. In these infants, the transverse colon may be obstructed by pressure from the SMA.

Subhepatic Cecum and Appendix

If the cecum adheres to the inferior surface of the liver when it returns to the abdomen (see Fig. 13 – 12*D*), it is drawn superiorly as the liver diminishes in size. As a result, the cecum remains in its fetal position (see Fig. 13 – 19*D*). Subhepatic cecum and appendix are more common in male infants than in female infants. Subhepatic cecum is not common in adults; however, when it occurs, it may create problems in the diagnosis of appendicitis and during surgical removal of the appendix (*appendectomy*).

Internal Hernia

In the case of an internal hernia, the small intestine passes into the mesentery of the midgut loop during the return of the intestines to the abdomen (see Fig. 13 – 19*E*). As a result, a hernialike sac forms. This very uncommon condition usually does not produce symptoms and is often detected at autopsy or during an anatomical dissection.

Midgut Volvulus

Midgut volvulus is an anomaly in which the small intestine fails to enter the abdominal cavity normally, and the mesenteries fail to undergo normal fixation. As a result, twisting of the intestines occurs (see Fig. 13 – 19*F*). Only two parts of the intestine — the duodenum and the proximal colon — are attached to the posterior abdominal wall. The small intestine hangs by a narrow stalk that contains the superior mesenteric artery and vein. These vessels are usually twisted in this stalk and become obstructed at or near the duodenojejunal junction. The circulation to the twisted intestine is often restricted; if the vessels are completely obstructed, gangrene develops.

Stenosis and Atresia of the Intestine

Partial occlusion (stenosis) and complete occlusion (atresia) of the intestinal lumen (see Fig. 13 – 5) account for about one third of cases of intestinal obstruction. The obstructive lesion occurs most often in the duodenum (25%) and ileum (50%). These anomalies result from failure of an adequate number of vacuoles to form during recanalization of the intestine. In some cases, a transverse diaphragm forms, producing a **diaphragmatic atresia.** Another possible cause of stenoses and atresias is interruption of the

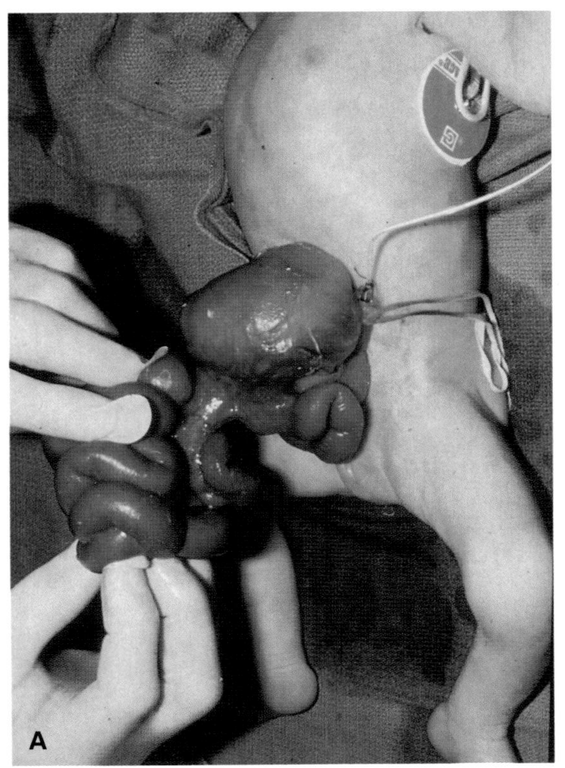

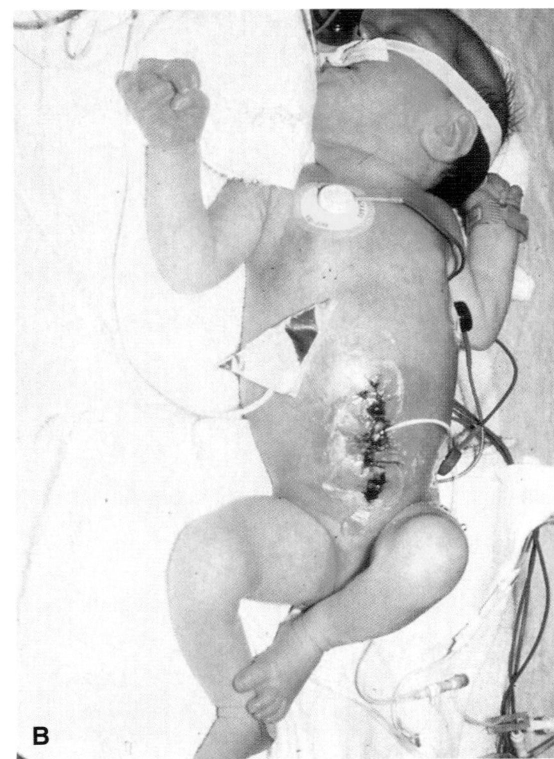

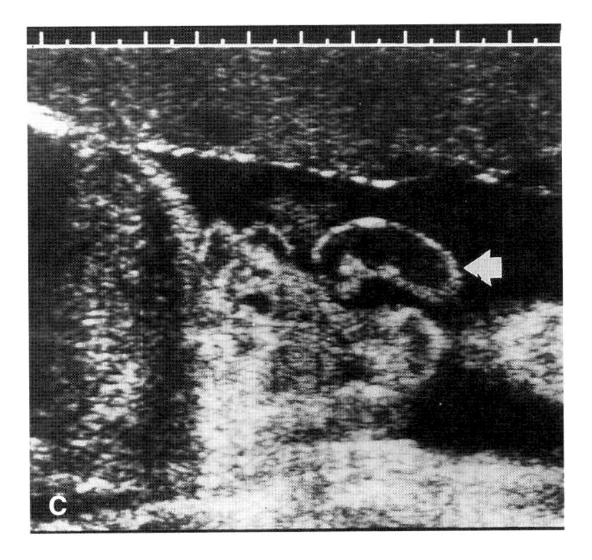

Figure 13 - 18. *A,* Photograph of a newborn infant with an anterior abdominal wall defect — (gastroschisis). The defect was relatively small (2 to 4 cm long) and involved all layers of the abdominal wall. It was located to the right of the umbilicus. *B,* Photograph of the same infant after the viscera were returned to the abdomen and the defect was surgically closed. *C,* Sonogram of a fetus (20 weeks' gestation) with gastroschisis. Loops of small bowel can be seen floating freely in the amniotic fluid (*arrow*) anterior to the fetal abdomen (*left*). (*A* and *B,* Courtesy of Dr. AE Chudley, Section of Genetics and Metabolism, Department of Pediatrics and Child Health, Children's Hospital, Winnipeg, Manitoba, Canada; *C,* Courtesy of Dr. CR Harman, Department of Obstetrics, Gynecology and Reproductive Services, Women's Hospital and University of Maryland, Baltimore, Maryland.)

blood supply to a loop of fetal intestine resulting from a **fetal vascular accident**; for example, an excessively mobile loop of intestine may become twisted, thereby interrupting its blood supply and leading to necrosis of the section of bowel involved. This necrotic segment later becomes a fibrous cord connecting the proximal and distal ends of normal intestine. Most atresias of the ileum are probably caused by infarction of the fetal bowel as the result of impairment of its blood supply secondary to volvulus. This impairment most likely occurs during the 10th week of development as the intestines return to the abdomen.

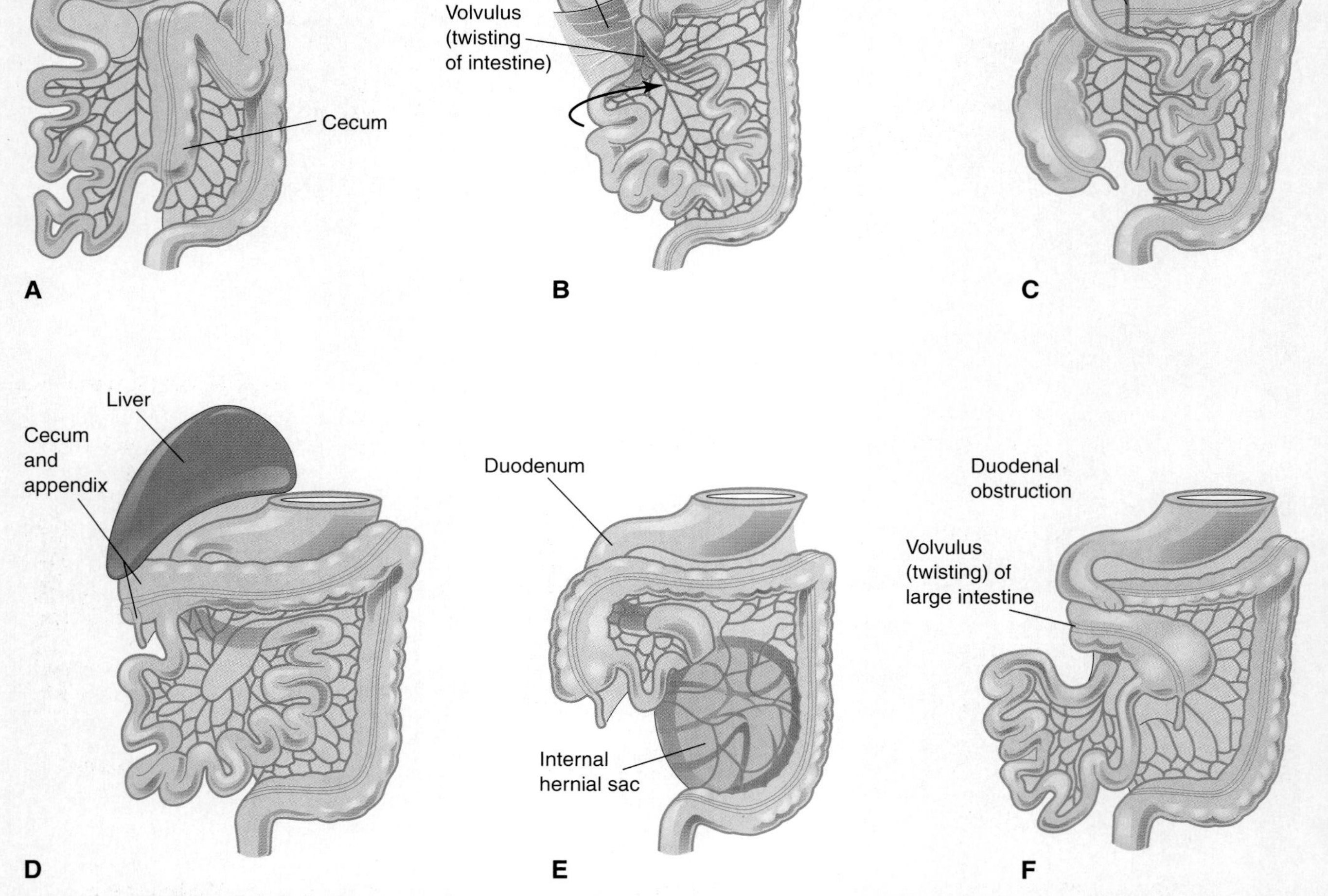

Figure 13 - 19. Drawings illustrating various abnormalities of midgut rotation. *A,* Nonrotation. *B,* Mixed rotation and volvulus. *C,* Reversed rotation. *D,* Subhepatic cecum and appendix. *E,* Internal hernia. *F,* Midgut volvulus.

Ileal Diverticulum and Other Yolk Stalk Remnants

Ileal diverticulum is one of the most common anomalies of the digestive tract (Fig. 13 – 20). A congenital ileal diverticulum (Meckel diverticulum) occurs in 2 to 4% of infants and is three to five times more prevalent in male infants than in female infants. *An ileal diverticulum is of clinical significance because it sometimes becomes inflamed* and causes symptoms that mimic appendicitis. The wall of the diverticulum contains all layers of the ileum and may contain small patches of gastric and pancreatic tissues. The gastric mucosa often secretes acid, producing ulceration and bleeding (Fig. 13 – 21*A*). An ileal diverticulum represents a remnant of the proximal portion of the yolk stalk. It typically appears as a fingerlike pouch about 3 to 6 cm long that *arises from the antimesenteric border of the ileum,* 40 to 50 cm from the ileocecal junction. An ileal diverticulum may be connected to the umbilicus by a fibrous cord or an **omphaloenteric (umbilicoileal) fistula** (see Fig. 13 – 21*B* and *C*); other possible remnants of the yolk stalk are illustrated in Figure 13 – 21*D* to F.

Duplication of the Intestine

Most intestinal duplications are cystic duplications or tubular duplications, of which the former are most common. Tubular duplications usually communicate with the intestinal lumen. Almost all duplications are caused by failure of normal recanalization of the intestine; as a result, two lumina form. The duplicated segment of the bowel lies on the mesenteric side of the intestine.

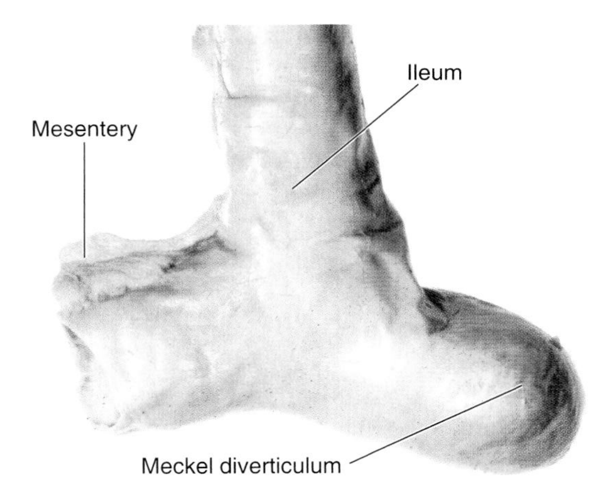

Figure 13 - 20. Photograph of a typical ileal diverticulum (cadaveric specimen), commonly referred to clinically as a Meckel diverticulum. Only a small percentage of these diverticula produce symptoms. Ileal diverticula are one of the most common anomalies of the digestive tract. They occur in 2 to 4% of individuals. The condition is three to four times more prevalent in males than in females. (From Moore KL, Persaud TVN, Shiota K: *Color Atlas of Clinical Embryology*, 2nd ed. Philadelphia, WB Saunders, 2000.)

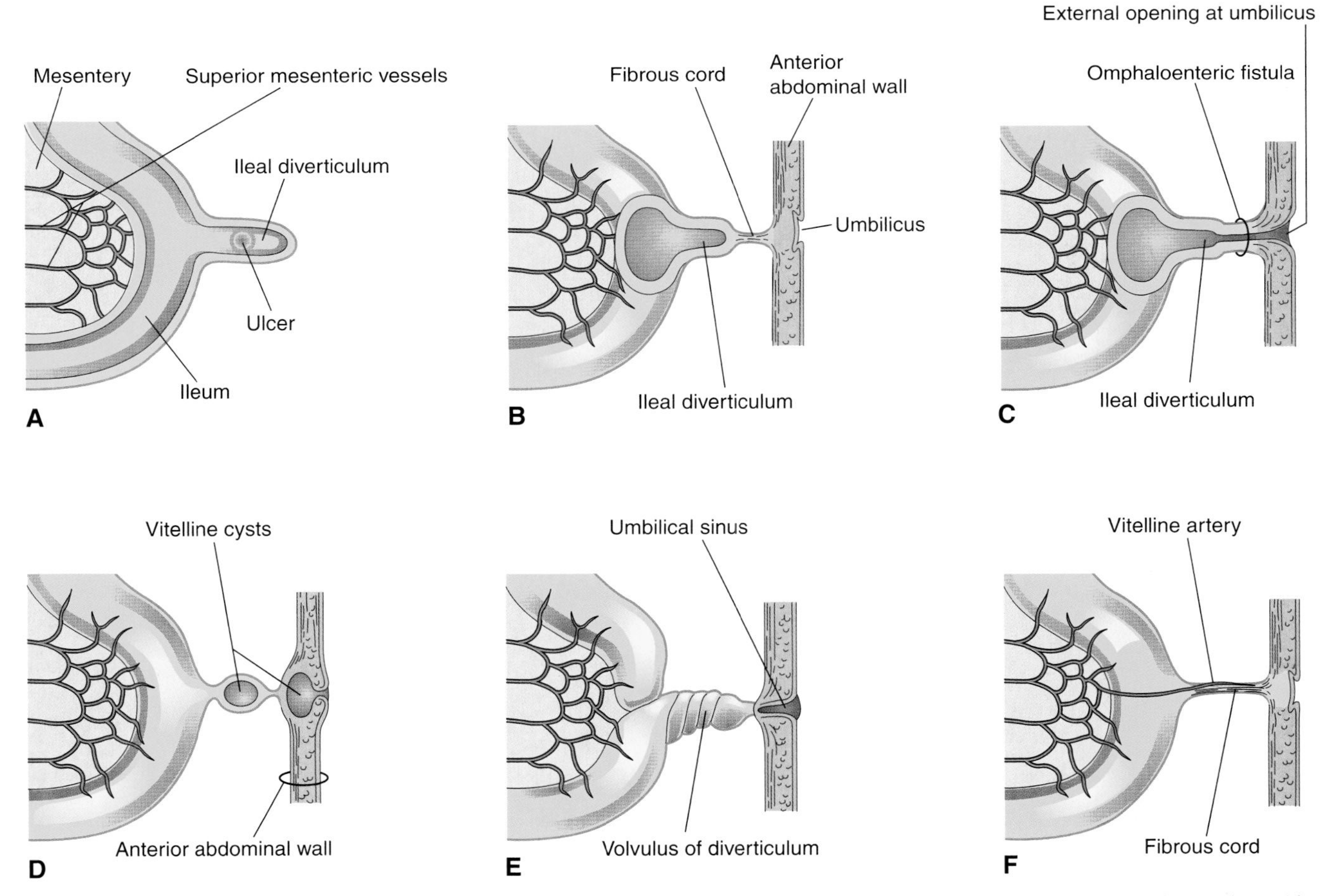

Figure 13 - 21. Drawings illustrating ileal (Meckel) diverticula and other remnants of the yolk stalk. *A,* Section of the ileum and a diverticulum with an ulcer. *B,* A diverticulum connected to the umbilicus by a fibrous cord. *C,* Omphaloenteric fistula resulting from persistence of the entire intra-abdominal portion of the yolk stalk. *D,* Vitelline cysts at the umbilicus and in a fibrous remnant of the yolk stalk. *E,* Umbilical sinus resulting from the persistence of the yolk stalk near the umbilicus. *F,* The yolk stalk has persisted as a fibrous cord connecting the ileum with the umbilicus. A persistent vitelline artery extends along the fibrous cord to the umbilicus.

The Hindgut

The derivatives of the hindgut are as follows:

- the left one-third to one half of the transverse colon, the descending colon and sigmoid colon, the rectum and the superior part of the anal canal
- the epithelium of the urinary bladder and most of the urethra

All of these hindgut derivatives are supplied by the **inferior mesenteric artery**, the artery of the hindgut. The junction between the segment of transverse colon derived from the midgut and that originating from the hindgut is indicated by the change in blood supply from a branch of the superior mesenteric artery (midgut artery) to a branch of the inferior mesenteric artery (hindgut artery). The descending colon becomes retroperitoneal as its mesentery fuses with the peritoneum on the left posterior abdominal wall and then disappears (see Fig. 13 - 14). The mesentery of the sigmoid colon is retained, but it is shorter than in the embryo.

The Cloaca

The terminal part of the hindgut, the cloaca, is an endoderm-lined chamber that is in contact with the surface ectoderm at the **cloacal membrane** (Fig. 13 - 22*A* and *B*). This membrane is composed of endoderm of the cloaca and ectoderm of the proctodeum or anal pit (Fig. 13 - 22*D*). The cloaca, the expanded terminal part of the hindgut, receives the **allantois** ventrally (Fig. 13 - 22*A*); the allantois is a fingerlike diverticulum of the yolk sac. (See Chapter 5 for a description of this rudimentary structure.)

Partitioning of the Cloaca

The cloaca is divided into dorsal and ventral parts by a wedge of mesenchyme — the **urorectal septum** — that develops in the angle between the allantois and hindgut. As the septum grows toward the cloacal membrane, it develops forklike extensions that produce infoldings of the lateral walls of the cloaca (see Fig. 13 - 22B_1). These folds grow toward each other and fuse, forming a partition that divides the cloaca into two parts (Fig. 13 - 22D_1 and F_1):

- the *rectum* and cranial part of the *anal canal* dorsally
- the *urogenital sinus* ventrally

By the seventh week of development, the urorectal septum has fused with the cloacal membrane, dividing it into a dorsal **anal membrane** and a larger ventral **urogenital membrane** (Fig. 13 - 22*E* and *F*). The area of fusion of the urorectal septum with the cloacal membrane is represented in the adult by the **perineal body,** the tendinous center of the perineum. Mesenchymal proliferations produce elevations of the surface ectoderm around the **anal membrane.** As a result, this membrane is soon located at the bottom of an ectodermal depression — the proctodeum or anal pit (see Fig. 13 - 22*E*). The anal membrane usually ruptures at the end of the eighth week of development, bringing the distal part of the digestive tract (anal canal) into communication with the amniotic cavity.

The Anal Canal

The superior two thirds (about 25 mm) of the adult anal canal are derived from the **hindgut**; the inferior one third (about 13 mm) develops from the **proctodeum** (Fig. 13 - 23). The junction of the epithelium derived from the ectoderm of the proctodeum and the endoderm of the hindgut is roughly indicated by an irregular **pectinate line,** located at the inferior limit of the anal valves. This line indicates the approximate former site of the anal membrane. At the anus, the epithelium is keratinized and continuous with the skin around the anus. The other layers of the wall of the anal canal are derived from splanchnic mesenchyme.

Because of its hindgut origin, the superior two-thirds of the anal canal are supplied mainly by the *superior rectal artery,* the continuation of the inferior mesenteric artery (hindgut artery). The venous drainage of this superior part occurs mainly via the *superior rectal vein,* a tributary of the inferior mesenteric vein. The lymphatic drainage of the superior part eventually courses to the *inferior mesenteric lymph nodes.* Its nerves are derived from the autonomic nervous system. Because of its origin from the proctodeum, the inferior one third of the anal canal is supplied mainly by the *inferior rectal arteries,* branches of the internal pudendal artery. The venous drainage is through the *inferior rectal vein,* a tributary of the internal pudendal vein that drains into the internal iliac vein. The lymphatic drainage of the inferior part of the anal canal courses to the *superficial inguinal lymph nodes.* Its nerve supply is from the *inferior rectal nerve*; hence, it is sensitive to pain, temperature, touch, and pressure.

The differences in blood supply, nerve supply, and venous and lymphatic drainage of the anal canal are important clinically, especially when considering the metastasis (spread) of cancer cells. The characteristics of carcinomas involving the two parts of the anal canal also differ. Tumors in the superior part are painless and arise from columnar epithelium, whereas those in the inferior part are painful and arise from squamous epithelium.

Anomalies of the Hindgut

Most anomalies of the hindgut are located in the anorectal region and result from abnormal development of the urorectal septum.

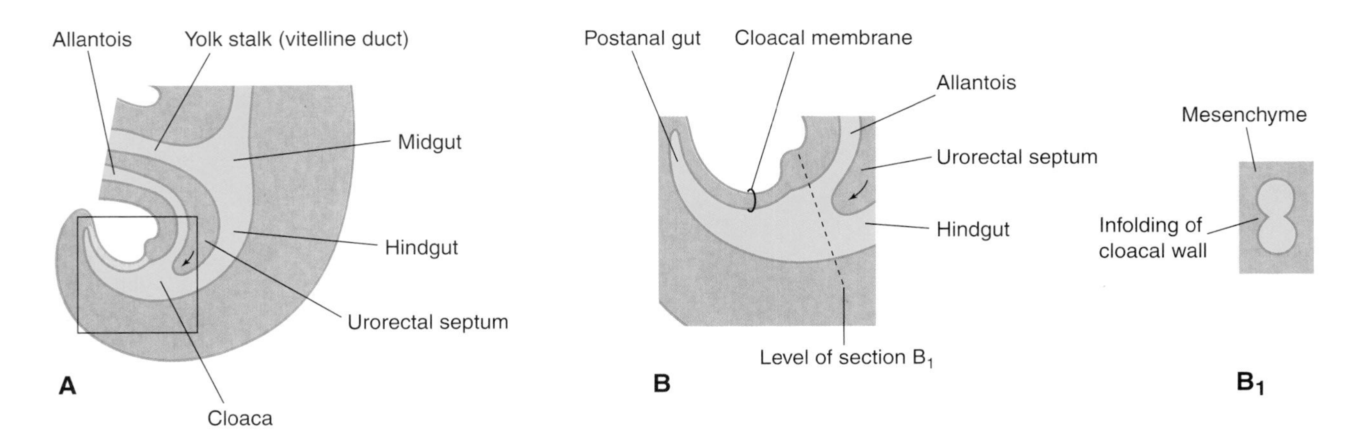

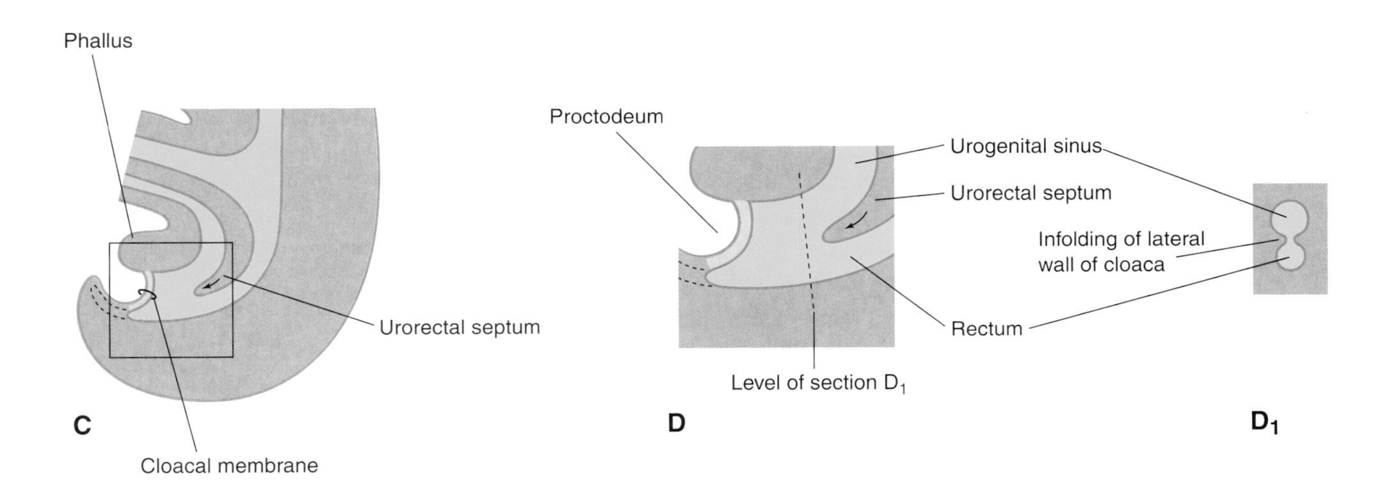

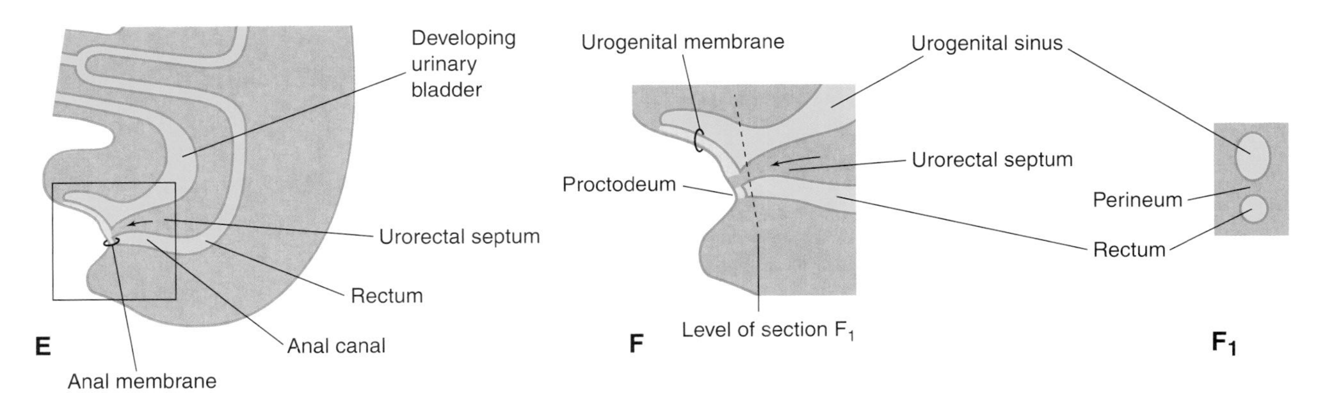

Figure 13 - 22. Drawings illustrating successive stages in the partitioning of the cloaca into the rectum and urogenital sinus by the urorectal septum. *A, C,* and *E,* Views from the left side at 4, 6, and 7 weeks, respectively. *B, D,* and *F,* Enlargements of the cloacal region. B_1, D_1, and F_1, Transverse sections of the cloaca at the levels shown in *B, D,* and *F,* respectively. Note that the postanal or tailgut (shown in *B*) degenerates and disappears as the rectum forms from the dorsal part of the cloaca (shown in *C*).

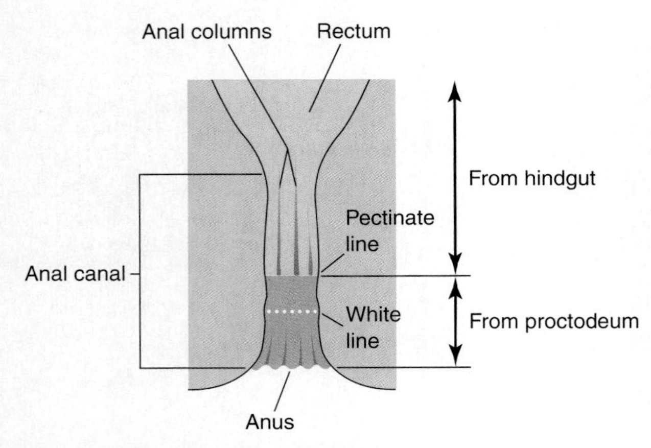

Figure 13 - 23. Sketch of the rectum and anal canal showing their developmental origins. Note that the superior two thirds of the anal canal are derived from the hindgut and are endodermal in origin, whereas the inferior one third of the anal canal is derived from the proctodeum and is ectodermal in origin. Because of their different embryologic origins, the superior and inferior parts of the anal canal are supplied by different arteries and nerves and have different venous and lymphatic drainages.

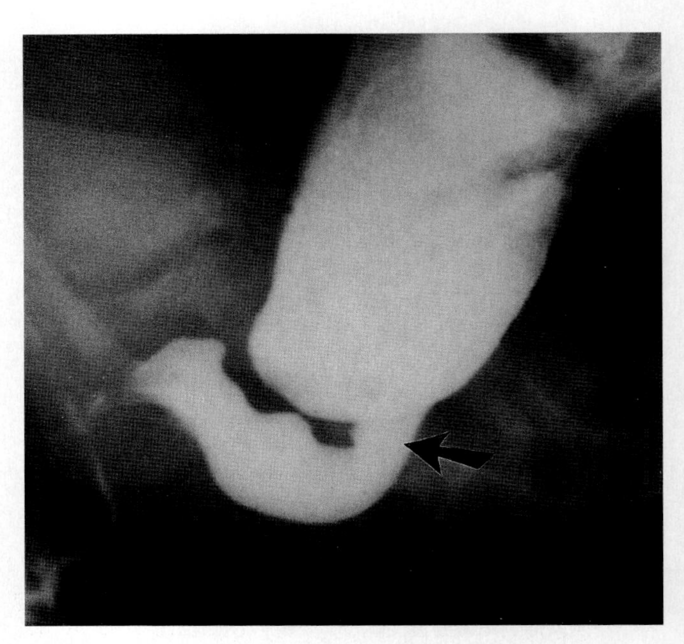

Figure 13 - 24. Radiograph of the colon after a barium enema in a 1-month-old infant with megacolon or Hirschsprung disease. The distal, aganglionic segment is narrow, with a dilated proximal colon full of fecal material above it. Note the transition zone (*arrow*). (Courtesy of Dr. Martin H. Reed, Department of Radiology, University of Manitoba and Children's Hospital, Winnipeg, Manitoba, Canada.)

Congenital Megacolon

In infants with congenital megacolon, or **Hirschsprung disease** (Fig. 13–24), a part of the colon is dilated because of the *absence of autonomic ganglion cells* in the myenteric plexus distal to the dilated segment of colon. The enlarged colon — termed **megacolon** (Gr. *megas*, big) — has the normal number of ganglion cells. The dilation results from failure of peristalsis in the aganglionic segment, which prevents movement of the intestinal contents. Male infants are affected more often than female infants (ratio of 4:1). Congenital megacolon results from failure of neural crest cells to migrate into the wall of the colon during the fifth to seventh weeks of development. This results in failure of parasympathetic ganglion cells to develop in the *Auerbach and Meissner plexuses.* Of the genes involved in the pathogenesis of Hirschsprung disease, the *RET* proto-oncogene accounts for most cases.

Imperforate Anus and Anorectal Anomalies

Imperforate anus occurs in about once in every 5000 newborn infants, and is most common in male infants (Figs. 13–25 and 13–26*C*). *Most anorectal anomalies result from abnormal development of the urorectal septum,* resulting in incomplete separation of the cloaca into urogenital and anorectal parts (see Fig. 13–26*A*). Lesions are classified as low or high, depending on whether the rectum ends superior or inferior to the puborectalis muscle.

Low Rectal Anomalies

Anal Agenesis, With or Without a Fistula

The anal canal may end blindly, or there may be an **ectopic anus** or an **anoperineal fistula** that opens into the perineum (see Fig. 13–26*D* and *E*). The abnormal canal

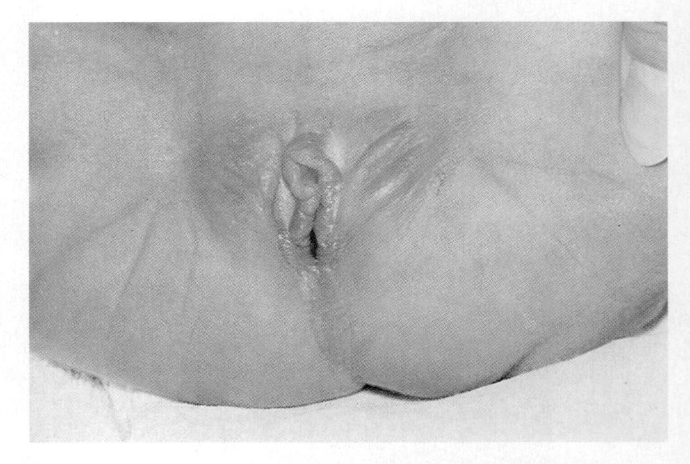

Figure 13 - 25. Female neonate with membranous anal atresia (imperforate anus). A tracheoesophageal fistula was also present. In most cases of anal atresia, a thin layer of tissue separates the anal canal from the exterior. This anomaly results from failure of the anal membrane to perforate at the end of the eighth week of development. Some form of imperforate anus occurs in about once in every 5000 neonates; it is most common in male infants. (Courtesy of Dr. AE Chudley, Section of Genetics and Metabolism, Department of Pediatrics and Child Health, Children's Hospital, Winnipeg, Manitoba, Canada.)

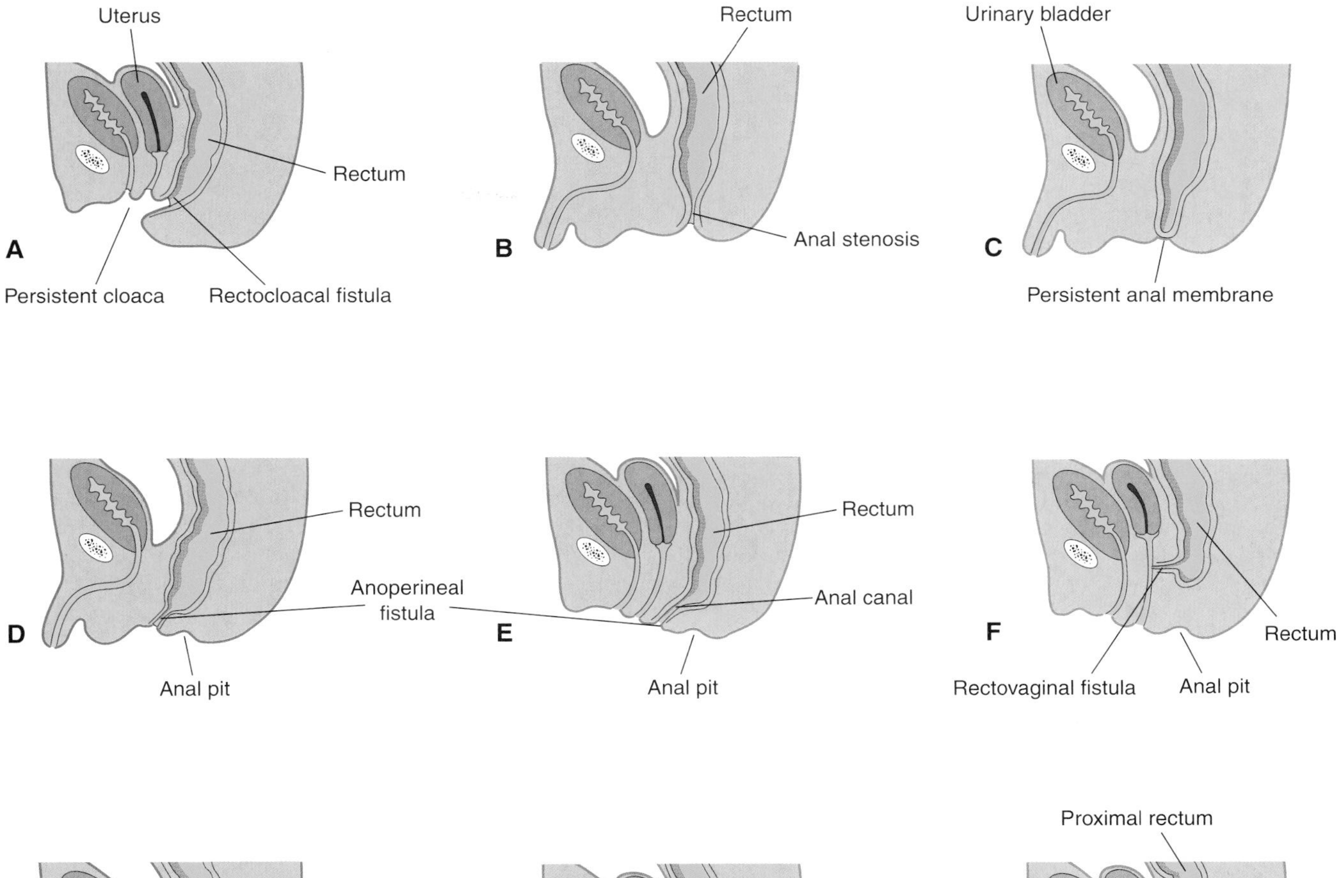

Figure 13-26. Drawings illustrating various types of anorectal anomaly. *A,* Persistent cloaca. Note the common outlet for the intestinal, urinary, and reproductive tracts. *B,* Anal stenosis. *C,* Membranous anal atresia (covered anus). *D* and *E,* Anal agenesis with a perineal fistula. *F,* Anorectal agenesis with a rectovaginal fistula. *G,* Anorectal agenesis with a rectourethral fistula. *H* and *I,* Rectal atresia.

may, however, open into the vagina in females or into the urethra in males (see Fig. 13-26*F* and *G*). Most low anorectal anomalies are associated with an external fistula. **Anal agenesis with a fistula** results from incomplete separation of the cloaca by the urorectal septum.

Anal Stenosis

In anal stenosis, the anus is in a normal position, but the anus and anal canal are narrow (see Fig. 13-26*B*). This anomaly is probably caused by a slight dorsal deviation of the urorectal septum as it grows caudally to fuse with the cloacal membrane. As a result, the anal canal and anal membrane are small. Sometimes, only a small probe can be inserted into the anal canal.

Membranous Atresia of the Anus

In membranous anal atresia, the anus is in the normal position, but a thin layer of tissue separates the anal canal from the exterior (see Figs. 13-25 and 13-26*C*). The anal membrane is thin enough to bulge on straining and appears blue from the presence of meconium superior to it. This anomaly results from failure of the anal membrane to perforate at the end of the eighth week of development.

High Anorectal Anomalies

Anorectal Agenesis, With or Without a Fistula

In anorectal agenesis, the rectum ends superior to the puborectalis muscle. This is the most common type of anorectal anomaly, which accounts for about two-thirds of anorectal defects. Although the rectum ends blindly, there is usually a fistula to the bladder (*rectovesical fistula*) or urethra (*rectourethral fistula*) in male individuals, or to the vagina (*rectovaginal fistula*) or the vestibule of the vagina (*rectovestibular fistula*) in female individuals (Fig. 13–26*F* and *G*). *Anorectal agenesis with a fistula* is the result of incomplete separation of the cloaca by the urorectal septum.

Rectal Atresia

With rectal atresia, the anal canal and rectum are present, but they are separated (Fig. 13–26*H* and *I*). Sometimes, the two segments of bowel are connected by a fibrous cord, the remnant of the atretic portion of the rectum. The cause of rectal atresia may be abnormal recanalization of the colon or, more likely, a defective blood supply, as discussed with atresia of the small intestine.

Summary of the Digestive System

The *primordial gut* forms during the fourth week of development from the part of the yolk sac that is incorporated into the embryo. The endoderm of the primordial gut gives rise to the epithelial lining of most of the digestive tract and biliary passages, together with the parenchyma of its glands, including the liver and pancreas. The epithelium at the cranial and caudal extremities of the digestive tract is derived from the ectoderm of the stomodeum and proctodeum, respectively. The muscular and connective tissue components of the digestive tract are derived from the splanchnic mesenchyme surrounding the primordial gut.

The **foregut** gives rise to the pharynx, lower respiratory system, esophagus, stomach, duodenum (proximal to the opening of the bile duct), liver, pancreas, and biliary apparatus. Because the trachea and esophagus have a common origin from the foregut (see Chapter 12), incomplete partitioning by the tracheoesophageal septum results in stenoses or atresias, with or without fistulas between them.

The **hepatic diverticulum**, the primordium of the liver, gallbladder, and biliary duct system, is an outgrowth of the endodermal epithelial lining of the foregut. The epithelial hepatic cords and primordia of the *biliary system,* which develop from the hepatic diverticulum, grow into the septum transversum. Between the layers of the *ventral mesentery*, which is derived from the septum transversum, these primordial cells differentiate into the liver and the lining of the ducts of the biliary system.

Congenital duodenal atresia results from failure of the vacuolization and recanalization process to occur following the normal solid stage of the duodenum. Usually, these epithelial cells degenerate and the lumen of the duodenum is restored. Obstruction of the duodenum can also be caused by an *anular pancreas.*

The **pancreas** develops from dorsal and ventral *pancreatic buds* that form from the endodermal lining of the foregut. When the duodenum rotates to the right, the **ventral pancreatic bud** moves dorsally and fuses with the dorsal pancreatic bud. The ventral pancreatic bud forms most of the head of the pancreas, including the uncinate process. The **dorsal pancreatic bud** forms the remainder of the pancreas. In some fetuses, the duct systems of the two buds fail to fuse, and an *accessory pancreatic duct* forms.

The **midgut** gives rise to the duodenum (distal to the bile duct), jejunum, ileum, cecum, appendix, ascending colon, and right half to two-thirds of the transverse colon. The midgut forms a U-shaped intestinal loop that herniates into the umbilical cord during the sixth week of development because there is no room for it in the abdomen. While in the umbilical cord, the **midgut loop** rotates counterclockwise 90 degrees. During the 10th week, the intestines rapidly return to the abdomen, rotating another 180 degrees during this process.

Omphaloceles, **malrotations**, and **abnormal fixation of the gut** result from failure of return or abnormal rotation of the intestine in the abdomen. Because the gut is normally occluded during the fifth and sixth weeks of development, owing to the rapid mitotic activity of its epithelium, *stenosis* (partial obstruction), *atresia* (complete obstruction), and *duplications* can result if recanalization fails to occur or occurs abnormally. Various remnants of the yolk stalk may persist. **Ileal diverticula** are common; however, only a few of them become inflamed and produce pain.

The **hindgut** gives rise to the left one-third to one-half of the transverse colon, the descending and sigmoid colon, the rectum, and the superior part of the anal canal. The inferior part of the anal canal develops from the **proctodeum**. The caudal part of the hindgut, known as the *cloaca,* is divided by the *urorectal septum* into the urogenital sinus and rectum. The urogenital sinus gives rise mainly to the urinary bladder and urethra (see Chapter 14). At first the rectum and the superior part of the anal canal are separated from the exterior by the *anal membrane,* but this membrane normally breaks down by the end of the eighth week of development.

Most **anorectal anomalies** result from abnormal partitioning of the cloaca by the urorectal septum into the rectum and anal canal posteriorly and the urinary bladder and urethra anteriorly. Arrested growth or deviation of the urorectal septum in a dorsal direction, or both, causes most anorectal abnormalities, such as rectal atresia and fistulas between the rectum and the urethra, urinary bladder, or vagina.

Clinically Oriented Questions

1. About 2 weeks after birth, an infant began to vomit shortly after feeding. Each time, the vomitus was propelled about 2 feet. The physician told the mother that the baby had a stomach tumor resulting in a narrow outlet from its stomach. Is there an embryologic basis for this anomaly? Is the tumor malignant?
2. Some say that infants with *Down syndrome* have an increased incidence of *duodenal atresia.* Is this true? Can the condition be corrected?
3. A patient claimed that his appendix was on his left side. Is this possible and, if so, how could this happen?
4. A nurse reported that a patient supposedly had two appendices and had had separate operations to remove them. Do people ever have two appendices?
5. What is *Hirschsprung disease*? Some sources state that it is a congenital condition resulting from large bowel obstruction. Is this correct? If so, what is its embryologic basis?
6. A nurse once stated that feces can sometimes be expelled from a baby's umbilicus. Was she kidding, or can this really happen? If so, what conditions would likely be present?

The answers to these questions are at the back of the book.

14

The Urogenital System

The urogenital system can be divided functionally into the *urinary* (*excretory*) *system* and the *genital* (*reproductive*) *system.* Embryologically, these systems are closely associated. The urogenital system develops from the intermediate mesoderm (Fig. 14-1*A* and *B*). During folding of the embryo (see Chapter 6), this mesoderm is carried ventrally and loses its connection with the somites (see Fig. 14-1*C*). A longitudinal elevation of mesoderm — the **urogenital ridge** — forms on each side of the dorsal aorta (Fig. 14-1*D*). It gives rise to parts of the urinary and genital systems. The part of the urogenital ridge giving rise to the urinary system is the **nephrogenic cord** or ridge (Fig. 14-1*C* and *F*); the part giving rise to the genital system is the **gonadal ridge** (see Fig. 14-19*C*).

Development of the Urinary System

Development of the Kidneys and Ureters

Three sets of excretory organs or kidneys develop in human embryos: the *pronephros*, *mesonephros*, and *metanephros*. The first set of kidneys — the *pronephroi* (pleural of *pronephros*) — are rudimentary and nonfunctional. The second set of kidneys — the *mesonephroi* — are well developed and function briefly. The third set of kidneys — the *metanephroi* — become the permanent kidneys.

Pronephroi. The transitory pronephroi appear in human embryos early in the fourth week of development. They are represented by a few cell clusters in the neck region (Fig. 14-2*A*). The pronephric ducts run caudally and open into the *cloaca* (see Fig. 14-2*B*). The rudimentary pronephroi soon degenerate; however, most of the pronephric ducts persist and are utilized by the next set of kidneys.

Mesonephroi. The mesonephroi, or interim kidneys, appear late in the fourth week of development and are located caudal to the rudimentary pronephroi (see Figs. 14-2 and 14-4*A*). They function until the permanent kidneys develop (Fig. 14-3). The mesonephric kidneys consist of glomeruli and mesonephric tubules. The tubules open into the **mesonephric ducts**, originally the pronephric ducts. The mesonephric ducts open into the cloaca. The mesonephroi degenerate toward the end of the first trimester; however, their tubules become the efferent ductules of the testes, and the mesonephric ducts have several adult derivatives in the male (Table 14-1).

Metanephroi. The metanephroi — the primordia of the permanent kidneys — begin to develop early in the fifth week and start to function about 4 weeks later. *Urine formation continues throughout fetal life.* Urine is excreted into the amniotic cavity and mixes with the amniotic fluid. The definitive kidneys develop from two sources:

- the *metanephric diverticulum* (ureteric bud)
- the *metanephric mass of intermediate mesoderm* (metanephrogenic blastema)

The metanephric diverticulum is an outgrowth from the mesonephric duct near its entrance into the cloaca, and the metanephric mass of intermediate mesoderm is derived from the caudal part of the nephrogenic cord (Fig. 14-4). Both primordia of the metanephros are of mesodermal origin. The **metanephric diverticulum** is the primordium of the *ureter, renal pelvis, calices,* and *collecting tubules* (see Fig. 14-4*C* to *E*). As it elongates, the diverticulum penetrates the **metanephric mass of intermediate mesoderm** (see Fig. 14-4*B*). The stalk of the metanephric diverticulum becomes the **ureter,** and its cranial end forms the renal pelvis.

The straight **collecting tubules** undergo repeated branching, forming successive generations of collecting tubules. The first four generations of tubules enlarge and become confluent to form the **major calices** (see Fig. 14-4*C* to *E*); the second four generations coalesce to form the **minor calices.** The remaining generations of tubules form the collecting tubules. The end of each arched collecting tubule induces clusters of mesenchymal cells in the metanephric mass of mesoderm to form small **metanephric vesicles** (Fig. 14-5*A*). These vesicles elongate and become **metanephric tubules** (see Fig. 14-5*B* and *C*). The proximal ends of these renal tubules are invaginated by glomeruli. The **renal corpuscle** (glomerulus and glomerular capsule) and its proximal convoluted tubule, nephron loop (of Henle), and distal convoluted tubule constitute a **nephron** (see Fig. 14-5*D*). Each distal convoluted tubule contacts an arched collecting tubule and the tubules become confluent, forming a uriniferous tubule. Consequently, each **uriniferous tubule** consists of two embryologically different parts (see Figs. 14-4 and 14-5):

- a *nephron* derived from the metanephric mass of intermediate mesoderm
- a *collecting tubule* derived from the metanephric diverticulum

Branching of the metanephric diverticulum depends on an inductive signal from the metanephric mesoderm and differentiation of the nephrons depends on induction by the collecting tubules. The **molecular aspects** of the reciprocal interactions between the metanephric mesenchyme and the collecting tubules are shown in Figure 14-6.

The **fetal kidneys** are subdivided into lobes. The lobulation usually disappears during infancy as the

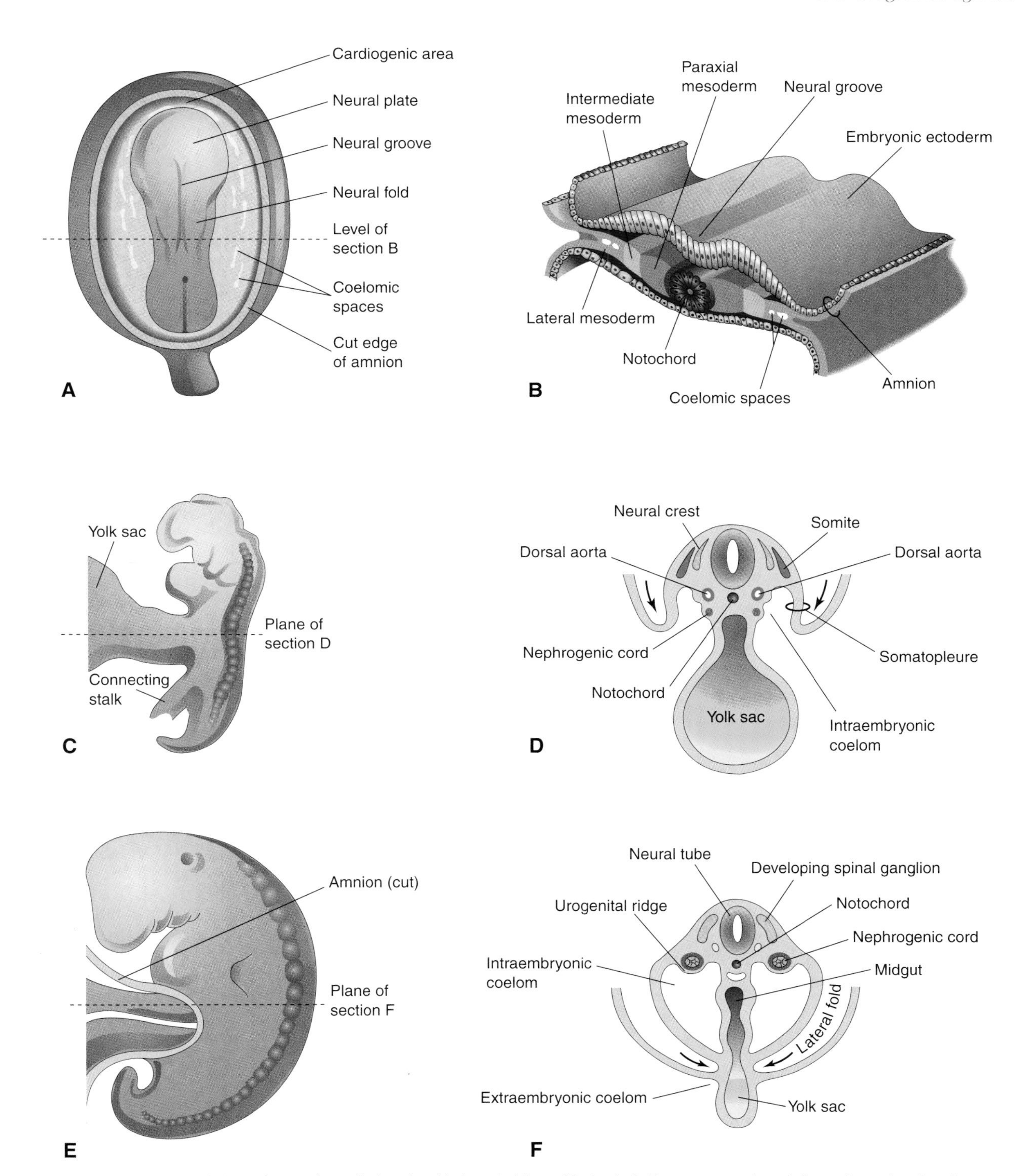

Figure 14 - 1. *A,* Dorsal view of an embryo during the third week (about 18 days). *B,* Transverse section of the embryo showing the position of the intermediate mesoderm before lateral folding of the embryo. *C,* Lateral view of an embryo during the fourth week (about 24 days). *D,* Transverse section of the embryo after the commencement of folding. Note the nephrogenic cords of mesoderm. *E,* Lateral view of an embryo later in the fourth week (about 26 days). *F,* Transverse section of the embryo showing the lateral folds meeting each other ventrally. Observe the position of the urogenital ridges and nephrogenic cords.

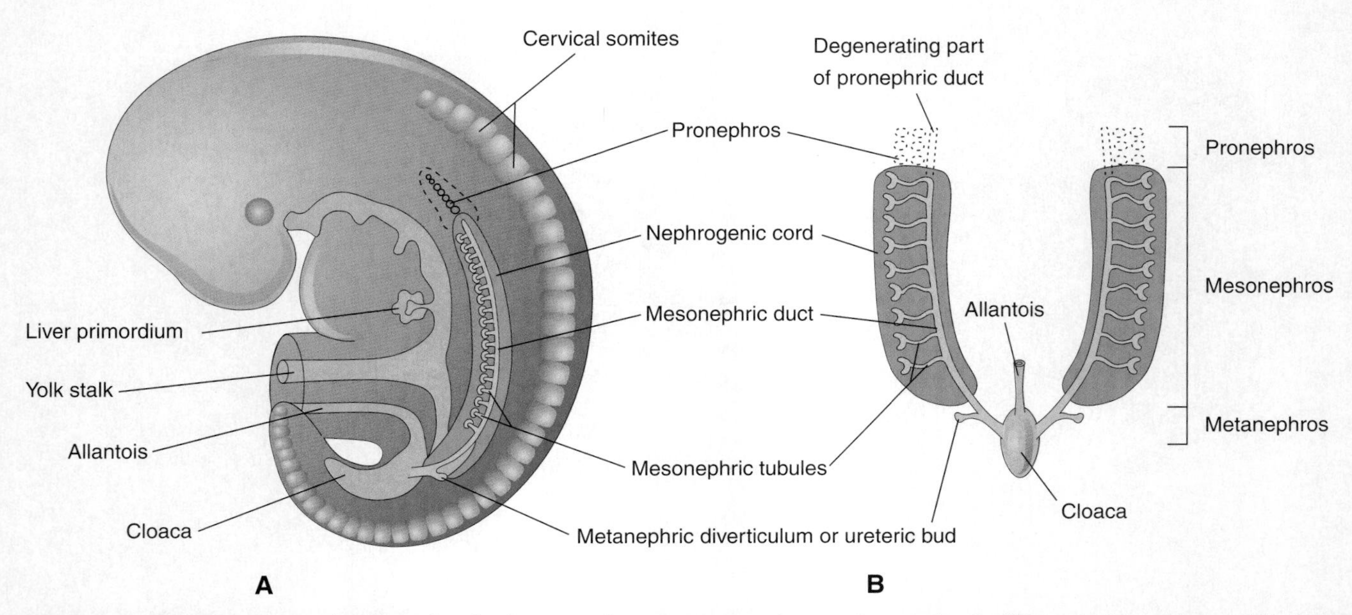

Figure 14-2. Diagrammatic sketches illustrating the three sets of excretory systems in an embryo during the fifth week. *A,* Lateral view. *B,* Ventral view. In this diagram, the mesonephric tubules have been pulled laterally; their normal position is shown in *A.*

nephrons increase and grow. At term, each kidney contains 800,000 to 1,000,000 nephrons. It is now believed that nephron formation is complete at birth except in premature infants. Functional maturation of the kidneys occurs after birth.

Positional Changes of the Kidneys

The metanephric kidneys lie close to each other in the pelvis (Fig. 14-7*A*). As the abdomen and pelvis grow, the kidneys gradually come to lie in the abdomen and move farther apart (see Fig. 14-7*B* and *C*). They attain their adult position by the ninth week of development (see Fig. 14-7*D*). This "migration" (relative ascent) results mainly from the growth of the embryo's body caudal to the kidneys. In effect, the caudal part of the embryo grows away from the kidneys so that the kidneys progressively occupy higher cranial levels. As the kidney "ascends," it rotates medially almost 90 degrees. By the ninth week, the hilum is directed anteromedially (see Fig. 14-7*C* and *D*).

Changes in the Blood Supply of the Kidneys

As the kidneys "ascend" from the pelvis, they receive their blood supply from vessels that are close to them. Initially, the renal arteries are branches of the common iliac arteries (see Fig. 14-7*A* and *B*). As they "ascend" further, the kidneys receive their blood supply from the distal end of the aorta. When they reach a higher level, they receive new branches from the aorta (see Fig. 14-7*C* and *D*). Normally, the caudal branches undergo involution and disappear. When the kidneys come into contact with the **suprarenal glands** in the ninth week, their "ascent" stops. The kidneys receive their most cranial arterial branches from the abdominal aorta; these branches become the permanent **renal arteries.**

Accessory Renal Arteries

The relatively common variations in the blood supply to the kidneys reflect the manner in which the blood supply continually changes during embryonic and early fetal life (see Fig. 14-7). A single renal artery to each kidney is present in about 75% of people. About 25% of adult kidneys have two to four renal arteries. Accessory (supernumerary) renal arteries usually arise from the aorta superior or inferior to the main renal artery and follow it to the hilum (Fig. 14-8*A* and *B*). Accessory renal arteries may enter the kidneys directly, usually into the superior or inferior poles. An accessory artery to the inferior pole may cross anterior to the ureter and obstruct it, causing hydronephrosis, or distention of the pelvis and calices with urine (see Fig. 14-8*B*). If the artery enters the inferior pole of the right kidney, it usually crosses anterior to the inferior vena cava and ureter. It is important to recognize that accessory renal arteries are end arteries; consequently, if an accessory artery is damaged or ligated, the part of the kidney supplied by it is likely to become ischemic. Accessory arteries are about twice as common as accessory veins (see Fig. 14-8*C* and *D*).

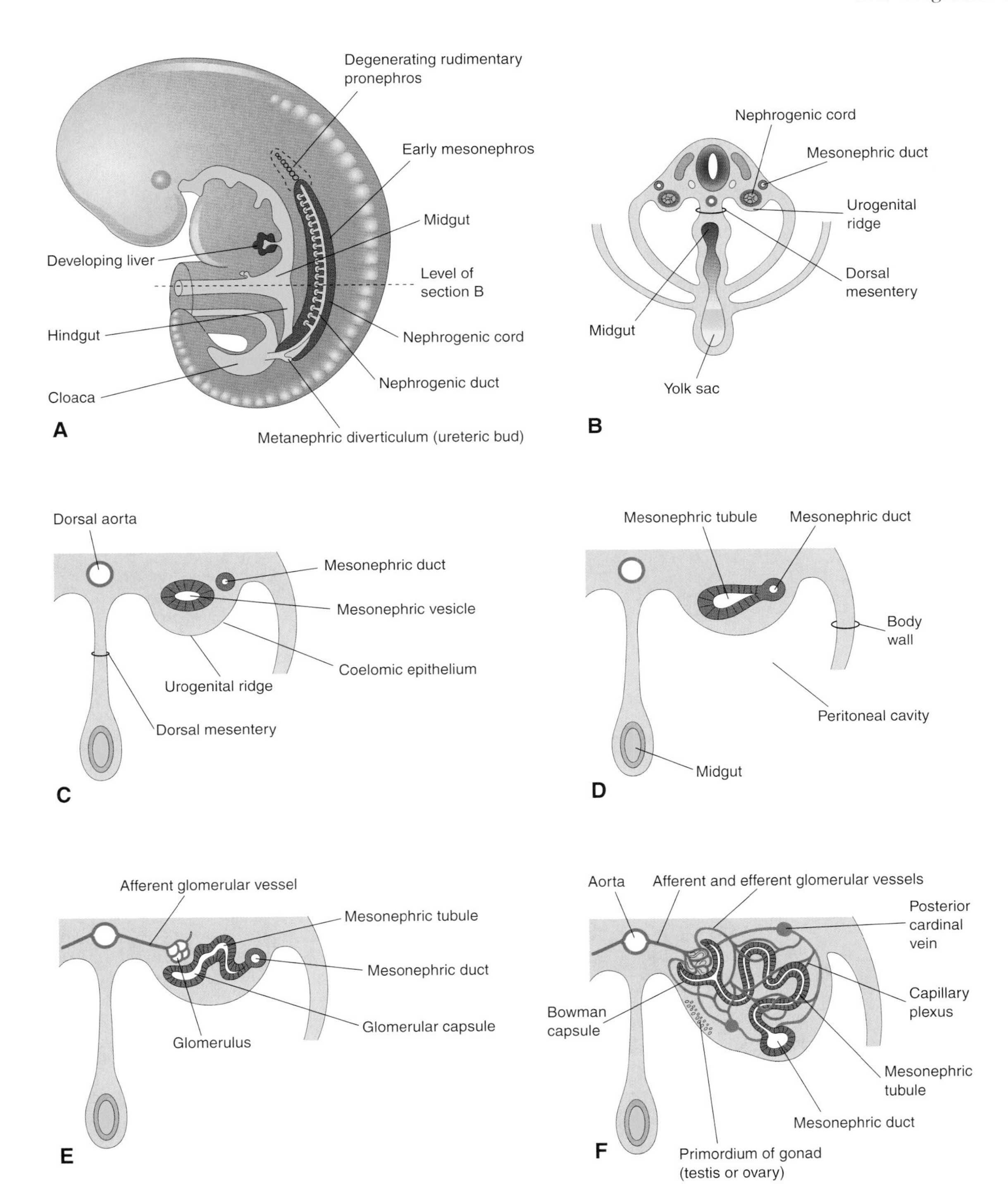

Figure 14 - 3. *A,* Sketch of a lateral view of a 5-week embryo showing the extent of the mesonephros and the primordium of the metanephros or permanent kidney. *B,* Transverse section of the embryo showing the nephrogenic cords from which the mesonephric tubules develop. *C* to *F,* Sketches of transverse sections showing successive stages in the development of a mesonephric tubule between the 5th and 11th weeks. Note that the mesenchymal cell cluster in the nephrogenic cord develops a lumen, thereby forming a mesonephric vesicle. The vesicle soon becomes an S-shaped mesonephric tubule and extends laterally to join the mesonephric duct. The expanded medial end of the mesonephric tubule is invaginated by blood vessels to form a glomerular capsule (Bowman capsule). The cluster of capillaries projecting into this capsule is the glomerulus.

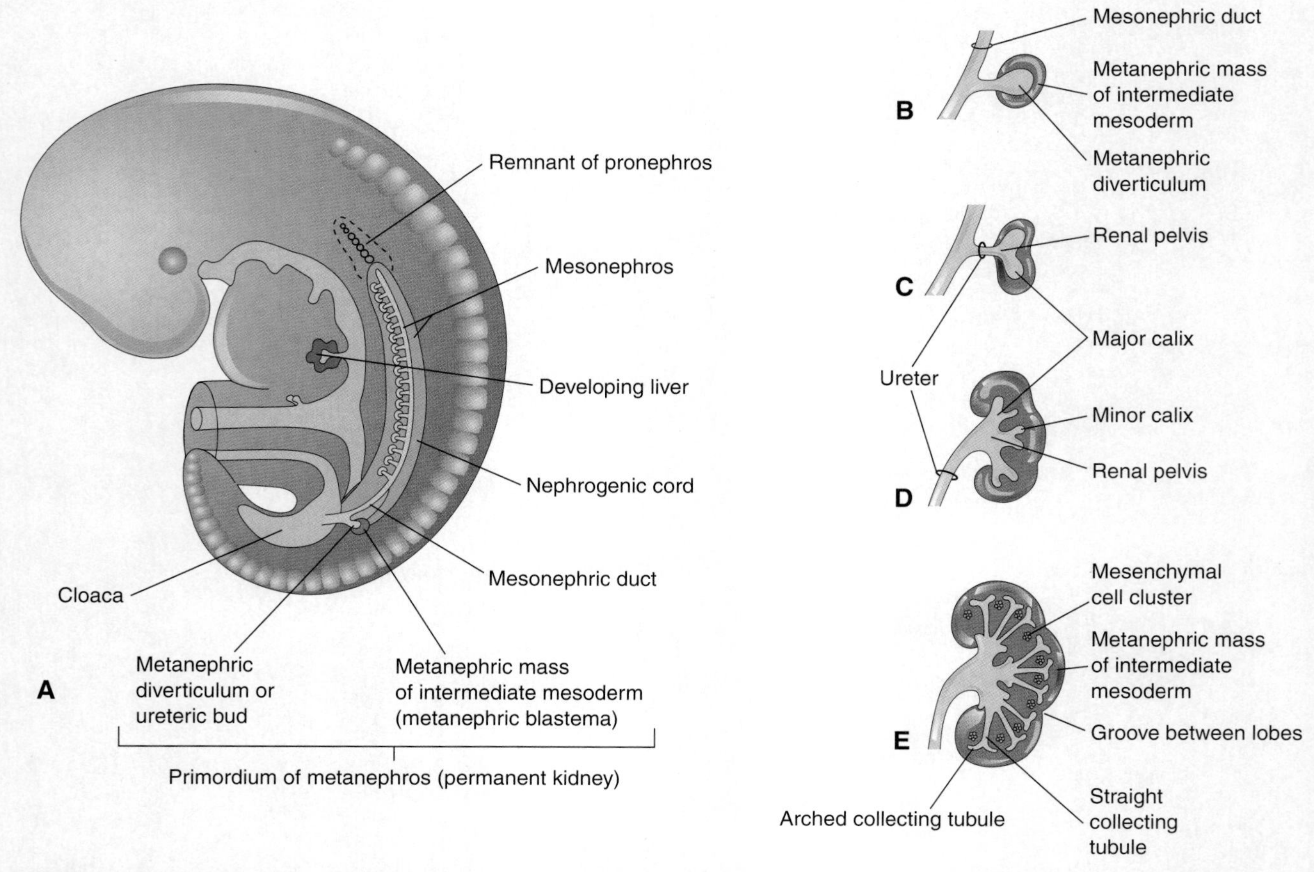

Figure 14-4. Development of the metanephros, the primordium of the permanent kidney. *A,* Sketch of a lateral view of a 5-week embryo, showing the primordium of the metanephros. *B* to *E,* Sketches showing successive stages in the development of the metanephric diverticulum or ureteric bud (fifth to eighth weeks). Observe the development of the ureter, renal pelvis, calices, and collecting tubules.

Congenital Anomalies of the Kidneys and Ureters

Developmental abnormalities of the kidneys and ureters are relatively common. Many of these abnormalities can be detected before birth by ultrasonography.

Renal Agenesis

Unilateral renal agenesis occurs in about 1 in every 1000 newborn infants (Fig. 14–9*A*). Male infants are affected more often than female infants, and the left kidney is usually the one that is absent. The other kidney usually undergoes compensatory hypertrophy and performs the function of the missing kidney. Unilateral renal agenesis should be suspected in infants with a *single umbilical artery* (see Chapter 8). **Bilateral renal agenesis** *is associated with oligohydramnios* (see Chapter 8) because little or no urine is excreted into the amniotic cavity. Decreased amniotic fluid volume in the absence of other causative factors, such as rupture of the fetal membranes, alerts the sonographer to search for urinary tract anomalies. Bilateral absence of the kidneys occurs in about 1 in every 3000 births and is incompatible with postnatal life. Most infants with bilateral renal agenesis die shortly after birth or during the first few months of life.

Absence of the kidneys results when the metanephric diverticula fail to develop or the primordia of the ureters degenerate. Failure of the metanephric diverticulum to penetrate the metanephric mesoderm results in an absence of renal development because no nephrons are induced by the collecting tubules to develop from the metanephric mass of intermediate mesoderm. Renal agenesis probably has a multifactorial cause.

Malrotation of the Kidneys

If a kidney fails to rotate, the hilum faces anteriorly; that is, the fetal kidney retains its embryonic position (see Figs. 14–7*A* and 14–9*C*). If the hilum faces posteriorly, rotation of the kidney has proceeded too far; if it faces laterally, then lateral instead of medial rotation has occurred. Abnormal rotation of the kidneys is often associated with ectopic kidneys.

Ectopic Kidneys

One or both kidneys may be in an abnormal position (see Fig. 14–9*B, E,* and *F*). Most ectopic kidneys are located

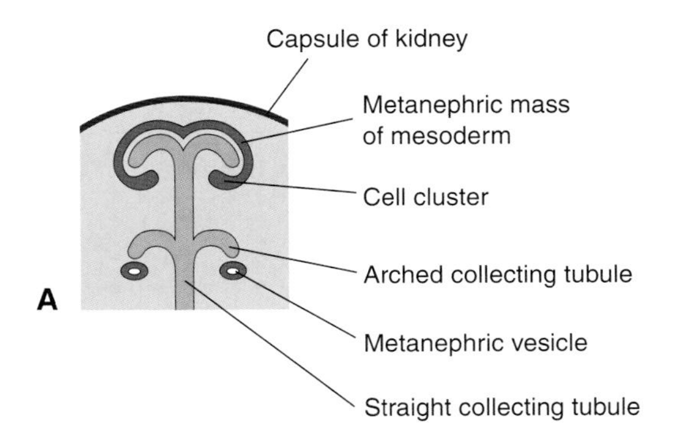

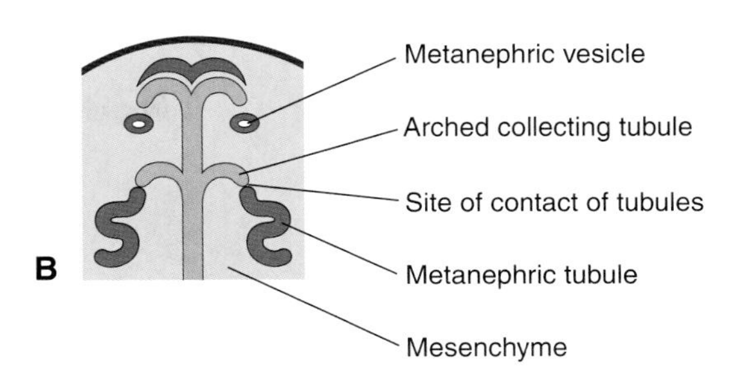

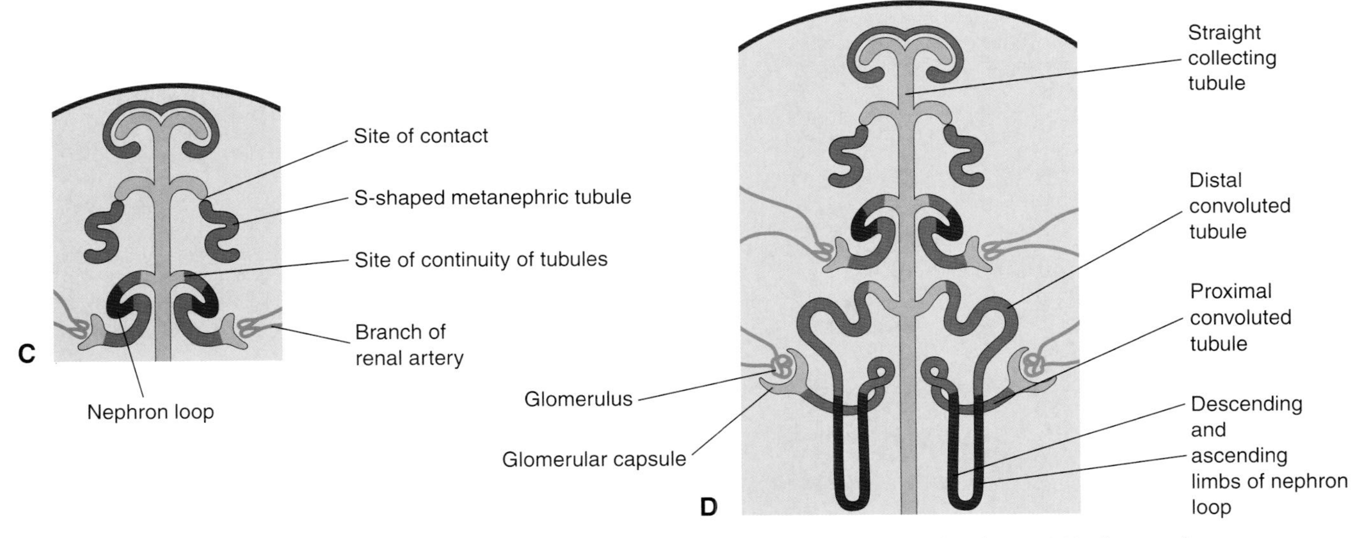

Figure 14–5. *A* to *D,* Diagrammatic sketches illustrating stages in nephrogenesis — the development of nephrons. *A,* Nephrogenesis commences around the beginning of the eighth week of development. *B* and *C,* Note that the metanephric tubules, the primordia of the nephrons, become continuous with the collecting tubules to form uriniferous tubules. *D,* The number of nephrons more than doubles from 20 weeks to 38 weeks. Observe that nephrons are derived from the metanephric mass of mesoderm, and that the collecting tubules are derived from the metanephric diverticulum.

in the pelvis, but some lie in the inferior part of the abdomen. **Pelvic kidneys** and other forms of ectopia result from failure of the kidneys to "ascend." Pelvic kidneys are close to each other and may fuse to form a **discoid kidney** — also called a "pancake kidney" (see Fig. 14–9*E*). Sometimes, a kidney crosses to the other side, resulting in **crossed renal ectopia** with or without fusion. An unusual type of abnormal kidney is *unilateral fused kidney* (see Fig. 14–9*D*). In such cases, the developing kidneys fuse while they are in the pelvis, and one kidney "ascends" to its normal position, carrying the other one with it.

In 1 in about 500 persons, the poles of the kidneys are fused; usually it is the inferior poles that fuse to form a **horseshoe kidney** (Fig. 14–10). Normal ascent of the fused kidneys is prevented because they are caught by the root of the inferior mesenteric artery.

Duplications of the Urinary Tract

Duplications of the abdominal part of the ureter and the renal pelvis are common, but a **supernumerary kidney** is

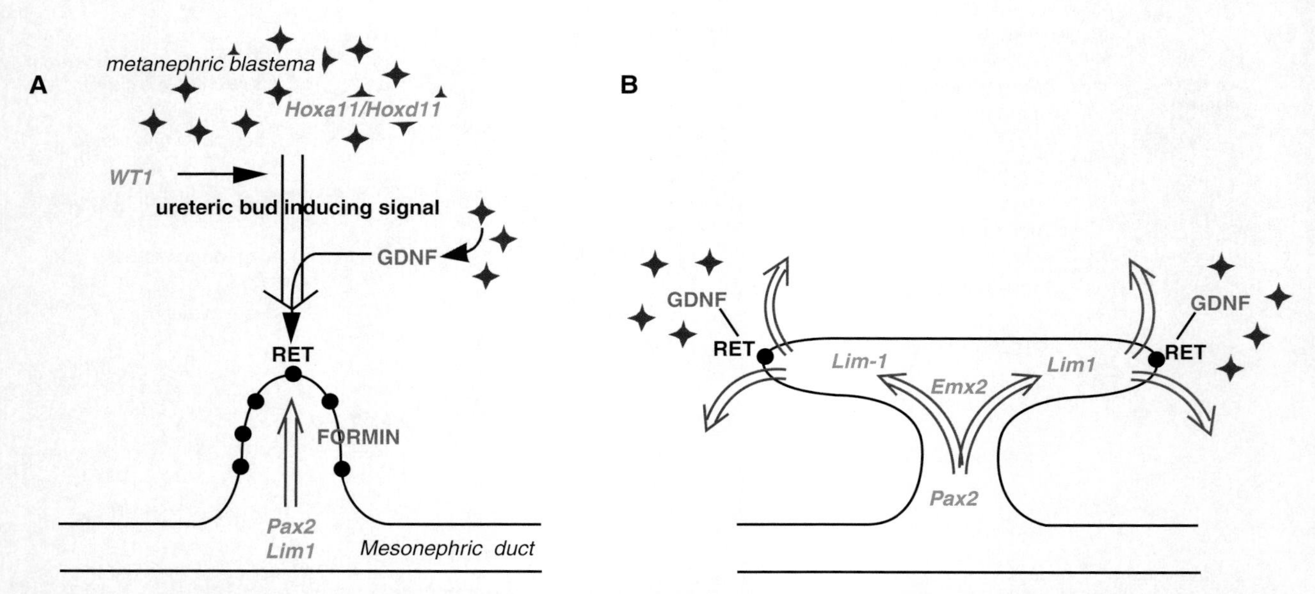

Figure 14 - 6. Molecular control of kidney development. *A,* Metanephric diverticulum or uretic bud outgrowth requires inductive signals derived from metanephric mesenchyme under control of transcription factors (orange text) such as *WT1* and signaling molecules (red text) including GDNF and its epithelial receptor, RET. Normal uretric bud response to these inductive signals is under the control of transcription factors such as *PAX2, LIM1,* and the *FORMIN* gene. *B,* Branching of the metanephric diverticulum is initiated and maintained by interaction with the mesenchyme under the regulation of genes such as *Emx2* and specified expression of *GDNF* and *RET* at the tips of the invading uretric bud. (From Piscione TD, Rosenblum ND: The malformed kidney: disruption of the glomerular and tubular development. *Clin Genet* 56: 341–356, 1999.)

rare (see Fig. 14–9*C* and *F*). These anomalies result from division of the metanephric diverticulum. The extent of the duplication depends on how complete the division of the diverticulum was. Incomplete division of the ureteric primordium results in a divided kidney with a bifid ureter (see Fig. 14–9*B*). Complete division results in a double kidney with a bifid ureter or separate ureters (Fig. 14–11). A supernumerary kidney with its own ureter

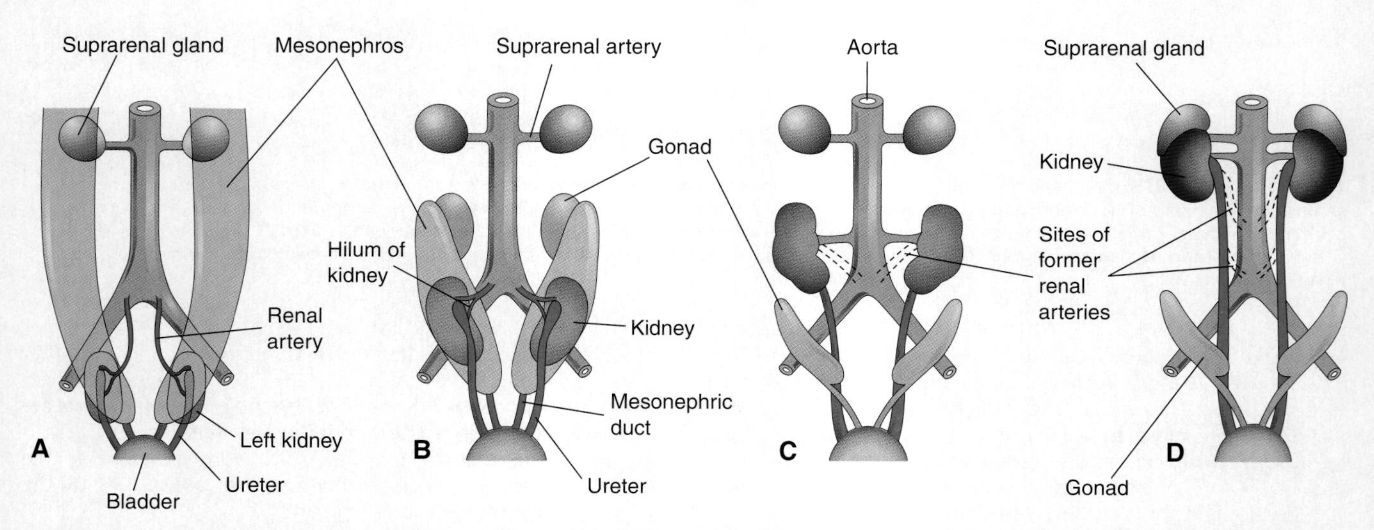

Figure 14 - 7. *A* to *D,* Diagrammatic ventral views of the abdominopelvic region of embryos and fetuses (sixth to ninth week of development) showing medial rotation and "ascent" of the kidneys from the pelvis to the abdomen. *A* and *B,* Observe also the size regression of the mesonephroi. *C* and *D,* Note that, as the kidneys "ascend," they are supplied by arteries at successively higher levels, and that the hilum of the kidney (where the vessels and nerves enter) is eventually directed anteromedially.

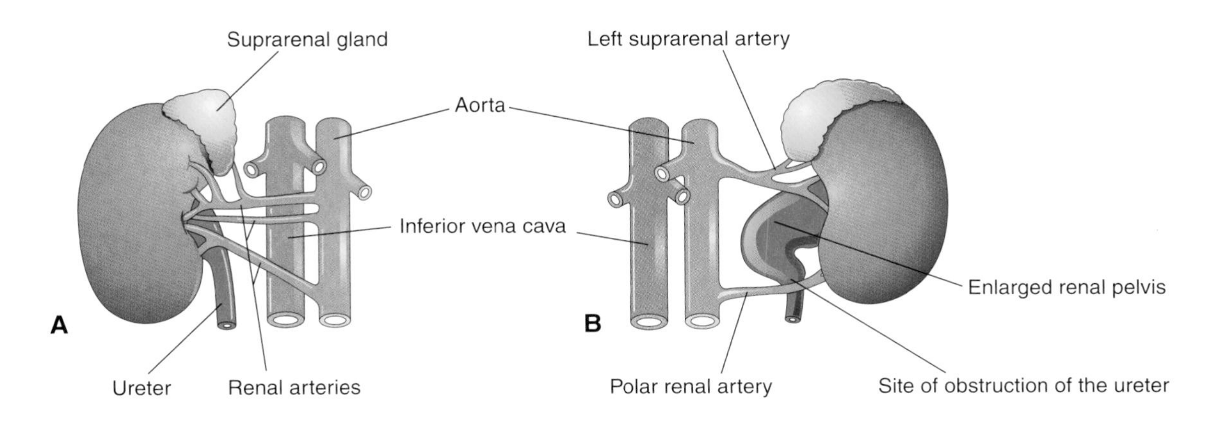

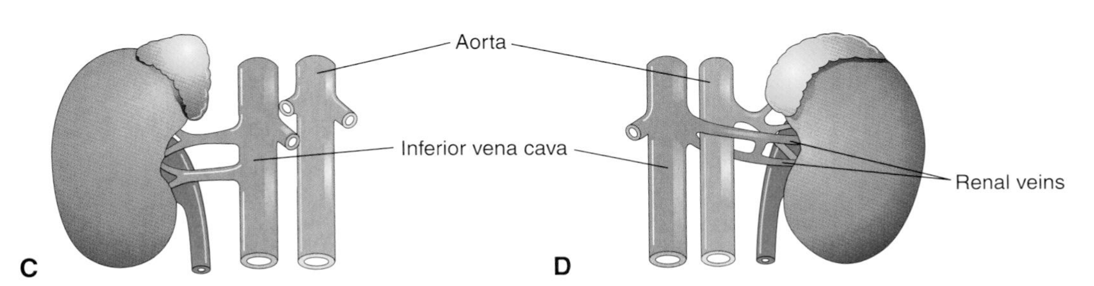

Figure 14-8. Drawings illustrating common variations of renal vessels. *A* and *B,* Multiple renal arteries. Note the accessory vessels entering the poles of the kidney. The polar renal artery illustrated in *B* has obstructed the ureter and produced an enlarged renal pelvis. *C* and *D,* Multiple renal veins are less common than supernumerary arteries.

probably results from the formation of two metanephric diverticula.

Ectopic Ureter

An ectopic ureter opens anywhere except into the urinary bladder. In males, an ectopic ureter usually opens into the neck of the bladder or into the prostatic part of the urethra, but it may enter the ductus deferens, prostatic utricle, or seminal gland. In females, an ectopic ureteric orifice may be located in the bladder neck, urethra, vagina, or vestibule. **Ureteric ectopia** results when the ureter is not incorporated into the posterior part of the urinary bladder; instead it is carried caudally with the mesonephric duct and is incorporated into the caudal portion of the vesical part of the urogenital sinus. When two ureters form on one side (see Fig. 14-11), they usually open into the urinary bladder. In some males, the extra ureter is carried caudally, and drains into the neck of the bladder or into the prostatic part of the urethra.

Cystic Kidney Disease

Polycystic kidney disease (*PKD*) is an autosomal recessive disorder (ARPKD) that is diagnosed at birth or in utero by ultrasonography. Both kidneys contain many hundreds of small cysts, resulting in renal insufficiency. Death of the infant usually occurs shortly after birth; however, an increasing number of these infants are surviving because of postnatal dialysis and kidney transplantation. *Multicystic dysplastic kidney* (*MDK*) *disease* results from dysmorphology during development of the renal system. The outcome for children with MDK is generally good since, in 75% of cases, the disease is unilateral. In MDK, fewer cysts are seen than in ARPKD, and these can range in size from a few millimeters to many centimeters in the same kidney. For many years, it was thought that the cysts were the result of failure of the metanephric vesicles to join the tubules derived from the metanephric mesoderm (Fig. 14-5*B*). It is now believed that the cystic structures are wide dilations of parts of the otherwise continuous nephrons, particularly the nephron loops. Gene mutations and faulty signaling have been implicated.

Development of the Urinary Bladder

Division of the cloaca by the **urorectal septum** (Fig. 14-12*A*) into a dorsal rectum and a ventral urogenital sinus is described in Chapter 13. For

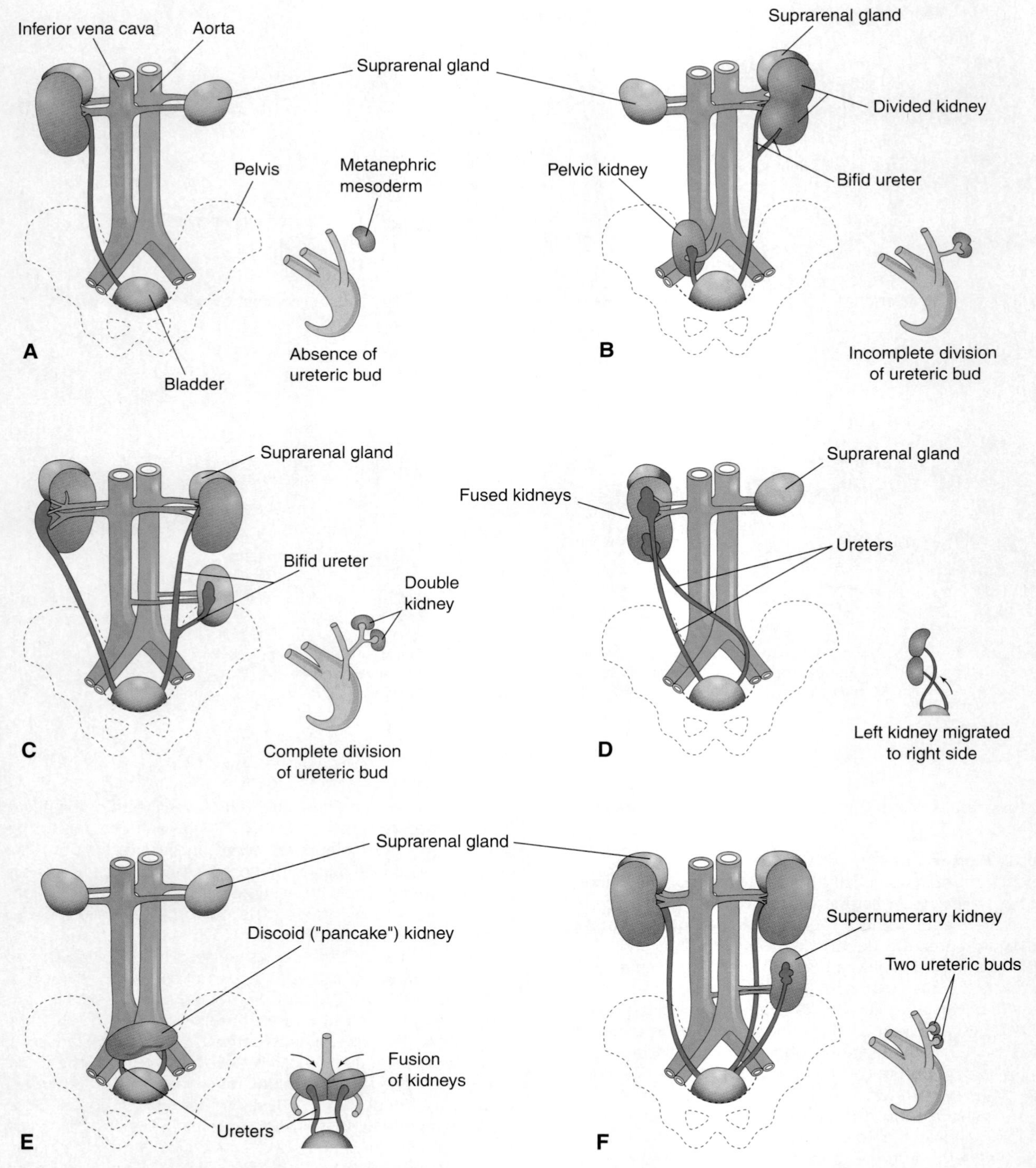

Figure 14 - 9. Drawings illustrating various anomalies of the urinary system. The small sketch to the lower right of each drawing illustrates the probable embryological basis of the anomaly. *A,* Unilateral renal agenesis. *B,* Right side of fetus, pelvic kidney; left side of fetus, divided kidney with a bifid ureter. *C,* Right side of fetus, malrotation of the kidney; left side of fetus, bifid ureter and supernumerary kidney. *D,* Crossed renal ectopia. The left kidney has crossed to the right side and fused with the right kidney. *E,* Discoid kidney resulting from fusion of the kidneys while they were in the pelvis. *F,* Supernumerary left kidney resulting from the development of two metanephric diverticula or ureteric buds.

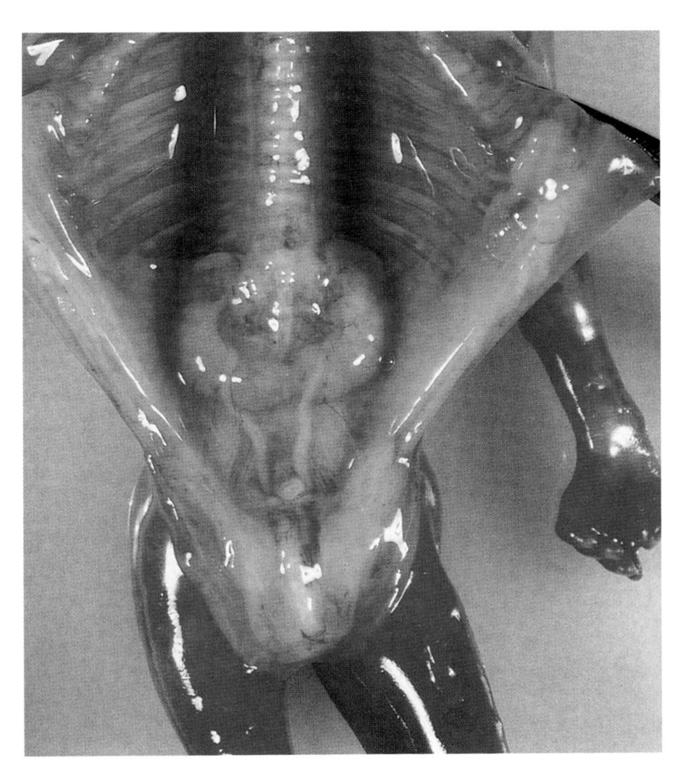

Figure 14 - 10. Horseshoe kidney in a female fetus (13 weeks). This anomaly resulted from fusion of the inferior poles of the kidneys while they were in the pelvis. (Courtesy of Dr. DK Kalousek, Department of Pathology, University of British Columbia, Children's Hospital, Vancouver, British Columbia, Canada.)

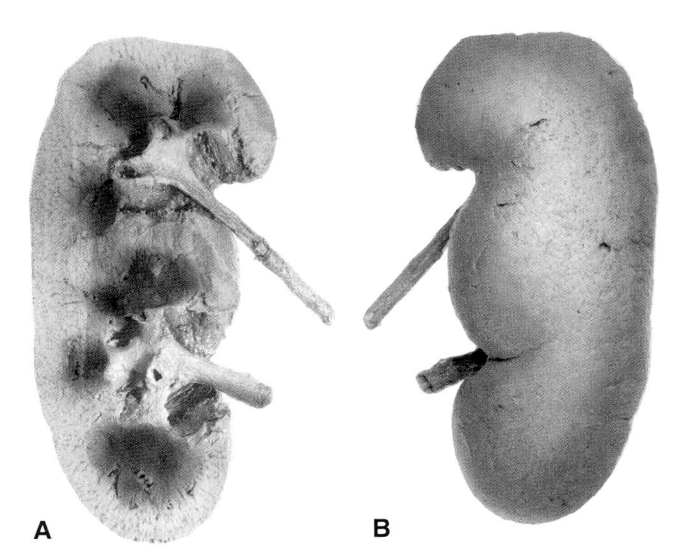

Figure 14 - 11. Photographs of a duplex kidney with two ureters and renal pelves. This congenital anomaly results from incomplete division of the metanephric diverticulum or ureteric bud. *A,* Longitudinal section through the kidney showing two renal pelves and calices. *B,* Anterior surface of the kidney.

descriptive purposes, the **urogenital sinus** is divided into three parts (see Fig. 14 - 12*A* and *C*):

- a cranial *vesical part* that forms most of the bladder and is continuous with the allantois
- a middle *pelvic part* that becomes the urethra in the bladder neck and the prostatic part of the urethra in males and the entire urethra in females
- a caudal *phallic part* that grows toward the genital tubercle — primoridium of the penis or clitoris

The bladder develops mainly from the vesical part of the urogenital sinus (see Fig. 14 - 12*D*), but its trigone region is derived from the caudal ends of the mesonephric ducts. The epithelium of the bladder is derived from the endoderm of the vesical part of the urogenital sinus. The other layers of its wall develop from adjacent splanchnic mesenchyme. Initially, the bladder is continuous with the **allantois,** a vestigial structure (see Fig. 14 - 12*C*). The allantois soon constricts and becomes a thick fibrous cord, the **urachus**, which extends from the apex of the bladder to the umbilicus (see Fig. 14 - 12*G*). In the adults, the urachus is represented by the *median umbilical ligament*. As the bladder enlarges, distal parts of the mesonephric ducts are incorporated into its dorsal wall (see Fig. 14 - 12*B* to *H*). These ducts contribute to the formation of the connective tissue in the *trigone of the bladder*, but the epithelium of the entire bladder is derived from the endoderm of the urogenital sinus. As the mesonephric ducts are absorbed, the ureters come to open separately into the urinary bladder (see Fig. 14 - 12*C* to *H*). The orifices of the mesonephric ducts move close together and enter the prostatic part of the urethra as the caudal ends of these ducts become the *ejaculatory ducts*. In females, the distal ends of the mesonephric ducts degenerate.

The apex of the urinary bladder in adults is continuous with the **median umbilical ligament**, which extends posteriorly along the posterior surface of the anterior abdominal wall; this ligament is the fibrous remnant of the urachus. The *median umbilical ligament* lies between the *medial umbilical ligaments*, which are the fibrous remnants of the umbilical arteries (see Chapter 15).

Urachal Anomalies

A remnant of the lumen usually persists in the inferior part of the urachus in infants, and in about 50% of cases, the lumen is continuous with the cavity of the bladder. Remnants of the epithelial lining of the urachus may give rise to **urachal cysts** (Fig. 14 - 13*A*). The patent inferior end of the urachus may dilate to form a **urachal sinus** that opens into the bladder. The lumen in the superior part of the urachus may

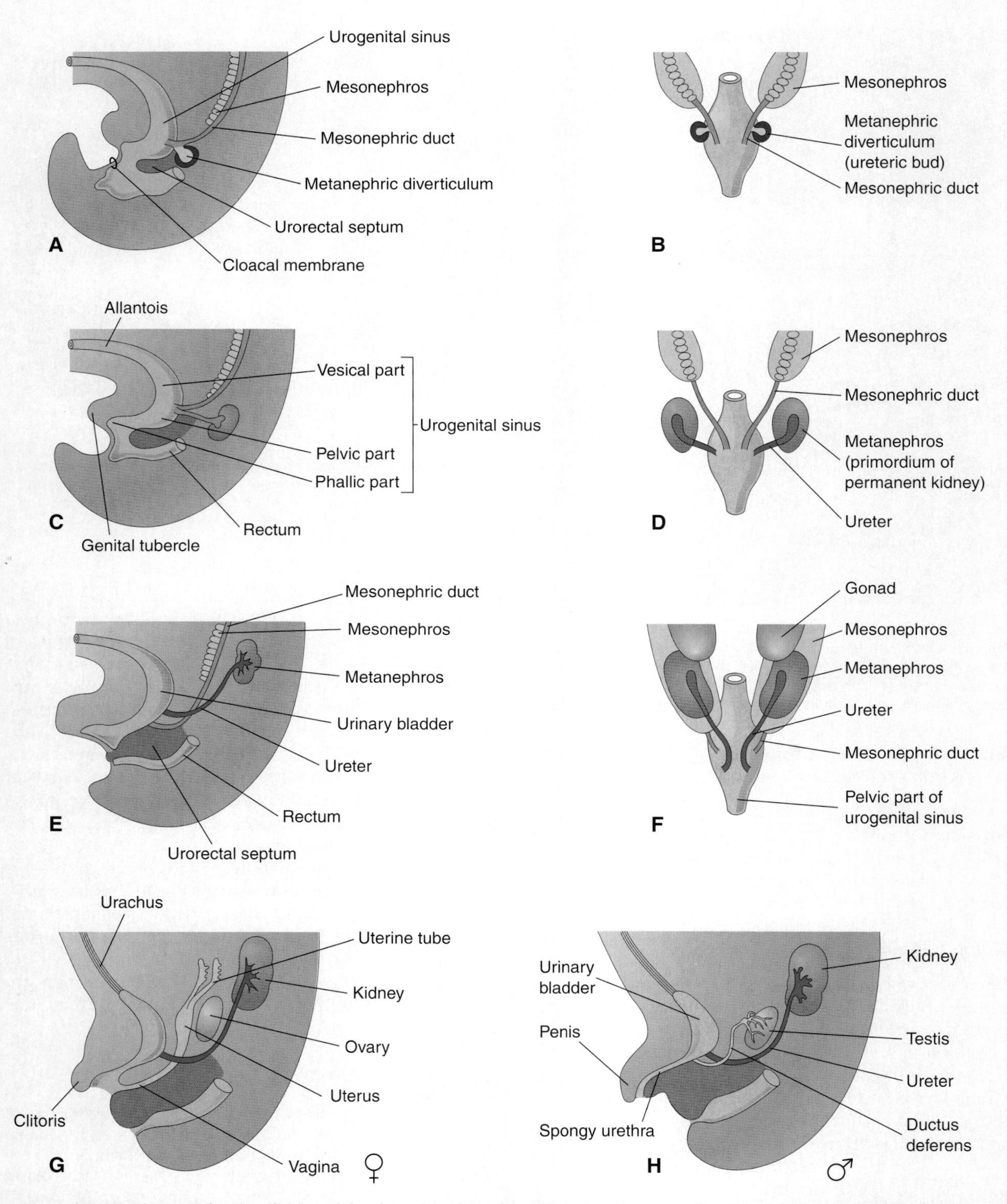

Figure 14 - 12. Diagrams showing division of the cloaca into the urogenital sinus and rectum; absorption of the mesonephric ducts; development of the urinary bladder, urethra, and urachus; and changes in the location of the ureters. *A,* Lateral view of the caudal half of a 5-week embryo. *B, D,* and *F,* Dorsal views. *C, E, G,* and *H,* Lateral views. The stages shown in *G* and *H* are reached by the 12th week of development.

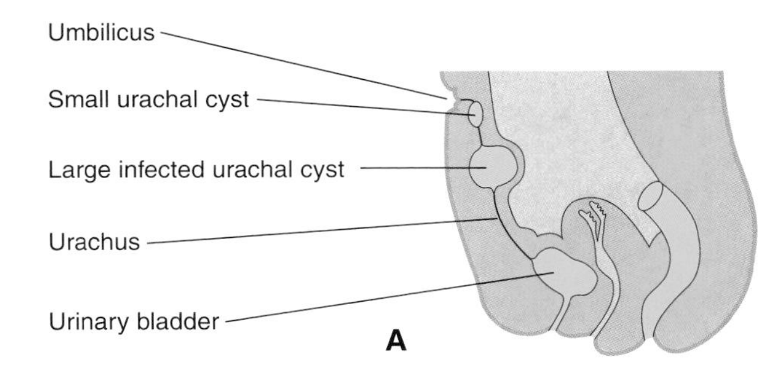

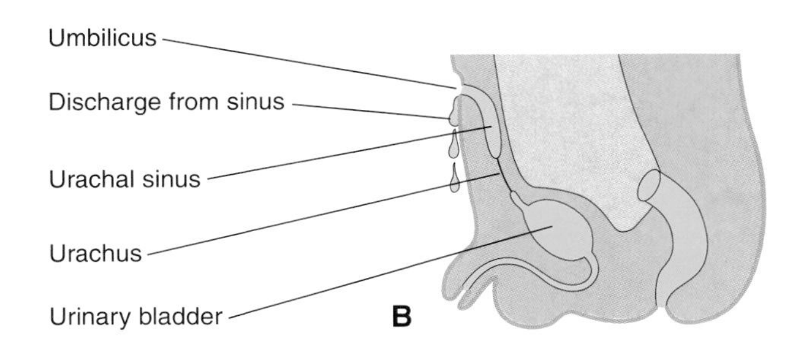

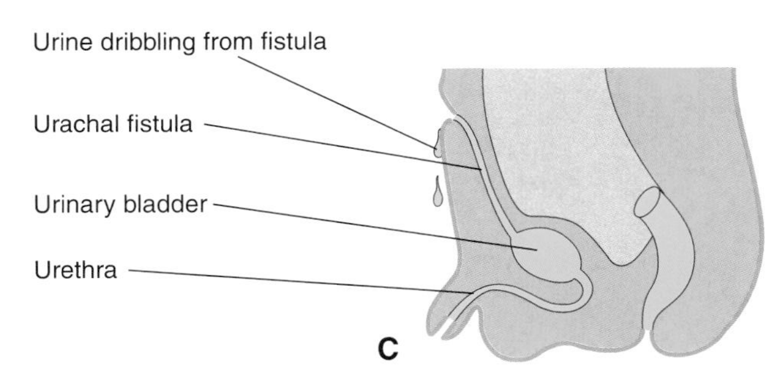

Figure 14 - 13. Diagrams illustrating urachal anomalies. *A,* Urachal cysts. The most common site for these cysts is in the superior end of the urachus, just inferior to the umbilicus. *B,* Two types of urachal sinus are illustrated: one that opens into the bladder and the other opens at the umbilicus. *C,* Patent urachus or urachal fistula connecting the bladder and the umbilicus.

also remain patent and form a urachal sinus that opens at the umbilicus (see Fig. 14 – 13*B*). Very rarely, the entire urachus remains patent and forms a **urachal fistula** that allows urine to escape from its umbilical orifice (see Fig. 14 – 13*C*).

Exstrophy of Bladder

Exstrophy of the bladder is a severe anomaly that occurs about once in every 10,000 to 40,000 births, predominantly affecting male infants (Fig. 14 – 14). *Exposure and protrusion of the posterior wall of the bladder* characterize this congenital anomaly. The trigone of the bladder and the ureteric orifices are exposed, and urine dribbles intermittently from the everted bladder. **Epispadias** (urethra opens on dorsum of penis) and wide separation of the pubic bones are associated with complete exstrophy of the bladder. In some cases, the penis is divided into two parts, and the scrotum is bifid (split). Exstrophy of the bladder is caused by incomplete median closure of the inferior part of the anterior abdominal wall (Fig. 14 – 15*A* to *F*). The defect involves the anterior abdominal wall and the anterior wall of the urinary bladder. The anomaly is the result of failure of mesenchymal cells to migrate between the ectoderm of the abdomen and cloaca during the fourth week of development (see Fig. 14 – 15*B* and *C*). As a result, no muscle and connective tissue form in the anterior abdominal wall over the urinary bladder. Later, the thin epidermis and anterior wall of the bladder rupture, causing wide communication between the exterior and the mucous membrane of the bladder.

Development of the Urethra

The epithelium of most of the male urethra and the entire female urethra is derived from endoderm of the urogenital sinus (Fig. 14 – 16; see also Fig. 14 – 12). The distal part of the urethra in the glans of the penis is derived from a solid cord of ectodermal cells that grows from the tip of the glans penis to meet the part of the spongy urethra derived from the phallic part of the urogenital sinus (see Fig. 14 – 16*A* to *C*). The ectodermal

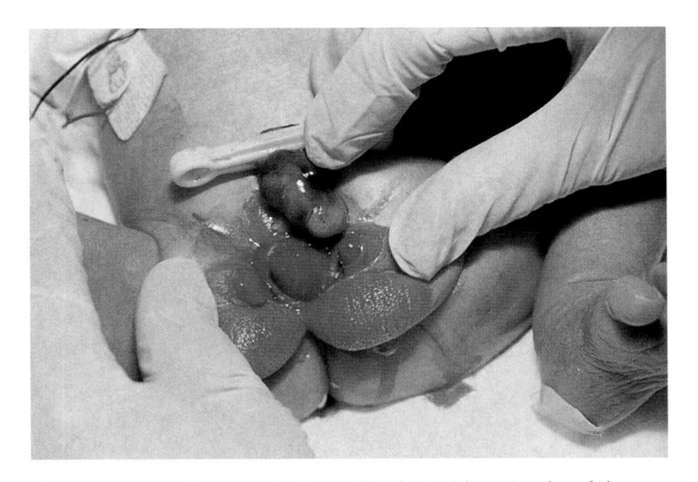

Figure 14 - 14. Photograph of a male infant with exstrophy of the bladder. Because of defective closure of the inferior part of the anterior abdominal wall and the anterior wall of the bladder, the urinary bladder appears as an everted, bulging mass inferior to the umbilicus. (Courtesy of Dr. AE Chudley, Department of Pediatrics and Child Health, University of Manitoba, Children's Hospital, Winnipeg, Manitoba, Canada.)

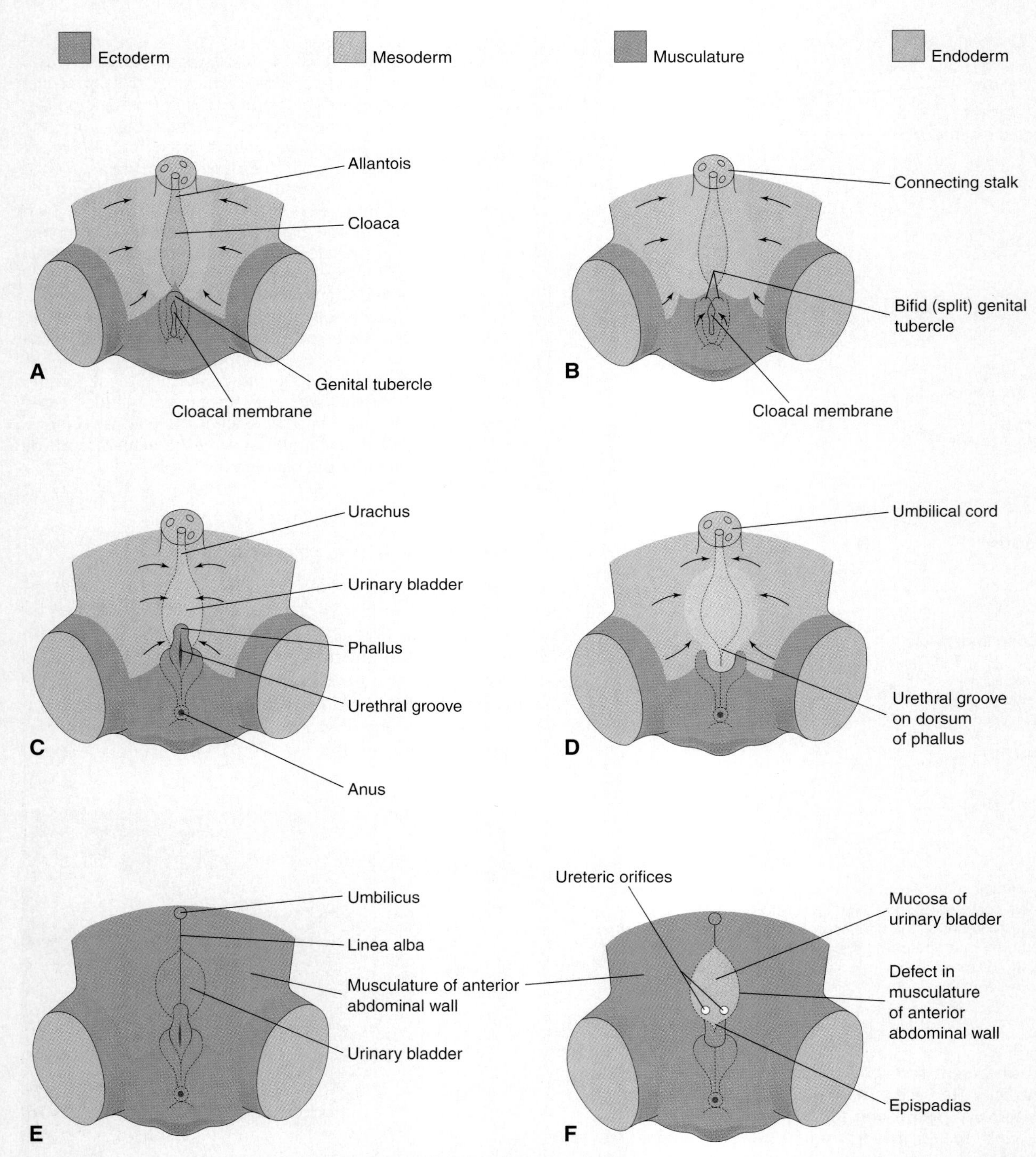

Figure 14 - 15. *A, C,* and *E,* Normal stages in the development of the infraumbilical abdominal wall and the penis during the fourth to eighth weeks. Note that mesoderm and (later) muscle reinforce the ectoderm of the developing anterior abdominal wall. *B, D,* and *F,* Probable stages in the development of exstrophy of the bladder and epispadias. In *B* and *D,* note that the mesenchyme (embryonic connective tissue) fails to extend into the anterior abdominal wall anterior to the urinary bladder. Also note that the genital tubercle is located in a more caudal position than usual, and that the urethral groove has formed on the dorsal surface of the penis. In *F,* the surface ectoderm and anterior wall of the bladder have ruptured, resulting in exposure of the posterior wall of the bladder. Note that the musculature of the anterior abdominal wall is present on each side of the defect. (Based on Patten BM, Barry A: The genesis of exstrophy of the bladder and epispadias. *Am J Anat* 90:35, 1952.)

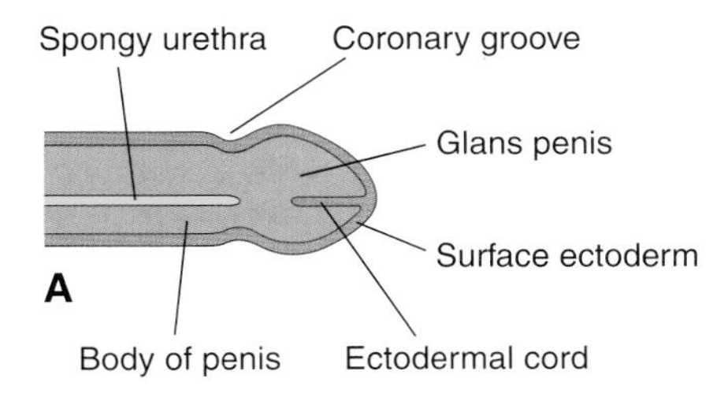

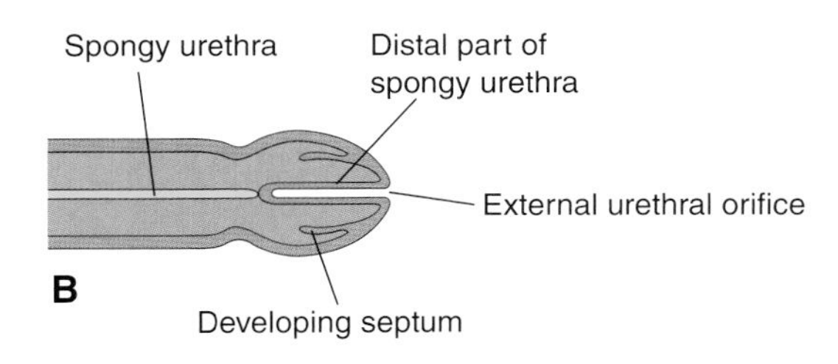

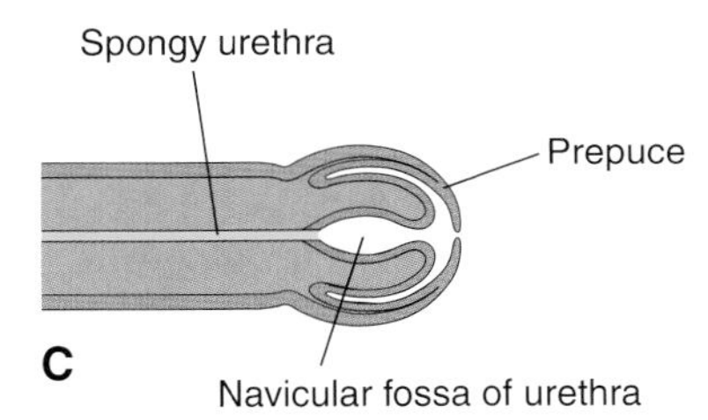

Figure 14-16. Schematic longitudinal sections of the distal part of the developing penis, illustrating development of the prepuce (foreskin) and the distal part of the spongy urethra. *A,* Eleven weeks. *B,* Twelve weeks. *C,* Fourteen weeks. The epithelium of the spongy urethra has a dual origin; most of it is derived from endoderm of the phallic part of the urogenital sinus. The distal part of the urethra lining the navicular fossa is derived from surface ectoderm (see also page 252).

cord canalizes and joins the rest of the spongy urethra; consequently, the epithelium of the terminal part of the urethra is derived from surface ectoderm. The connective tissue and smooth muscle of the urethra in both sexes are derived from splanchnic mesenchyme.

Development of the Suprarenal Glands

The cortex and medulla of the suprarenal (adrenal) glands have different origins (Fig. 14-17*A* to *H*). The **cortex** develops from **mesoderm**, whereas the **medulla** differentiates from **neural crest cells.** The cortex first becomes evident during the sixth week of development by an aggregation of mesenchymal cells on each side, between the root of the dorsal mesentery and the developing gonad (see Fig. 14-19*C*). The cells that form the *fetal cortex* are derived from the mesothelium lining the posterior abdominal wall. The cells that form the medulla are derived from an adjacent *sympathetic ganglion,* which is derived from the neural crest cells (see Fig. 14-17*B*). These cells differentiate into the *secretory cells* of the suprarenal medulla. Mesenchymal cells from the mesothelium enclose the fetal cortex and give rise to the permanent cortex (see Fig. 14-17*C*). Differentiation of the characteristic suprarenal cortical zones begins during the late fetal period. The *zona glomerulosa* and *zona fasciculata* are present at birth, but the *zona reticularis* is not recognizable until the end of the third year (see Fig. 14-17*H*). The suprarenal glands of the human fetus are 10 to 20 times larger than the adult glands relative to body weight and are large compared with the kidneys (see Fig. 14-6). These large glands result from the extensive size of the fetal cortex. The suprarenal medulla remains relatively small until after birth (see Fig. 14-17*F*). The suprarenal glands rapidly become smaller as the fetal cortex regresses during the first year (see Fig. 14-17*G*).

Congenital Adrenal Hyperplasia

An abnormal increase in the cells of the suprarenal cortex results in excessive androgen production during the fetal period. In females, this usually causes masculinization of the external genitalia and enlargement of the clitoris (Fig. 14-18). Affected male infants have normal external genitalia and may remain undiagnosed in early infancy. Later in childhood in both sexes, androgen excess leads to rapid growth and accelerated skeletal maturation. The **adrenogenital syndrome** associated with congenital adrenal hyperplasia (CAH) manifests itself in various clinical forms that can be correlated with enzymatic deficiencies of cortisol biosynthesis. CAH represents a group of *autosomal recessive disorders* that result in virilization of female fetuses. It is caused by a genetically determined mutation in the cytochrome P450c21-steroid 21-hydroxylase gene, which causes a deficiency of suprarenalcortical enzymes that are necessary for the biosynthesis of various steroid hormones. The reduced hormone output results in an increased release of adrenocorticotropic hormone, which causes adrenal hyperplasia and overproduction of androgens by the hyperplastic suprarenal glands.

Development of the Genital System

Although the chromosomal and genetic sex of an embryo is determined at fertilization by the kind of

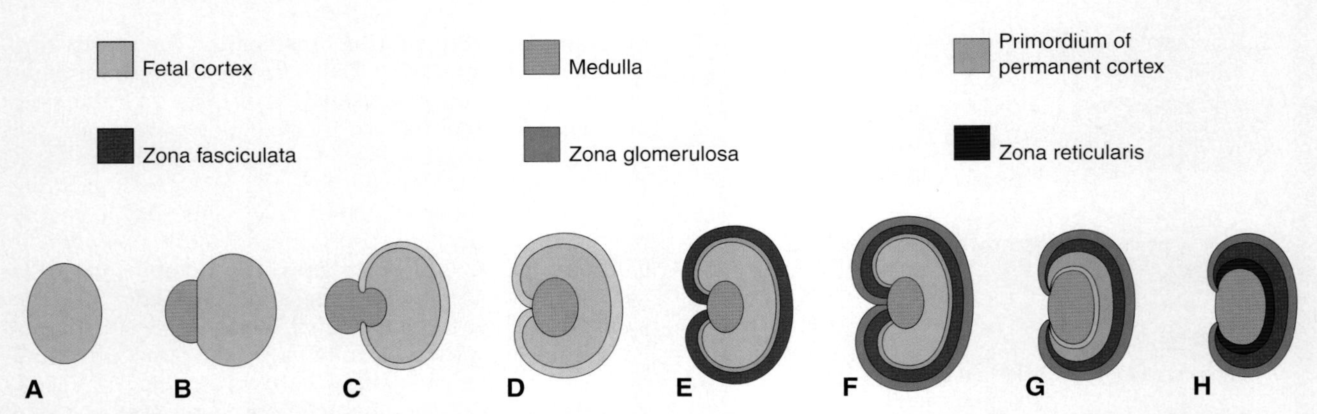

Figure 14–17. Schematic drawings illustrating development of the suprarenal glands. *A,* Six weeks, showing the mesodermal primordium of the fetal cortex. *B,* Seven weeks, showing the addition of neural crest cells. *C,* Eight weeks, showing the fetal cortex and the early permanent cortex beginning to encapsulate the medulla. *D* and *E,* Later stages of encapsulation of the medulla by the cortex. *F,* Neonatal period showing the fetal cortex and two zones of the permanent cortex. *G,* One year of age. Note that the fetal cortex has almost disappeared. *H,* Four years of age. Note the adult pattern of cortical zones. Observe that the fetal cortex has disappeared and that the gland is smaller than it was at birth *(F).*

sperm that fertilizes the ovum (see Chapter 3), male and female morphological characteristics do not begin to develop until the seventh week. The early genital systems in the two sexes are similar; therefore, the initial period of genital development is referred to as the *indifferent state of sexual development.*

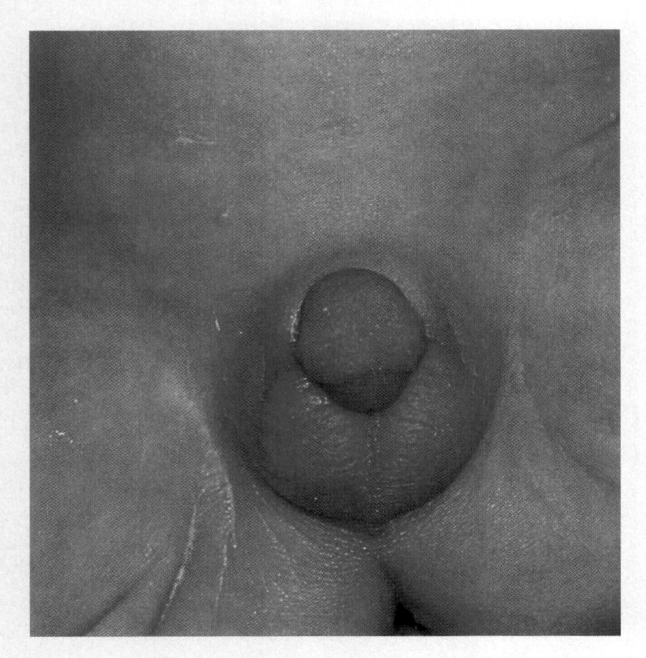

Figure 14–18. External genitalia of a newborn female infant with congenital adrenal hyperplasia. The virilization was caused by excessive androgens produced by the suprarenal (adrenal) glands during the fetal period. Note the enlarged clitoris and fusion of the labia majora. (Courtesy of Dr. Heather Dean, Department of Pediatrics and Child Health, University of Manitoba, Winnipeg, Manitoba, Canada.)

Development of the Gonads

The gonads (testes and ovaries) are derived from three sources (Fig. 14–19*A* to *C*):

- the *mesothelium* (mesodermal epithelium) lining the posterior abdominal wall
- the underlying *mesenchyme* (embryonic connective tissue)
- the *primordial germ cells*

Indifferent Gonads

Gonadal development begins during the fifth week when a thickened area of mesothelium develops on the medial side of the mesonephros (see Fig. 14–19*A* to *C*). Proliferation of this epithelium and the underlying mesenchyme produces a bulge on the medial side of the mesonephros, termed the **gonadal ridge** (see Fig. 14–19*B* and *C*). Fingerlike epithelial cords, known as the **gonadal cords**, soon grow into the underlying mesenchyme (see Fig. 14–19*D*). The **indifferent gonad** now consists of an external *cortex* and an internal *medulla.* In embryos with an XX sex chromosome complex, the cortex of the indifferent gonad differentiates into an ovary and the medulla regresses. In embryos with an XY sex chromosome complex, the medulla differentiates into a testis and the cortex regresses, except for vestigial remnants (see Table 14–1).

Primordial Germ Cells

The primordial germ cells originate in the wall of the yolk sac and migrate by amoeboid movements along the dorsal mesentery of the gut to the gonadal ridges. During the sixth week of development, the primordial germ cells enter the underlying mesenchyme and are incorporated

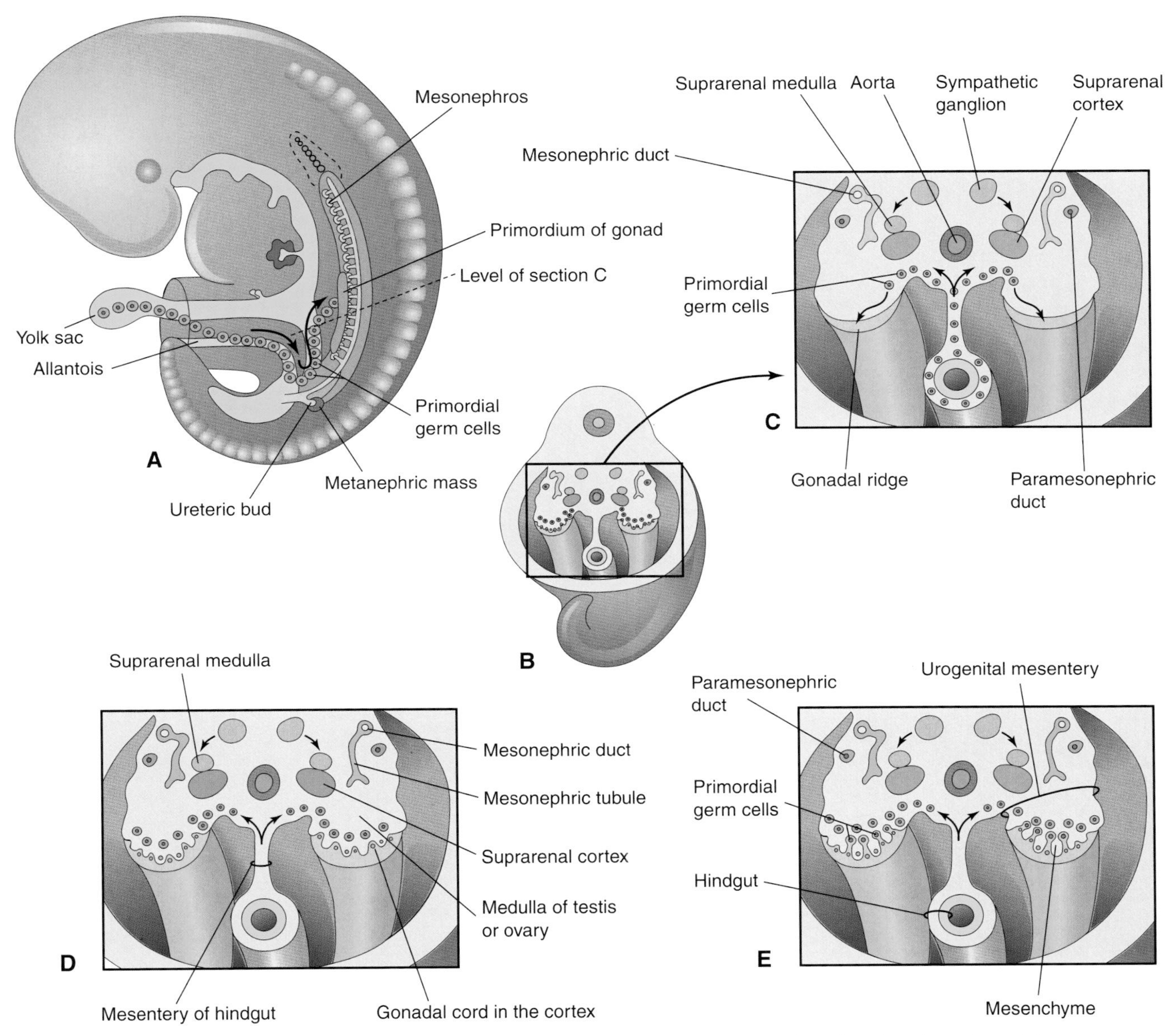

Figure 14-19. *A,* Sketch of a 5-week embryo illustrating the migration of primordial germ cells from the yolk sac into the embryo. *B,* Three-dimensional sketch of the caudal region of a 5-week embryo, showing the location and extent of the gonadal ridges. *C,* Transverse section showing the primordium of the suprarenal (adrenal) glands, the gonadal ridges, and the migration of primordial germ cells into the developing gonads. *D,* Transverse section of a 6-week embryo showing the gonadal cords. *E,* Similar section at a later stage showing the indifferent gonads and paramesonephric ducts.

in the *gonadal cords* (see Fig. 14-19*E*). They eventually differentiate into ova and sperms.

Sex Determination

Chromosomal sex, established at fertilization, depends upon whether an X-bearing sperm or a Y-bearing sperm fertilizes the X-bearing ovum. The type of gonads that develop is determined by the sex chromosome complex of the embryo (XX or XY). Before the seventh week of development, the gonads of the two sexes are identical in appearance and are called **indifferent gonads** (Fig. 14-20). Development of the male phenotype requires a Y chromosome, but only the short arm of this chromosome is critical for sex determination. The *SRY gene* for a **testis-determining factor (TDF)** has been localized in the sex-determining region of the Y

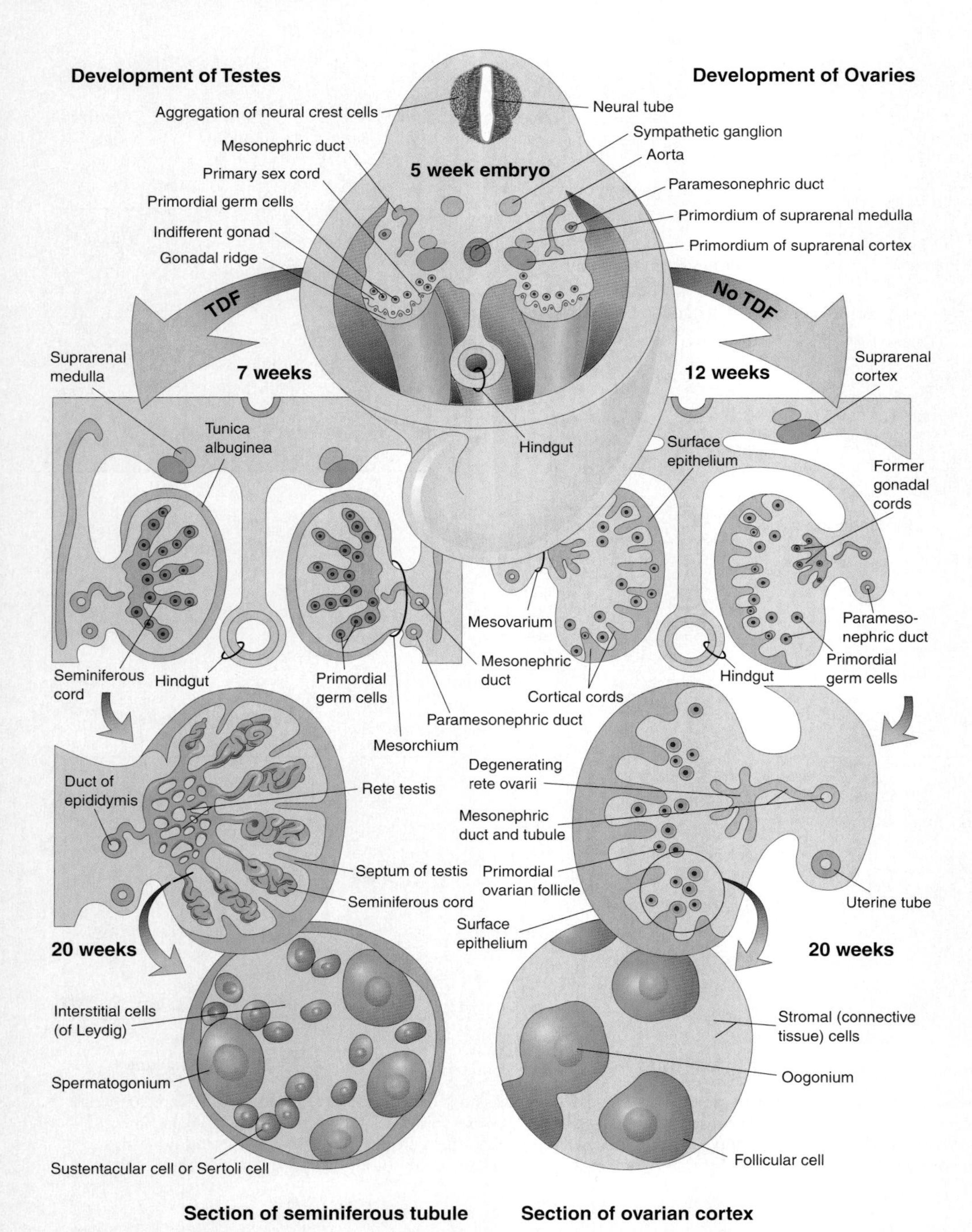

Figure 14 – 20. Schematic illustrations showing differentiation of the indifferent gonads of a 5-week embryo (*top*) into ovaries or testes. The left side of the illustration shows the development of testes resulting from the effects of the testis-determining factor (TDF) located on the Y chromosome. Note that the gonadal cords become seminiferous cords, the primordia of the seminiferous tubules. The parts of the gonadal cords that enter the medulla of the testis form the rete testis. In the section of the testis at the bottom left, observe that there are two kinds of cells: spermatogonia, which are derived from the primordial germ cells, and sustentacular or Sertoli cells, which are derived from mesenchyme. The right side of the figure shows the development of ovaries in the absence of TDF. Cortical cords have extended from the surface epithelium of the gonad, and primordial germ cells have entered them. They are the primordia of the oogonia. Follicular cells are derived from the surface epithelium of the ovary.

chromosome. Two X chromosomes are required for the development of the female phenotype. A number of genes and regions of the X chromosome have special roles in sex determination.

The *Y chromosome has a testis-determining effect* on the medulla of the indifferent gonad. Expression of the transcription factor *SOX9* is essential for testicular determination. Under the influence of TDF, the gonadal cords differentiate into **seminiferous cords** —primorida of seminiferous tubules (see Fig. 14–20). The absence of a Y chromosome (i.e., an XX sex chromosome complement) results in the formation of an ovary. Consequently, the type of sex chromosome complex established at fertilization determines the type of gonad that differentiates from the indifferent gonad. The type of gonads present then determines the type of sexual differentiation that occurs in the genital ducts and external genitalia. **Testosterone,** produced by the fetal testes, determines maleness. Primary female sexual differentiation in the fetus does not depend on hormones; it occurs even if the ovaries are absent and apparently is unaffected by hormonal changes.

Abnormal Sex Chromosome Complexes

In embryos with abnormal sex chromosome complexes, such as XXX or XXY, the number of X chromosomes appears to be unimportant in sex determination. If a *normal* Y chromosome is present, the embryo develops as a male. If no Y chromosome is present or the testis-determining region of the Y chromosome has been lost, female development occurs. The loss of an X chromosome does not appear to interfere with the migration of primordial germ cells to the gonadal ridges because some germ cells have been observed in the fetal gonads of 45,X females with Turner syndrome. Two X chromosomes are needed, however, to bring about complete ovarian development.

Development of the Testes

Embryos with a Y chromosome in their sex chromosome complement usually develop testes. A coordinated sequence of genes induces the development of testes. *The SRY gene for the testis-determining factor (TDF) on the short arm of the Y chromosome acts as the switch that directs development of the indifferent gonad into a testis.* TDF induces the gonadal cords to condense and extend into the medulla of the indifferent gonad, where they branch and anastomose to form the **rete testis.** The connection of the prominent gonadal cords — the seminiferous cords — with the surface epithelium is lost when a thick fibrous capsule, the tunica albuginea, develops (see Fig. 14–20). The dense **tunica albuginea** is a characteristic feature of testicular development in the fetus. Gradually, the testis separates from the degenerating mesonephros and becomes suspended by its own mesentery, the **mesorchium.** The seminiferous cords develop into the seminiferous tubules, straight tubules (tubuli recti), and rete testis.

The **seminiferous tubules** are separated by mesenchyme, giving rise to the **interstitial cells** (of Leydig). By the eighth week of development, these cells secrete androgenic hormones — *testosterone* and *androstenedione* — which induce masculine differentiation of the mesonephric ducts and the external genitalia. Testosterone production is stimulated by **human chorionic gonadotropin** (hCG), which reaches peak amounts during the 8- to 12-week period. The fetal testes also produce a glycoprotein known as **antimüllerian hormone (AMH)** or *müllerian-inhibiting substance* (MIS). AMH is produced by the sustentacular Sertoli cells, which are present until puberty, at which time the levels of AMH decrease. AMH suppresses development of the paramesonephric (müllerian) ducts, which form the uterus and uterine tubes. The seminiferous tubules remain solid (i.e., without lumina) until puberty, at which time lumina begin to develop. The walls of the seminiferous tubules are composed of two kinds of cells (see Fig. 14–20):

- *Sertoli cells,* supporting cells derived from the surface epithelium of the testis
- *Spermatogonia,* primordial sperm cells derived from primordial germ cells

Sertoli cells constitute most of the seminiferous epithelium in the fetal testis (see Fig. 14–20). The **rete testis** becomes continuous with 15 to 20 mesonephric tubules that become **efferent ductules**. These ductules are connected with the mesonephric duct, which becomes the **ductus epididymis** (see Figs. 14–20 and 14–21*A*).

Development of the Ovaries

Gonadal development occurs slowly in female embryos. The X chromosomes bear genes for ovarian development, and an autosomal gene also appears to play a role in ovarian organogenesis. The ovary is not identifiable by histological examination until about the 10th week of development. **Gonadal cords** do not become prominent, but they extend into the medulla and form a rudimentary *rete ovarii.* This structure and the gonadal cords normally degenerate and disappear (see Fig. 14–20). **Cortical cords** extend from the surface epithelium of the developing ovary into the underlying mesenchyme during the early fetal period. This epithelium is derived from the mesothelium. As the cortical cords increase in size, **primordial germ cells** are incorporated into them. At about 16 weeks, these cords begin to break up into isolated cell clusters — called **primordial follicles** — each of which consists of an **oogonium** (derived from a primordial germ cell), surrounded by a single layer of

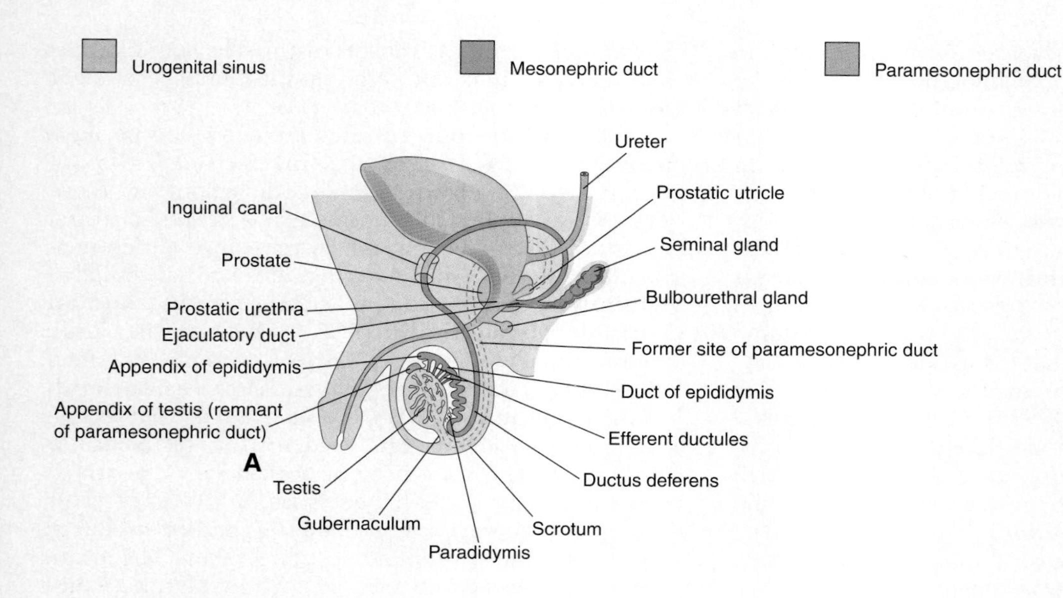

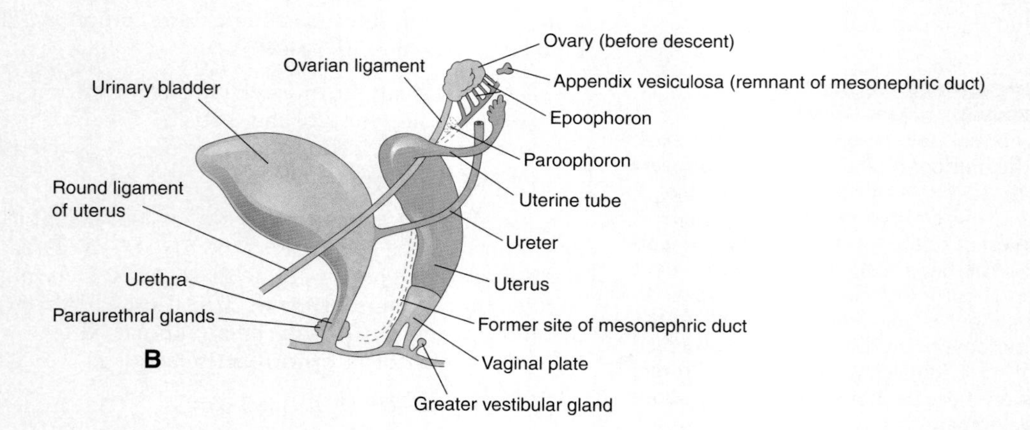

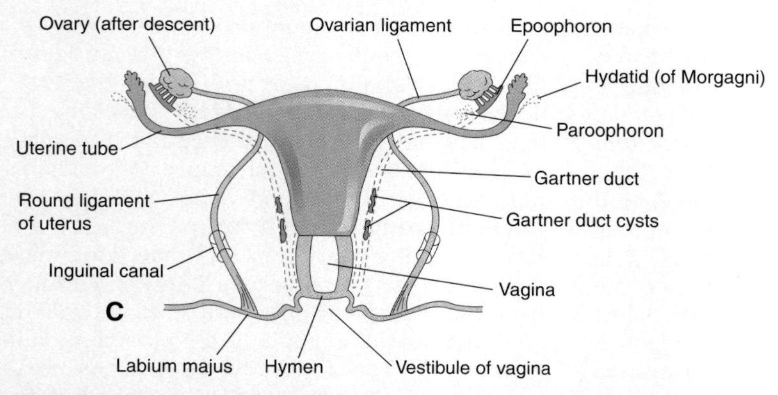

Figure 14–21. Schematic drawings illustrating the development of the male and female reproductive systems from the genital ducts and urogenital sinus. Vestigial structures are also shown. *A,* Reproductive system in a newborn male. *B,* Female reproductive system in a 12-week, female fetus. *C,* Reproductive system in a newborn female.

follicular cells derived from the cortical cord — derived from the surface epithelium (see Fig. 14 – 20). Active mitosis of oogonia occurs during fetal life, producing thousands of these primordial follicles.

No oogonia form postnatally. Although many oogonia degenerate before birth, the 2 million or so that remain enlarge to become primary oocytes before birth. After birth, the surface epithelium of the ovary flattens to a single layer of cells that is continuous with the mesothelium of the peritoneum at the hilum of the ovary. The surface epithelium was formerly called the "germinal epithelium," but this has since been recognized as an inappropriate term because it is now known that the germ cells differentiate from the primordial germ cells (see Fig. 14 – 20). The surface epithelium becomes separated from the follicles in the cortex by a thin fibrous capsule, the **tunica albuginea.** As the ovary separates from the regressing mesonephros, it is suspended by the **mesovarium,** which is its mesentery.

Development of the Genital Ducts

Both male and female embryos have two pairs of genital ducts. The mesonephric ducts (wolffian ducts) play an important part in the development of the male reproductive system, whereas the paramesonephric ducts (müllerian ducts) have a leading role in the development of the female reproductive system. During the fifth and sixth weeks of development, the genital system is in an indifferent state, when male and female genital ducts are present.

The **mesonephric ducts,** which drain urine from the mesonephric kidneys, play an essential role in the development of the **male reproductive system** (Fig. 14 – 21*A*). Under the influence of testosterone produced by the fetal testes in the eighth week of development, the proximal part of each mesonephric duct becomes highly convoluted to form the **epididymis.** The remainder of this duct forms the **ductus deferens** and **ejaculatory duct.** In female fetuses, the mesonephric ducts almost completely disappear, leaving only a few nonfunctional remnants (see Fig. 14 – 21*B* and *C*; Table 14 – 1).

The **paramesonephric ducts,** which develop lateral to the gonads and mesonephric ducts (see Fig. 14 – 20), play an essential role in the development of the **female reproductive system.** The paramesonephric ducts form on each side from longitudinal invaginations of the mesothelium on the lateral aspects of the mesonephroi. The funnel-shaped cranial ends of these ducts open into the peritoneal cavity (see Fig. 14 – 21*A* to *C*). The paramesonephric ducts pass caudally, parallel to the mesonephric ducts, until they reach the future pelvic region of the embryo. There, they cross ventral to the mesonephric ducts, approach each other in the median plane, and fuse to form a Y-shaped **uterovaginal primordium** (Fig. 14 – 22*A*). This tubular structure projects into the dorsal wall of the urogenital sinus and produces an elevation known as the **sinus (müller) tubercle** (see Fig. 14 – 22*B*).

Development of the Male Genital Ducts and Glands

The fetal testes produce *masculinizing hormones* and a müllerian inhibiting substance (MIS). **Testosterone,** the production of which is stimulated by hCG, stimulates the mesonephric ducts to form male genital ducts; **MIS** causes the paramesonephric ducts to disappear by epithelial-mesenchymal transformation. Laboratory studies have shown that the *molecular mechanisms* of MIS-mediated paramesonephric duct regression involve overexpression of betacatenin and lymphoid enhancer factor 1. As the mesonephros degenerates, some mesonephric tubules persist and are transformed into efferent ductules (see Fig. 14 – 21*A*). These ductules open into the mesonephric duct, which has been transformed into the **duct of the epididymis** (L. *ductus epididymis*) in this region. Distal to the epididymis, the mesonephric duct acquires a thick investment of smooth muscle and becomes the **ductus deferens**. A lateral outgrowth from the caudal end of each mesonephric duct gives rise to the **seminal gland** (vesicle). The secretions of this pair of glands nourish the sperms. The part of the mesonephric duct between the duct of this gland and the urethra becomes the **ejaculatory duct**.

Prostate. Multiple endodermal outgrowths arise from the prostatic part of the urethra and grow into the surrounding mesenchyme (Fig. 14 – 23). The glandular epithelium of the prostate differentiates from these endodermal cells, and the associated mesenchyme differentiates into the dense stroma and smooth muscle of the prostate. Expression of *sonic hedgehog* (*Shh*) and *Hox genes* in the urogenital sinus appears to be essential for prostate development.

Bulbourethral Glands. The bulbourethral glands are the pea-sized structures that develop from paired outgrowths from the spongy part of the urethra (see Fig. 14 – 21*A*). The smooth muscle fibers and the stroma differentiate from the adjacent mesenchyme. The secretions of these glands contribute to the semen.

Development of the Female Genital Ducts and Glands

In female embryos, the mesonephric ducts regress because of lack of testosterone, and the paramesonephric ducts develop because of the absence of MIS. Female sexual development does not depend on the presence of

Table 14 - 1. **Adult Derivatives and Vestigial Remains of Embryonic Urogenital Structures***

Male	Embryonic	Structure Female
Testis	***Indifferent Gonad***	*Ovary*
Seminiferous tubules	***Cortex***	*Ovarian follicles*
Rete testis	***Medulla***	*Rete ovarii*
Gubernaculum	***Gubernaculum***	*Ovarian ligament*
		Round ligament of uterus
Efferent ductules of testis	***Mesonephric Tubules***	*Epoophoron*
Paradidymis		*Paroophoron*
Appendix of epididymis	***Mesonephric Duct***	*Appendix vesiculosa*
Duct of epididymis		*Duct of epoophoron*
Ductus deferens		*Duct of Gartner*
Ureter, pelvis, calices and collecting tubules		*Ureter, pelvis, calices and collecting tubules*
Ejaculatory duct and seminal gland		
Appendix of testis	***Paramesonephric Duct***	*Hydatid (of Morgagni)*
		Uterine tube
		Uterus
Urinary bladder	***Urogenital Sinus***	*Urinary bladder*
Urethra (except navicular fossa)		*Urethra*
Prostatic utricle		*Vagina*
Prostate		*Urethral and paraurethral glands*
Bulbourethral glands		*Greater vestibular glands*
Seminal colliculus	***Sinus Tubercle***	*Hymen*
Penis	***Phallus***	*Clitoris*
Glans of penis		*Glans of clitoris*
Corpora cavernosa of penis		*Corpora cavernosa of clitoris*
Corpus spongiosum of penis		*Bulb of vestibule*
Ventral aspect of penis	***Urogenital Folds***	Labia minora
Scrotum	***Labioscrotal Swellings***	Labia majora

*Functional derivatives are in italics.

ovaries or hormones. The paramesonephric ducts form most of the female genital tract. The **uterine tubes** develop from the unfused cranial parts of the paramesonephric ducts (see Fig. 14 - 21*B* and *C*). The caudal fused portions of these ducts form the **uterovaginal primordium.** As the name of this structure indicates, it gives rise to the uterus and vagina (superior part). The endometrial stroma and myometrium are derived from splanchnic mesenchyme. Fusion of the paramesonephric ducts also brings together two peritoneal folds that form the right and left **broad ligament** and two peritoneal compartments, the **rectouterine pouch** and **vesicouterine pouch** (Fig. 14 - 24*B* to *D*). Along the sides of the uterus, between the layers of the broad ligament, the mesenchyme proliferates and differentiates into cellular tissue — the *parametrium* — which is composed of loose connective tissue and smooth muscle.

Development of the Vagina

The vaginal epithelium is derived from the endoderm of the urogenital sinus, and the fibromuscular wall of the vagina develops from the surrounding mesenchyme.

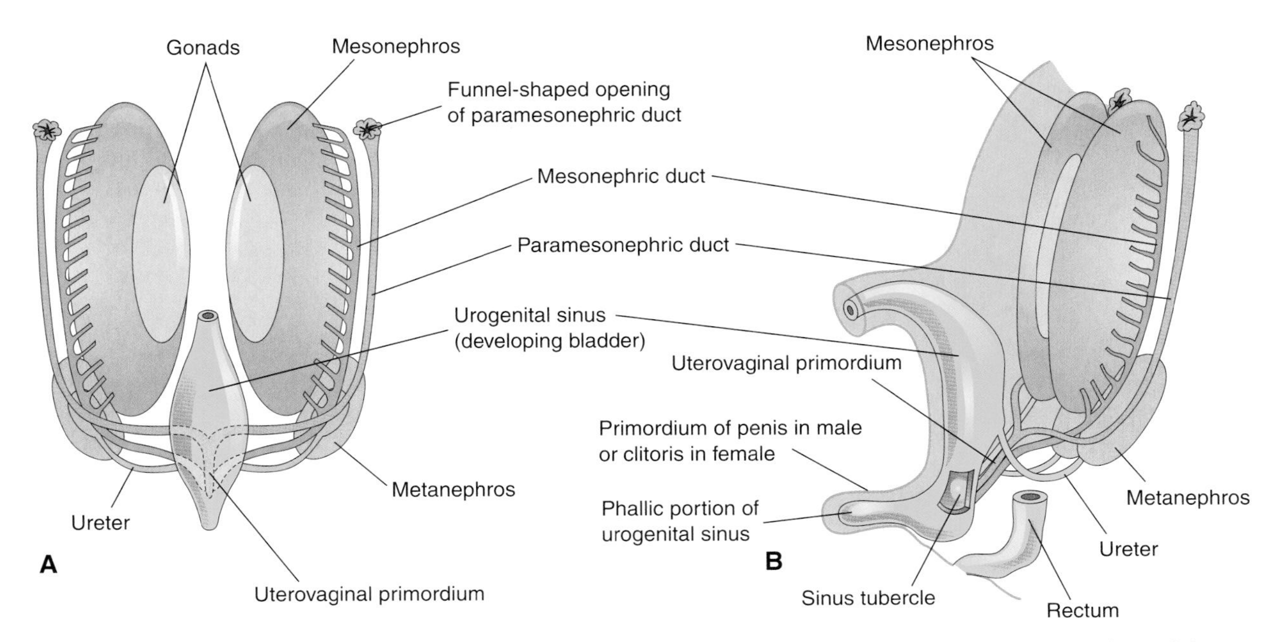

Figure 14 - 22. *A,* Sketch of a ventral view of the posterior abdominal wall of a 7-week embryo showing the two pairs of genital ducts present during the indifferent state of sexual development. *B,* Lateral view of a 9-week fetus showing the sinus tubercle (müller tubercle) on the posterior wall of the urogenital sinus. It becomes the hymen in females and the seminal colliculus in males. The colliculus is an elevated part of the urethral crest on the posterior wall of the prostatic urethra.

Contact of the uterovaginal primordium with the urogenital sinus, forming the **sinus tubercle** (see Fig. 14 - 22*B*), induces the formation of paired endodermal outgrowths called **sinovaginal bulbs** (see Fig. 14 - 24*A*). They extend from the urogenital sinus to the caudal end of the uterovaginal primordium. The sinovaginal bulbs fuse to form a **vaginal plate** (see Fig. 14 - 21*B*). Later, the central cells of this plate break down, forming the lumen of the vagina. Its peripheral cells form the vaginal epithelium (see Fig. 14 - 21*C*). The lining of the vagina — vaginal epithelium — is derived from the vaginal plate. Until late fetal life, the lumen of the vagina is separated from the cavity of the urogenital sinus by a membrane called the **hymen** (Fig. 14 - 25*H;* see also Fig. 14 - 21*C*). The hymen is formed by invagination of the posterior wall of the urogenital sinus, resulting from expansion of the caudal end of the vagina. The hymen usually ruptures during the perinatal period and remains as a thin fold of mucous membrane just within the **vaginal orifice** (entrance to the vagina).

Female Auxiliary Genital Glands

Buds grow from the urethra into the surrounding mesenchyme, forming the **urethral glands** and **paraurethral glands** (of Skene). These glands correspond to the prostate in the male. Outgrowths from the urogenital sinus form the **greater vestibular glands** (of Bartholin), which are homologous to the bulbourethral glands in the male (see Table 14 - 1).

Vestigial Structures Derived from Embryonic Genital Ducts

During conversion of the mesonephric and paramesonephric ducts into adult structures, parts of them

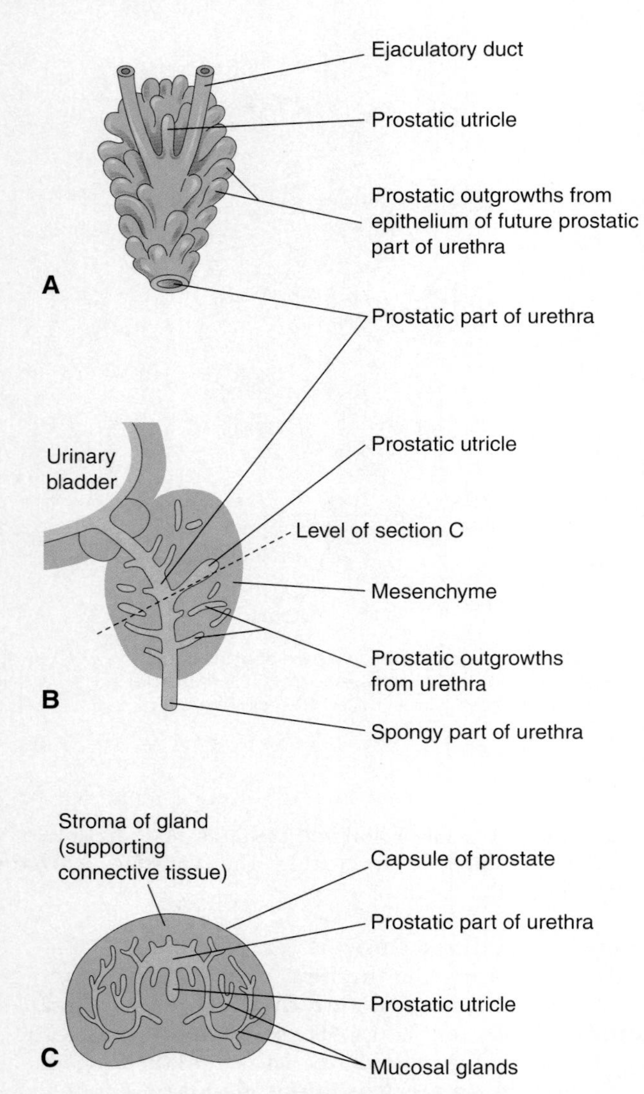

Figure 14 - 23. *A,* Dorsal view of the developing prostate in an 11-week fetus. *B,* Sketch of a median section of the developing urethra and prostate. Note the numerous endodermal outgrowths from the prostatic urethra. The vestigial prostatic utricle is also shown. *C,* Section of the prostate (at 16 weeks) at the level shown in *B.*

remain as vestigial structures (see Table 14 - 1 and Fig. 14 - 21*A* to *C*). These vestiges are rarely seen unless pathologic changes develop in them.

Development of the External Genitalia

Early development of the external genitalia is similar in both sexes. Distinguishing sexual characteristics begin to appear during the ninth week of development, but the external genitalia are not fully differentiated until the 12th week. From the fourth to the early part of the seventh week, the external genitalia are sexually undifferentiated (see Fig. 14 - 25*A* and *B*). Early in the fourth week, proliferating mesenchyme produces a **genital tubercle** in both sexes at the cranial end of the cloacal membrane. **Labioscrotal swellings** and **urogenital folds** soon develop on each side of the cloacal membrane. The genital tubercle soon elongates to form a **primordial phallus**. When the urorectal septum fuses with the cloacal membrane at the end of the sixth week, it divides the cloacal membrane into a dorsal anal membrane and a ventral urogenital membrane (see Fig. 14 - 25*B*). The **urogenital membrane** lies in the floor of a median cleft, the **urogenital groove**, which is bounded by the urogenital folds. The anal and urogenital membranes rupture a week or so later, forming the **anus** and **urogenital orifice**, respectively. In the female fetus, the urethra and vagina open into a common cavity, the **vestibule**.

Development of Male External Genitalia

Masculinization of the indifferent external genitalia is induced by **testosterone** produced by the fetal testes (see Fig. 14 - 25*C, E,* and *G*). As the phallus enlarges and elongates to become the penis, the urogenital folds form the lateral walls of the **urethral groove** on the ventral surface of the penis. This groove is lined by a proliferation of endodermal cells, the **urethral plate** (see Fig. 14 - 25), which extends from the phallic portion of the urogenital sinus. The **urogenital folds** fuse with each other along the ventral surface of the penis to form the **spongy urethra** (see Fig. 14 - 25E_1 to E_3). The surface ectoderm fuses in the median plane of the penis, forming the **penile raphe** and enclosing the spongy urethra within the penis. At the tip of the glans of the penis, an ectodermal ingrowth forms a cellular **ectodermal cord**, the which grows toward the root of the penis to meet the spongy urethra (see Fig. 14 - 16*A*). This cord canalizes and joins the previously formed spongy urethra. This completes the terminal part of the urethra and moves the external urethral orifice to the tip of the glans of the penis (see Fig. 14 - 16*C*). During the 12th week of development, a circular ingrowth of ectoderm occurs at the periphery of the glans (see Fig. 14 - 16*B*). When this ingrowth breaks down, it forms the **prepuce** (foreskin), a covering fold of skin (see Fig. 14 - 16*C*). For some time, the prepuce is adherent to the glans and is usually not retractable at birth. Breakdown of the adherent surfaces normally occurs during infancy. The **corpora cavernosa** penis and **corpus spongiosum** penis develop from mesenchyme in the phallus. The **labioscrotal swellings** grow toward each other and fuse to form the **scrotum** (Fig. 14 - 25*E* and *G*). The line of fusion of these folds is clearly visible as the **scrotal raphe** (see Fig. 14 - 25*G*).

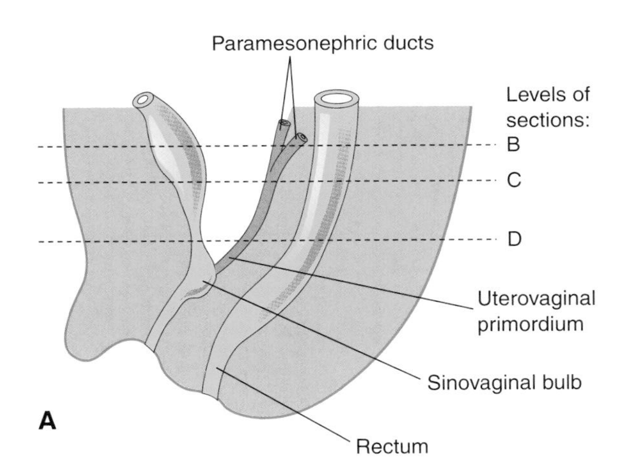

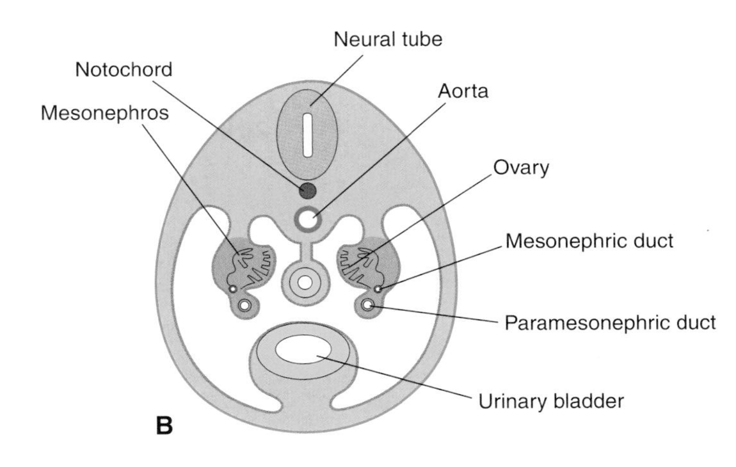

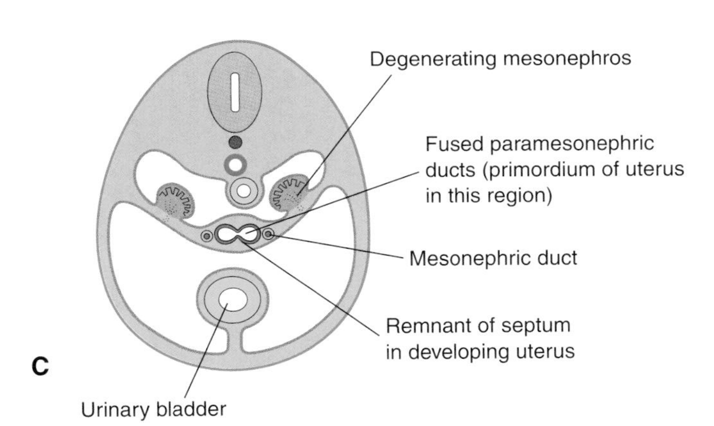

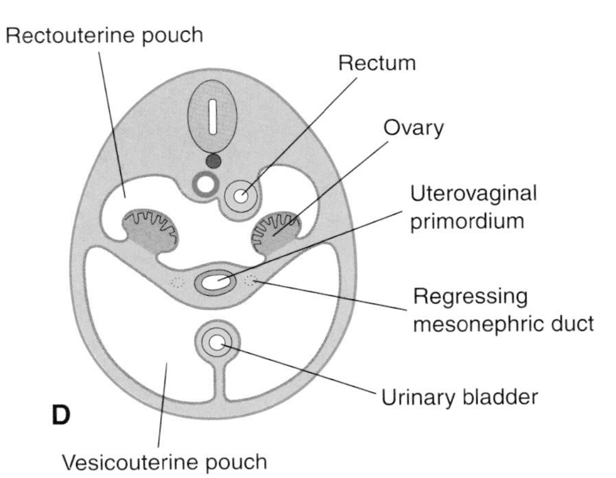

Figure 14 - 24. Early development of the ovaries and uterus. *A,* Schematic drawing of a sagittal section of the caudal region of an 8-week female embryo. *B,* Transverse section showing the paramesonephric ducts approaching each other. *C,* Similar section at a more caudal level illustrating fusion of the paramesonephric ducts. A remnant of the septum that initially separated them is shown. *D,* Similar section showing the uterovaginal primordium, broad ligament, and pouches in the pelvic cavity. Note that the mesonephric ducts have regressed.

Development of Female External Genitalia

Feminization of the indifferent external genitalia is not clearly understood, but estrogens produced by the placenta and fetal ovaries appear to be involved (see Fig. 14 - 25*D, F,* and *H*). Growth of the phallus gradually ceases, and it becomes the **clitoris**. The clitoris develops like the penis but the urogenital folds do not fuse, except posteriorly, where they join to form the **frenulum of the labia minora**. The unfused parts of the urogenital folds form the **labia minora**. The labioscrotal folds fuse posteriorly to form the **posterior labial commissure** and anteriorly to form the *anterior labial commissure* and *mons pubis* (see Fig. 14 - 25*D, F,* and *H*). Most parts of the **labioscrotal folds** remain unfused and form two large folds of skin, the **labia majora**, which are homologous to the scrotum.

Determination of Fetal Sex

Visualization of the external genitalia during ultrasonography is clinically important for several reasons, such as detection of fetuses at risk for severe X-linked disorders. Careful examination of the perineum may detect **ambiguous genitalia.** Only documentation of testes in the scrotum provides 100% gender determination, which is not possible in utero until 22 to 36 menstrual weeks. Because early embryos have the potential to develop as either males or females, errors in sex determination and differentiation result in intermediate sex, a condition known as intersexuality or **hermaphroditism.** Hermaphroditism implies a discrepancy between the morphology of the gonads (testes/ovaries) and the appearance of the external genitalia. A person with ambiguous external genitalia is called an **intersex** or

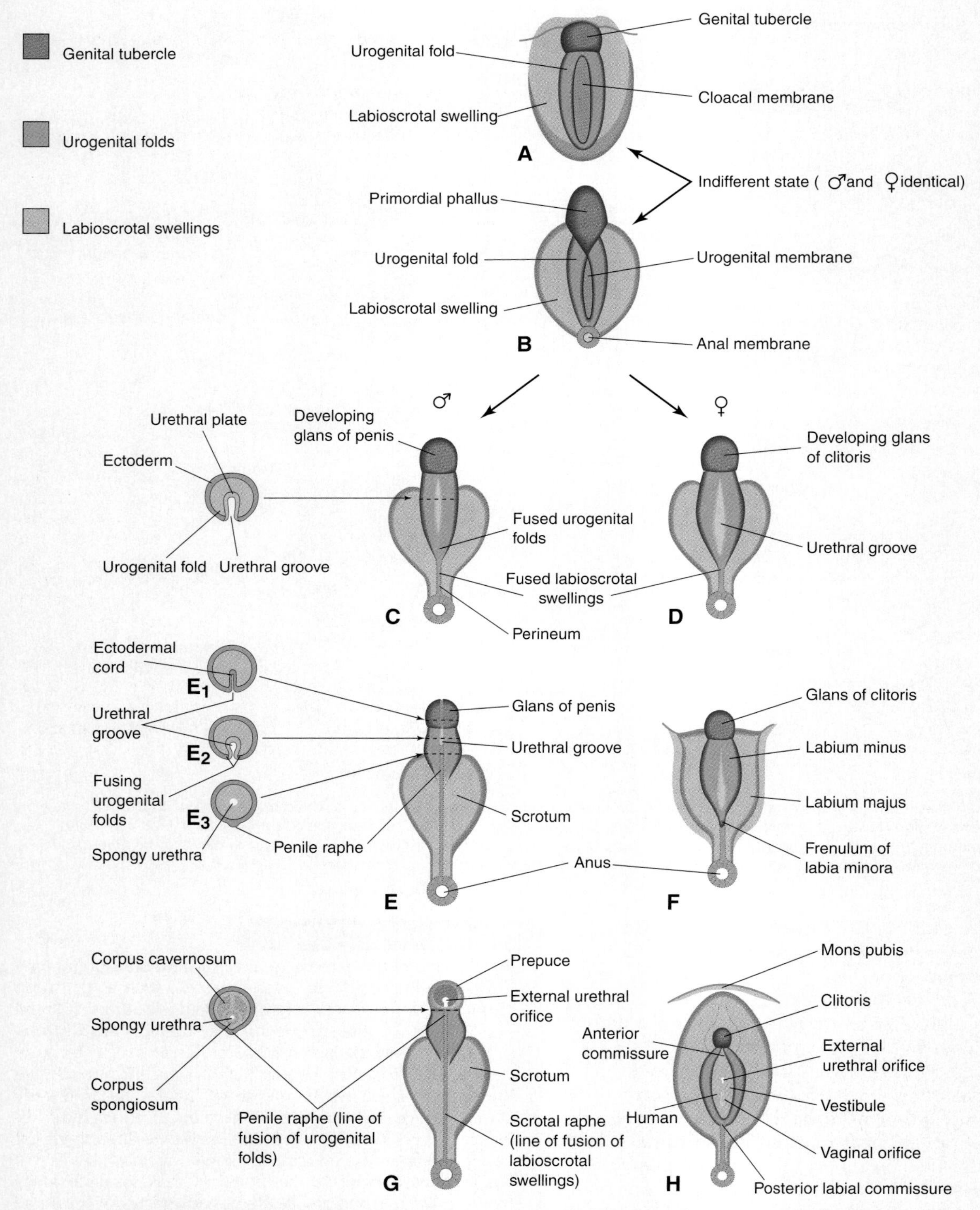

Figure 14 - 25. Development of the external genitalia. *A* and *B,* Diagrams illustrating the appearance of the genitalia during the indifferent state (fourth to seventh weeks). *C, E,* and *G,* Stages in the development of male external genitalia at 9, 11, and 12 weeks, respectively. To the left of the drawings are schematic transverse sections of the developing penis, illustrating formation of the spongy urethra. *D, F,* and *H,* Stages in the development of female external genitalia at 9, 11, and 12 weeks, respectively.

hermaphrodite. Intersexual conditions are classified according to the histologic appearance of the gonads:

- *True hermaphrodites* have ovarian and testicular tissue either in the same or in opposite gonads.
- *Female pseudohermaphrodites* have ovaries.
- *Male pseudohermaphrodites* have testes.

True Hermaphroditism

Persons with the extremely rare intersexual condition—true hermaphroditism—usually have a 46,XX sex chromosome constitution. *True hermaphroditism results from an error in sex determination.* The phenotype may be male or female, but the external genitalia are always ambiguous.

Female Pseudohermaphroditism

Persons with the intersexual condition known as female pseudohermaphroditism have *chromatin-positive nuclei* and a 46,XX chromosome constitution. This anomaly results from exposure of the female fetus to excessive androgens, the principal effect of which is virilization of the external genitalia (clitoral enlargement and labial fusion [Fig. 14–26]). The common cause of female pseudohermaphroditism is caused by congenital hyperplasia (CAH). There is no ovarian abnormality, but the excessive production of androgens by the fetal suprarenal glands causes masculinization of the external genitalia, varying from enlargement of the clitoris to almost masculine genitalia. Commonly, there is **clitoral hypertrophy**, partial fusion of the labia majora, and a persistent urogenital sinus are noted.

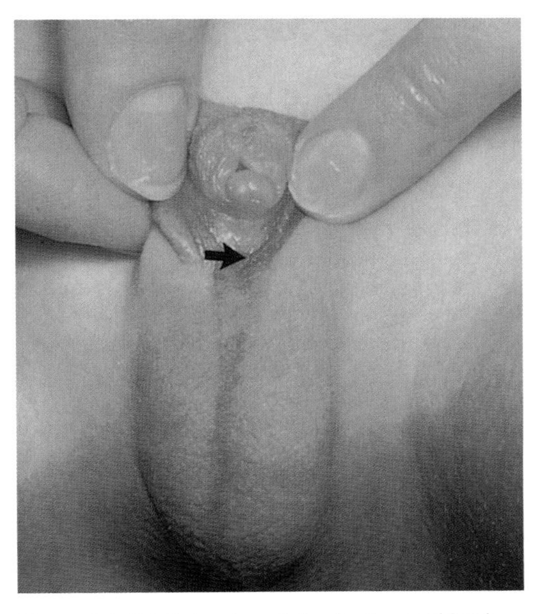

Figure 14–26. External genitalia of a 6-year-old girl showing a scrotum-like structure formed by an enlarged clitoris and fused labia majora. The arrow indicates the opening into the urogenital sinus. This extreme masculinization is the result of congenital adrenal hyperplasia.

Male Pseudohermaphroditism

Persons with this uncommon intersexual condition — male pseudohermaphroditism — have *chromatin-negative nuclei* (i.e., nuclei that do not contain sex chromatin) and a 46,XY chromosome constitution. The external and internal genitalia are variable, owing to varying degrees of development of the external genitalia and paramesonephric ducts. These anomalies are caused by inadequate production of testosterone and müllerian-inhibiting factor by the fetal testes. Testicular development in these males ranges from rudimentary to normal.

Androgen Insensitivity Syndrome

Androgen-insensitivity syndrome — also called **testicular feminization syndrome** — occurs in 1 in 20,000 live births. Individuals with this unusual condition are normal-appearing females, despite the presence of testes and a 46,XY chromosome constitution. The external genitalia are female, but the vagina usually ends in a blind pouch, and the uterus and uterine tubes are absent or rudimentary. At puberty, there is normal development of breasts and female characteristics, but menstruation does not occur and pubic hair is scanty or absent. The psychosexual orientation of women with androgen insensitivity syndrome is entirely female, and medically, legally, and socially, they are considered to be females. The testes are usually situated in the abdomen or the inguinal canals, but in some individuals, they may descend into the labia majora. The failure of masculinization to occur in these individuals results from a resistance to the action of testosterone at the cellular level in the genital tubercle and labioscrotal and urogenital folds.

Hypospadias

Hypospadias is the most common anomaly involving the penis. In 1 of every 300 male infants, the external urethral orifice is on the ventral surface of the glans of the penis or on the ventral surface of the body of the penis **(penile hypospadias)**. Usually, the penis is underdeveloped and curved ventrally, a condition known as **chordee**. There are four types of hypospadias: hypospadias of the glans, penile hypospadias, penoscrotal hypospadias, and perineal hypospadias. The glans and penile types of hypospadias constitute about 80% of the cases (Fig. 14–27*A* and *B*). In **penoscrotal hypospadias**, the urethral orifice is at the junction of the penis and scrotum. In **perineal hypospadias**, the labioscrotal folds fail to fuse and the external urethral orifice is located between the unfused halves of the scrotum. Hypospadias results from inadequate production of androgens by the fetal testes and/or inadequate receptor sites for these hormones, or both.

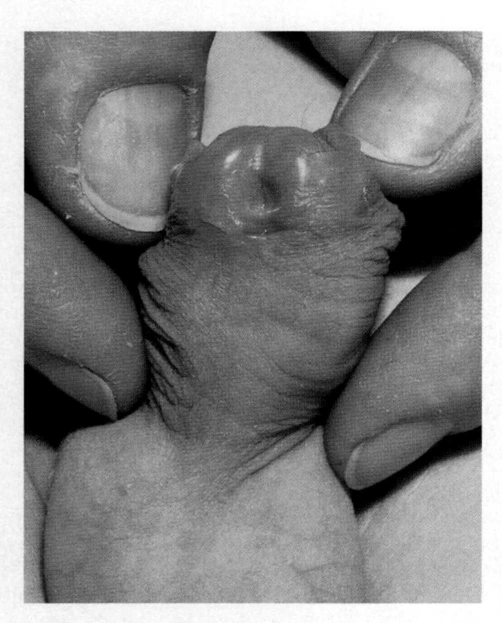

Figure 14-27. Hypospadias of the glans in an infant. This is the most common form of hypospadias. The external urethral orifice is on the ventral aspect of the glans of the penis. There is a shallow pit in the glans penis at the usual site of the urethral orifice. (Courtesy of Dr. AE Chudley, Department of Pediatrics and Child Health, University of Manitoba, Children's Hospital, Winnipeg, Manitoba, Canada.)

Epispadias

In rare cases, the urethra opens on the dorsal surface of the penis, a condition known as **epispadias.** Although epispadias may occur as a separate entity, it is *often associated with exstrophy of the bladder* (see Fig. 14-14). Epispadias may result from inadequate ectodermal-mesenchymal interactions during development of the genital tubercle. As a consequence, the genital tubercle develops more dorsally than in normal embryos. Consequently, when the urogenital membrane ruptures, the urogenital sinus opens on the dorsal surface of the penis. Urine is expelled at the root of the malformed penis.

Anomalies of the Female Genital Tract

Various types of uterine duplication and vaginal anomalies result from developmental arrest of the uterovaginal primordium during the eighth week of development (Fig. 14-28*B* to *G*). These developmental anomalies may include:

- incomplete fusion of the paramesonephric ducts
- incomplete development of a paramesonephric duct
- failure of parts of one or both paramesonephric ducts to develop
- incomplete canalization of the vaginal plate that forms the vagina

A **double uterus** (uterus didelphys) results from failure of fusion of the inferior parts of the paramesonephric ducts. It may be associated with a double or a single vagina (Fig. 14-28*A* to *C*). In some cases, the uterus is divided internally by a thin septum (Fig. 14-28*F*). If the duplication involves only the superior part of the body of the uterus, the condition is called **bicornuate uterus** (Fig. 14-28*D* and *E*). If one paramesonephric duct is retarded in its growth and does not fuse with the other one, a **bicornuate uterus with a rudimentary horn** (L. *cornu*) develops (see Fig. 14-28*E*). The rudimentary horn may not communicate with the cavity of the uterus. A **unicornuate uterus** develops when one paramesonephric duct fails to develop; this results in a uterus with one uterine tube (see Fig. 14-28*G*).

Agenesis of the vagina results from failure of the sinovaginal bulbs to develop and form the vaginal plate (see Fig. 14-21*B*). When the vagina is absent, the uterus is usually absent also, because the developing uterus (uterovaginal primordium) induces the formation of sinovaginal bulbs, which fuse to form the vaginal plate. Failure of canalization of the vaginal plate results in blockage of the vagina. Failure of the inferior end of the vaginal plate to perforate results in an **imperforate hymen.**

Development of the Inguinal Canals

The inguinal canals form pathways for the testes to descend from their intra-abdominal position through the anterior abdominal wall into the scrotum. *Inguinal canals develop in both sexes* because of the morphologically indifferent state of sexual development. As the mesonephros degenerates, a ligament called the **gubernaculum** descends on each side of the abdomen from the inferior pole of the gonad (Fig. 14-29*A*). The gubernaculum passes obliquely through the developing anterior abdominal wall at the site of the future inguinal canal. The gubernaculum attaches caudally to the internal surface of the *labioscrotal swellings* (future halves of the scrotum or labium majora).

The **processus vaginalis,** or vaginal process, an evagination of peritoneum, develops ventral to the gubernaculum and herniates through the abdominal wall along the path formed by the gubernaculum (see Fig. 14-29*B* to *E*). The vaginal process carries extensions of the layers of the abdominal wall before it, which form the walls of the inguinal canal. In males, these layers also form the coverings of the spermatic cord and testis (see Fig. 14-29*E* and *F*). The opening in the transversalis fascia produced by the vaginal process becomes the **deep inguinal ring,** and the opening created in the external oblique aponeurosis forms the **superficial inguinal ring.**

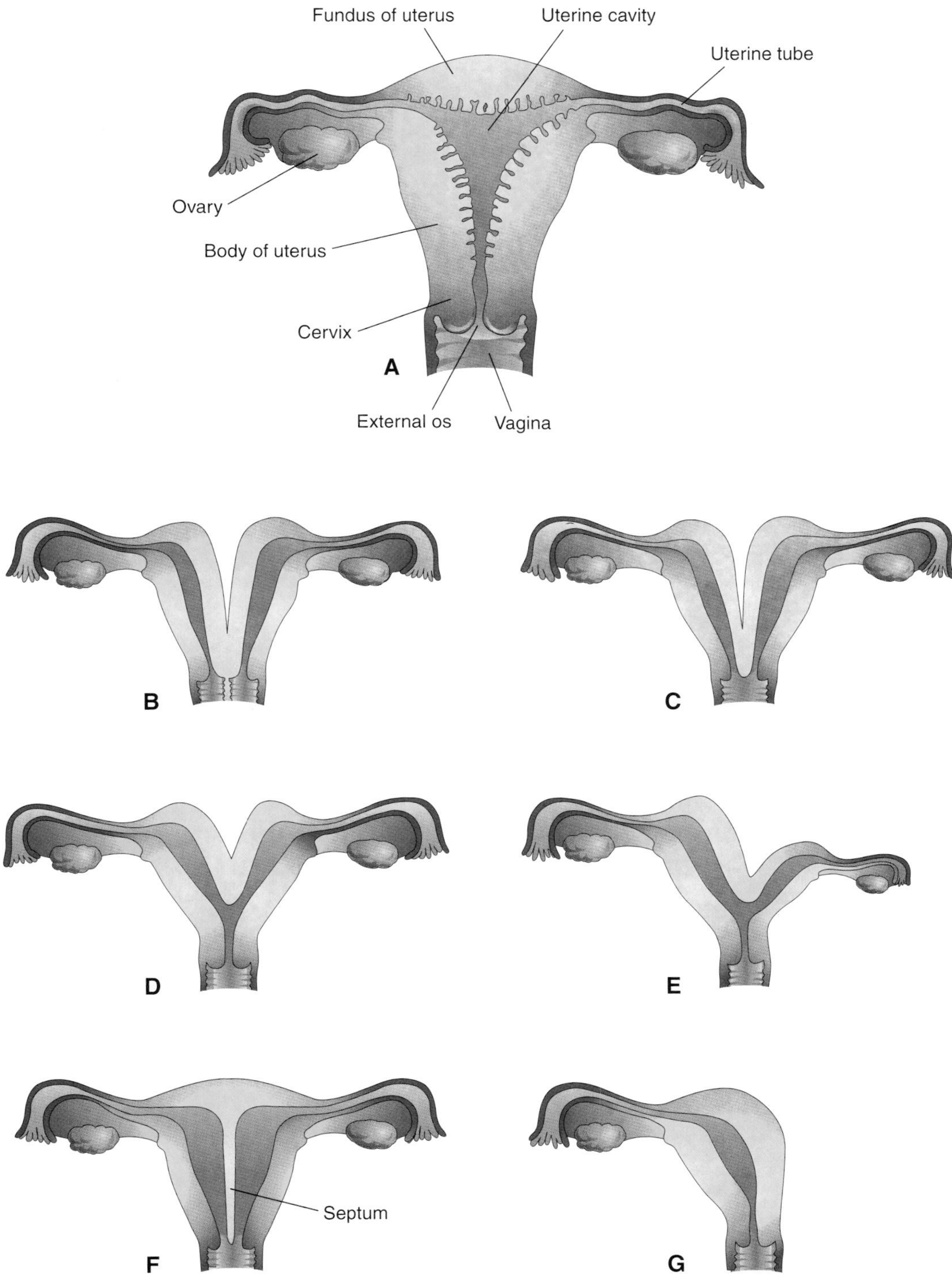

Figure 14 - 28. Drawings illustrating various types of uterine anomaly. *A,* Normal uterus and vagina. *B,* Double uterus (uterus didelphys) and double vagina. *C,* Double uterus with single vagina. *D,* Bicornuate uterus. *E,* Bicornuate uterus with a rudimentary left horn. *F,* Septate uterus. *G,* Unicornuate uterus.

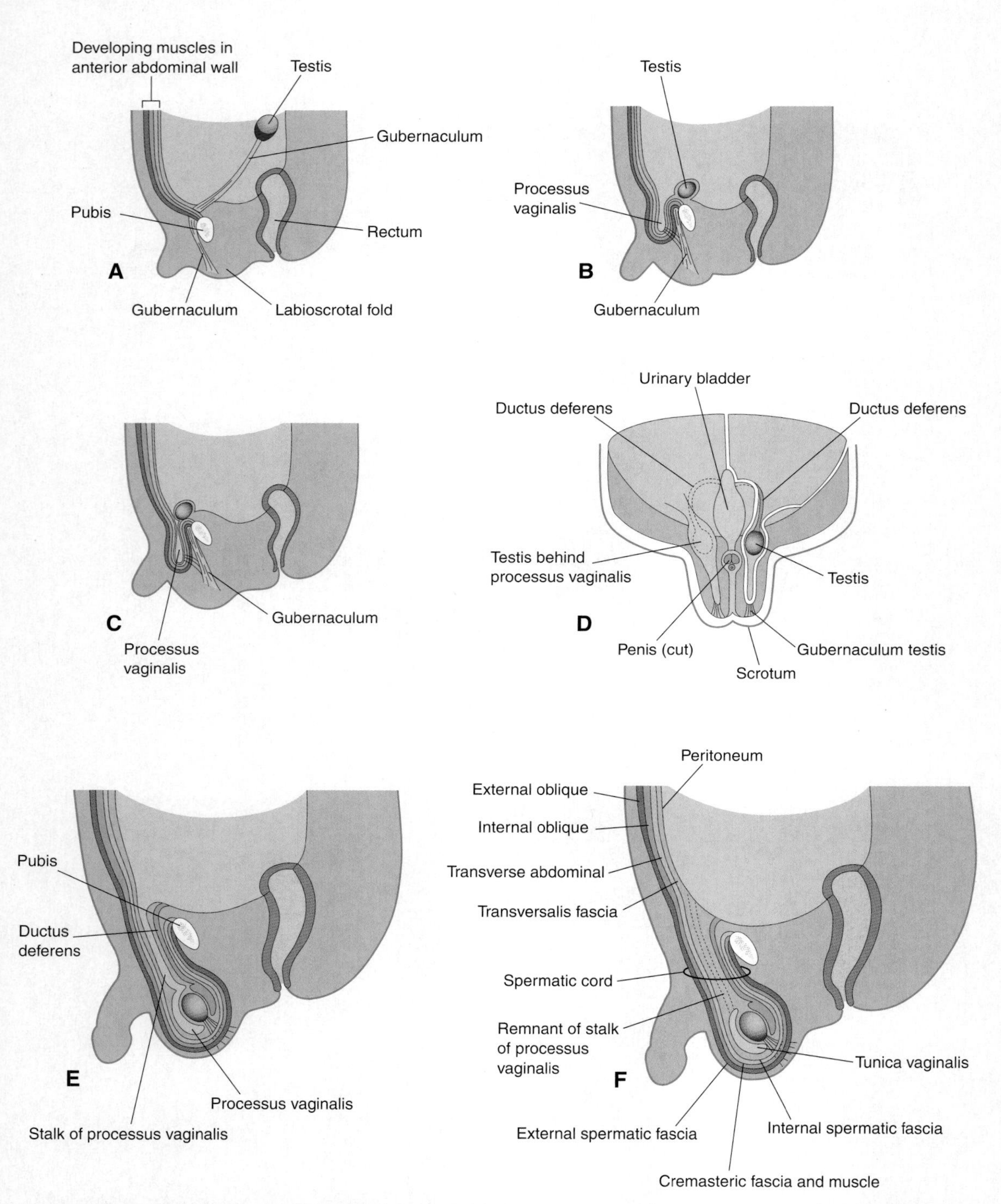

Figure 14 - 29. Schematic diagrams illustrating formation of the inguinal canals and descent of the testes. *A,* Sagittal section of a 7-week embryo showing the testis before its descent from the dorsal abdominal wall. *B* and *C,* Similar sections at about 28 weeks showing the processus vaginalis and the testis beginning to pass through the inguinal canal. Note that the processus vaginalis carries fascial layers of the abdominal wall before it. *D,* Frontal section of a fetus about 3 days later illustrating descent of the testis posterior to the processus vaginalis. The processus vaginalis has been cut away on the left side to show the testis and ductus deferens. *E,* Sagittal section of a newborn infant showing the processus vaginalis communicating with the peritoneal cavity by a narrow stalk. *F,* Similar section of a 1-month-old infant after obliteration of the stalk of the processus vaginalis. Note that the extended fascial layers of the abdominal wall now form the coverings of the spermatic cord.

Descent of the Testes

Testicular descent is associated with:

- enlargement of the testes and atrophy of the mesonephroi (mesonephric kidneys) allow movement of the testes caudally along the posterior abdominal wall
- atrophy of the paramesonephric ducts induced by the MIS, which enables the testes to move transabdominally to the deep inguinal rings
- enlargement of the processus vaginalis, which guides the testis through the inguinal canal into the scrotum

By 26 weeks, the testes have descended retroperitoneally (external to the peritoneum) from the posterior abdominal wall to the deep inguinal rings (see Fig. 14 – 29*B* and *C*). This change in position occurs as the fetal pelvis enlarges and the trunk of the embryo elongates. Transabdominal movement of the testes is largely a relative movement that results from growth of the cranial part of the abdomen away from the caudal part (future pelvic region).

Testicular descent through the inguinal canals and into the scrotum is controlled by androgens (e.g., testosterone) produced by the fetal testes. The gubernaculum appears to guide the testes in their descent. Descent of the testes through the inguinal canals and into the scrotum usually begins during the 26th week of development and takes 2 to 3 days. In more than 97% of full-term newborn boys, both testes are in the scrotum. During the first 3 months after birth, most undescended testes descend into the scrotum. When the testis descends, it carries its ductus deferens and vessels with it. As the testis and ductus deferens descend, they are ensheathed by the fascial extensions of the abdominal wall, as described (see Fig. 14 – 29*F*):

- The extension of the transversalis fascia becomes the **internal spermatic fascia.**
- The extensions of the internal oblique muscle and fascia become the **cremasteric muscle** and **fascia.**
- The extension of the external oblique aponeurosis becomes the **external spermatic fascia.**

Within the scrotum, the testis projects into the distal end of the processus vaginalis. During the perinatal period, the connecting stalk of the process is normally obliterated, isolating the **tunica vaginalis** as a peritoneal sac related to the testis (see Fig. 14 – 29*F*).

Descent of the Ovaries

The ovaries also descend from the posterior abdominal wall to the pelvis, just inferior to the pelvic brim. The gubernaculum is attached to the uterus near the attachment of the uterine tube. The cranial part of the gubernaculum becomes the **ovarian ligament,** and the caudal part forms the round ligament of the uterus (see Fig. 14 – 21*C*). The **round ligaments** pass through the inguinal canals and terminate in the labia majora. The relatively small processus vaginalis in the female is usually obliterated, and disappears long before birth.

Cryptorchidism or Undescended Testes

Cryptorchidism (Gr. *kryptos,* hidden) occurs in up to 30% of premature male infants and in about 3 to 4% of full-term male infants. Cryptorchidism may be unilateral or bilateral. In most cases, the testes descend into the scrotum by the end of the first year. If both testes remain within or just outside the abdominal cavity, they fail to mature and sterility is common. **Cryptorchid testes** may be in the abdominal cavity or anywhere along the usual path of descent of the testis, but they are usually in the inguinal canal (Fig. 14 – 30*A*). The cause of most cases of cryptorchidism is unknown, but a deficiency of androgen production by the fetal testes is an important factor.

Ectopic Testes

After traversing the inguinal canal, the testis may deviate from its usual path of descent and lodge in various abnormal locations (see Fig. 14 – 30*B*):

- interstitial (external to the aponeurosis of the external oblique muscle)
- in the proximal part of the medial thigh
- dorsal to the penis
- on the opposite side (crossed ectopia)

All types of ectopic testis are rare, but **interstitial ectopia** occurs most frequently. Ectopic testis occurs when a part of the gubernaculum passes to an abnormal location and the testis follows it.

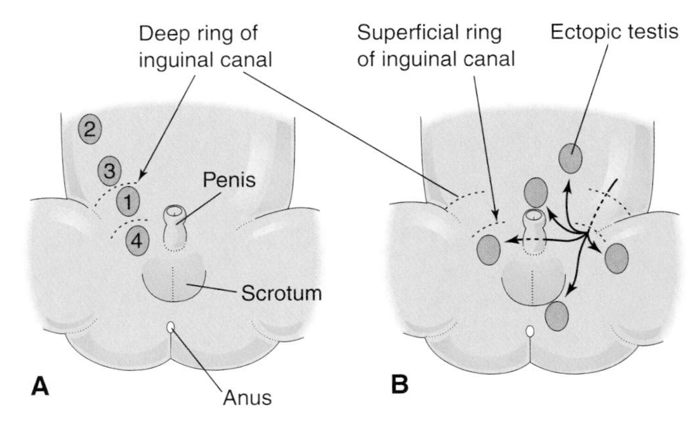

Figure 14 – 30. Diagrams showing the possible sites of cryptorchid and ectopic testes. *A,* Positions of cryptorchid testes, numbered in order of frequency. *B,* Usual locations of ectopic testes.

Congenital Inguinal Hernia

If the communication between the tunica vaginalis and the peritoneal cavity fails to close, a **persistent processus vaginalis** exists. A loop of intestine may herniate through it into the scrotum or labium majus (Fig. 14 – 31*A* and *B*). Embryonic remnants resembling the ductus deferens or epididymis are often found in inguinal hernial sacs. Congenital inguinal hernia is much more common in male infants than in female infants, and it is often associated with cryptorchidism and, in females, with the androgen insensitivity syndrome.

Hydrocele

Occasionally, the abdominal end of the processus vaginalis remains open but is too small to permit herniation of intestine (see Fig. 14 – 31*D*). In such cases, peritoneal fluid passes into the patent processus vaginalis and forms a **hydrocele of the testis.** If the middle part of the processus vaginalis remains open, fluid may accumulate, and give rise to a **hydrocele of the spermatic cord** (see Fig. 14 – 31*C*).

Summary of the Urogenital System

The urogenital system develops from:

- intermediate mesoderm
- mesothelium lining the abdominal cavity
- endoderm of the urogenital sinus

The urinary system begins to develop about 3 weeks before the genital system is evident. Three successive kidney systems develop:

- the *pronephroi,* which are vestigial and nonfunctional
- the *mesonephroi,* which serve as temporary excretory organs
- the *metanephroi,* which become the permanent kidneys

The **metanephroi** (*primordia of the permanent kidneys*) develop from two sources:

- the *metanephric diverticulum,* which gives rise to the ureter, renal pelvis, calices, and collecting tubules

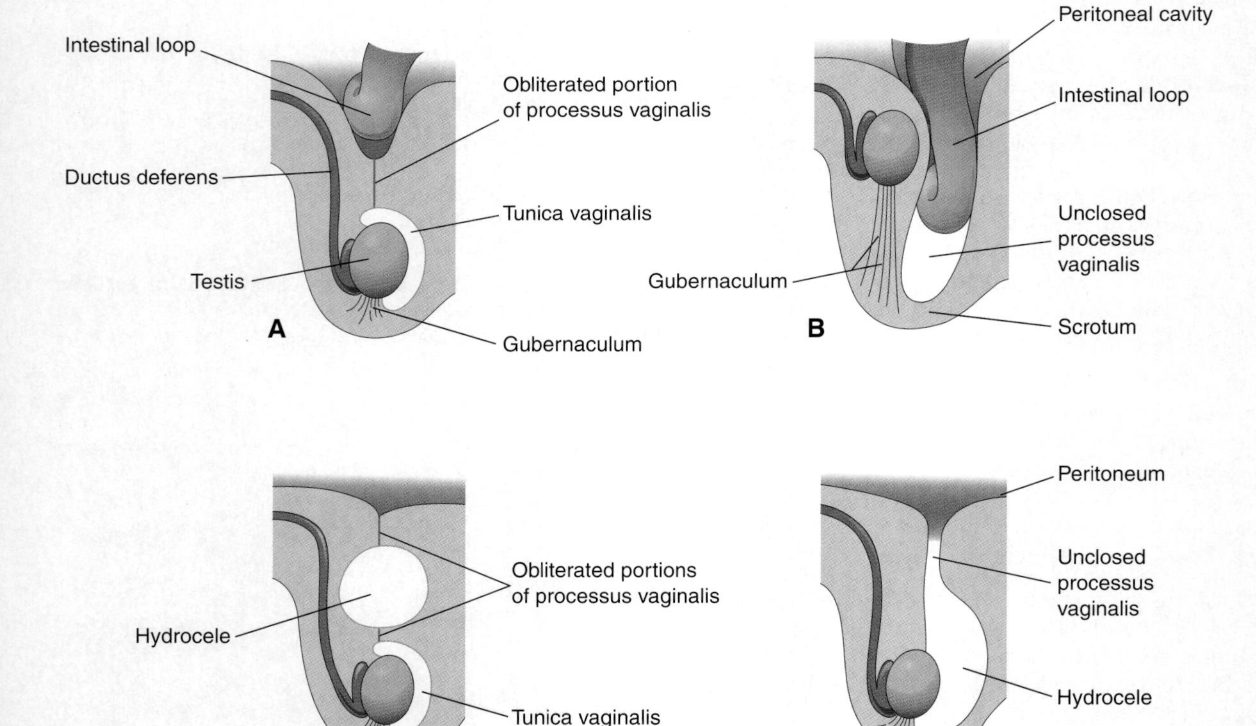

Figure 14 – 31. Diagrams of sagittal sections illustrating conditions resulting from failure of closure of the processus vaginalis. *A,* Incomplete congenital inguinal hernia resulting from persistence of the proximal part of the processus vaginalis. *B,* Complete congenital inguinal hernia into the scrotum resulting from persistence of the processus vaginalis. Cryptorchidism, a commonly associated condition, is also illustrated. *C,* Large cyst or hydrocele that arose from an unobliterated portion of the processus vaginalis. *D,* Hydrocele of the testis and spermatic cord resulting from peritoneal fluid passing into a patent processus vaginalis.

- the *metanephric mass of intermediate mesoderm,* which gives rise to the nephrons

At first, the kidneys are located in the pelvis, but they gradually "ascend" to the abdomen. This apparent migration results from disproportionate growth of the fetal lumbar and sacral regions. Developmental abnormalities of the kidneys and ureters are common. Incomplete division of the metanephric diverticulum results in a double ureter and supernumerary kidney. Failure of the kidney to "ascend" from its embryonic position in the pelvis results in an ectopic kidney that is abnormally rotated.

The **urinary bladder** develops from the urogenital sinus and the surrounding splanchnic mesenchyme. The female urethra and almost all of the male urethra have a similar origin. *Exstrophy of the bladder* results from a rare ventral body wall defect through which the posterior wall of the urinary bladder protrudes onto the abdominal wall. In males, *epispadias* is a common associated anomaly.

The **genital system** develops in close association with the urinary or excretory system. Genetic sex is established at fertilization, but the gonads do not begin to attain sexual characteristics until the seventh week of development. *Primordial germ cells* form in the wall of the yolk sac during the fourth week and migrate into the developing gonads, where they differentiate into germ cells (oogonia/spermatogonia). The external genitalia do not acquire distinct masculine or feminine characteristics until the 12th week of development. The reproductive organs develop from primordia that are identical in both sexes. During this *indifferent state,* an embryo has the potential to develop into either a male or a female.

Gonadal sex is determined by the testis-determining factor (TDF) on the Y chromosome. TDF is located on the sex-determining region of the short arm of the Y chromosome and directs testicular differentiation. The Leydig cells produce the testosterone that stimulates development of the mesonephric ducts into male genital ducts. These androgens also stimulate development of the indifferent external genitalia into the penis and scrotum. A *müllerian-inhibiting substance (MIS)* produced by the Sertoli cells of the testes inhibits development of the paramesonephric ducts.

In the absence of a Y chromosome and in the presence of two X chromosomes, ovaries develop, the mesonephric ducts regress, the paramesonephric ducts develop into the uterus and uterine tubes, the vagina develops from the vaginal plate derived from the urogenital sinus, and the indifferent external genitalia develop into the clitoris and labia (majora and minora).

Persons with *true hermaphroditism,* an extremely rare intersexual condition, have both ovarian and testicular tissue and variable internal and external genitalia. *Male pseudohermaphroditism* results from failure of the fetal testes to produce adequate amounts of masculinizing hormones or from the tissue insensitivity of the sexual structures. *Female pseudohermaphroditism* usually results from congenital adrenal hyperplasia (CAH), a disorder of the fetal suprarenal (adrenal) glands that causes excessive production of androgens and masculinization of the external genitalia.

Most anomalies of the female genital tract, such as *double uterus,* result from incomplete fusion of the paramesonephric ducts. *Cryptorchidism* and *ectopic testes* result from abnormalities of testicular descent. *Congenital inguinal hernia* and *hydrocele* result from persistence of the processus vaginalis. Failure of the urogenital folds to fuse in males results in various types of *hypospadias.*

Clinically Oriented Questions

1. Does a horseshoe kidney usually function normally? What sort of problems may occur with this anomaly, and how can they be corrected?
2. A patient has been told that he has two kidneys on one side and none on the other. How did this abnormality probably happen? Are there likely to be any problems associated with this condition?
3. Do true hermaphrodites ever marry? Are they ever fertile?
4. When a baby is born with ambiguous external genitalia, how long does it take to assign the appropriate sex? What does the physician tell the parents? How is the appropriate sex determined?
5. What is the most common type of disorder producing ambiguity of the external genitalia? Will masculinizing or androgenic hormones given during the fetal period of development cause ambiguity of external genitalia in female fetuses?

The answers to these questions are at the back of the book.

15

The Cardiovascular System

The cardiovascular system is the first major system to function in the embryo. The primordial heart and vascular system appear in the middle of the third week of embryonic development. *The heart starts to function at the beginning of the fourth week.* This precocious development is necessary because the rapidly growing embryo can no longer satisfy its nutritional and oxygen requirements by diffusion alone.

Early Development of the Heart and Vessels

The earliest sign of the heart is the appearance of paired endothelial strands — **angioblastic cords** — in the cardiogenic mesoderm during the third week of development (Fig. 15-1*B* and *C*). These cords canalize to form **heart tubes** that fuse to form the tubular heart late in the third week (see Fig. 15-5). The heart begins to beat at 22 to 23 days (Fig. 15-2). An inductive influence from the anterior endoderm stimulates early formation of the heart. Cardiac morphogenesis is controlled by a cascade of *regulatory genes and transcription factors* (see the list of References at the back of the book).

Development of the Veins Associated with the Heart

Three paired veins drain into the tubular heart of a 4-week embryo (see Fig. 15-2):

- *Vitelline veins* return poorly oxygenated blood from the yolk sac.
- *Umbilical veins* carry well-oxygenated blood from the chorion—the primordial placenta; only the left umbilical vein persists.
- *Common cardinal veins* return poorly oxygenated blood from the body of the embryo.

After passing through the septum transversum, the **vitelline veins** enter the venous end of the heart, known as the **sinus venosus** (Fig. 15-3*A* and *B;* see also Fig. 15-2). As the liver primordium grows into the septum transversum, the *hepatic cords* anastomose around pre-existing endothelium-lined spaces. These spaces, the primordia of the **hepatic sinusoids**, later become linked to the vitelline veins. The **hepatic veins** form from the remains of the right vitelline vein in the region of the developing liver. The **portal vein** develops from an anastomotic network of vitelline veins around the duodenum (Fig. 15-4*B*). The fate of the umbilical veins may be summarized as follows (see Fig. 15-4):

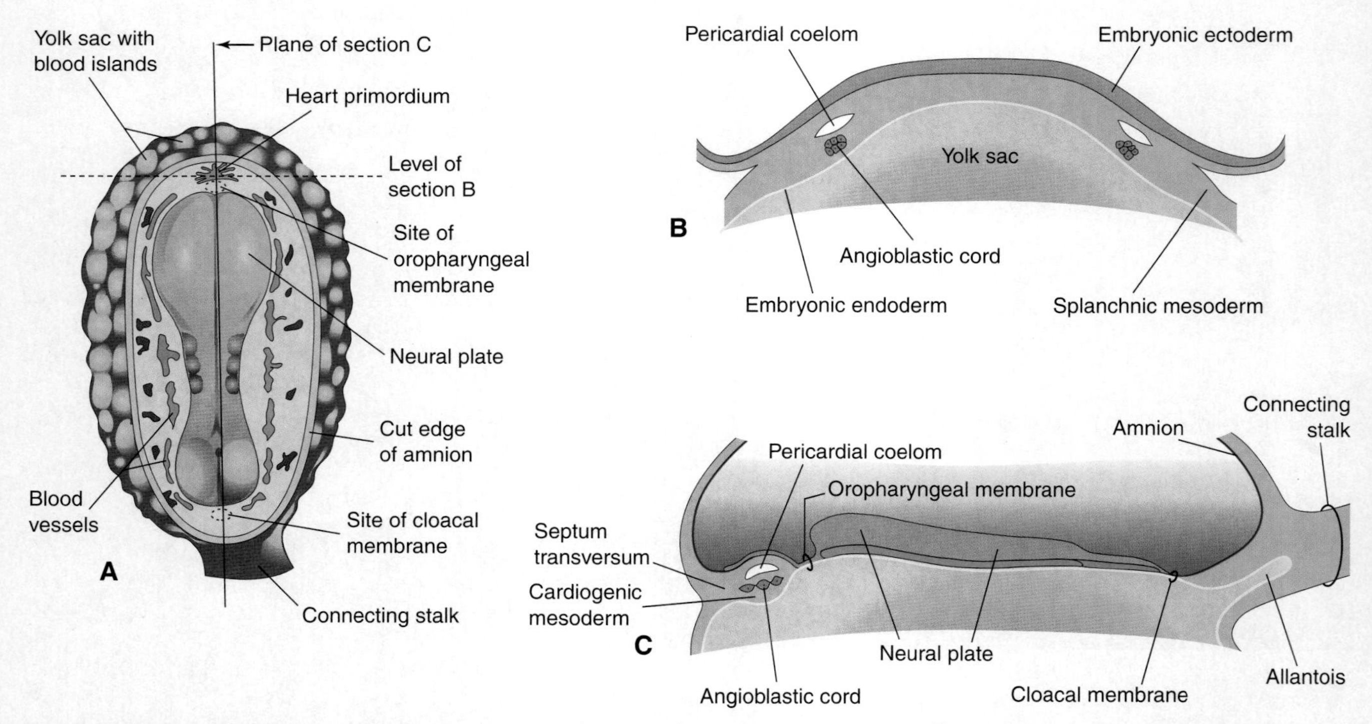

Figure 15-1. Early development of the heart. *A,* Drawing of a dorsal view of an embryo (about 18 days). *B,* Transverse section of the embryo demonstrating angioblastic cords and their relationship to the pericardial coelom. *C,* Longitudinal section through the embryo illustrating the relationship of the angioblastic cords to the oropharyngeal membrane, pericardial coelom, and septum transversum.

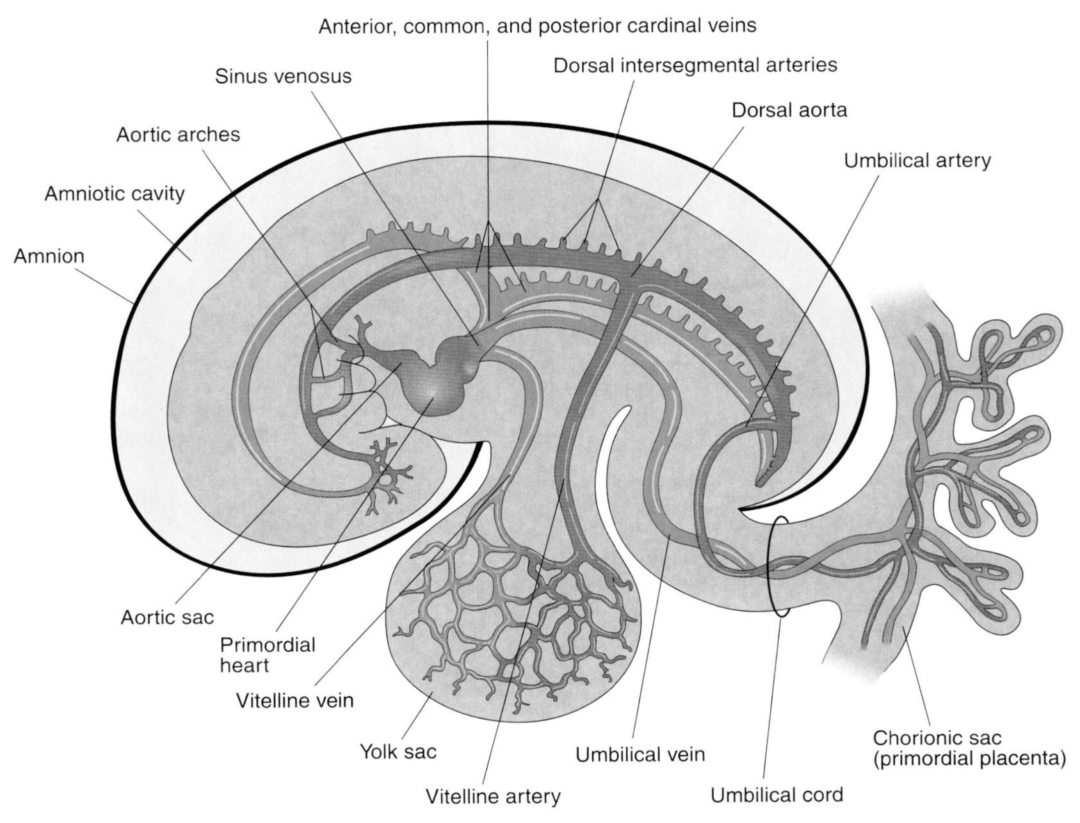

Figure 15 - 2. Drawing of the embryonic cardiovascular system (at about 26 days) showing vessels on the left side only. The umbilical vein carries well-oxygenated blood and nutrients from the chorion (the embryonic part of the placenta) to the embryo. The umbilical arteries carry poorly oxygenated blood and waste products from the embryo to the chorion (primordium of the placenta).

- The right umbilical vein and the caudal part of the left umbilical vein between the liver and the sinus venosus degenerate.
- The persistent caudal part of the left umbilical vein becomes the **umbilical vein**, which carries the well-oxygenated blood from the placenta to the embryo.
- A large venous shunt — the **ductus venosus** — develops within the liver (Fig. 15 - 4*B*) and connects the umbilical vein with the inferior vena cava. The ductus venosus forms a bypass through the liver, enabling most of the blood from the placenta to pass directly to the heart without passing through the capillary networks of the liver.

The **cardinal veins** (see Figs. 15 - 2 and 15 - 3*A*) constitute the main venous drainage system of the embryo. The anterior and posterior cardinal veins drain the cranial and caudal parts of the embryo, respectively (see Fig. 15 - 3*B*). The anterior and posterior cardinal veins join the **common cardinal veins**, which enter the **sinus venosus** (see Fig. 15 - 2). During the eighth week of development, the **anterior cardinal veins** are connected by an oblique anastomosis (see Fig. 15 - 4*B*) that shunts blood from the left to the right anterior cardinal vein. This anastomotic shunt becomes the **left brachiocephalic vein** when the caudal part of the left anterior cardinal vein degenerates (see Figs. 15 - 3*D* and 15 - *C*). The **superior vena cava** forms from the right anterior cardinal vein and the right common cardinal vein. The only adult derivatives of the *posterior cardinal veins* are the *root of the azygos vein* and the *common iliac veins*.

The subcardinal and supracardinal veins gradually replace and supplement the posterior cardinal veins. The **subcardinal veins** appear first (see Fig. 15 - 3*A*) and form the stem of the left renal vein, the suprarenal veins, the gonadal veins (testicular and ovarian), and a segment of the inferior vena cava (see Fig. 15 - 3*D*). The **supracardinal veins** become disrupted in the region of the kidneys (Fig. 15 - 3*C*). Cranial to this, they become united by an anastomosis that forms the **azygos** and the

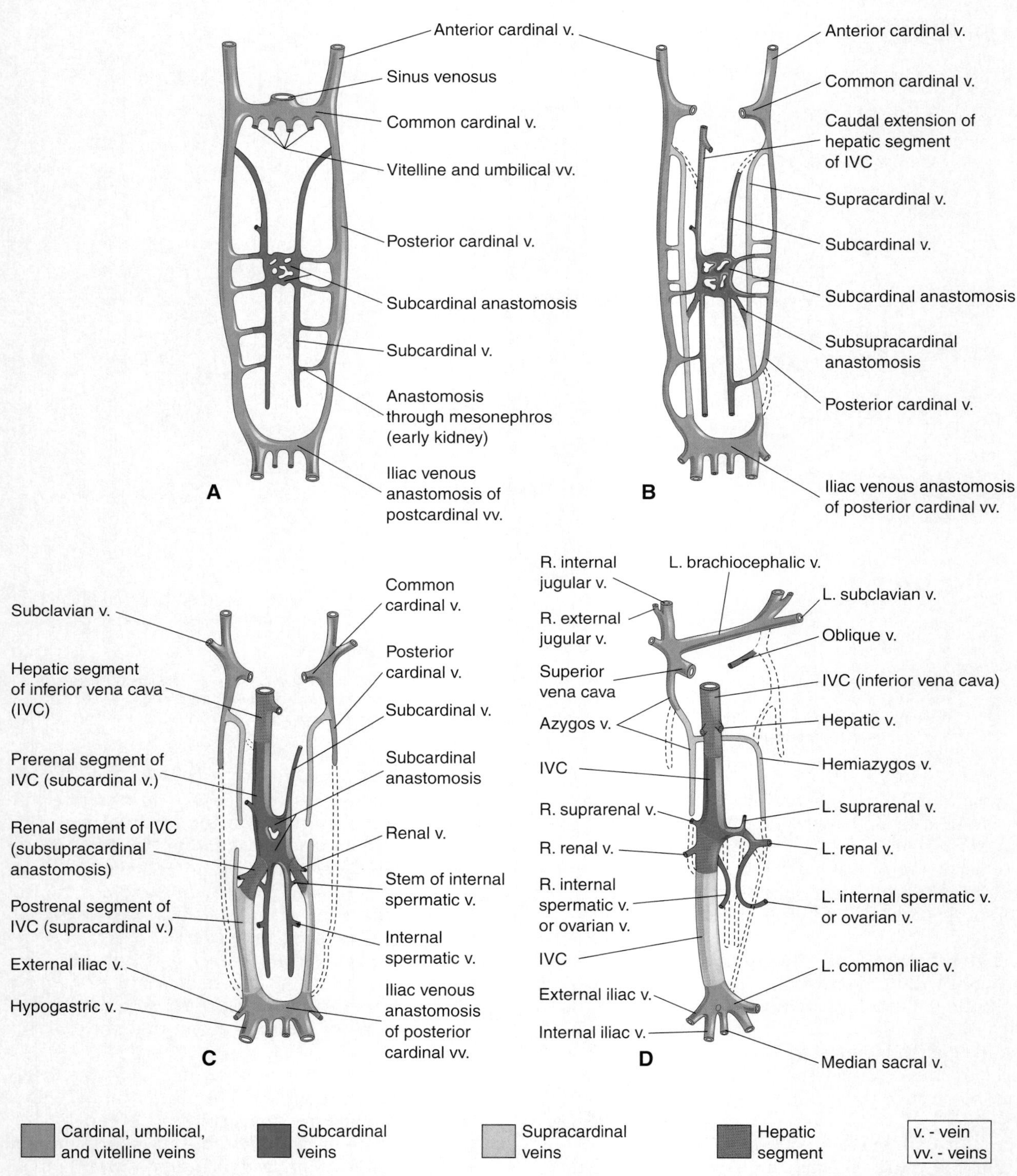

Figure 15 - 3. Drawings illustrating the primordial veins of the trunk in the human embryo (ventral views). Initially, three systems of veins are present: the umbilical veins from the chorion, the vitelline veins from the yolk sac, and the cardinal veins from the body of the embryo. Next, the subcardinal veins appear, and finally, the supracardinal veins develop. *A,* Six weeks. *B,* Seven weeks. *C,* Eight weeks. *D,* Adult. This drawing illustrates the transformations that produce the adult venous pattern. R., right; L., left. (Modified from Arey LB: *Developmental Anatomy*, rev. 7th ed. Philadelphia, WB Saunders, 1974.)

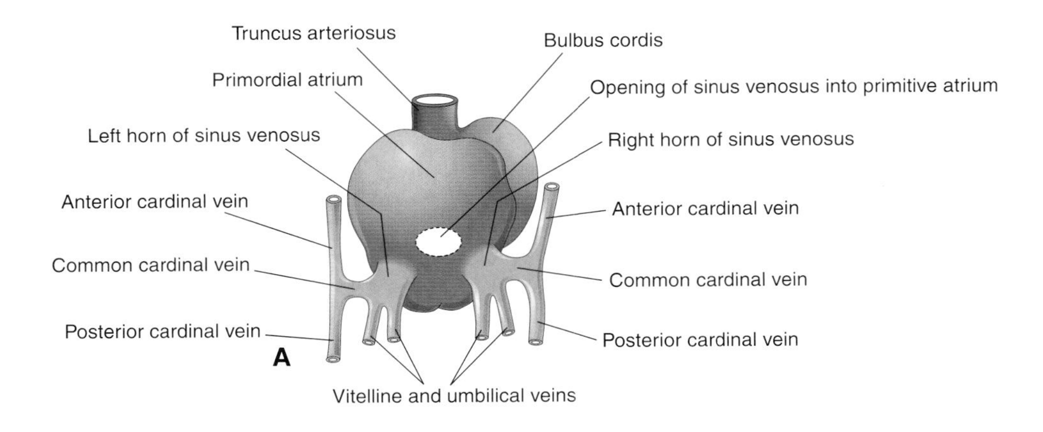

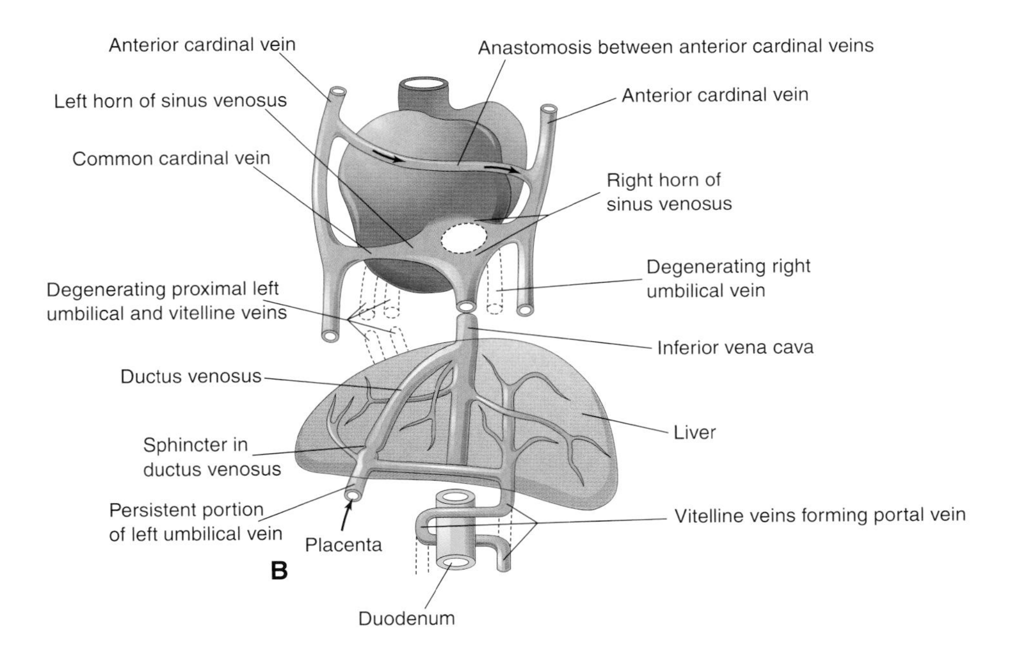

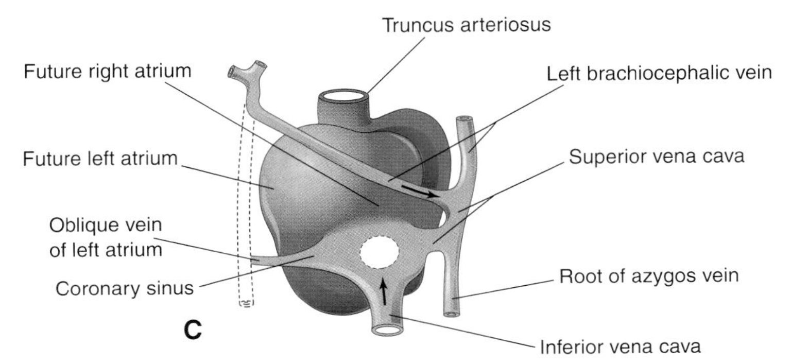

Figure 15-4. Dorsal views of the developing heart. *A,* During the fourth week of development (about 24 days), the primordial atrium and sinus venosus, as well as the veins draining into them, are evident. *B,* At seven weeks, the right sinual horn is enlarged and the venous circulation through the liver is established. (The organs are not drawn to scale.) *C,* Eight weeks, illustrating the adult derivatives of the cardinal veins.

hemiazygos veins (see Figs. 15 - 3*D* and 15 - 4*C*). Caudal to the kidneys, the left supracardinal vein degenerates, but the right supracardinal vein becomes the inferior part of the IVC (see Fig. 15 - 3*D*). The **IVC** forms as blood returning from the caudal part of the embryo is shifted from the left to the right side of the body. The IVC is composed of four main segments (see Fig. 15 - 3*C*):

- a *hepatic segment* derived from the hepatic vein (proximal part of the right vitelline vein) and hepatic sinusoids
- a *prerenal segment* derived from the right subcardinal vein
- a *renal segment* derived from the subcardinal-supracardinal anastomosis
- a *postrenal segment* derived from the right supracardinal vein

Anomalies of the Venae Cavae

The most common anomaly of the venae cavae is a persistent left superior vena cava (SVC), which drains into the right atrium through the enlarged orifice of the *coronary sinus*. The most common anomaly of the IVC is interruption of its abdominal course; as a result, blood drains from the lower limbs, abdomen, and pelvis to the heart through the azygos system of veins.

The Aortic Arches and Other Branches of the Dorsal Aorta

As the *pharyngeal arches* form during the fourth and fifth weeks of development, they are supplied by arteries — the **aortic arches** — that arise from the **aortic sac** and terminate in the dorsal aortae (see Fig. 15 - 2). Initially, the paired dorsal aortae run through the entire length of the embryo, but they soon fuse to form a single **dorsal aorta**, just caudal to the pharyngeal arches.

Intersegmental Arteries

Thirty or so branches of the dorsal aorta, collectively known as the **intersegmental arteries**, pass between and carry blood to the somites and their derivatives (see Fig. 15 - 2). The dorsal intersegmental arteries in the neck join to form a longitudinal artery on each side, the **vertebral artery**. Most of the original connections of the intersegmental arteries to the dorsal aorta eventually disappear. In the thorax, the dorsal intersegmental arteries persist as **intercostal arteries**. Most of the dorsal intersegmental arteries in the abdomen become **lumbar arteries**; however, the fifth pair of lumbar intersegmental arteries remains as the **common iliac arteries** (see Fig. 15 - 3*D*). In the sacral region, the intersegmental arteries form the **lateral sacral arteries**. The caudal end of the dorsal aorta becomes the median sacral artery.

Fate of the Vitelline and Umbilical Arteries

The unpaired ventral branches of the dorsal aorta supply the yolk sac, allantois, and chorion (see Fig. 15 - 2). The **vitelline arteries** supply the yolk sac and, later, the primordial gut, which forms from the incorporated part of the yolk sac. Three vitelline arteries remain and become:

- the *celiac trunk* to the foregut
- the *superior mesenteric artery* to the midgut
- the *inferior mesenteric artery* to the hindgut

The paired **umbilical arteries** pass through the connecting stalk (later, the *umbilical cord*) and give rise to vessels in the chorion. The umbilical arteries carry poorly oxygenated blood to the placenta (see Fig. 15 - 2). The proximal parts of these arteries become the *internal iliac arteries* and *superior vesical arteries*, whereas the distal parts are obliterated after birth and become the *medial umbilical ligaments*.

Further Development of the Heart

As lateral embryonic folding occurs, the heart tubes approach each other and fuse to form a single tube (see Figs. 15 - 5*B* and *C*; Fig. 15 - 6*D*). Fusion of the heart tubes begins at the cranial end of the developing heart and extends caudally. As the heart tubes fuse, the **primordial myocardium** is formed from splanchnic mesoderm surrounding the pericardial coelom (see Fig. 15 - 6*B* and *C*). At this stage, the developing heart is composed of a thin endothelial tube that is separated from a thick muscular tube (the primordial myocardium) by gelatinous connective tissue called **cardiac jelly** (see Fig. 15 - 6*C* and *D*). The endothelial tube becomes the internal endothelial lining of the heart, the **endocardium**, and the primordial myocardium becomes the muscular wall of the heart, the **myocardium**. The visceral pericardium, or **epicardium**, is derived from mesothelial cells that arise from the external surface of the sinus venosus (see Fig. 15 - 6*F*).

As folding of the head region occurs, the heart and pericardial cavity come to lie ventral to the foregut and caudal to the oropharyngeal membrane (Fig. 15 - 7*A* to *C*). Concurrently, the tubular heart elongates and develops alternate dilations and constrictions (see Fig. 15 - 5*C* to *E*): the truncus arteriosus, bulbus cordis, ventricle, atrium, and sinus venosus.

The tubular **truncus arteriosus** is continuous cranially with the **aortic sac** (Fig. 15 - 8*A*), from which the *aortic arches* arise. The **sinus venosus** receives the umbilical, vitelline, and common cardinal veins from

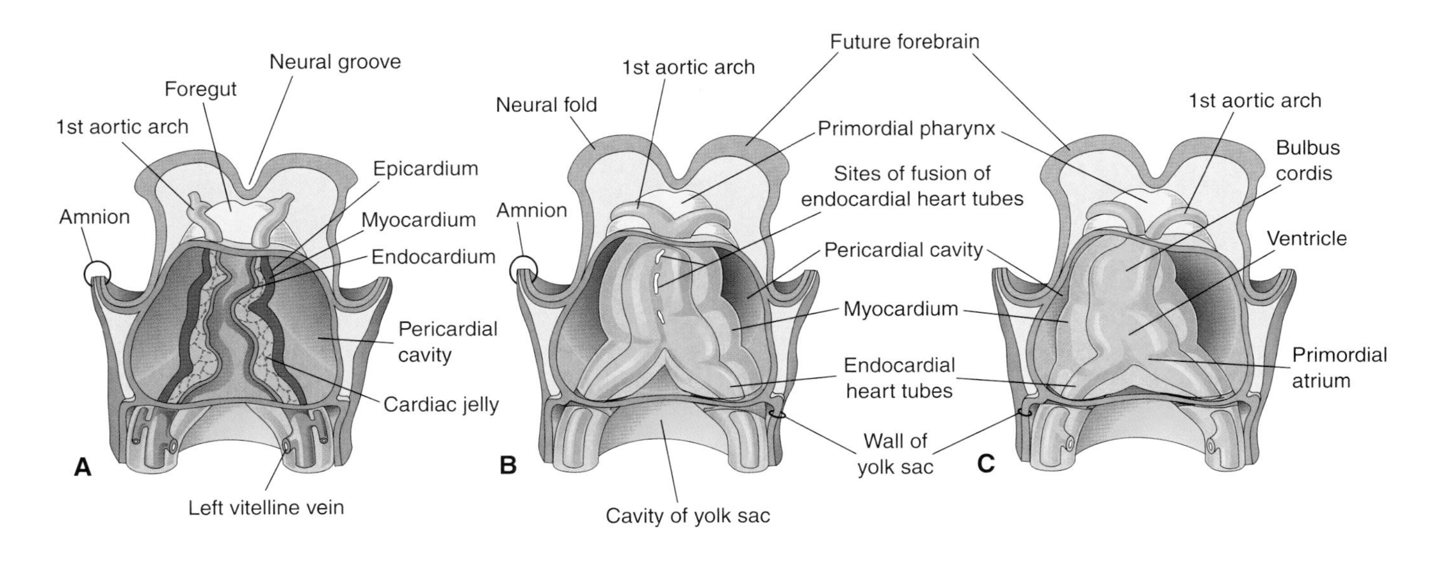

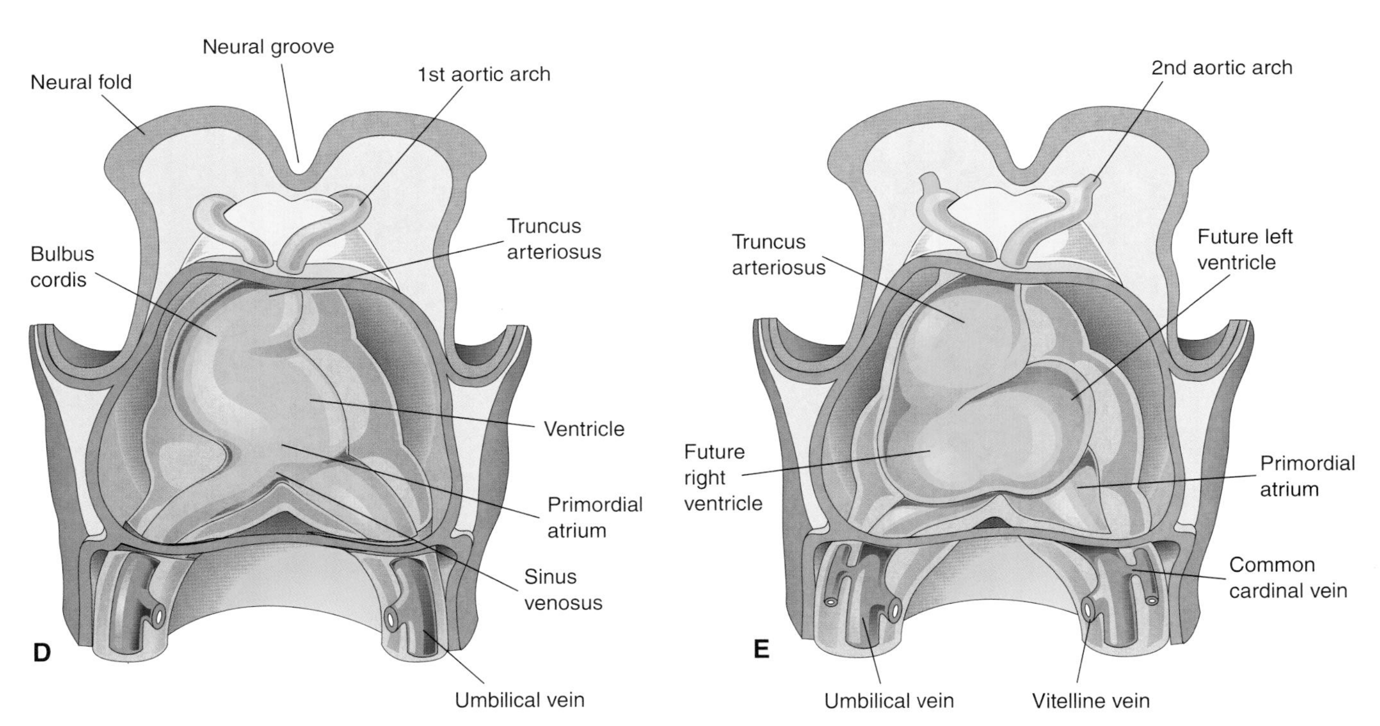

Figure 15-5. *A* to *C*, Drawings of ventral views of the developing heart and pericardial region (22 to 35 days after fertilization). The ventral pericardial wall has been removed to show the developing myocardium and fusion of the two heart tubes to form a single heart tube. The fusion begins at the cranial ends of the tubes and extends caudally until a single tubular heart is formed. *D* and *E*, As the heart elongates, it bends upon itself, forming an S-shaped heart.

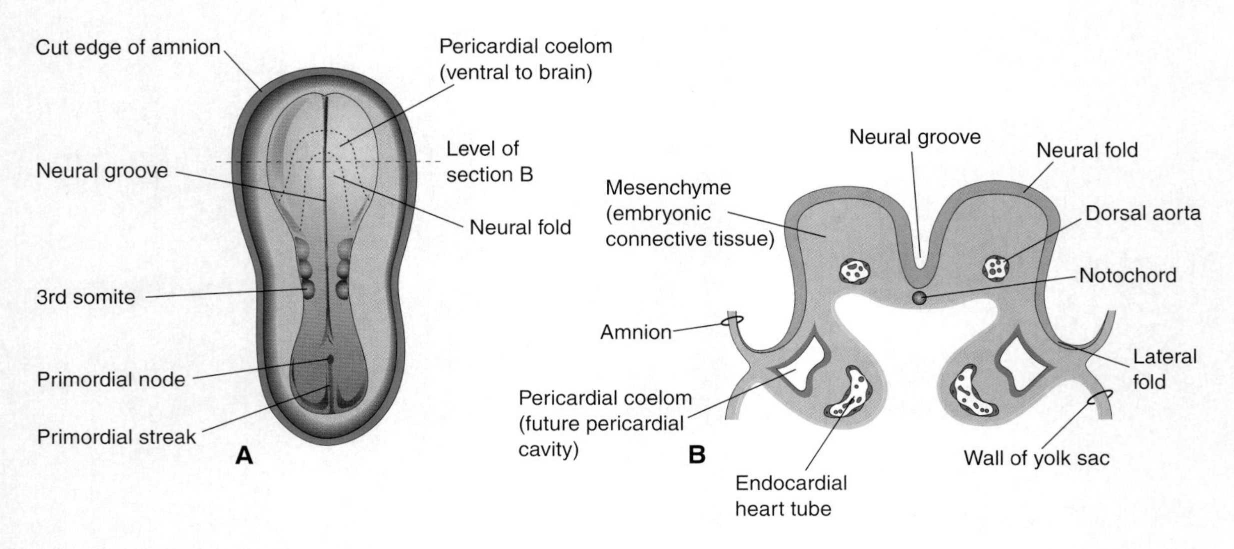

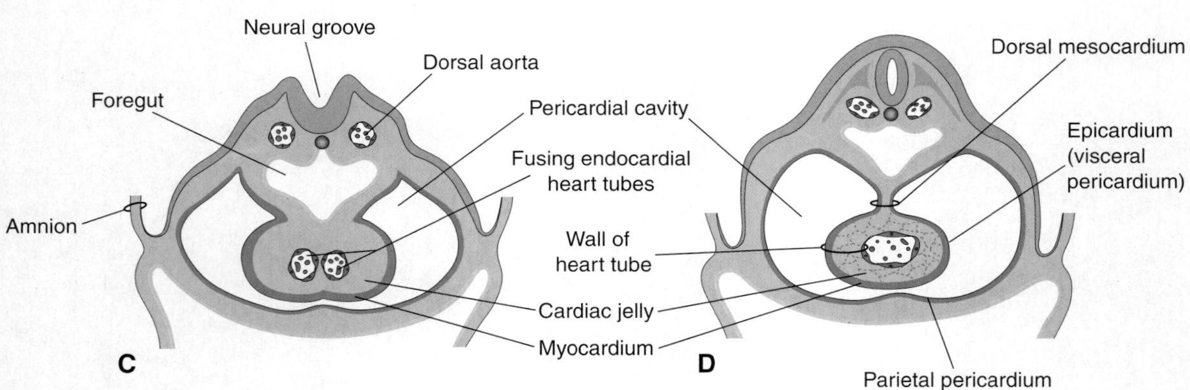

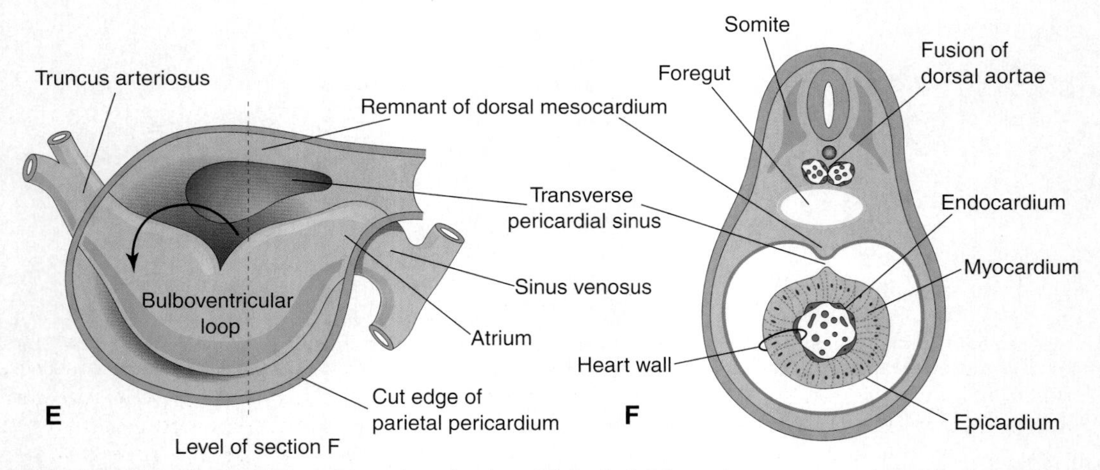

Figure 15 - 6. *A*, Drawing of a dorsal view of an embryo (at about 20 days). *B*, Schematic transverse section of the heart region of the embryo illustrated in *A*, showing the two endocardial heart tubes and the lateral folds of the body. *C*, Transverse section of a slightly older embryo, showing the formation of the pericardial cavity and the fusing heart tubes. *D*, Similar section (about 22 days) showing the single heart tube suspended by the dorsal mesocardium. *E*, Schematic drawing of the heart (about 28 days) showing degeneration of the central part of the dorsal mesocardium and formation of the transverse sinus of the pericardium. *F*, Transverse section of the embryo at the level shown in *E*, showing the layers of the heart wall.

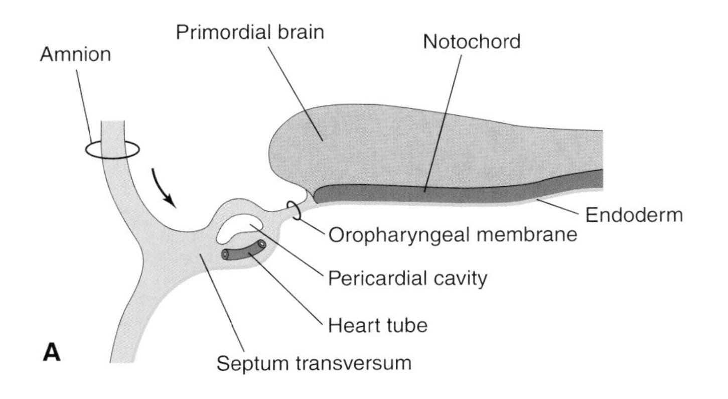

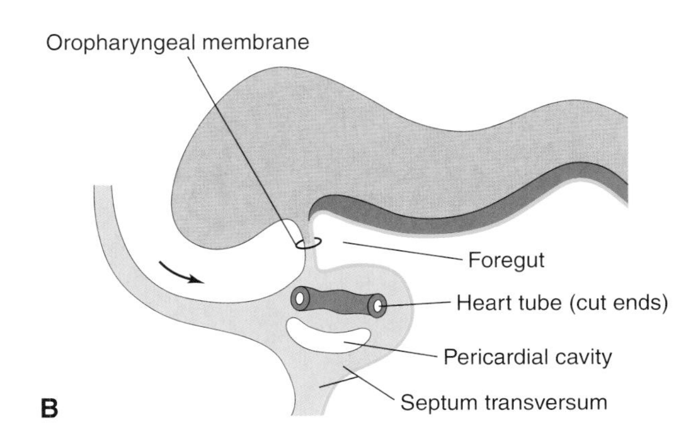

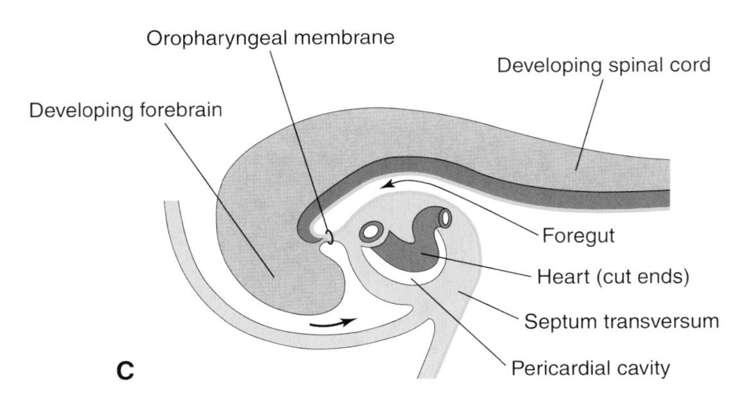

Figure 15 - 7. Drawings of longitudinal sections through the cranial half of human embryos during the fourth week of development. The illustrations show the effect of the head fold (*arrow*) on the position of the heart and other structures. *A* and *B*, As the head fold develops, the heart tube and pericardial cavity come to lie ventral to the foregut and caudal to the oropharyngeal membrane. *C*, Note that the positions of the pericardial cavity and septum transversum have reversed with respect to each other. The septum transversum now lies posterior to the pericardial cavity, where it will form the central tendon of the diaphragm.

the chorion, yolk sac, and embryo, respectively (see Fig. 15-8*B*). The arterial and venous ends of the heart are fixed by the pharyngeal arches and septum transversum, respectively. Because the **bulbus cordis** and ventricle grow faster than other regions, the heart bends on itself, forming a U-shaped **bulboventricular loop** (see Fig. 15-6*E*). As the primordial heart bends, the atrium and sinus venosus come to lie dorsal to the truncus arteriosus, bulbus cordis, and ventricle (see Fig. 15-8*A* and *B*). By this stage, the sinus venosus has developed lateral expansions, the right and left **horns of the sinus venosus**. As the heart develops, it gradually invaginates the **pericardial cavity** (see Figs. 15-6*C* and *D* and 15-7*C*). The heart is initially suspended from the dorsal wall by a mesentery, the **dorsal mesocardium**. However, the central part of this mesentery soon degenerates, forming a communication called the **transverse pericardial sinus** between the right and left sides of the pericardial cavity (see Fig. 15-6*E* and *F*). At this stage, the heart is attached only at its cranial and caudal ends.

Circulation Through the Primordial Heart

The initial contractions of the heart originate in muscle; that is, they are of myogenic origin. Blood enters the sinus venosus from a number of sites (see Fig. 15-8*A* and *B*), including the:

- embryo through the common cardinal veins
- placenta through the umbilical veins
- yolk sac through the vitelline veins

Blood from the sinus venosus enters the primordial atrium; flow from it is controlled by **sinuatrial valves** (see Figs. 15-8*A*). The blood then passes through the **atrioventricular canal** into the **primordial ventricle**. When the ventricle contracts, blood is pumped through the **bulbus cordis** and **truncus arteriosus** into the **aortic sac**, from which it is distributed to the aortic arches in the pharyngeal arches (see Fig. 15-8*C*). The blood then passes into the dorsal aortae for distribution to the embryo, yolk sac, and placenta. Partitioning of the atrioventricular canal, primordial atrium, and ventricle

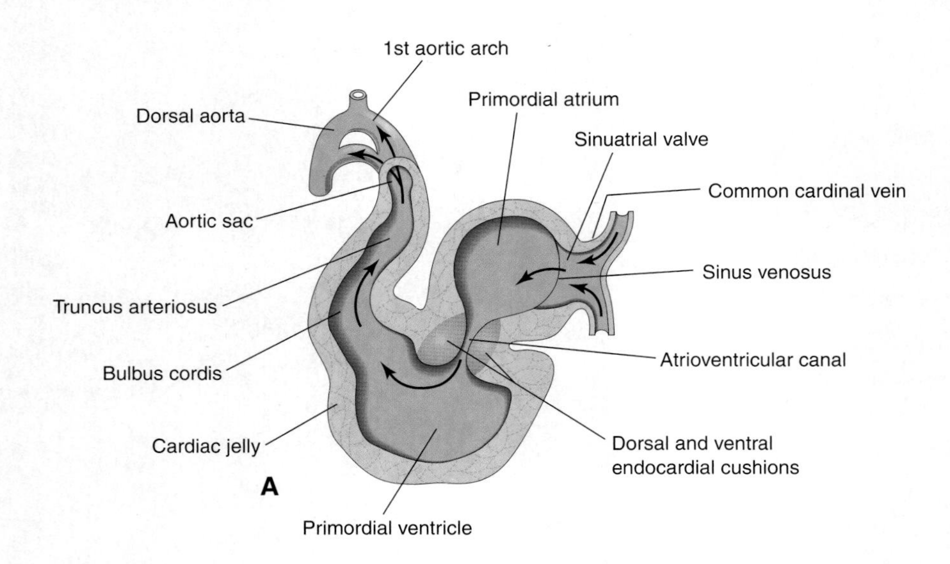

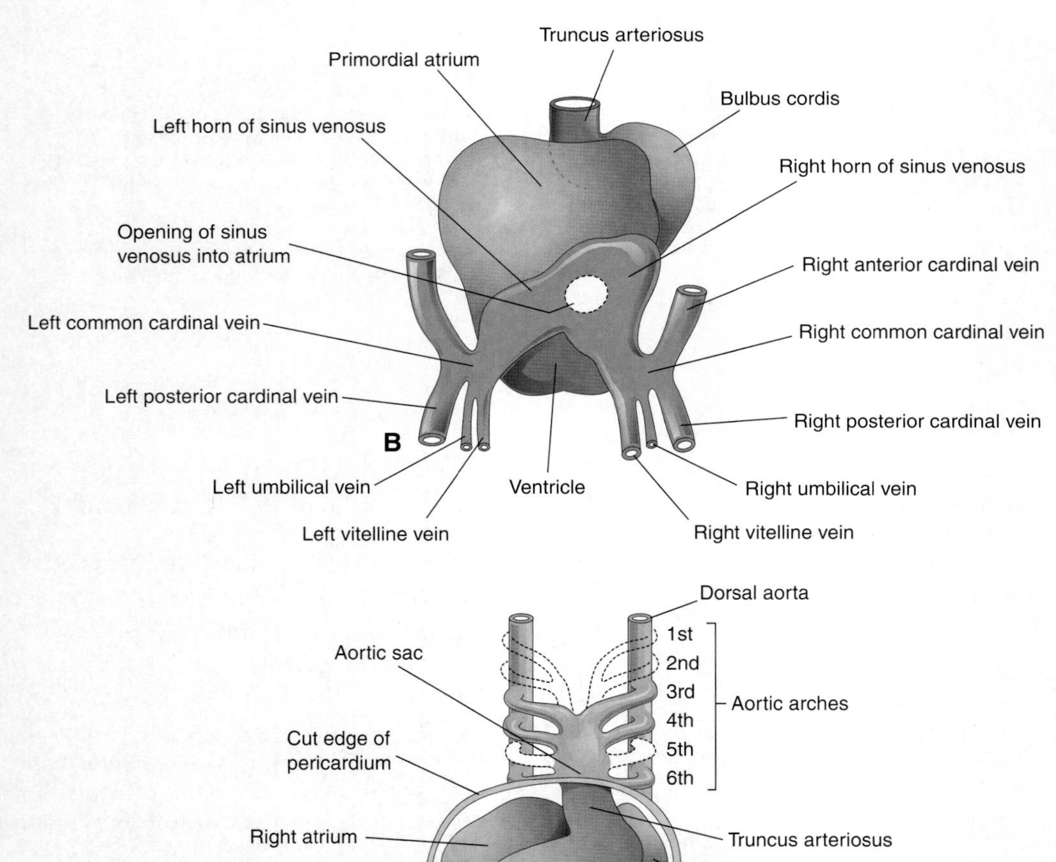

Figure 15 – 8. *A*, Sagittal section of the primordial heart (at about 24 days), showing blood flowing through it (*arrows*). *B*, Dorsal view of the heart (at about 26 days), illustrating the horns of the sinus venosus and the dorsal location of the primordial atrium. *C*, Ventral view of the heart and aortic arches (at about 35 days). The ventral wall of the pericardial sac has been removed to show the heart in the pericardial cavity.

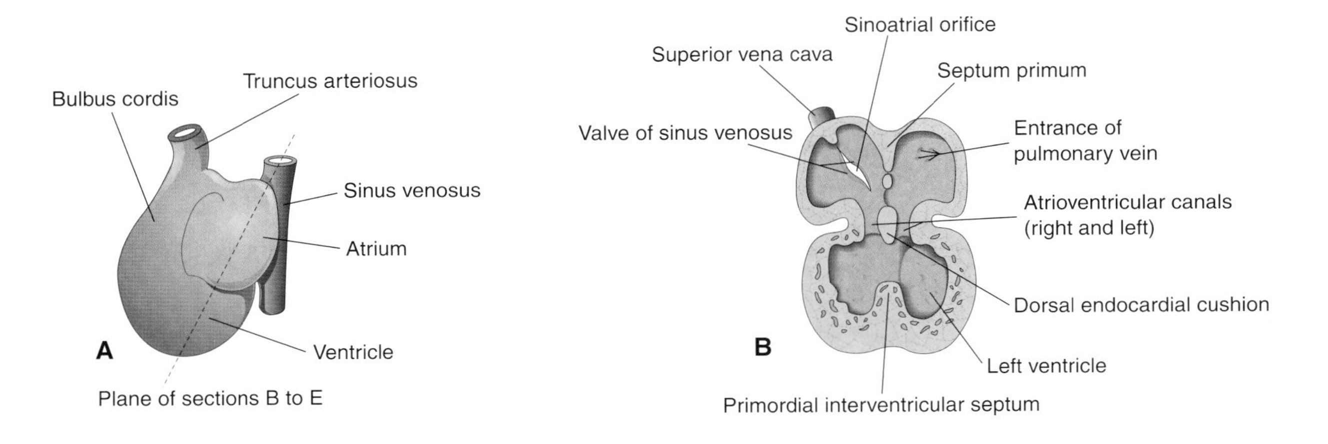

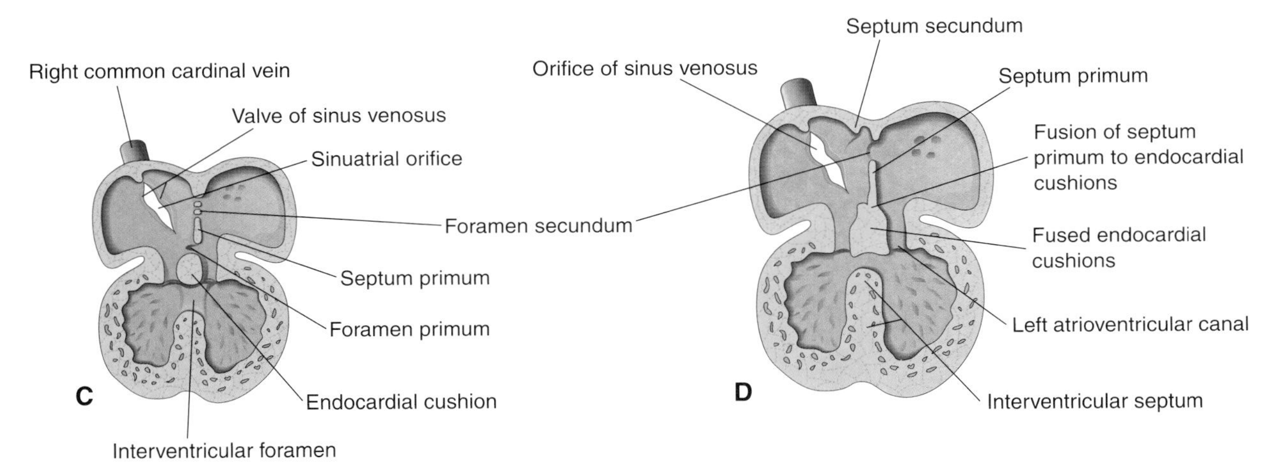

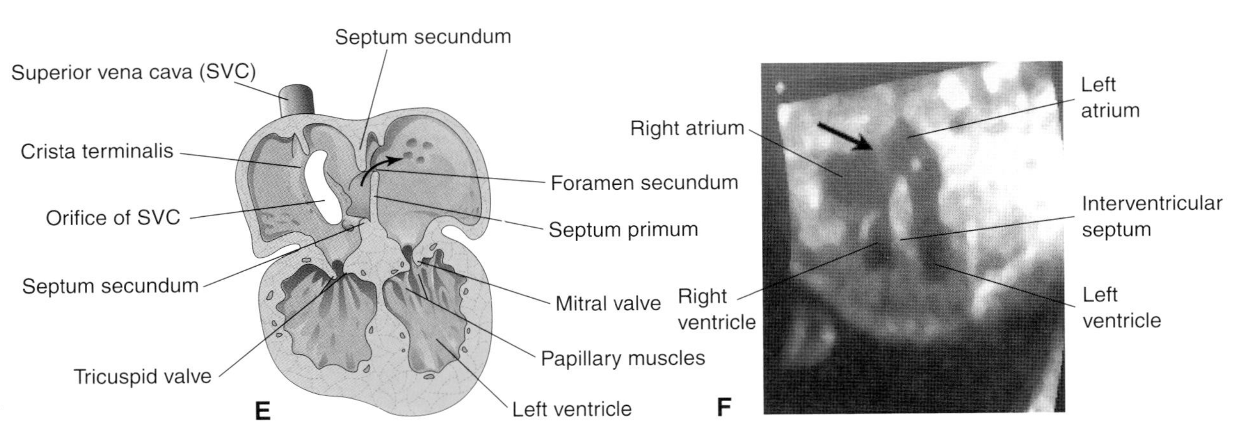

Figure 15 - 9. Drawings of the developing heart showing partitioning of the atrioventricular canal, primordial atrium, and ventricle. *A*, Sketch showing the plane of the sections (*B* to *E*). *B*, Fourth week of development (at about 28 days), showing the early appearance of the septum primum, interventricular septum, and dorsal endocardial cushion. *C*, Frontal section of the heart (at about 32 days) showing perforations in the dorsal part of the septum primum. *D*, Frontal section of the heart (at about 35 days), showing the foramen secundum. *E*, About 8 weeks after fertilization, the heart is partitioned into four chambers. The arrow indicates the flow of well-oxygenated blood from the right to the left atrium. *F*, Sonogram of a second-trimester fetus showing the four chambers of the heart. Note the septum secundum (*arrow*) and the descending aorta. (*F*, Courtesy of Dr. Greg Reid, Women's Hospital and the Department of Obstetrics, Gynecology, and Reproductive Sciences, University of Manitoba, Winnipeg, Manitoba, Canada.)

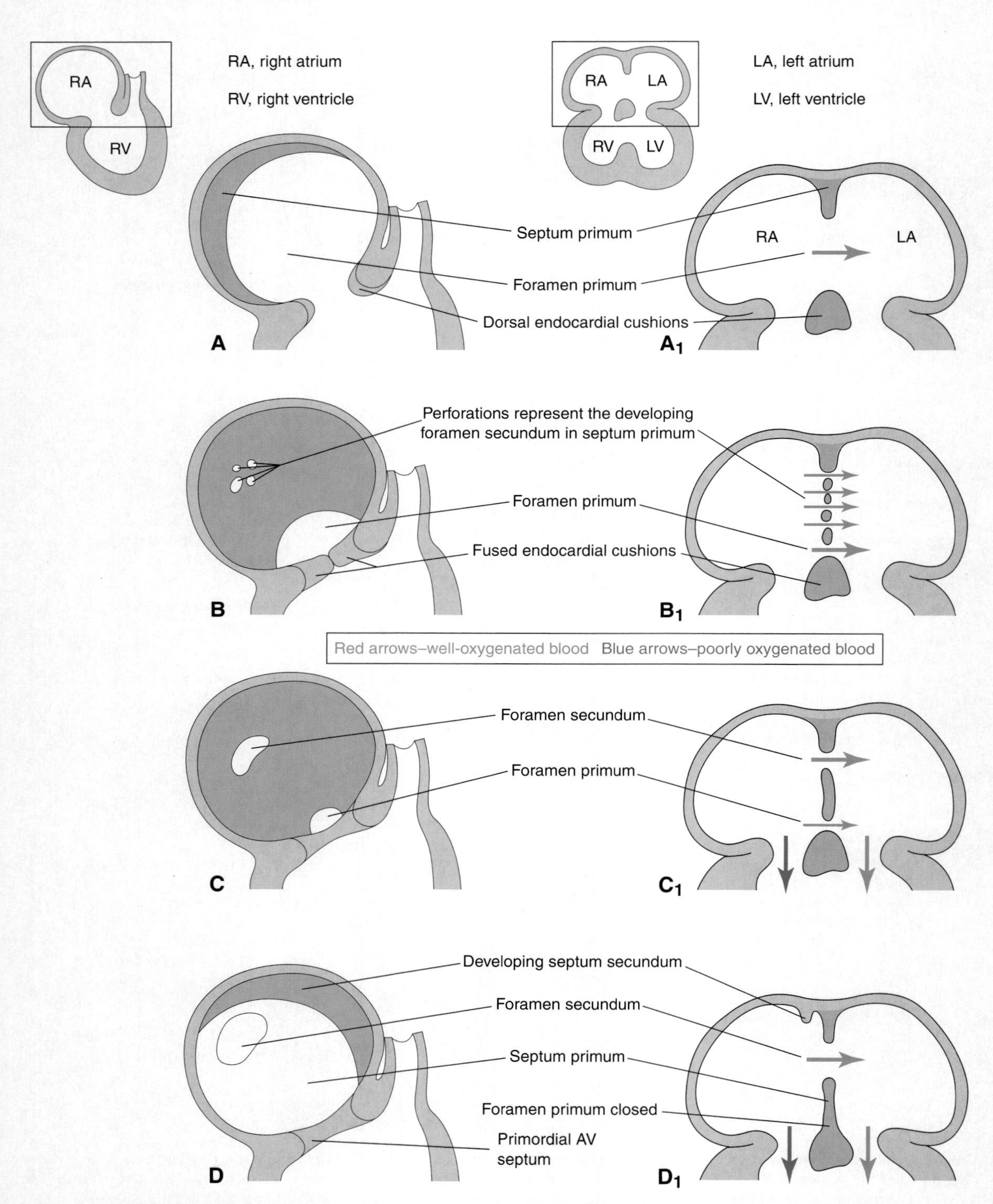

Figure 15 - 10. Diagrams illustrating progressive stages in the partitioning of the primordial atrium. *A* to *H,* are views of the developing interatrial septum as viewed from the right side. A_1 to H_1, are frontal (coronal) sections of the developing interatrial septum. As the septum secundum grows, note that it overlaps the opening in the septum primum (foramen secundum). Observe the valve of the oval foramen in G_1 and H_1. When pressure in the right atrium exceeds that in the left atrium, blood passes from the right to the left side of the heart. When the pressures are equal or higher in the left atrium, the valve formed by the septum primum closes the oval foramen.

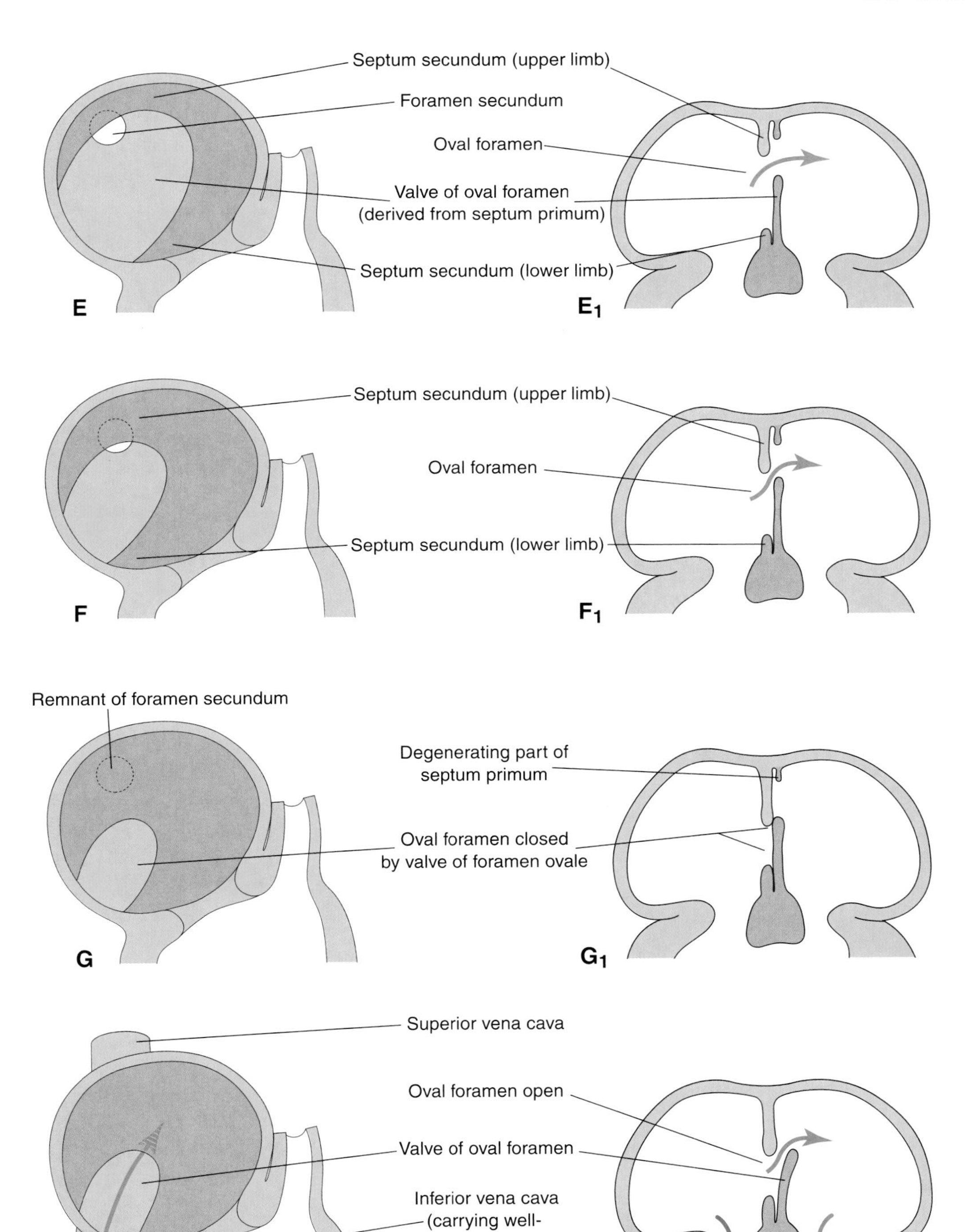

Figure 15 - 10. Continued

begins around the middle of the fourth week of development and is essentially completed by the end of the fifth week.

Partitioning of the Atrioventricular Canal

Endocardial cushions form on the dorsal and ventral walls of the atrioventricular (AV) canal. As these tissue masses are invaded by mesenchymal cells (see Fig. 15 – 8*A*), the AV cushions approach each other and fuse, dividing the AV canal into right and left AV canals (Fig. 15 – 9*B* and *D*). These canals partially separate the primordial atrium from the ventricle, and the cushions function as AV valves. The **endocardial cushions** develop from a specialized matrix related to the myocardium. *Its formation is associated with the expression of TGFβ2 and bone morphogenetic factors* (*BMP2A and PMP4*).

Partitioning of the Primordial Atrium

The primordial atrium is divided into right and left atria by the formation and subsequent modification and fusion of two septa, the septum primum and septum secundum (see Fig. 15 – 9*A* to *E*; Fig. 15 – 10).

The **septum primum**, a thin, crescent-shaped membrane, grows toward the fusing endocardial cushions from the roof of the primordial atrium, partially dividing the atrium into right and left halves. As this curtainlike septum develops, a large opening, called the interatrial **foramen primum**, forms between its free edge and the endocardial cushions (see Figs. 15 – 9*C* and 15 – 10*A* to *C*). The foramen primum serves as a shunt, enabling oxygenated blood to pass from the right to the left atrium. The foramen primum becomes progressively smaller and disappears as the septum primum fuses with the fused endocardial cushions to form the **primordial AV septum** (see Fig. 15 – 10*D* and D_1). Before the foramen primum disappears, perforations, produced by programmed cell death, appear in the central part of the septum primum. As the septum fuses with the endocardial cushions, the perforations coalesce to form another opening, the **foramen secundum** (Fig. 15 – 10*C*). Concurrently, the free edge of the septum primum fuses with the left side of the fused endocardial cushions, obliterating the foramen primum (see Figs. 15 – 9*D* and 15 – 10*D*). The foramen secundum ensures a continuous flow of oxygenated blood from the right to the left atrium.

The **septum secundum**, a crescentic muscular membrane, grows from the ventrocranial wall of the atrium, immediately to the right of the septum primum (see Fig. 15 – 10D_1). As this thick septum grows during the fifth and sixth weeks of development, it gradually overlaps the foramen secundum in the septum primum (see Fig. 15 – 10*E* and *F*). The septum secundum forms an incomplete partition between the atria; the opening in the foramen secundum is called the **oval foramen** (L. foramen ovale). The cranial part of the septum primum, initially attached to the roof of the left atrium, gradually disappears (see Fig. 15 – 10G_1 and H_1). The remaining part of the septum primum, attached to the endocardial cushions, forms the **valve of the oval foramen**.

Before birth, the oval foramen allows most of the oxygenated blood entering the right atrium from the IVC to pass into the left atrium (Fig. 15 – 11*A*). It also prevents the passage of blood in the opposite direction because the septum primum closes against the relatively rigid septum secundum (see Fig. 15 – 11*B*). *After birth*, the oval foramen normally closes, and the valve of the oval foramen fuses with the septum secundum. As a result, the interatrial septum becomes a complete partition between the atria.

Changes in the Sinus Venosus

Initially, the sinus venosus opens into the posterior wall of the primordial atrium, and its right and left horns are about the same size. By the end of the fourth week of development, the right sinual horn becomes larger than the left (Fig. 15 – 12*A* and *B*). As this occurs, the sinuatrial orifice moves to the right and opens in the part of the primordial atrium that will become the adult right atrium (see Fig. 15 – 12*C*). The results of the two left-to-right venous shunts are:

- The left sinual horn decreases in size and importance.
- The right sinual horn enlarges and receives all the blood from the head and neck through the SVC, and from the placenta and caudal regions of the body through the IVC.

As heart development proceeds, the left sinual horn becomes the **coronary sinus**, and the right sinual horn becomes incorporated into the wall of the right atrium (see Fig. 15 – 12*B* and *C*).

Because it is derived from the sinus venosus, the smooth part of the wall of the right atrium is called the **sinus venarum** (see Fig. 15 – 12*B* and *C*). The remainder of the internal surface of the wall of the right atrium, as well as the muscular pouch called the **auricle**, has a rough, trabeculated appearance. These two parts are derived from the primordial atrium. The smooth part (sinus venarum) and the rough part (primordial atrium) are demarcated internally in the right atrium by a vertical ridge, termed the **crista terminalis** or terminal crest (see Fig. 15 – 12*C*), and externally by a shallow inconspicuous groove, the **sulcus terminalis** or terminal groove (see Fig. 15 – 12*B*). The crista terminalis represents the cranial part of the right **sinuatrial valve** (see Fig. 15 – 12*C*); the caudal part of

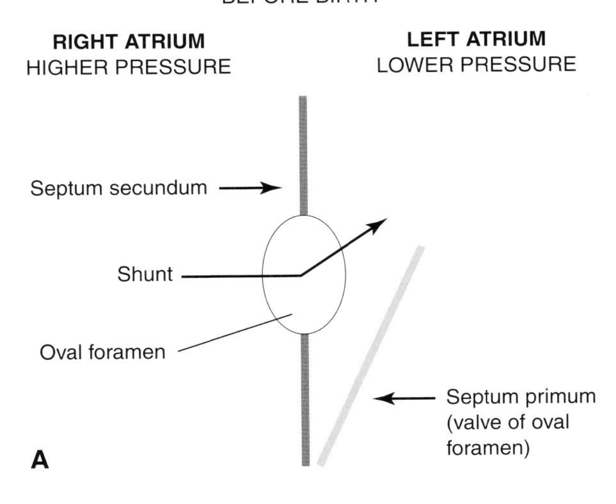

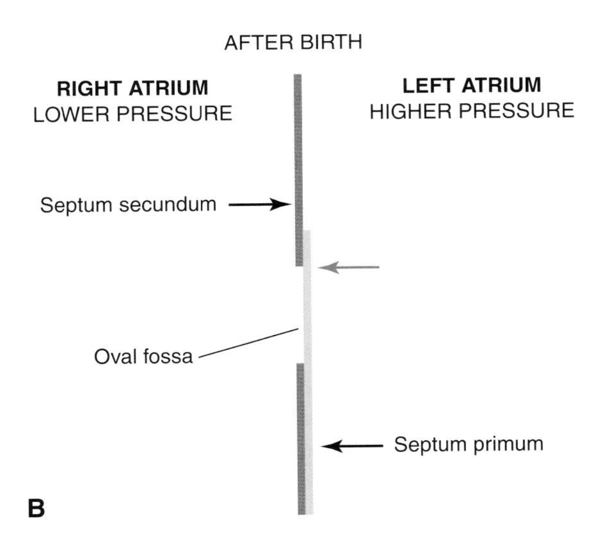

Figure 15 - 11. Diagrams illustrating the relationship of the septum primum to the oval foramen and septum secundum. *A,* Before birth, well-oxygenated blood is shunted from the right atrium through the oval foramen into the left atrium when the pressure rises. When the pressure decreases in the right atrium, the flaplike valve of the oval foramen is pressed against the relatively rigid septum secundum; this closes the oval foramen. *B,* After birth, the pressure in the left atrium increases as the blood returns from the lungs, which are now functioning. Eventually, the septum primum is pressed against the septum secundum and adheres to it, permanently closing the oval foramen and forming the oval fossa.

this valve forms the valves of the IVC and coronary sinus. The left sinuatrial valve fuses with the septum secundum and is incorporated with it into the interatrial septum.

Primordial Pulmonary Vein and Formation of the Left Atrium

Most of the wall of the left atrium is smooth because it is formed by incorporation of the primordial pulmonary vein (Fig. 15 - 13*A*). This vein develops as an outgrowth of the dorsal atrial wall, just to the left of the septum primum. As the atrium expands, the primordial pulmonary vein and its main branches are gradually incorporated into the wall of the left atrium (see Fig. 15 - 13*B*); as a result, four pulmonary veins are formed (see Fig. 15 - 13*C* and *D*). The small left auricle is derived from the primordial atrium; its internal surface has a rough, trabeculated appearance.

Partitioning of the Primordial Ventricle

Division of the primordial ventricle into two ventricles is first indicated by a median muscular ridge — the **primordial interventricular (IV) septum** — in the floor of the ventricle near its apex (see Fig. 15 - 9*B*). This thick, crescentic fold has a concave free edge (Fig. 15 - 14*A*). Initially, most of its increase in height results from dilation of the ventricles on each side of the IV septum (see Fig. 15 - 14*B*). The medial walls of the ventricles approach each other and fuse to form the primordium of the **muscular part of the IV septum**. Later, active

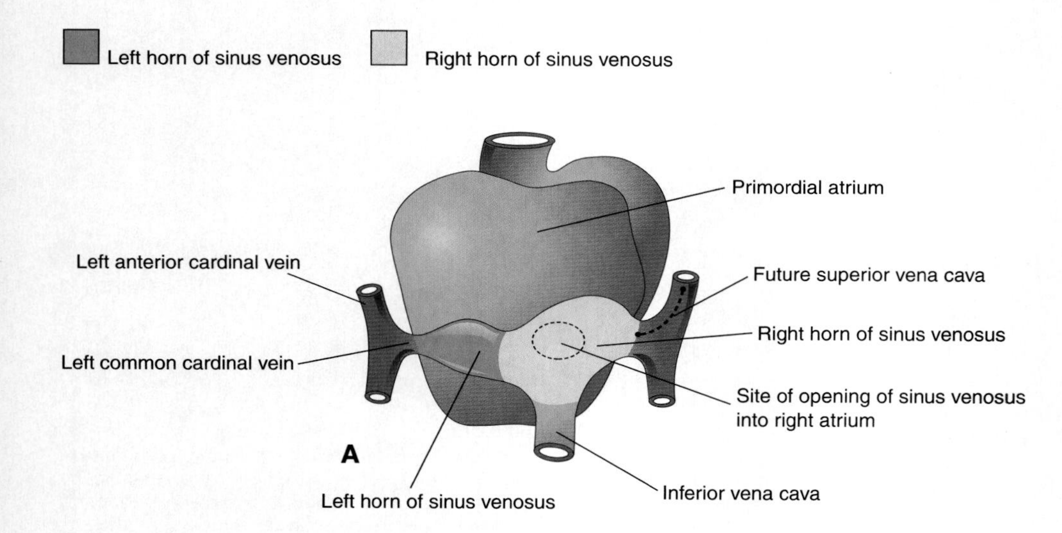

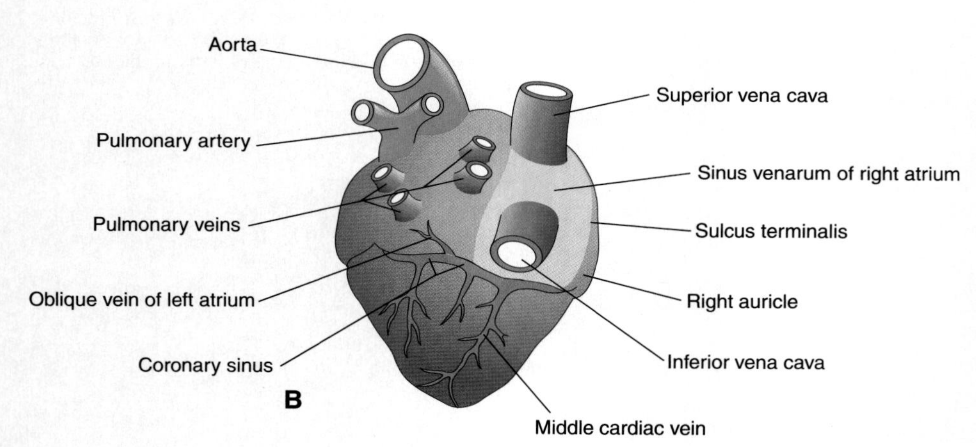

Figure 15 - 12. Diagrams illustrating the fate of the sinus venosus. *A*, Dorsal view of the heart (at about 26 days) showing the primordial atrium and sinus venosus. *B*, Dorsal view at 8 weeks after incorporation of the right sinual horn into the right atrium. The left sinual horn has become the coronary sinus. *C*, Internal view of the fetal right atrium showing the smooth part of the wall of the right atrium (sinus venarum) derived from the right sinual horn, and the crista terminalis and the valves of the inferior vena cava and coronary sinus derived from the right sinuatrial valve. The primordial right atrium becomes the right auricle, a conical muscular pouch.

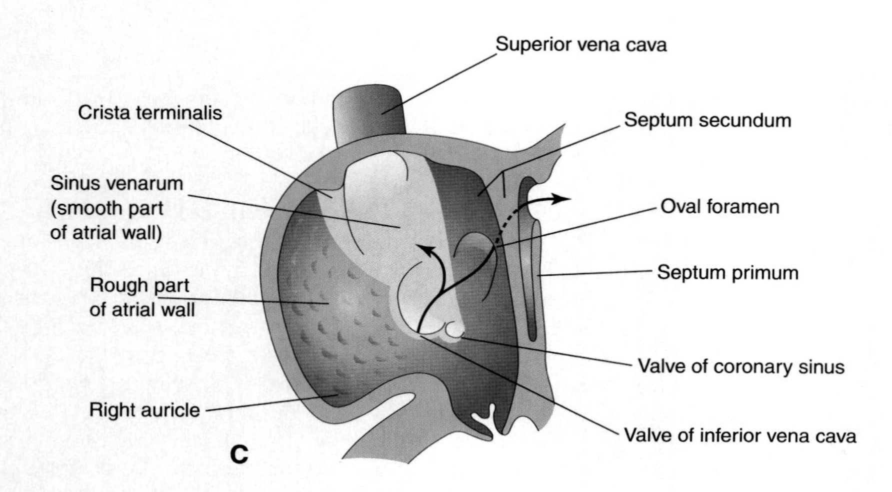

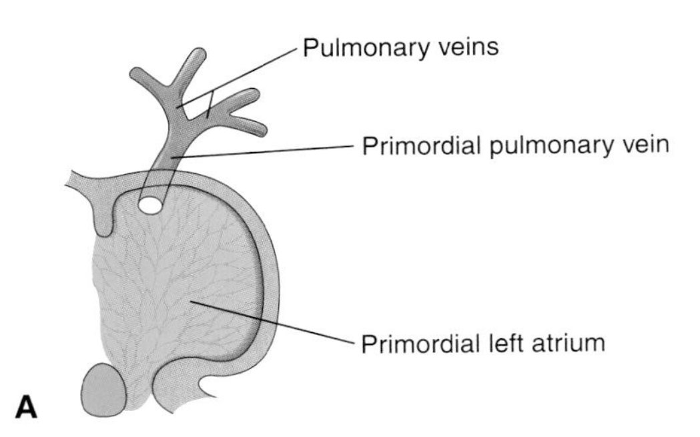

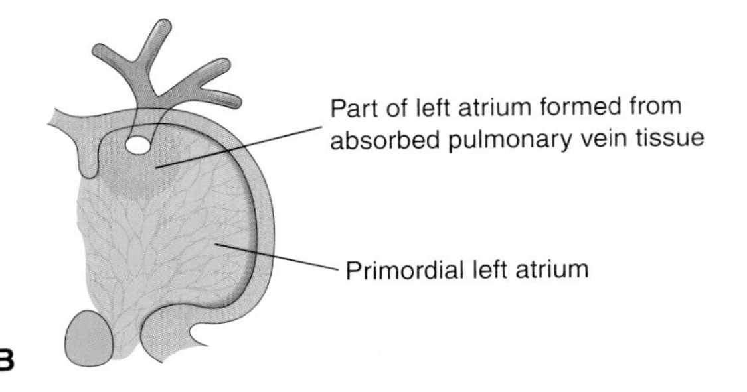

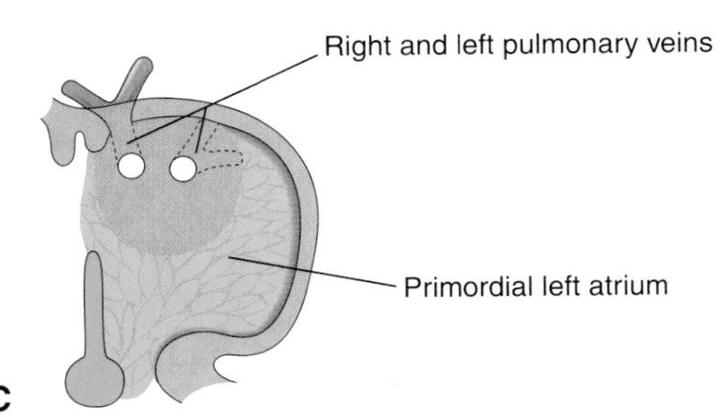

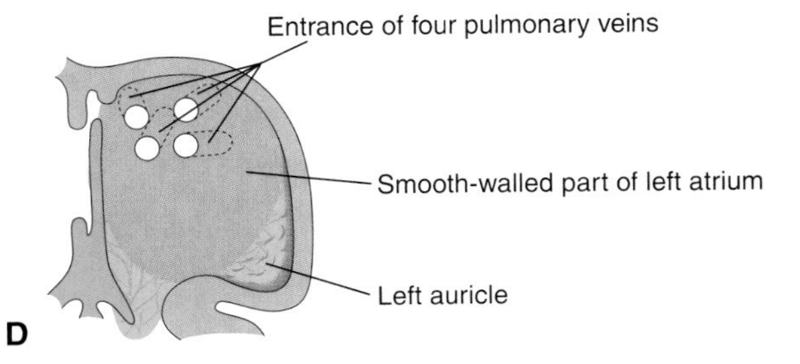

Figure 15 - 13. Diagrams illustrating absorption of the pulmonary vein into the left atrium. *A*, Five weeks, showing the common pulmonary vein opening into the primordial left atrium. *B*, Later stage, showing partial absorption of the common pulmonary vein. *C*, Six weeks, showing the openings of two pulmonary veins into the left atrium as a result of absorption of the common pulmonary vein. *D*, Eight weeks, showing four pulmonary veins with separate atrial orifices. The primordial left atrium becomes the left auricle, a tubular pouch of the atrium. Most of the left atrium is formed by absorption of the primordial pulmonary vein and its branches.

proliferation of myoblasts in the septum increases its size. Until the seventh week of development, there is a crescent-shaped **IV foramen** between the free edge of the IV septum and the fused endocardial cushions (see Fig. 15-16*A* and *B*). The IV foramen permits communication between the right and left ventricles (Fig. 15-15*B;* see also Fig. 15-14). The IV foramen usually closes by the end of the seventh week as the bulbar ridges fuse with the endocardial cushion (see Fig. 15-15*C* to *E*). **Closure of the IV foramen** and formation of the membranous part of the IV septum result from the fusion of tissues from three sources:

- the right bulbar ridge
- the left bulbar ridge
- the endocardial cushion

The **membranous part of the IV septum** is derived from an extension of tissue from the right side of the endocardial cushion to the muscular part of the IV septum. This tissue merges with the aorticopulmonary septum and the thick muscular part of the IV septum (Fig. 15-16*D*). After closure of the IV foramen and formation of the membranous part of the IV septum, the pulmonary trunk communicates with the right ventricle, and the aorta communicates with the left ventricle (see Fig. 15-15*E*). Cavitation of the ventricular walls forms a spongelike mass of muscular bundles. Some bundles remain as the **trabeculae carneae** (muscular bundles on the ventricular walls). Other bundles become the **papillary muscles** and tendinous cords (L. **chordae tendineae**). The tendinous cords run from the papillary muscles to the AV valves (see Fig. 15-16*D*).

Fetal Cardiac Ultrasonography

Echocardiography and Doppler ultrasonography have made it possible for sonographers to recognize normal and abnormal fetal cardiac anatomy. Most studies are performed at between 18 and 22 weeks' gestation because the heart is large enough to examine easily; however, real-time ultrasound images of the fetal heart can be obtained at 16 weeks.

Partitioning of the Bulbus Cordis and Truncus Arteriosus

During the fifth week of development, active proliferation of mesenchymal cells in the walls of the bulbus cordis results in the formation of **bulbar ridges** (Fig. 15-17*B* and *C;* see also Fig. 15-15*C* and *D*). Similar ridges form in the truncus arteriosus; these are continuous with the bulbar ridges. The bulbar and **truncal ridges** are derived mainly from neural crest mesenchyme. **Neural crest**

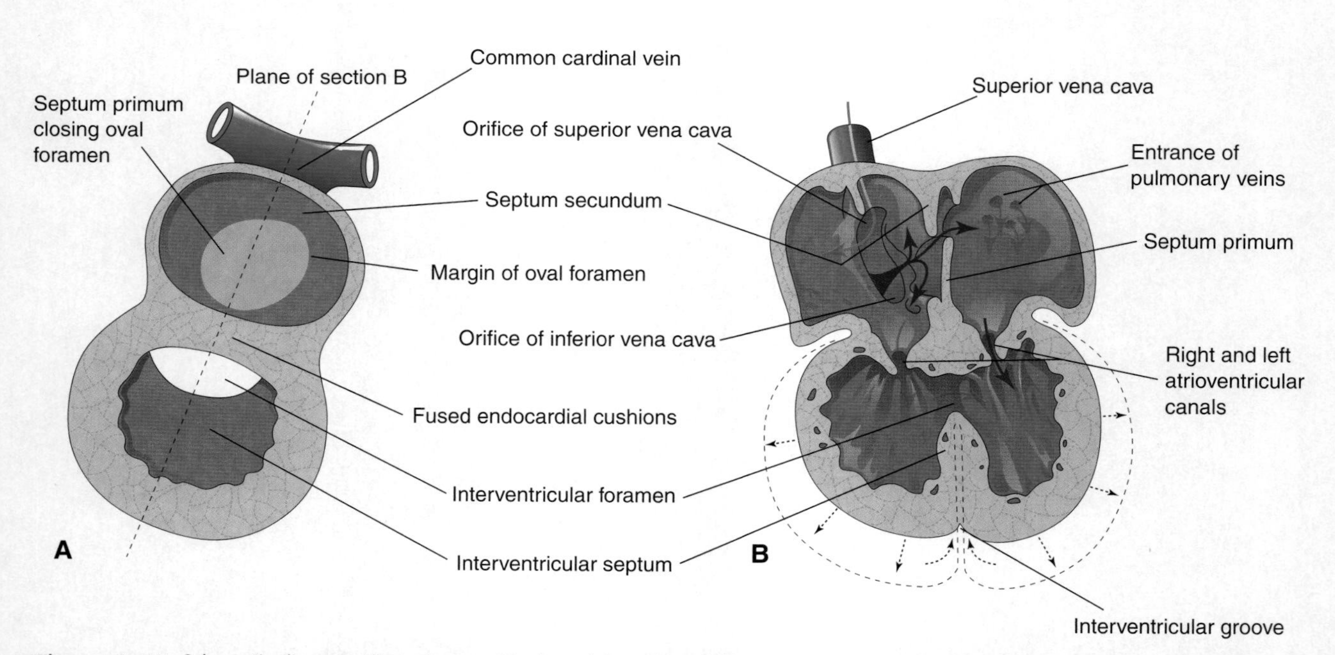

Figure 15-14. Schematic diagrams illustrating partitioning of the primordial heart. *A,* A sagittal section late in the fifth week shows the cardiac septa and foramina. *B,* A frontal section at a slightly later stage illustrates the directions of blood flow through the heart and the expansion of the ventricles.

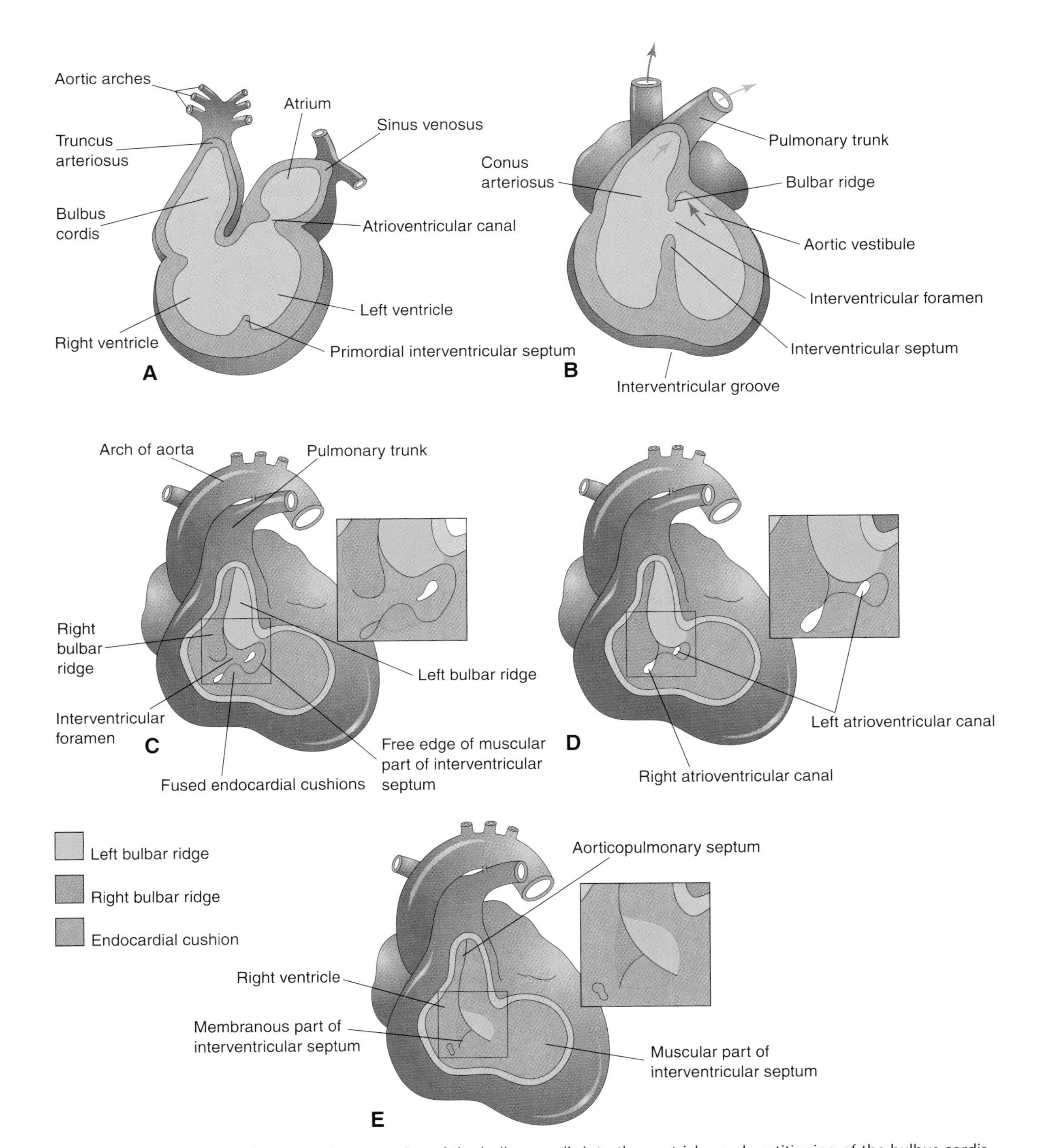

Figure 15 – 15. Sketches illustrating incorporation of the bulbus cordis into the ventricles and partitioning of the bulbus cordis and truncus arteriosus into the aorta and pulmonary trunk. *A*, A sagittal section at 5 weeks shows the bulbus cordis as one of the chambers of the primordial heart. *B*, Schematic coronal section at 6 weeks, after the bulbus cordis has been incorporated into the ventricles to become the conus arteriosus (infundibulum) of the right ventricle and the aortic vestibule of the left ventricle. *C* to *E*, Schematic drawings illustrating closure of the interventricular (IV) foramen and formation of the membranous part of the IV septum. The walls of the truncus arteriosus, bulbus cordis, and right ventricle have been removed. *C*, Five weeks, showing the bulbar ridges and fused endocardial cushions. *D*, Six weeks, showing how proliferation of subendocardial tissue diminishes the IV foramen. *E*, Seven weeks, showing the fused bulbar ridges, the membranous part of the IV septum formed by extensions of tissue from the right side of the endocardial cushions, and closure of the IV foramen.

cells migrate through the primordial pharynx and pharyngeal arches to reach the ridges. *Bone morphogenetic protein (BMP) and other signaling systems, such as Wnt and fibroblast growth factor (FGF), have been implicated in the induction and migration of neural crest cells.* As this occurs, the bulbar and truncal ridges undergo an 180-degree spiraling. The spiral orientation of the bulbar and truncal ridges, possibly caused by the streaming of blood from the ventricles, results in the formation of a spiral **aorticopulmonary septum** when the ridges fuse (see

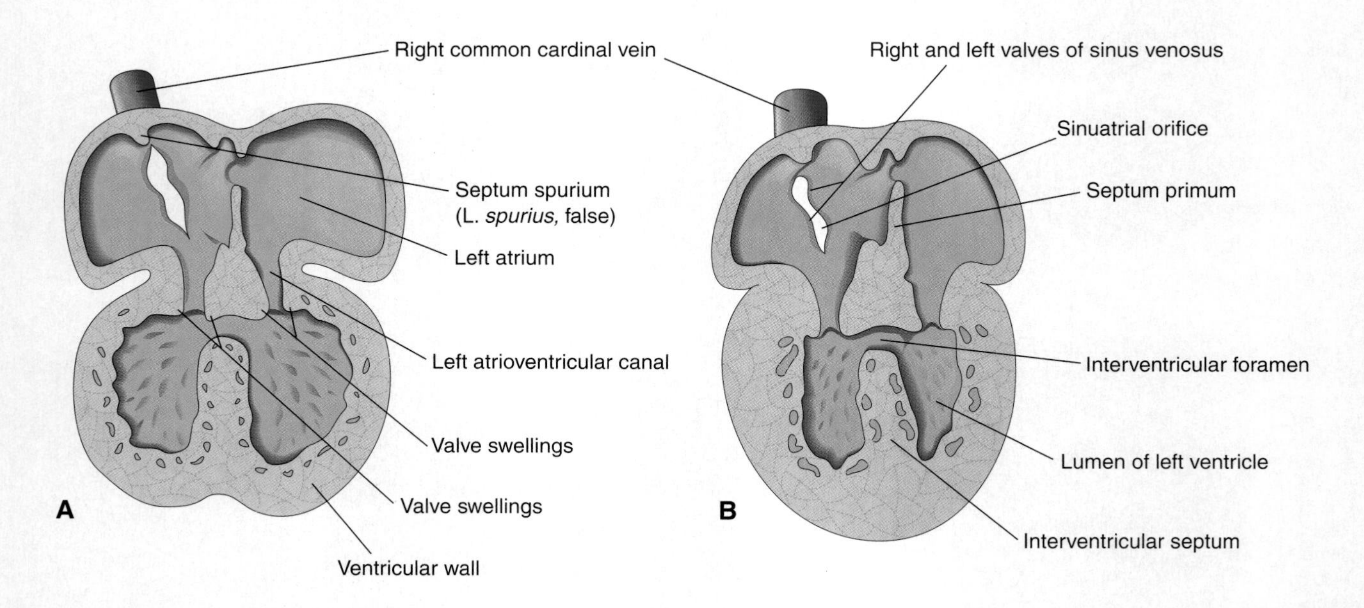

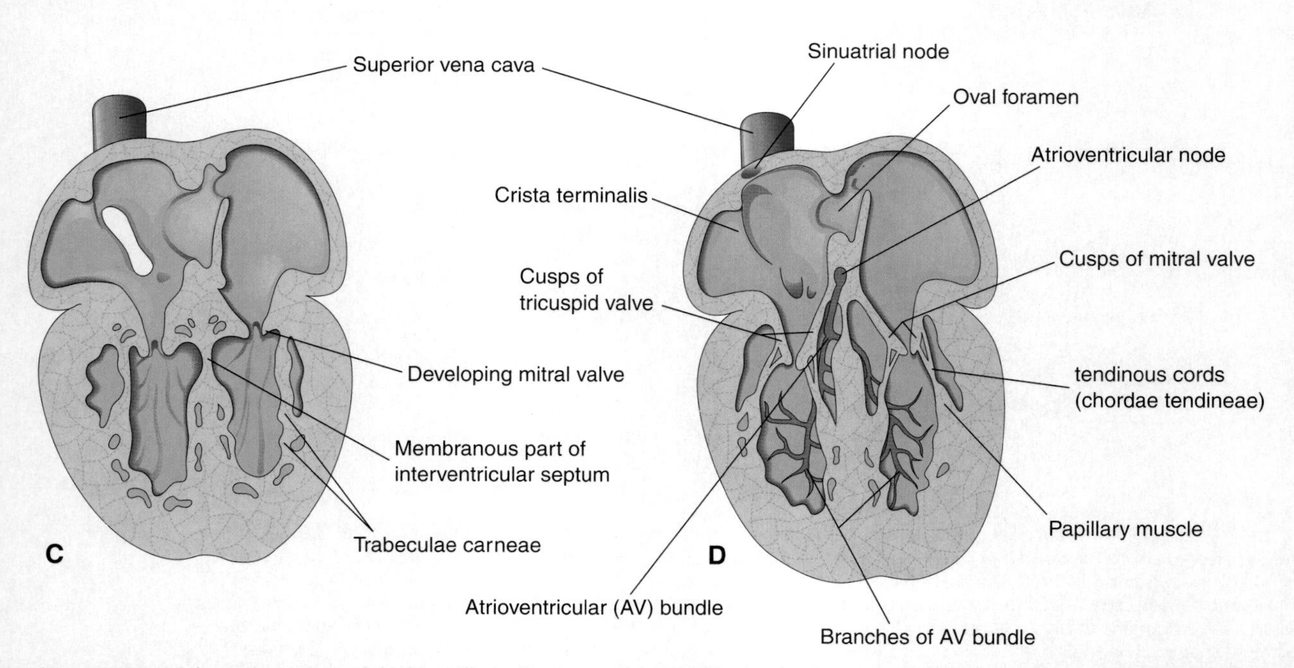

Figure 15 - 16. Schematic sections of the heart illustrating successive stages in the development of the atrioventricular (AV) valves, tendinous cords (L. chordae tendineae), and papillary muscles. *A*, Five weeks. *B*, Six weeks. *C*, Seven weeks. *D*, Twenty weeks, showing the conducting system of the heart.

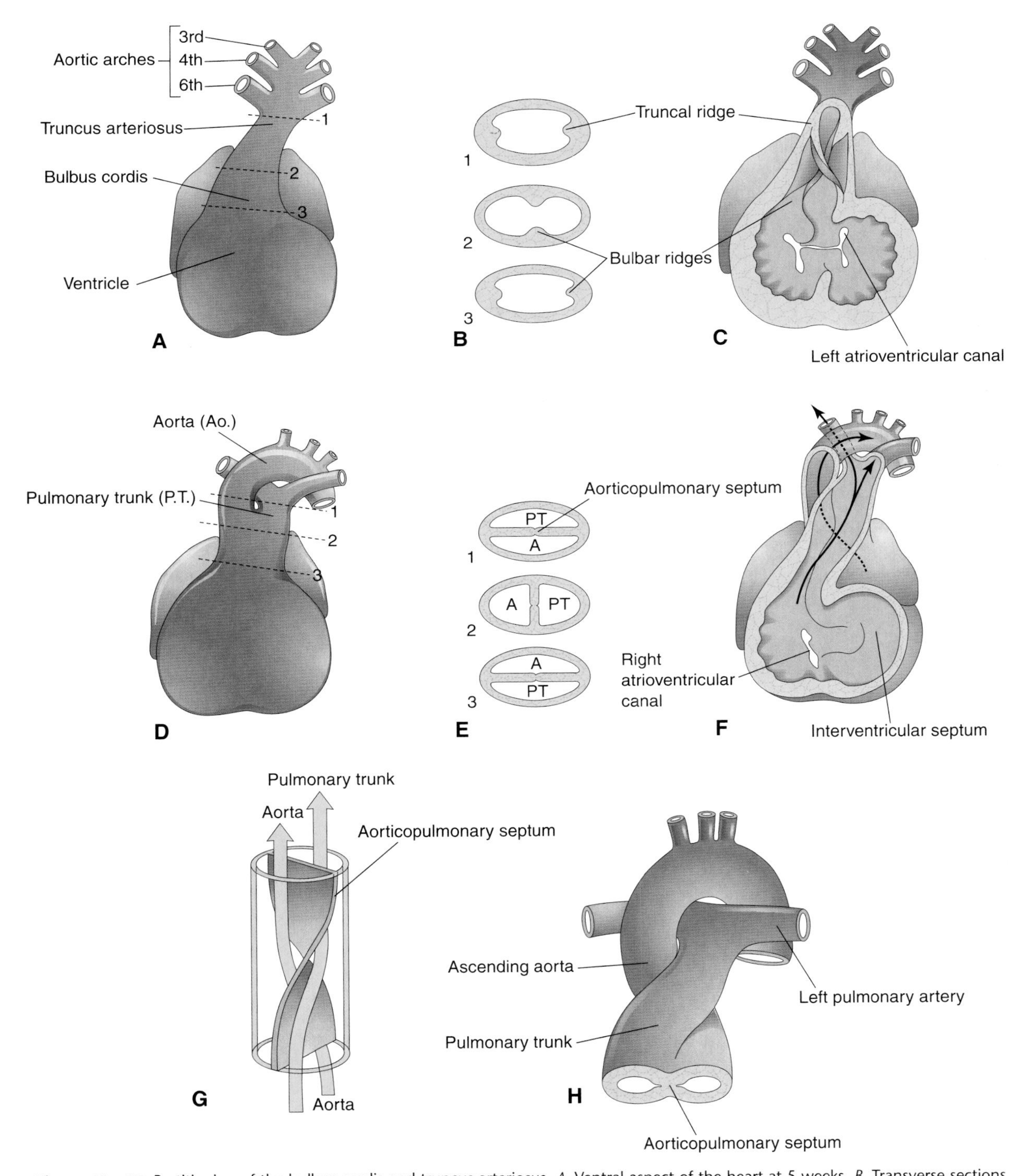

Figure 15 – 17. Partitioning of the bulbus cordis and truncus arteriosus. *A*, Ventral aspect of the heart at 5 weeks. *B*, Transverse sections of the truncus arteriosus and bulbus cordis, illustrating the truncal and bulbar ridges. *C*, The ventral wall of the heart and truncus arteriosus has been removed to demonstrate these ridges. *D*, Ventral aspect of the heart after partitioning of the truncus arteriosus. *E*, Sections through the newly formed aorta (A) and pulmonary trunk (PT) show the aorticopulmonary septum. *F*, Six weeks. The ventral wall of the heart and pulmonary trunk has been removed to show the aorticopulmonary septum. *G*, Diagram illustrating the spiral form of the aorticopulmonary septum. *H*, Drawing showing the great arteries twisting around each other as they leave the heart.

Fig. 15 – 17*D* to *G*). This septum divides the bulbus cordis and truncus arteriosus into two arterial channels, the **aorta** and **pulmonary trunk**. Because of the spiraling of the aorticopulmonary septum, the pulmonary trunk twists around the ascending aorta (see Fig. 15 – 17*H*).

The **bulbus cordis** is incorporated into the walls of the definitive ventricles in several ways (see Fig. 15 – 15*A* and *B*):

- In the right ventricle, the bulbus cordis is represented by the **conus arteriosus** (infundibulum), which gives origin to the pulmonary trunk.
- In the left ventricle, the bulbus cordis forms the walls of the **aortic vestibule**, the part of the ventricular cavity just inferior to the aortic valve.

Development of the Cardiac Valves

The **semilunar valves** develop from three swellings of subendocardial tissue around the orifices of the aorta and pulmonary trunk (Fig. 15 – 18*B* to *F*). These swellings are hollowed out and reshaped to form three, thin-walled cusps. The **AV valves** (tricuspid and mitral valves) develop similarly from localized proliferations of tissue around the AV canals.

Conducting System of the Heart

Initially, the muscle layers of the atrium and ventricle are continuous. The primordial atrium acts as the interim pacemaker of the heart, but the sinus venosus soon takes over this function. The **sinuatrial (SA) node** develops during the fifth week. The SA node is located in the right atrium, near the entrance of the SVC (see Fig. 15 – 16D). After incorporation of the sinus venosus, cells from its left wall are found in the base of the interatrial septum near the opening of the coronary sinus. Together with cells from the AV region, they form the **AV node and bundle**, which are located just superior to the endocardial cushions. The fibers arising from the **AV bundle** pass from the atrium into the ventricle and split into right and left **bundle branches**, which are distributed throughout the ventricular myocardium (see Fig. 15 – 16*D*). The SA node, AV node, and AV bundle are richly supplied by nerves; however, the primordial conducting system is developed before these nerves enter the heart.

Abnormalities of the Conducting System

Abnormalities of the conducting tissue may cause unexpected death during infancy, as in "crib death," or **sudden infant death syndrome (SIDS)**. There remains a lack of consensus that a single mechanism is responsible for the sudden and unexpected deaths of apparently healthy infants. A **brain stem developmental abnormality** or maturational delay related to neuroregulation of cardiorespiratory control appears to be the most compelling hypothesis relating to etiology.

Anomalies of the Heart and Great Vessels

Congenital heart defects (CHDs) are common, occurring with a frequency of 6 to 8 cases per 1000 births. Some cases of CHD are caused by single-gene or chromosomal mechanisms; others result from exposure to teratogens, such as the *rubella virus.* Most CHDs are thought to be caused by multiple factors, both genetic and environmental (i.e., **multifactorial inheritance**). Recent technology, such as real-time two-dimensional echocardiography, has permitted detection of fetal CHDs as early as the 17th or 18th weeks of gestation.

Dextrocardia

If the heart tube bends to the left instead of to the right (Fig. 15 – 19), the heart is displaced to the right and there is transposition whereby the heart and its vessels are reversed, left to right, as in a mirror image. **Dextrocardia** is the most frequent positional abnormality of the heart. In **dextrocardia with situs inversus** (transposition of viscera, such as the liver), the incidence of accompanying cardiac defects is low. In **isolated dextrocardia**, the abnormal position of the heart is not accompanied by displacement of other viscera. The TGF-β factor **Nodal** is involved in looping of the heart tube. Its role in dextrocardia is uncertain.

Ectopia Cordis

In ectopia cordis, an extremely rare condition, the heart is in an abnormal location outside the thoracic cavity. The most common thoracic form of ectopia cordis results from faulty development of the sternum and pericardium secondary to incomplete fusion of the lateral folds in the formation of the thoracic wall during the fourth week of development. Death occurs in most cases during the first few days after birth, usually from infection, cardiac failure, or hypoxemia. If no severe cardiac defects are present, surgical therapy usually consists of covering the heart with skin.

Atrial Septal Defects

Atrial septal defects (**ASDs**) are common congenital heart anomalies, occurring more frequently in females than in males. The most common form of ASD is a **patent oval foramen** (Figs. 15 – 20*A* and 15 – 21*A* to *D*). A small, isolated, patent oval foramen is of no hemodynamic significance. However, if other defects are present (e.g., pulmonary stenosis or atresia), blood is shunted through the oval foramen into the left atrium, producing **cyanosis**. This condition is characterized by a dark bluish or purplish coloration of the skin and mucous membranes secondary to deficient oxygenation of the blood. A **probe-patent oval foramen** is present in up to 25% of people (see Fig. 15 – 20*A* and *B*). With this abnormality, a probe can be passed from one atrium to the other through

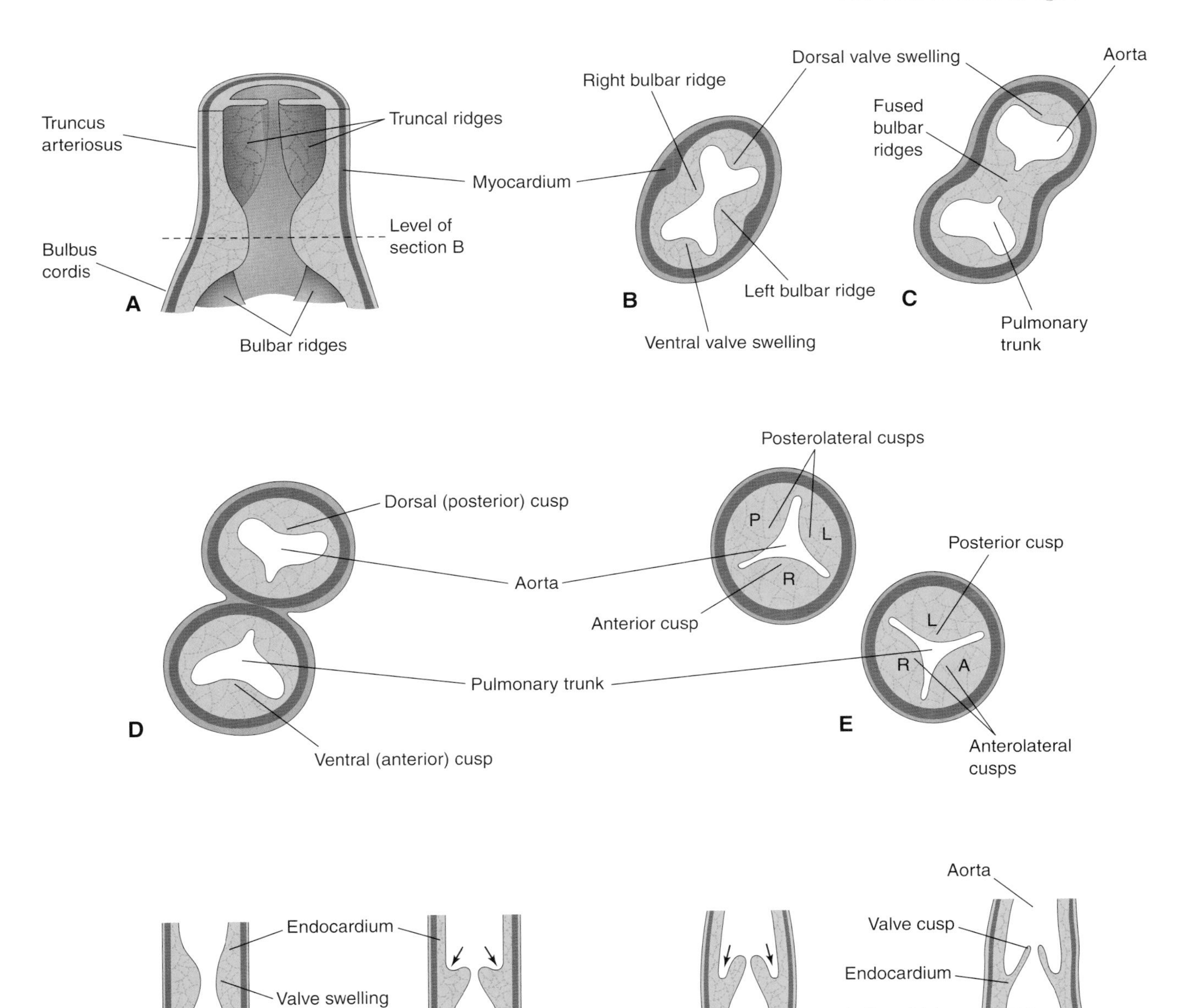

Figure 15 - 18. Development of the semilunar valves of the aorta and pulmonary trunk. *A*, Sketch of a section of the truncus arteriosus and bulbus cordis, showing the valve swellings. *B*, Transverse section of the bulbus cordis. *C*, Similar section after fusion of the bulbar ridges. *D*, Formation of the walls and valves of the aorta and pulmonary trunk. *E*, Rotation of the vessels has established the adult positions of the valves in relation to each other. *P*, posterior; *L*, left; and *R*, right. *F*, Longitudinal sections of the aorticoventricular junction, illustrating successive stages in the hollowing (*arrows*) and thinning of the valve swellings to form the valve cusps.

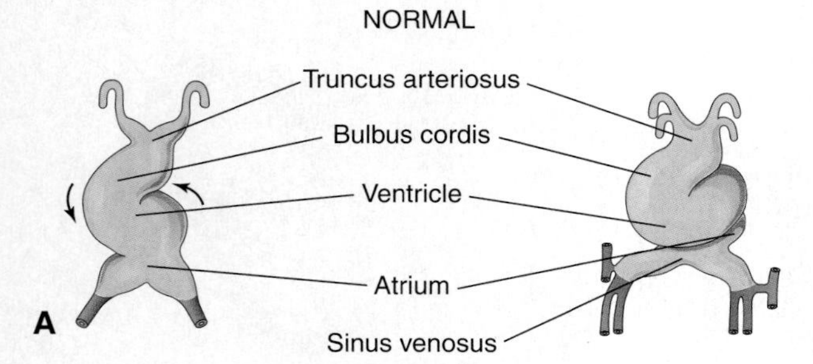

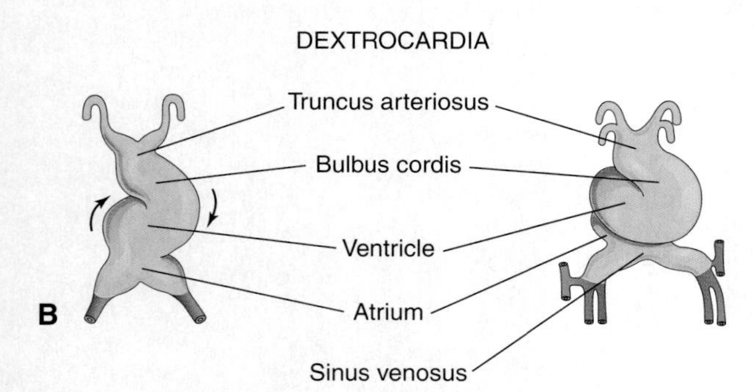

Figure 15 – 19. The primordial heart during the fourth week. *A*, Normal bending to the right. *B*, Abnormal bending to the left.

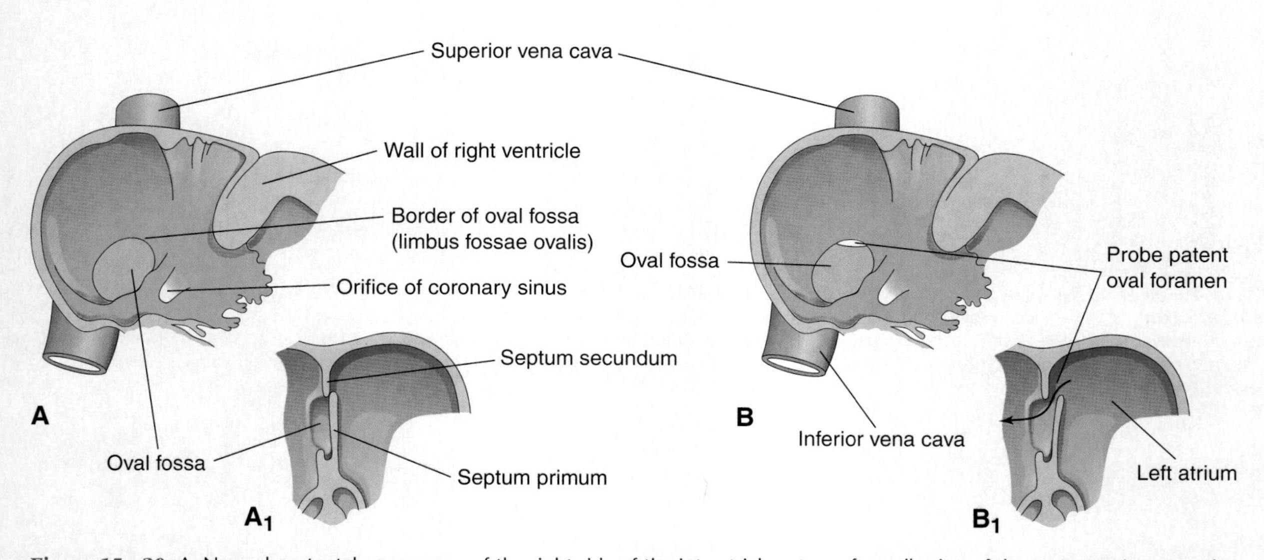

Figure 15 – 20. A, Normal postnatal appearance of the right side of the interatrial septum after adhesion of the septum primum to the septum secundum. A_1, Section of the interatrial septum illustrating formation of the oval fossa in the interatrial septum. Note that the floor of this fossa is formed by the septum primum. *B* and B_1, Similar views of a probe-patent oval foramen resulting from incomplete adhesion of the septum primum to the septum secundum.

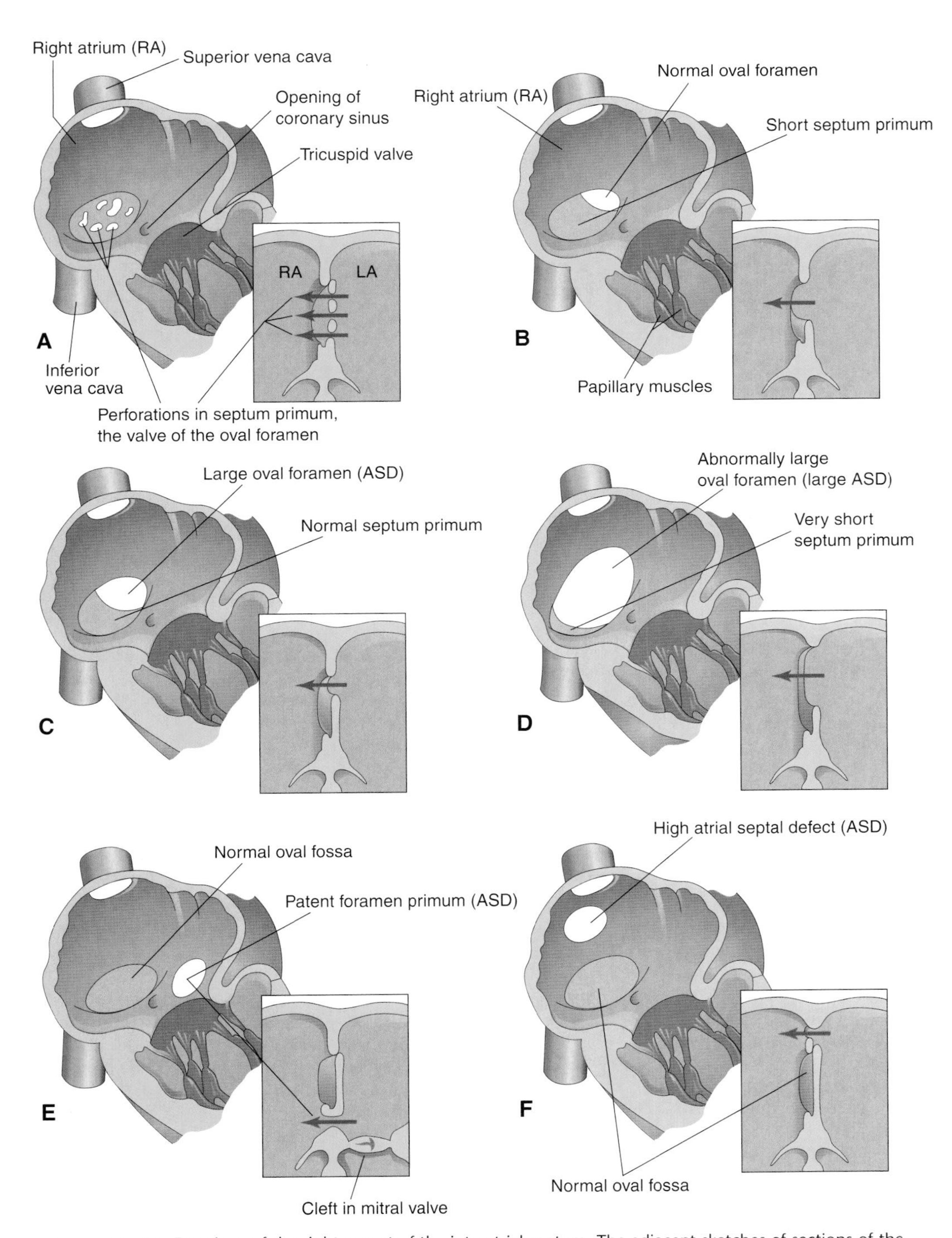

Figure 15 - 21. Drawings of the right aspect of the interatrial septum. The adjacent sketches of sections of the septa illustrate various types of atrial septal defect (ASD). *A,* Patent oval foramen resulting from resorption of the septum primum in abnormal locations. *B,* Patent oval foramen caused by excessive resorption of the septum primum ("short flap defect"). *C,* Patent oval foramen resulting from an abnormally large oval foramen. *D,* Patent oval foramen resulting from an abnormally large oval foramen and excessive resorption of the septum primum. *E,* Endocardial cushion defect with primum-type ASD. The adjacent section shows the cleft in the anterior cusp of the mitral valve. *F,* Sinus venosus ASD. The high septal defect resulted from abnormal absorption of the sinus venosus into the right atrium. In *E* and *F,* note that the oval fossa has formed normally.

the superior part of the floor of the oval fossa. This defect, which is usually small, is not clinically significant. A probe-patent oval foramen results from incomplete adhesion between the original flap of the valve of the oval foramen and the septum secundum after birth.

There are four clinically significant types of ASD (see Fig. 15 – 21), of which the first two types are relatively common:

- ostium secundum defect
- endocardial cushion defect with a foramen primum (ostium primum) defect
- sinus venosus defect
- common atrium

Ostium secundum ASDs (see Fig. 15 – 21*A* to *D*) occur in the area of the oval fossa and include defects of both the septum primum and the septum secundum. Females with ASDs outnumber males 3 to 1. Ostium secundum ASDs are one of the most common types of CHD. The patent oval foramen usually results from abnormal resorption of the septum primum during the formation of the foramen secundum. If resorption occurs in abnormal locations, the septum primum is fenestrated or netlike (see Fig. 15 – 21*A*). If excessive resorption of the septum primum occurs, the resulting short septum primum does not close the oval foramen (see Fig. 15 – 21*B*). If an abnormally large oval foramen develops as a result of defective development of the septum secundum, a normal septum primum does not close the abnormal oval foramen at birth (see Fig. 15 – 21*C*). Large ostium secundum ASDs may occur because of a combination of excessive resorption of the septum primum and a large oval foramen (see Fig. 15 – 21*D*). Closure of the ASD is accomplished by open heart surgery.

Endocardial cushion defects with a patent foramen primum (see Fig. 15 – 21*E*) are less common forms of ASD. Several cardiac abnormalities are grouped together under this heading because they result from the same developmental defect, a deficiency of the endocardial cushions and the AV septum. The septum primum does not fuse with the endocardial cushions, resulting in a **patent foramen primum**. Usually there is also a cleft in the anterior cusp of the mitral valve.

Sinus venosus ASDs are located in the superior part of the interatrial septum, close to the entry of the SVC (see Fig. 15 – 21*F*). These defects result from incomplete absorption of the sinus venosus into the right atrium or abnormal development of the septum secundum, or both.

A **common atrium** is a rare cardiac defect in which the AV septum is absent.

Ventricular Septal Defects

Ventricular septal defects (VSDs) are the most common type of CHD, accounting for about 25% of such defects. VSDs occur more frequently in males than in females. Most VSDs involve the membranous part of the IV septum (Fig. 15 – 22*B*). Many small VSDs close spontaneously, most frequently during the first year of life. Most patients with a large VSD have a massive left-to-right shunt of blood. **Muscular VSD** is a less common type of defect that may appear anywhere in the muscular part of the IV septum. **Transposition of the great arteries** (Fig. 15 – 23) and a rudimentary outlet chamber are present in most infants with this severe CHD.

Truncus Arteriosus

Truncus arteriosus (TA) or **persistent TA** results from failure of the truncal ridges and aorticopulmonary septum to develop normally and to divide the truncus arteriosus into the aorta and pulmonary trunk (see Fig. 15 – 22). The most common type of persistent TA is a single arterial vessel that branches to form the pulmonary trunk and ascending aorta (see Fig. 15 – 22*A* and *B*). In this anomaly, a single arterial trunk, the TA, arises from the heart and supplies the systemic, pulmonary, and coronary circulations. A VSD is always present with the TA anomaly, and the TA overrides the VSD (see Fig. 15 – 22*B*). The cause of this condition is unknown.

Transposition of the Great Arteries

Transposition of the great arteries (TGA) is the most common cause of **cyanotic heart disease** in newborn infants (see Fig. 15 – 23). In typical cases, the aorta lies anterior and to the right of the pulmonary trunk and arises anteriorly from the morphologic right ventricle, whereas the pulmonary trunk arises from the morphologic left ventricle. There is also an *ASD* with or without an associated *patent ductus arteriosus* (PDA) and VSD. This defect is thought to result from failure of the conus arteriosus to develop normally during incorporation of the bulbus cordis into the ventricles. Defective neural crest cell migration may also be involved.

Unequal Division of the Truncus Arteriosus

Unequal division of the TA (Figs. 15 – 24*A* and *B;* see also Fig. 15 – 21) results when partitioning of the TA superior to the valves is disparate, producing one large great artery and the other a small one. As a result, the aorticopulmonary septum is not aligned with the IV septum, and a VSD results. The larger vessel (aorta or pulmonary trunk) usually straddles (overrides) the VSD (see Fig. 15 – 24*A* and *B*). In **pulmonary valve stenosis**, the cusps of the pulmonary valve are fused together to form a dome with a narrow central opening (Fig. 15 – 25*D*). In **infundibular stenosis**, the conus arteriosus of the right ventricle is underdeveloped. The two types of pulmonary stenosis may be concurrent. Depending upon the degree of obstruction to blood flow, there is a variable degree of hypertrophy of the right ventricle (see Fig. 15 – 24*B*).

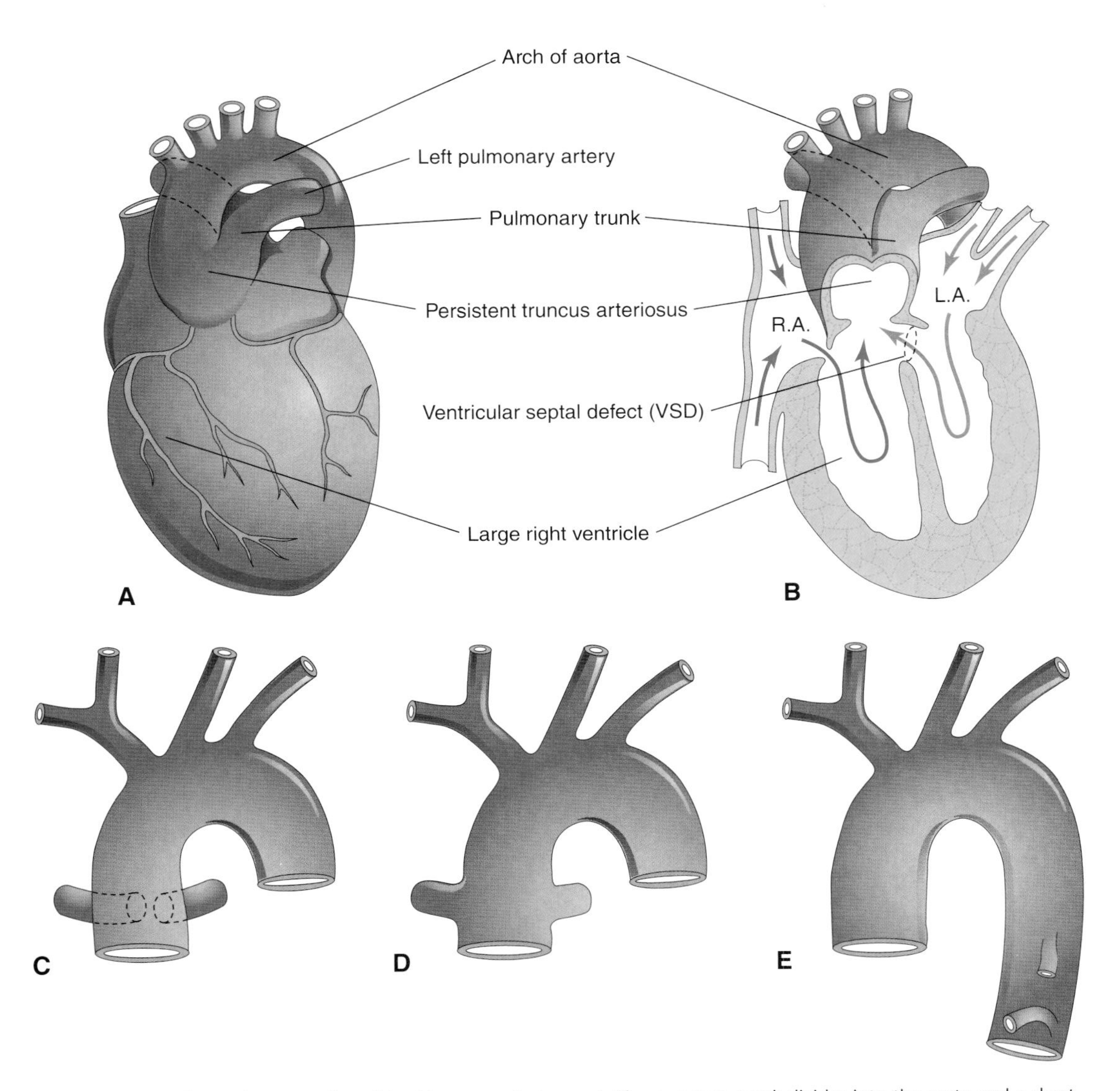

Figure 15 – 22. The main types of persistent truncus arteriosus. *A,* The common trunk divides into the aorta and a short pulmonary trunk. *B,* Frontal section of the heart shown in *A.* Observe the circulation in this heart (*arrows*) and the VSD. R.A., right aorta; L.A., left aorta. *C,* The right and left pulmonary arteries arise close together from the truncus arteriosus. *D,* The pulmonary arteries arise independently from the sides of the truncus arteriosus. *E,* No pulmonary arteries are present.

Tetralogy of Fallot

The classic group of four cardiac defects known as tetralogy of Fallot (see Fig. 15 – 24*A* and *B*) consists of:

- pulmonary stenosis (obstructed right ventricular outflow)
- ventricular septal defect (VSD)
- dextroposition of aorta (overriding aorta)
- right ventricular hypertrophy

The pulmonary trunk is usually small, and there may be varying degrees of pulmonary artery stenosis as well.

Aortic Stenosis and Atresia

In **aortic valve stenosis**, the edges of the valve are usually fused to form a dome with a narrow opening (see Fig. 15 – 25). This anomaly may be present at birth (congenital), or it may develop after birth (acquired form). The valvular stenosis causes extra work for the heart and results in hypertrophy of the left ventricle and abnormal heart sounds **(heart murmurs)**. In *subaortic stenosis*, there is often a band of fibrous tissue just inferior to the aortic valve. The narrowing of the aorta results from persistence of tissue that normally degenerates as the valve forms. Aortic atresia is present when obstruction of the aorta or its valve is complete.

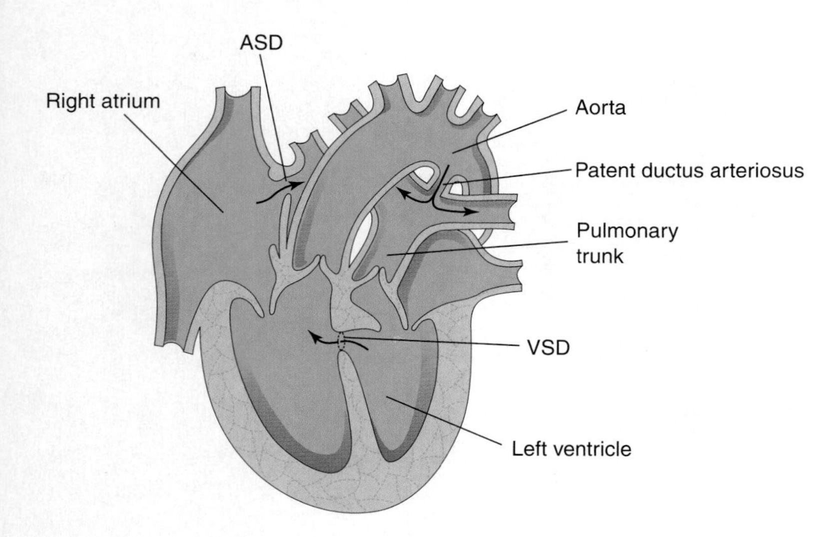

Figure 15 – 23. Frontal section of a malformed heart illustrating transposition of the great arteries. The ventricular septal defect (VSD) and the atrial septal defect (ASD) allow mixing of the arterial and venous blood. Transposition of the great arteries is the most common single cause of cyanotic heart disease in newborn infants. As is shown here, it is often associated with other cardiac anomalies (VSD and ASD).

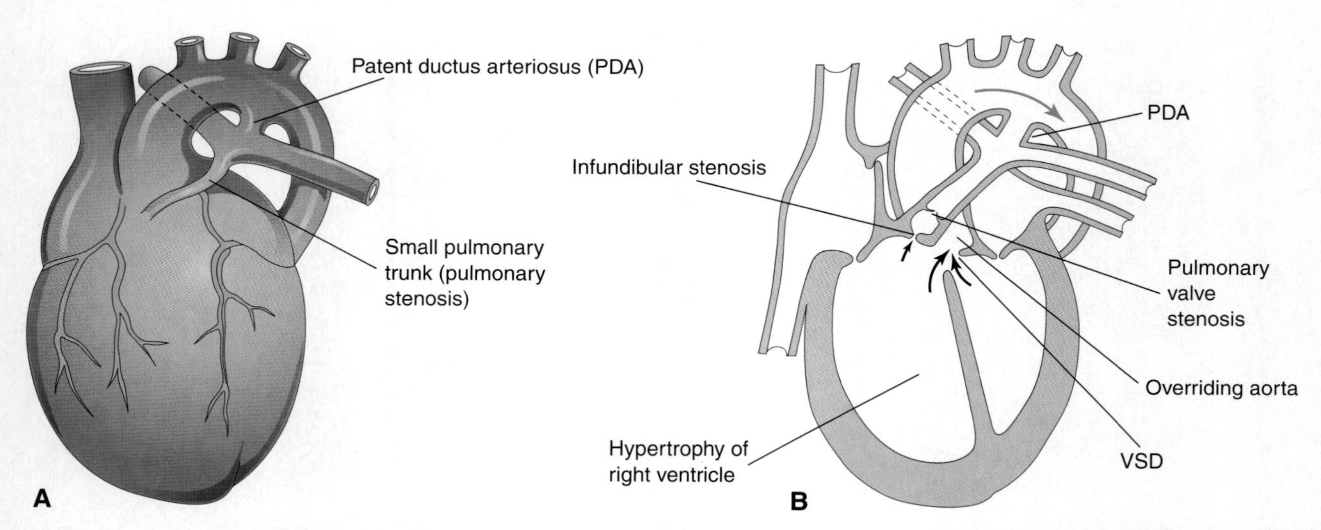

Figure 15 – 24. *A,* Drawing of an infant's heart showing a small pulmonary trunk (pulmonary stenosis) and a large aorta resulting from unequal partitioning of the truncus arteriosus. There is also hypertrophy of the right ventricle and a patent ductus arteriosus (PDA). *B,* Frontal section of a heart illustrating tetralogy of Fallot. Observe the four cardiac defects: pulmonary valve stenosis, ventricular septal defect, overriding aorta, and hypertrophy of the right ventricle. In this case, infundibular stenosis is also shown.

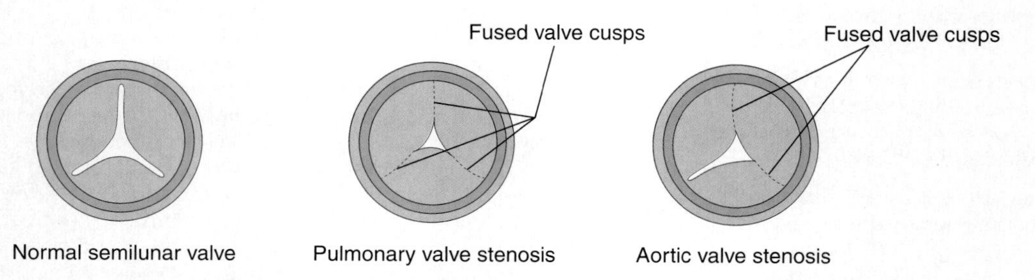

Figure 15 – 25. Sketches illustrating a normal semilunar valve and stenotic pulmonary and aortic valves.

Aortic Arch Derivatives

As the pharyngeal arches develop during the fourth week (Fig. 15-26*A*), they are supplied by arteries — the **aortic arches** — from the *aortic sac*, the homolog of the ventral aorta in other mammals (see Fig. 15-26*B*). The aortic arches terminate in the dorsal aorta on the ipsilateral side. Although six pairs of aortic arches usually develop, they are not all present at the same time. By the time the sixth pair of aortic arches has formed, the first two pairs have disappeared (see Fig. 15-26*C*).

Derivatives of the First Pair of Aortic Arches

The first pair of aortic arches (arteries) largely disappears, but remnants of these vessels form parts of the **maxillary arteries**, which supply the ears, teeth, and muscles of the eyes and face. These arteries may contribute to parts of the **external carotid arteries**.

Derivatives of the Second Pair of Aortic Arches

Dorsal parts of the second pair of aortic arches persist, and form the stems of the **stapedial arteries.** These are small vessels that run through the ring of the stapes, a small ear bone.

Derivatives of the Third Pair of Aortic Arches

Proximal parts of the third pair of aortic arches form the **common carotid arteries,** which supply structures in the head. Distal parts of the third pair of aortic arches

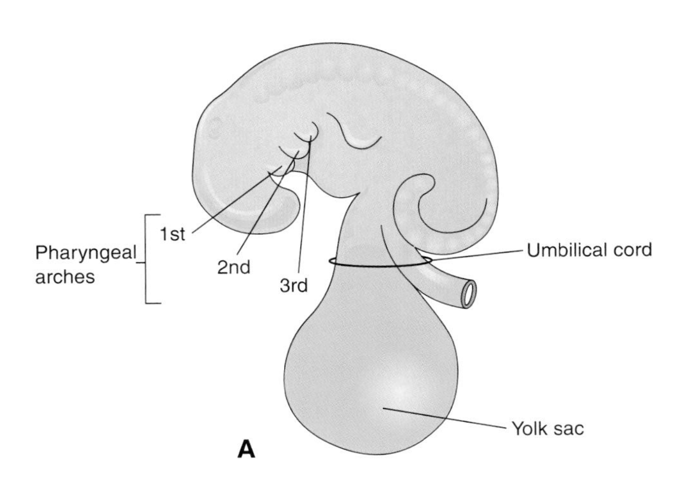

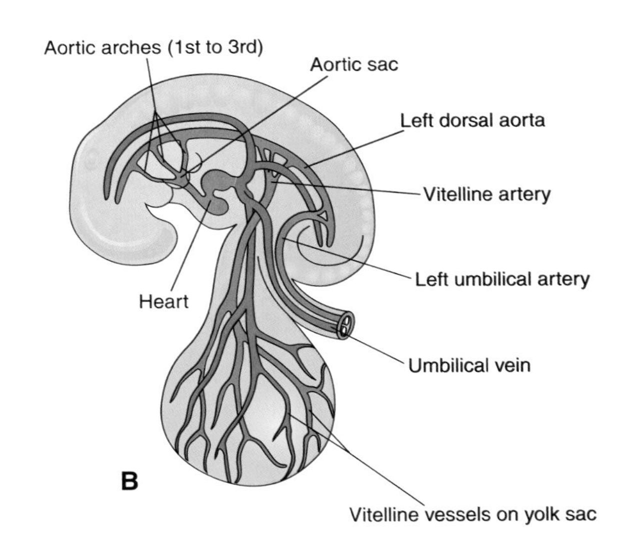

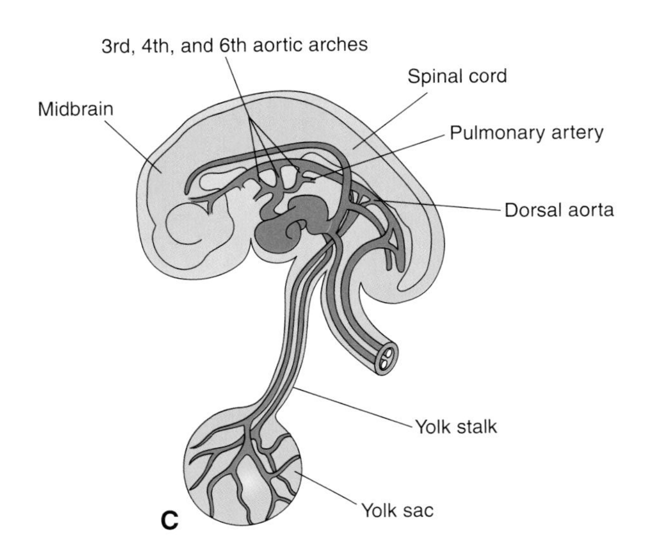

Figure 15-26. The pharyngeal and aortic arches. *A,* Left side of an embryo (at about 26 days). *B,* Schematic drawing of this embryo showing the left aortic arches arising from the aortic sac, running through the pharyngeal arches, and terminating in the left dorsal aorta. *C,* An embryo (at about 37 days). Note the single dorsal aorta and the mostly degenerated first two pairs of aortic arches.

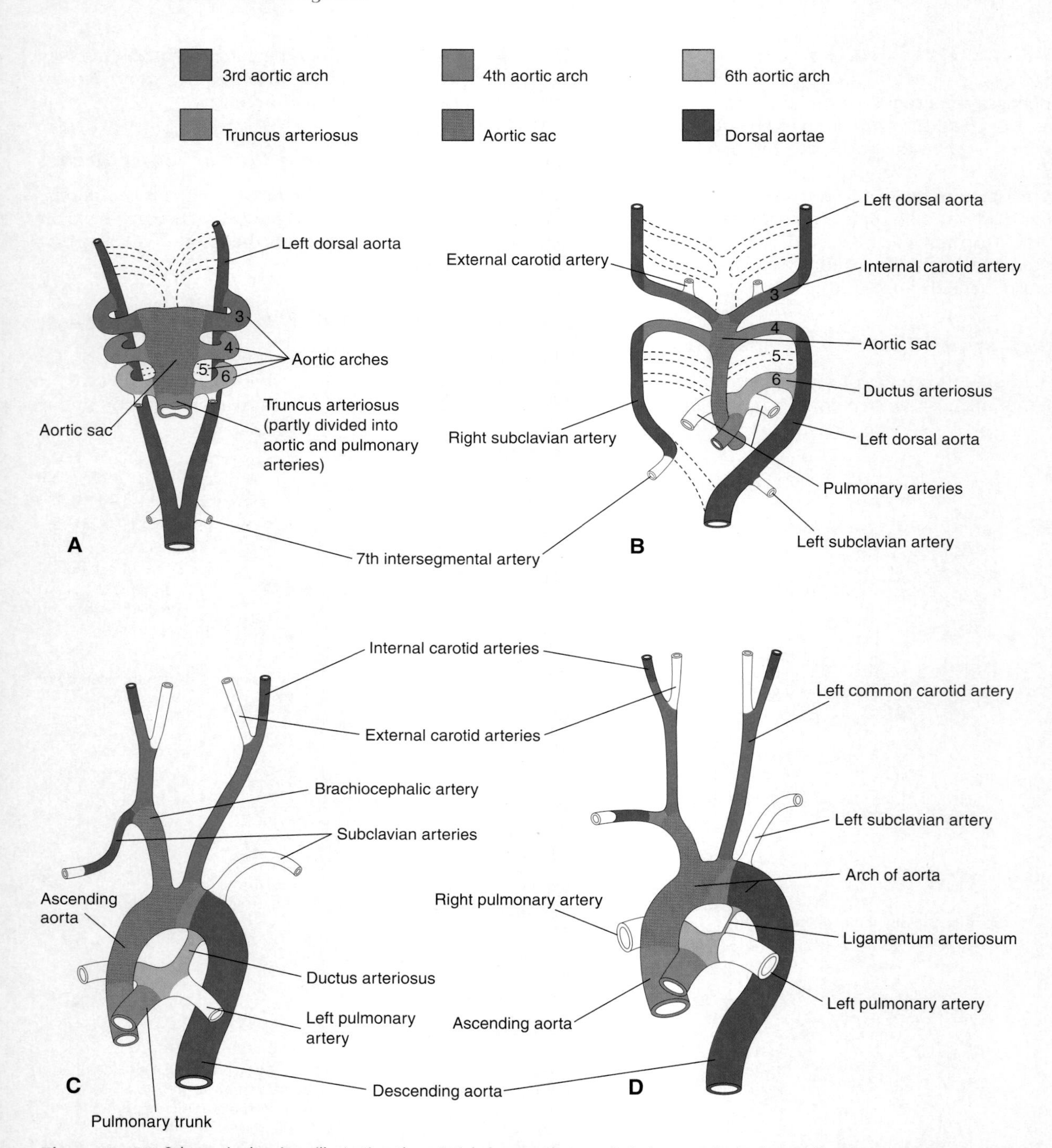

Figure 15-27. Schematic drawings illustrating the arterial changes that result during transformation of the truncus arteriosus, aortic sac, aortic arches, and dorsal aortae into the adult arterial pattern. The vessels that are not colored are not derived from these structures. *A,* Aortic arches at 6 weeks. By this stage, the first two pairs of aortic arches have largely disappeared. *B,* Aortic arches at 7 weeks. The parts of the dorsal aortae and aortic arches that normally disappear are indicated with broken lines. *C,* Arterial arrangement at 8 weeks. *D,* Sketch of the arterial vessels of a 6-month-old infant. Note that the ascending aorta and pulmonary arteries are considerably smaller in *C* than in *D.* This represents the relative flow through these vessels at the different stages of development. Observe the large size of the ductus arteriosus (DA) in *C;* it is essentially a direct continuation of the pulmonary trunk. The DA normally becomes functionally closed within the first few days after birth. Eventually, the DA becomes the ligamentum arteriosum, as shown in *D.*

join with the dorsal aortae to form the **internal carotid arteries,** which supply the ears, orbits, and brain and its meninges (protective membranes for the brain).

Derivatives of the Fourth Pair of Aortic Arches

The **left fourth aortic arch** forms part of the arch of the aorta (Fig. 15 – 27*C* and *D*). The proximal part of the arch develops from the aortic sac, whereas the distal part is derived from the left dorsal aorta. The **right fourth aortic arch** becomes the proximal part of the **right subclavian artery**. The distal part of the subclavian artery forms from the right dorsal aorta and right seventh intersegmental artery. The left subclavian artery is not derived from an aortic arch; it forms from the left seventh intersegmental artery (see Fig. 15 – 27*A*). As development proceeds, differential growth shifts the origin of the left subclavian artery cranially; consequently, it comes to lie close to the origin of the left common carotid artery (see Fig. 15 – 27*D*).

Fate of the Fifth Pair of Aortic Arches

In about 50% of embryos, the fifth pair of aortic arches comprises rudimentary vessels that soon degenerate, leaving no vascular derivatives. In other embryos, these arteries do not develop.

Derivatives of the Sixth Pair of Aortic Arches

The **left sixth aortic arch** develops as follows (see Fig. 15 – 27*B* and *C*):

- The proximal part of the arch persists as the proximal part of the **left pulmonary artery**.
- The distal part of the arch passes from the left pulmonary artery to the dorsal aorta to form a prenatal shunt, the **ductus arteriosus (DA)**.

The **right sixth aortic arch** develops as follows:

- The proximal part of the arch persists as the proximal part of the **right pulmonary artery**.
- The distal part of the arch degenerates.

The transformation of the sixth pair of aortic arches explains why the course of the **recurrent laryngeal nerves** differs on the two sides. These nerves supply the sixth pair of pharyngeal arches and hook around the sixth pair of aortic arches on their way to the developing larynx (Fig. 15 – 28*A*). **On the right**, because the distal part of the right sixth aortic arch degenerates, the right recurrent laryngeal nerve moves superiorly and hooks around the proximal part of the right subclavian artery, the derivative of the fourth aortic arch (see Fig. 15 – 28*B*). **On the left**, the left recurrent laryngeal nerve hooks around the DA formed by the distal part of the sixth aortic arch. When this arterial shunt involutes after birth, the nerve hooks around the **ligamentum arteriosum** (remnant of the DA) and the arch of the aorta (see Fig. 15 – 28*C*).

Aortic Arch Anomalies

Because of the many changes involved in transformation of the embryonic pharyngeal arch system of arteries into the adult arterial pattern, it is understandable why anomalies may occur. Most irregularities result from the persistence of parts of aortic arches that usually disappear, or from disappearance of parts that normally persist.

Coarctation of the Aorta

Aortic coarctation (constriction) occurs in about 10% of children and adults with CHDs. Coarctation is characterized by a constriction of varying length of the aorta (Fig. 15 – 29). Most constrictions occur distal to the origin of the left subclavian artery at the entrance of the DA **(juxtaductal coarctation)**. A classification system of preductal and postductal coarctations is commonly used; however, in most instances, the coarctation is directly opposite the DA. Coarctation of the aorta occurs twice as often in males as in females and is associated with a bicuspid aortic valve in 70% of cases. Coarctation of the aorta is caused by genetic or environmental factors, or both.

Double Aortic Arch

Double aortic arch is a rare anomaly that is characterized by a **vascular ring** around the trachea and esophagus (Fig. 15 – 30). The vascular ring results from failure of the distal part of the right dorsal aorta to disappear (see Fig. 15 – 30*A*); as a result, right and left arches form. Usually, the right arch of the aorta is the larger one and passes posterior to the trachea and esophagus (see Fig. 15 – 30*B*).

Right Arch of the Aorta

When the entire right dorsal aorta persists (Fig. 15 – 31*A*) and the distal part of the left dorsal aorta involutes, a right aortic arch results. There are two main types:

- *Right arch of the aorta without a retroesophageal component* (see Fig. 15 – 31*B*). The DA (or ligamentum arteriosum) passes from the right pulmonary artery to the right arch of the aorta.
- *Right arch of the aorta with a retroesophageal component* (see Fig. 15 – 31*C*). Originally, a small left arch of the aorta probably involuted, leaving the right arch of the aorta posterior to the esophagus. The DA (or ligamentum arteriosum) attaches to the distal part of the arch of the aorta and forms a ring, which may constrict the esophagus and trachea.

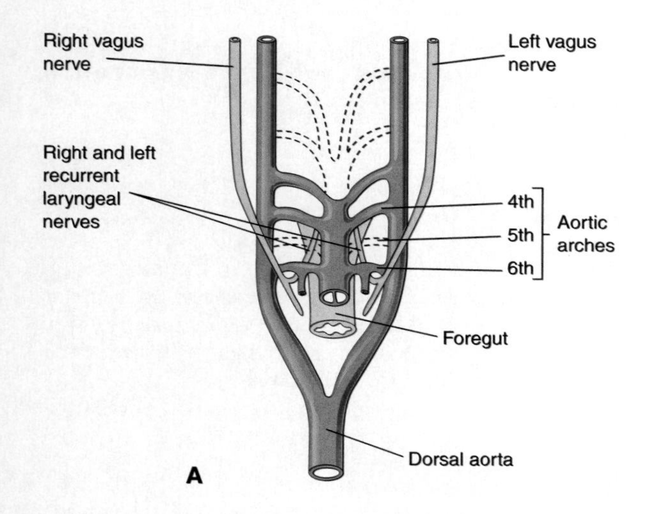

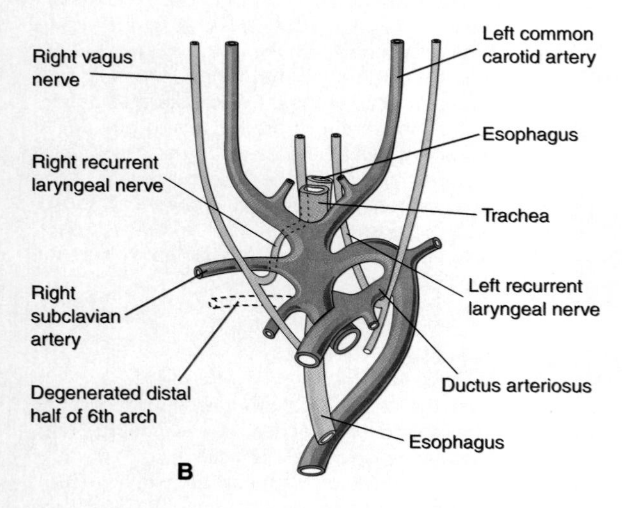

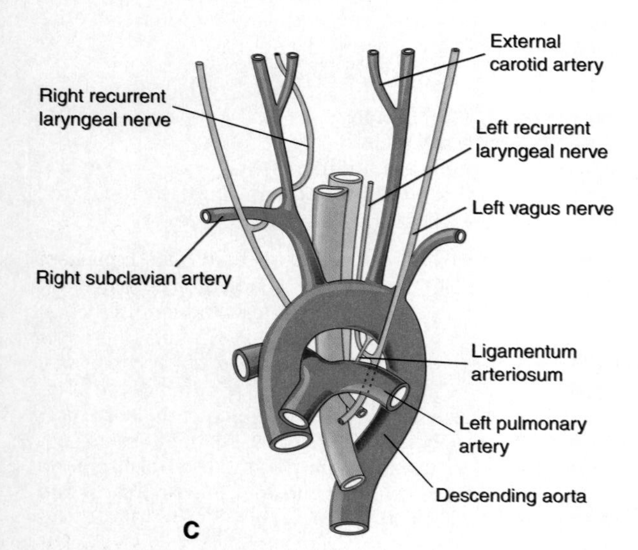

Figure 15–28. The relation of the recurrent laryngeal nerves to the aortic arches. *A,* Six weeks, showing that the recurrent laryngeal nerves are hooked around the sixth pair of aortic arches. *B,* Eight weeks, showing that the right recurrent laryngeal nerve is hooked around the right subclavian artery, and the left recurrent laryngeal nerve is hooked around the ductus arteriosus and arch of the aorta. *C,* In a child, the left recurrent laryngeal nerve is hooked around the ligamentum arteriosum and arch of the aorta.

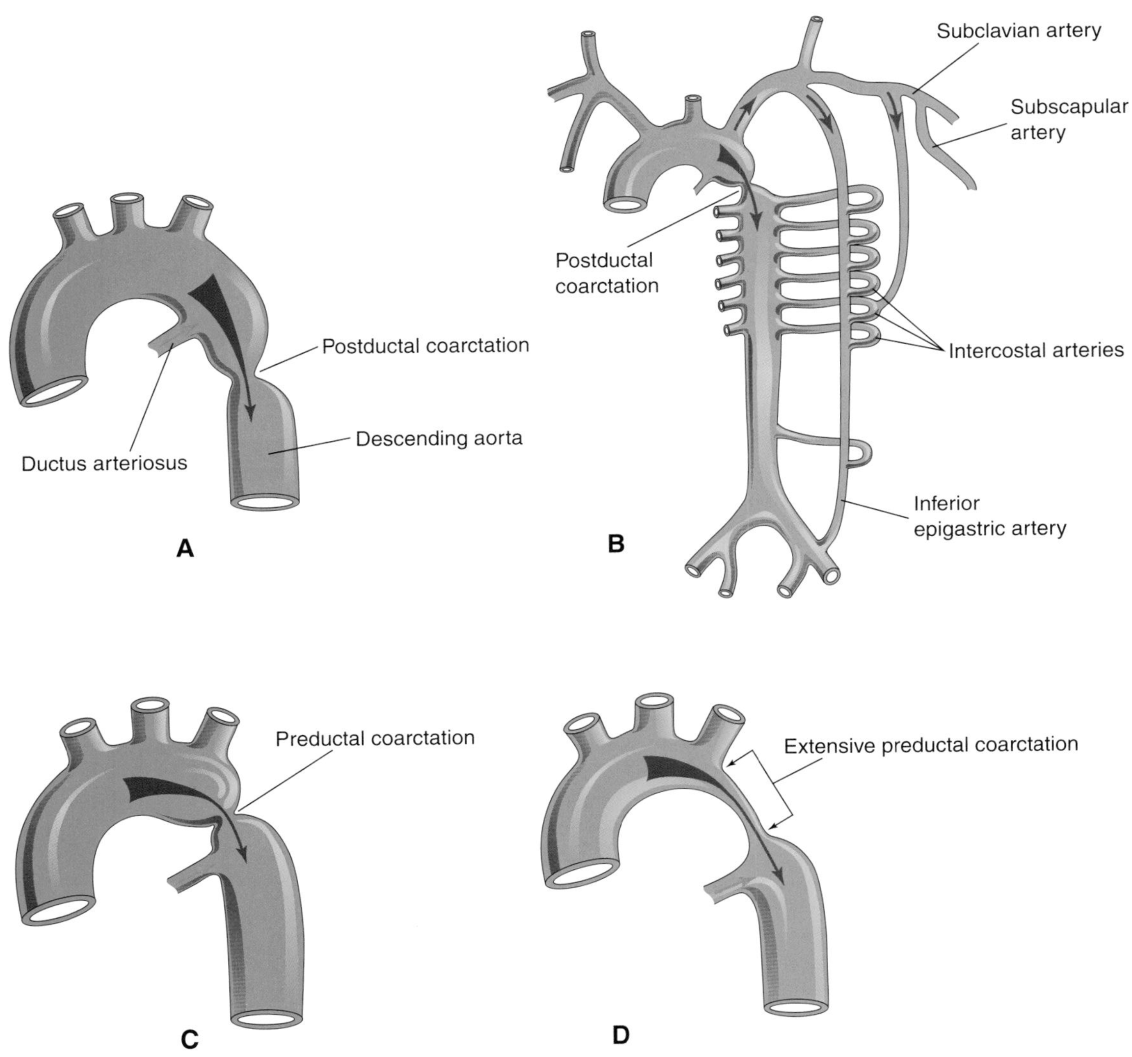

Figure 15-29. *A,* Postductal coarctation of the aorta. *B,* Diagrammatic representation of the common routes of collateral circulation that develop in association with postductal coarctation of the aorta. *C* and *D,* Preductal coarctation.

Anomalous Right Subclavian Artery

The right subclavian artery arises from the distal part of the arch of the aorta and passes posterior to the trachea and esophagus to supply the right upper limb (Fig. 15–32). A **retroesophageal right subclavian artery** occurs when the right fourth aortic arch and the right dorsal aorta disappear cranial to the seventh intersegmental artery. As a result, the right subclavian artery forms from the right seventh intersegmental artery and the distal part of the right dorsal aorta. As development proceeds, differential growth shifts the origin of the right subclavian artery cranially until it comes to lie close to the origin of the left subclavian artery.

Fetal and Neonatal Circulation

The fetal circulation (Fig. 15–33) is designed to serve prenatal needs and permit modifications at birth that establish the neonatal pattern (Fig. 15–34). Prenatally, the lungs do not provide gas exchange and the pulmonary vessels are vasoconstricted. Three shunts are essential in the transitional circulation: the ductus venosus, oval foramen, and ductus arteriorsus (DA).

Fetal Circulation

Highly oxygenated, nutrient-rich blood returns from the placenta in the **umbilical vein** (see Fig. 15–33). On

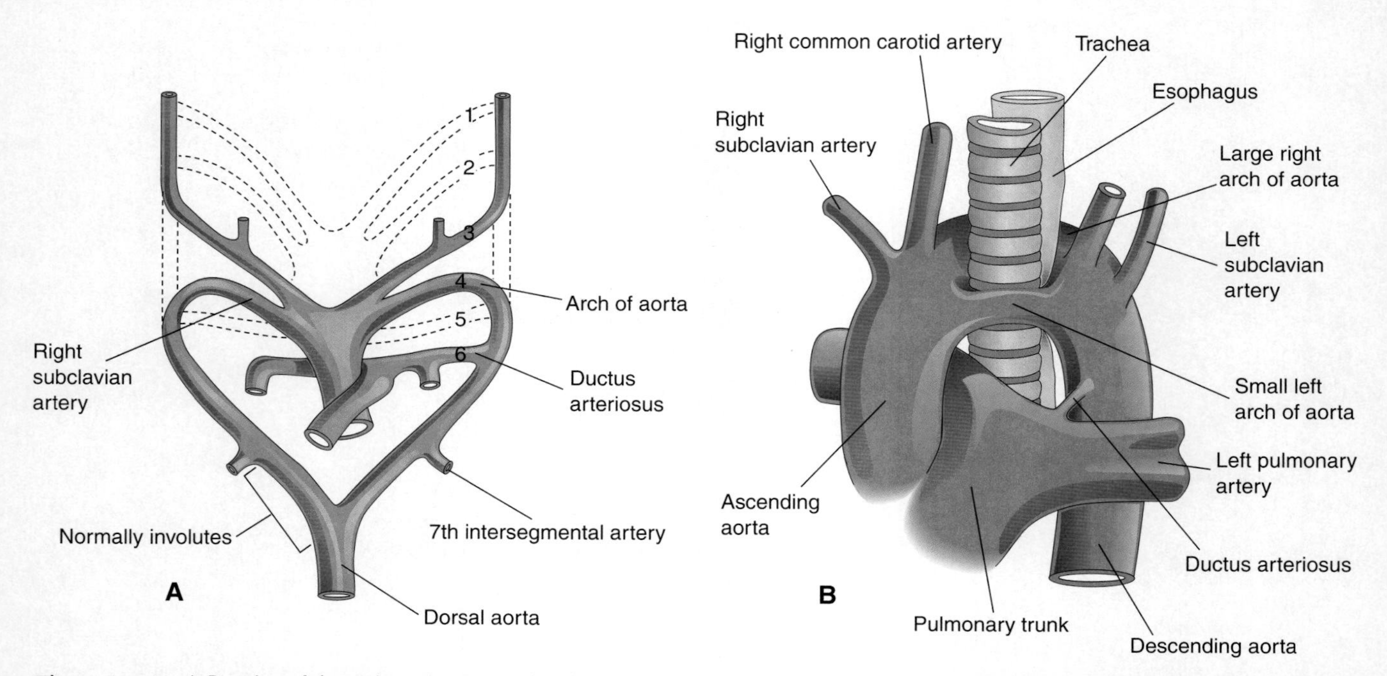

Figure 15 - 30. *A,* Drawing of the embryonic aortic arches illustrating the embryological basis of a double aortic arch. The distal portion of the right dorsal aorta persists and forms a right aortic arch. *B,* A large right aortic arch and a small left aortic arch arise from the ascending aorta and form a vascular ring around the trachea and esophagus. Note the compression of the esophagus and trachea. The right common carotid and subclavian arteries arise separately from the large right arch of the aorta.

approaching the liver, about half of the blood under high pressure passes directly into the **ductus venosus (DV)**, a fetal vessel connecting the umbilical vein to the IVC; consequently, the blood bypasses the liver. The other half of the blood in the umbilical vein flows into the *sinusoids of the liver* and enters the IVC through the **hepatic veins**. Blood flow through the DV is regulated by a *sphincter mechanism* close to the umbilical vein. After a short course in the IVC, the blood enters the right atrium of the heart. Most blood from the IVC is directed by the inferior border of the septum secundum, the **crista dividens**, through the **oval foramen** and into the left atrium. There, it mixes with the relatively small amount of poorly oxygenated blood returning from the lungs through the pulmonary veins. The fetal lungs extract oxygen from the blood instead of providing it. From the left atrium, the blood passes to the left ventricle and leaves through the ascending aorta.

The arteries to the heart, head, neck, and upper limbs receive well-oxygenated blood. The liver also receives well-oxygenated blood from the umbilical vein. The small amount of well-oxygenated blood from the IVC that remains in the right atrium mixes with poorly oxygenated blood from the SVC and coronary sinus and passes into the right ventricle. This blood, with a medium oxygen content, leaves through the pulmonary trunk. About 10% of the blood goes to the lungs, but most of it passes through the DA into the aorta to the fetal body. It then returns to the placenta through the umbilical arteries (see Fig. 15 - 33). Because of the high pulmonary vascular resistance in fetal life, pulmonary blood flow is low. Only a small volume of blood from the ascending aorta (about 10% of cardiac output) enters the descending aorta. About 65% of the blood in the descending aorta passes into the umbilical arteries and is returned to the placenta for reoxygenation. The remaining 35% of the blood supplies the viscera and the inferior part of the body.

Transitional Neonatal Circulation

Important circulatory adjustments occur at birth when the circulation of fetal blood through the placenta ceases and the infant's lungs expand and begin to function (see Fig. 15 - 34). *As soon as the baby is born, the oval foramen, DA, DV, and umbilical vessels are no longer needed.* The sphincter in the DA constricts, so that all blood entering the liver passes through the hepatic sinusoids. Occlusion

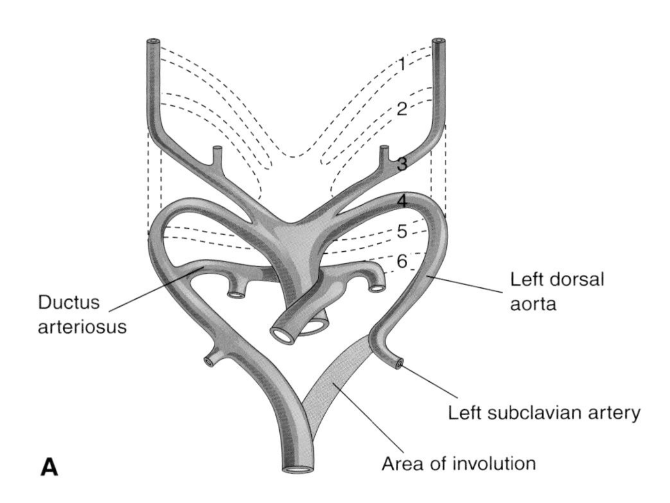

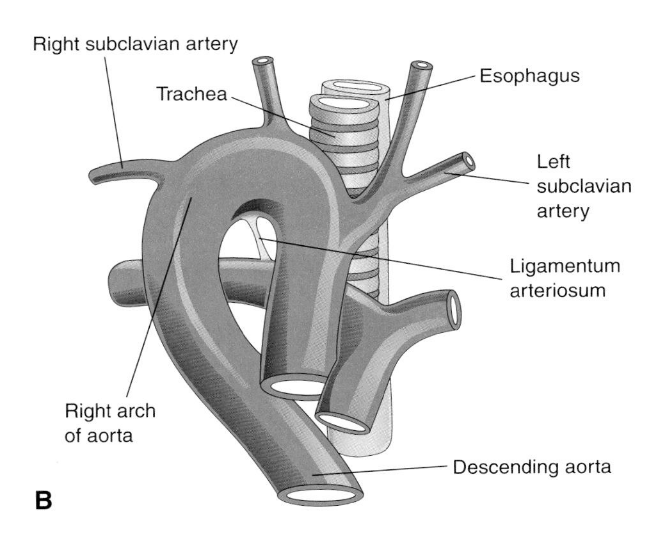

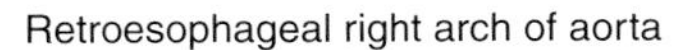

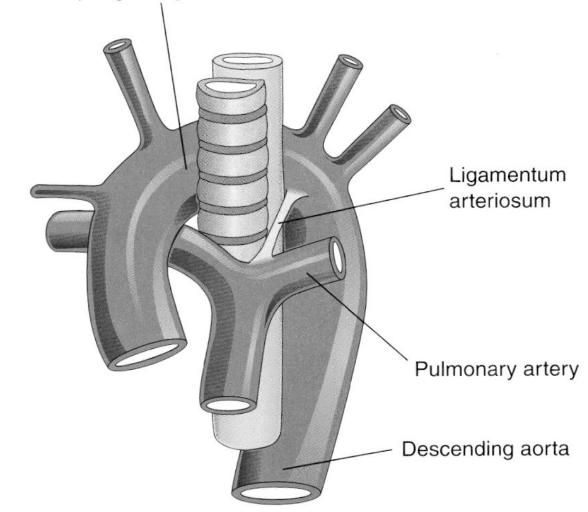

Figure 15 - 31. *A,* Sketch of the aortic arches showing abnormal involution of the distal portion of the left dorsal aorta. There is also persistence of the entire right dorsal aorta and the distal part of the right sixth aortic arch artery. *B,* Right aortic arch without a retroesophageal component. *C,* Right aortic arch with a retroesophageal component. The abnormal right aortic arch and the ligamentum arteriosum (postnatal remnant of the ductus arteriosus) form a vascular ring that compresses the esophagus and trachea.

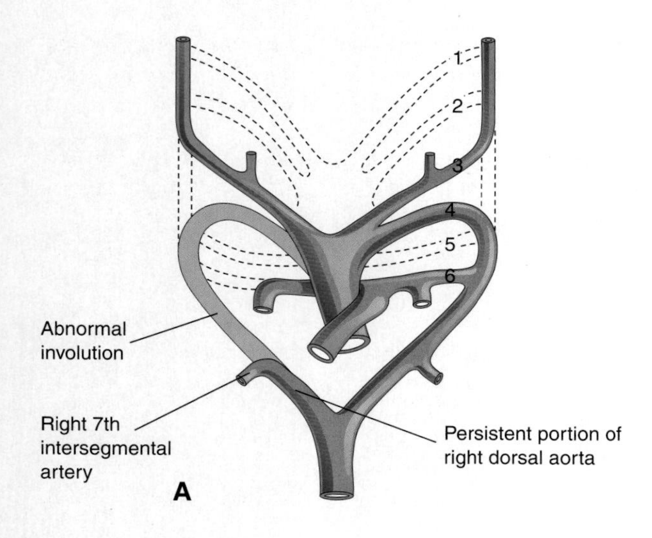

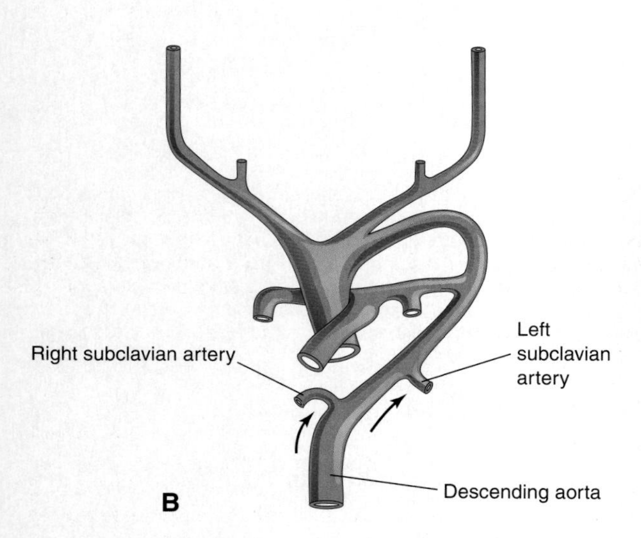

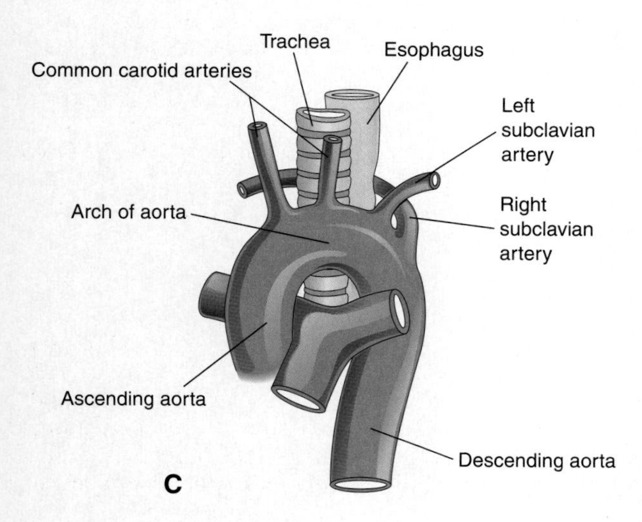

Figure 15 - 32. Sketches illustrating the possible embryological basis for an abnormal origin of the right subclavian artery. *A,* The right fourth aortic arch and cranial part of the right dorsal aorta have involuted. As a result, the right subclavian artery forms from the right seventh intersegmental artery and the distal segment of the right dorsal aorta. *B,* As the arch of the aorta forms, the right subclavian artery is carried cranially (*arrows*) with the left subclavian artery. *C,* The abnormal right subclavian artery arises from the aorta and passes posterior to the trachea and esophagus.

of the placental circulation causes an immediate decrease in blood pressure in the IVC and right atrium.

The oval foramen closes at birth. Because of increased pulmonary blood flow, the pressure in the left atrium is higher than in the right atrium. The increased left atrial pressure closes the oval foramen by pressing the valve of the foramen against the septum secundum (see Fig. 15 - 34). The output from the right ventricle then flows entirely into the pulmonary circulation. Because pulmonary vascular resistance is lower than systemic vascular resistance, blood flow in the DA reverses, passing from the aorta to the pulmonary trunk.

The **DA constricts at birth**, but for a few days, there is often a small shunt of blood from the aorta to the pulmonary in healthy, full-term infants. In premature infants and in those with persistent hypoxia, the DA may remain open much longer. Oxygen is the most important factor in controlling closure of the DA in full-term infants. Closure of the DA appears to be mediated by *bradykinin* and *prostaglandins*.

The umbilical arteries constrict at birth, preventing loss of the infant's blood. The umbilical cord is not tied for a minute or so; consequently, blood flow through the umbilical vein continues, transferring fetal blood from the placenta to the infant. *The change from the fetal to the adult pattern of blood circulation is not a sudden occurrence*. Some changes occur with the first breath; others take place over hours and days. During the transitional stage, a right-to-left flow may occur through the oval foramen. The closure of fetal vessels and the oval foramen is initially a functional change. Later, anatomical closure results from proliferation of endothelial and fibrous tissues.

Adult Derivatives of Fetal Vascular Structures

Because of the changes in the cardiovascular system at birth, certain vessels and structures are no longer required postnatally.

Umbilical Vein and Ligamentum Teres

The intra-abdominal part of the *umbilical vein* eventually becomes the *round ligament of the liver* (*ligamentum teres*) (see Fig. 15 - 34), which passes from the umbilicus to the liver; there, it is attached to the left branch of the portal vein. The umbilical vein remains patent for a considerable period and may be used for *exchange transfusions of blood* during early infancy. These transfusions are done to prevent brain damage and death of anemic erythroblastotic infants.

Ductus Venosus and Ligamentum Venosum

The DA becomes the *ligamentum venosum*; however, its closure is more prolonged than that of the DA. The ligamentum venosum passes through the liver from the left branch of the portal vein to the IVC, to which it is attached (Fig. 15 - 35).

Umbilical Arteries and Abdominal Ligaments

Most of the intra-abdominal parts of the umbilical arteries become the **medial umbilical ligaments** (see Fig. 15 - 34); the proximal parts of these vessels persist as the **superior vesical arteries**, which supply the urinary bladder.

Oval Foramen and Oval Fossa

The oval foramen normally closes functionally at birth. Anatomical closure occurs by the third month and results from tissue proliferation and adhesion of the septum primum (valve of the oval foramen) to the left margin of the septum secundum. The septum primum forms the floor of the oval fossa. The inferior edge of the septum secundum forms a rounded fold, the border of the oval fossa (L. *limbus fossae ovalis*), which marks the former boundary of the oval foramen (see Fig. 15 - 20).

Ductus Arteriosus and Ligamentum Arteriosum

Functional closure of the DA is usually completed 10 to 15 hours after birth (see Fig. 15 - 35*A*). It passes from the left pulmonary artery to the arch of the aorta. Anatomical closure of the DA and formation of the ligamentum arteriosum usually occurs by the 12th postnatal week (see Fig. 15 - 35*C*).

Patent Ductus Arteriosus

PDA is a common anomaly that occurs two to three times more frequently in females than in males (see Fig. 15 - 35*B*). Functional closure of the DA usually occurs soon after birth; however, if it remains patent, aortic blood is shunted into the pulmonary artery. PDA is the most common congenital anomaly associated with maternal rubella infection during early pregnancy. *Premature infants and infants born at high altitude may have a PDA*; the patency is the result of hypoxia and immaturity. The embryological basis of PDA is failure of the DA to involute after birth and form the ligamentum arteriosum. Failure of contraction of the muscular wall of the DA after birth is the primary cause of patency.

Development of the Lymphatic System

The lymphatic system begins to develop at the end of the sixth week of development. Lymphatic vessels develop in a manner similar to that previously described for blood vessels, and they make connections with the venous system. The early lymphatic capillaries join each other to form a network of lymphatics. **Six primary lymph**

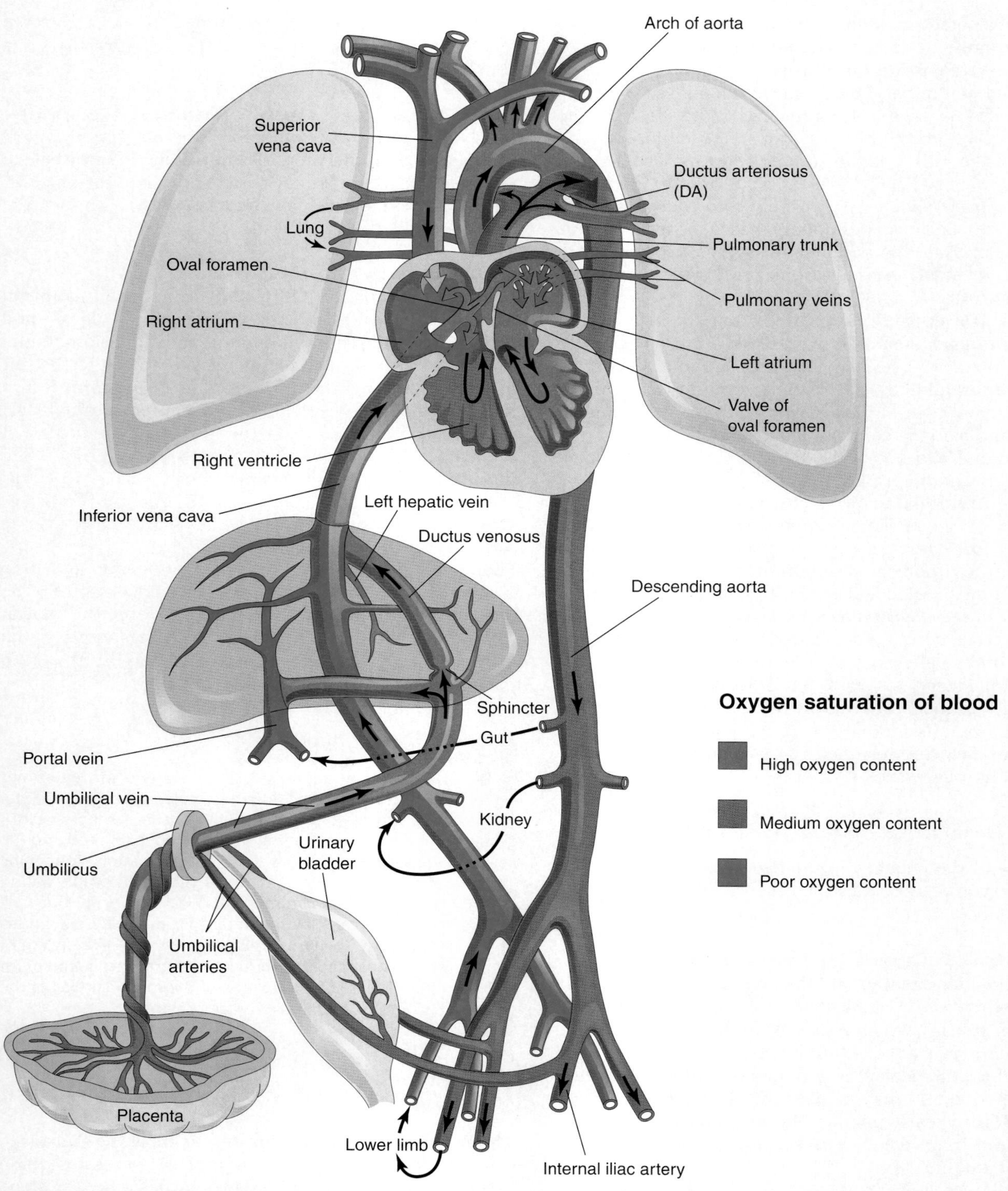

Figure 15 - 33. Fetal circulation. The colors indicate the oxygen saturation of the blood, and the arrows show the course of the blood from the placenta to the heart. The organs are not drawn to scale. Observe that three shunts permit most of the blood to bypass the liver and lungs: (1) ductus venosus, (2) oval foramen, and (3) ductus arteriosus. The poorly oxygenated blood returns to the placenta for oxygen and nutrients through the umbilical arteries.

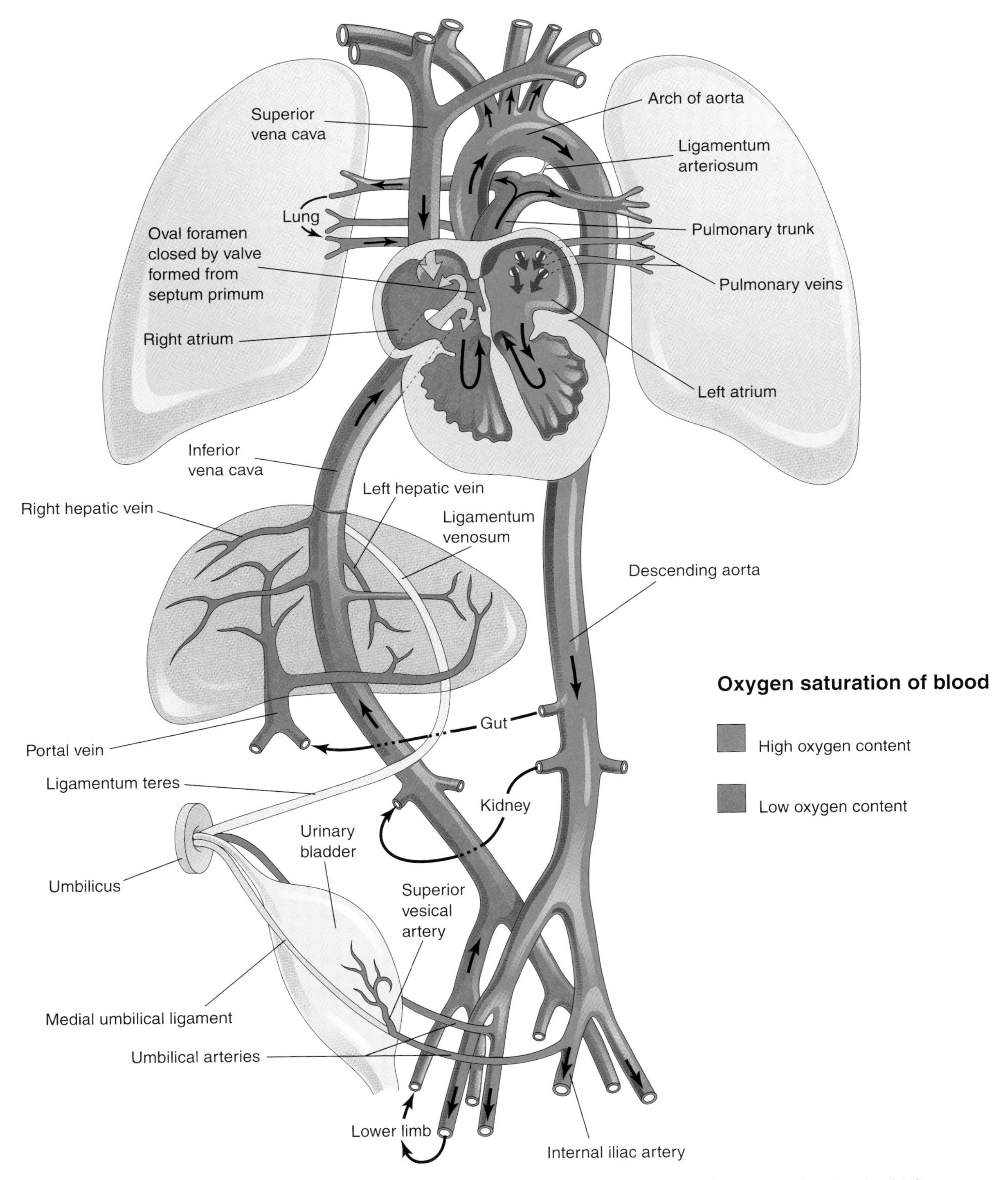

Figure 15 - 34. Neonatal circulation. The adult derivatives of the fetal vessels and structures that become nonfunctional at birth are shown. The arrows indicate the course of the blood in the infant. The organs are not drawn to scale. After birth, the three shunts that short-circuited the blood during fetal life cease to function, and the pulmonary and systemic circulations become separated.

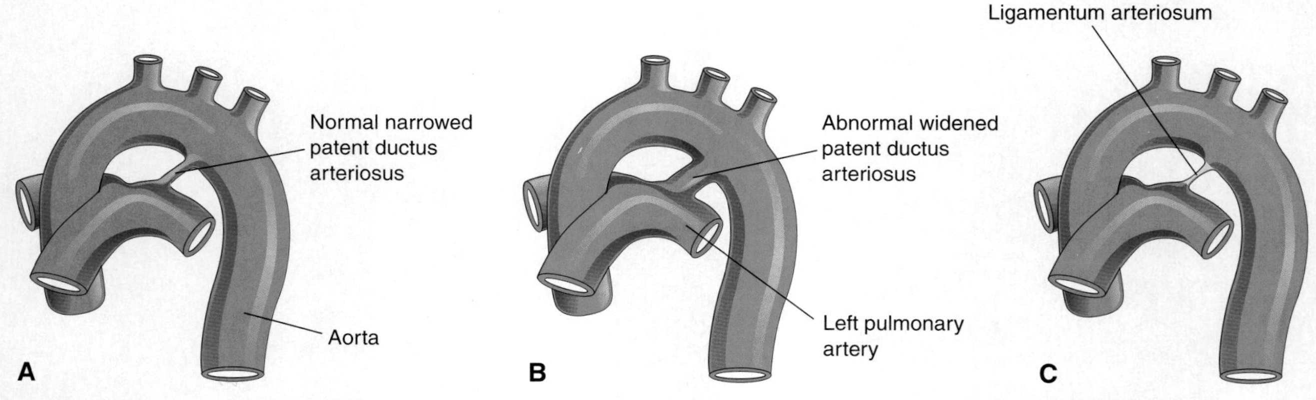

Figure 15 - 35. Closure of the ductus arteriosus (DA). *A,* The DA of a newborn infant. *B,* Abnormal patent DA in a 6-month-old infant. *C,* The ligamentum arteriosum in a 6-month-old infant.

sacs exist at the end of the embryonic period (Fig. 15 - 36*A*):

- two *jugular lymph sacs* near the junction of the subclavian veins with the anterior cardinal veins (the future internal jugular veins)
- two *iliac lymph sacs* near the junction of the iliac veins with the posterior cardinal veins
- one *retroperitoneal lymph sac* in the root of the mesentery on the posterior abdominal wall
- one chyle cistern (L. *cisterna chyli*) located dorsal to the retroperitoneal lymph sac

Lymphatic vessels soon join the lymph sacs, and accompany main veins (1) to the head, neck, and upper limbs from the jugular lymph sacs; (2) to the lower trunk and lower limbs from the iliac lymph sacs; and (3) to the primordial gut from the retroperitoneal lymph sac and the **chyle cistern**. Two large channels (right and left thoracic ducts) connect the jugular lymph sacs with the chyle cistern. Soon, a large anastomosis forms between these channels (see Fig. 15 - 36*B*). The **thoracic duct** develops from:

- the caudal part of the right thoracic duct
- the anastomosis between the thoracic ducts and the cranial part of the left thoracic duct

The **right lymphatic duct** is derived from the cranial part of the right thoracic duct (see Fig. 15 - 36*C*). The thoracic duct and right lymphatic duct connect with the venous system at the venous angle between the internal jugular and subclavian veins. The superior part of the embryonic chyle cistern persists.

Development of the Lymph Nodes

Except for the superior part of the chyle cistern, the lymph sacs are transformed into groups of lymph nodes during the early fetal period. Mesenchymal cells invade each lymph sac and form a network of lymphatic channels, the primordia of the *lymph sinuses*. Other mesenchymal cells give rise to the capsule and connective tissue framework of the lymph node. The lymphocytes are derived originally from primordial stem cells in the yolk sac mesenchyme and later from the liver and spleen. The lymphocytes eventually enter the bone marrow, where they divide to form *lymphoblasts*. The lymphocytes that appear in lymph nodes before birth are derived from the *thymus*, a derivative of the third pair of pharyngeal pouches (see Chapter 11). Small lymphocytes leave the thymus and circulate to other lymphoid organs. Later, some mesenchymal cells in the lymph nodes differentiate into lymphocytes. Lymph nodules do not appear in the lymph nodes until just before or after birth.

Development of the Spleen and Tonsils

The spleen develops from an aggregation of mesenchymal cells in the dorsal mesentery of the stomach (see Chapter 13). The **palatine tonsils** develop from the second pair of pharyngeal pouches. The **tubal tonsils** develop from aggregations of lymph nodules around the pharyngeal openings of the pharyngotympanic (auditory) tubes. The **pharyngeal tonsils** (adenoids) develop from an aggregation of lymph nodules in the wall of the nasopharynx. The **lingual tonsil** develops from an aggregation of lymph nodules in the root of the tongue. Lymph nodules also develop in the mucosa of the respiratory and digestive systems.

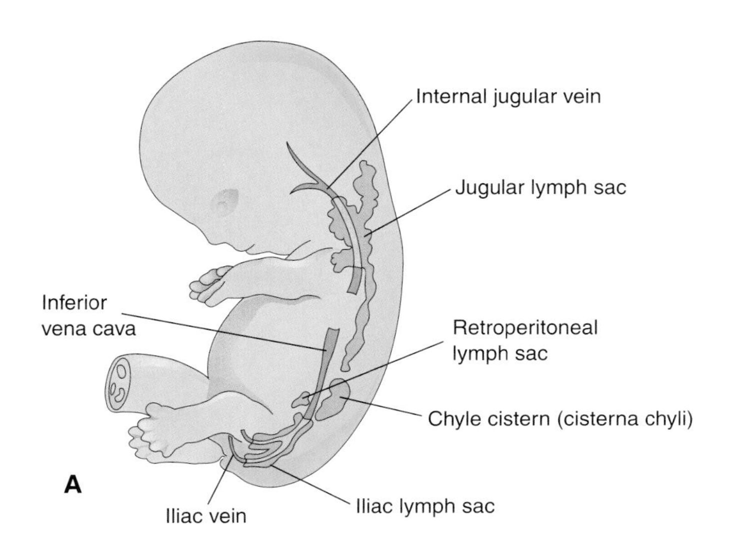

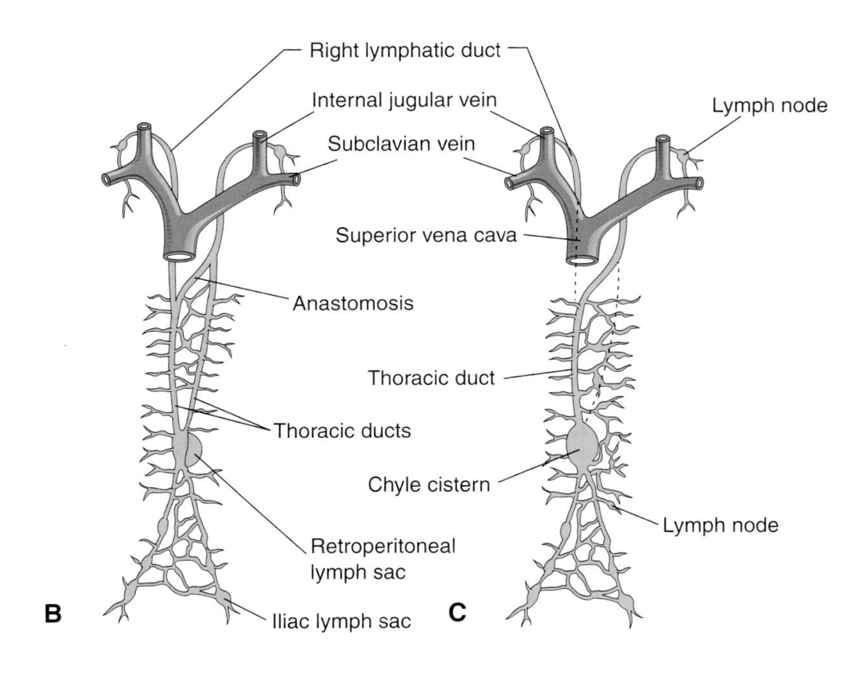

Figure 15 - 36. Development of the lymphatic system. *A,* Left side of an 8-week embryo showing the primary lymph sacs. *B,* Ventral view of the lymphatic system at 9 weeks, showing the paired thoracic ducts. *C,* Later in the fetal period, illustrating formation of the thoracic duct and right lymphatic duct.

Anomalies of the Lymphatic System

Congenital anomalies of the lymphatic system are uncommon. There may be diffuse swelling of a part of the body, termed **congenital lymphedema**. This condition may result from dilation of primordial lymphatic channels or from congenital hypoplasia of lymphatic vessels. **Cystic hygromas** are large swellings, usually appearing in the inferolateral part of the neck, that consist of large, single or multilocular, fluid-filled cavities. Hygromas may be present at birth, but they often enlarge and become evident during later infancy.

Hygromas are believed to arise from parts of a jugular lymph sac that are pinched off, or from lymphatic spaces that fail to establish connections with the main lymphatic channels.

Summary of the Cardiovascular System

The cardiovascular system begins to develop at the end of the third week, and the heart starts to beat at the beginning of the fourth week of development. Mesenchymal cells derived from the splanchnic mesoderm proliferate and form isolated cell clusters, which soon develop into **heart tubes** that fuse to form the primordial heart. Splanchnic mesoderm surrounding the heart tube forms the **primordial myocardium**. The primordium of the heart consists of four chambers:

- bulbus cordis
- ventricle
- atrium
- sinus venosus

The **truncus arteriosus** (TA; primordium of the ascending aorta and pulmonary trunk) is continuous caudally with the **bulbus cordis**, which becomes part of the ventricles. As the heart grows, it bends and soon acquires the general external appearance of the adult heart. The heart becomes partitioned into four chambers between the fourth and seventh weeks of development. Three systems of paired veins drain into the primordial heart:

- the *vitelline system*, which becomes the *portal system*
- the *cardinal veins*, which form the *caval system*
- the *umbilical system*, which involutes after birth

As the pharyngeal arches form during the fourth and fifth weeks of development, they are penetrated by arteries — the **aortic arches** — that arise from the **aortic sac**. During the sixth to eighth weeks of development, the aortic arches are transformed into the adult arterial arrangement of the carotid, subclavian, and pulmonary arteries.

The critical period of heart development extends from day 20 to day 50 after fertilization. Because partitioning of the primordial heart results from complex cellular and molecular processes, defects of the cardiac septa are relatively common, particularly VSDs. Some congenital anomalies result from abnormal transformation of the aortic arches into the adult arterial pattern (e.g., right aortic arch). Because the lungs are nonfunctional during prenatal life, the **fetal cardiovascular system** is structurally designed so that the blood is oxygenated in the placenta and largely bypasses the lungs. The modifications that establish the postnatal circulatory pattern are not abrupt, but extend into infancy. Failure of these circulatory changes to occur at birth is the cause of two of the most common congenital anomalies of the heart and great vessels: patent oval foramen and patent ductus arteriosus (PDA).

The **lymphatic system** begins to develop late in the sixth week in close association with the venous system. Six primary *lymph sacs* develop; these later become interconnected by lymphatic vessels. **Lymph nodes** develop along the network of lymphatic vessels; lymph nodules do not appear until just before or after birth. Sometimes, a part of a jugular lymph sac becomes pinched off, giving rise to a mass of dilated lymphatic spaces called a *cystic hygroma*.

Clinically Oriented Questions

1. The pediatrician diagnosed a heart murmur in a newborn baby. What does this mean? What causes this condition and what does it indicate?
2. Are congenital anomalies of the heart common? What is the most common congenital heart defect in children?
3. What are the causes of congenital anomalies of the cardiovascular system? Can drugs taken by the mother during pregnancy cause congenital cardiac defects? One mother who drank heavily during her pregnancy had a child with a heart defect. Could her drinking have caused her infant's heart defect?
4. Can viral infections cause congenital heart disease? Is it true that if a mother has measles during pregnancy, her baby will have an abnormality of the cardiovascular system? Is it true that women can be given a vaccine that will protect their babies against certain viruses?
5. A baby had its aorta arising from the right ventricle and its pulmonary artery arising from the left ventricle. The baby died during the first week. What is this anomaly called, and how common is this disorder? Can the condition be corrected surgically? If so, how is this done?
6. During a routine examination of identical twin sisters in their forties, it was found that one of them had a *reversed heart*. Is this a serious heart anomaly? How common is this among identical twins, and what causes this condition to develop?

The answers to these questions are at the back of the book.

16

The Skeletal System

The skeletal system develops from mesodermal and neural crest cells. As the notochord and neural tube form, the *intraembryonic mesoderm* lateral to these structures thickens to form two longitudinal columns of **paraxial mesoderm** (Fig. 16 – 1*A* to *C*). Toward the end of the third week, these columns become segmented into blocks of mesoderm — the **somites**. Each somite differentiates into two parts (see Fig. 16 – 1*D* and *E*):

- The ventromedial part is the **sclerotome**; its cells form the vertebrae and ribs.
- The dorsolateral part is the **dermomyotome**; cells from its *myotome region* form myoblasts (primordial muscle cells), whereas those from its *dermatome region* form the dermis.

Mesodermal cells give rise to loosely organized embryonic connective tissue called **mesenchyme**. Considerable mesenchyme in the head region is also derived from the neural crest. **Neural crest cells** migrate into the pharyngeal arches and form the bones and connective tissue of craniofacial structures. These cells are important for the complex patterning of the head and face. *Homeobox* (*Hox*) *genes* play an essential role in the migration and subsequent differentiation of the neural crest cells.

Development of Bone and Cartilage

Bones first appear as condensations of mesenchymal cells that form bone models. Condensation marks the beginning of selective gene activity, which precedes cell differentiation. Most flat bones develop in mesenchyme within pre-existing membranous sheaths; this type of osteogenesis is **intramembranous bone formation**. Mesenchymal models of most limb bones are transformed into cartilaginous bone models, which later become ossified by **endochondral bone formation**. *Bone morphogenetic proteins (BMP 5 and BMP 7) and growth and differentiation factor 5 (Gdf5), members of the TGF-β superfamily, as well as other signaling molecules, have been implicated as endogenous regulators of skeletal development.*

Histogenesis of Cartilage

Cartilage develops from mesenchyme and first appears in embryos during the fifth week. In areas where cartilage is to develop, the mesenchyme condenses to form **chondrification centers**. The mesenchymal cells proliferate and become rounded. Cartilage-forming cells — **chondroblasts** — secrete collagenous fibrils and the ground substance of the matrix. Subsequently, collagenous and/or elastic fibers are deposited in the intercellular substance or matrix. *Three types of cartilage are distinguishable according to the type of matrix that is formed:*

- *hyaline cartilage,* the most widely distributed type (e.g., in joints)
- *fibrocartilage* (e.g., in intervertebral disks)
- *elastic cartilage* (e.g., in the auricle of ear)

Histogenesis of Bone

Bone develops in two types of connective tissue: mesenchyme and cartilage. Like cartilage, bone consists of cells and an organic intercellular substance, called **bone matrix**, which comprises collagen fibrils embedded in an amorphous component.

Intramembranous Ossification

This type of bone formation occurs in mesenchyme that has formed a membranous sheath (Fig. 16 – 2), hence the name *intramembranous ossification.* The mesenchyme condenses and becomes highly vascular; some cells differentiate into **osteoblasts** (bone-forming cells) and begin to deposit matrix, termed **osteoid tissue**, or prebone. The osteoblasts are almost completely separated from one another, with contact being maintained by only a few tiny processes. Calcium phosphate is then deposited in the osteoid tissue as it is organized into bone. Bone osteoblasts are trapped in the matrix and become **osteocytes**. At first, new bone has no organized pattern. Spicules of bone soon become organized and coalesce into lamellae (layers). Concentric lamellae develop around blood vessels, forming **haversian systems**. Some osteoblasts remain at the periphery of the bone and continue to lay down layers, forming plates of compact bone on the surfaces. Between the surface plates, the intervening bone remains spiculated or spongy. This spongy environment is somewhat accentuated by the action of cells with a different origin — **osteoclasts** — which absorb bone. In the interstices of spongy bone, the mesenchyme differentiates into **bone marrow**. During fetal and postnatal life, continuous remodeling of bone occurs by the simultaneous action of osteoclasts and osteoblasts. Studies of the cellular and molecular events during embryonic bone formation suggest that osteogenesis and chondrogenesis are programmed early in development and are independent events under the influence of vascular factors.

Intracartilaginous Ossification

Intracartilaginous ossification is a type of bone formation that occurs in preexisting cartilaginous models (Fig. 16 – 3*A* to *E*). In a long bone, for example, the **primary center of ossification** appears in the **diaphysis** — the part of the bone between its ends — which forms the **shaft** of the bone. Here, the cartilage

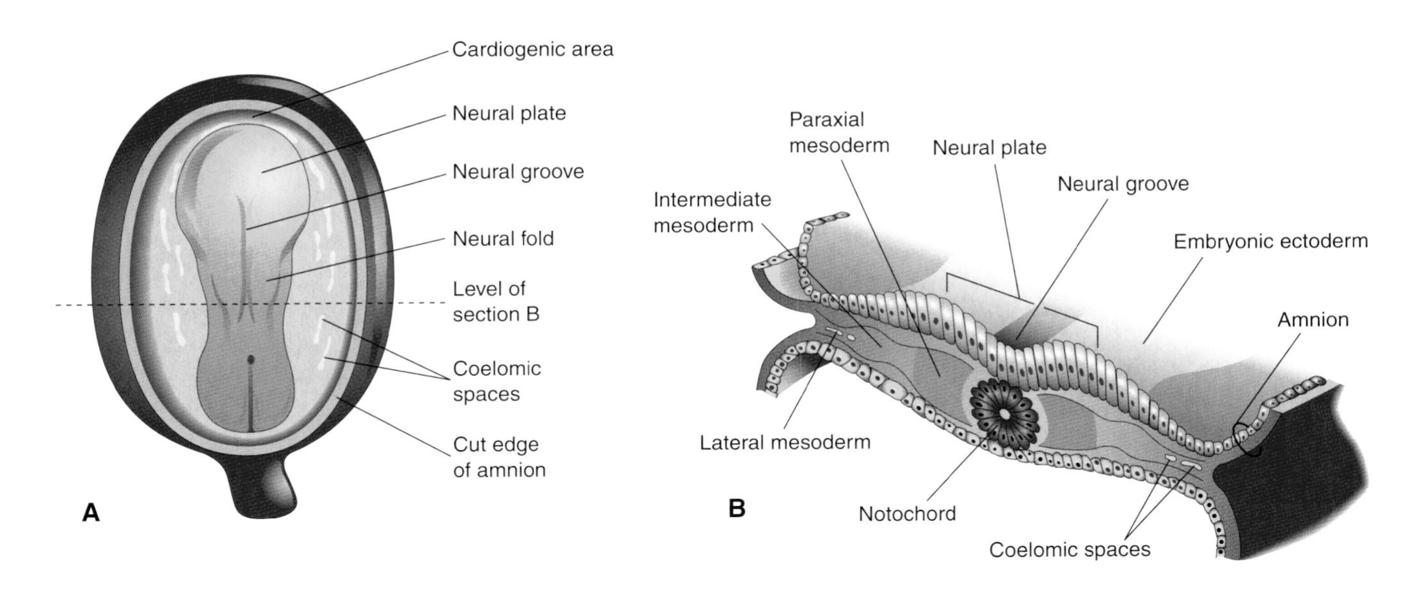

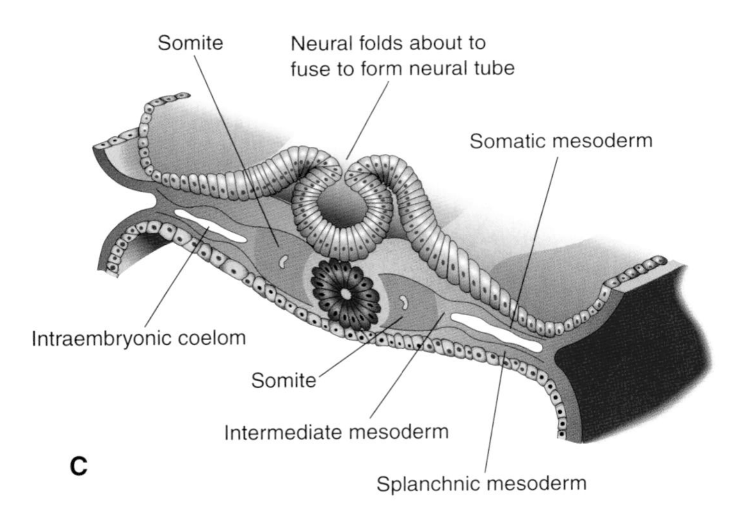

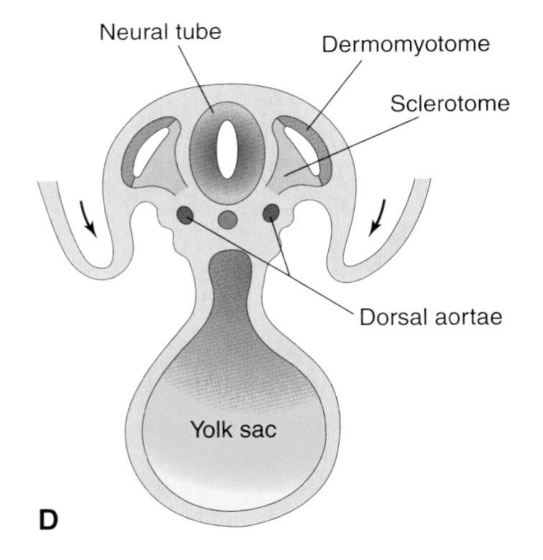

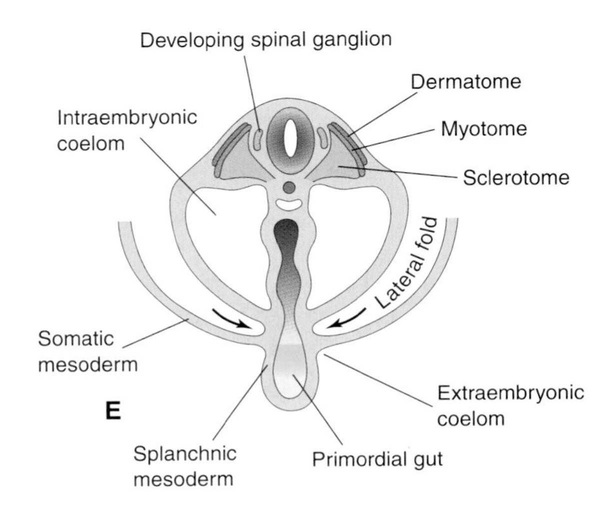

Figure 16 - 1. Drawings illustrating formation and early differentiation of somites. *A*, Dorsal view of a presomite embryo (about 18 days). *B*, Transverse section of the embryo shown in *A*, illustrating the paraxial mesoderm from which the somites are derived. *C*, Transverse section of an embryo at about 22 days, at which time early somites have appeared. *D*, Transverse section of an embryo at about 24 days. Note the folding of the embryo in the horizontal plane (*arrows*). The dermomyotome region of the somite gives rise to the dermatome and myotome. *E*, Transverse section of an embryo at about 26 days, showing the dermatome, myotome, and sclerotome regions of the somite.

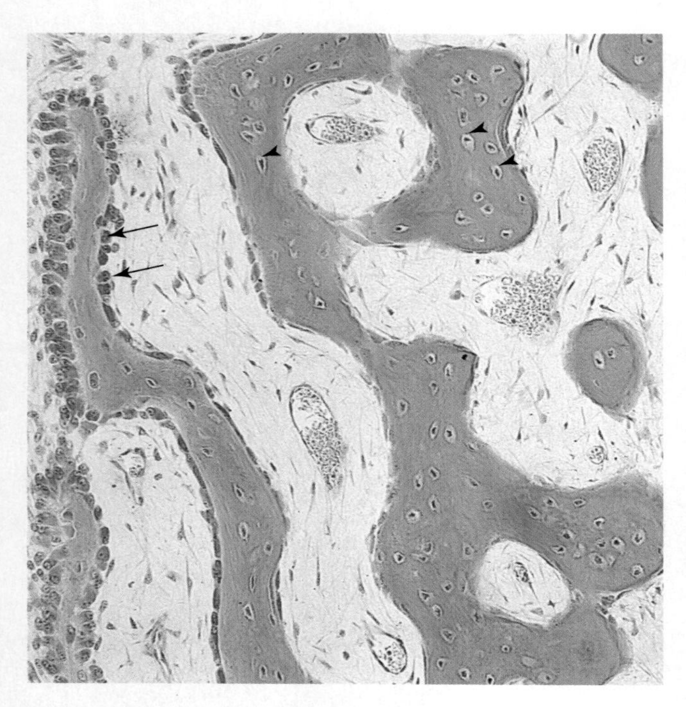

Figure 16 - 2. Light micrograph of intramembranous ossification (× 132). Trabeculae of bone are being formed by osteoblasts lining their surface (*arrows*). Observe that osteocytes are trapped in lacunae (*arrowheads*) and that primordial osteons are beginning to form. The osteons (canals) contain blood capillaries. (From Gartner LP, Hiatt JL: *Color Textbook of Histology*, 2nd ed. Philadelphia, WB Saunders, 2001.)

cells increase in size (hypertrophy), the matrix becomes calcified, and the cells die. Concurrently, a thin layer of bone is deposited under the **perichondrium** surrounding the diaphysis; thus, the perichondrium becomes the **periosteum**. Invasion of vascular connective tissue from the periosteum breaks up the cartilage. Some invading cells differentiate into **hemopoietic cells**, which are responsible for the formation of blood cells in the bone marrow. Other invading cells differentiate into osteoblasts that deposit bone matrix on the spicules of calcified cartilage. This process continues toward the **epiphyses**, or ends of the bone. The spicules of bone are remodeled by the action of osteoclasts and osteoblasts. *Lengthening of long bones occurs at the diaphysial-epiphysial junction.* The lengthening of bone depends on the **epiphysial cartilage plates** (growth plates), whose chondrocytes proliferate and participate in endochondral bone formation (see Fig. 16 - 3*D* and *E*). Cartilage cells in the diaphysial-epiphysial region proliferate by mitosis. Toward the diaphysis, the cartilage cells hypertrophy and the matrix becomes calcified and broken up into spicules by vascular tissue from the marrow or **medullary cavity**. Bone is deposited on these spicules; absorption of this bone keeps the spongy bone masses relatively constant in length and enlarges the medullary cavity.

Ossification of limb bones begins at the end of the embryonic period; and thereafter the process makes demands on the maternal supply of calcium and phosphorus. The region of bone formation at the center of the shaft (body) of a long bone is the **primary ossification center** (see Fig. 16 - 3*B* and *E*). At birth, the *diaphyses* are largely ossified, but most of the ends or *epiphyses* are still cartilaginous. Most **secondary ossification centers** appear in the epiphyses during the first few years after birth. The epiphysial cartilage cells hypertrophy, and there is invasion by vascular connective tissue. Ossification spreads in all directions, and only the articular cartilage and a transverse plate of cartilage, the **epiphysial cartilage plate**, remain cartilaginous. Upon completion of growth, this plate is replaced by spongy bone, the epiphyses and diaphysis are united, and no further elongation of the bone occurs. In most bones, the epiphyses have fused with the diaphysis by about the age of 20 years. Growth in the diameter of a bone results from deposition of bone at the periosteum and from absorption on the medullary surface. The rate of deposition and absorption is balanced to regulate the thickness of the compact bone and the size of the medullary cavity. The internal reorganization of bone continues throughout life.

Rickets

Rickets is a disease that occurs in children with vitamin D deficiency. Calcium absorption by the intestine is impaired, which causes disturbances in ossification of the epiphysial cartilage plates and disorientation of cells at the metaphysis. The limbs are shortened and deformed, with severe bowing of the bones.

Development of Joints

Joints begin to develop during the sixth embryonic week (Fig. 16 - 4*A*), and by the end of the eighth week, they resemble adult joints. Joints are classified as:

- fibrous joints
- cartilaginous joints
- synovial joints

Joints with little or no movement are classified according to the type of material holding the bones together.

Fibrous Joints

During development of fibrous joints, the interzonal mesenchyme between the developing bones differentiates into dense fibrous tissue (see Fig. 16 - 4*D*). The sutures of the cranium are an example of fibrous joints.

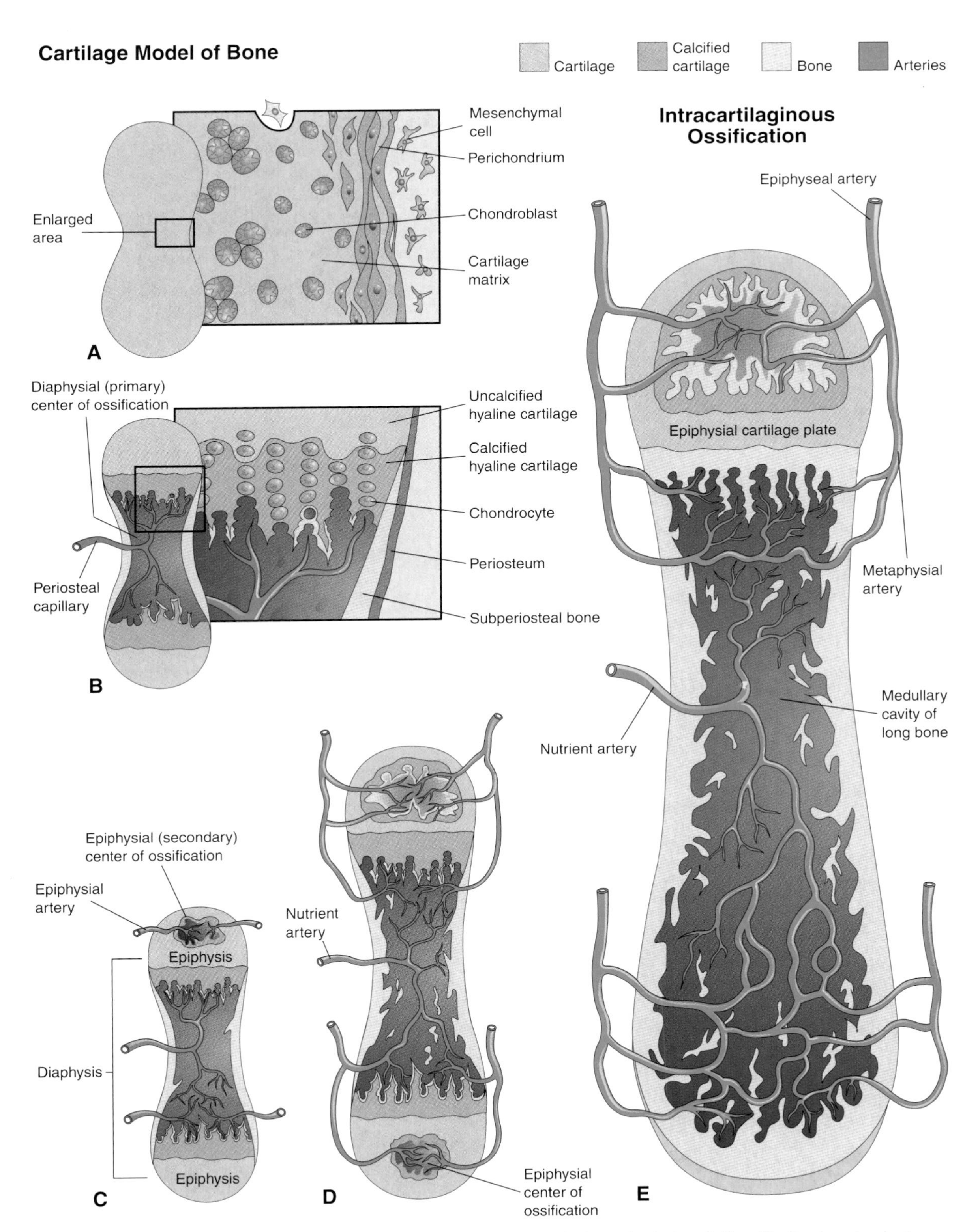

Figure 16-3. *A* to *E*, Schematic longitudinal sections illustrating intracartilaginous (endochondral) ossification in a developing long bone.

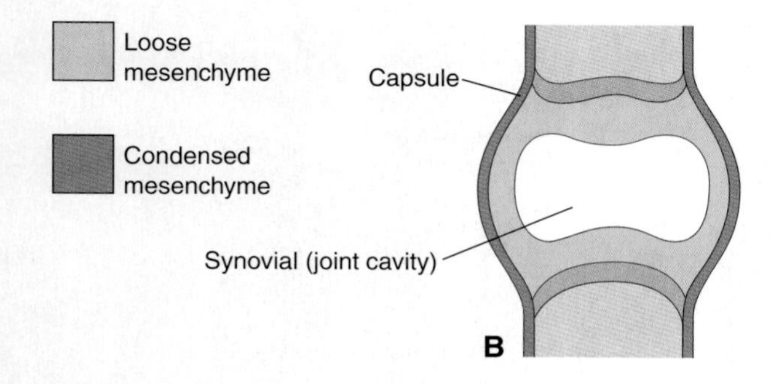

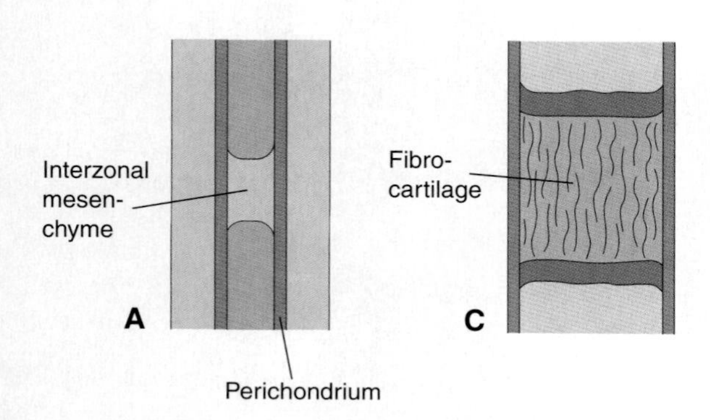

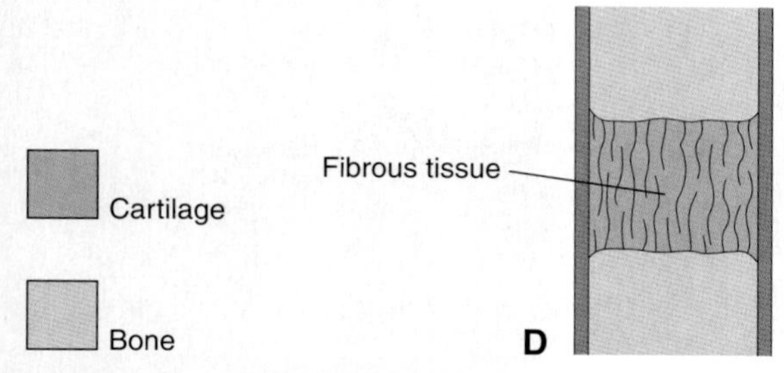

Figure 16 - 4. Drawings illustrating the development of joints during the sixth and seventh embryonic weeks. *A,* Condensed mesenchyme continues across the gap or interzone between the developing bones, enclosing some interzonal mesenchyme between them. This primordial joint may differentiate into a synovial joint *(B)*, a cartilaginous joint *(C)*, or a fibrous joint *(D)*.

Cartilaginous Joints

During the development of cartilaginous joints, the interzonal mesenchyme between the developing bones differentiates into hyaline cartilage (e.g., costochondral joints) or fibrocartilage (see Fig. 16 - 4*C*). An example of this type of joint is the *pubic symphysis* between the bodies of the pubic bones.

Synovial Joints

During the development of the relatively common synovial joints (e.g., the knee joint), the interzonal mesenchyme between the developing bones differentiates as follows (see Fig. 16 - 4*B*):

- Peripherally, it forms the capsular and other ligaments.
- Centrally, it disappears, and the resulting space becomes the synovial (joint) cavity.
- Where it lines the fibrous capsule and articular surfaces, it forms the synovial membrane.

Development of the Axial Skeleton

The axial skeleton is composed of the cranium (skull), vertebral column, ribs, and sternum. During formation of this part of the skeleton, the cells in the sclerotomes of the somites change their position (see Fig. 16 - 1). During the fourth week of development, they surround the neural tube (primordium of the spinal cord) and the notochord, the structure around which the primordia of the vertebrae develop. This positional change of the sclerotomal cells is effected by differential growth of the surrounding structures, not by active migration of sclerotomal cells.

Development of the Vertebral Column

During the precartilaginous stage, mesenchymal cells from the sclerotomes are found in three main areas (Fig. 16 - 5*A*):

- around the notochord
- surrounding the neural tube
- in the body wall

In a frontal section of a 4-week embryo, the sclerotomes appear as paired condensations of mesenchymal cells around the notochord (see Fig. 16 - 5*B*). Each sclerotome consists of loosely arranged cells cranially and densely packed cells caudally. Some densely packed cells move cranially, opposite the center of the myotome, where they form the **intervertebral (IV) disc** (see Fig. 16 - 5*C* and *D*). These cells express *Pax-1*, a paired box gene. The remaining densely packed cells fuse with the loosely arranged cells of the immediately caudal sclerotome to form the mesenchymal **centrum**, the primordium of the body of a vertebra. Thus, each centrum develops from two adjacent sclerotomes and becomes an intersegmental structure. The nerves now lie in close relationship to the IV discs, and the **intersegmental arteries** lie on each side of the vertebral bodies. In the thorax, the dorsal intersegmental arteries become the **intercostal arteries**. Laboratory studies suggest that the development of the vertebral column is regulated by homeobox and paired box genes.

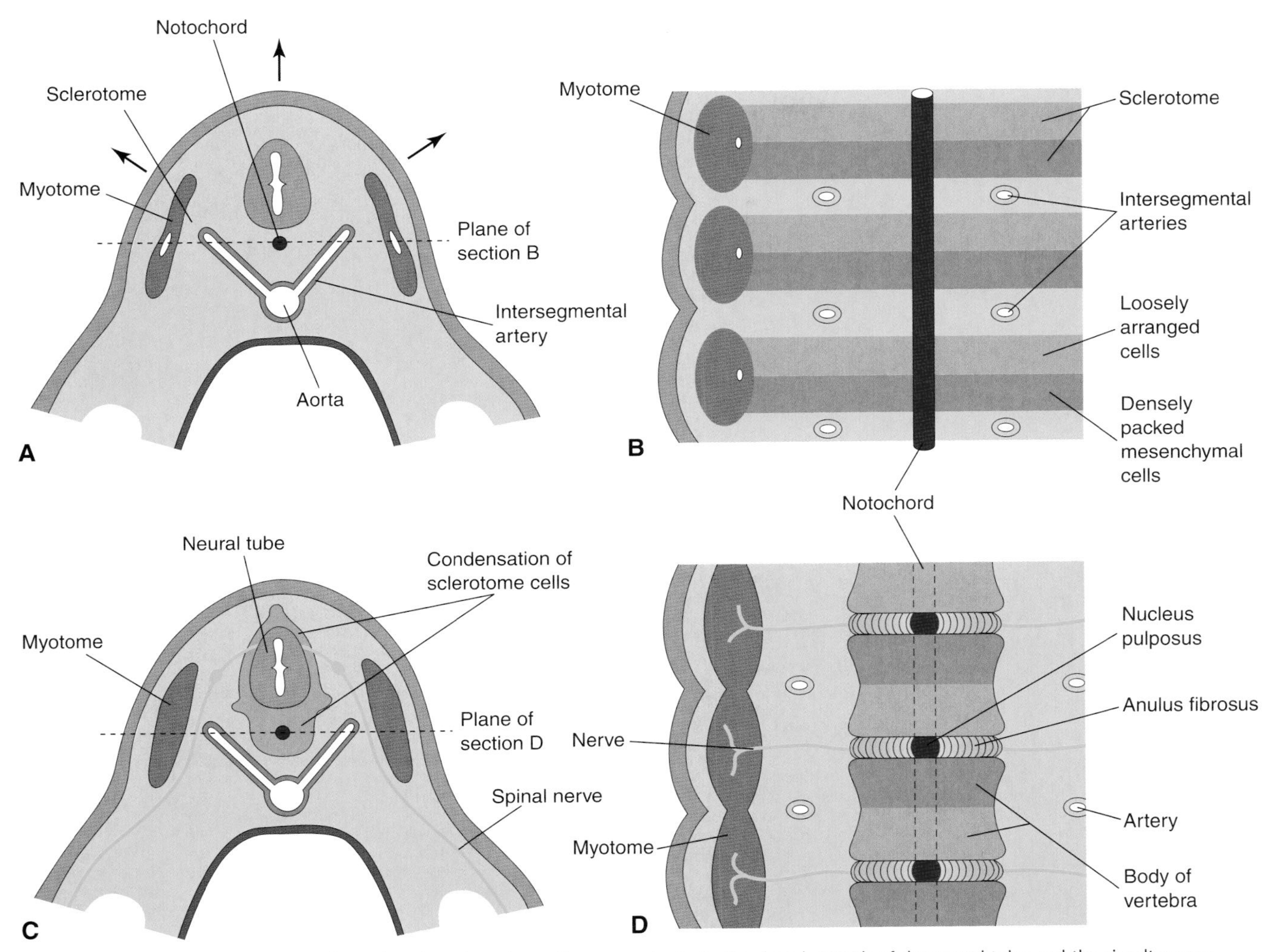

Figure 16 - 5. *A*, Transverse section through a 4-week embryo. The arrows indicate the dorsal growth of the neural tube and the simultaneous dorsolateral movement of the somite remnant, which leaves behind a trail of sclerotomal cells. *B*, A diagram of the frontal section of the same embryo as in *A* showing that the condensation of sclerotomal cells around the notochord consists of a cranial area of loosely packed cells and a caudal area of densely packed cells. *C*, Transverse section through a 5-week embryo. Note the condensation of sclerotomal cells around the notochord and neural tube, which forms a mesenchymal vertebra. *D*, Diagram of the frontal section illustrating that the vertebral body forms from the cranial and caudal halves of two successive sclerotomal masses. The intersegmental arteries now cross the bodies of the vertebrae, and the spinal nerves lie between the vertebrae. The notochord is degenerating except in the region of the intervertebral disk, where it forms the nucleus pulposus.

Where it is surrounded by the developing vertebral bodies, the **notochord** degenerates and disappears. Between the vertebrae, the notochord expands to form the gelatinous center of the intervertebral disc, the **nucleus pulposus** (see Fig. 16 - 5*D*). This nucleus is later surrounded by circularly arranged fibers that form the **anulus fibrosus.** The nucleus pulposus and anulus fibrosus together constitute the intervertebral disc. The mesenchymal cells that surround the neural tube form the vertebral (neural) arch (Fig. 16 - 6*D*). The mesenchymal cells in the body wall form the **costal processes**, which form ribs in the thoracic region.

Chordoma

Remnants of the notochord may persist and give rise to a **chordoma.** About a third of these slowly growing, malignant tumors involve the base of the cranium and extend to the nasopharynx. They infiltrate bone and are difficult to remove. Few patients with chordoma survive longer than 5 years. Chordomas may also develop in the lumbosacral region.

Cartilaginous Stage of Vertebral Development

During the sixth week of development, chondrification centers appear in each mesenchymal vertebra (see

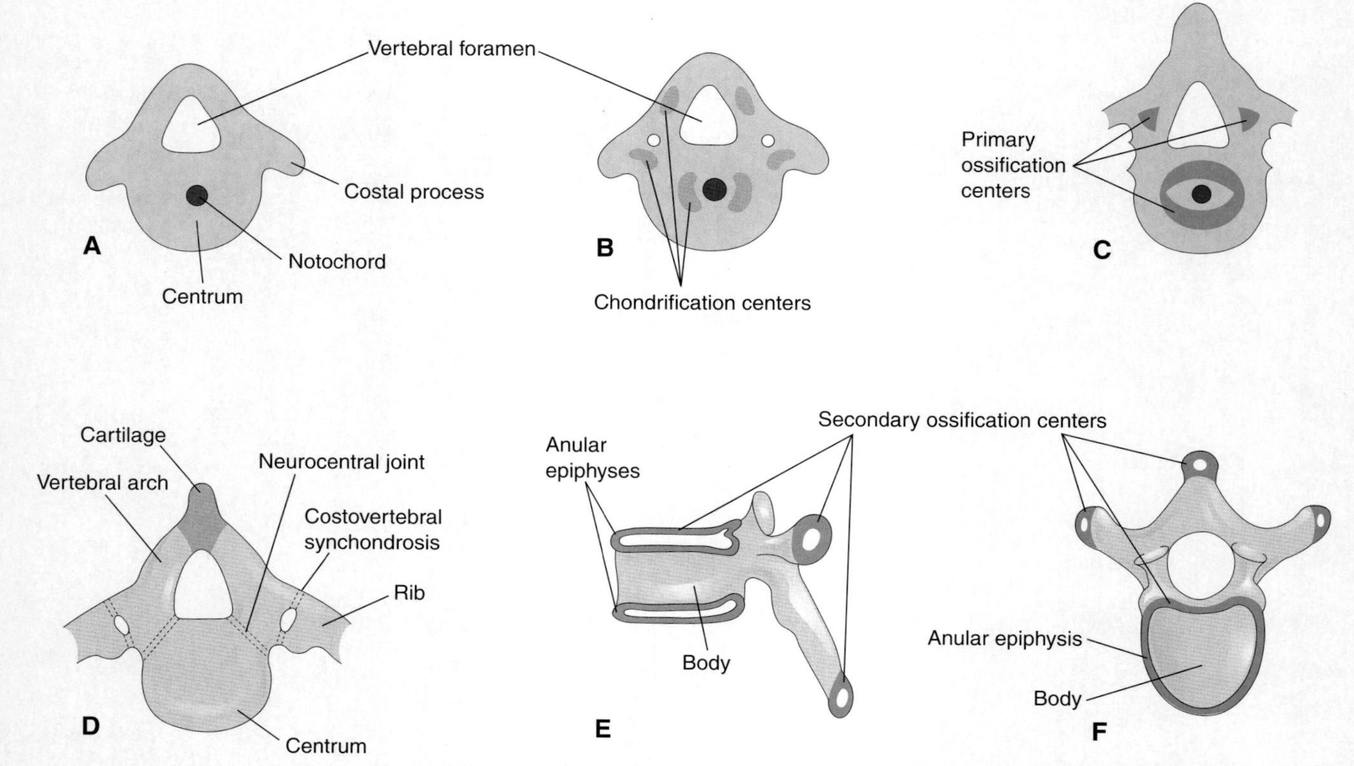

Figure 16-6. The stages of vertebral development. *A*, Mesenchymal vertebra at 5 weeks. *B*, Chondrification centers in a mesenchymal vertebra at 6 weeks. *C*, Primary ossification centers in a cartilaginous vertebra at 7 weeks. *D*, Thoracic vertebra, consisting of three bony parts, at birth. Note the cartilage between the halves of the vertebral arch and between the arch and centrum (neurocentral joint). *E* and *F*, Two views of a typical thoracic vertebra at puberty showing the location of the secondary centers of ossification.

Fig. 16-6*A* and *B*). At the end of the embryonic period, the two centers in each centrum fuse to form a cartilaginous centrum. Concomitantly, the centers in the vertebral arches fuse with each other and the centrum. The spinous and transverse processes develop from extensions of chondrification centers in the vertebral arch. Chondrification spreads until a cartilaginous vertebral column is formed.

Bony Stage of Vertebral Development

Ossification of typical vertebrae begins during the embryonic period. There are two primary ossification centers — ventral and dorsal — for the centrum (Fig. 16-6*C*). These **primary ossification centers** soon fuse to form one center. Three primary centers are present by the end of the embryonic period:

- one in the centrum
- one in each half of the vertebral arch

Ossification becomes evident in the vertebral arches during the eighth week of development. At birth, each vertebra consists of three bony parts connected by cartilage (see Fig. 16-6*D*). The bony halves of the vertebral arch usually fuse during the first 3 to 5 years of life. The arches first unite in the lumbar region, and union progresses cranially. The vertebral arch articulates with the centrum at cartilaginous **neurocentral joints**. These articulations permit the vertebral arches to grow as the spinal cord enlarges. These joints disappear when the vertebral arch fuses with the centrum during the third to sixth years. Five **secondary ossification centers** appear in the vertebrae after puberty:

- one at the tip of the spinous process
- one at the tip of each transverse process
- two *anular epiphyses*, one on the superior and one on the inferior rim of the vertebral body (Fig. 16-6*E* and *F*)

The **vertebral body** is a composite of the anular epiphyses and the mass of bone between them. The vertebral body includes the centrum, parts of the vertebral arch, and the facets for the heads of the ribs. All secondary centers unite with the rest of the vertebra at around 25 years of age. Exceptions to the typical ossification of vertebrae occur in the atlas (C1), axis (C2), C7, lumbar vertebrae, sacrum, and coccyx.

Variations in the Number of Vertebrae

About 95% of people have 7 cervical, 12 thoracic, 5 lumbar, and 5 sacral vertebrae. About 3% of people have one or two more vertebrae, and about 2% have one less. To determine the number of vertebrae, it is necessary to examine the entire vertebral column because an apparent extra (or absent) vertebra in one segment of the column may be compensated for by an absent (or extra) vertebra in an adjacent segment (e.g., 11 thoracic-type vertebrae with 6 lumbar-type vertebrae).

Development of the Ribs

The ribs develop from the mesenchymal costal processes of the thoracic vertebrae (see Fig. 16–6*A*). They become cartilaginous during the embryonic period and ossify during the fetal period. The original site of union of the **costal processes** with the vertebrae is replaced by **costovertebral joints**. These are the plane type of synovial joint (Fig. 16–6*D*). Seven pairs of ribs (1 to 7), the so-called **true ribs**, attach by their own cartilages to the sternum. Five pairs of ribs (8 to 12), the so-called **false ribs**, attach to the sternum through the cartilage of another rib or ribs. The last two pairs of ribs (11 and 12) do not attach to the sternum; they are called **floating ribs**.

Development of the Sternum

A pair of vertical mesenchymal bands, called **sternal bars**, develop ventrolaterally in the body wall. *Chondrification* occurs in these bars as they move medially. They fuse craniocaudally in the median plane to form cartilaginous models of the manubrium, sternebrae (segments of the sternal body), and xiphoid process. Fusion at the inferior end of the sternum is sometimes incomplete; as a result, the xiphoid process in these infants is bifid or perforated. Centers of ossification appear craniocaudally in the sternum before birth, except the ossification center for the xiphoid process, which appears during childhood.

Development of the Cranium

The cranium (skull) develops from mesenchyme around the developing brain. The cranium consists of:

- the **neurocranium**, a protective case for the brain
- the **viscerocranium**, the skeleton of the face

Cartilaginous Neurocranium

At six weeks, the cartilaginous neurocranium, or **chondrocranium**, consists of the cartilaginous base of the developing cranium, which forms by fusion of several cartilages (Fig. 16–7*A* to *D*). Later, endochondral ossification of the chondrocranium forms the bones of the base of the cranium. The ossification pattern of these bones has a definite sequence, beginning with the occipital bone, body of the sphenoid, and ethmoid bone. The **parachordal cartilage**, or **basal plate**, forms around the cranial end of the notochord (Fig. 16–7*A*) and fuses with the cartilages derived from the sclerotome regions of the occipital somites. This cartilaginous mass contributes to the base of the occipital bone; later, extensions grow around the cranial end of the spinal cord and form the boundaries of the foramen magnum (Fig. 16–7*C*). The **hypophysial cartilage** forms around the developing pituitary gland and fuses to form the body of the sphenoid bone. The *trabeculae cranii* fuse to form the body of the ethmoid bone, and the *ala orbitalis* forms the lesser wing of the sphenoid bone. **Otic capsules** develop around the otic vesicles, the primordia of the internal ears (see Chapter 20), and form the petrous and mastoid parts of the temporal bone. **Nasal capsules** develop around the nasal sacs (see Chapter 11) and contribute to the formation of the ethmoid bone.

Membranous Neurocranium

Intramembranous ossification occurs in the mesenchyme at the sides and top of the brain, forming the **calvaria** (cranial vault). During fetal life, the flat bones of the calvaria are separated by dense connective tissue membranes that form fibrous joints, the **sutures** (Fig. 16–8). Six large fibrous areas — known as the **fontanelles** — are present where several sutures meet. The softness of the bones and their loose connections at the sutures enable the calvaria to undergo changes of shape (termed molding) during birth. During **molding of the fetal cranium** (adaptation of the fetal head to the pelvic cavity during birth), the frontal bones become flat, the occipital bone is drawn out, and one parietal bone slightly overrides the other one. Within a few days after birth, the shape of the calvaria returns to normal.

Cartilaginous Viscerocranium

The parts of the fetal cranium that constitute the cartilaginous viscerocranium are derived from the cartilaginous skeleton of the first two pairs of pharyngeal arches (see Chapter 11).

- The dorsal end of the *first arch cartilage* forms two middle ear bones, the malleus and incus.
- The dorsal end of the *second arch cartilage* forms the stapes of the middle ear and the styloid process

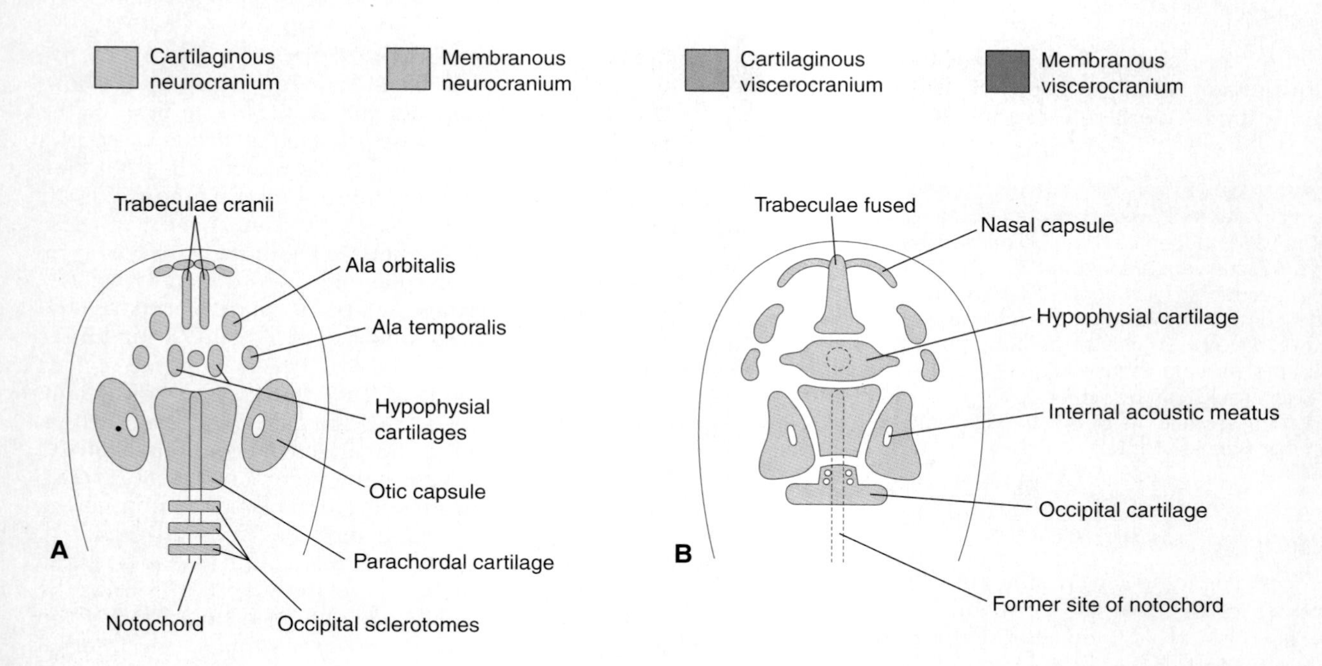

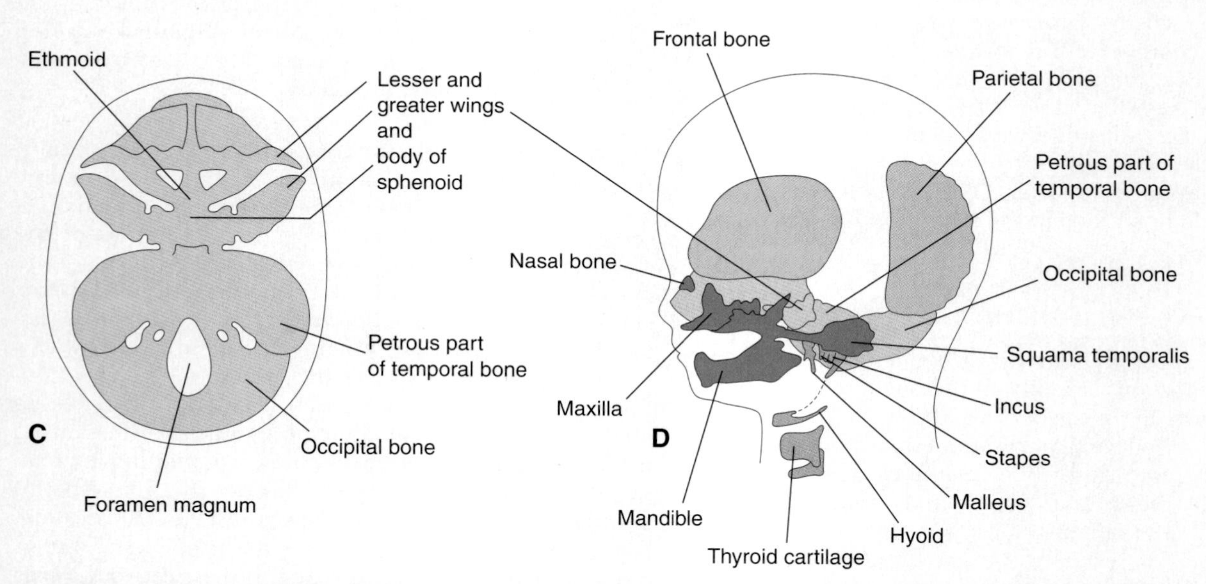

Figure 16 – 7. Diagrams illustrating stages in the development of the cranium. Views of the base of the developing cranium, viewed superiorly (*A* to *C*), and laterally (*D*), are illustrated. *A*, Six weeks, showing the various cartilages that will fuse to form the chondrocranium. *B*, Seven weeks, after fusion of some of the paired cartilages. *C*, Twelve weeks, showing the cartilaginous base of the cranium or chondrocranium formed by the fusion of various cartilages. *D*, Twenty weeks, indicating the derivation of the bones of the fetal cranium.

of the temporal bone. Its ventral end ossifies to form the lesser horn (cornu) and superior part of the body of the hyoid bone.

- The *third, fourth, and sixth arch cartilages* form only in the ventral parts of the arches. The *third arch cartilages* give rise to the greater horns and inferior part of the body of the hyoid bone.
- The *fourth and sixth arch cartilages* fuse to form the laryngeal cartilages, except for the epiglottis.

Membranous Viscerocranium

Intramembranous ossification occurs in the maxillary prominence of the first pharyngeal arch (see Chapter 11) and subsequently forms the squamous temporal,

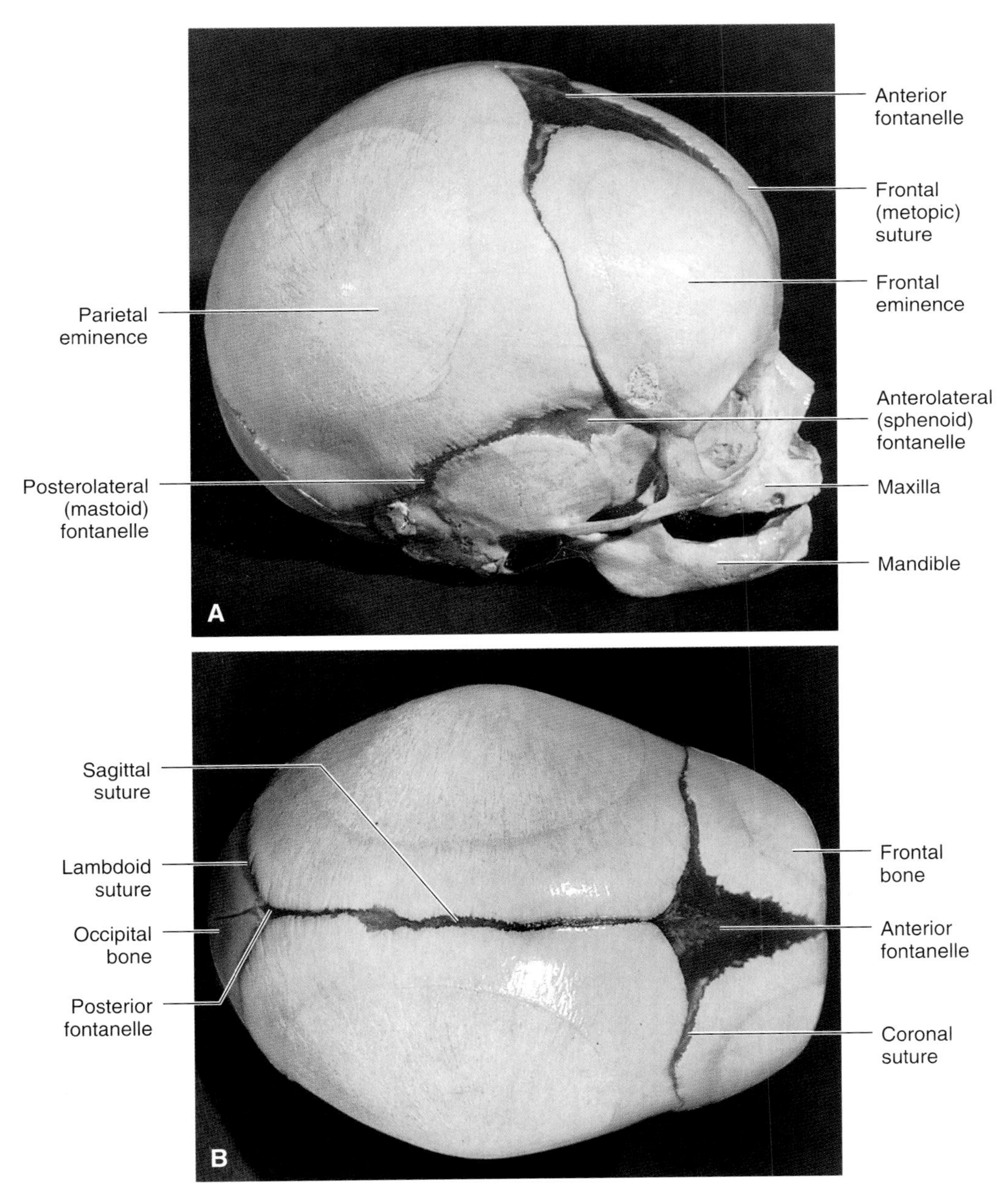

Figure 16 – 8. Photographs of a fetal cranium showing the bones, fontanelles, and connecting sutures. *A,* Lateral view. *B,* Superior view. The posterior and anterolateral fontanelles disappear within 2 or 3 months after birth because of growth of surrounding bones, but they remain as sutures for several years. The posterolateral fontanelles disappear in a similar manner by the end of the first year, and the anterior fontanelle disappears by the end of the second year. The halves of the frontal bone normally begin to fuse during the second year, and the frontal (metopic) suture is usually obliterated by the eighth year. The other sutures disappear during adult life.

maxillary, and zygomatic bones. The squamous temporal bones become part of the neurocranium. The mesenchyme in the mandibular prominence of the first arch condenses around its cartilage and undergoes intramembranous ossification to form the mandible. Some endochondral ossification occurs in the median plane of the chin and in the mandibular condyle.

Newborn Cranium

After recovering from molding, the newborn's cranium is rather round and its bones are thin. Like the fetal cranium (see Fig. 16 – 8), it is large in proportion to the rest of the skeleton, and the face is relatively small compared with the calvaria. The small facial region of the cranium is the result of several factors:

- the small size of the jaws
- the virtual absence of paranasal (air) sinuses
- underdevelopment of the facial bones at birth

Postnatal Growth of the Cranium

The fibrous sutures of the newborn's calvaria permit the brain to enlarge during infancy and childhood. The increase in the size of the calvaria is greatest during the first 2 years of life, the period of most rapid postnatal growth of the brain. The capacity of the calvaria normally increases until about 16 years of age. After this time, its size usually increases slightly for 3 to 4 years because of thickening of its bones. There is also rapid growth of the face and jaws, coinciding with eruption of the primary (deciduous) teeth. These facial changes are more marked after the secondary (permanent) teeth erupt (see Chapter 21). There is concurrent enlargement of the frontal and facial regions, associated with the increase in the size of the paranasal sinuses. Most paranasal sinuses are rudimentary or absent at birth. Growth of these sinuses is important in altering the shape of the face and in adding resonance to the voice.

Klippel-Feil Syndrome (Brevicollis)

The main features of the Klippel-Feil syndrome are a short neck, low hairline, and restricted neck movements. In most cases, the number of cervical vertebral bodies is fewer than normal. In some cases, there is a lack of segmentation of several elements of the cervical region of the vertebral column. The number of cervical nerve roots may be normal, but they are small, as are the intervertebral foramina. Persons with this syndrome are often otherwise normal, but the association of this anomaly with other congenital anomalies is not uncommon.

Spina Bifida

Failure of the halves of the vertebral arch to fuse results in a major defect called spina bifida. **Spina bifida occulta** is a relatively minor anomaly of the vertebral column that usually causes no clinical symptoms. It can be diagnosed in utero by sonography. Spina bifida occulta of the first sacral vertebra occurs in about 20% of vertebral columns that are examined radiographically. The spinal cord and spinal nerves are usually normal, and neurological symptoms are commonly absent. The skin over the bifid vertebral arch is intact, and there may be no external evidence of the vertebral defect. Sometimes, the anomaly is indicated by a tuft of hair. **Spina bifida cystica**, a severe type of spina bifida involving the spinal cord and meninges, is discussed in Chapter 19. Neurological symptoms are present in these infants.

Accessory Ribs

Accessory ribs, usually rudimentary, result from the development of the costal processes of the cervical or lumbar vertebrae (see Fig. 16–6*A*). These processes form ribs in the thoracic region. The most common type of accessory rib is a **lumbar rib**, but it usually causes no problems. **Cervical ribs** occur in 0.5 to 1% of people (Fig. 16–9*A*). A cervical rib is attached to the seventh cervical vertebra and may be unilateral or bilateral. Pressure of a cervical rib on the brachial plexus or the subclavian artery often produces neurovascular symptoms.

Fused Ribs

Fusion of ribs occasionally occurs posteriorly when two or more ribs arise from a single vertebra. Fused ribs are often associated with a hemivertebra.

Hemivertebra

The developing vertebral bodies have two chondrification centers that soon unite. A hemivertebra results from failure of one of the chondrification centers to appear and subsequent failure of half of the vertebra to form (see Fig. 16–9*B*). These vertebral defects produce **scoliosis** (lateral curvature) of the vertebral column.

Rachischisis

The term *rachischisis* (cleft vertebral column) refers to vertebral abnormalities in a complex group of anomalies (*axial dysraphic disorders*) that primarily affect axial structures (Fig. 16–10). In affected infants, the neural folds fail to fuse, either because of faulty induction by the underlying notochord or because of the action of teratogenic agents on the neuroepithelial cells in the neural folds. The neural and vertebral defects may be extensive, or they may be restricted to a small area.

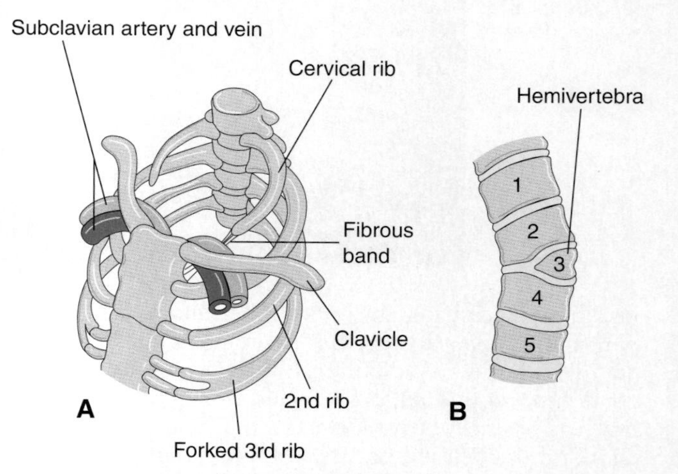

Figure 16–9. Vertebral and rib abnormalities. *A*, Cervical and forked ribs. Observe that the left cervical rib has a fibrous band that passes posterior to the subclavian vessels and attaches to the sternum. *B*, Anterior view of the vertebral column showing a hemivertebra. The right half of the third thoracic vertebra is absent. Note the associated lateral curvature (scoliosis) of the vertebral column.

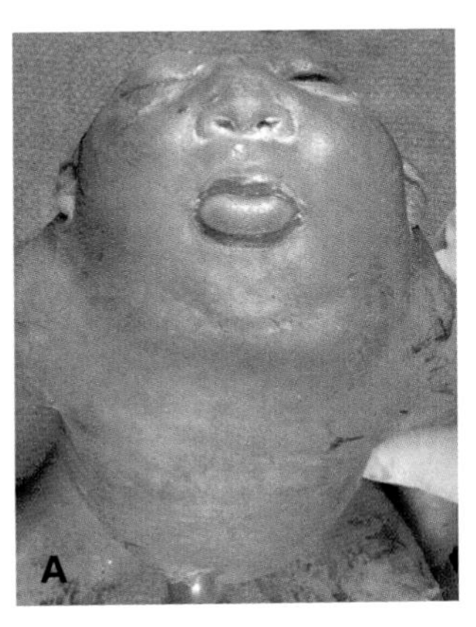

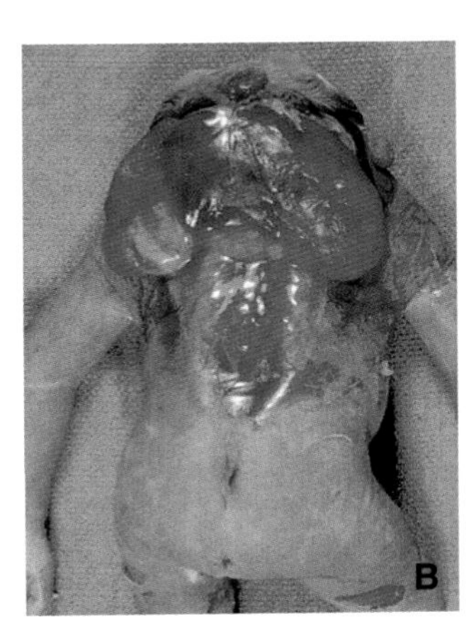

Figure 16 - 10. Photographs depicting anterior *(A)* and posterior *(B)* views of a 20-week fetus with severe defects, including acrania (absence of calvaria), cervical rachischisis (extensive clefts in vertebral arches), cerebral regression (meroanencephaly), and iniencephaly (defect in occiput — back of cranium), and a sacral dimple. (Courtesy of R. Del Bigio, Department of Pathology [Neuropathology], University of Manitoba, Winnipeg, Manitoba Canada.)

Acrania

In acrania, the calvaria is absent, and extensive defects of the vertebral column are often present (Fig. 16–10). Acrania associated with **meroanencephaly** or **anencephaly** (partial absence of the brain) occurs in about one in every 1000 births and is incompatible with life. Meroanencephaly results from failure of the cranial end of the neural tube to close during the fourth week of development. This anomaly causes subsequent failure of the calvaria to form.

Craniosynostosis

Several cranial deformities result from premature closure of the cranial sutures. Prenatal closure results in the most severe abnormalities. The cause of craniosynostosis is unclear, but genetic factors appear to be important. *Homeobox gene (MSX2 and ALX4) mutations have been implicated in cases of craniosynostosis and other cranial defects.* These abnormalities are much more common in males than in females, and they are often associated with other skeletal anomalies. The type of deformed cranium produced depends upon which sutures close prematurely. If the sagittal suture closes early, the cranium becomes elongated and wedge-shaped, a condition known as **scaphocephaly** or *dolichocephaly* (Fig. 16–11*A*). This type of cranial deformity constitutes about half the cases of craniosynostosis. Another 30% of cases involve premature closure of the coronal suture, which results in a high, towerlike cranium, a condition known as **oxycephaly** or *turricephaly* (Fig. 16–11*B*). If the coronal or lambdoid suture closes prematurely on one side only, the cranium is twisted and asymmetric, resulting in a condition known as **plagiocephaly.** Premature closure of the metopic suture results in a keel-shaped deformity of the frontal bone and other anomalies, known collectively as *trigonocephaly.*

Microcephaly

Infants with microcephaly are born with a normal-sized or slightly small calvaria. The fontanelles close during early infancy, and the sutures close during the first year. This anomaly is not caused by premature closure of sutures. Rather, microcephaly is the result of abnormal development of the central nervous system (CNS) that causes the brain, and consequently, the cranium to fail to grow. Usually, microcephalics are severely mentally retarded. This CNS anomaly is discussed further in Chapter 19 (p. 364).

Development of the Appendicular Skeleton

The appendicular skeleton consists of the pectoral and pelvic girdles and the limb bones. Mesenchymal bones

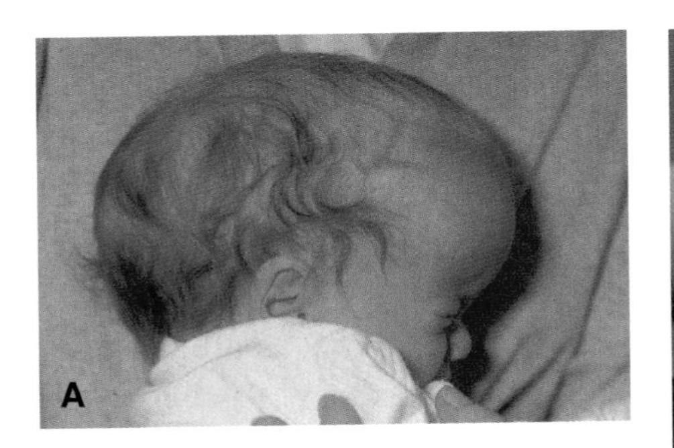

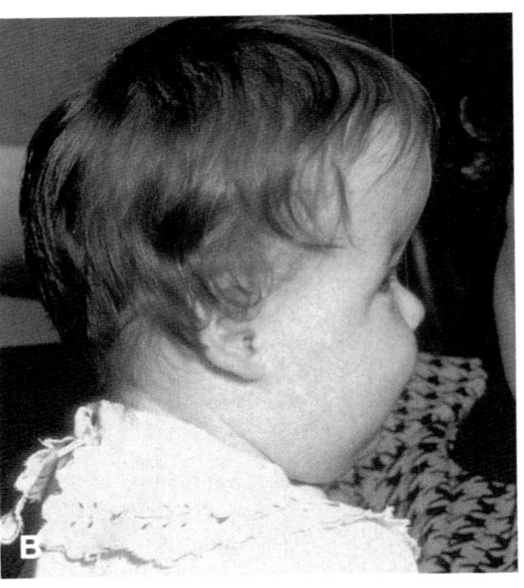

Figure 16 - 11. Craniosynostosis. *A,* Photograph of an infant with an elongated, wedge-shaped cranium (*scaphocephaly* or *dolichocephaly*) resulting from premature closure of the sagittal suture. *B,* Photograph of an infant with bilateral premature closure of the coronal suture (*oxycephaly* or *brachycephaly*) resulting in a high, towerlike forehead. (Courtesy of Dr. John A. Jane, Sr., David D. Weaver Professor of Neurosurgery, Department of Neurological Surgery, University of Virginia Health System, Charlottesville, Virginia.)

form during the fifth week of development as condensations of mesenchyme appear in the limb buds. During the sixth week, the mesenchymal bone models undergo chondrification to form hyaline cartilage bone models (Fig. 16-12*D* and *E*). The clavicle initially develops by intramembranous ossification; and later, growth cartilages form at both ends. The models of the pectoral girdle and upper limb bones appear slightly before those of the pelvic girdle and lower limbs; the bone models appear in a proximodistal sequence. Patterning in the developing limbs is regulated by **homeobox-containing (Hox) genes**. The molecular mechanisms of these Hox genes as they relate to limb morphogenesis remain uncertain.

Ossification begins in the long bones by the eighth week of development and initially occurs in the diaphyses of the bones from **primary centers of ossification** (see Fig. 16-3*B* and *C*). By 12 weeks, primary ossification centers have appeared in nearly all limb bones (Fig. 16-13). The clavicles begin to ossify before any other bones in the body. The femora are the next bones to show traces of ossification. The first indication of ossification in the cartilaginous model of a long bone is visible near the center of the future shaft; this is the primary center of ossification. Primary centers appear at different times in different bones; however, most of them appear between the 7th and 12th weeks of development. Virtually all primary centers of ossification are present at birth. The part of a bone ossified from a primary center is the **diaphysis**.

The secondary ossification centers of the bones at the knee are the first to appear. The centers for the distal end of the femur and the proximal end of the tibia usually appear during the last month of intrauterine life (34 to

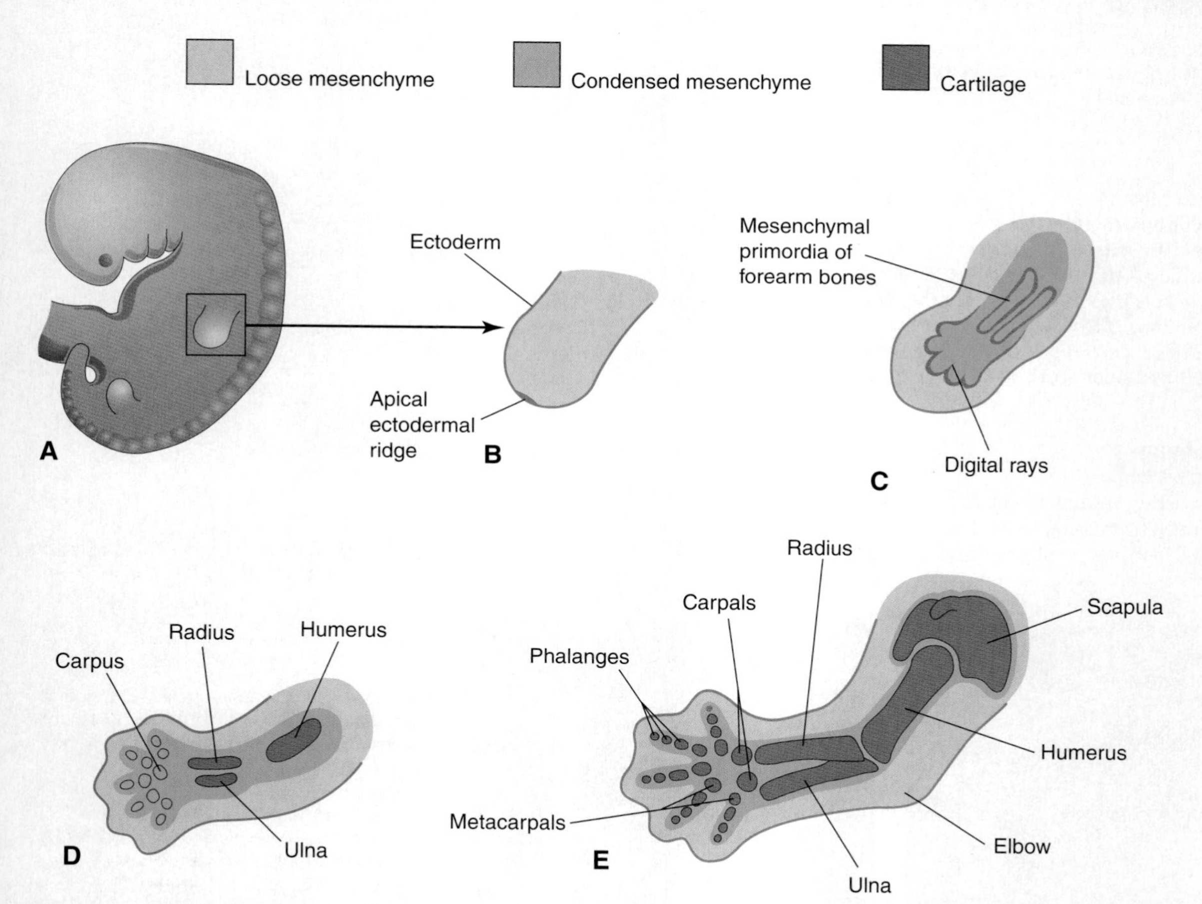

Figure 16-12. *A,* Embryo at about 28 days, showing the early appearance of a limb bud. *B,* Longitudinal section through an upper limb bud. The apical ectodermal ridge has an inductive influence on the mesenchyme in the limb bud; it promotes growth of mesenchyme and appears to give it the ability to form specific cartilaginous elements. *C,* Similar sketch of an upper limb bud at about 33 days, showing the mesenchymal primordia of the limb bones. The digital rays are mesenchymal condensations that undergo chondrification and ossification to form the bones of the hand. *D,* Upper limb at 6 weeks showing the cartilage models of the bones. *E,* Later in the sixth week. Note the completed cartilaginous models of the bones of the upper limb.

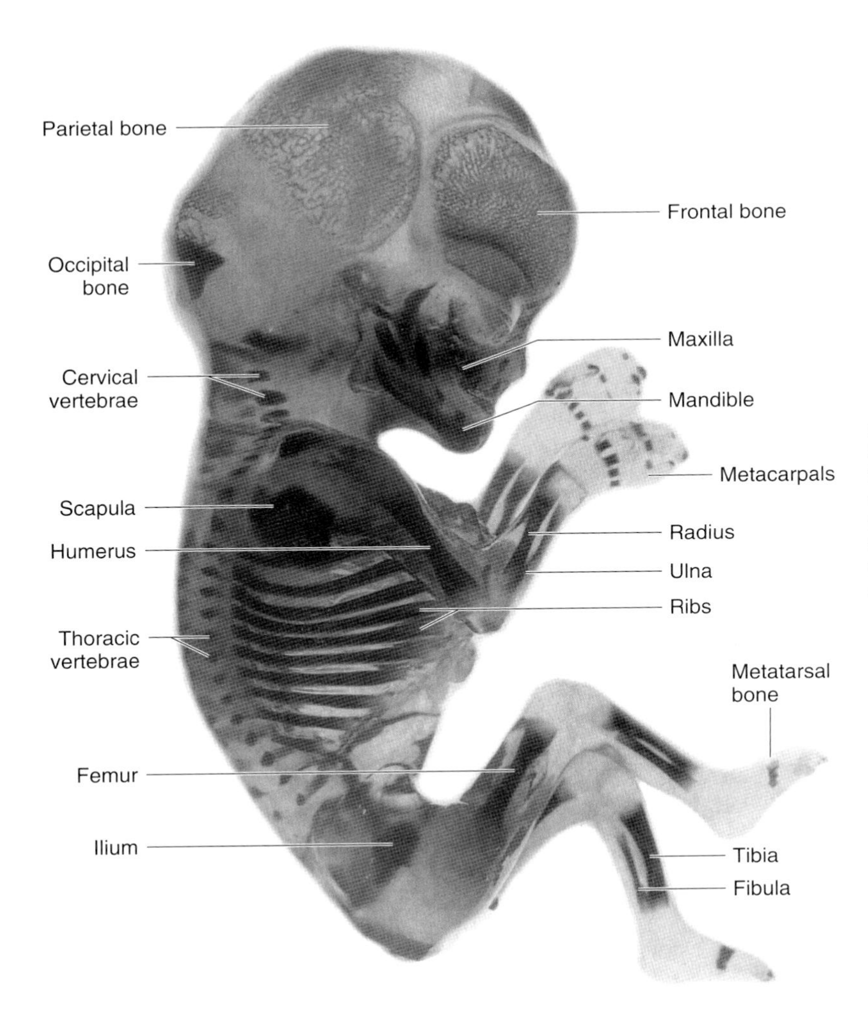

Figure 16-13. Alizarin-stained 12-week human fetus. Observe the degree of progression of ossification from the primary centers of ossification, which are endochondral in the appendicular and axial parts of the skeleton except for most of the cranial bones. Note that the carpus and tarsus are wholly cartilaginous at this stage, as are the epiphyses of all long bones. (Courtesy of Dr. Gary Geddes, Lake Oswego, OR.)

38 weeks). Consequently, they are usually present at birth; however, secondary centers of other bones appear after birth. The part of a bone ossified from a secondary center is the epiphysis. The bone formed from the primary center in the diaphysis does not fuse with that formed from the secondary centers in the epiphyses until the bone grows to its adult length. This delay enables lengthening of the bone to continue until the final size is reached. During bone growth, the epiphysial cartilage plate intervenes between the diaphysis and the epiphysis (see Fig. 16-3). The **epiphysial plate** is eventually replaced by bone development on each of its two sides, diaphysial and epiphysial. When this occurs, growth of the bone ceases.

Bone Age

Bone age is a good index of general maturation. Utilizing radiographic studies to determine the number, size, and fusion of epiphysial centers is a method commonly used to establish bone age. A radiologist can determine the bone age of a person by assessing the ossification centers using two criteria:

- The appearance of calcified material in the diaphysis and/or epiphysis is specific for each diaphysis and epiphysis and for each bone and sex.
- The disappearance of the dark line representing the epiphysial cartilage plate indicates that the epiphysis has fused with the diaphysis.

Fusion of the epiphysial centers, which occurs at specific times for each epiphysis, happens 1 to 2 years earlier in females than in males. Real-time ultrasonography is now increasingly used for the evaluation and measurement of fetal bones.

Generalized Skeletal Malformations

Achondroplasia is a common cause of **dwarfism**, or shortness of stature (see Chapter 9). It occurs in about one in every 15,000 births. The limbs are bowed and short because of disturbance of endochondral ossification at the epiphysial cartilage plates, particularly of the long bones (Fig. 16 – 14), during fetal life. The trunk is usually short, and the head is enlarged with a bulging forehead and "scooped-out" nose (flat nasal bone). Achondroplasia is an *autosomal dominant disorder,* and about 80% of cases arise from new mutations; the rate increases with paternal age.

Hyperpituitarism

Congenital infantile hyperpituitarism, which causes abnormally rapid growth in infancy is rare. This condition may result in **gigantism** (excessive height and body proportions) or *acromegaly* in the adult (enlargement of the soft tissues, visceral organs, and bones of the face, hands, and feet). In **acromegaly**, the epiphysial and diaphysial centers of the long bones fuse, thereby preventing elongation of the bones. Both gigantism and acromegaly result from an excessive secretion of growth hormone.

Hypothyroidism and Cretinism

A severe deficiency of fetal thyroid hormone production results in cretinism, a condition characterized by growth retardation, mental deficiency, skeletal abnormalities, and auditory and neurological disorders. Bone age in an affected individual appears as less than chronological age because epiphysial development is delayed. Cretinism is very rare except in areas where iodine is lacking in the soil and water. Agenesis of the thyroid gland also results in cretinism.

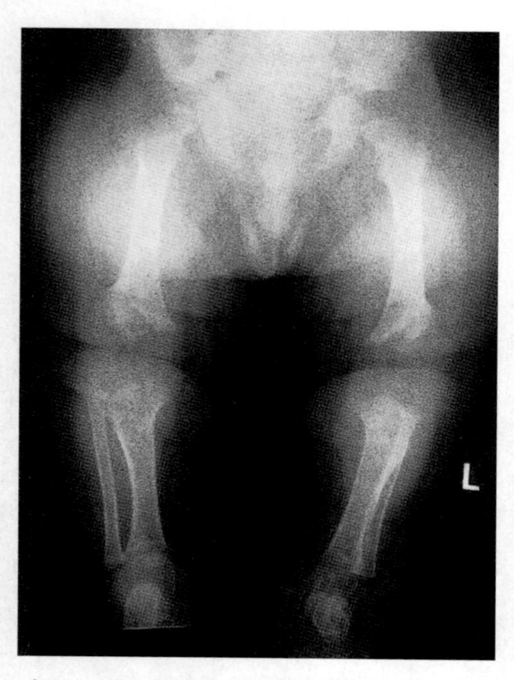

Figure 16 – 14. Achondroplasia in a 1-year-old male infant. The radiograph shows proximal shortening of the lower limbs. (Courtesy of Dr. Prem Sahni, Department of Radiology, Children's Hospital, Winnipeg, Manitoba, Canada.)

Summary of the Skeletal System

The skeletal system develops from mesenchyme derived from mesoderm and the neural crest. In most bones, such as the long bones in the limbs, the condensed mesenchyme undergoes chondrification to form cartilage bone models. Ossification centers appear in these models by the end of the embryonic period, and the bones ossify later by **endochondral ossification**. Some bones, the flat bones of the cranium, for example, develop by **intramembranous ossification**. The vertebral column and ribs develop from mesenchymal cells derived from the sclerotomes of the somites. Each vertebra is formed by fusion of a condensation of the caudal half of one pair of *sclerotomes* with the cranial half of the subjacent pair of sclerotomes.

The developing cranium consists of a neurocranium and a viscerocranium, each of which has membranous and cartilaginous components. The *neurocranium* forms the calvaria and the *viscerocranium* forms the facial skeleton. The *appendicular skeleton* develops from endochondral ossification of the bone models, which form from mesenchyme in the developing limbs.

Joints develop from interzonal mesenchyme between the primordia of bones. In a fibrous joint, the intervening mesenchyme soon differentiates into dense, fibrous connective tissue. In a cartilaginous joint, the mesenchyme between the bones differentiates into cartilage. In a synovial joint, a *synovial cavity* is formed within the intervening mesenchyme by breakdown of the cells. The mesenchyme also gives rise to the synovial membrane and the capsular and other ligaments of the joint.

Although there are numerous types of skeletal anomaly, most of them, except for spina bifida occulta and accessory ribs, are uncommon.

Clinically Oriented Questions

1. What is the most common congenital anomaly of the vertebral column? Where is the defect usually located? Does this congenital anomaly usually cause symptoms (e.g., back problems)?
2. Occasionally, rudimentary ribs are associated with the seventh cervical vertebra and the first lumbar vertebra. Are these accessory ribs of clinical importance? What is the embryological basis of accessory ribs?

3. What vertebral defect can produce scoliosis? Define this condition. What is the embryological basis of the vertebral defect?
4. What is meant by the term *craniosynostosis*? What results from this developmental abnormality? Give a common example and describe it.
5. A child presented with characteristics of the Klippel-Feil syndrome. What are the main features of this condition? What vertebral anomalies are usually present?

The answers to these questions are at the back of the book.

17

The Muscular System

The muscular system develops from **mesoderm**, except for the muscles of the iris, which develop from **neuroectoderm**. Myoblasts, which are embryonic muscle cells, are derived from mesenchyme (embryonic connective tissue). Much of the mesenchyme in the head is derived from the **neural crest**, particularly the tissues derived from the pharyngeal arches; however, the original mesenchyme in the arches gives rise to the musculature of the face and neck (see Table 11 – 1).

Development of Skeletal Muscle

The myoblasts that form the skeletal muscles of the trunk are derived from mesoderm in the myotome regions of the somites. The limb muscles develop from myogenic precursor cells in the limb buds. Studies show that these cells originate from the ventral dermamyotome of somites in response to *molecular signals* from nearby tissues (Fig. 17 – 1). The myogenic precursor cells migrate into the limb buds, where they undergo epitheliomesenchymal transformation. The first indication of **myogenesis** (muscle formation) is the elongation of the nuclei and cell bodies of mesenchymal cells as they differentiate into **myoblasts**. These primordial muscle cells soon fuse to form elongated, multinucleated, cylindrical structures called **myotubes.** At the *molecular level*, these events are preceded by gene activation and expression of the MyoD family of muscle-specific basic helix-loop-helix transcription factors (MyoD, myogenin, Myf-5, and MRF4) in the precursor myogenic cells. It has been suggested that signaling molecules from the ventral neural tube (Shh), notochord (Shh), dorsal neural tube (Wnts, BMP-4), and the overlying ectoderm (Wnts, BMP-4) regulate the beginning of myogenesis and the induction of the myotome.

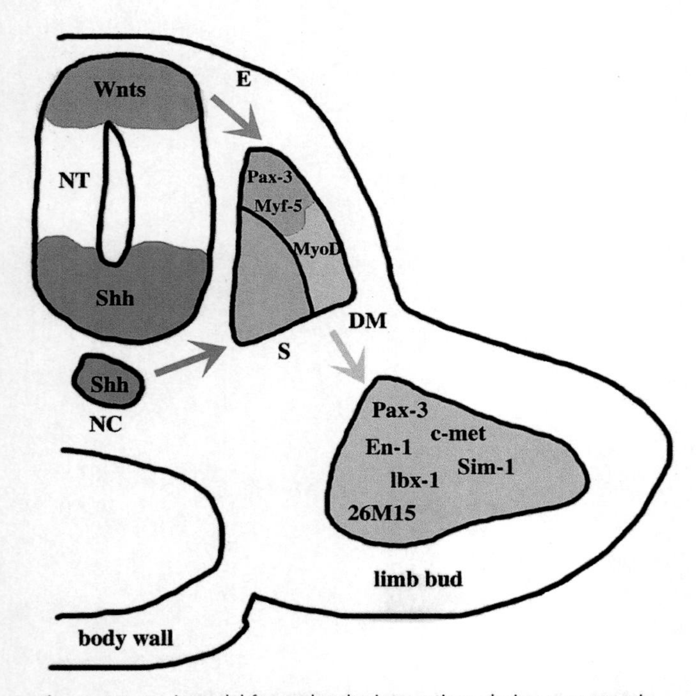

Figure 17 – 1. A model for molecular interactions during myogenesis. Shh and Wnts, produced by the neural tube (NT) and notochord (NC), induce Pax-3 and Myf-5 in the somites. Either of them can activate the initiation of MyoD transcription and myogenesis. Surface ectoderm (E) is also capable of inducing Myf-5 and MyoD. In addition, Pax-3 regulates the expression of c-met, which is necessary for the migratory ability of myogenic precursor cells, which also express En-1, Sim-1, lbx-1, and 26M15. DM, dermamyotome; S, sclerotome. (From Kablar B, Rudnicki MA: Skeletal muscle development in the mouse embryo. *Histol Histopathol* 15:649, 2000.)

Muscle growth during development results from the ongoing fusion of myoblasts and myotubes. **Myofilaments** develop in the cytoplasm of the myotubes during or after fusion of the myoblasts. Soon after that, myofibrils and other organelles characteristic of striated muscle cells develop. Because muscle cells are long and narrow, they are called **muscle fibers**. As the myotubes differentiate, they become invested with external laminae, which segregate them from the surrounding connective tissue. Fibroblasts produce the perimysium and epimysium layers of the fibrous sheath; the endomysium is formed by the external lamina, which is derived from the muscle fiber, and reticular fibers. Most skeletal muscle develops before birth, and almost all remaining muscles are formed by the end of the first year. The increase in the size of a muscle after the first year results from an increase in the diameter of the fibers because of the formation of more myofilaments. Muscles increase in length and width in order to grow with the skeleton.

Myotomes

Typically, each myotome part of a somite divides into a dorsal *epaxial division* and a ventral *hypaxial division* (Fig. 17 – 2*B*). Each developing **spinal nerve** also divides and sends a branch to each division, with the **dorsal primary ramus** supplying the epaxial division and the **ventral primary ramus** supplying the hypaxial division. Some muscles — the intercostal muscles, for example — remain segmentally arranged like the somites, but most myoblasts migrate away from the myotome and form nonsegmented muscles. *Gene-targeting studies* in the mouse embryo suggest that MyoD and Myf-5 are essential for the development of the hypaxial and epaxial muscles, respectively. Both genes are involved in the development of the abdominal and intercostal muscles.

Derivatives of Epaxial Divisions of Myotomes

Myoblasts from the epaxial divisions of the myotomes form the segmental muscles of the main body axis, the

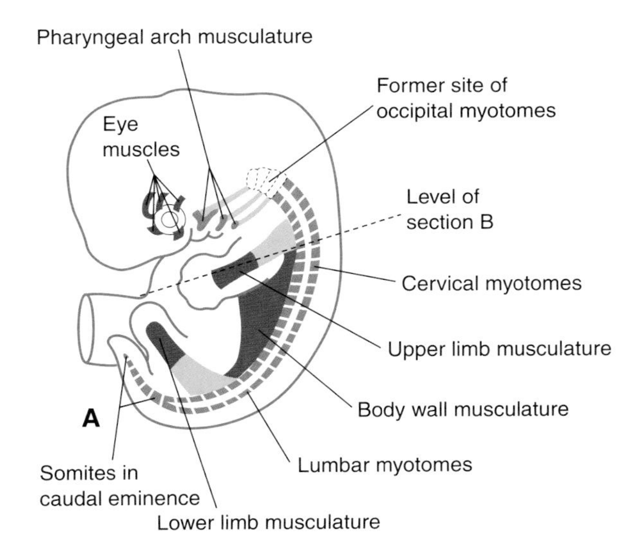

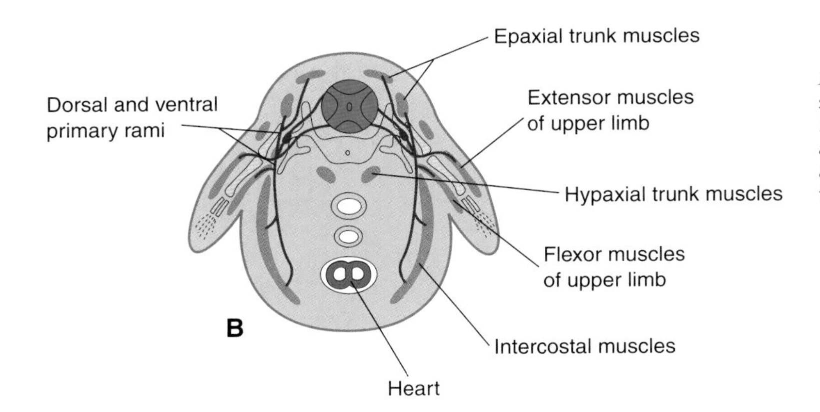

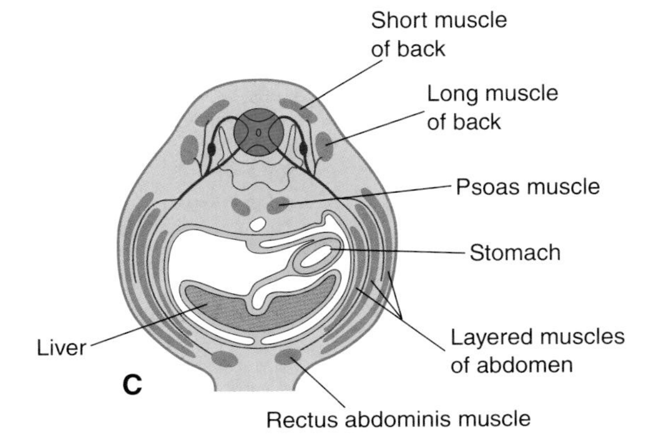

Figure 17-2. *A,* Sketch of an embryo (about 41 days old), showing the myotomes and developing muscular system. *B,* Transverse section of the embryo, illustrating the epaxial and hypaxial derivatives of a myotome. *C,* Similar section of a 7-week embryo, showing the muscle layers formed from the myotomes.

extensor muscles of the neck and vertebral column (Fig. 17-3). The embryonic extensor muscles that are derived from the sacral and coccygeal myotomes degenerate; their adult derivatives are the dorsal sacrococcygeal ligaments.

Derivatives of Hypaxial Divisions of Myotomes

Myoblasts from the hypaxial divisions of the cervical myotomes form the scalene, prevertebral, geniohyoid, and infrahyoid muscles (see Fig. 17-3). The thoracic

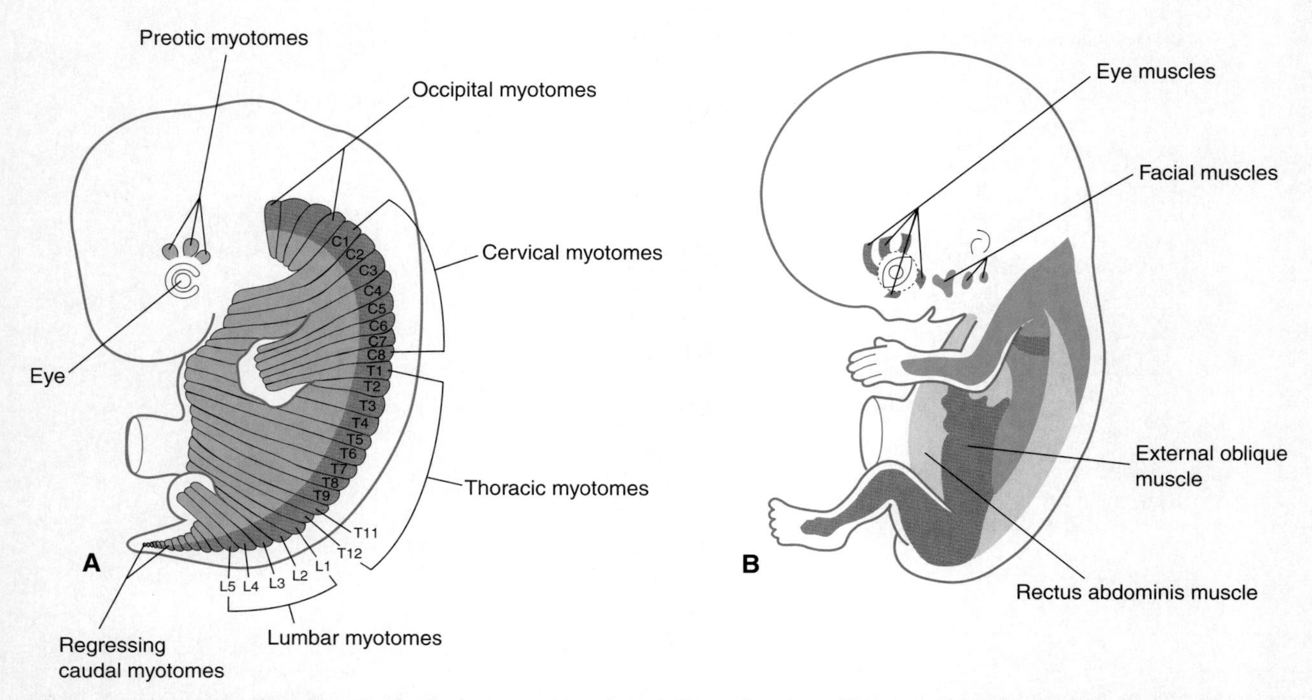

Figure 17–3. Drawings illustrating the developing muscular system. *A,* Six-week embryo. The myotome regions of the somites give rise to most skeletal muscles. *B,* Eight-week embryo showing the developing trunk and limb musculature.

myotomes form the lateral and ventral flexor muscles of the vertebral column, whereas the lumbar myotomes form the quadratus lumborum muscle. The muscles of the limbs, the intercostal muscles, and the abdominal muscles are also derived from the hypaxial division of myotomes. The sacrococcygeal myotomes form the muscles of the pelvic diaphragm and probably the striated muscles of the anus and sex organs.

Pharyngeal Arch Muscles

The migration of myoblasts from the pharyngeal arches to form the muscles of mastication, facial expression, pharynx, and larynx is described in Chapter 11. These muscles are innervated by pharyngeal arch nerves.

Ocular Muscles

The origin of the extrinsic eye muscles is unclear, but it is thought that they may be derived from mesenchymal cells near the prechordal plate (Figs. 17–2 and 17–3). The mesoderm in this area is thought to give rise to three *preotic myotomes.* Myoblasts differentiate from mesenchymal cells derived from these myotomes. Groups of myoblasts, each supplied by its own cranial nerve (CN III, CN IV, or CN VI), form the extrinsic muscles of the eye.

Tongue Muscles

Initially, there are four *occipital* (*postotic*) *myotomes;* however, the first pair disappears. Myoblasts from the remaining myotomes form the tongue muscles, which are innervated by the hypoglossal nerve (CN XII).

Limb Muscles

The musculature of the limbs develops from myoblasts surrounding the developing bones (see Fig. 17–2). Grafting and gene-targeting studies in birds and mammals have demonstrated that the precursor myogenic cells in the limb buds originate from the somites. These cells are first located in the ventral part of the dermamyotome, and they are epithelial in nature (see Fig. 16–1*D*). Following epitheliomesenchymal transformation, the cells migrate into the primordium of the limb. *Molecular signals* from the neural tube and notochord induce Pax-3 and Myf-5 in the somites. Pax-3 regulates the expression of c-met, a migratory peptide growth factor, in the limb bud that is involved in precursor myogenic cell migration.

Development of Smooth Muscle

Smooth muscle fibers differentiate from splanchnic mesenchyme surrounding the endoderm of the primordial gut and its derivatives (see Fig. 16–1). The smooth muscle in the walls of many blood and lymphatic vessels arises from somatic mesoderm. The muscles of the iris (sphincter and dilator pupillae) and the myoepithelial cells in mammary and sweat glands are thought to be derived from mesenchymal cells that originate from ectoderm. The first sign of differentiation of smooth muscle is the development of elongated nuclei in spindle-shaped myoblasts. During early development, new myoblasts continue to differentiate from mesenchymal cells but do not fuse; they remain mononucleated. During later development, division of existing myoblasts gradually replaces the differentiation of new myoblasts in the production of new smooth muscle tissue. Filamentous but nonsarcomeric contractile elements develop in their cytoplasm, and the external surface of each differential cell acquires a surrounding external lamina. As smooth muscle fibers develop into sheets or bundles, they receive autonomic innervation; fibroblasts and muscle cells synthesize and lay down collagenous, elastic, and reticular fibers.

Development of Cardiac Muscle

Cardiac muscle develops from the lateral splanchnic mesoderm, which gives rise to the mesenchyme surrounding the developing heart tube (see Chapter 15). **Cardiac myoblasts** differentiate from this primordial myocardium. Heart muscle is recognizable in the fourth week of development and likely develops through expression of cardiac-specific genes. **Cardiac muscle fibers** arise by differentiation and growth of single cells, unlike striated skeletal muscle fibers, which develop by fusion of cells. Growth of cardiac muscle fibers results from the formation of new **myofilaments**. The myoblasts adhere to each other as in developing skeletal muscle, but the intervening cell membranes do not disintegrate; these areas of adhesion give rise to **intercalated discs**. Late in the embryonic period, special bundles of muscle cells develop with relatively few myofibrils and relatively larger diameters than typical cardiac muscle fibers. These atypical cardiac muscle cells — **Purkinje fibers** — form the conducting system of the heart (see Chapter 15).

Anomalies of Muscles

Any muscle in the body may occasionally be absent; common examples are the sternocostal head of the pectoralis major, the palmaris longus, trapezius, serratus anterior, and quadratus femoris. Absence of the pectoralis major, often its sternal part, is usually associated with syndactyly (fusion of digits). These anomalies are part of the *Poland syndrome*. Absence of the pectoralis major is occasionally associated with absence of the mammary gland and/or hypoplasia of the nipple.

Some muscular anomalies are of a more vital nature, such as **congenital absence of the diaphragm**, which is usually associated with *pulmonary atelectasis* (incomplete expansion of the lungs or part of a lung) and pneumonitis (pneumonia).

Variations in Muscles

Certain muscles are functionally vestigial, such as those of the external ear and scalp. Some muscles that are present in other primates appear only in some humans (e.g., the sternalis muscle). Variations in the form, position, and attachments of muscles are common and are usually functionally insignificant.

The sternocleidomastoid muscle (SCM) is sometimes injured at birth, resulting in **congenital torticollis**. There is fixed rotation and tilting of the head because of fibrosis, and shortening of the SCM on one side (Fig. 17–4). Some cases of torticollis (wryneck) result from tearing of fibers of the SCM during childbirth. Although birth trauma is commonly considered a cause of congenital torticollis, genetic factors may also be involved in some cases.

Accessory Muscles

Accessory muscles occasionally develop, and some are clinically significant. For example, an *accessory soleus muscle* is present in about 6% of the population. It has been suggested that the primordium of the soleus muscle undergoes early splitting to form an accessory soleus.

Summary of the Muscular System

Skeletal muscle is derived from the myotome regions of somites. Some head and neck muscles are derived from pharyngeal arch mesoderm. The limb muscles develop from myogenic precursor cells, which are derived from somites. Cardiac muscle and most smooth muscle are derived from splanchnic mesoderm. Absence or variation of some muscles is common and is usually of little consequence.

Clinically Oriented Questions

1. A newborn infant was born with the prune-belly syndrome caused by failure of the abdominal musculature to

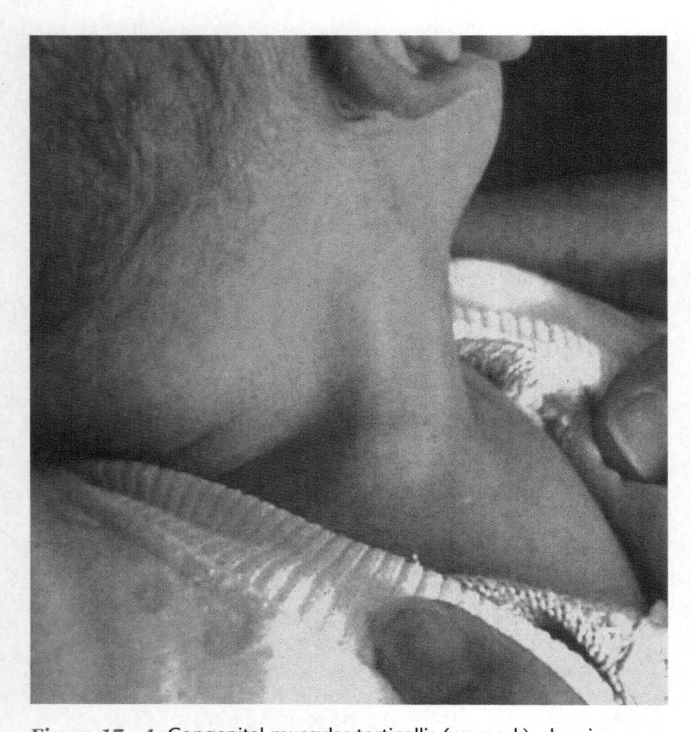

Figure 17 – 4. Congenital muscular torticollis (wryneck), showing extensive involvement of the left sternocleidomastoid muscle, in an infant who was examined at the age of 2 months. (Courtesy of Professor Jack CY Cheng, Department of Orthopaedics & Traumatology, Chinese University of Hong Kong, Hong Kong.)

develop normally. What do you think would cause this congenital anomaly? What urinary anomaly results from abnormal development of the anterior abdominal wall?

2. A boy asked his mother why one of his nipples was much lower than the other one. She was unable to explain this anomaly to her son. How would you explain the abnormally low position of the nipple?
3. An 8-year-old girl asked her doctor why the muscle on one side of her neck was so prominent. What would you tell her? What would happen if this muscle was not treated?
4. After strenuous exercise, a young athlete complained of pain on the posteromedial aspect of his ankle. He was told he had an accessory calf muscle. Is this possible? If so, what is the embryological basis of this anomaly?

The answers to these questions are at the back of the book.

18
The Limbs

The general features of limb development are described and illustrated in Chapter 6. The development of limb bones is described in Chapter 16, and the formation of limb musculature is outlined in Chapter 17. The purpose of this chapter is to consolidate this material and provide additional information about limb development.

Early Stages of Limb Development

The **limb buds** first appear as small elevations of the ventrolateral body wall toward the end of the fourth week of development (Fig. 18 – 1*A*). Limb development begins with the activation of a group of mesenchymal cells in the lateral mesoderm. The upper limb buds are visible by day 26 or 27, whereas the lower limb buds appear a day or two later. Each limb bud consists of a mass of mesenchyme covered by ectoderm. The mesenchyme is derived from the somatic layer of lateral mesoderm. The limb buds elongate by the proliferation of the mesenchyme. The early stages of limb development are alike for the upper and lower limbs; however, development of the upper limb buds precedes that of the lower limb buds by about 2 days (see Fig. 18 – 1*A* and *B*). In addition, there are distinct differences between the development of the hands and feet because of their form and function. The upper limb buds develop opposite the caudal cervical segments, whereas the lower limb buds form opposite the lumbar and upper sacral segments.

At the apex of each limb bud, the ectoderm thickens to form an **apical ectodermal ridge** (AER). The **AER**, a multilayered epithelial structure (Fig. 18 – 2), interacts with the mesenchyme in the limb bud, promoting outgrowth of the bud. Expression of endogenous fibroblast growth factors (FGFs) in the AER is involved in this process. *The AER exerts an inductive influence on the limb mesenchyme that initiates growth and development of the limbs in a proximodistal (proximal-distal) axis.* Mesenchymal cells aggregate at the posterior margin of the limb bud to form a **zone of polarizing activity** (ZPA). FGFs from the AER activate the ZPA, which causes expression of the *sonic hedgehog gene* (Shh). *Molecular studies* show that Shh secretions control the patterning of the limb along the anteroposterior axis. Expression of Wnt7 from the dorsal epidermis of the limb bud and engrailed-1 (En-1) from the ventral aspect are involved in specifying the dorsoventral axis. Of interest, the AER itself is maintained by inductive signals from Shh and Wnt7. The mesenchyme adjacent to the AER consists of undifferentiated, rapidly proliferating cells, whereas mesenchymal cells proximal to it differentiate into blood vessels and cartilage bone models. Expression of *homeobox genes* is essential for the normal patterning of the limbs. The distal ends of the flipper-like limb buds eventually flatten into paddle-like hand and foot plates (Fig. 18 – 3).

By the end of the sixth week of development, mesenchymal tissue in the **hand plates** has condensed to form **digital rays** (see Figs. 18 – 3 and 18 – 4*A* to *C*). These mesenchymal condensations — *finger buds* — outline the pattern of the digits (fingers). During the seventh week, similar condensations of mesenchyme form digital rays — *toe buds* — in the foot **plates** (see Fig. 18 – 4*E* to *I*). At the tip of each digital ray, a part of the AER induces development of the mesenchyme into the mesenchymal primordia of the bones (phalanges) in the digits. The intervals between the digital rays are occupied by loose mesenchyme. Soon the intervening regions of mesenchyme break down, forming **notches between the digital rays** (see Figs. 18 – 3 and 18 – 4*D* and *J*). As this tissue breakdown progresses, separate digits are produced by the end of the eighth week of development (Fig. 18 – 3). **Programmed cell death** (apoptosis) is responsible for the tissue breakdown in the interdigital regions, and it is probably mediated by **bone morphogenetic proteins** (BMPs), signaling molecules of the transforming growth factor (TGF-β) superfamily. Blocking of these cellular and molecular events could account for webbing or fusion of the fingers or toes, a condition known as **syndactyly** (see Fig. 18 – 11*C* and *D*).

Final Stages of Limb Development

As the limbs elongate during the early part of the fifth week of development, mesenchymal models of the bones are formed by cellular aggregations (Fig. 18 – 5*A* and *B*). **Chondrification centers** appear later in the fifth week. By the end of the sixth week, the entire limb skeleton is cartilaginous (see Fig. 18 – 5*C* and *D*). **Osteogenesis of the long bones** begins in the seventh week from primary ossification centers in the middle of the cartilaginous models of the long bones. **Primary ossification centers** are present in all long bones by the 12th week. Ossification of the carpal (wrist) bones begins during the first year after birth.

As the long bones form, myoblasts aggregate and form a large muscle mass in each limb bud (see Fig. 17 – 2). In general, this muscle mass separates into dorsal (extensor) and ventral (flexor) components. The mesenchyme in the limb bud gives rise to bones, ligaments, and blood vessels (see Fig. 18 – 5). From the dermamyotome regions of the somites, myogenic precursor cells also migrate into the limb bud and later differentiate into **myoblasts**, the precursors of muscle cells. The cervical and lumbosacral myotomes contribute to the muscles of the pectoral and pelvic girdles, respectively.

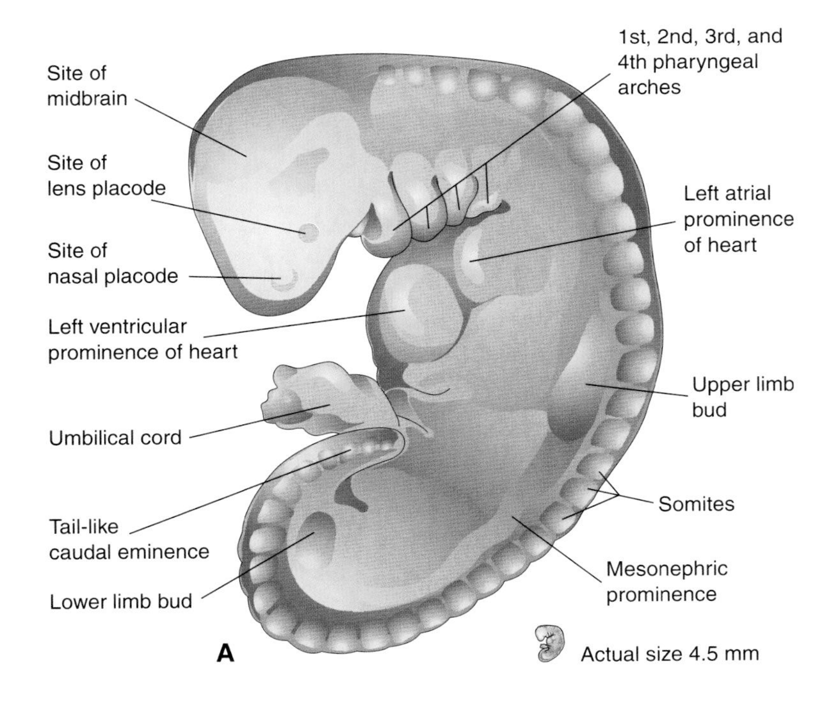

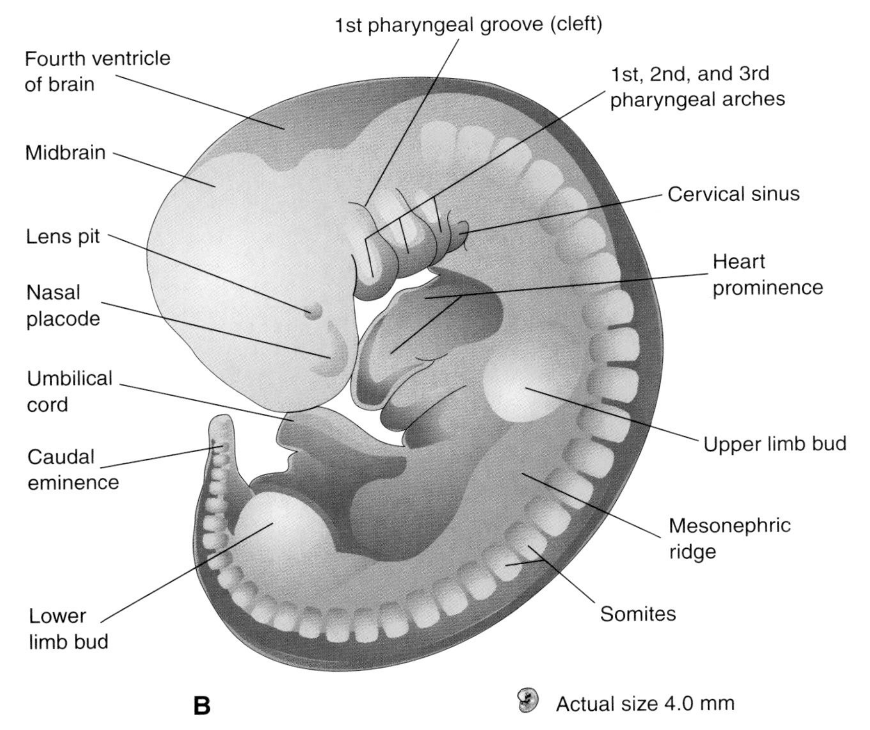

Figure 18 - 1. *A,* Lateral view of a human embryo, about 28 days. The upper limb buds appear as swellings on the ventrolateral body wall. The lower limb buds are not as well developed. *B,* Lateral view of an embryo, about 32 days. The upper limb buds are paddle-shaped, whereas the lower limb buds are flipper-like. (Modified from Nishimura H, Semba R, Tanimura T, Tanaka O: *Prenatal Development of the Human with Special Reference to Craniofacial Structures: An Atlas.* Washington, DC, National Institute of Health, 1977.)

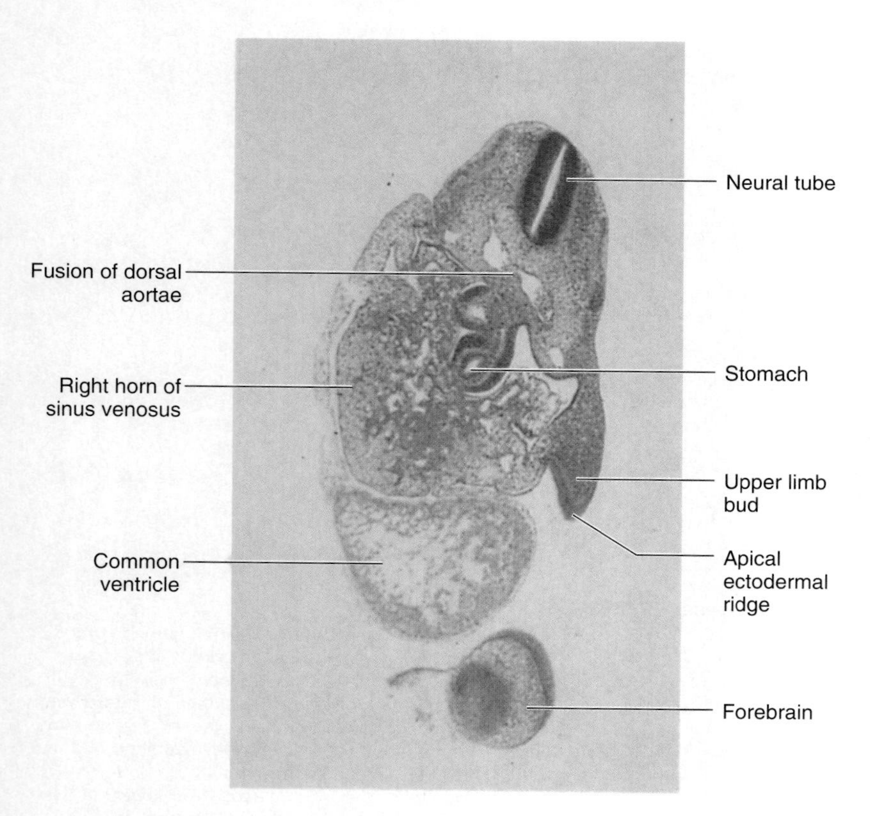

Figure 18 - 2. Oblique section of an embryo at about 28 days. Observe the flipper-like upper limb bud lateral to the embryonic heart. (From Moore KL, Persaud TVN, Shiota K: *Color Atlas of Clinical Embryology,* 2nd ed. Philadelphia, WB Saunders, 2000.)

Early in the seventh week, the limbs extend ventrally. The developing upper and lower limbs rotate in opposite directions and to different degrees (Fig. 18 - 6*A* to *D*):

- *The upper limbs rotate laterally through 90 degrees on their longitudinal axes*; thus the future elbows point dorsally and the extensor muscles lie on the lateral and posterior aspects of the limb.
- *The lower limbs rotate medially through almost 90 degrees*; thus the future knees face ventrally and the extensor muscles lie on the anterior aspect of the lower limb.

The radius and the tibia are homologous bones, as are the ulna and fibula, just as the thumb and great toe are homologous digits. Originally, the flexor aspect of the limbs is ventral and the extensor aspect dorsal, and the preaxial and postaxial borders are cranial and caudal, respectively (Fig. 18 - 7*A* and *D*). **Synovial joints** appear at the beginning of the fetal period, coinciding with functional differentiation of the limb muscles and their innervation.

Cutaneous Innervation of the Limbs

Because of its relationship to the growth and rotation of the limbs, the cutaneous segmental nerve supply of the limbs is considered in this chapter rather than in Chapter 19 on the nervous system. **Motor axons** arising from the spinal cord enter the limb buds during the fifth week of development and grow into the dorsal and ventral muscle masses. **Sensory axons** enter the limb buds after the motor axons and use them for guidance. **Neural crest cells**, the precursors of neurolemma (Schwann) cells, surround the motor and sensory nerve fibers in the limbs and form the *neurolemmal* and *myelin sheaths* (see Chapter 19).

A **dermatome** is the area of skin supplied by a single spinal nerve and its spinal ganglion. During the fifth week of development, the peripheral nerves grow from the developing limb plexuses (brachial and lumbosacral) into the mesenchyme of the limb buds (see Fig. 18 - 7*A, B,* and *E*). The spinal nerves are distributed in segmental bands, supplying both dorsal and ventral surfaces of the limb buds. As the limbs elongate, the cutaneous distribution of the spinal nerves migrates along the limbs and no longer reaches the surface in the distal part of the limbs. Although the original dermatomal pattern changes during growth of the limbs, an orderly sequence of distribution can still be recognized in the adult (see Fig. 18 - 7*C* and *F*). In the upper limb, the areas supplied by C5 and C6 adjoin the areas supplied by T2, T1, and C8, but

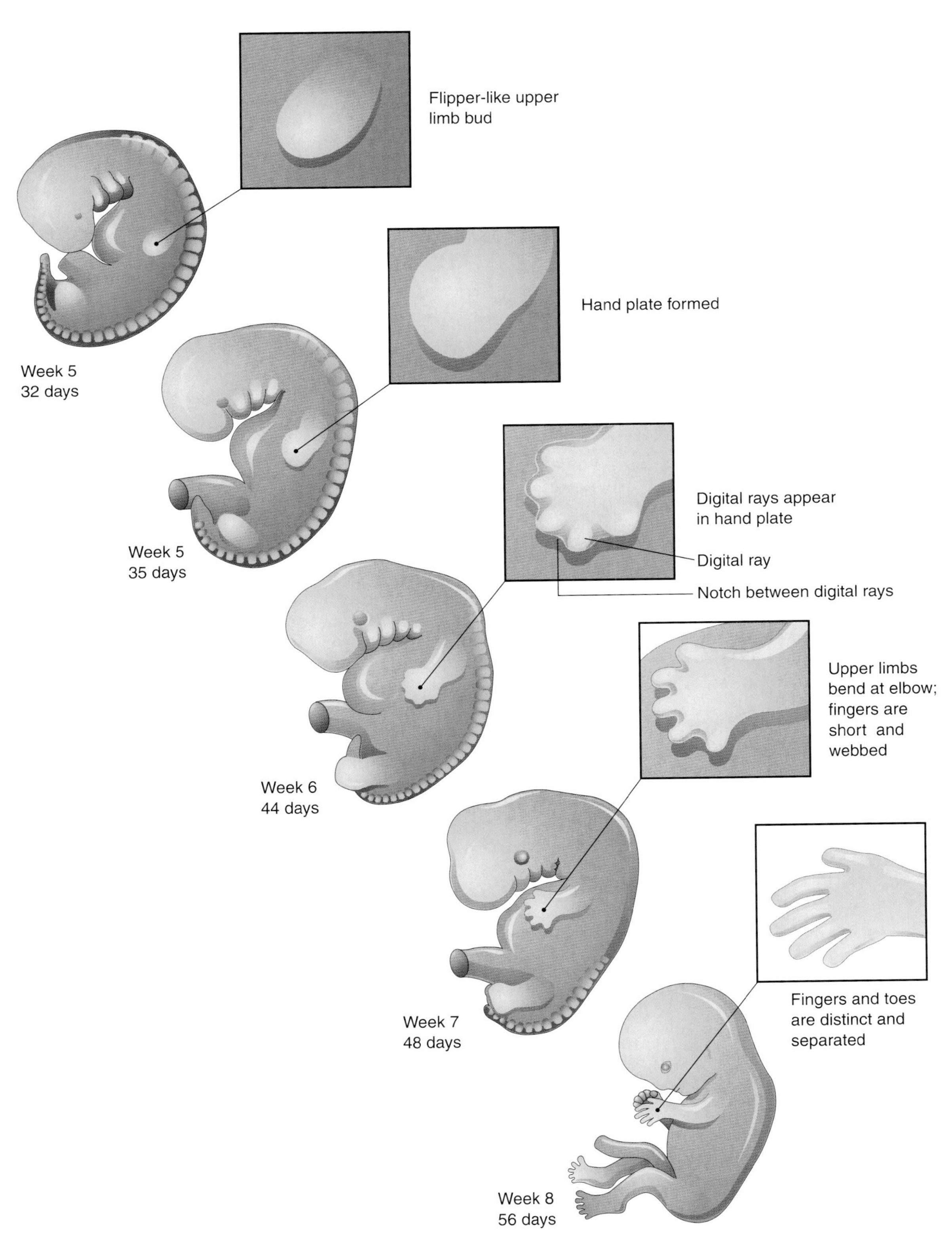

Figure 18 - 3. Drawings illustrating development of the limbs (32 to 56 days).

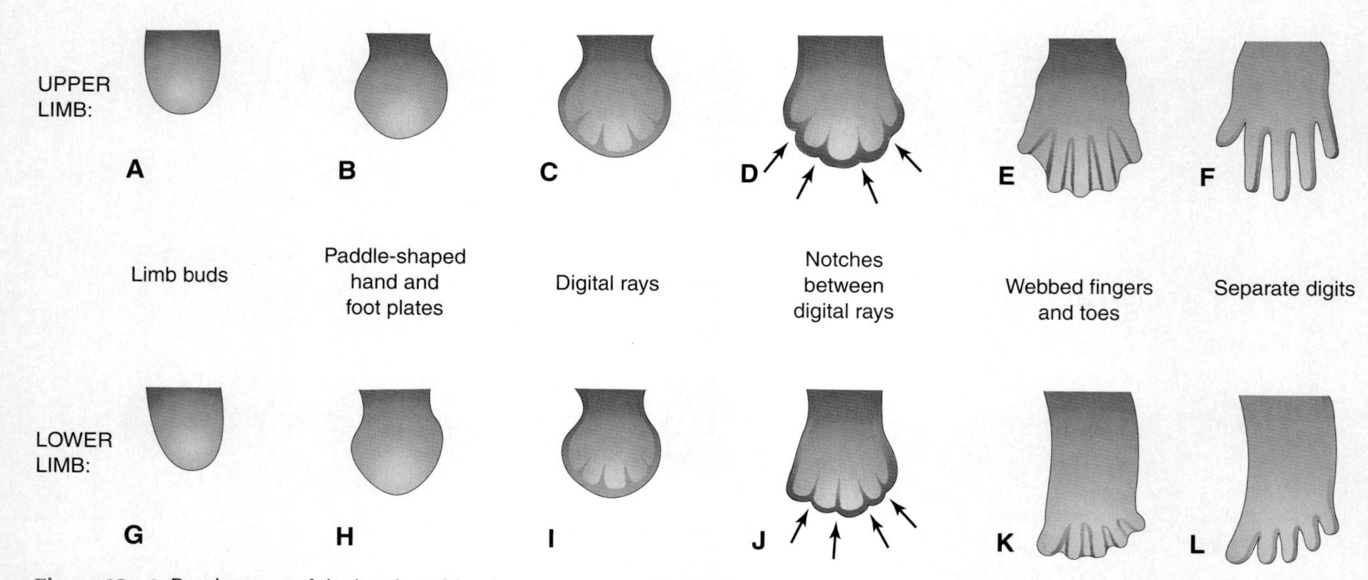

Figure 18 - 4. Development of the hands and feet between the fourth and eighth weeks. The early stages of limb development are similar except that development of the hands precedes that of the feet by a day or so. *A*, 27 days; *B*, 32 days; *C*, 41 days; *D*, 46 days; *E*, 50 days; *F*, 52 days; *G*, 28 days; *H*, 36 days; *I*, 46 days; *J*, 49 days; *K*, 52 days; *L*, 56 days.

the overlap between them is minimal at the *ventral axial line*.

A **cutaneous nerve area** is the area of skin supplied by a peripheral nerve. Cutaneous nerve areas and dermatomes show considerable overlapping. If the dorsal root supplying the area is cut, the dermatomal patterns indicate that there may be a slight deficit in the area indicated. Because there is overlapping of dermatomes, a particular area of skin is not exclusively innervated by a single segmental nerve. The limb dermatomes may be traced progressively down the lateral aspect of the upper limb and back up its medial aspect. A comparable distribution of dermatomes occurs in the lower limbs, which may be traced down the ventral aspect and then up the dorsal aspect of the lower limb. When the limbs descend, they carry their nerves with them; this explains the oblique course of the nerves arising from the brachial and lumbosacral plexuses.

Blood Supply to the Limbs

The limb buds are supplied by branches of the **dorsal intersegmental arteries** (Fig. 18 - 8*A*), which arise from the dorsal aorta and form a fine capillary network throughout the mesenchyme. The primordial vascular pattern consists of a **primary axial artery** and its branches (see Fig. 18 - 8*B*), which drain into a peripheral marginal sinus. Blood in the sinus drains into a peripheral vein. The vascular pattern changes as the limbs develop, chiefly as a result of vessels sprouting from existing vessels. The new vessels coalesce with other sprouts to form new vessels. The primary axial artery becomes the **brachial artery** in the arm and the **common interosseous artery** in the forearm, which has anterior and posterior interosseous branches. The **ulnar and radial arteries** are terminal branches of the brachial artery. As the digits form, the marginal sinus breaks up and the final venous pattern, represented by the basilic and cephalic veins and their tributaries, develops. In the thigh, the primary axial artery is represented by the **deep artery of the thigh** (profunda femoris artery). In the leg, the primary axial artery is represented by the anterior and posterior tibial arteries.

Anomalies of the Limbs

Minor limb anomalies are relatively common. Although these anomalies are usually of no serious medical consequence, they may indicate the presence of more severe anomalies, and they may be part of a recognizable pattern of birth defects. *The most critical period of limb development is from 24 to 36 days after fertilization.* This statement is based on clinical studies of infants exposed to thalidomide, a potent human teratogen during the embryonic period that produces **limb defects** and other anomalies (Fig. 18 - 9*A* and *B*). Exposure to a potent teratogen before day 33 may cause severe anomalies, such as absence of the limbs and hands. Exposure to a teratogen from days 34 to 36 produces absence or hypoplasia of the digits. Consequently, a teratogen that

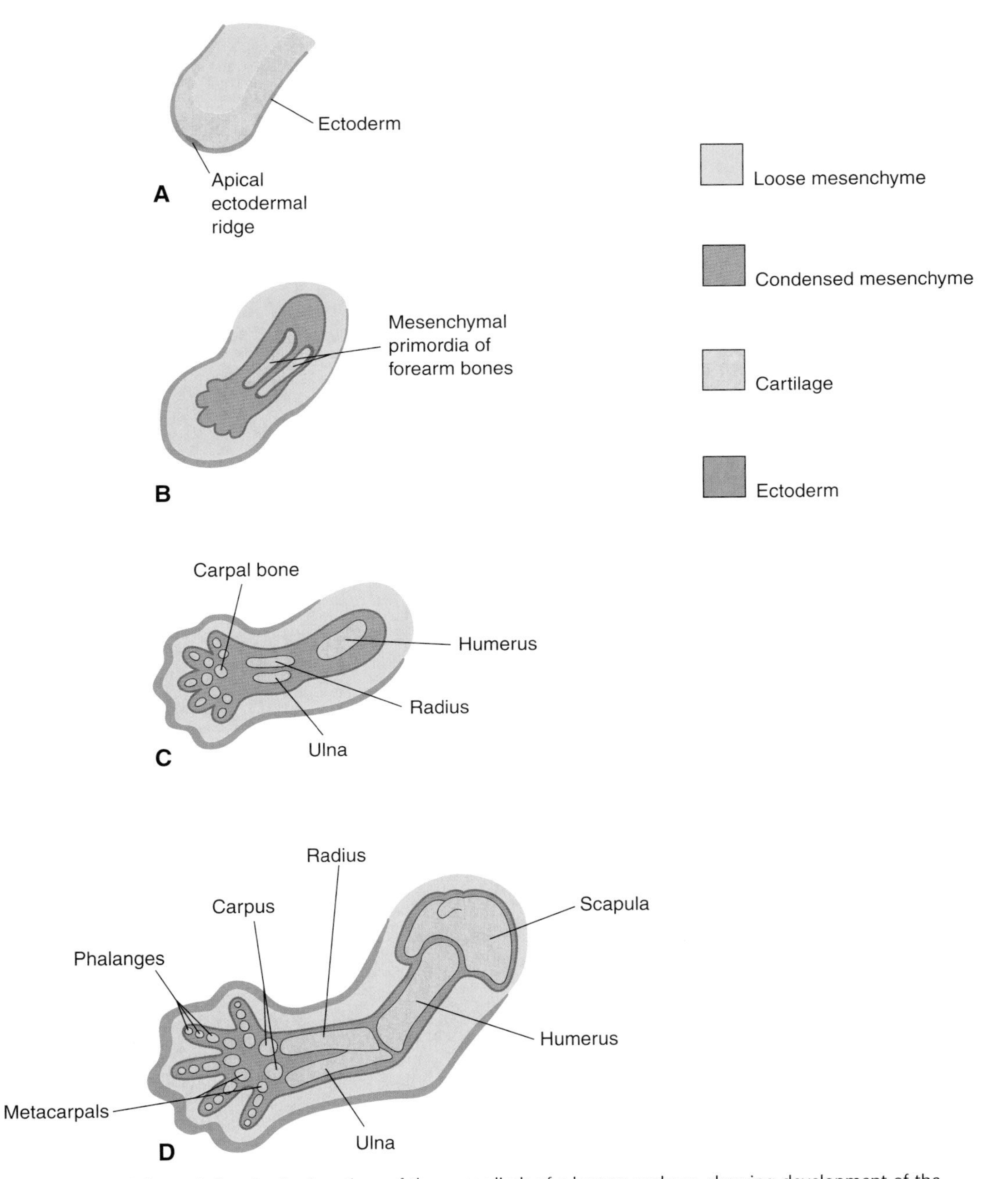

Figure 18 - 5. Schematic longitudinal sections of the upper limb of a human embryo, showing development of the cartilaginous bones. *A*, 28 days. *B*, 44 days. *C*, 48 days. *D*, 56 days.

could cause absence of the limbs or parts of them must act before the end of the critical period of limb development. Many severe limb anomalies occurred from 1957 to 1962 as a result of maternal ingestion of **thalidomide** (see Chapter 9).

Limb Defects

The terminology used to describe limb defects in this book follows the international nomenclature, in which only two basic descriptive terms are used:

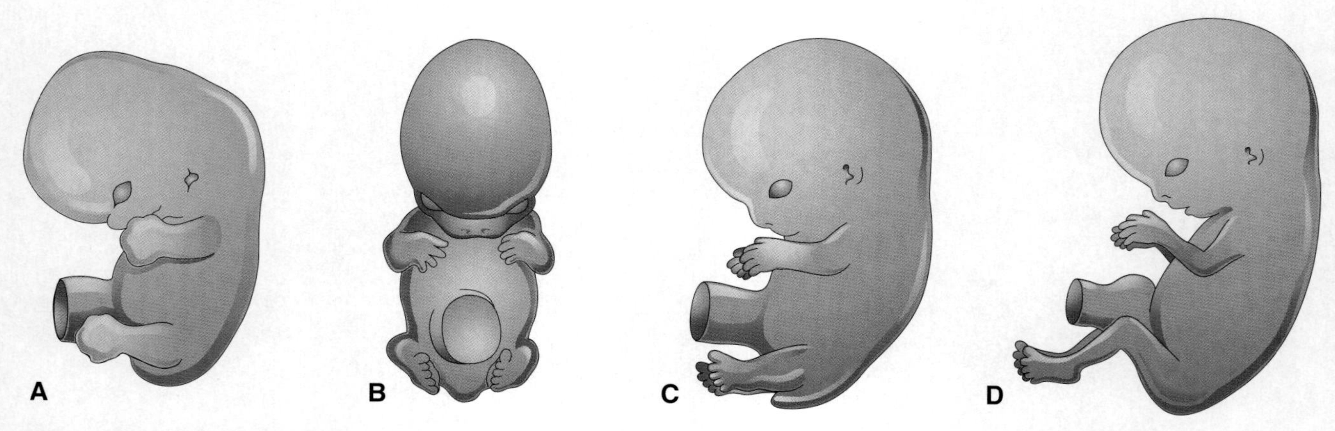

Figure 18 - 6. Positional changes of the developing limbs of human embryos. *A,* About 48 days, showing the limbs extending ventrally and the hand and foot plates facing each other. *B,* About 51 days, showing the upper limbs bent at the elbows and the hands curved over the thorax. *C,* About 54 days, showing the soles of the feet facing medially. *D,* About 56 days. Note that the elbows now point caudally and the knees cranially.

- *amelia*, or complete absence of a limb
- *meromelia* (Gr. *meros*, part, and *melos*, extremity), or partial absence of a limb

Terms such as *hemimelia*, *peromelia*, *ectromelia*, and *phocomelia*, are not used in current nomenclature because of their imprecision.

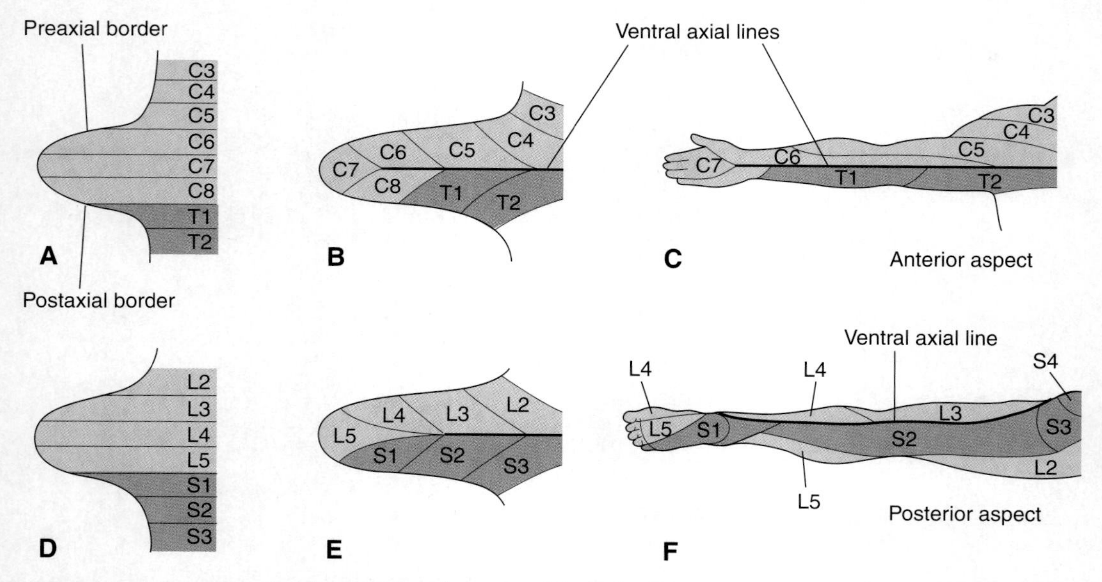

Figure 18 - 7. Development of the dermatomal patterns of the limbs. The axial lines indicate where no sensory overlap is present. *A* and *D,* Ventral aspect of the limb buds early in the fifth week. At this stage, the dermatomal patterns show the primordial segmental arrangement. *B* and *E,* Similar views later in the fifth week, showing the modified arrangement of dermatomes. *C* and *F,* The dermatomal patterns in the adult upper and lower limbs. The primordial dermatomal pattern has disappeared, but an orderly sequence of dermatomes can still be recognized. In *F,* note that most of the original ventral surface of the lower limb lies on the back of the adult limb. This results from the medial rotation of the lower limb that occurs toward the end of the embryonic period. In the upper limb, the ventral axial line extends along the anterior surface of the arm and forearm. In the lower limb, the ventral axial line extends along the medial side of the thigh and knee to the posteromedial aspect of the leg to the heel.

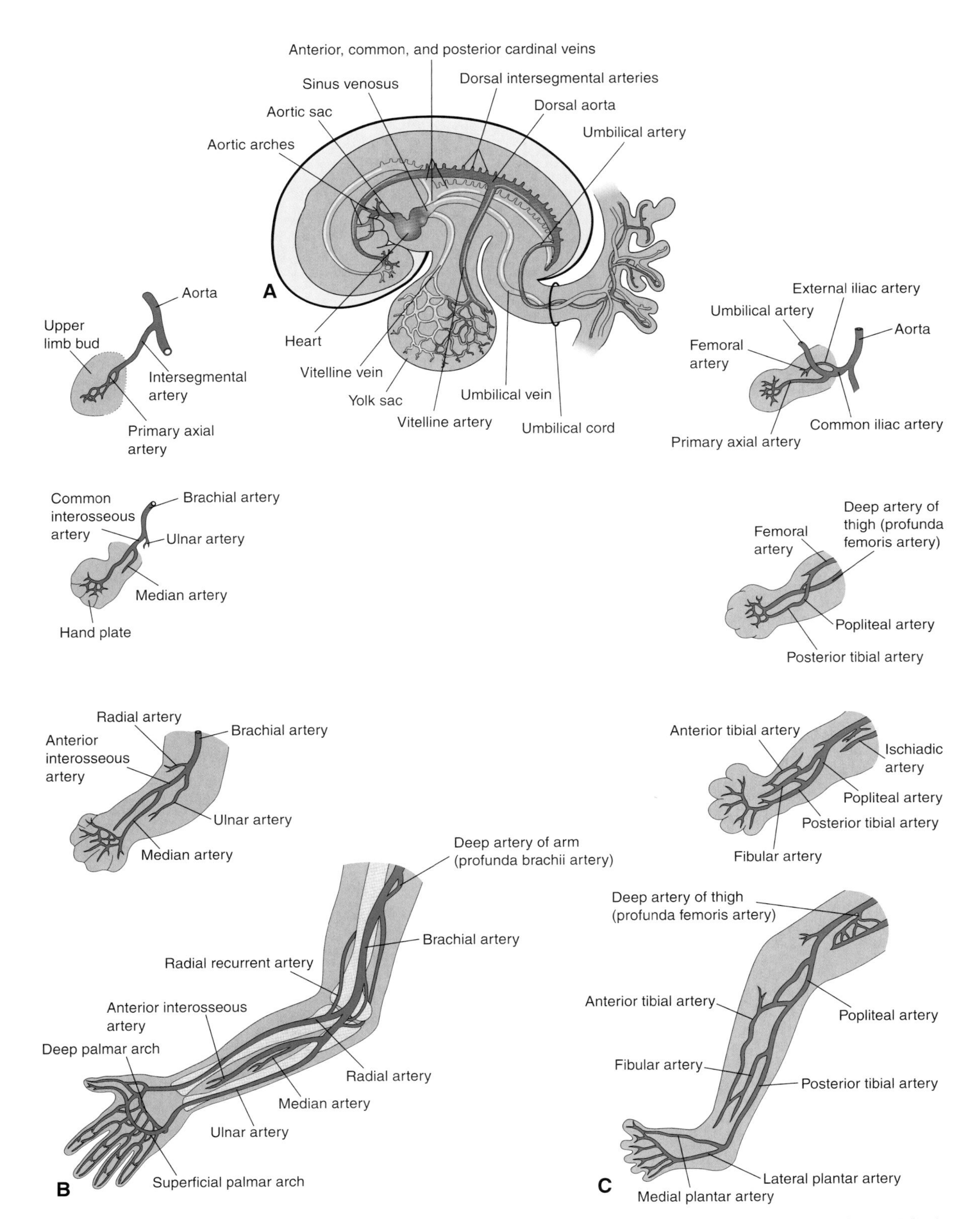

Figure 18 - 8. Development of limb arteries. *A,* Sketch of the primordial cardiovascular system in a 4-week embryo (about 26 days). *B,* Development of arteries in the upper limb. *C,* Development of arteries in the lower limb.

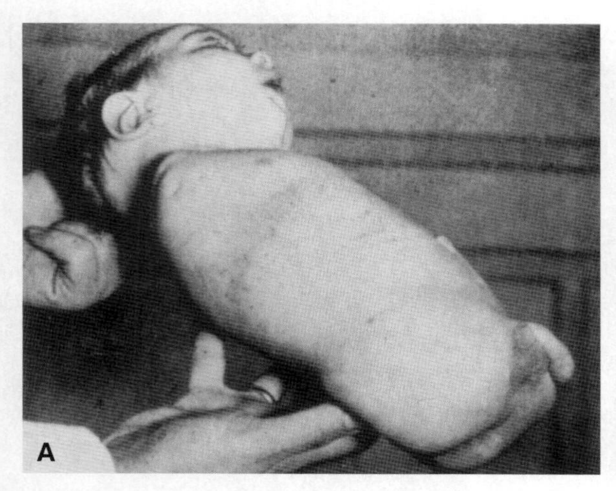

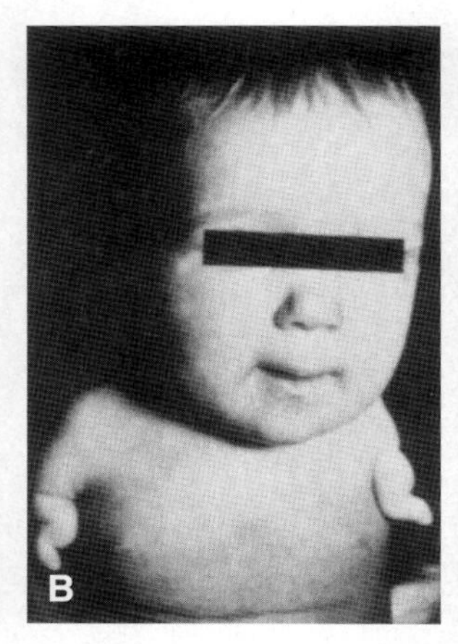

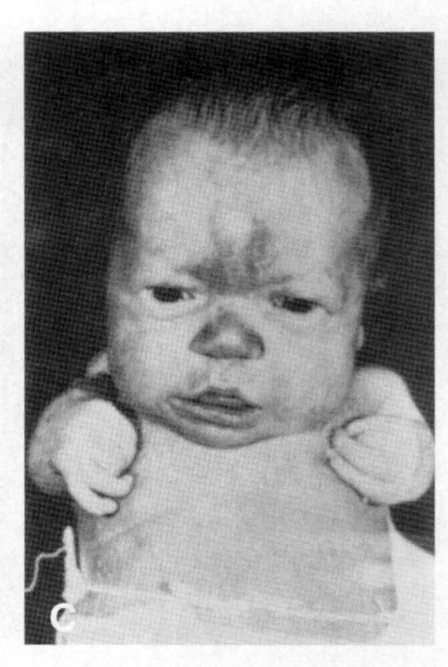

Figure 18 - 9. Limb anomalies caused by thalidomide. *A,* Quadruple amelia (absence of the upper and lower limbs). *B,* Meromelia of the upper limbs. The limbs are represented by rudimentary stumps. *C,* Meromelia in which the rudimentary upper limbs are attached directly to the trunk. (From Lenz W, Knapp K: Foetal malformations due to thalidomide. *Ger Med Mon* 7:253, 1962.)

Cleft Hand and Cleft Foot

In the rare "lobster-claw deformities" known as cleft hand or foot, one or more central digits are absent ectodactylyl, resulting from failure of one or more digital rays to develop (Fig. 18–10*A* and *B*). The hand or foot is divided into two parts that oppose each other like lobster claws. The remaining digits are partially or completely fused (syndactyly).

Congenital Absence of the Radius

In some individuals, the radius is partially or completely absent. The hand deviates laterally (radially), and the ulna bows with the concavity on the lateral side of the forearm. This anomaly results from failure of the mesenchymal primordium of the radius to form during the fifth week of development. Absence of the radius is usually caused by genetic factors.

Brachydactyly

Shortness of the digits (fingers or toes) is uncommon and is the result of a reduction in the length of the phalanges. This anomaly is usually inherited as a dominant trait and is often associated with shortness of stature.

Polydactyly

Supernumerary digits are common (Fig. 18–11*A* and *B*). Often, the extra digit is incompletely formed and lacks proper muscular development, rendering it useless. If the hand is affected, the extra digit is most commonly medial or lateral rather than central. In the foot, the extra toe is usually on the lateral side. Polydactyly is inherited as a dominant trait.

Syndactyly

Syndactyly occurs with a frequency of about 1 in 2200 births. Cutaneous syndactyly (simple webbing of the digits) is the most common limb anomaly (see Fig. 18–11*C*). It occurs more frequently in the foot than in the hand (see Fig. 18–11*C* and *D*). Syndactyly is most frequently observed between the third and fourth fingers and between the

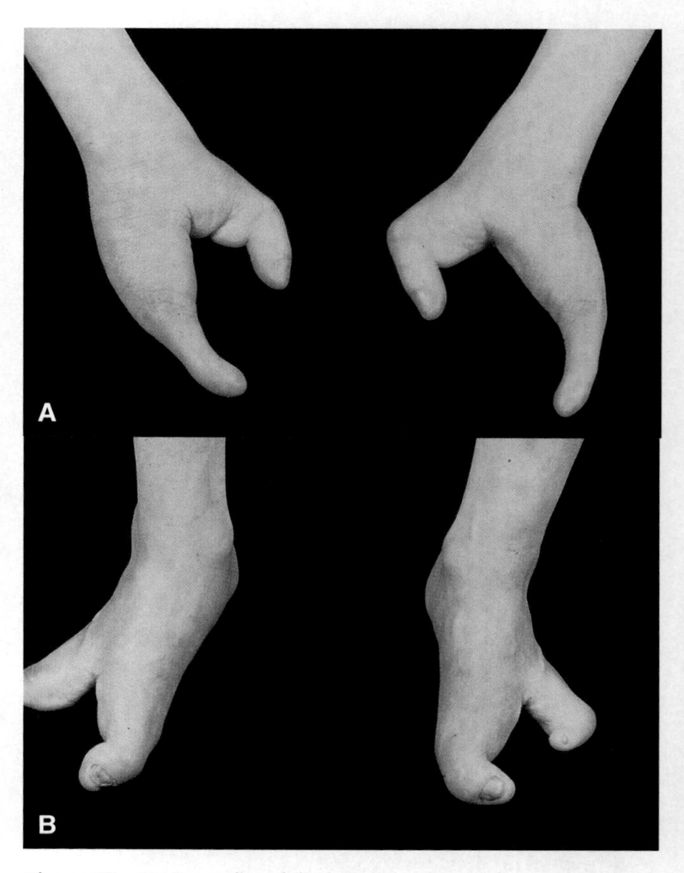

Figure 18 - 10. Anomalies of the hands and feet. *A,* Ectrodactyly in a child. Note the absence of the central digits of the hands, resulting in split hands (lobster-claw anomaly). *B,* A similar type of defect involving the feet. These limb defects can be inherited in an autosomal dominant pattern. (Courtesy of Dr. AE Chudley, Department of Pediatrics and Child Health, University of Manitoba, Children's Hospital, Winnipeg, Manitoba, Canada.)

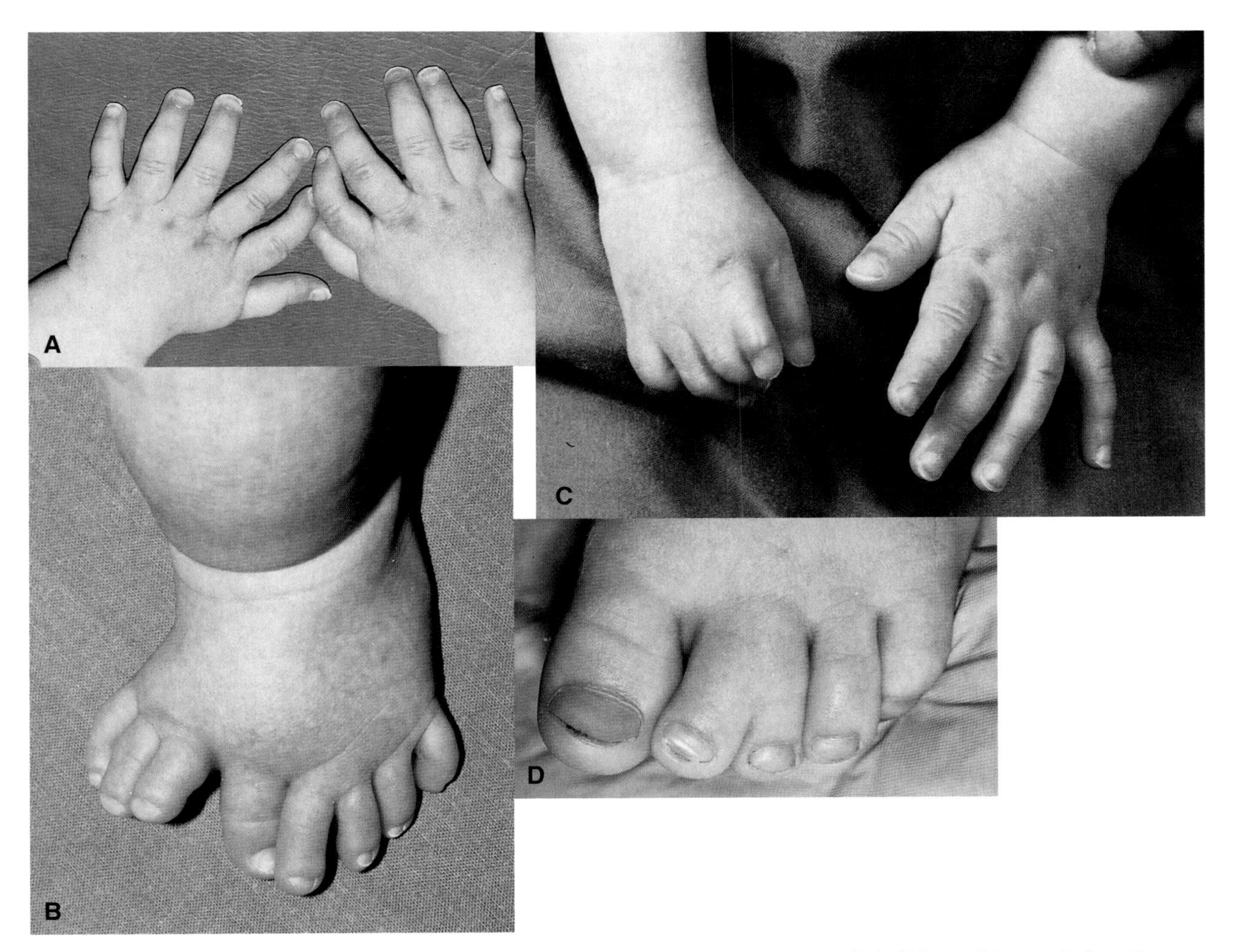

Figure 18 – 11. Various types of limb defects. *A* and *B,* Polydactyly of the hands and foot, respectively. This condition results from the formation of one or more extra digital rays during the embryonic period. *C* and *D,* Various forms of syndactyly involving the fingers and toes. Cutaneous syndactyly *(C),* the most common form of this condition, is probably caused by incomplete programmed cell death (apoptosis) in the tissues between the digital rays during embryonic life. Syndactyly of the second and third toes is shown in *D.* In osseous syndactyly, the digital rays merge as a result of excessive cell death, causing fusion of the bones. (Courtesy of Dr. AE Chudley, Department of Pediatrics and Child Health, University of Manitoba, Children's Hospital, Winnipeg, Manitoba, Canada.)

second and third toes (Fig. 18 – 11*D*). It is inherited as a simple dominant or simple recessive trait. **Cutaneous syndactyly** results from failure of the webs to degenerate between two or more digits. In some cases, there is fusion of the bones (synostosis). **Osseous syndactyly** occurs when the notches between the digital rays fail to develop during the seventh week; as a result, separation of the digits does not occur.

Congenital Clubfoot

Any deformity of the foot involving the talus (ankle bone) is called clubfoot or talipes (L. *talus*, heel, ankle, and *pes*, foot). Clubfoot occurs at a rate of about 1 in every 1000 births. **Talipes equinovarus**, the most common type of clubfoot (Fig. 18 – 12), occurs about twice as frequently in males than in females. The sole of the foot is turned medially and the foot is inverted. There is much uncertainty about the cause of clubfoot. Hereditary factors are involved in some cases, and it appears that environmental factors are involved in most cases. Clubfoot appears to follow a **multifactorial pattern of inheritance**; hence, any intrauterine position that results in abnormal positioning of the feet may cause clubfeet if the fetus is genetically predisposed to this deformity.

Congenital Dislocation of the Hip

Congenital dislocation of the hip occurs with a frequency of about 1 in every 1500 newborn infants and is more common

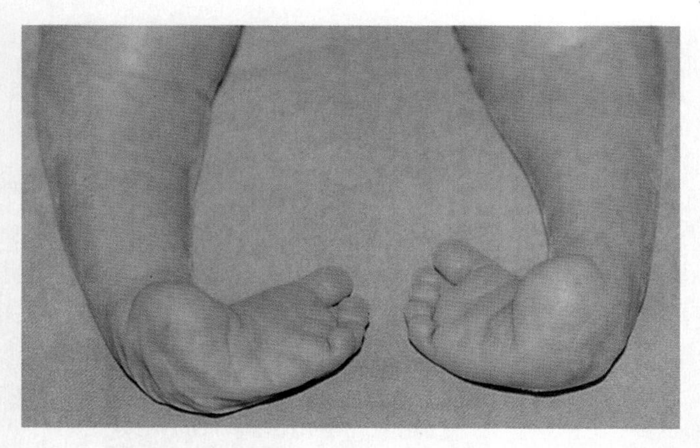

Figure 18 - 12. Neonate with bilateral talipes equinovarus deformities (clubfeet). This is the classic type of this anomaly, characterized by sharp and tight hyperextension and incurving of the feet. (Courtesy of Dr. AE Chudley, Department of Pediatrics and Child Health, University of Manitoba, Children's Hospital, Winnipeg, Manitoba, Canada.)

in females than in males. The fibrous capsule of the hip joint is very relaxed at birth, and the acetabulum of the hip bone and the head of the femur are underdeveloped. The actual dislocation almost always occurs after birth. The two causative factors commonly implicated are abnormal development of the acetabulum and generalized joint laxity.

Causes of Limb Anomalies

Anomalies of the limbs originate at different stages of development. Suppression of limb bud development during the early part of the fourth week results in *absence of the limbs,* a condition known as amelia (see Fig. 18–9*A*). Arrest or disturbance of differentiation or growth of the limbs during the fifth week results in partial absence of limbs, a condition known as **meromelia**. Like other congenital anomalies, some limb defects are caused by the following:

- genetic factors, such as chromosomal abnormalities associated with trisomy 18 (see Chapter 9)
- mutant genes, as in brachydactyly or osteogenesis imperfecta (*Molecular studies* have implicated gene mutation [Hox gene, BMP, Shh, Wnt7, En-1, and others] in some cases of limb defects.)
- environmental factors, such as teratogens like thalidomide
- a combination of genetic and environmental factors (*multifactorial inheritance*), as in congenital dislocation of the hip
- vascular disruption and ischemia, as in limb reduction defects

Summary of Limb Development

Limb buds appear toward the end of the fourth week of development as slight elevations of the ventrolateral body wall. Upper limb buds develop about 2 days before the lower limb buds. The tissues of the limb buds are derived from two main sources: mesoderm and ectoderm. The **apical ectodermal ridge** (AER) exerts an inductive influence on the limb mesenchyme, promoting growth and development of the limbs. The limb buds elongate by proliferation of the mesenchyme within them. *Programmed cell death* is an important mechanism in limb development, such as in the formation of notches between the digital rays. Limb muscles are derived from mesenchyme (myogenic precursor cells) originating in the somites. The muscle-forming cells (myoblasts) form dorsal and ventral muscle masses. Nerves grow into the limb buds after the muscle masses have formed. Arteries of the limbs arise as buds arise from the dorsal aorta.

Initially, the developing limbs are directed caudally; later, they project ventrally, and finally, they rotate on their longitudinal axes. The upper and lower limbs rotate in opposite directions and to different degrees. Most limb anomalies are caused by genetic factors; however, many limb abnormalities probably result from an interaction of genetic and environmental factors (multifactorial inheritance). Relatively few congenital anomalies of the limbs can be attributed to specific environmental teratogens, except those resulting from thalidomide.

Clinically Oriented Questions

1. A baby presented with short limbs. His trunk was normally proportioned, but his head was slightly larger than normal. Both parents had normal limbs, and these problems had never occurred in either of their families. Could the mother's ingestion of drugs during pregnancy have caused these abnormalities? If not, what would be the probable cause of these skeletal disorders? Could they occur again if the couple had more children?
2. A woman is interested in marrying a man who happens to have very short fingers (*brachydactyly*). He says that two of his relatives have exhibited short fingers, but none of his brothers or sisters has them. The woman has normal digits and so does everyone else in her family. Clearly, heredity is involved, but what are the chances that the couple's children would have brachydactyly if they were to marry?
3. About a year ago, a woman was reported to have given birth to a child with no right hand. She had taken a drug called *Bendectin* to alleviate nausea during the 10th week of her pregnancy (8 weeks after fertilization). The woman is instituting legal proceedings against the drug company that makes the drug. Does this drug cause limb defects? If it does, could it have caused failure of the child's hand to develop?

4. A baby presented with *syndactyly* of the left hand and absence of the left sternal head of the pectoralis major muscle. The baby seemed normal except that the nipple on the left side was about 2 inches lower than the other one. What is the cause of these anomalies? Can they be corrected?

5. What is the most common type of clubfoot? How common is it? What is the appearance of the feet of infants born with this anomaly?

The answers to these questions are at the back of the book.

19

The Nervous System

The nervous system consists of three main parts:

- the *central nervous system* (CNS), which includes the brain and spinal cord
- the *peripheral nervous system* (PNS), which includes neurons (nerve cells) outside the CNS and cranial and spinal nerves that connect the brain and spinal cord with peripheral structures
- The *autonomic nervous system* (ANS), which has parts in both the CNS and PNS and which consists of neurons that innervate smooth muscle, cardiac muscle, or glandular epithelium or combinations of these tissues

Origin of the Nervous System

The nervous system develops from the **neural plate** (Fig. 19 – 1*A*). The notochord and paraxial mesoderm induce the overlying ectoderm to differentiate into the neural plate. Experimental studies show that induction of the neural plate begins with signaling molecules (*chordin, noggin, follistatin, Xhr3, cerberus*) that inhibit expression of bone morphogenetic proteins (BMPs) in the epiblast. For the rostrocaudal patterning of the neural plate, complex molecular pathways, involving *fibroblast growth factors* (FGFs), homeobox (Hox) genes, Wnt, retinoic acid, and other signaling factors, have been implicated. Formation of the neural folds, neural tube, and neural crest from the neural plate is illustrated in Figure 19 – 1*B* to *F.*

- The **neural tube** differentiates into the CNS.
- The **neural crest** gives rise to cells that form most of the PNS and ANS.

Formation of the neural tube, a process known as **neurulation**, begins during the early part of the fourth week of development (22 to 23 days). Fusion of the neural folds proceeds in cranial and caudal directions until only small areas remain open at both ends (Fig. 19 – 2*A* and *B*). At these sites, the lumen of the neural tube — the **neural canal** — communicates freely with the amniotic cavity (see Fig. 19 – 2*C*). The cranial opening, called the *rostral neuropore*, closes on about the 25th day; the *caudal neuropore* closes 2 days later (see Fig. 19 – 2*D*). It has been suggested that multiple sites of neural tube closure exist in humans. **Closure of the neuropores** coincides with the establishment of a blood vascular circulation for the neural tube. The walls of the neural tube thicken to form the brain and spinal cord (Fig. 19 – 3). The neural canal is converted into the *ventricular system of the brain* and the *central canal* of the spinal cord. The dorsoventral patterning of the neural tube appears to involve sonic hedgehog (Shh), Pax genes, BMPs, and dorsalin, a transforming growth factor(TGF-β) growth factor.

Development of the Spinal Cord

The neural tube caudal to the fourth pair of somites develops into the spinal cord (see Figs. 19 – 3 and 19 – 4). The lateral walls of the neural tube thicken and gradually reduce the size of the neural canal to a small **central canal** (see Fig. 19 – 4*A* to *C*). Initially, the wall of the neural tube is composed of a thick, pseudostratified, columnar neuroepithelium (see Fig. 19 – 4*D*). These neuroepithelial cells constitute the **ventricular zone** (ependymal layer), which gives rise to all neurons and macroglial cells in the spinal cord (Fig. 19 – 5). Macroglial cells are the large types of neuroglial cells (e.g., astrocytes and oligodendrocytes). Soon, a **marginal zone** composed of the outer parts of the neuroepithelial cells becomes recognizable (see Fig. 19 – 4*E*). This zone gradually becomes the *white substance (matter) of the spinal cord* as axons grow into it from nerve cell bodies in the spinal cord, spinal ganglia, and brain. Some dividing neuroepithelial cells in the ventricular zone differentiate into primordial neurons called **neuroblasts**. These embryonic cells form an **intermediate zone** (mantle layer) between the ventricular and marginal zones. Neuroblasts become neurons as they develop cytoplasmic processes (see Fig. 19 – 5). The primordial supporting cells of the CNS — the **glioblasts** (spongioblasts) — differentiate from neuroepithelial cells, mainly after neuroblast formation has ceased. The glioblasts migrate from the ventricular zone into the intermediate and marginal zones. Some glioblasts become **astroblasts** and later **astrocytes**, whereas others become **oligodendroblasts** and eventually **oligodendrocytes** (see Fig. 19 – 5). When neuroepithelial cells cease producing neuroblasts and glioblasts, they differentiate into *ependymal cells,* which form the **ependyma** lining the central canal of the spinal cord. Ependymal cells appear to be stem cells.

Microglial cells (microglia), scattered throughout the gray and white substance, are small cells that are derived from **mesenchymal cells** (see Fig. 19 – 5). Microglial cells invade the CNS rather late in the fetal period, after it has been penetrated by blood vessels. Microglia originate in the bone marrow and are part of the mononuclear phagocytic cell population. Proliferation and differentiation of neuroepithelial cells in the primordial spinal cord produce thick walls and thin roof and floor plates (see Fig. 19 – 4*B*). Differential thickening of the lateral walls of the spinal cord soon produces a shallow, longitudinal groove on each side, the **sulcus limitans** (see Figs. 19 – 4*B* and 19 – 6). This groove separates the dorsal part, the **alar plate** (lamina), from the ventral part, the **basal plate** (lamina). The alar and basal plates produce longitudinal bulges extending through most of the length of the developing spinal cord. This regional separation is of fundamental

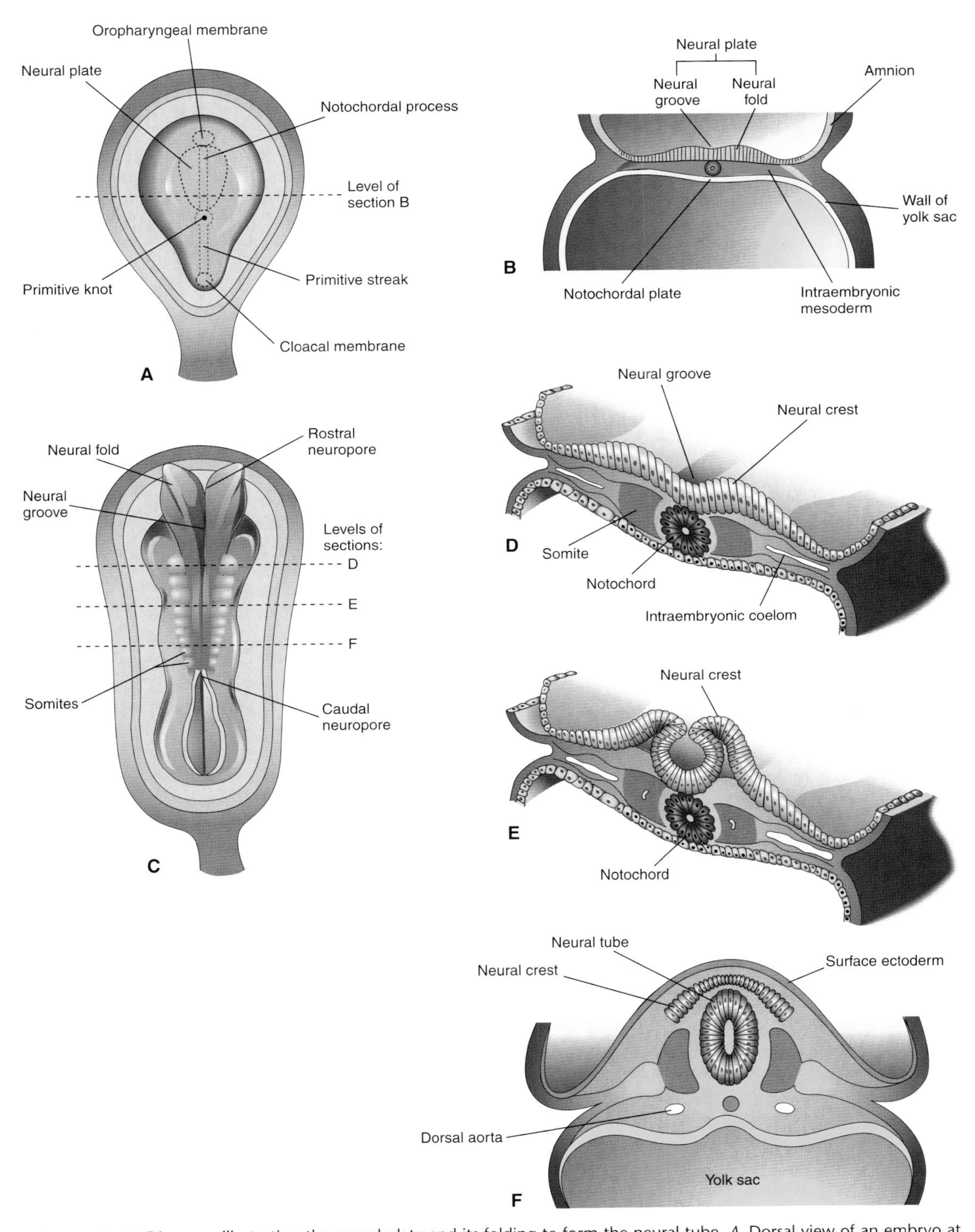

Figure 19 – 1. Diagrams illustrating the neural plate and its folding to form the neural tube. *A,* Dorsal view of an embryo at about 18 days, exposed by removing the amnion. *B,* Transverse section of the embryo showing the neural plate and early development of the neural groove. *C,* Dorsal view of an embryo at about 22 days. The neural folds have fused opposite the fourth to sixth somites but are spread apart at both ends. *D* to *F,* Transverse sections of this embryo at the levels shown in *C,* illustrating formation of the neural tube and its detachment from the surface ectoderm. Note that some neuroectodermal cells are not included in the neural tube but remain between it and the surface ectoderm as the neural crest.

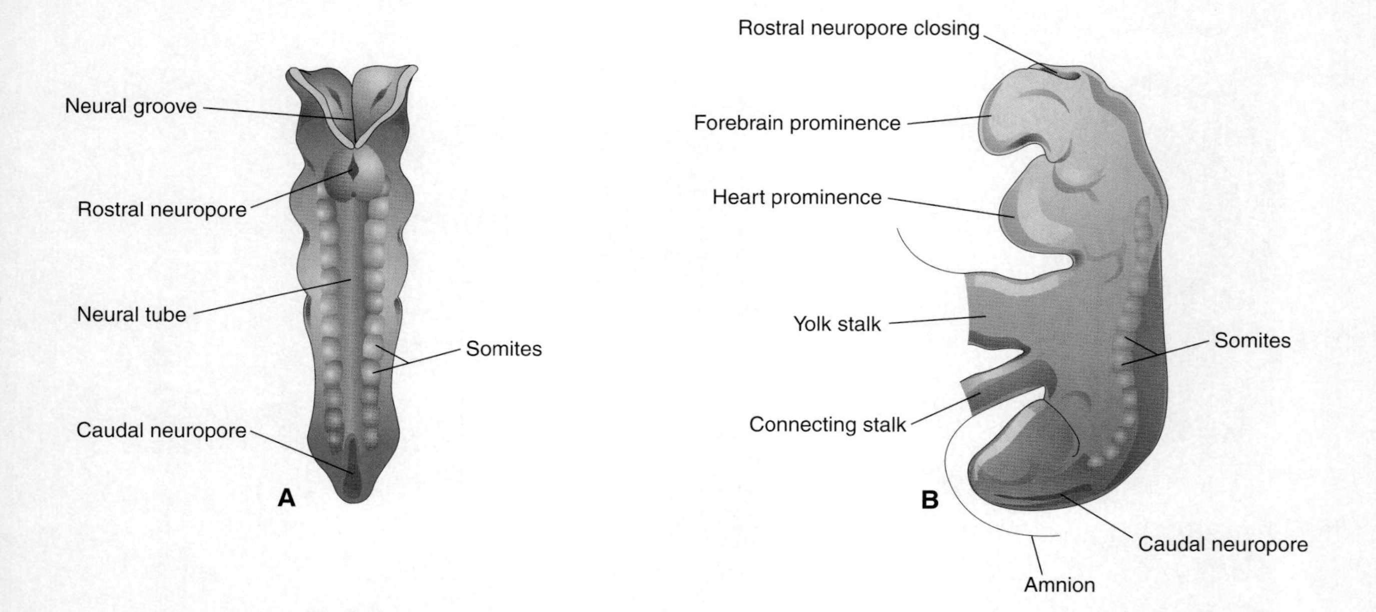

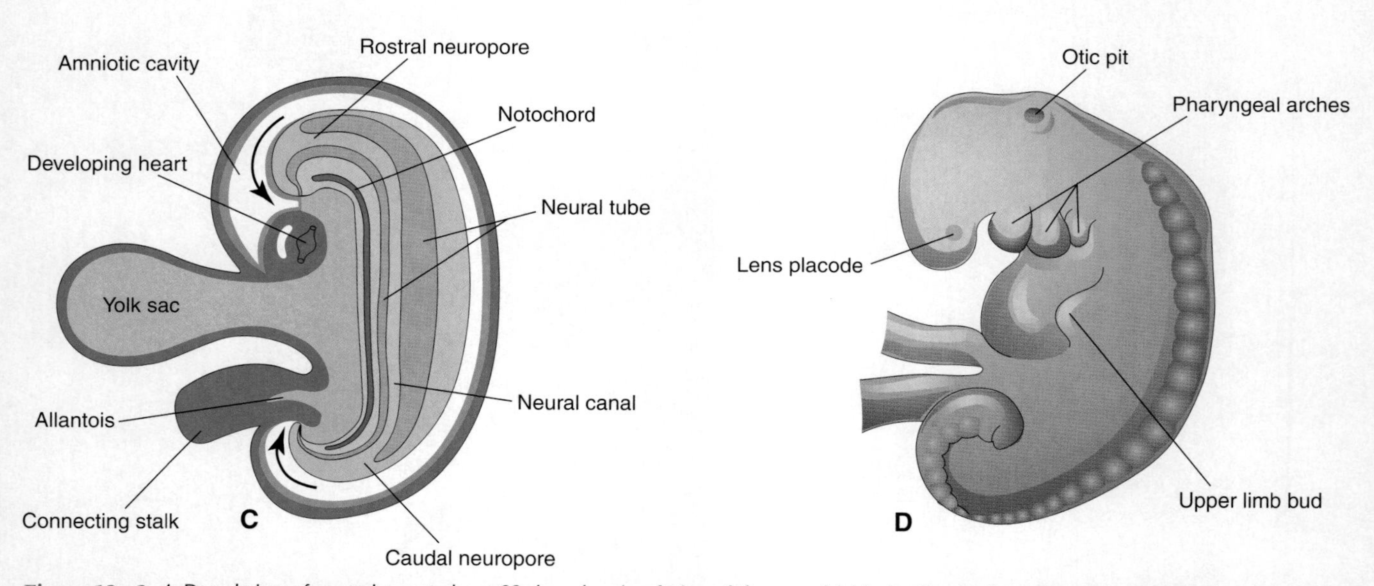

Figure 19-2. *A,* Dorsal view of an embryo at about 23 days showing fusion of the neural folds, leading to formation of the neural tube. *B,* Lateral view of an embryo at about 24 days, showing the forebrain prominence and closing of the rostral neuropore. *C,* Sagittal section of this embryo showing the transitory communication of the neural canal with the amniotic cavity (*arrows*). *D,* Lateral view of an embryo at about 27 days. Note that the neuropores shown in *B* are closed.

importance because the alar and basal plates are later associated with afferent and efferent functions, respectively.

Cell bodies in the alar plates form the dorsal gray columns that extend the length of the spinal cord. In transverse sections, these columns are the **dorsal horns** (Fig. 19-7). Neurons in these columns constitute afferent nuclei, and groups of these nuclei form the **dorsal gray columns.** As the alar plates enlarge, the *dorsal median septum* forms. Cell bodies in the basal plates form the ventral and lateral gray columns. In transverse sections of the spinal cord, these columns are the

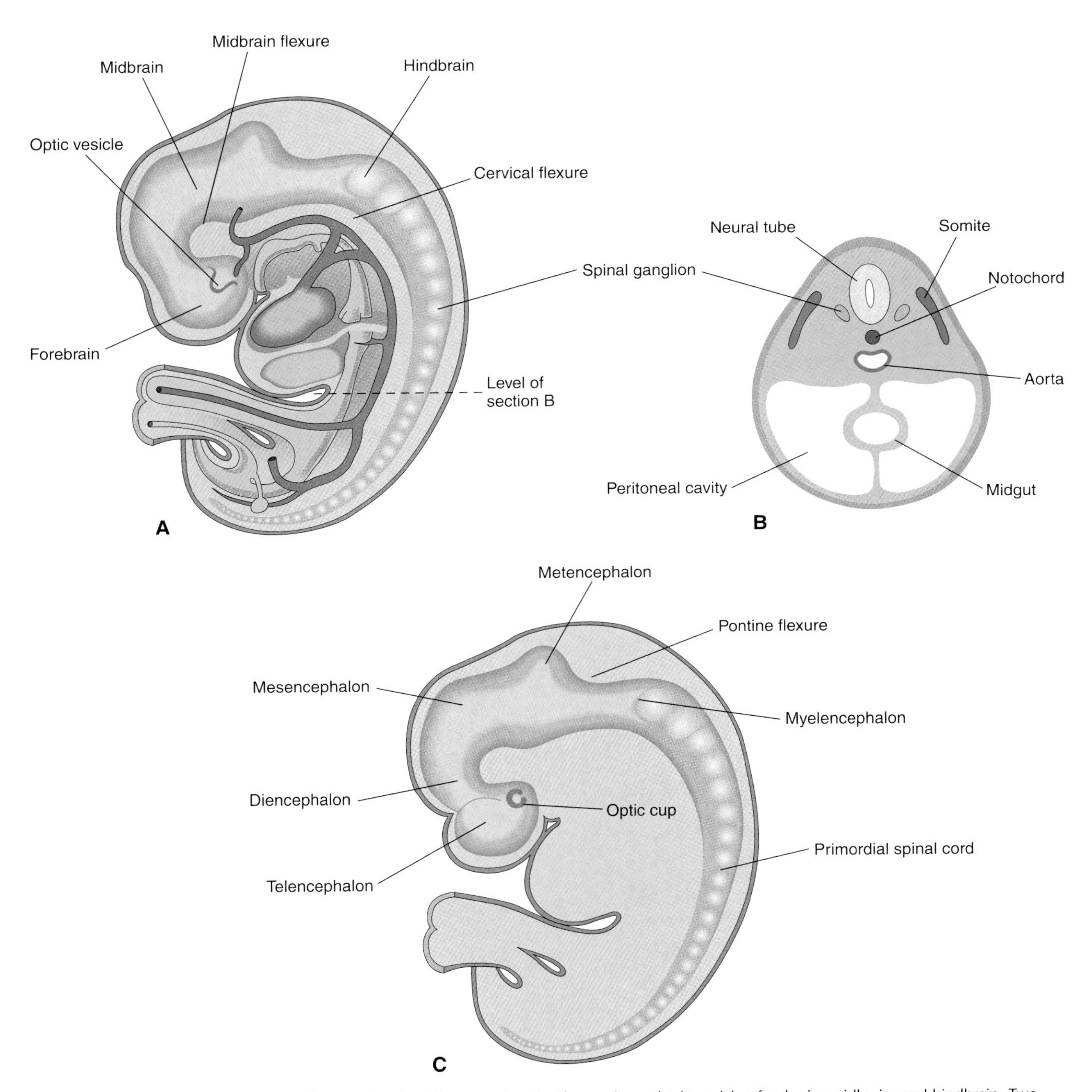

Figure 19 – 3. *A,* Lateral view of an embryo at about 28 days showing the three primary brain vesicles: forebrain, midbrain, and hindbrain. Two flexures demarcate the primary divisions of the brain. *B,* Transverse section of this embryo showing the neural tube that will develop into the spinal cord in this region. The spinal ganglia derived from the neural crest are also shown. *C,* Schematic lateral view of the central nervous system of a 6-week embryo showing the secondary brain vesicles and pontine flexure. The flexure (bend) occurs as the brain grows rapidly.

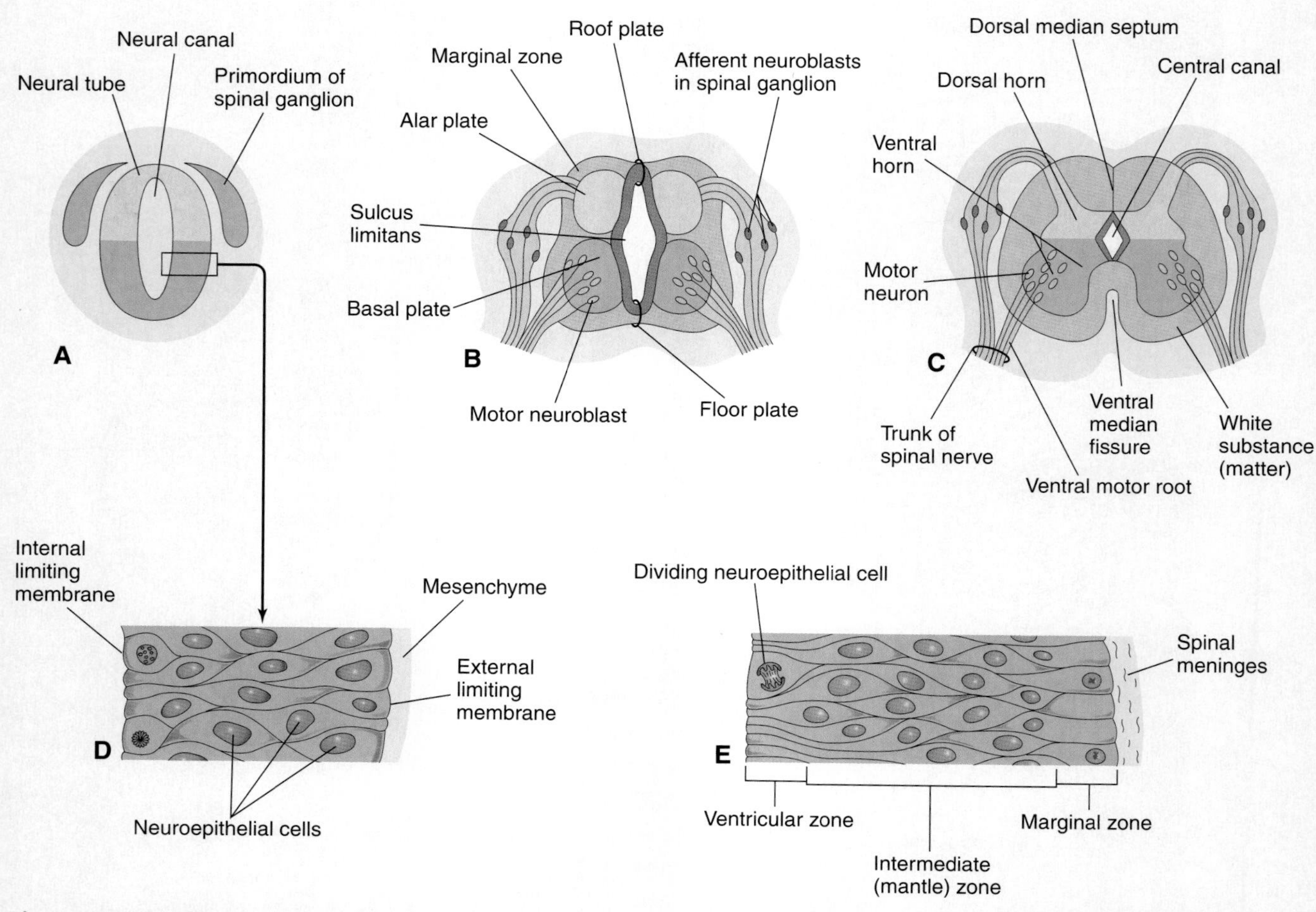

Figure 19 – 4. Development of the spinal cord. *A,* Transverse section of the neural tube of an embryo at about 23 days. *B* and *C,* Similar sections in 6- and 9-weeks embryos, respectively. *D,* Section of the wall of the neural tube shown in *A. E,* Section of the wall of the developing spinal cord showing its three zones. In *A* to *C,* note that the neural canal of the neural tube is converted into the central canal of the spinal cord.

ventral horns and **lateral horns**, respectively. Axons of ventral horn cells grow out of the cord and form the **ventral roots of the spinal nerves** (see Fig. 19 – 7). As the basal plates enlarge, they bulge ventrally on each side of the median plane. As this occurs, the **ventral median septum** forms and a deep longitudinal groove – the **ventral median fissure** – develops on the ventral surface of the cord.

Development of the Spinal Ganglia

The unipolar neurons in the spinal ganglia (dorsal root ganglia) are derived from **neural crest cells** (see Fig. 19 – 7). The peripheral processes of **spinal ganglion cells** pass in the spinal nerves to sensory endings in somatic or visceral structures (see Fig. 19 – 7). The central processes enter the spinal cord, constituting the **dorsal roots of the spinal nerves**.

Development of the Spinal Meninges

The mesenchyme surrounding the neural tube (Fig. 19 – 4*E*) condenses to form the *primordial meninx* (membrane). The external layer of this membrane gives rise to the **dura mater** (Fig. 19 – 8). The internal layer forms the **pia mater** and **arachnoid mater**; together, these layers constitute the leptomeninges. *Neural crest cells* migrate into the **leptomeninges**, and appear to be involved in the formation of the pia mater. Fluid-filled spaces appear within the leptomeninges that soon coalesce to form the **subarachnoid space** (Fig. 19 – 9*A*). Embryonic **cerebrospinal fluid** (CSF), which may provide a

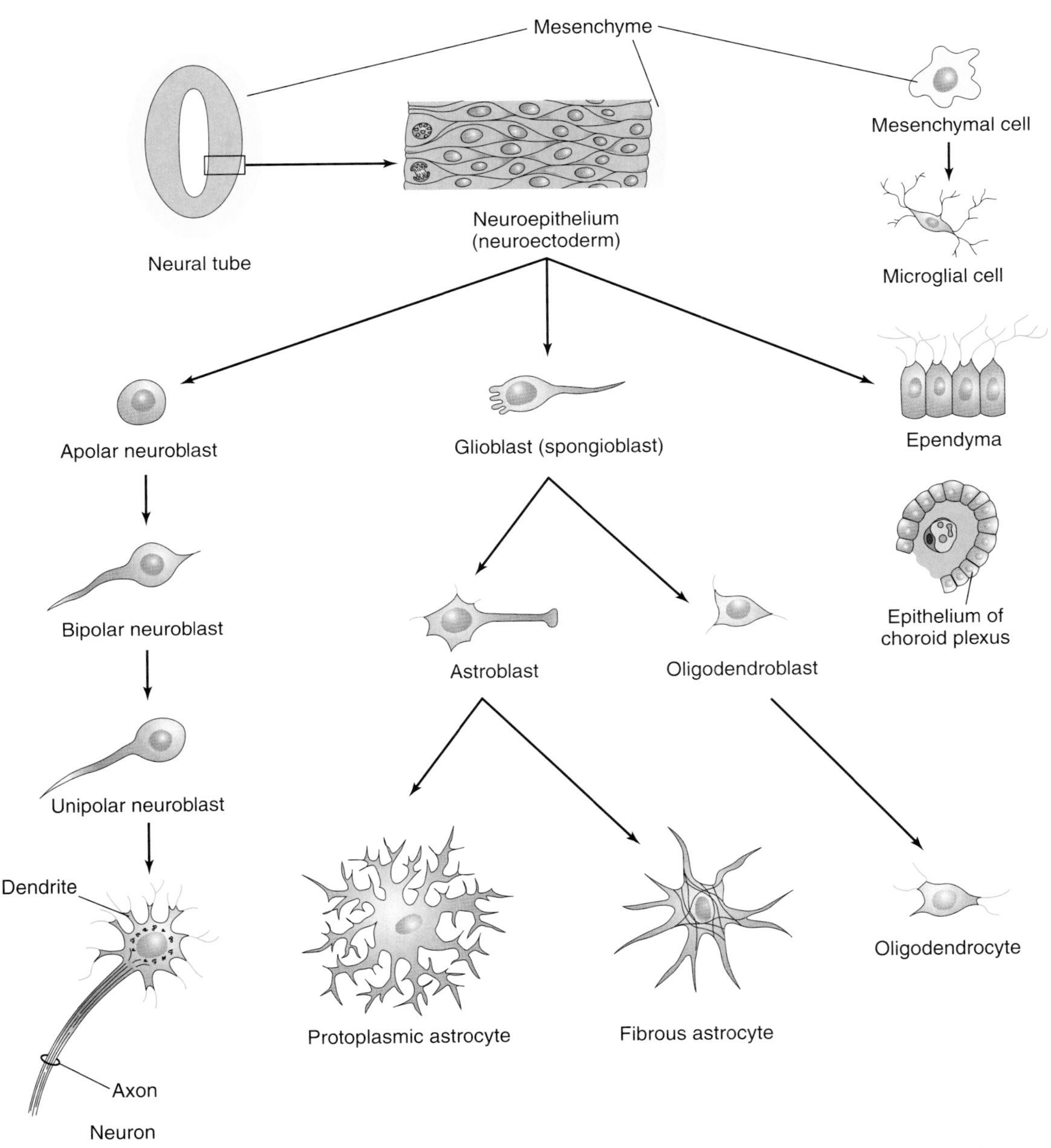

Figure 19 - 5. Histogenesis of cells in the central nervous system. After further development, the multipolar neuroblast (*lower left*) becomes a nerve cell or neuron. Neuroepithelial cells give rise to all neurons and macroglial cells. Microglial cells are derived from mesenchymal cells that invade the developing nervous system with the blood vessels.

nutrient medium for the epithelial neural tissues, begins to form during the fifth week of development.

Positional Changes of the Spinal Cord

The spinal cord in the embryo extends the entire length of the vertebral canal (see Fig. 19 - 8*A*). The spinal nerves pass through the intervertebral foramina near their levels of origin. Because the vertebral column and dura mater grow more rapidly than the spinal cord, this relationship does not persist. The caudal end of the spinal cord gradually comes to lie at relatively higher levels. At 6 months, it lies at the level of the first sacral vertebra (see Fig. 19 - 8*B*). The spinal cord in the newborn infant terminates at the level of the second or third lumbar vertebra (see Fig. 19 - 8*C*). The spinal cord in the adult usually terminates at the inferior border of the first lumbar vertebra (see Fig. 19 - 8*D*). As a result,

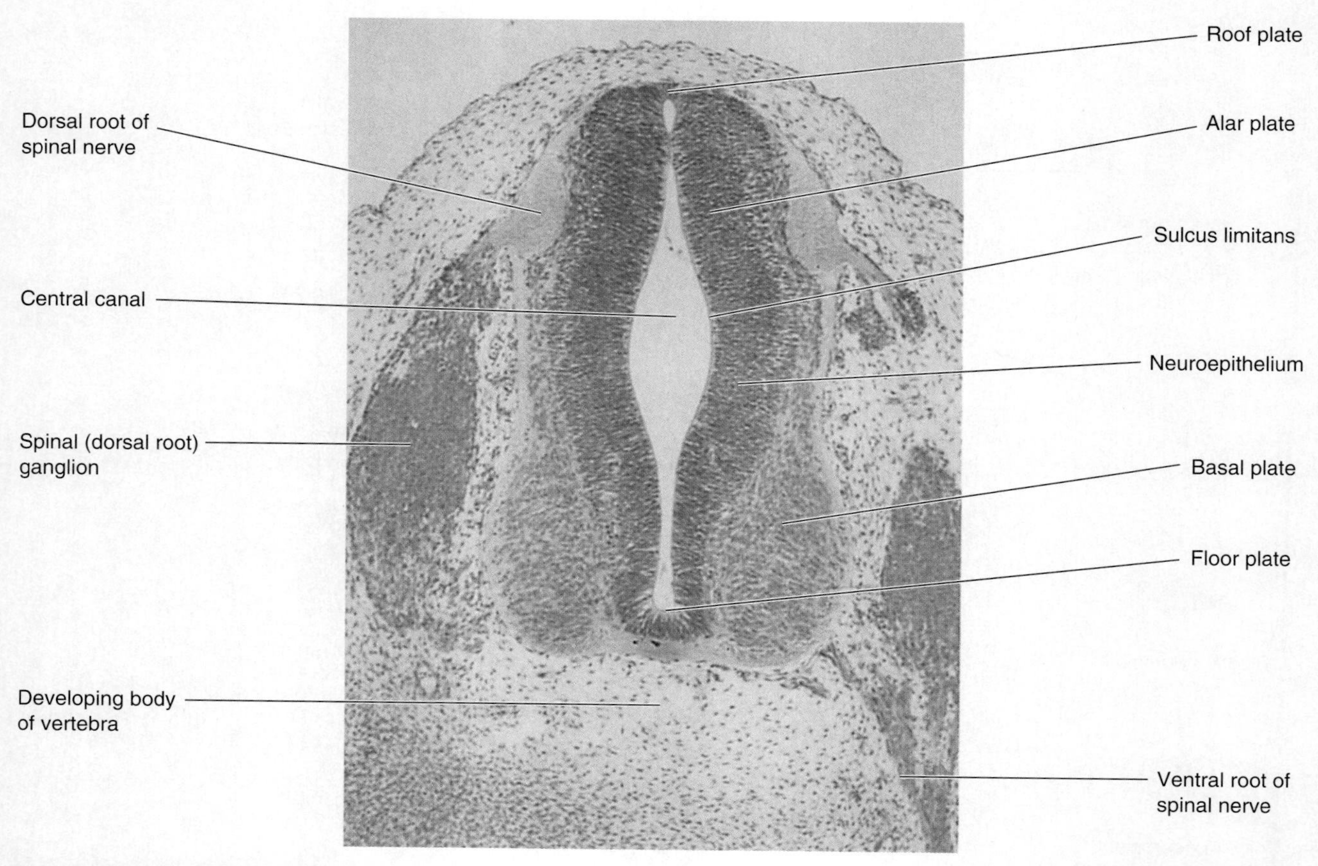

Figure 19 - 6. Transverse section of an embryo (× 100) at 40 days. The ventral root of the spinal nerve is composed of nerve fibers arising from neuroblasts in the basal plate, whereas the dorsal root is formed by nerve processes arising from neuroblasts in the spinal ganglion.

the spinal nerve roots, especially those of the lumbar and sacral segments, run obliquely from the spinal cord to the corresponding level of the vertebral column. The nerve roots inferior to the end of the cord — the **medullary cone** (L. *conus medullaris*) form a sheaf of nerve roots, the **cauda equina** (L., horse's tail). Although the dura mater and arachnoid mater usually end at the S2 vertebra in adults, the pia mater does not. Distal to the caudal end of the spinal cord, the pia mater forms a long fibrous thread, the **terminal filum** (L. *filum terminale*), which indicates the line of regression of the caudal end of the embryonic spinal cord (see Fig. 19 - 8*C* and *D*). The terminal filum extends from the medullary cone to the periosteum of the first coccygeal vertebra.

Myelination of Nerve Fibers

Myelin sheaths in the spinal cord begin to form during the late fetal period and continue to form during the first postnatal year. In general, fiber tracts become myelinated at about the time they become functional. The **myelin sheaths** surrounding nerve fibers within the spinal cord are formed by **oligodendrocytes.** The myelin sheaths around the axons of peripheral nerve fibers are formed by the plasma membranes of **neurolemma (Schwann) cells.** These neuroglial cells are derived from **neural crest cells** that migrate peripherally and wrap themselves around the axons of somatic motor neurons and presynaptic autonomic motor neurons as they pass out of the CNS (see Fig. 19 - 7). These cells also wrap themselves around both the central and peripheral processes of somatic and visceral sensory neurons, as well as around the axons of postsynaptic autonomic motor neurons. Motor roots are myelinated before sensory roots.

Congenital Anomalies of the Spinal Cord

Most congenital anomalies of the spinal cord result from defective closure of the neural tube during the fourth

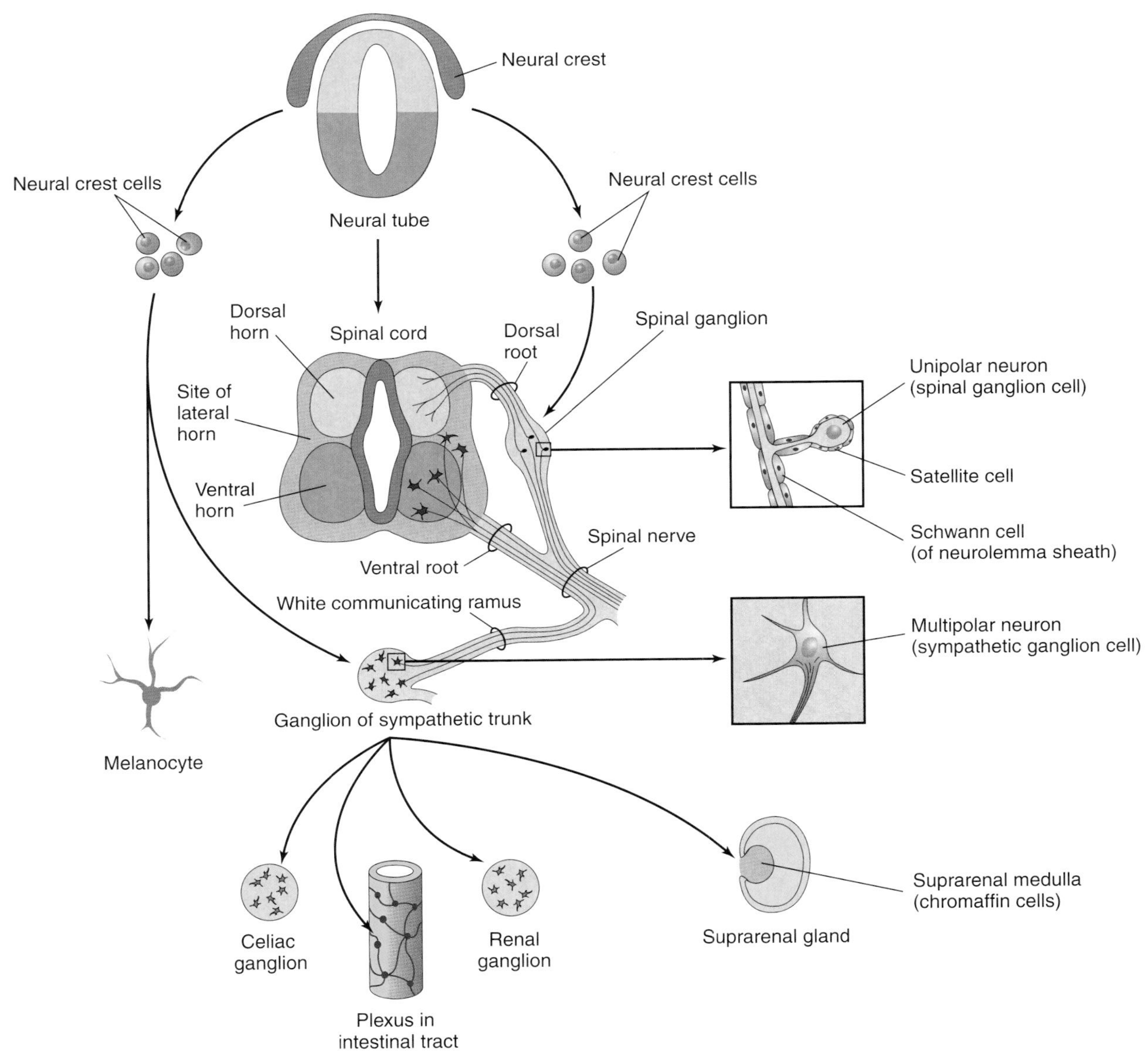

Figure 19 - 7. Derivatives of the neural crest. Neural crest cells also differentiate into the cells in the afferent ganglia of cranial nerves and many other structures. The formation of a spinal nerve is also illustrated.

week of development. Mutations in the human Pax-3 gene have been implicated in some **neural tube defects** (NTDs). These defects affect the tissues overlying the spinal cord, including the meninges, vertebral arch, muscles, and skin (Fig. 19 – 9*B* to *D*). Anomalies involving the vertebral arches are referred to as **spina bifida**. This term denotes *nonfusion of the primordial halves of the vertebral arches*, which is common to all types of spina bifida. Severe anomalies also involve the spinal cord and meninges.

Spina Bifida Occulta

Spina bifida occulta results from failure of the embryonic halves of the arch to grow normally and fuse in the median plane (see Fig. 19 – 9*A*). Spina bifida occulta occurs in L5 or S1 vertebrae in about 10% of otherwise normal people. In its most minor form, the only evidence of its presence may be a small dimple with a tuft of hair arising from it (see Fig. 19 – 10). Spina bifida occulta usually produces no clinical symptoms.

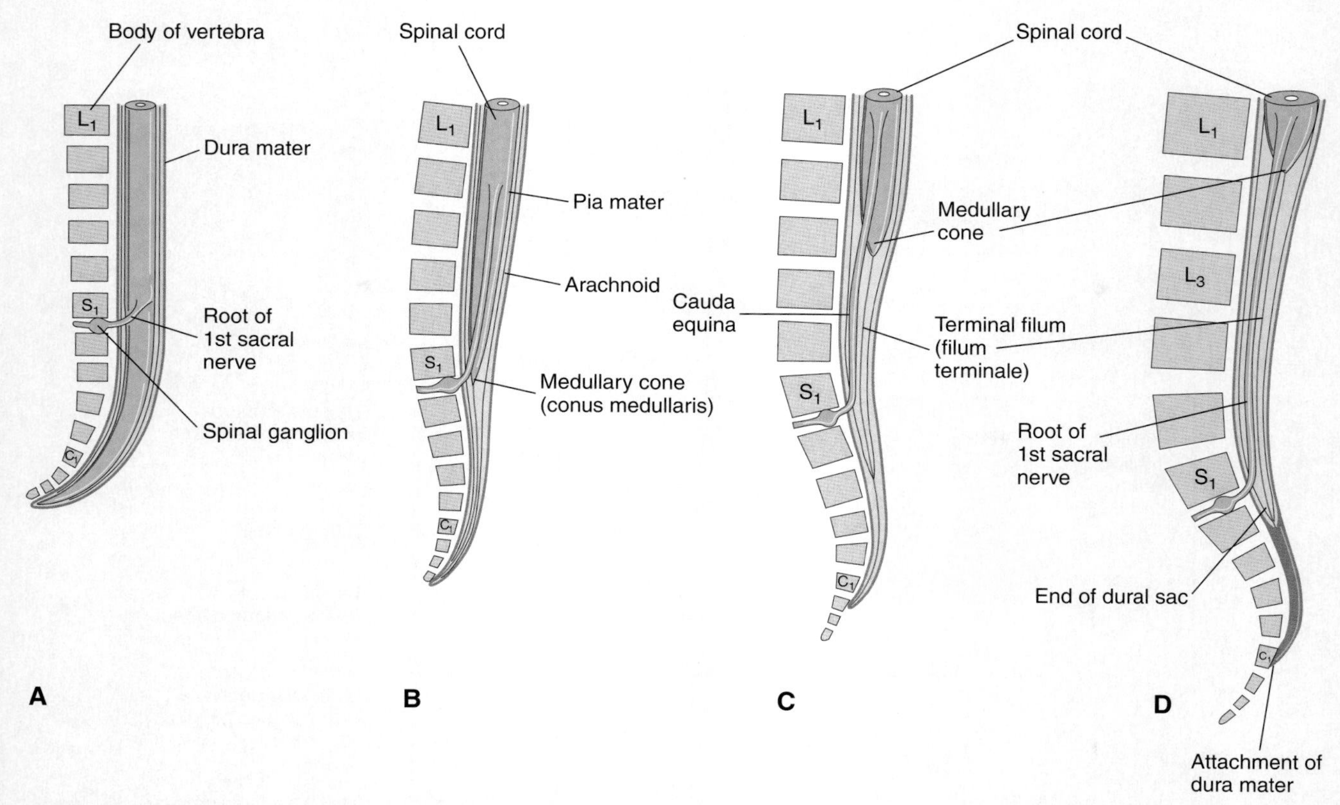

Figure 19-8. The position of the caudal end of the spinal cord in relation to the vertebral column and meninges at various stages of development. The increasing inclination of the root of the first sacral nerve is also illustrated. *A,* Eight weeks. *B,* Twenty-four weeks. *C,* Newborn infant. *D,* Adult.

Spina Bifida Cystica

Severe types of spina bifida, involving protrusion of the spinal cord and/or meninges through the defect in the vertebral arches, are referred to collectively as *spina bifida cystica* because of the cystlike sac that is associated with these anomalies (see Figs. 19-9*B* to *D* and 19-11). Spina bifida cystica occurs in about 1 in every 1000 births. When the sac contains meninges and CSF, the anomaly is called **spina bifida with meningocele** (see Fig. 19-9*B*). The spinal cord and spinal roots are in their normal position, but spinal cord abnormalities may be present. If the spinal cord and/or nerve roots are included in the sac, the anomaly is called **spina bifida with meningomyelocele** (see Figs. 19-9*C* and 19-11). Meningoceles are rare compared with meningomyeloceles. Spina bifida with meningomyelocele involving several vertebrae is often associated with partial absence of the brain, termed **meroanencephaly** or **anencephaly** (Fig. 19-12). Spina bifida cystica is associated with varying degrees of neurologic deficit, depending on the position and extent of the lesion. Spina bifida cystica and/or meroanencephaly is strongly suspected in utero when the level of *alpha-fetoprotein* (*AFP*) in the amniotic fluid is high (see Chapter 8). Alpha-fetoprotein levels may also be elevated in the maternal blood serum.

Causes of Neural Tube Defects

Genetic, nutritional, and environmental factors play a role in the production of NTDs. Studies have shown that vitamins and folic acid supplements taken prior to conception reduce the incidence of NTDs. Certain drugs increase the risk of meningomyelocele (see Chapter 9, valproic acid). For example, valproic acid, an anticonvulsant, causes NTDs in 1 to 2% of pregnant women if given during the fourth week of development when the neural folds are fusing.

Development of the Brain

The neural tube cranial to the fourth pair of somites develops into the brain. Even before the neural folds are completely fused, three distinct vesicles are recognizable in the rostral end of the developing neural tube. From rostral to caudal, these three **primary brain vesicles** (Fig. 19-13) form the *forebrain* (prosencephalon), *midbrain* (mesencephalon), and *hindbrain* (rhombencephalon). By the beginning of the fourth week, the forebrain partly divides into two **secondary brain vesicles,** the *telencephalon* and *diencephalon*; the midbrain remains

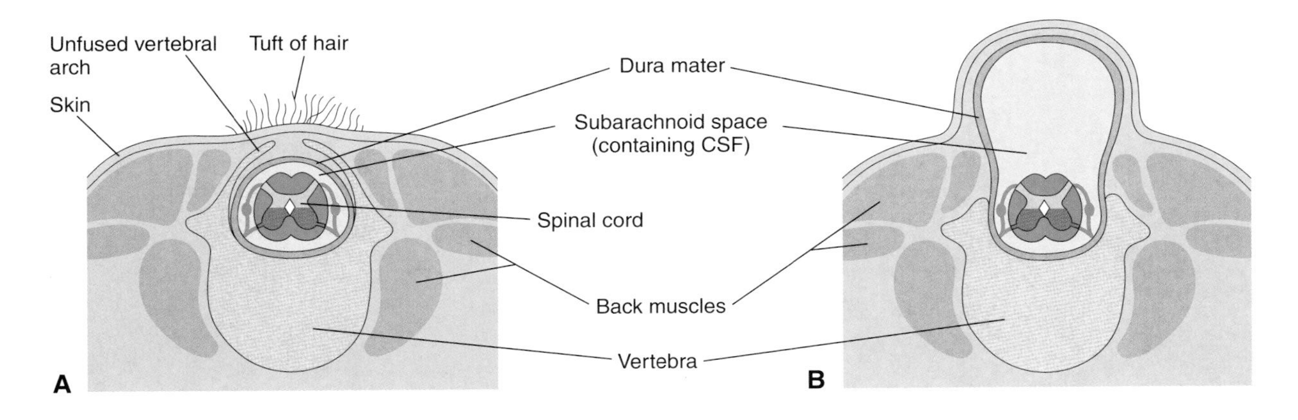

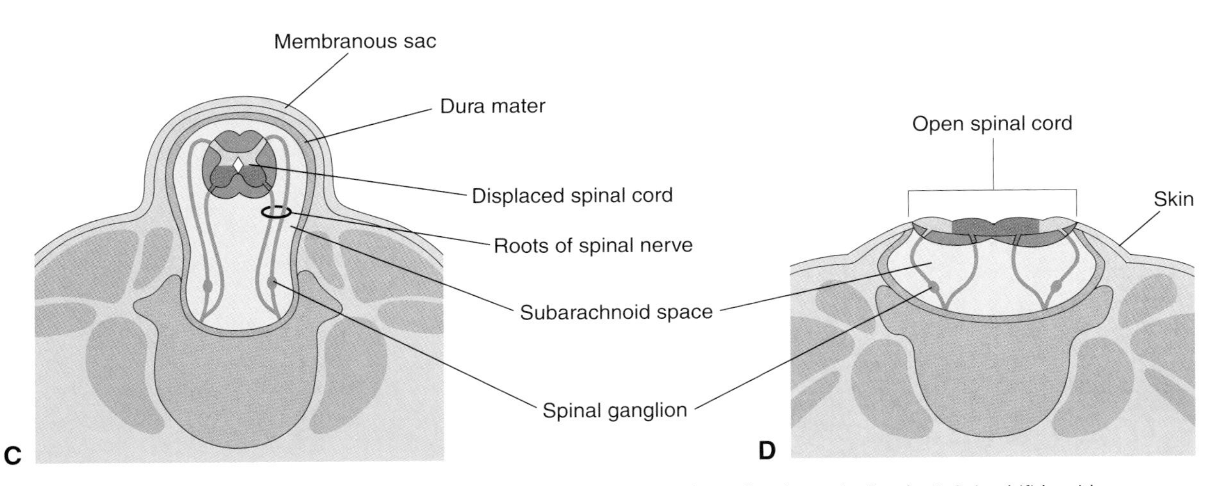

Figure 19 - 9. Various types of spina bifida. *A,* Spina bifida occulta. Observe the unfused vertebral arch. *B,* Spina bifida with meningocele. *C,* Spina bifida with meningomyelocele. *D,* Spina bifida with myeloschisis. This severe type of spina bifida results from failure of the caudal neuropore to close at the end of the fourth week of development. The types illustrated in *B* to *D* are referred to collectively as spina bifida cystica because of the cystlike sac that is associated with them.

undivided and the hindbrain gives rise to the *metencephalon* and *myelencephalon*. By the fifth week, there are five secondary brain vesicles (see Fig. 19 - 13).

Brain Flexures

The embryonic brain grows rapidly and bends ventrally with the head fold. This produces the **midbrain flexure** in the midbrain region and the **cervical flexure** at the junction of the hindbrain and spinal cord (Fig. 19 - 14). Later, unequal growth of the brain between these flexures produces the **pontine flexure** in the opposite direction. This flexure results in thinning of the roof of the hindbrain. The **sulcus limitans** extends cranially to the junction of the midbrain and forebrain, and the alar and basal plates are recognizable only in the midbrain and hindbrain.

Hindbrain

The *cervical flexure* demarcates the hindbrain from the spinal cord (see Fig. 19 - 14*A*). The *pontine flexure*, located in the future pontine region, divides the hindbrain into caudal (myelencephalon) and rostral (metencephalon) parts. The myelencephalon becomes the **medulla oblongata** or medulla, whereas the metencephalon becomes the **pons** and **cerebellum**. The cavity of the hindbrain becomes the fourth ventricle and the central canal in the caudal part of the medulla (see Fig. 19 - 14*B*).

Myelencephalon

Neuroblasts from the alar plates in the myelencephalon migrate into the marginal zone and form isolated areas of gray substance (matter) known as the **gracile nuclei** medially and the **cuneate nuclei** laterally (see

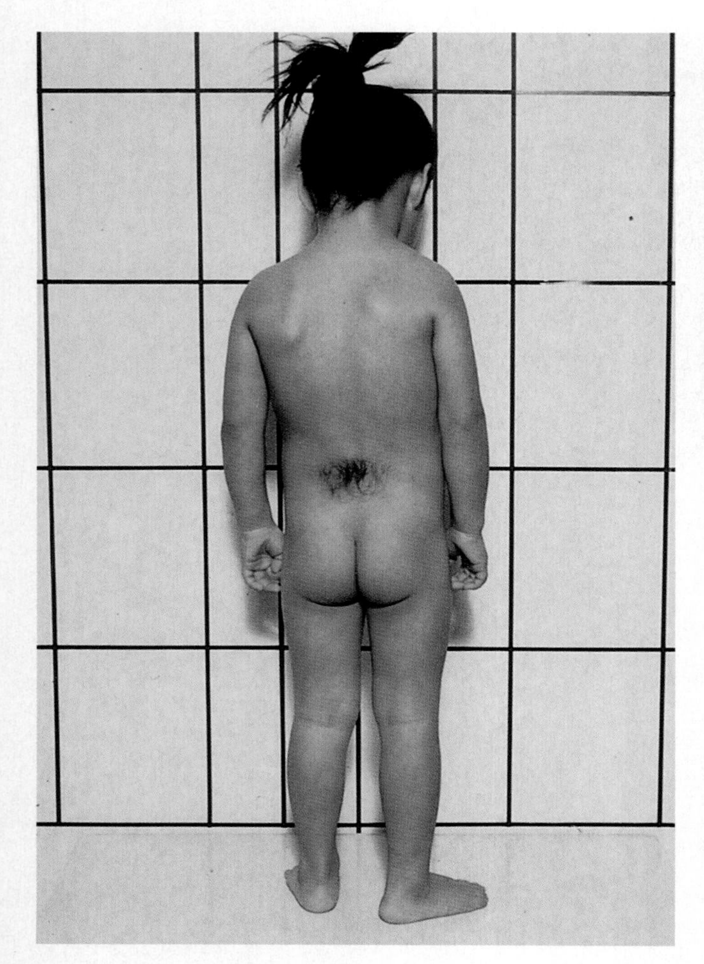

Figure 19 – 10. A female child with a hairy patch in the lumbosacral region indicating the site of a spina bifida occulta. (Courtesy of AE Chudley, MD, Section of Genetics and Metabolism, Department of Pediatrics and Child Health, Children's Hospital and University of Manitoba, Winnipeg, Manitoba, Canada.)

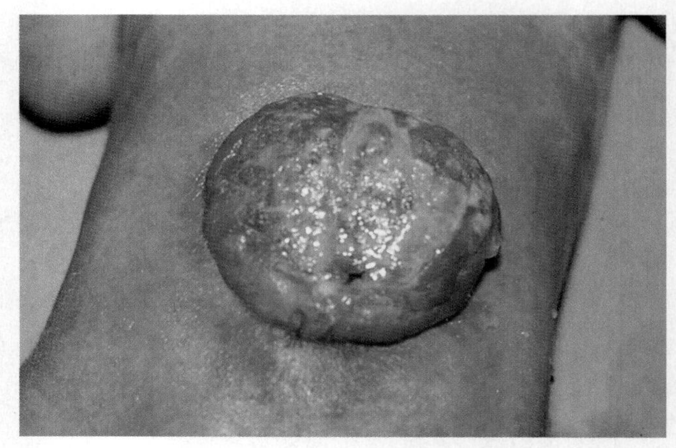

Figure 19 – 11. The back of a newborn infant with a large lumbar meningomyelocele (myelomeningocele). The NTD is covered with a thin membrane. (Courtesy of AE Chudley, MD, Section of Genetics and Metabolism, Department of Pediatrics and Child Health, Children's Hospital and University of Manitoba, Winnipeg, Manitoba, Canada.)

Fig. 19 – 14*B*). These nuclei are associated with correspondingly named tracts that enter the medulla from the spinal cord. The ventral area of the medulla contains a pair of fiber bundles, called **pyramids**, which consist of corticospinal fibers descending from the developing cerebral cortex. The rostral part of the myelencephalon (the "open" part of the medulla) is wide and rather flat, especially opposite the pontine flexure (see Fig. 19 – 14*C* and *D*). As the pontine flexure is formed, the walls of the medulla move laterally and the alar plates come to lie lateral to the basal plates. As the positions of the plates change, the motor nuclei generally develop medial to the sensory nuclei (see Fig. 19 – 14*C*). Neuroblasts in the basal plates of the medulla, like those in the spinal cord, develop into motor neurons. In the medulla, the neuroblasts form nuclei (groups of nerve cells) and organize into three cell columns on each side (see Fig. 19 – 14*D*). From medial to lateral, they are the:

- *general somatic efferent*, represented by neurons of the hypoglossal nerve
- *special visceral efferent*, represented by neurons innervating muscles derived from the pharyngeal arches (see Chapter 11)
- *general visceral efferent*, represented by some neurons of the vagus and glossopharyngeal nerves

Neuroblasts of the alar plates form neurons that are arranged in four columns on each side. From medial to lateral, they are the:

- *general visceral afferent*, receiving impulses from the viscera
- *special visceral afferent*, receiving taste fibers
- *general somatic afferent*, receiving impulses from the surface of the head
- *special somatic afferent*, receiving impulses from the ear

Some neuroblasts from the alar plates migrate ventrally and form the neurons in the **olivary nuclei** (see Fig. 19 – 14*C* and *D*).

Metencephalon

The walls of the metencephalon form the pons and cerebellum, and its cavity forms the superior part of the fourth ventricle (Fig. 19 – 15*A*). As in the rostral part of the myelencephalon, the pontine flexure causes divergence of the lateral walls of the pons, which spreads the gray substance in the floor of the fourth ventricle. As in the

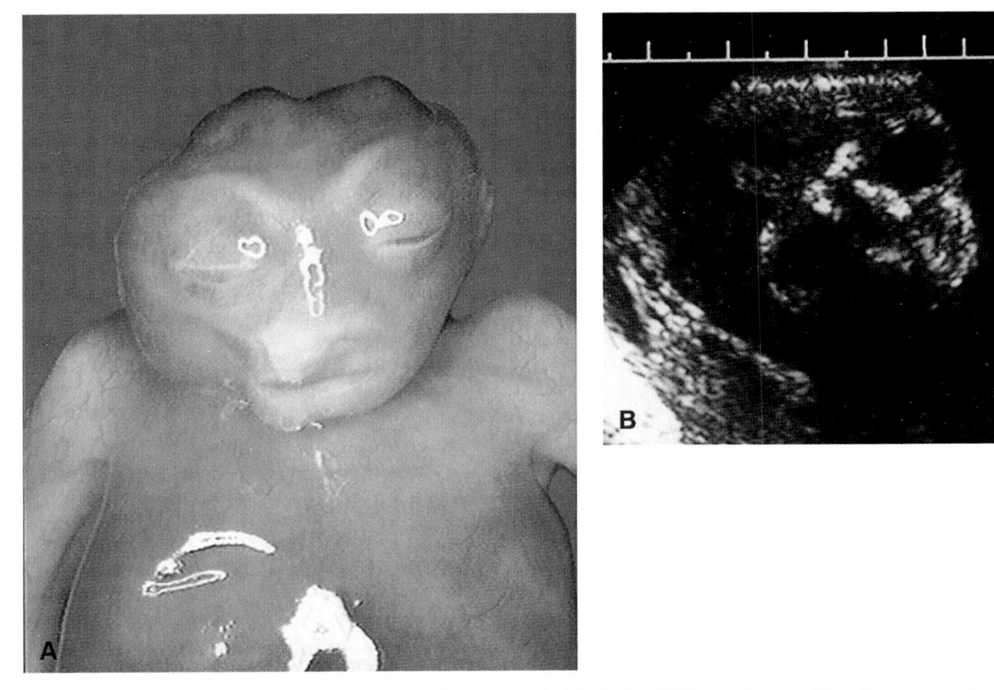

Figure 19–12. *A,* A fetus with meroanencephaly (anencephaly). *B,* The NTD was detected by ultrasonography at 18 weeks' gestation. Note the absence of the calvaria. (Courtesy of Wesley Lee, MD, Division of Fetal Imaging, Department of Obstetrics and Gynecology, William Beaumont Hospital, Royal Oak, Michigan.)

myelencephalon, neuroblasts in each basal plate develop into motor nuclei and organize into three columns on each side.

The **cerebellum** develops from dorsal parts of the alar plates (rhombomere 1) and the rhombic lip. Initially, the cerebellar swellings project into the fourth ventricle (see Fig. 19–15*A* and *B*). As the swellings enlarge and fuse in the median plane, they overgrow the rostral half of the fourth ventricle and overlap the pons and medulla (see Fig. 19–15*D*). Some neuroblasts in the

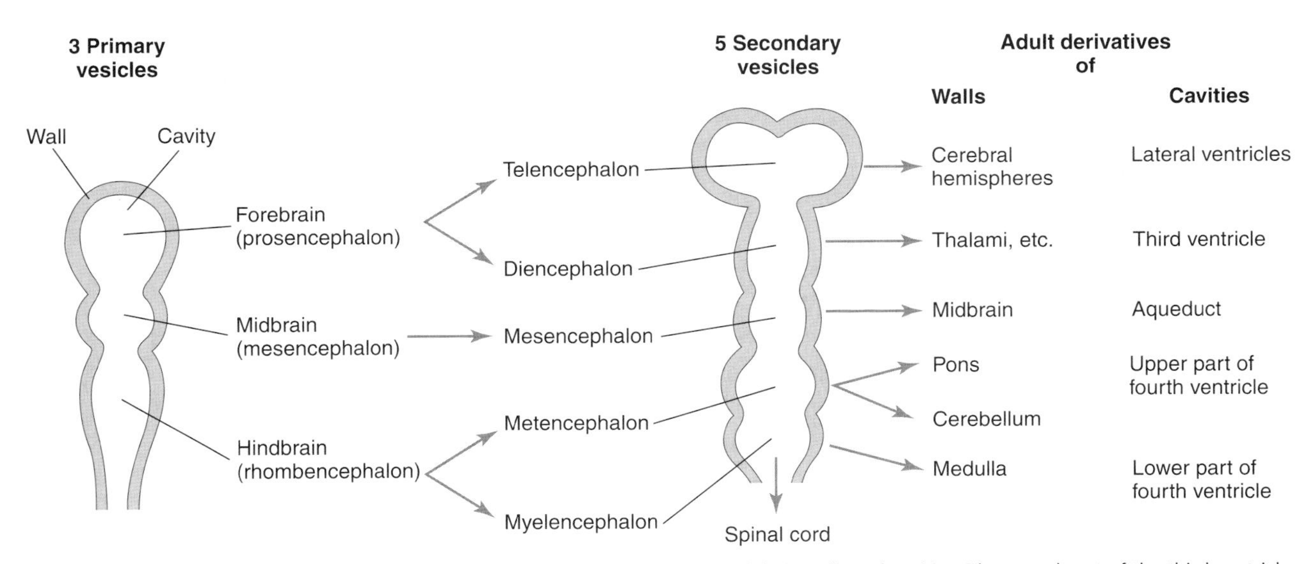

Figure 19–13. Diagrams of the brain vesicles, indicating the adult derivatives of their walls and cavities. The rostral part of the third ventricle forms from the cavity of the telencephalon; most of this ventricle is derived from the cavity of the diencephalon.

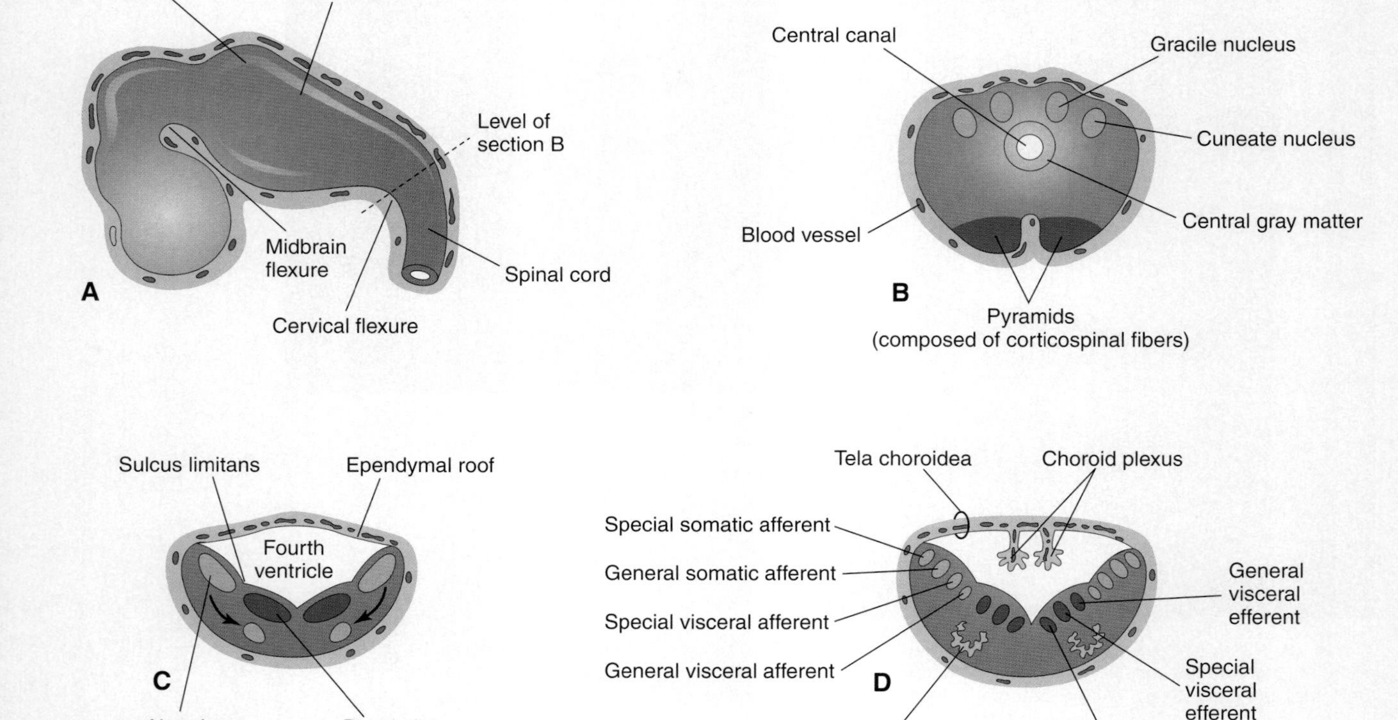

Figure 19 - 14. *A,* The developing brain at the end of the fifth week, showing three primary divisions of the brain and the brain flexures. *B,* Transverse section of the caudal part of the myelencephalon. *C* and *D,* Similar sections of the rostral part of the myelencephalon, showing the position and successive stages of differentiation of the alar and basal plates. The arrows in *C* show the pathway taken by neuroblasts from the alar plates to form the olivary nuclei.

intermediate zone of the alar plates migrate to the marginal zone and differentiate into the neurons of the **cerebellar cortex**. Other neuroblasts from these plates give rise to central nuclei, the largest of which is the **dentate nucleus**. Cells from the alar plates also give rise to the pontine nuclei, the cochlear and vestibular nuclei, and the sensory nuclei of the trigeminal nerve.

Nerve fibers connecting the cerebral and cerebellar cortices with the spinal cord pass through the marginal layer of the ventral region of the metencephalon. This region of the brain stem is the **pons** (L., bridge; see Fig. 19 - 15*C* and *D*).

Choroid Plexuses and Cerebrospinal Fluid

The thin ependymal roof of the fourth ventricle is covered externally by *pia mater* (see Fig. 19 - 15*C* and *D*). This vascular pia mater, together with the ependymal roof, forms the **tela choroidea**. Because of the active proliferation of the pia mater, the tela choroidea invaginates the fourth ventricle, where it differentiates into the **choroid plexus**. Similar choroid plexuses develop in the roof of the third ventricle and in the medial walls of the lateral ventricles. *The choroid plexuses secrete cerebrospinal fluid (CSF*). The thin roof of the fourth ventricle evaginates in three locations. These outpouchings rupture to form openings, the **median** and **lateral apertures** (foramen of Magendie and foramina of Luschka, respectively). These apertures permit CSF to enter the **subarachnoid space** from the fourth ventricle. The main site of absorption of CSF into the venous system is through protrusions of the arachnoid, called **arachnoid villi**.

Midbrain

The midbrain (mesencephalon) undergoes less change than any other part of the developing brain, except for the caudal part of the hindbrain. The neural canal narrows and becomes the **cerebral aqueduct** (see Fig. 19 - 15*D*),

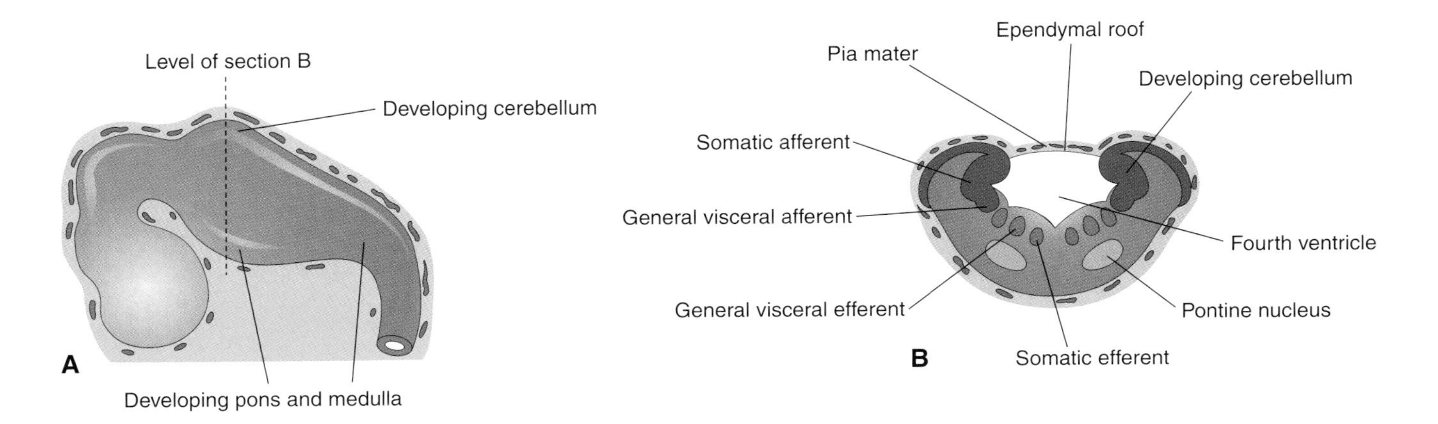

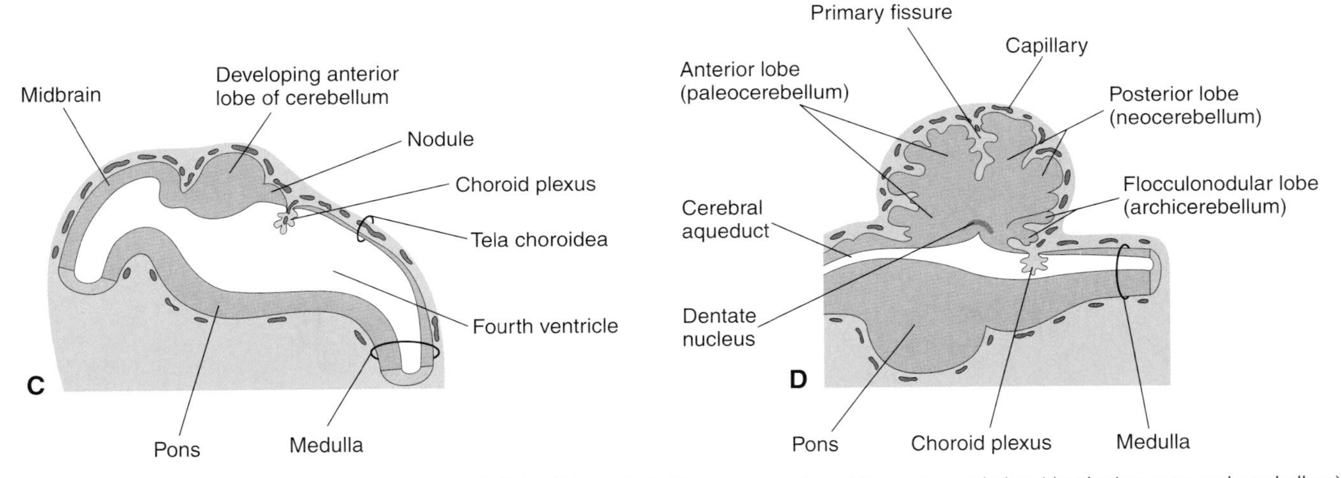

Figure 19-15. *A,* The developing brain at the end of the fifth week. *B,* Transverse section of the metencephalon (developing pons and cerebellum) showing the derivatives of the alar and basal plates. *C* and *D,* Sagittal sections of the hindbrain at 6 and 17 weeks, respectively, showing successive stages in the development of the pons and cerebellum.

a canal that connects the third and fourth ventricles. Neuroblasts migrate from the alar plates of the midbrain into the *tectum* (roof), where they aggregate to form four large groups of neurons — the paired superior and inferior colliculi (Fig. 19-16*B*), which are concerned with visual and auditory reflexes, respectively. Neuroblasts from the basal plates may give rise to groups of neurons in the ***tegmentum*** (red nuclei, nuclei of the third and fourth cranial nerves, and the reticular nuclei). The **substantia nigra**, a broad layer of gray matter adjacent to the cerebral peduncle (see Fig. 19-16*D* and *E*), may also differentiate from the basal plate, but some authorities believe that it is derived from cells in the alar plate that migrate ventrally. Fibers growing from the cerebrum form the cerebral peduncles anteriorly (see Fig. 19-16B). The **cerebral peduncles** become progressively more prominent as additional descending fiber groups (corticopontine, corticobulbar, and corticospinal) pass through the developing midbrain on their way to the brain stem and spinal cord.

Forebrain

As closure of the rostral neuropore occurs, two lateral outgrowths, called **optic vesicles**, appear (see Fig. 19-3*A*), one on each side of the forebrain. The optic vesicles are the primordia of the retinae and *optic nerves* (see Chapter 20). A second pair of diverticula soon arises more dorsally and rostrally, representing the **telencephalic vesicles** (see Fig. 19-16*C*). They are the primordia of the **cerebral hemispheres**, and their cavities become the **lateral ventricles** (Fig. 19-17*A* and *B*). The rostral or anterior part of the forebrain, including the primordia of the cerebral hemispheres, is

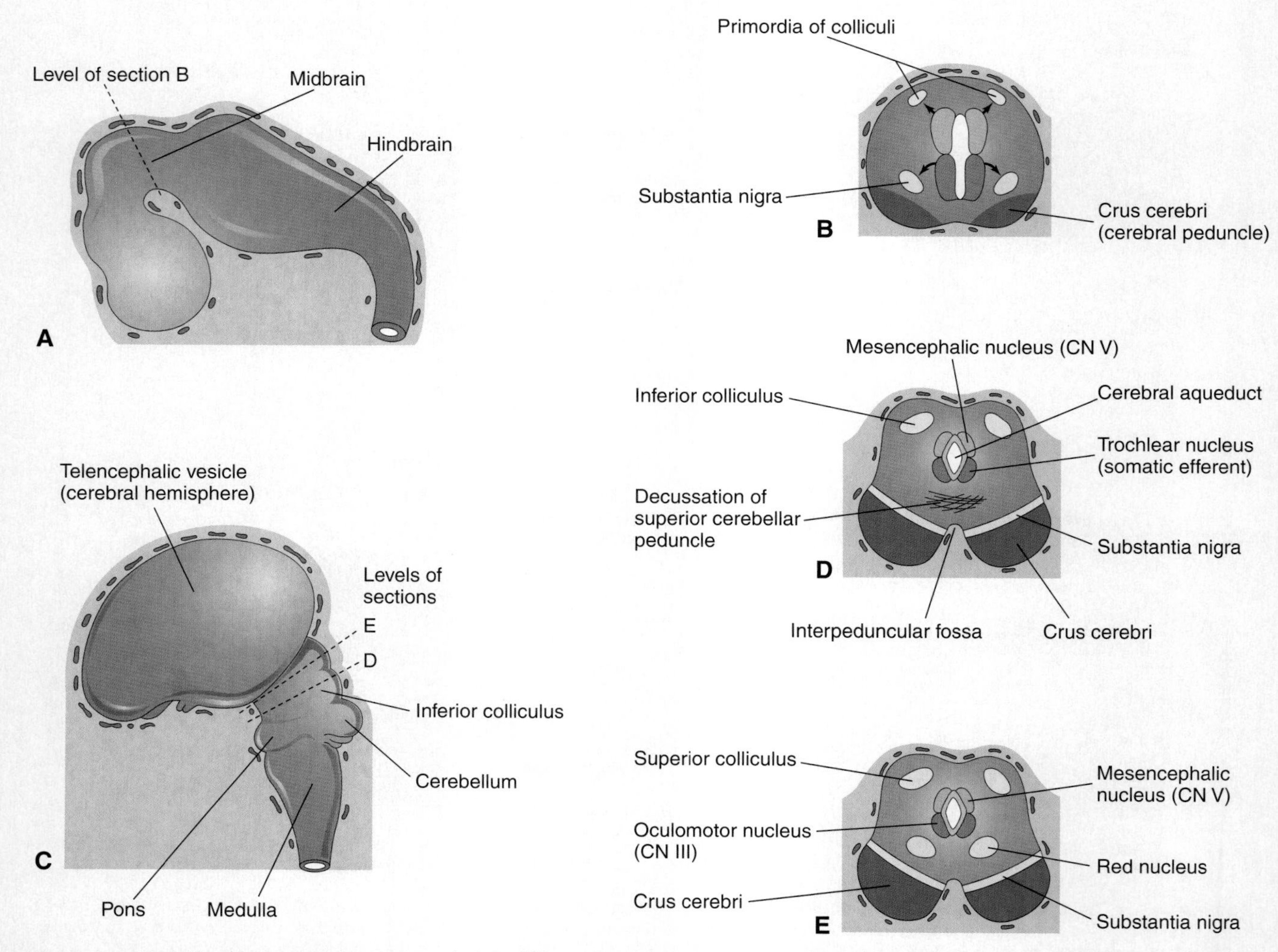

Figure 19 - 16. *A,* The developing brain at the end of the fifth week. *B,* Transverse section of the developing midbrain showing the early migration of cells from the basal and alar plates. *C,* Sketch of the developing brain at 11 weeks. *D* and *E,* Transverse sections of the developing midbrain at the level of the inferior and superior colliculi, respectively.

known as the **telencephalon**, whereas the caudal or posterior part of the forebrain is called the **diencephalon**. The cavities of the telencephalon and diencephalon contribute to the formation of the **third ventricle** (see Fig. 19 - 17*C*). *Molecular studies* show that sonic hedgehog (Shh) and Notch signals play a role in the modulations of neural precursor cells that give rise to the forebrain.

Diencephalon

Three swellings develop in the lateral walls of the third ventricle; these later become the epithalamus, thalamus, and hypothalamus (see Fig. 19 - 17*C* to *E*). The **thalamus** is separated from the epithalamus by the *epithalamic sulcus* and from the hypothalamus by the *hypothalamic sulcus*. The thalamus develops rapidly on each side and bulges into the cavity of the third ventricle, eventually reducing it to a narrow cleft. The **hypothalamus** arises by proliferation of neuroblasts in the intermediate zone of the diencephalic walls, ventral to the hypothalamic sulci. A pair of nuclei, the **mammillary bodies**, form pea-sized swellings on the ventral surface of the hypothalamus (see Fig. 19 - 17*C*). The **epithalamus** develops from the roof and dorsal part of the lateral wall of the diencephalon. Initially, the epithalamic swellings are large, but later, they become relatively small. The **pineal gland** (body) develops as a median diverticulum of the caudal part of the roof of the diencephalon (see Fig. 19 - 17*C* and *D*). Proliferation of the cells in its walls soon converts it into a solid, cone-shaped gland.

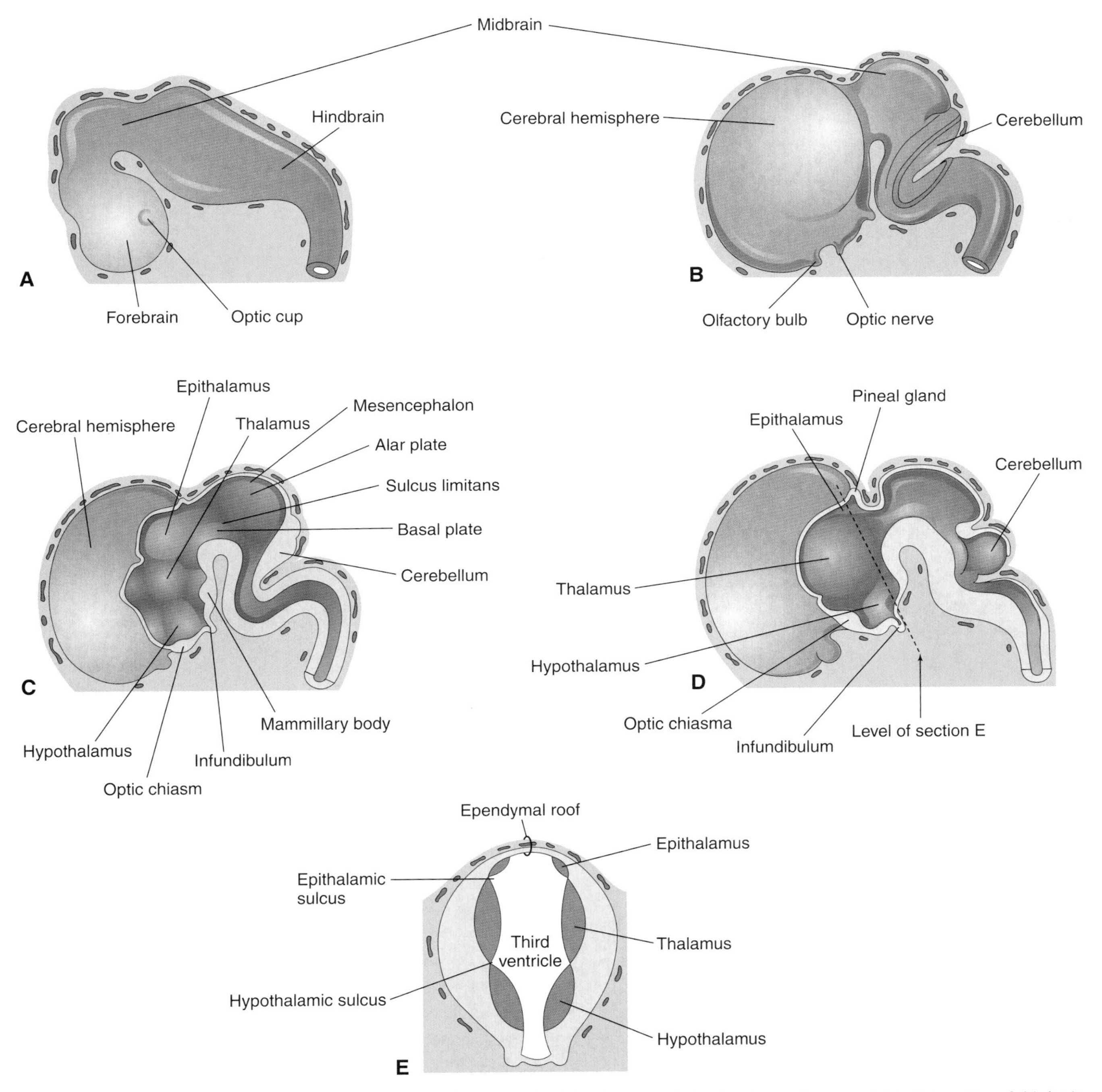

Figure 19-17. *A,* External view of the brain at the end of the fifth week of development. *B,* Similar view at 7 weeks. *C,* Median section of this brain showing the medial surface of the forebrain and midbrain. *D,* Similar section at 8 weeks. *E,* Transverse section of the diencephalon showing the epithalamus dorsally, the thalamus laterally, and the hypothalamus ventrally.

The **pituitary gland** (L. hypophysis) (Fig. 19-18; Table 19-1) is ectodermal in origin. It develops from two sources:

- an upgrowth from the ectodermal roof of the stomodeum called the **hypophysial diverticulum**

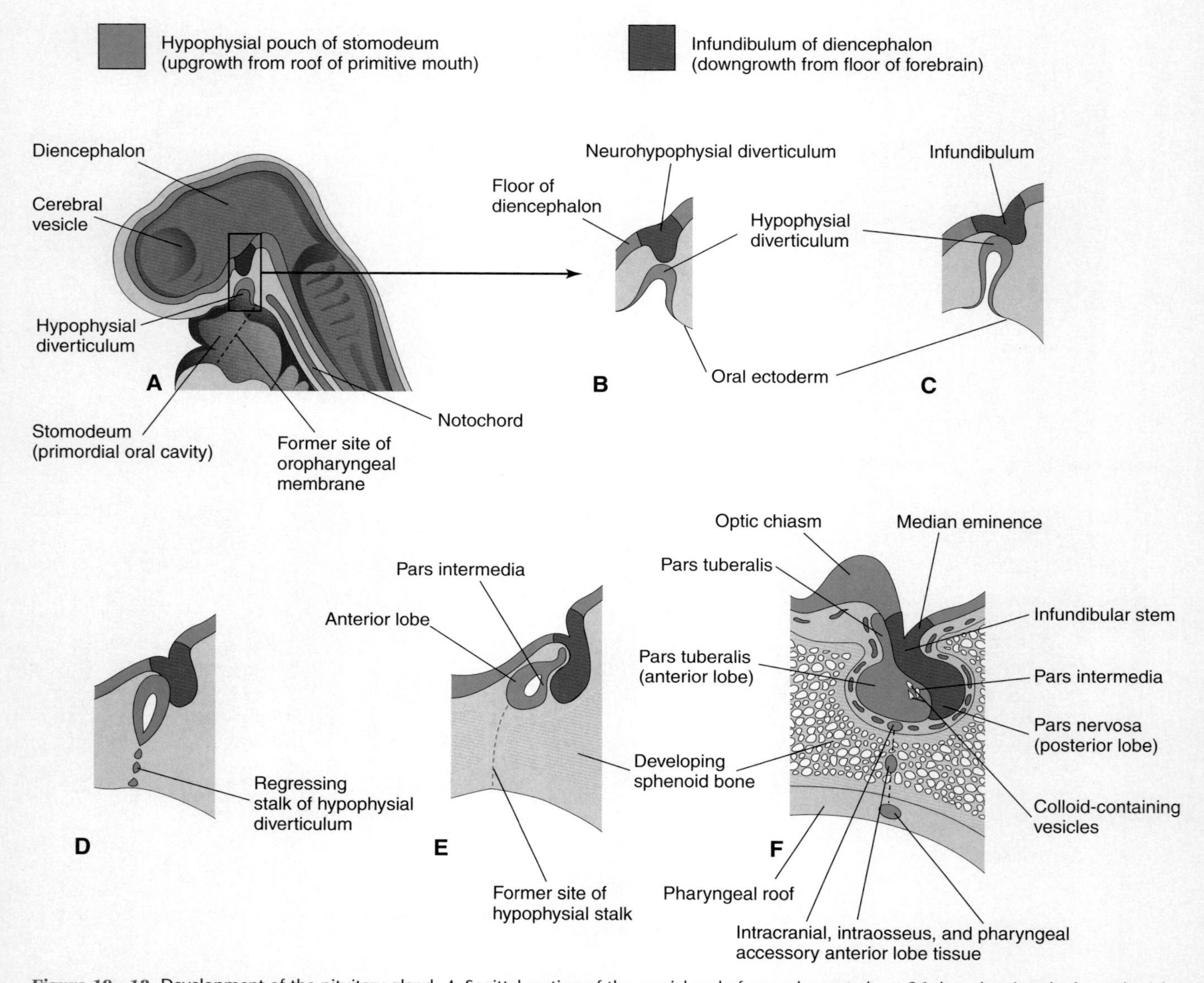

Figure 19 - 18. Development of the pituitary gland. *A,* Sagittal section of the cranial end of an embryo at about 36 days showing the hypophysial diverticulum (pouch), an upgrowth from the stomodeum, and the neurohypophysial diverticulum, a downgrowth from the forebrain. *B* to *D,* Successive stages of the developing pituitary gland. By 8 weeks, the diverticulum loses its connection with the oral cavity and is in close contact with the infundibulum and the posterior lobe (neurohypophysis) of the pituitary gland. *E* and *F,* Later stages, showing proliferation of the anterior wall of the hypophysial diverticulum to form the anterior lobe (adenohypophysis) of the pituitary gland.

Table 19 - 1. **Derivation and Terminology of the Pituitary Gland**

Origin		Derivative	Parts	Lobe
Oral Ectoderm (Hypophysial diverticulum from roof of stomodeum)	⟶	Adenohypophysis (glandular portion)	Pars distalis Pars tuberalis Pars intermedia	Anterior lobe (Pars distalis, Pars tuberalis)
Neuroectoderm (Neurohypophysial diverticulum from floor of diencephalon)	⟶	Neurohypophysis (nervous portion)	Pars nervosa Infundibular stem Median eminence	Posterior lobe (Pars intermedia, Pars nervosa, Infundibular stem, Median eminence)

- a downgrowth from the neuroectoderm of the diencephalon called the **neurohypophysial diverticulum**

This double embryonic origin explains why the pituitary gland is composed of two completely different types of tissue.

- The **adenohypophysis** (glandular part), or anterior lobe, arises from oral ectoderm.
- The **neurohypophysis** (nervous part), or posterior lobe, originates from neuroectoderm.

During the fourth week of development, a **hypophysial diverticulum** — (Rathke pouch) — projects from the roof of the stomodeum and lies adjacent to the floor (ventral wall) of the diencephalon (see Fig. 19 - 18*A* and *B*). By the fifth week, this pouch has elongated and has become constricted at its attachment to the oral epithelium, giving it a nipplelike appearance (see Fig. 19 - 18*C*). By this stage, it has come into contact with the **infundibulum** (derived from the neurohypophysial diverticulum), a ventral downgrowth of the diencephalon (see Fig. 19 - 18*C*). The parts of the pituitary gland that develop from the ectoderm of the stomodeum — pars anterior, pars intermedia, and pars tuberalis — form the **adenohypophysis** (see Table 19 - 1).

Pharyngeal Hypophysis and Craniopharyngioma

A remnant of the stalk of the hypophysial diverticulum may persist and form a *pharyngeal hypophysis* in the roof of the oropharynx (see Fig. 19 - 18*E* and *F*). Occasionally, **craniopharyngiomas** develop in the pharynx or in the basisphenoid (posterior part of the sphenoid bone) from remnants of the stalk of the hypophysial diverticulum, but most often, they form in or superior to the sella turcica of the cranium, or both.

Cells of the anterior wall of the hypophysial diverticulum give rise to the **pars distalis** of the pituitary gland. Later, a small extension, the **pars tuberalis**, grows around the infundibular stem. The extensive proliferation of the anterior wall of the hypophysial diverticulum reduces its lumen to a narrow cleft (see Fig. 19 - 18*E*). Cells in the posterior wall of the hypophysial diverticulum do not proliferate; they give rise to the thin, poorly defined **pars intermedia** (see Fig. 19 - 18*F*). The part of the pituitary gland that develops from the neuroectoderm of the brain (infundibulum) is the **neurohypophysis** (see Table 19 - 1). The infundibulum gives rise to the *median eminence, infundibular stem,* and *pars nervosa.*

Telencephalon

The telencephalon consists of a median part and two lateral diverticula, the **cerebral vesicles** (see Fig. 19 - 18*A*). These vesicles give rise to the **cerebral hemispheres**, which are identifiable at 7 weeks. The cavity of the median part of the telencephalon forms the extreme anterior part of the third ventricle. At first, the cerebral hemispheres are in wide communication with the cavity of the third ventricle through the **interventricular foramina** (Fig. 19 - 19*B*). As the cerebral hemispheres expand, they cover successively the diencephalon, midbrain, and hindbrain. The hemispheres eventually meet each other in the midline, flattening their medial surfaces.

The **corpus striatum** appears during the sixth week as a prominent swelling in the floor of each cerebral hemisphere (Fig. 19 - 20*B*). The floor of each hemisphere expands more slowly than its thin cortical wall because it contains the rather large corpus striatum; consequently, the cerebral hemispheres become C-shaped (Fig. 19 - 21). The growth and curvature of the hemispheres also affect the shape of the lateral ventricles. They become roughly C-shaped cavities filled with CSF. The caudal end of each cerebral hemisphere turns ventrally and then rostrally, forming the temporal lobe; in so doing, it carries the ventricle (forming the temporal horn) and **choroid fissure** with it (see Fig. 19 - 21). Here, the thin medial wall of the hemisphere is invaginated along the choroid fissure by vascular pia mater to form the *choroid plexus of the temporal horn* of the lateral ventricle (see Figs. 19 - 20*B* and 19 - 21*B*).

As the cerebral cortex differentiates, fibers passing to and from it pass through the **corpus striatum** and divide it into the *caudate* and *lentiform nuclei*. This fiber pathway — the **internal capsule** (see Fig. 19 - 20*C*) — becomes C-shaped as the hemisphere assumes this form. The **caudate nucleus** becomes elongated and C-shaped, conforming to the outline of the lateral ventricle (see Fig. 19 - 21*A* to *C*). Its pear-shaped head and elongated body lie in the floor of the frontal horn and body of the lateral ventricle; its tail makes a U-shaped turn to gain the roof of the temporal horn.

Cerebral Commissures

As the cerebral cortex develops, groups of fibers— called **commissures** — connect corresponding areas of the cerebral hemispheres with one another (see Fig. 19 - 20). The most important commissures cross in the **lamina terminalis**, the rostral end of the forebrain. This lamina extends from the roof plate of the diencephalon to the optic chiasm. The **anterior commissure** connects the olfactory bulb and related brain areas of one hemisphere with those of the opposite side. The **hippocampal commissure** connects the hippocampal formations. The anterior and hippocampal commissures connect phylogenetically older parts of the brain and are the first to form. The **corpus callosum** (see Fig. 19 - 20*A*), the

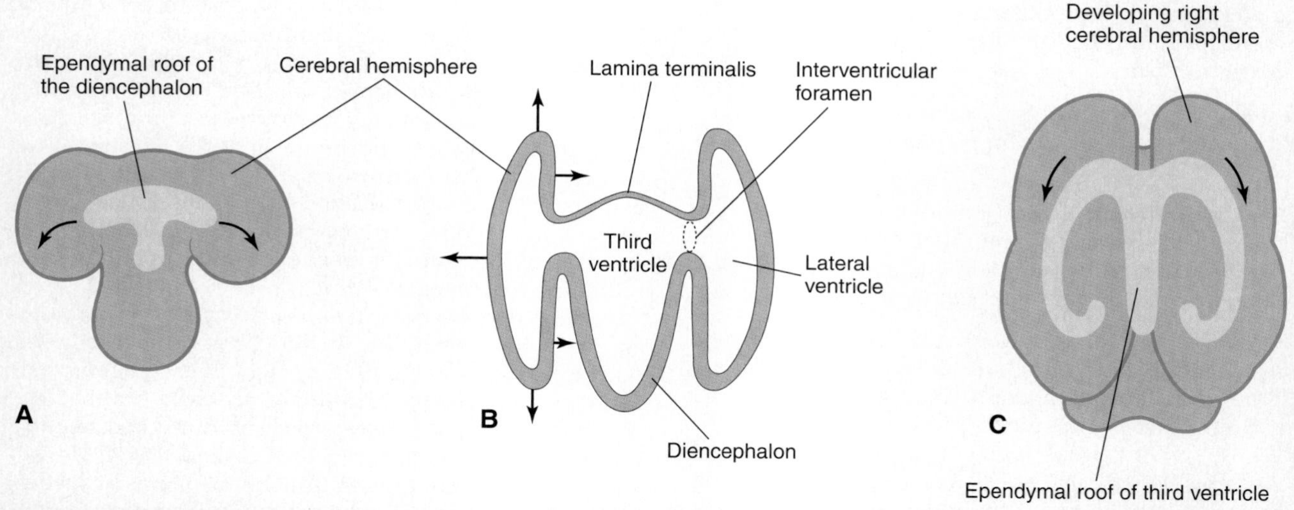

Figure 19–19. *A,* The dorsal surface of the forebrain, showing how the ependymal roof of the diencephalon is carried out to the dorsomedial surface of the cerebral hemispheres. *B,* Diagrammatic section of the forebrain showing how the developing cerebral hemispheres grow from the lateral walls of the forebrain and expand in all directions until they cover the diencephalon. The arrows indicate some of the directions in which the hemispheres expand. The rostral wall of the forebrain, the *lamina terminalis,* is very thin. *C,* Sketch of the forebrain showing how the ependymal roof is finally carried into the temporal lobes as a result of the C-shaped growth pattern of the cerebral hemispheres.

largest cerebral commissure, connects the neocortical areas. The corpus callosum initially lies in the lamina terminalis, but fibers are added to it as the cortex enlarges; as a result, it gradually extends beyond the lamina terminalis. The rest of the **lamina terminalis** lies between the corpus callosum and the fornix. It becomes stretched to form the **septum pellucidum**, a thin plate of brain tissue. At birth, the corpus callosum extends over the roof of the diencephalon. The **optic chiasm**, which develops in the ventral part of the lamina terminalis (see Fig. 19–20*A*), consists of fibers from the medial halves of the retinae that cross to join the optic tract of the opposite side.

The walls of the developing cerebral hemispheres initially show the three typical zones of the neural tube (ventricular, intermediate, and marginal); later, a fourth zone, the subventricular zone, appears. Cells of the intermediate zone migrate into the marginal zone and give rise to the cortical layers. The gray substance is thus located peripherally, and axons from its cell bodies pass centrally to form the large volume of white substance, the **medullary center**.

Initially, the surface of the hemispheres is smooth (Fig. 19–22*A*); however, as growth proceeds, **sulci** (grooves) and **gyri** (convolutions) develop (see Fig. 19–22*B* to *C*). The sulci and gyri permit a considerable increase in the surface area of the cerebral cortex without requiring an extensive increase in cranial size. As each cerebral hemisphere grows, the cortex covering the external surface of the corpus striatum grows relatively slowly and is soon overgrown. This buried cortex, hidden from view in the depths of the lateral sulcus (fissure) of the cerebral hemisphere, is known as the **insula** (L., island). *Molecular studies* suggest that homeogenes of the Emx family may play a role in the patterning of the cerebral cortex.

Congenital Anomalies of the Brain

Because of the complexity of its embryological history, abnormal development of the brain is common (occurring with a frequency of about 3 per 1000 births). Most major congenital anomalies of the brain, such as meroanencephaly (anencephaly) and meningoencephalocele, result from defective closure of the rostral neuropore during the fourth week of development (see Figs. 19–22 and 19–23) and involve the overlying tissues (meninges and calvaria). The factors causing NTDs are genetic, nutritional, or environmental in nature, or a combination of the three. Molecular studies have implicated many genes and signaling mechanisms.

Cranium Bifidum

Defects in the formation of the cranium (*cranium bifidum*) are often associated with congenital anomalies of the brain or meninges, or both. Defects of the cranium usually involve the median plane of the calvaria. The defect is often in the

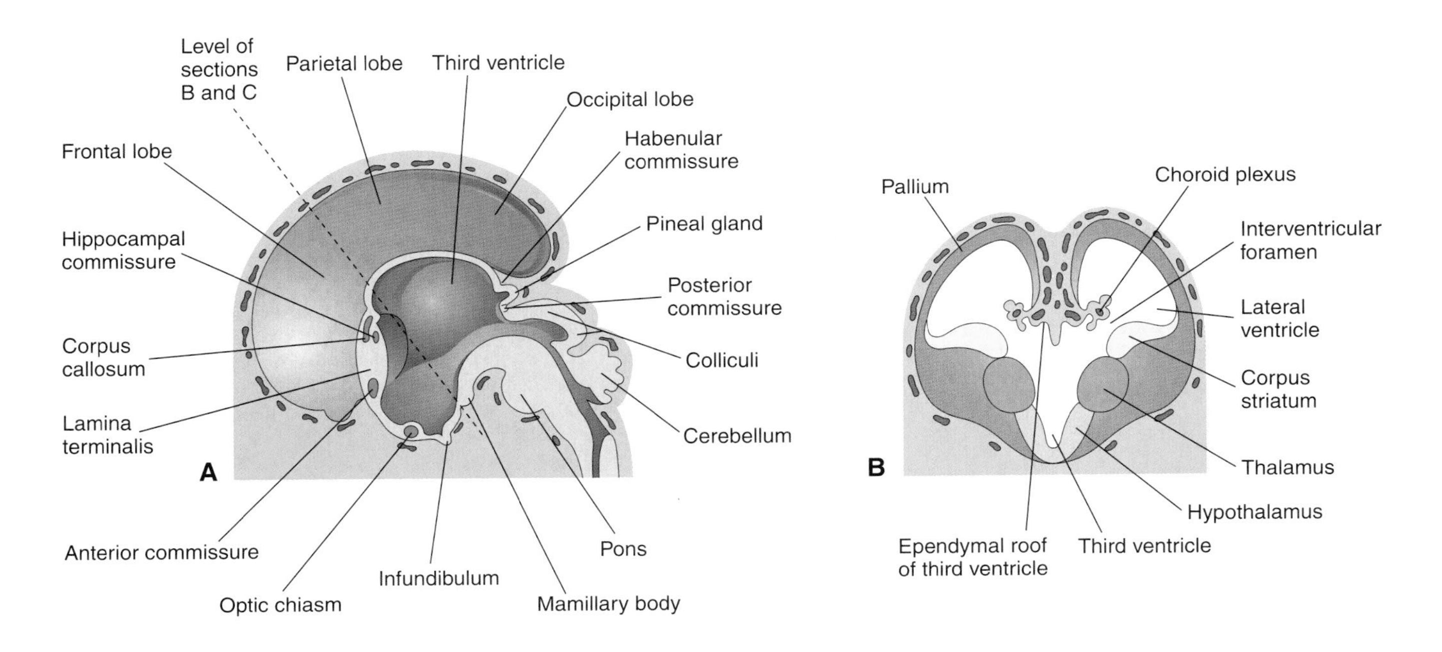

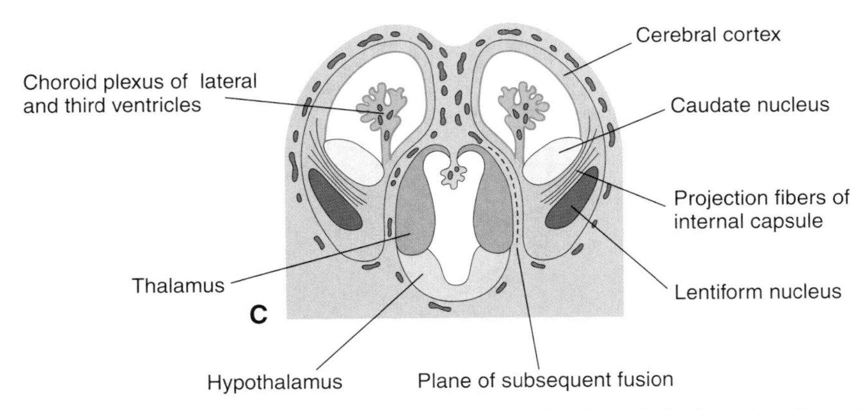

Figure 19–20. *A,* The medial surface of the forebrain of a 10-week embryo showing the diencephalic derivatives, the main commissures, and the expanding cerebral hemispheres. *B,* Transverse section of the forebrain at the level of the interventricular foramina, showing the corpus striatum and choroid plexuses of the lateral ventricles. *C,* Similar section at about 11 weeks showing division of the corpus striatum into the caudate and lentiform nuclei by the internal capsule. The developing relationship of the cerebral hemispheres to the diencephalon is also illustrated.

squamous part of the occipital bone and may include the posterior part of the foramen magnum. When the defect is small, usually only the meninges herniate, and the anomaly is called a **cranial meningocele**, or cranium bifidum with meningocele. Cranium bifidum associated with herniation of the brain and/or its meninges occurs in about 1 in every 2000 births. When the cranial defect is large, the meninges and part of the brain herniate, forming a **meningoencephalocele** (Fig. 19–23*A* to *C*). If the protruding brain contains part of the ventricular system, the anomaly is called a **meningohydroencephalocele**.

Meroanencephaly and Exencephaly

Meroanencephaly is a severe anomaly of the brain that results from failure of the rostral neuropore to close during the fourth week of development. As a result, the forebrain primordium is abnormal, and development of the calvaria is defective (see Fig. 19–12). In such cases, most of the infant's brain extrudes from the cranium, a condition known as **exencephaly**. Because of the abnormal structure and vascularization of the embryonic exencephalic brain, the nervous tissue undergoes degeneration, leaving a spongy, vascular mass consisting mostly of hindbrain structures.

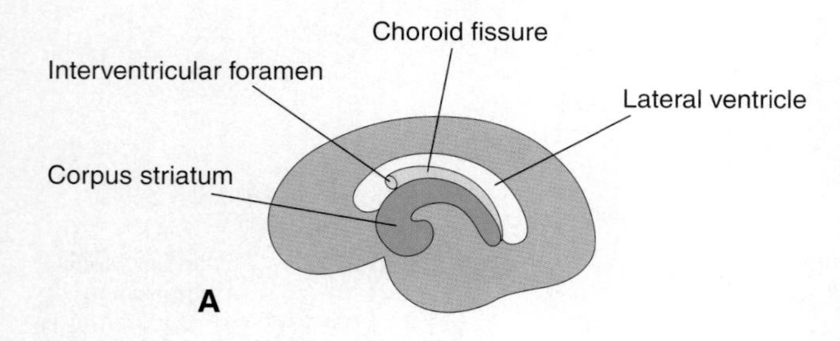

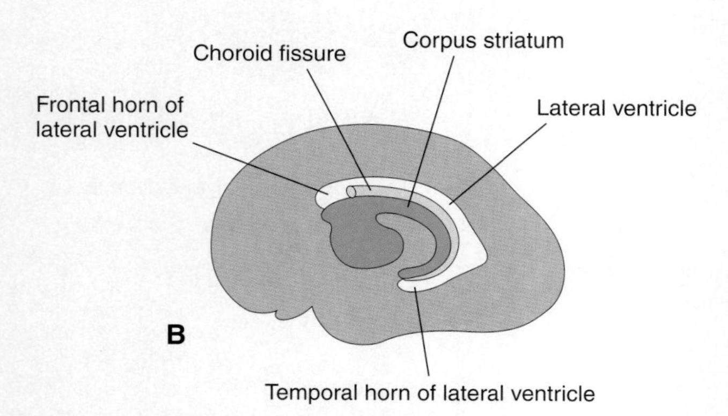

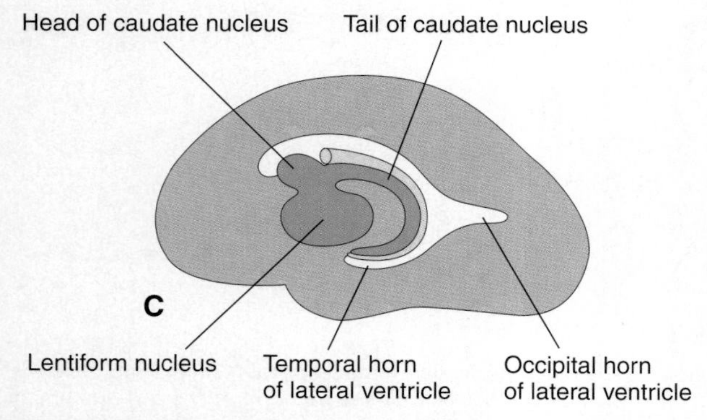

Figure 19 – 21. Drawings of the medial surface of the developing right cerebral hemisphere, showing the development of the lateral ventricle, choroid fissure, and corpus striatum. *A,* At 13 weeks. *B,* At 21 weeks. *C,* At 32 weeks.

Although this NTD is often called **anencephaly** (Gr. *an,* without + *enkephalos,* brain), a rudimentary brain stem and functioning neural tissue are always present in living infants. For this reason, meroanencephaly (Gr. *meros,* part) is a better name for this anomaly. **Meroanencephaly** is a common *lethal anomaly,* occurring at a rate of at least 1 in every 1000 births. It is more common in females than in males. *Meroanencephaly is usually associated with a multifactorial pattern of inheritance.* An excess of amniotic fluid **(polyhydramnios)** is often associated with meroanencephaly, possibly because the fetus lacks the neural control for swallowing amniotic fluid.

Microcephaly

In microcephaly, the calvaria and brain are small but the face is of normal size. Affected infants are *grossly mentally retarded* because the brain is underdeveloped (microencephaly). Some cases of microcephaly appear to be genetic in origin (autosomal recessive); others are caused by environmental factors, such as a cytomegalovirus infection in utero (see Chapter 9). Exposure during the fetal period to large amounts of ionizing radiation, infectious agents, and certain drugs (as with maternal alcohol abuse) is a contributing factor in some cases.

Hydrocephalus

Hydrocephalus results from impaired circulation and absorption of CSF or, in unusual cases, from increased production of CSF by a choroid plexus adenoma. An **excess of CSF** is present in the ventricular system of the brain (Fig. 19–24). Impaired circulation of CSF often results from **congenital aqueductal stenosis** (narrow cerebral aqueduct). *Blockage of CSF circulation results in dilation of the ventricles proximal to the obstruction and in pressure on the cerebral hemispheres.* This squeezes the brain between the ventricular fluid and the calvaria. In infants, the internal pressure results in an accelerated rate of expansion of the brain and calvaria because the fibrous sutures of the calvaria are not fused. *Hydrocephalus usually refers to obstructive or noncommunicating hydrocephalus,* in which all or part of the ventricular system is enlarged. All ventricles are enlarged if the apertures of the fourth ventricle or the subarachnoid spaces are blocked, whereas only the lateral and third ventricles are dilated when the cerebral aqueduct is obstructed. Hydrocephalus resulting from obliteration of the subarachnoid cisterns or malfunction of the arachnoid villi is known as *nonobstructive* or *communicating hydrocephalus.* Although hydrocephalus may be associated with spina bifida cystica, enlargement of the head may not be obvious at birth.

Holoprosencephaly

Holoprosencephaly results from a failure of cleavage of the forebrain. It occurs in about 1 in 15,000 to 20,000 live births, and most affected infants die within 6 months. Infants with this condition have a small, undivided forebrain and often a large, single, fused ventricle. Defects in forebrain development usually induce facial anomalies. This severe spectrum of congenital anomalies is attributable to genetic and environmental factors that cause extensive cell death in the median plane of the embryonic disc during the third

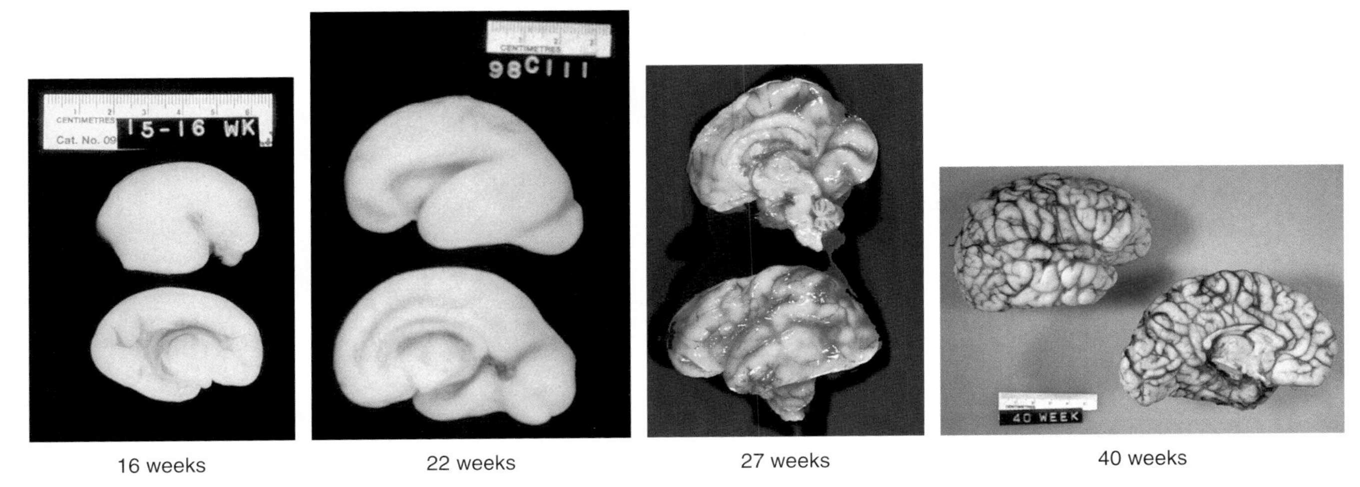

Figure 19 - 22. Photographs showing lateral and medial surfaces of human fetal brains at the gestational ages of 16, 22, 27, and 40 weeks. As the brain enlarges (see scales), the gyral pattern of the cerebral hemispheres becomes more complex. Note that, in the younger brains, the leptomeninges and associated blood vessels are not shown because they are easily stripped away. (Courtesy of Dr. Marc R. Del Bigio, Department of Pathology (Neuropathology), University of Manitoba and Health Sciences Centre, Winnipeg, Manitoba, Canada.)

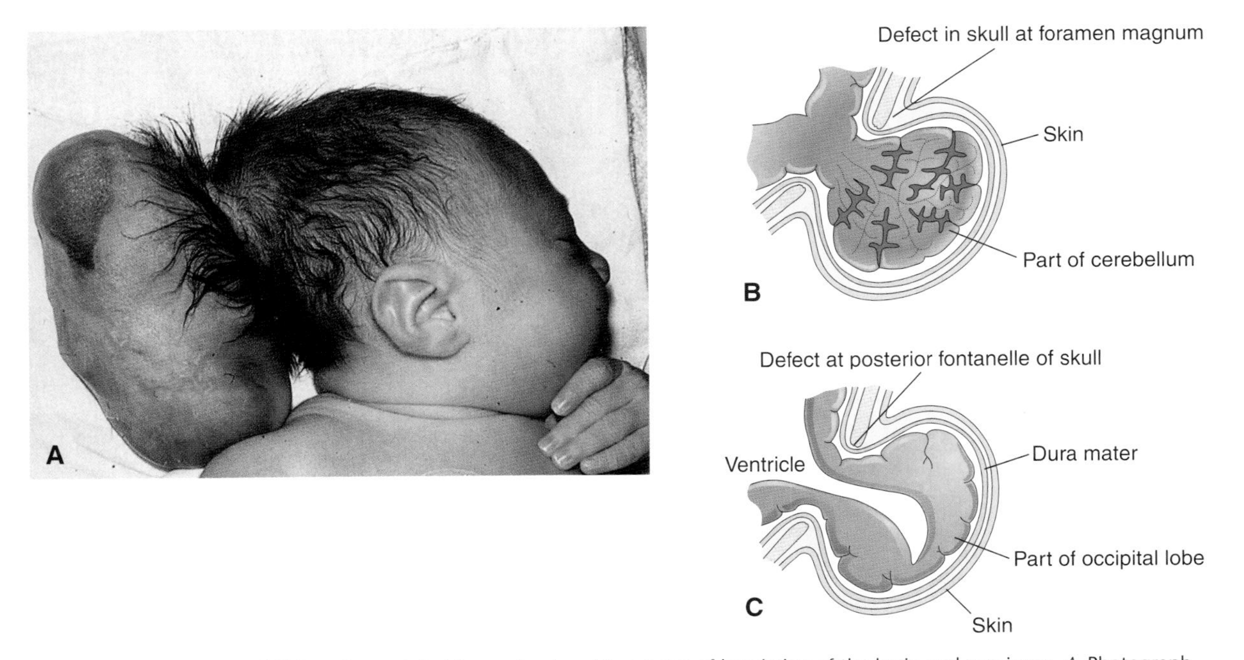

Figure 19 - 23. Cranium bifidum (bony defect in cranium) and two types of herniation of the brain and meninges. *A,* Photograph of an infant with a large meningoencephalocele in the occipital area. (Courtesy of Dr. AE Chudley, Section of Genetics and Metabolism, Department of Pediatrics and Child Health, Children's Hospital and University of Manitoba, Winnipeg, Manitoba, Canada.) *B,* Meningoencephalocele consisting of a protrusion of part of the cerebellum that is covered by meninges and skin. *C,* Meningohydroencephalocele consisting of a protrusion of part of the occipital lobe that contains part of the posterior horn of a lateral ventricle.

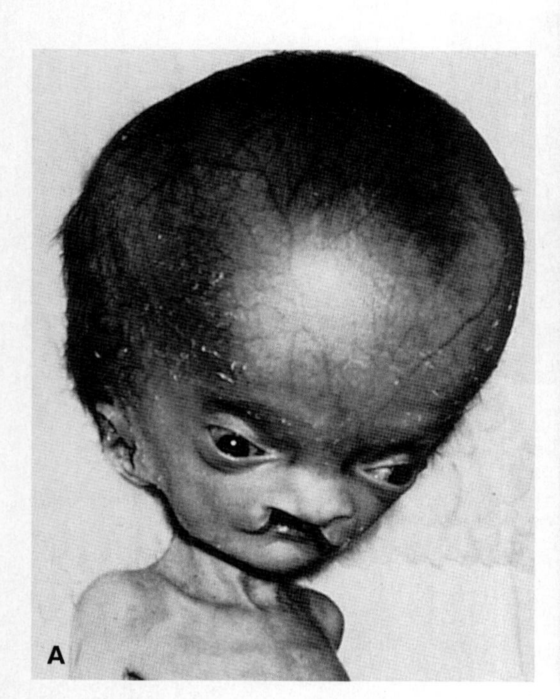

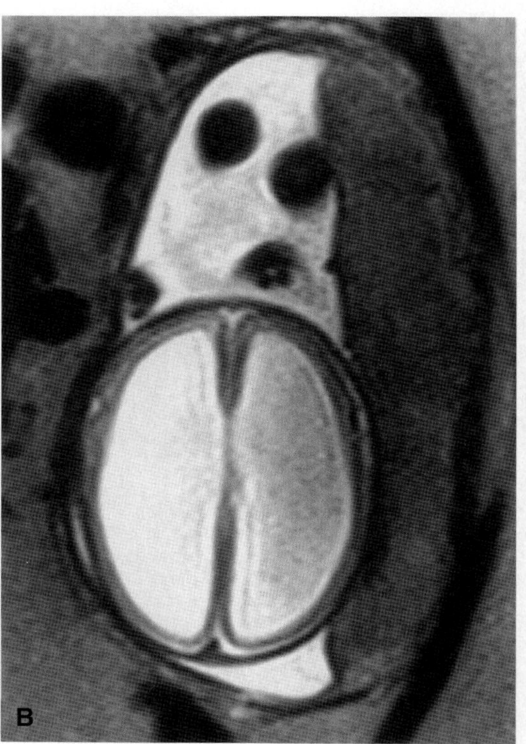

Figure 19 - 24. *A,* An infant with hydrocephalus and bilateral cleft palate. Hydrocephalus often produces thinning of the bones of the calvaria, prominence of the forehead, and atrophy of the cerebral cortex and white substance. *B,* Axial MRI (transverse section through the brain) of a fetus with X-linked hydrocephalus at about 29 weeks' gestation. (Courtesy of Dr. EH Whitby et al., Magnetic Resonance Imaging Unit, University of Sheffield, U.K.)

week of development. *Several genes (Shh, Zic, SIX3, and TG1F) have been implicated in the development of holoprosencephaly in humans.*

Arnold-Chiari Malformation

Arnold-Chiari malformation is the most common congenital anomaly involving the cerebellum (Fig. 19 – 25*B*). A tongue-like projection of the medulla and *inferior displacement of the vermis of the cerebellum herniates through the foramen magnum into the vertebral canal.* The condition results in a type of communicating hydrocephalus in which there is interference with the absorption of CSF; as a result, the entire ventricular system is distended. The Arnold-Chiari malformation occurs in about 1 in 1000 births and is frequently associated with spina bifida with meningomyelocele, spina bifida with myeloschisis, and hydrocephalus. The cause of the Arnold-Chiari malformation is uncertain; however, the posterior cranial fossa is abnormally small in some infants.

Development of the Peripheral Nervous System

The peripheral nervous system (PNS) consists of cranial, spinal, and visceral nerves and cranial, spinal, and autonomic ganglia. The PNS develops from various sources, mostly from the *neural crest.* All sensory cells (somatic and visceral) of the PNS are derived from **neural crest cells**. The cell bodies of these sensory cells are located outside the CNS. The cell body of each afferent neuron is closely invested by a capsule of modified Schwann cells, called satellite cells (see Fig. 19 – 7), which are derived from neural crest cells. This capsule is continuous with the neurolemma sheath of Schwann cells that surrounds the axons of afferent neurons.

Neural crest cells in the developing brain migrate to form sensory ganglia only in relation to the trigeminal (CN V), facial (CN VII), vestibulocochlear (CN VIII), glossopharyngeal (CN IX), and vagus (CN X) nerves. Neural crest cells also differentiate into multipolar neurons of the *autonomic ganglia* (see Fig. 19 – 7), including

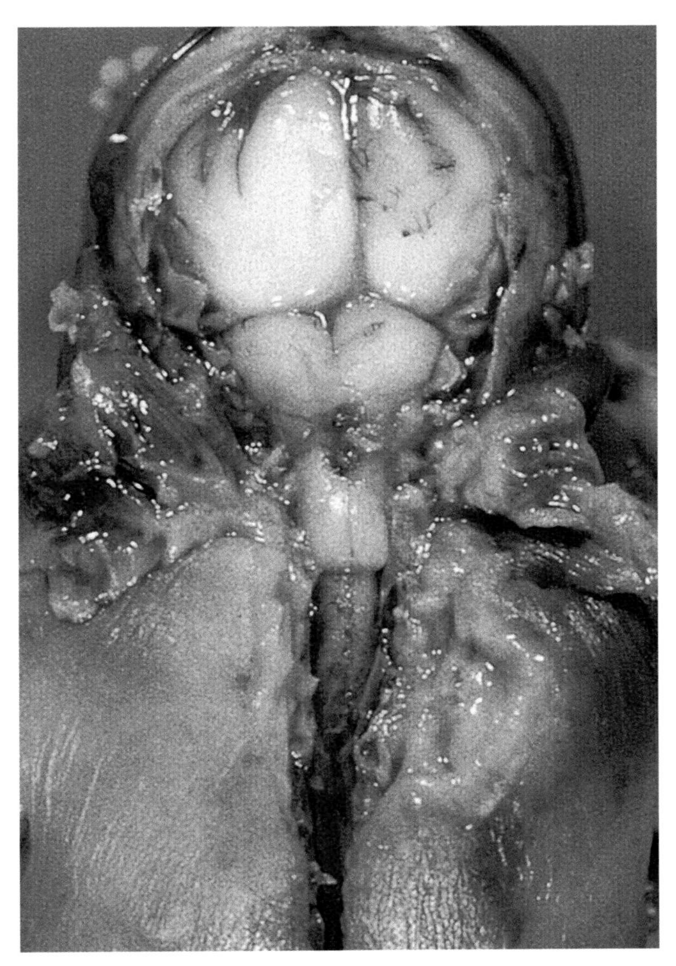

Figure 19 - 25. Photograph showing Arnold-Chiari type II malformation in a 23-week (gestational age) fetus. In situ exposure of the hindbrain reveals cerebellar tissue well below the foramen magnum. (Courtesy of Dr. Marc R. Del Bigio, Department of Pathology [Neuropathology], University of Manitoba, Winnipeg, Manitoba, Canada.)

ganglia of the sympathetic trunks that lie along the sides of the vertebral bodies; collateral or prevertebral ganglia in plexuses of the thorax and abdomen (e.g., the cardiac, celiac, and mesenteric plexuses); and parasympathetic or terminal ganglia in or near the viscera (e.g., the submucosal or Meissner plexus). Cells of the paraganglia – **chromaffin cells** – are also derived from the neural crest. The term *paraganglia* includes several widely scattered groups of cells that are similar in many ways to medullary cells of the suprarenal (adrenal) glands. The cell groups largely lie retroperitoneally, often in association with sympathetic ganglia. The carotid and aortic bodies also have small islands of chromaffin cells associated with them. These widely scattered groups of chromaffin cells constitute the **chromaffin system**. Neural crest cells also give rise to melanoblasts (precursors of *melanocytes*) and cells of the medulla of the suprarenal gland.

Spinal Nerves

Motor nerve fibers arising from the spinal cord begin to appear at the end of the fourth week of development (see Figs. 19 - 4, 19 - 6, and 19 - 7). The nerve fibers arise from cells in the *basal plates* of the developing spinal cord and emerge as a continuous series of rootlets along its ventrolateral surface. The fibers destined for a particular developing muscle group become arranged in a bundle, forming a **ventral nerve root**. The nerve fibers of the **dorsal nerve root** are formed by axons derived from neural crest cells that migrate to the dorsolateral aspect of the spinal cord, where they differentiate into the cells of the **spinal ganglion** (see Fig. 19 - 7). The central processes of neurons in the spinal ganglion form a single bundle that grows into the spinal cord, opposite the apex of the dorsal horn of gray substance (see Fig. 19 - 4*B* and *C*). The distal processes of spinal ganglion cells grow toward the ventral nerve root and eventually join it to form a **spinal nerve**.

As each limb bud develops, the nerves from the spinal cord segments opposite to it elongate and grow into the limb. The nerve fibers are distributed to its muscles, which differentiate from myogenic cells that originate from the somites (see Chapter 17). The skin of the developing limbs is also supplied in a segmental manner.

Cranial Nerves

Twelve pairs of cranial nerves (CNs) form during the fifth and sixth weeks of development. They are classified into three groups according to their embryological origins.

Somatic Efferent Cranial Nerves

The trochlear (CN IV), the abducent (CN VI), the hypoglossal (CN XII), and the greater part of the oculomotor (CN III) nerves are homologous with the ventral roots of spinal nerves (Fig. 19 - 26). The cells of origin of these nerves are located in the *somatic efferent column* (derived from the basal plates) of the brainstem. Their axons are distributed to the muscles derived from the head myotomes (preotic and occipital; see Fig. 17 - 2).

The **hypoglossal nerve** (CN XII) develops by fusion of the ventral root fibers of three or four occipital nerves (see Fig. 19 - 26*A*). Sensory roots, corresponding to the dorsal roots of spinal nerves, are absent. The somatic motor fibers originate from the *hypoglossal nucleus*. These fibers leave the ventrolateral wall of the

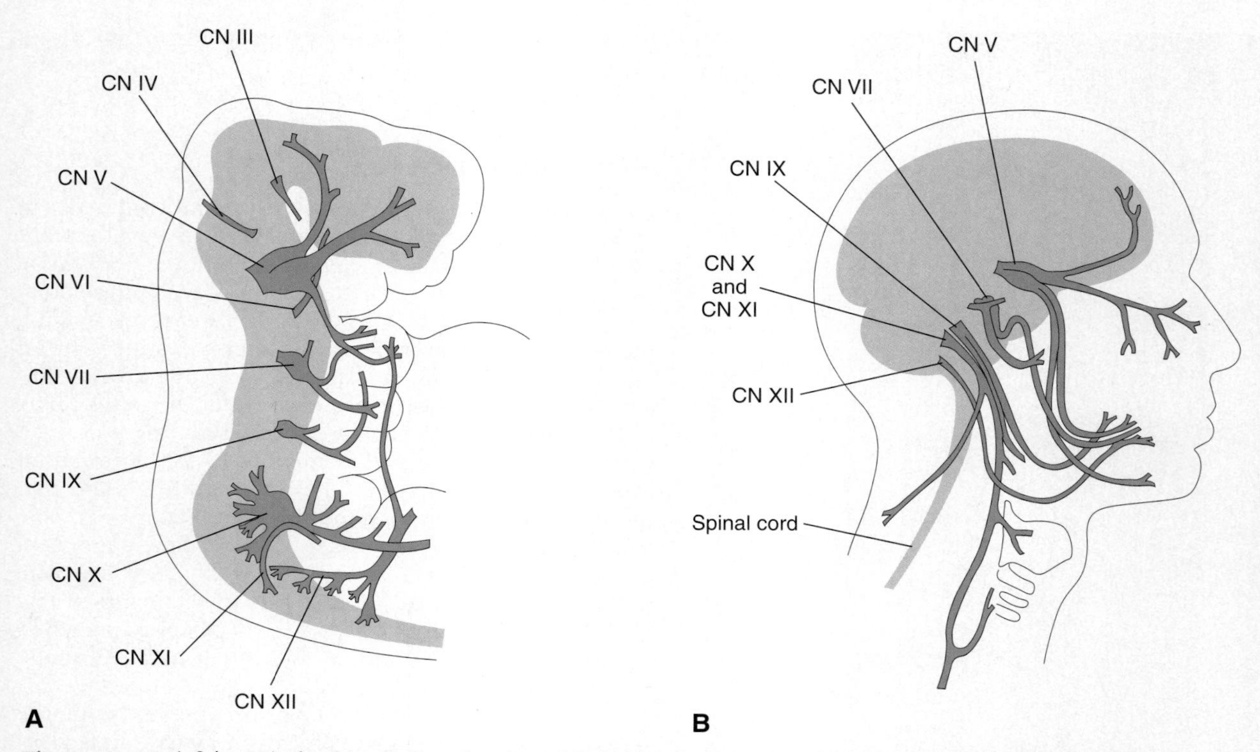

Figure 19 - 26. *A,* Schematic drawing of a 5-week embryo showing the distribution of most of the cranial nerves, especially those supplying the pharyngeal arches. *B,* Schematic drawing of the head and neck of an adult, showing the general distribution of most of the cranial nerves.

medulla in several groups — the *hypoglossal nerve roots* — converging to form the common trunk of CN XII (see Fig. 19 - 26*B*). They grow rostrally and eventually innervate the muscles of the tongue, which are derived from the occipital myotomes (see Fig. 17 - 2*A*).

The **abducent nerve** (CN VI) arises from nerve cells in the basal plates of the metencephalon. It passes from its ventral surface to the posterior of the three preotic myotomes from which the lateral rectus muscle of the eye is thought to originate.

The **trochlear nerve** (CN IV) arises from nerve cells in the somatic efferent column in the posterior part of the midbrain. Although a motor nerve, it emerges from the brain stem dorsally and then passes ventrally to supply the superior oblique muscle of the eye.

The **oculomotor nerve** (CN III) supplies most of the muscles of the eye (i.e., the superior, inferior, and medial recti) and inferior oblique muscles. The latter are thought to be derived from the first preotic myotomes.

Nerves of the Pharyngeal Arches

Cranial nerves V, VII, IX, and X supply the embryonic pharyngeal arches; thus, the structures that develop from these arches are innervated by these cranial nerves (see Fig. 19 - 26*A* and Table 11 - 1).

The **trigeminal nerve** (CN V) is the nerve of the first pharyngeal arch, but it has an ophthalmic division that is not a pharyngeal arch component. *Cranial nerve V is chiefly sensory and is the principal sensory nerve for the head.* The cells of the large **trigeminal ganglion** are derived from the most anterior part of the neural crest. The central processes of cells in this ganglion form the large sensory root of CN V, which enters the lateral part of the pons. The peripheral processes of cells in this ganglion separate into three large divisions (ophthalmic, maxillary, and mandibular nerves). Their sensory fibers supply the skin of the face as well as the lining of the mouth and nose. The *motor fibers of CN V* arise from cells in the most anterior part of the *special visceral efferent column* in the metencephalon. These fibers pass to the muscles of mastication and to other muscles that develop in the mandibular prominence of the first pharyngeal arch (see Table 11 - 1). The mesencephalic nucleus of CN V differentiates from cells in the midbrain.

The **facial nerve** (CN VII) is the nerve of the second pharyngeal arch. It consists mostly of motor fibers that arise principally from a nuclear group in the *special visceral efferent column* in the caudal part of the pons. These fibers are distributed to the *muscles of facial expression* and to other muscles that develop in the mesenchyme of the second pharyngeal arch (see Table 11 - 1). The small general visceral efferent component of CN VII terminates in the peripheral autonomic ganglia of the head. The sensory fibers of CN VII arise from the cells of the *geniculate ganglion*. The central processes of these cells enter the pons; the peripheral processes pass to the greater superficial petrosal nerve and, via the chorda tympani nerve, to the taste buds in the anterior two thirds of the tongue.

The **glossopharyngeal nerve** (CN IX) is the nerve of the third pharyngeal arch. Its motor fibers arise from the special and, to a lesser extent, general visceral efferent columns of the anterior part of the myelencephalon. Cranial nerve IX forms from several rootlets that arise from the medulla just caudal to the developing internal ear. All the fibers from the special visceral efferent column are distributed to the stylopharyngeus muscle, which is derived from mesenchyme in the third pharyngeal arch (see Table 11 - 1). The general efferent fibers are distributed to the otic ganglion from which postsynaptic fibers pass to the parotid and posterior lingual glands. The *sensory fibers of CN IX* are distributed as general sensory and special visceral afferent fibers (taste fibers) to the posterior part of the tongue.

The **vagus nerve** (CN X) is formed by fusion of the nerves of the fourth and sixth pharyngeal arches (see Table 11 - 1). The nerve of the fourth pharyngeal arch becomes the **superior laryngeal nerve**, which supplies the cricothyroid muscle and constrictor muscles of the pharynx. The nerve of the sixth pharyngeal arch becomes the **recurrent laryngeal nerve**, which supplies various laryngeal muscles.

The **accessory nerve** (CN XI) has two separate origins (see Fig. 19 - 26). The cranial root is a posterior extension of CN X, whereas the spinal root arises from the cranial five or six cervical segments of the spinal cord. The fibers of the cranial root emerge from the lateral surface of the medulla, where they join the vagus nerve and supply the muscles of the soft palate and intrinsic muscles of the larynx. The fibers of the spinal root supply the sternocleidomastoid and trapezius muscles.

Special Sensory Nerves

The **olfactory nerve** (CN I) arises from the olfactory bulb. The olfactory cells are bipolar neurons that differentiate from cells in the epithelial lining of the primordial nasal sac. The axons of the olfactory cells are collected into 18 to 20 bundles around which the *cribriform plate* of the ethmoid bone develops. These unmyelinated nerve fibers end in the olfactory bulb.

The **optic nerve** (CN II) is formed by more than a million nerve fibers that grow into the brain from neuroblasts in the primordial retina. Because the optic nerve develops from the evaginated wall of the forebrain, it actually represents a fiber tract of the brain. Development of the optic nerve is described in Chapter 20.

The **vestibulocochlear nerve** (CN VIII) consists of two kinds of sensory fiber in two bundles; these fibers are known as the vestibular and cochlear nerves. The **vestibular nerve** originates in the semicircular ducts, whereas the **cochlear nerve** proceeds from the cochlear duct, in which the **spiral organ** (of Corti) develops. The bipolar neurons of the vestibular nerve have their cell bodies in the vestibular ganglion. The central processes of these cells terminate in the *vestibular nuclei* in the floor of the fourth ventricle. The bipolar neurons of the *cochlear nerve* have their cell bodies in the spiral ganglion. The central processes of these cells end in the ventral and dorsal *cochlear nuclei* in the medulla.

Development of the Autonomic Nervous System

Functionally, the autonomic system can be divided into sympathetic (thoracolumbar) and parasympathetic (craniosacral) parts.

Sympathetic Nervous System

During the fifth week of development, *neural crest cells* in the thoracic region migrate along each side of the spinal cord, where they form paired cellular masses (ganglia) dorsolateral to the aorta (see Fig. 19 - 7). All of these segmentally arranged **sympathetic ganglia** are connected in a bilateral chain by longitudinal nerve fibers. These ganglionated cords, called **sympathetic trunks**, are located on each side of the vertebral bodies. Some neural crest cells migrate ventral to the aorta and form neurons in the **preaortic ganglia**, such as the celiac and mesenteric ganglia (see Fig. 19 - 7). Other neural crest cells migrate to the area of the heart, lungs, and gastrointestinal tract, where they form terminal ganglia in sympathetic organ plexuses, located near or within these organs.

After the sympathetic trunks have formed, axons of sympathetic neurons located in the **intermediolateral cell column** (lateral horn) of the thoracolumbar segments of the spinal cord pass through the ventral root of a spinal nerve and a **white communicating ramus** (L. white ramus communicans) to a paravertebral ganglion (see Fig. 19 - 7). There, they may synapse with neurons or ascend or descend in the sympathetic trunk to synapse at other levels. Other presynaptic fibers pass through the paravertebral ganglia without synapsing, forming splanchnic nerves to the viscera. The postsynaptic fibers course through a **gray communicating ramus**, passing from a sympathetic ganglion into a spinal nerve; hence, the sympathetic trunks are composed of ascending and descending fibers.

Parasympathetic Nervous System

The presynaptic parasympathetic fibers arise from neurons in nuclei of the brain stem and in the sacral region of the spinal cord. The fibers from the brain stem leave through the oculomotor (CN III), facial (CN VII), glossopharyngeal (CN IX), and vagus (CN X) nerves. The postsynaptic neurons are located in peripheral ganglia or in plexuses near or within the structure being innervated (e.g., the pupil of the eye and salivary glands).

Congenital Aganglionic Megacolon

Congenital aganglionic megacolon, or **Hirschsprung disease**, results from absence of ganglion cells in the wall of the large intestine, extending proximally and continuously from the anus for a variable distance. Hirschsprung disease is the most common cause of lower intestinal obstruction in the neonate, with an overall incidence of 1 in 5000 births. The absence of innervation of the colon results from failure of enteric neuronal precursors to migrate into the wall of the lower bowel. The affected segment of colon is paralyzed in a constricted state, which results in distention of the proximal, normally innervated bowel. The aganglionic segment is limited to the rectosigmoid colon in most cases. The clinical symptoms of Hirschsprung disease usually begin within 48 hours of birth with the delayed passage of meconium (fetal feces). Males are affected more often than females (4:1). *Molecular analysis* shows that mutations of the endothelin-B receptor and endothelin-3 genes cause Hirschsprung disease.

Summary of the Nervous System

The CNS develops from a dorsal thickening of ectoderm — the **neural plate** — that appears around the middle of the third week of development. The neural plate is induced by the underlying **notochord** and paraxial mesoderm. The neural plate becomes infolded to form a **neural groove** that has **neural folds** on each side. When the neural folds begin to fuse during the fourth week to form the **neural tube**, some neuroectodermal cells are not included in it but remain between the neural tube and surface ectoderm as the **neural crest**. The cranial end of the neural tube forms the brain, the primordia of which are the forebrain, midbrain, and hindbrain. The **forebrain** gives rise to the cerebral hemispheres and diencephalon. The embryonic **midbrain** becomes the adult midbrain, and the **hindbrain** gives rise to the pons, cerebellum, and medulla oblongata. The remainder of the neural tube becomes the spinal cord. The **neural canal**, the lumen of the neural tube, becomes the ventricles of the brain and the **central canal** of the spinal cord. The walls of the neural tube thicken by proliferation of its neuroepithelial cells. These cells give rise to all nerve and macroglial cells in the CNS. The microglia differentiate from mesenchymal cells that enter the CNS with the blood vessels.

Cells in the cranial, spinal, and autonomic ganglia are derived from **neural crest cells**, which originate in the neural crest. Neurolemma cells, which myelinate the axons external to the spinal cord, also arise from neural crest cells. Similarly, most of the autonomic nervous system and all chromaffin tissue, including the suprarenal medulla, develop from neural crest cells.

Congenital anomalies of the CNS are common, occurring at a rate of about 3 per 1000 births. Defects in the closure of the neural tube account for most severe anomalies (e.g., spina bifida cystica). Some anomalies of the CNS are caused by genetic factors (e.g., numerical chromosomal abnormalities, such as trisomy 21); others result from environmental factors, such as exposure to infectious agents, drugs, and metabolic disease. However, *most CNS anomalies are caused by a combination of genetic and environmental factors*. Gross congenital anomalies (e.g., meroanencephaly) are incompatible with life. Other severe anomalies (e.g., spina bifida with meningomyelocele) cause functional disability (e.g., muscle paralysis of the lower limbs). Severe abnormalities of the CNS also result from congenital anomalies of the ventricular system of the brain. **There are two main types of hydrocephalus:**

- *obstructive or noncommunicating hydrocephalus* (blockage of CSF flow in the ventricular system)
- *nonobstructive or communicating hydrocephalus* (blockage of CSF flow in the subarachnoid space)

Clinically Oriented Questions

1. Are neural tube defects (NTDs) hereditary? A mother had a baby with spina bifida cystica and her daughter had an infant with meroanencephaly. Is the daughter likely to have another child with a NTD? Can meroanencephaly and spina bifida be detected early in fetal life?
2. A baby was born with no cerebral hemispheres but a normal-appearing head. The baby exhibited excessive sleepiness, continuous crying when awake, and feeding problems. What is this condition called? What is its embryologic basis? Do affected children usually survive?
3. Some say that pregnant women who are heavy drinkers may have babies who exhibit mental and growth retardation. Is this true? There are reports of women getting drunk during pregnancy, yet their babies seem to be normal. Is there a safe threshold for alcohol consumption during pregnancy?
4. A woman was told that cigarette smoking during her pregnancy probably caused the slight mental retardation of her baby. She is not a heavy smoker. Was the woman correctly informed?
5. Do all types of spina bifida cause loss of motor function in the lower limbs? Which type of spina bifida cystica is more common and serious? What treatments are there for infants with spina bifida cystica?

The answers to these questions are at the back of the book.

20

The Eye and Ear

Development of the Eye and Related Structures

Early eye development results from a series of inductive signals. The eyes are derived from four sources:

- neuroectoderm of the forebrain
- surface ectoderm of the head
- mesoderm between the neuroectoderm and the surface ectoderm
- neural crest cells

The neuroectoderm of the forebrain differentiates into the retina, the posterior layers of the iris, and the optic nerve. The surface ectoderm of the head forms the lens of the eye and the corneal epithelium. The mesoderm between the neuroectoderm and surface ectoderm gives rise to the fibrous and vascular coats of the eye. Mesenchymal cells are derived from mesoderm, but neural crest cells migrate into the mesenchyme and differentiate into the choroid, sclera, and corneal endothelium.

Eye development is first evident at the beginning of the fourth week of development. **Optic grooves** (sulci) appear in the neural folds at the cranial end of the embryo (Fig. 20 – 1*A* and *B*). As the neural folds fuse to form the **forebrain**, the optic grooves evaginate to form hollow diverticula, known as **optic vesicles**, which project from the wall of the forebrain into the adjacent mesenchyme (see Fig. 20 – 1*C*). The cavities of the optic vesicles are continuous with the cavity of the forebrain. Formation of the optic vesicles is induced by the mesenchyme adjacent to the developing brain, probably through a chemical mediator. As the optic vesicles grow, their distal ends expand and their connections with the forebrain constrict to form hollow **optic stalks** (see Fig. 20 – 1*D*). The optic vesicles soon come in contact with the surface ectoderm.

Concurrently, the surface ectoderm adjacent to the optic vesicles thickens to form **lens placodes**, the primordia of the lenses (see Fig. 20 – 1*C*). Formation of the lens placodes is induced by the optic vesicles after the surface ectoderm has been conditioned by the underlying mesenchyme. An inductive message passes from the optic vesicles, stimulating the surface ectodermal cells to form the lens primordia. The lens placodes invaginate as they sink deep to the surface ectoderm, forming **lens pits** (see Figs. 20 – 1*D* and 20 – 2). The edges of the lens pits approach each other and fuse to form spherical **lens vesicles** (see Fig. 20 – 1*F* and *H*), which soon lose their connection with the surface ectoderm. Further development of the lenses is described after formation of the eyeball is discussed.

As the lens vesicles are developing, the optic vesicles invaginate to form double-walled **optic cups** (see Figs. 20 – 1*H* and 20 – 2). The opening of each optic cup is large at first, but its rim is infolded around the lens (Fig. 20 – 3*A*). By this stage, the lens vesicles have entered the cavities of the optic cups (Fig. 20 – 4). Linear grooves — retinal (optic) fissures — develop on the ventral surface of the optic cups and along the optic stalks (see Figs. 20 – 1*E* to *H* and 20 – 3*A* to *D*). The **retinal fissures** contain vascular mesenchyme from which the hyaloid blood vessels develop. The **hyaloid artery**, a branch of the *ophthalmic artery,* supplies the inner layer of the optic cup, the lens vesicle, and the mesenchyme in the optic cup (see Figs. 20 – 1*H* and 20 – 3). The **hyaloid vein** returns blood from these structures. As the edges of the retinal fissure fuse, the hyaloid vessels are enclosed within the **primordial optic nerve** (see Fig. 20 – 3*C* to *F*). Distal parts of the hyaloid vessels eventually degenerate, but proximal parts persist as the **central artery and vein of the retina** (Fig. 20 – 5*D*).

Development of the Retina

The retina develops from the walls of the optic cup (see Figs. 20 – 1 and 20 – 2). The outer, thinner layer of the optic cup becomes the **retinal pigment epithelium**, whereas the inner, thicker layer differentiates into the **neural retina**. During the embryonic and early fetal periods, the two retinal layers are separated by an **intraretinal space** (see Fig. 20 – 4), which is the original cavity of the optic cup. This space gradually disappears as the two layers of the retina fuse (see Fig. 20 – 5*D*), but this fusion is never firm; hence, when an adult eyeball is dissected, the neural retina is often separated from the retinal pigment epithelium. Because the optic cup is an outgrowth of the forebrain, the layers of the optic cup are continuous with the wall of the brain (see Fig. 20 – 1*H*). Under the influence of the developing lens, the inner layer of the optic cup proliferates to form a thick **neuroepithelium** (see Fig. 20 – 4). Subsequently, the cells of this layer differentiate into the **neural retina**, the light-sensitive region of the optic part of the retina. This region contains photoreceptors (*rods* and *cones*) and the cell bodies of neurons (e.g., bipolar and ganglion cells). Because the optic vesicle invaginates as it forms the optic cup, the neural retina is "inverted"; that is, light-sensitive parts of the photoreceptor cells are adjacent to the retinal pigment epithelium. As a result, light must pass through most of the retina before reaching the receptors; however, because the retina is thin and transparent, it does not form a barrier to light. The axons of ganglion cells in the superficial layer of the neural retina grow proximally in the wall of the optic stalk to the brain (see Figs. 20 – 3 and 20 – 4). The cavity of the optic stalk is gradually obliterated as the axons of the many ganglion cells form the **optic nerve** (see Fig. 20 – 3*F*).

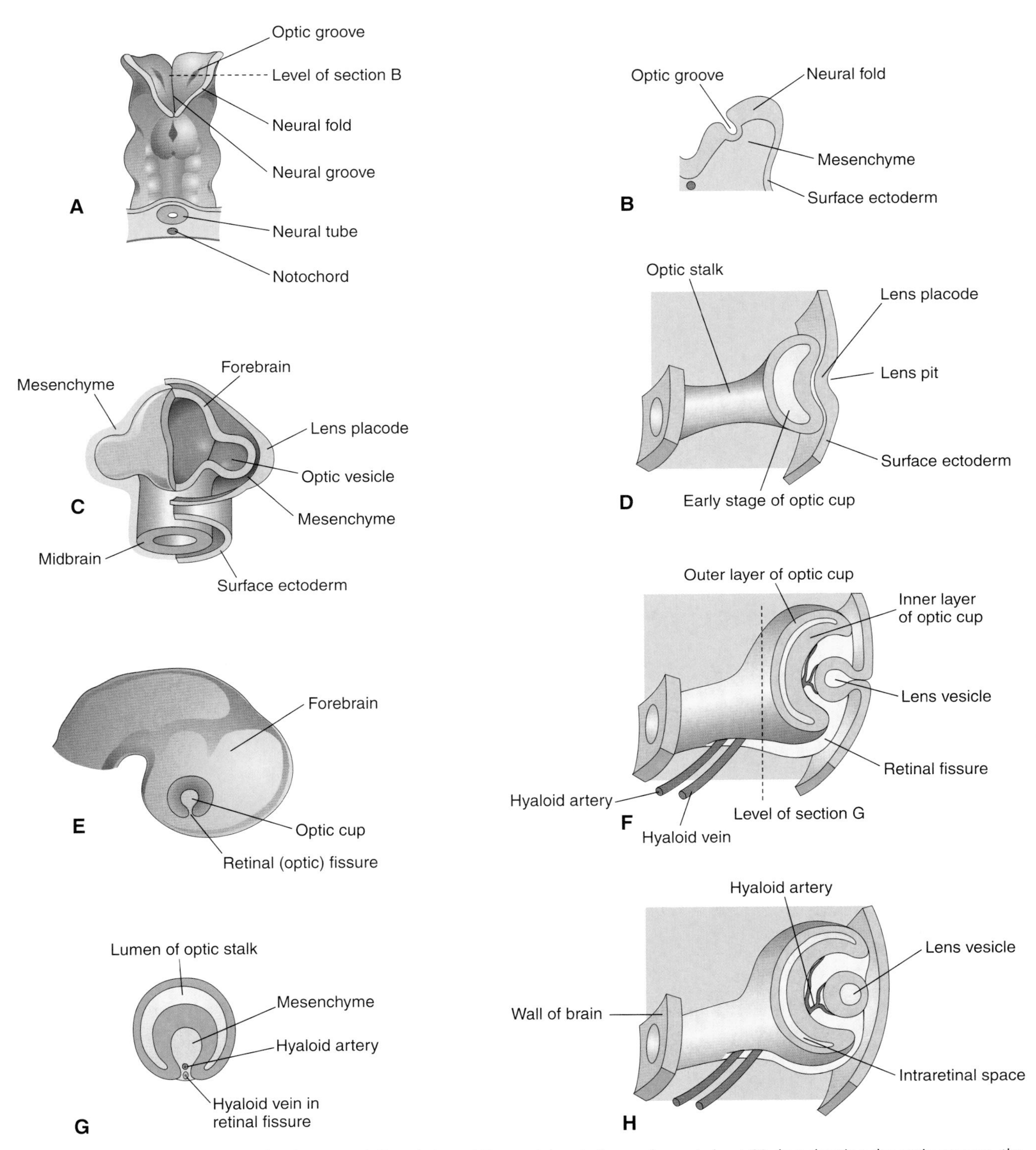

Figure 20 - 1. Early stages of eye development. *A,* Dorsal view of the cranial end of an embryo at about 22 days showing the optic grooves, the first indication of eye development. *B,* Transverse section of a neural fold showing the optic groove. *C,* Schematic drawing of the forebrain of an embryo at about 28 days. *D, F,* and *H,* Schematic sections of the developing eye illustrating successive stages in the development of the optic cup and lens vesicle. *E,* Lateral view of an embryo at about 32 days, showing the external appearance of the optic cup. *G,* Transverse section of the optic stalk showing the retinal fissure and its contents.

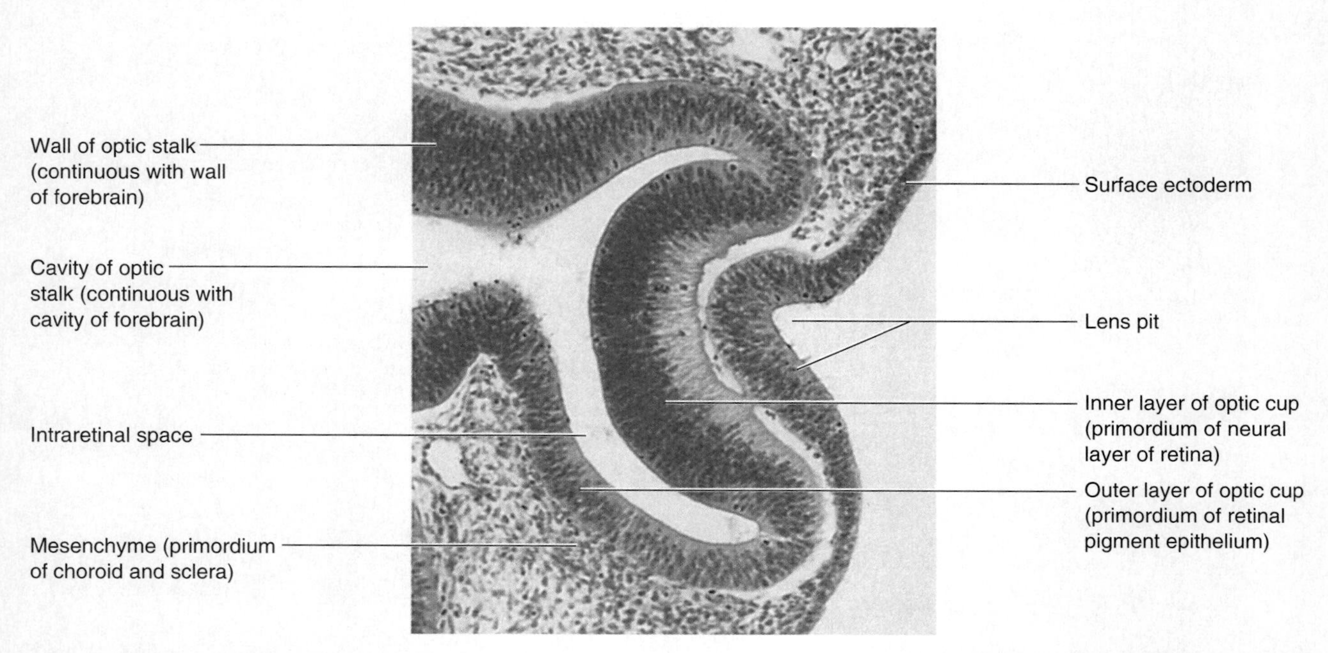

Figure 20 – 2. Photomicrograph of a sagittal section of the eye of an embryo (× 200), at about 32 days. Observe the primordium of the lens (invaginated lens placode), the walls of the optic cup (primordium of the retina), and the optic stalk, the primordium of the optic nerve. (From Moore KL, Persaud TVN, Shiota K: *Color Atlas of Clinical Embryology,* 2nd ed. Philadelphia, WB Saunders, 2000.)

Myelination of optic nerve fibers is incomplete at birth. After the eyes have been exposed to light for about 10 weeks, myelination is complete, but the process normally stops short of the optic disc, which is insensitive to light. Normal newborn infants can see, but not too well; they are able to respond to changes in illumination and to fixate points of contrast. Visual acuity has been estimated to be in the range of 20/400. At 2 weeks of age, the infants show a more sustained interest in large objects.

Congenital Anomalies of the Eye

Because of the complexity of eye development from several sources, including the diencephalon, many anomalies occur, but most of them are uncommon. The type and severity of the anomaly depend upon the embryonic stage when development is disrupted. Several environmental teratogens can cause congenital eye defects (see Chapter 9). Most common eye anomalies result from *defects in closure of the retinal fissure.*

Congenital Retinal Detachment

Congenital detachment of the retina occurs when the inner and outer layers of the optic cup fail to fuse during the fetal period to form the retina and obliterate the intraretinal space (see Figs. 20 – 3 and 20 – 5). The separation of the neural and pigmented layers may be partial or complete. Retinal detachment may result from unequal rates of growth of the two retinal layers; as a result, the layers of the optic cup are not in perfect apposition. Knowledge about eye development makes it clear that, where there is a detached retina, it is not a detachment of the entire retina because the retinal pigment epithelium remains firmly attached to the choroid. The detachment is at the site of adherence of the outer and inner layers of the optic cup. Although separated from the retinal pigment epithelium, the neural retina retains its blood supply (central artery of retina). Normally, the retinal pigment epithelium becomes firmly fixed to the choroid, but its attachment to the neural retina is not firm; hence, a retinal detachment may follow a blow to the eyeball, as may occur during a boxing match. In such cases, fluid accumulates between the layers and vision is impaired.

Coloboma of the Retina

Retinal coloboma is a defect that is characterized by a localized gap in the retina, usually inferior to the optic disc. The defect is bilateral in most cases. A typical coloboma results from defective closure of the retinal fissure.

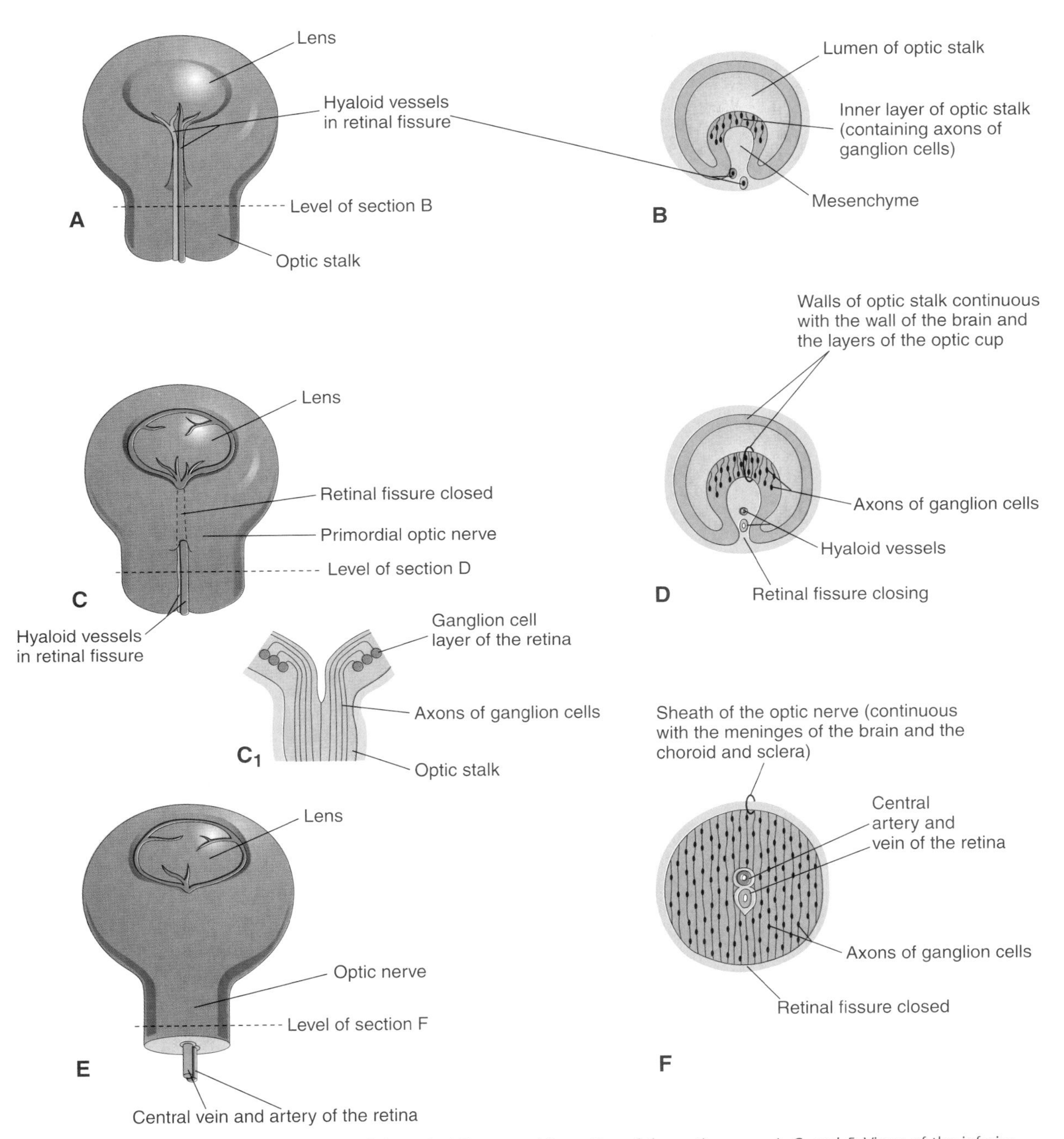

Figure 20 - 3. Diagrams illustrating closure of the retinal fissure and formation of the optic nerve. *A, C,* and *E,* Views of the inferior surface of the optic cup and stalk, showing progressive stages in the closure of the retinal fissure. C_1, Schematic sketch of a longitudinal section of a part of the optic cup and stalk, showing axons of ganglion cells of the retina growing through the optic stalk to the brain. *B, D,* and *F,* Transverse sections of the optic stalk, showing successive stages in closure of the retinal fissure and formation of the optic nerve. Note that the lumen of the optic stalk is gradually obliterated as axons of ganglion cells accumulate in the inner layer of the optic stalk with formation of the optic nerve.

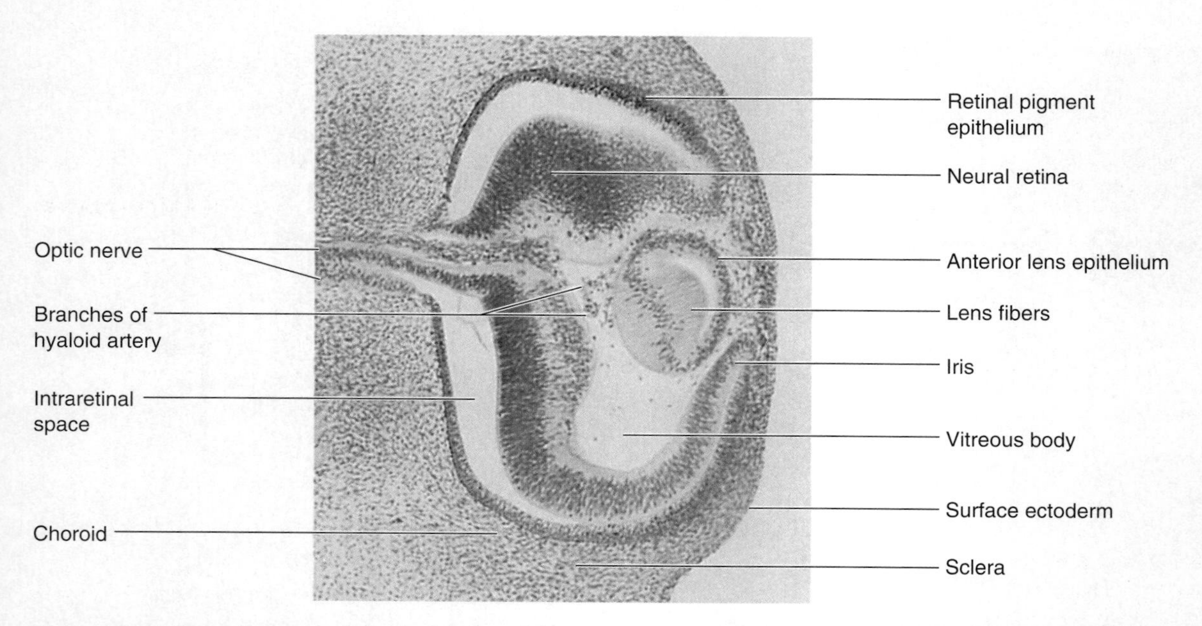

Figure 20 – 4. Photomicrograph of a sagittal section of the eye of an embryo (× 100), at about 44 days. Observe that the posterior wall of the lens vesicle forms the lens fibers. The anterior wall does not change appreciably as it becomes the anterior lens epithelium. (From Nishimura H [ed]: *Atlas of Human Prenatal Histology.* Tokyo, Igaku-Shoin, 1983.)

Development of the Ciliary Body

The ciliary body is the wedge-shaped extension of the choroid (Fig. 20 – 5*C* and *D*). Its medial surface projects toward the lens, forming **ciliary processes**. The pigmented part of the ciliary epithelium is derived from the outer layer of the optic cup and is continuous with the retinal pigment epithelium. The nonpigmented part of the ciliary epithelium represents the anterior prolongation of the neural retina, in which no neural elements differentiate. The smooth **ciliary muscle** that is responsible for focusing the lens and the connective tissue in the ciliary body develops from mesenchyme located at the edge of the optic cup in the region between the anterior scleral condensation and the ciliary pigment epithelium.

Development of the Iris

The iris develops from the rim of the optic cup, which grows inward and partially covers the lens (Fig. 20 – 5*D*). The two layers of the optic cup remain thin in this area. The epithelium of the iris represents both layers of the optic cup; it is continuous with the double-layered epithelium of the ciliary body and with the retinal pigment epithelium and neural retina. The connective tissue framework of the iris is derived from neural crest cells that migrate into the iris. The **dilator pupillae** and **sphincter pupillae muscles** of the iris are *derived from neuroectoderm of the optic cup.* They appear to arise from the anterior epithelial cells of the iris. These smooth muscles result from a transformation of epithelial cells into smooth muscle cells.

Color of the Iris

The iris is typically light blue or gray in most newborn infants. It acquires its definitive color as pigmentation occurs during the first 6 to 10 months. It is the concentration and distribution of pigment-containing cells, called *chromatophores,* in the loose vascular connective tissue of the iris that determines eye color. If the melanin pigment is confined to the pigmented epithelium on the posterior surface of the iris, the eye appears blue. If melanin is also distributed throughout the connective tissue of the iris, the eye appears brown.

Coloboma of the Iris

In infants with coloboma of the iris, there is a defect in the inferior sector of the iris or the pupillary margin that gives the pupil a keyhole appearance (Fig. 20 – 6). The coloboma may be limited to the iris, or it may extend deeper and involve the ciliary body and retina. A typical coloboma *results from failure of closure of the retinal fissure* during the sixth week. The defect may be genetically determined, or it may be caused by environmental factors. A simple coloboma of the iris is frequently hereditary and is transmitted as an autosomal dominant characteristic.

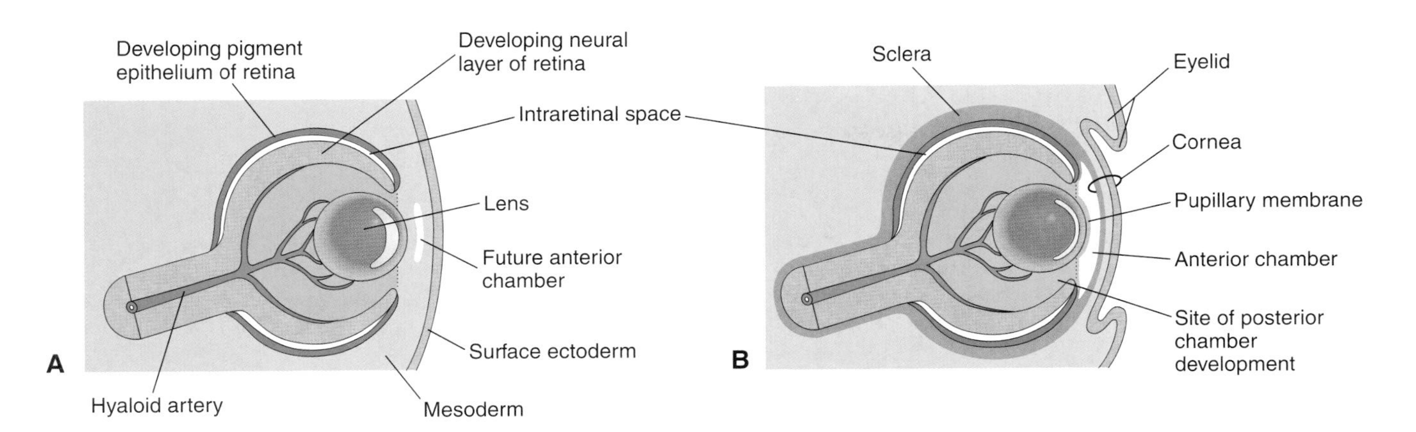

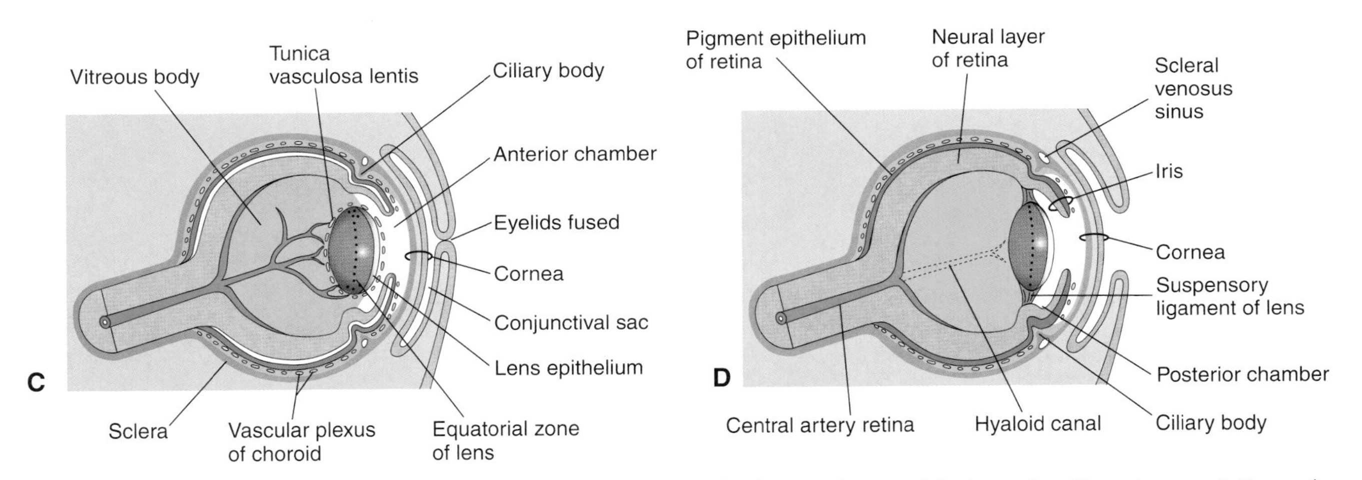

Figure 20 - 5. Diagrams of sagittal sections of the eye, showing the successive developmental stages of the lens, retina, iris, and cornea. *A,* Five weeks. *B,* Six weeks. *C,* Twenty weeks. *D,* Newborn infant. Note that the layers of the optic cup fuse to form the retinal pigment epithelium and neural retina and that they extend anteriorly as the double epithelium of the ciliary body and its iris. The retina and optic nerve are formed from the optic cup and optic stalk (outgrowths of the brain).

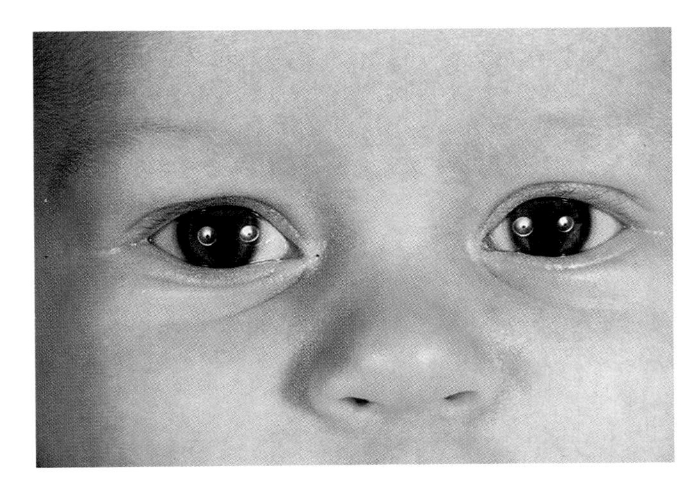

Figure 20 - 6. Bilateral coloboma of iris. Observe the defect in the inferior part of the iris (at 6-o'clock position). The defect is in the area of closure of the retinal fissure. (Courtesy of AE Chudley, MD, Section of Genetics and Metabolism, Department of Pediatrics and Child Health, Children's Hospital, University of Manitoba, Winnipeg, Manitoba, Canada.)

Development of the Lens

The lens develops from the **lens vesicle**, a derivative of the surface ectoderm (see Fig. 20 - 1). Lens formation involves the expression of L-Maf (lens-specific Maf) and other transcription factors in the lens placode and vesicle. The anterior wall of the lens vesicle, composed of cuboidal epithelium, becomes the subcapsular **lens epithelium** (see Fig. 20 - 5*C*). The nuclei of the tall columnar cells forming the posterior wall of the lens vesicle undergo dissolution. These cells lengthen considerably to form highly transparent epithelial cells, the **primary lens fibers**. As these fibers grow, they gradually obliterate the cavity of the lens vesicle (see Figs. 20 - 5*A* to *C* and 20 - 7). The rim of the lens is known as the **equatorial zone** because it is located midway between the anterior and posterior poles of the lens. The cells in the equatorial zone are cuboidal; as they elongate, they lose their nuclei and become **secondary lens fibers** (Fig. 20 - 8). These fibers are added to the external sides of the primary lens fibers. Although secondary lens fibers continue to form during adulthood and the lens increases in diameter, the primary lens fibers must last a lifetime. The developing lens is supplied by the distal part of the **hyaloid artery** (see Figs. 20 - 4 and 20 - 5); however, it becomes avascular in the fetal period when this part of the artery degenerates. After this occurs, the lens depends on diffusion from the aqueous humor in the anterior chamber of the eye, which bathes its anterior surface, and from the vitreous humor in other parts. The developing lens is invested by a vascular mesenchymal layer, the **tunica vasculosa lentis**. The anterior part of this capsule is the **pupillary membrane** (see Fig. 20 - 5*B* and *C*). The part of the hyaloid artery that supplies the tunica vasculosa lentis disappears during the late fetal period. As a result, the tunica vasculosa lentis and pupillary membrane degenerate (see Fig. 20 - 5*D*); however, the **lens capsule** produced by the anterior lens epithelium and the lens fibers persists. The lens capsule represents a greatly thickened basement membrane and has a lamellar structure. The former site of the hyaloid artery is indicated by the **hyaloid canal** in the vitreous body (see Fig. 20 - 5*D*); this canal is usually inconspicuous in the living eye.

The **vitreous body** forms within the cavity of the optic cup (see Fig. 20 - 5*C*). It is composed of **vitreous humor**, an avascular mass of transparent, gel-like, intercellular substance. The **primary vitreous humor** is derived from mesenchymal cells of neural crest origin. The primary vitreous humor does not increase, but it is surrounded by a gelatinous **secondary vitreous humor**, the origin of which is uncertain.

Persistence of the Hyaloid Artery

The distal part of the hyaloid artery normally degenerates as its proximal part becomes the central artery of the retina. If part of the artery persists distally, it may appear as a freely moving, nonfunctional vessel or as a wormlike structure projecting from the optic disc. Sometimes, the hyaloid artery remnant may appear as a fine strand traversing the vitreous body. In other cases, the remnant may form a cyst.

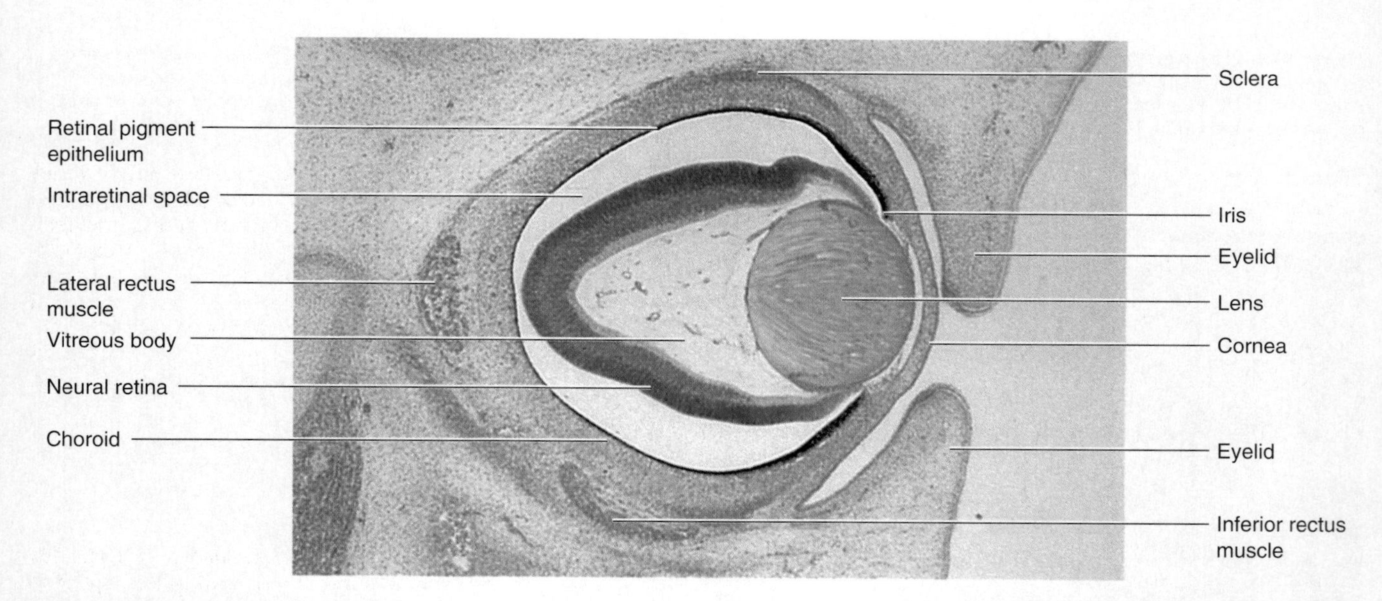

Figure 20 - 7. Photomicrograph of a sagittal section of the eye of an embryo (× 50), at about 56 days. Observe the developing neural retina and the retinal pigment epithelium. The intraretinal space normally disappears as these two layers of the retina fuse. (From Moore KL, Persaud TVN, Shiota K: *Color Atlas of Clinical Embryology,* 2nd ed. Philadelphia, WB Saunders, 2000.)

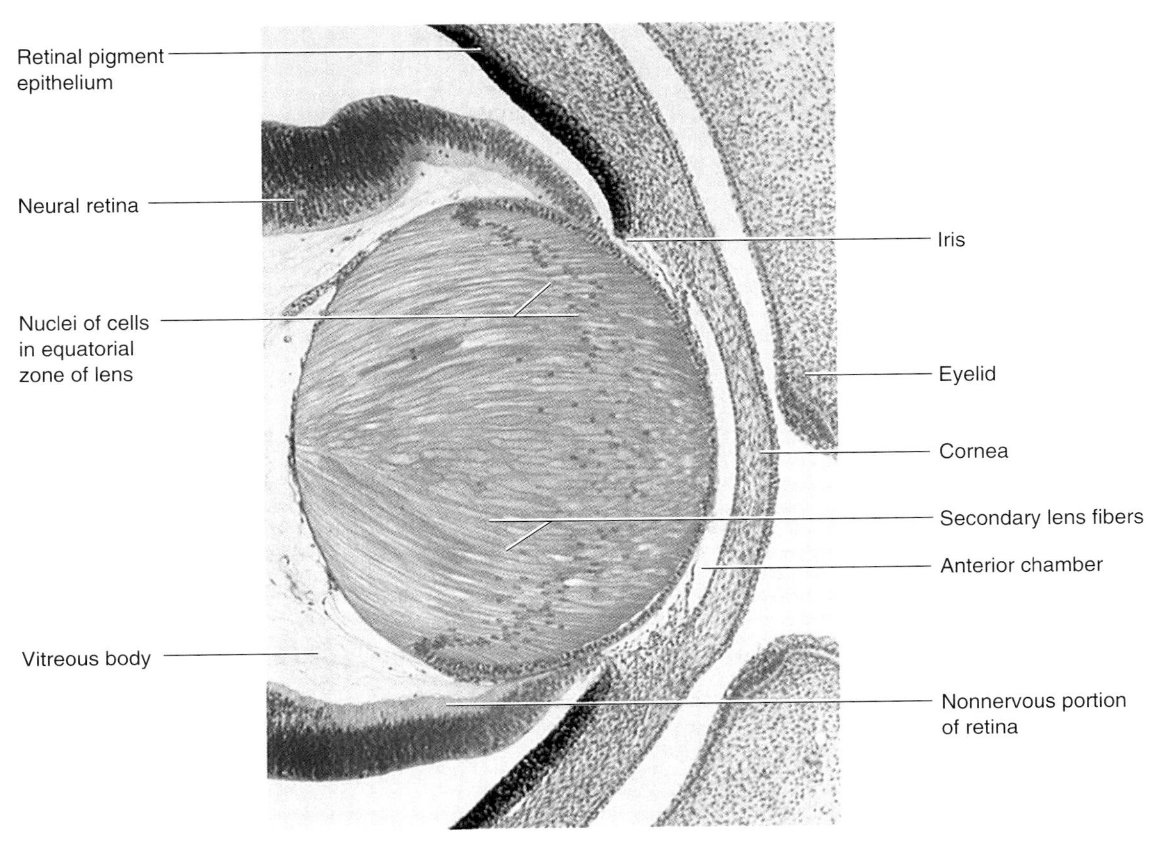

Figure 20 - 8. Photomicrograph of a sagittal section of a portion of the developing eye of an embryo (× 280), at about 56 days. Observe that the lens fibers have elongated and obliterated the cavity of the lens vesicle. Note that the inner layer of the optic cup has thickened greatly to form the neural retina and that the outer layer is heavily pigmented (retinal pigment epithelium). (From Moore KL, Persaud TVN, Shiota K: *Color Atlas of Clinical Embryology,* 2nd ed. Philadelphia, WB Saunders, 2000.)

Development of the Aqueous Chambers

The **anterior chamber of the eye** develops from a cleftlike space that forms in the mesenchyme located between the developing lens and cornea (see Fig. 20 - 5*A*). The mesenchyme superficial to this space forms the substantia propria of the cornea and the mesothelium of the anterior chamber. After the lens is established, it induces the surface ectoderm to develop into the epithelium of the cornea and the conjunctiva. The **posterior chamber of the eye** develops from a space that forms in the mesenchyme posterior to the developing iris and anterior to the developing lens (see Fig. 20 - 5*D*). When the pupillary membrane disappears and the pupil forms, the anterior and posterior chambers of the eye are able to communicate with each other through a circumferential **scleral venous sinus** (L. sinus venosus sclerae). This sinus encircles the anterior chamber and is the *outflow site of aqueous humor* from the anterior chamber of the eye to the venous system.

Congenital Glaucoma

Abnormal elevation of intraocular pressure in newborn infants usually results from abnormal development of the drainage mechanism of aqueous humor during the fetal period (see Fig. 9 - 16*B*). *Intraocular tension* rises because of an imbalance between the production of aqueous humor and its outflow. This imbalance may result from abnormal development of the *scleral venous sinus* (see Fig. 20 - 5*D*). Congenital glaucoma is usually genetically heterogeneous, but the condition may result from a rubella infection during early pregnancy (see Chapter 9).

Congenital Cataracts

In congenital cataracts, the lens is opaque and frequently appears grayish white. Blindness results. Many lens opacities are inherited, with dominant transmission being more common than recessive or sex-linked transmission. Some congenital cataracts are caused by teratogenic agents (particularly the *rubella virus* ([see Fig. 9–16*A*]), that affect early development of the lenses. The lenses are vulnerable to **rubella virus** between the fourth and seventh weeks when primary lens fibers are forming. Physical agents, such as **radiation**, can also damage the lens and produce cataracts.

Development of the Cornea

The cornea is formed from three sources:

- The external corneal epithelium is derived from **surface ectoderm**.
- The embryonic connective tissue or mesenchyme is derived from **mesoderm**.
- **Neural crest cells** migrate from the lip of the optic cup through the embryonic connective tissue and are transformed into the corneal endothelium.

Formation of the cornea is induced by the lens vesicle, which acts on the surface ectoderm. The inductive influence results in transformation of the surface ectoderm into the transparent avascular cornea.

Development of the Choroid and Sclera

The mesenchyme surrounding the optic cup (largely of neural crest origin) reacts to the inductive influence of the retinal pigment epithelium by differentiating into an inner vascular layer, the **choroid**, and an outer fibrous layer, the **sclera** (see Fig. 20–5*C*). The sclera develops from a condensation of the mesenchyme external to the choroid and is continuous with the stroma (connective tissue framework) of the cornea. Toward the rim of the optic cup, the choroid becomes modified to form the cores of the **ciliary processes**, consisting chiefly of capillaries supported by delicate connective tissue. The first choroidal blood vessels appear during the 15th week of development; by the 22nd week, arteries and veins can be distinguished.

Development of the Eyelids

The eyelids develop during the sixth week of development from mesenchyme derived from the neural crest and from two folds of skin that grow over the cornea (see Fig. 20–5*B*). The eyelids adhere to one another by the beginning of the 10th week, and remain adherent until the 26th to 28th weeks (see Fig. 20–5*C*). While the eyelids are adherent, there is a closed **conjunctival sac** anterior to the cornea; when the eyelids begin to open, the **bulbar conjunctiva** is reflected over the anterior part of the sclera and the surface epithelium of the cornea. The **palpebral conjunctiva** lines the inner surface of the eyelids. The eyelashes and glands in the eyelids are derived from the surface ectoderm in a manner similar to that described for other parts of the integument (see Chapter 21). The connective tissue and tarsal plates develop from mesenchyme in the developing eyelids. The *orbicularis oculi muscle* is derived from the mesenchyme in the second pharyngeal arch (see Chapter 11) and is supplied by its nerve (cranial nerve VII).

Congenital Ptosis of the Eyelid

Drooping of the superior (upper) eyelids (ptosis) at birth is relatively common. Congenital ptosis (Gr., a falling) may result from failure of normal development of the *levator palpebrae superioris muscle*. Ptosis may also result from prenatal injury or abnormal development of the superior division of the **oculomotor nerve** (CN III) that supplies this muscle. *Congenital ptosis is hereditary*; an isolated defect is usually transmitted as an autosomal dominant trait.

Coloboma of the Eyelid

Large defects of the eyelid **(palpebral coloboma)** are uncommon. Most colobomas are characterized by a small notch in the superior eyelid, but the defect may involve almost the entire lid. Coloboma of the inferior (lower) eyelid is rare. Palpebral colobomas appear to result from local developmental disturbances in the formation and growth of the eyelids.

Development of the Lacrimal Glands

The lacrimal glands develop at the superolateral angles of the orbits; they are derived from a number of solid buds from the surface ectoderm. The buds branch and become canalized to form the ducts and alveoli of the glands. The lacrimal glands are small at birth and do not function fully until about 6 weeks; hence, the newborn infant does not produce tears when it cries. Tears are often not present with crying until 1 to 3 months of age.

Development of the Ear

The ear is composed of three anatomical parts:

- the *external ear*, consisting of the auricle (pinna), the external acoustic meatus, and the external layer of the tympanic membrane (eardrum)
- the *middle ear*, consisting of three auditory ossicles (small ear bones) that connect the internal layer of the tympanic membrane to the oval window of the internal ear

- the *internal ear*, consisting of the vestibulocochlear organ, which plays a role in both hearing and balance

The external and middle parts of the ear regulate the transference of sound waves from the exterior to the internal ear; this process converts the sound waves into nerve impulses and registers changes in equilibrium.

Development of the Internal Ear

The internal ear is the first of the three parts of the ear to begin to develop. Early in the fourth week, a thickening of surface ectoderm — the **otic placode** — appears on each side of the myelencephalon, the caudal part of the hindbrain (Fig. 20 - 9*A* and *B*). Inductive influences from the notochord and paraxial mesoderm stimulate the surface ectoderm to form the otic placodes. Experimental studies have shown that fibroblast growth factor 3 (FGF-3) may play a role in this process. Each otic placode soon invaginates and sinks deep to the surface ectoderm into the underlying mesenchyme. In so doing, it forms an **otic pit** (see Fig. 20 - 9*C* and *D*). The edges of the otic pit soon come together and fuse to form an **otic vesicle**, the primordium of the *membranous labyrinth* (see Fig. 20 - 9*E* to *G*). The otic vesicle soon loses its connection with the surface ectoderm, and a diverticulum grows from the otic vesicle and elongates to form the **endolymphatic duct** and **sac** (Fig. 20 - 10*A* to *E*). Two regions of the otic vesicle are now recognizable:

- a dorsal **utricular part** from which the endolymphatic duct, utricle, and semicircular ducts arise
- a ventral **saccular part**, which gives rise to the saccule and cochlear duct in which the spiral organ (of Corti) is located

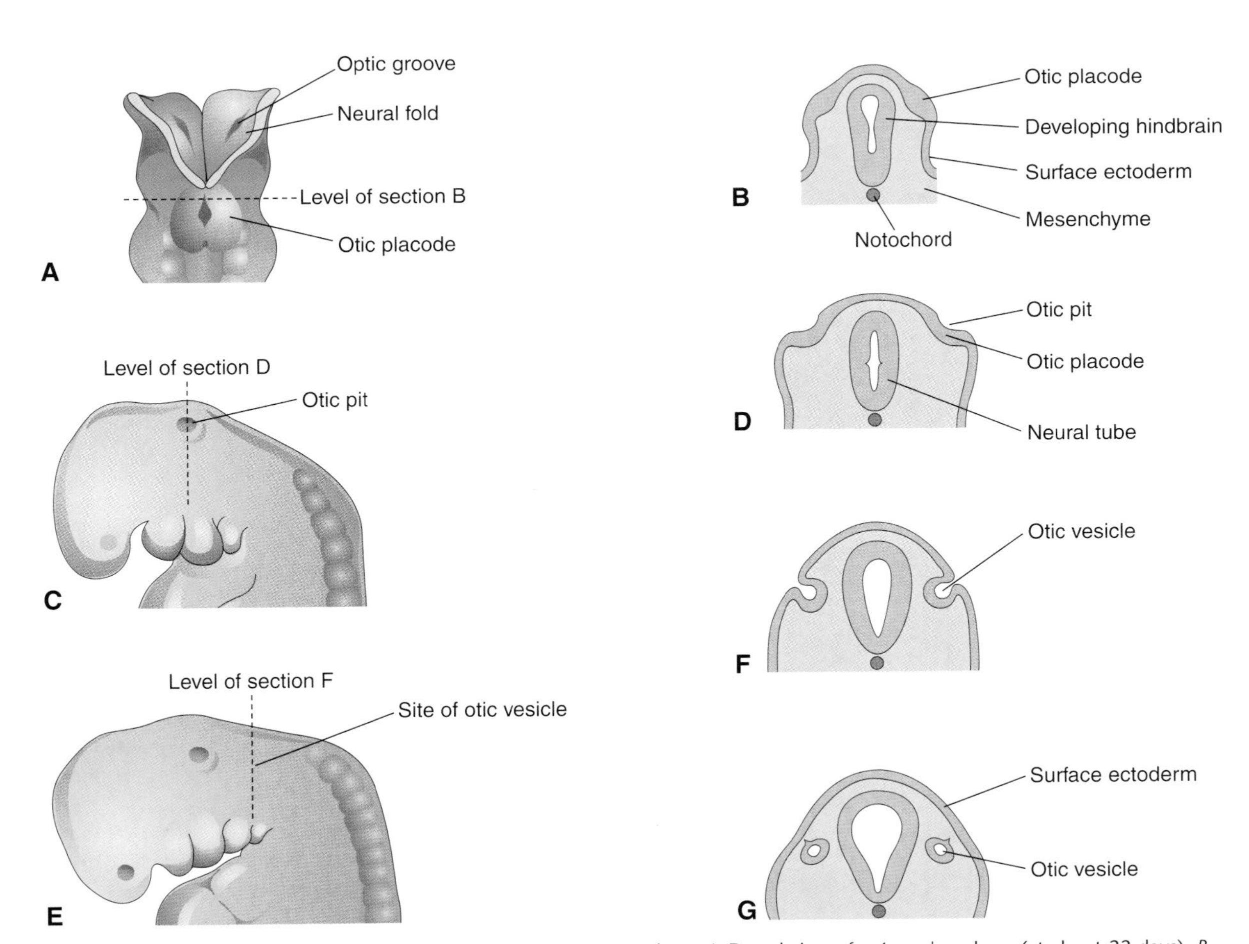

Figure 20 - 9. Drawings illustrating early development of the internal ear. *A,* Dorsal view of a 4-week embryo (at about 22 days). *B, D, F,* and *G,* Schematic coronal sections illustrating successive stages in the development of otic vesicles. *C* and *E,* Lateral views of the cranial region of embryos, at about 24 and 28 days, respectively.

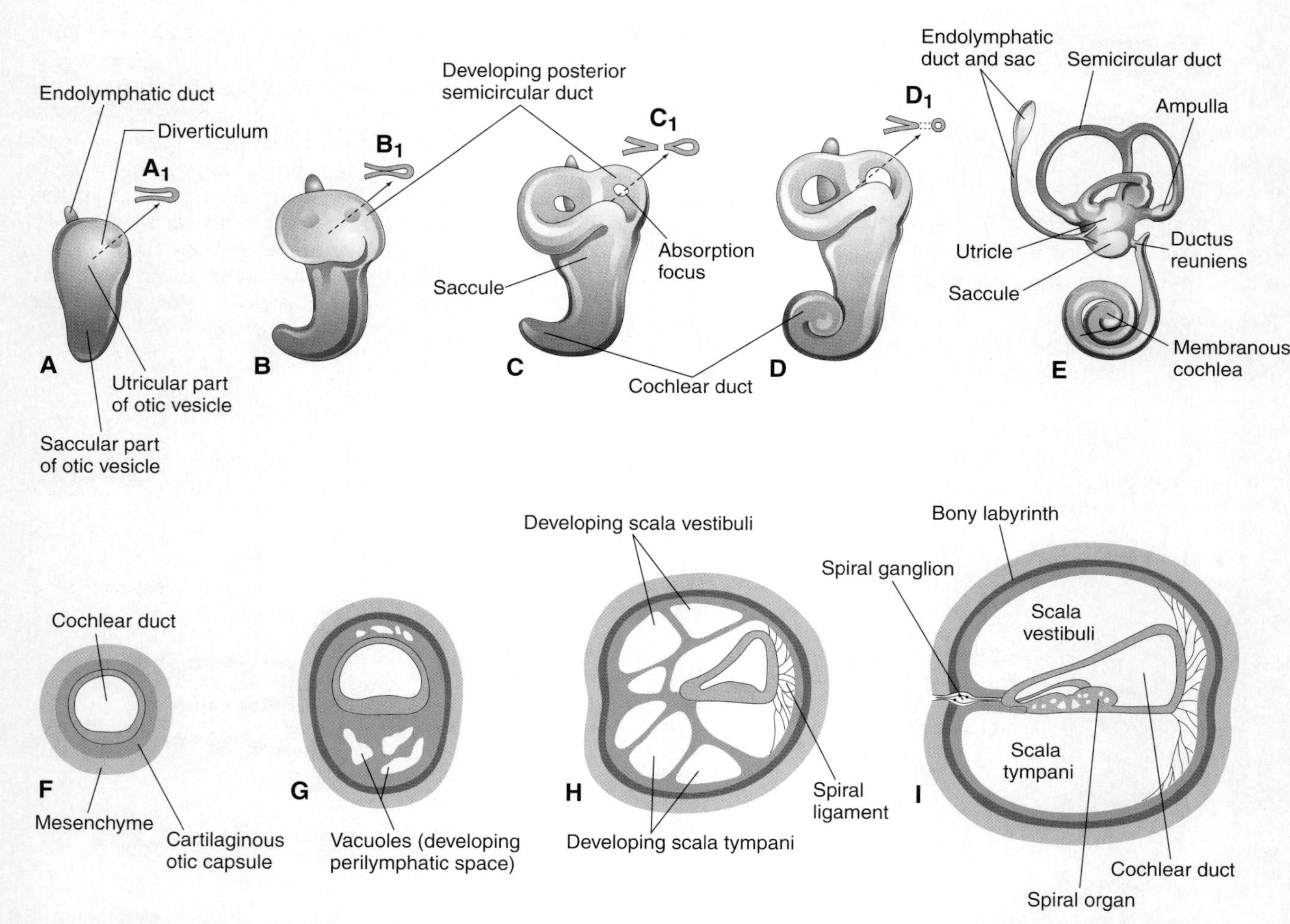

Figure 20 - 10. Drawings of the otic vesicle showing development of the membranous and bony labyrinths of the internal ear. *A* to *E,* Lateral views showing successive stages in the development of the otic vesicle into the membranous labyrinth from the fifth to eighth weeks. A_1 to D_1, Diagrams illustrating development of a semicircular duct. *F* to *I,* Sections through the cochlear duct showing successive stages in the development of the spiral organ and the perilymphatic space from the 8th to the 20th weeks.

Three disklike diverticula grow out from the utricular part of the primordial **membranous labyrinth**. Soon, the central parts of these diverticula fuse and disappear (see Fig. 20 - 10*B* to *E*). The peripheral unfused parts of the diverticula become the **semicircular ducts**, which are attached to the utricle and are later enclosed in the **semicircular canals** of the **bony labyrinth**. Localized dilatations, the **ampullae**, develop at one end of each semicircular duct. Specialized receptor areas, called cristae ampullares, differentiate in these ampullae and in the utricle and saccule (maculae utriculi and sacculi).

From the ventral saccular part of the otic vesicle, a tubular diverticulum — the **cochlear duct** — grows and coils to form the **membranous cochlea** (see Fig. 20 - 10*C* to *E*). A connection of the cochlea with the saccule, the **ductus reuniens**, soon forms. The **spiral organ** (of Corti) differentiates from cells in the wall of the cochlear duct (see Fig. 20 - 10*F* to *I*). Ganglion cells of CN VIII migrate along the coils of the membranous cochlea and form the **spiral ganglion** (cochlear ganglion). Nerve processes extend from this ganglion to the spiral organ, where they terminate on *hair cells.* The cells in the spiral ganglion retain their embryonic bipolar condition; that is, they do not become unipolar like spinal ganglion cells.

Inductive influences from the otic vesicle stimulate the mesenchyme around the otic vesicle to differentiate into a **cartilaginous otic capsule** (see Fig. 20 - 10*F*). Transforming growth factor-β_1 (TGF-β_1) may play a role in modulating epithelial-mesenchymal interaction in the internal ear and in directing the formation of the otic capsule. As the **membranous labyrinth** enlarges, vacuoles appear in the cartilaginous otic capsule that soon coalescing to form the **perilymphatic space**. The membranous labyrinth is now suspended in **perilymph** (fluid in the perilymphatic space). The perilymphatic space related to the cochlear duct develops two divisions, the **scala tympani** and **scala vestibuli** (see Fig. 20 - 10*H* and *I*). The cartilaginous otic capsule later ossifies to form the **bony labyrinth** of the internal ear. The internal ear reaches its adult size and shape by the middle of the fetal period (20 to 22 weeks).

Development of the Middle Ear

Development of the **tubotympanic recess** (Fig. 20 - 11*B*) from the first pharyngeal pouch is described in Chapter 11. The proximal part of the tubotympanic recess forms the **pharyngotympanic tube** (auditory tube). The distal part of the tubotympanic recess expands and becomes the **tympanic cavity** (Fig. 20 - 11*C*), which gradually envelops the **auditory ossicles** (malleus, incus, and stapes), their tendons and ligaments, and the chorda tympani nerve. All these structures receive a more or less complete epithelial investment. An epithelial-type organizer, located at the tip of the tubotympanic recess, probably plays a role in the early development of the middle ear cavity by inducing programmed cell death — apoptosis. During the late fetal period, expansion of the tympanic cavity gives rise to the **mastoid antrum**, located in the temporal bone. The mastoid antrum is almost adult size at birth; however, *no mastoid cells are present in newborn infants*. By 2 years of age, the mastoid cells are well developed and produce conical projections of the temporal bones, the **mastoid processes**. The middle ear continues to grow through puberty. Development of the **auditory ossicles** is described in Chapter 11. The *tensor tympani*, the muscle attached to the malleus, is derived

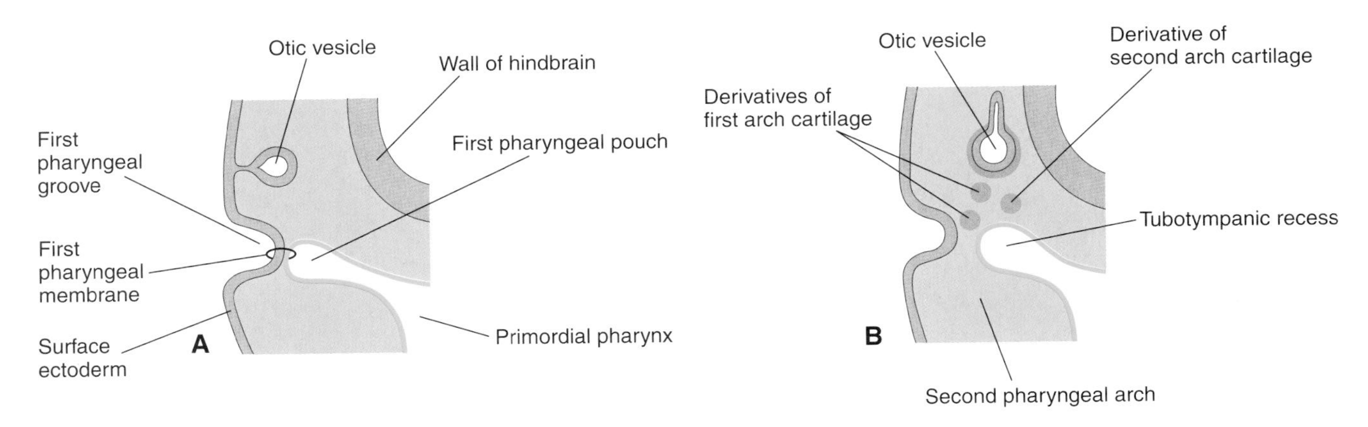

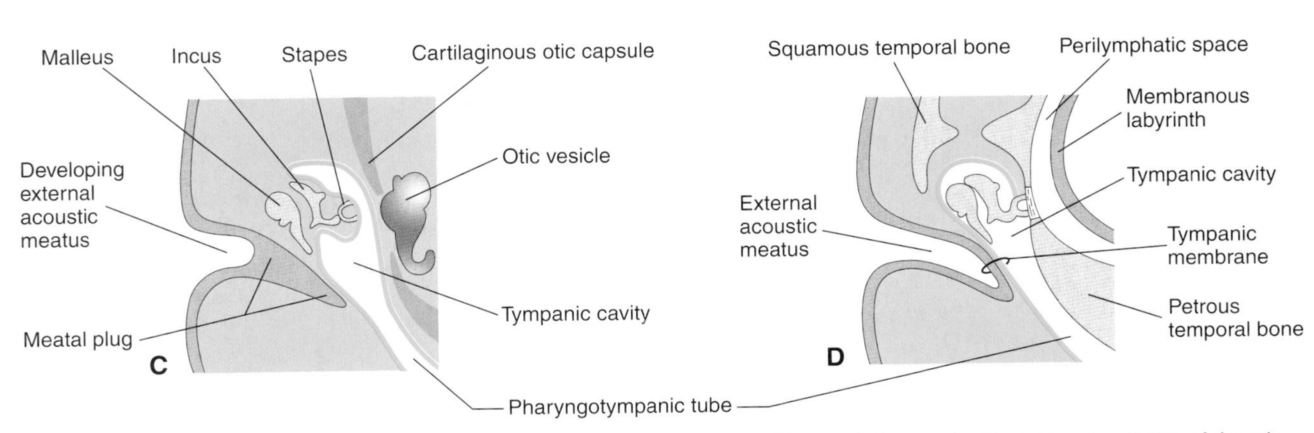

Figure 20 - 11. Schematic drawings illustrating development of the external and middle ear. *A,* Four weeks, illustrating the relation of the otic vesicle to the pharyngeal apparatus. *B,* Five weeks, showing the tubotympanic recess and pharyngeal arch cartilages. *C,* Later stage, showing the tubotympanic recess (future tympanic cavity and mastoid antrum) beginning to envelop the ossicles. *D,* Final stage of ear development, showing the relationship of the middle ear to the perilymphatic space and the external acoustic meatus. Note that the tympanic membrane develops from three germ layers.

from mesenchyme in the first pharyngeal arch and is innervated by CN V, the nerve of this arch. The *stapedius muscle* is derived from the second pharyngeal arch and is supplied by CN VII, the nerve of that arch.

Development of the External Ear

The **external acoustic meatus** develops from the dorsal part of the first pharyngeal groove. The ectodermal cells at the bottom of this tube proliferate to form a solid epithelial plate, the **meatal plug** (see Fig. 20 - 11*C*). Late in the fetal period, the central cells of this plug degenerate, forming a cavity that becomes the internal part of the external acoustic meatus (see Fig. 20 - 11*D*). The external acoustic meatus attains its adult length around the ninth year. The primordium of the **tympanic membrane** is the first pharyngeal membrane, which separates the first pharyngeal groove from the first pharyngeal pouch (see Fig. 20 - 11*A*). As development proceeds, mesenchyme grows between the two parts of the pharyngeal membrane and differentiates into the collagenic fibers in the tympanic membrane. The external covering of the tympanic membrane is derived from the surface ectoderm, whereas its internal lining is derived from the endoderm of the tubotympanic recess.

The **auricle** develops from six mesenchymal proliferations in the first and second pharyngeal arches. Prominences, called **auricular hillocks**, surround the first pharyngeal groove (Fig. 20 - 12*A*). As the auricle grows, the contribution by the first arch is reduced (Fig. 20 - 12*B* to *D*). The lobule (earlobe) is the last part to develop. The auricles begin to develop at the base of the neck (Fig. 20 - 12*A* and *B*). As the mandible develops, the auricles move to their normal position at the side of the head (Fig. 20 - 12*D*).

Congenital Deafness

Because formation of the internal ear is independent of development of the middle and external ears, congenital impairment of hearing may be the result of maldevelopment of the sound-conducting apparatus of the middle and external ears, or of the neurosensory structures in the internal ear. About 1 in every 1000 newborn infants has significant hearing loss. Most types of congenital deafness are caused by genetic factors; in many cases, the genes involved have been identified. In **deaf-mutism**, the ear defect is usually perceptive in type. Congenital deafness may be associated with several other head and neck anomalies as a part of the *first arch syndrome* (see Chapter 11). Abnormalities of the malleus and incus are often associated with this syndrome. A **rubella infection** during the critical period of development of the internal ear, particularly the seventh and eighth weeks, can cause maldevelopment of the spiral organ and deafness. **Congenital fixation of the stapes** results in conductive deafness in an otherwise normal ear. Failure of differentiation of the *anular ligament,* which attaches the base of the stapes to the oval windows (L. fenestra) vestibuli, results in fixation of the stapes to the bony labyrinth.

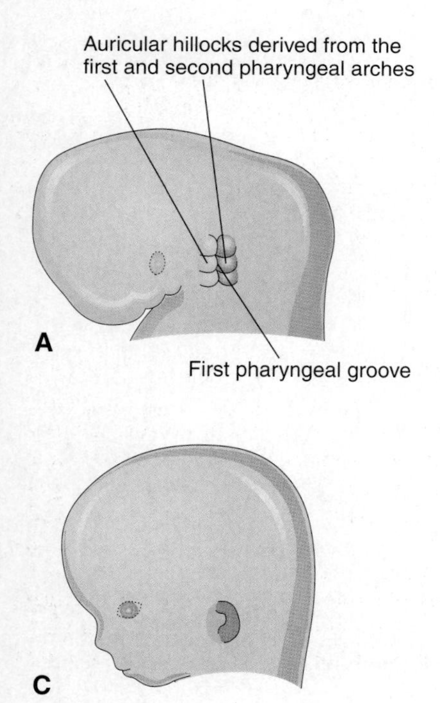

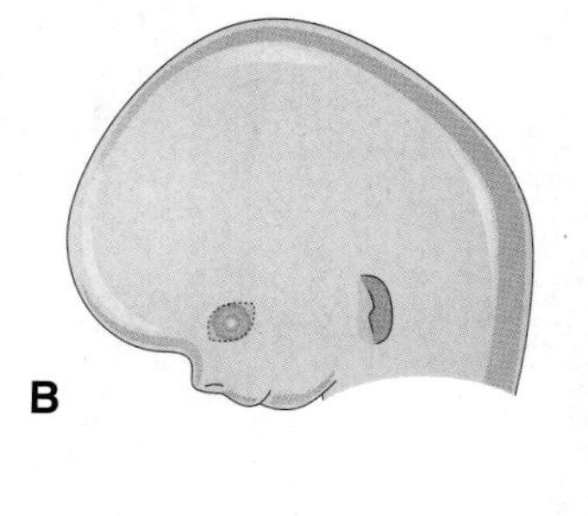

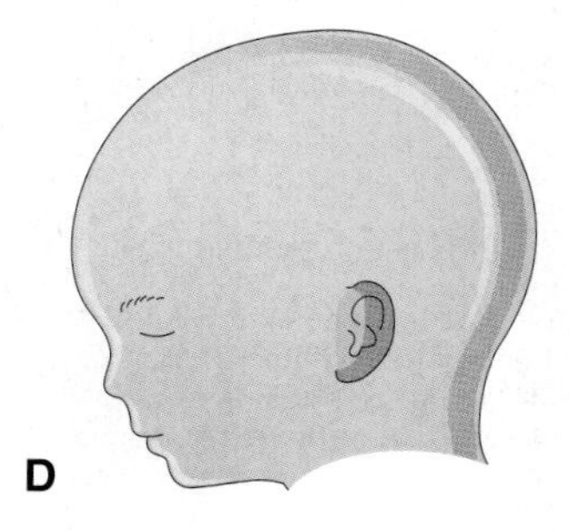

Figure 20 - 12. Drawings illustrating development of the auricle. *A,* Six weeks. Note that three auricular hillocks are located on the first pharyngeal arch and three are situated on the second arch. *B,* Eight weeks. *C,* Ten weeks. *D,* Thirty-two weeks. As the mandible and teeth develop, the auricles move from the neck to the side of the head.

Auricular Anomalies

Minor auricular deformities are common. A wide variation exists in the shape of the auricle. Minor anomalies of the auricles may serve as indicators of a specific pattern of congenital anomalies. For example, the auricles are often abnormal in shape and low-set in infants with chromosomal syndromes, such as trisomy 18 (see Chapter 9) and in infants affected by maternal ingestion of certain drugs (e.g., trimethadione).

Auricular Appendages

Auricular appendages (skin tags) are common and result from the development of accessory auricular hillocks (Fig. 20 – 13). The appendages usually appear anterior to the auricle, more often unilaterally than bilaterally. The appendages, often with narrow pedicles, consist of skin, but they may also contain some cartilage.

Microtia

Microtia (small auricle) results from suppressed development of the auricular hillocks (see Fig. 20 – 13). This anomaly often serves as an indicator of associated anomalies, such as atresia of the external acoustic meatus and middle ear anomalies.

Preauricular Sinuses

Pitlike cutaneous depressions or shallow sinuses are commonly located in a triangular area anterior to the auricle. The sinuses are usually shallow pits that have pinpoint external openings. Some sinuses contain a vestigial cartilaginous mass. The embryological basis for auricular sinuses is uncertain, but in some cases, the defect is related to abnormal development of the auricular hillocks and defective closure of the dorsal part of the first pharyngeal groove. Other auricular sinuses appear to represent ectodermal folds that are sequestered during formation of the auricle. Preauricular sinuses are familial and are frequently bilateral. *Auricular fistulas* (narrow canals) connecting the exterior with the tympanic cavity or the tonsillar sinus (fossa) are extremely rare.

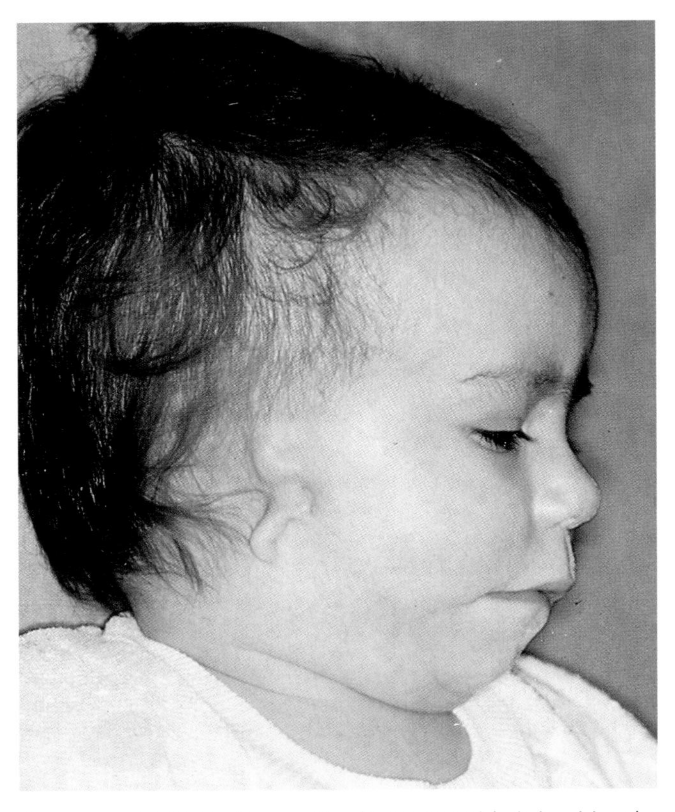

Figure 20 – 13. Child with a small rudimentary auricle (microtia) and preauricular tag. The external acoustic meatus is also absent. (Courtesy of AE Chudley, MD, Section of Genetics and Metabolism, Department of Pediatrics and Child Health, Children's Hospital, University of Manitoba, Winnipeg, Manitoba, Canada.)

Atresia of the External Acoustic Meatus

Blockage of the external acoustic canal results from failure of the meatal plug to canalize (see Fig. 20 – 11*C*). Usually the deep part of the meatus is open but the superficial part is blocked by bone or fibrous tissue. Most cases are associated with the *first arch syndrome* (see Chapter 11). Often, abnormal development of both the first and second pharyngeal arches is involved. Usually, the auricle is also severely affected and anomalies of the middle or internal ear, or both, are sometimes present. Atresia of the external acoustic meatus can occur bilaterally or unilaterally and usually results from autosomal dominant inheritance.

Absence of the External Acoustic Meatus

Absence of the external acoustic meatus is rare (see Fig. 20 – 13); often, the auricle is normal. This anomaly results from failure of inward expansion of the first pharyngeal groove and failure of the meatal plug to disappear (see Fig. 20 – 13).

Summary of Development of the Eye

The first indication of the eye is the **optic groove**, which forms at the beginning of the fourth week of development. The groove deepens to form an **optic vesicle** that projects from the forebrain. The optic vesicle contacts the surface ectoderm and induces development of the **lens placode**, the primordium of the lens. As the lens placode invaginates to form a **lens pit** and **lens vesicle**, the optic vesicle invaginates to form an **optic cup**. The retina forms from the two layers of the optic cup.

- The retina, optic nerve fibers, muscles of the iris, and epithelium of the iris and ciliary body are derived from the *neuroectoderm* of the forebrain. The sphincter and dilator muscles of the iris develop from the ectoderm at the rim of the optic cup.
- The *surface ectoderm* gives rise to the lens and the epithelium of the lacrimal glands, eyelids, conjunctiva, and cornea.
- The *mesoderm* gives rise to the eye muscles, except those of the iris, and to all connective and vascular tissues of the cornea, iris, ciliary body, choroid, and sclera.

There are many **ocular anomalies**, but most of them are uncommon. The eye is very sensitive to the teratogenic effects of infectious agents (e.g., cytomegalovirus and rubella virus). The most serious defects result from disturbances of development during the fourth to sixth weeks of development, but defects of sight may also result from infection of tissues and organs by certain microorganisms during the fetal period (e.g., rubella virus). Most ocular anomalies are caused by defective closure of the retinal fissure during the sixth week (e.g., coloboma of iris). *Congenital cataract* and *glaucoma* may result from intrauterine infections, but most congenital cataracts are inherited.

Summary of Development of the Ear

The **otic placodes** and the **otic vesicle** develop from the surface ectoderm during the fourth week. The vesicle develops into the **membranous labyrinth** of the internal ear. The otic vesicle divides into two parts:

- a dorsal utricular part, which gives rise to the utricle, semicircular ducts, and endolymphatic duct
- a ventral saccular part, which gives rise to the saccule and cochlear duct

The cochlear duct gives rise to the **spiral organ**. The **bony labyrinth** develops from the mesenchyme adjacent to the membranous labyrinth. The epithelium lining the tympanic cavity, mastoid antrum, and **pharyngotympanic tube** is derived from the endoderm of the **tubotympanic recess**, which develops from the first pharyngeal pouch. The **auditory ossicles** (malleus, incus, and stapes) develop from the dorsal ends of the cartilages in the first two pharyngeal arches. The epithelium of the **external acoustic meatus** develops from the ectoderm of the first pharyngeal groove. The tympanic membrane is derived from three sources:

- endoderm of the first pharyngeal pouch
- ectoderm of the first pharyngeal groove
- mesoderm between the layers listed above

The auricle develops from six **auricular hillocks** that form from mesenchymal prominences around the margins of the first pharyngeal groove. These hillocks fuse to form the auricle.

Congenital deafness may result from abnormal development of the membranous or bony labyrinth, or both, as well as from abnormalities of the auditory ossicles. *Recessive inheritance is the most common cause of congenital deafness*, but a rubella virus infection near the end of the embryonic period is a major environmental factor known to cause abnormal development of the spiral organ and defective hearing. There are many minor anomalies of the auricle that alert the clinician to the possible presence of associated major anomalies (e.g., defects of the middle ear). Low-set, severely malformed ears are often associated with chromosomal abnormalities, particularly trisomy 13 and trisomy 18.

Clinically Oriented Questions

1. If a woman has rubella (German measles) during the first trimester of her pregnancy, what are the chances that the eyes and ears of the fetus will be affected? What is the most common manifestation of late fetal rubella infection? If a pregnant woman is exposed to rubella, can it be determined if she is immune to the infection?
2. Some believe that a good way of preventing the congenital anomalies caused by German measles is by the purposeful exposure of young girls to rubella. Is this the best way for a woman to avoid having a blind and deaf baby secondary to a rubella infection during pregnancy? If not, what can be done to provide immunization against rubella infection?
3. It has been reported that deafness and tooth defects occurring during childhood can result from a condition called "fetal syphilis." Is this true? If so, how could this happen? Can these congenital defects be prevented?
4. There are reports that blindness and deafness can result from herpes virus infections. Is this true? If so, which herpes viruses are involved? What are the affected infant's chances of normal development?
5. An article in the newspaper reported that methyl mercury exposure in utero can cause mental retardation, deafness, and blindness. The article cited the eating of contaminated fish as the cause of the abnormalities. Can you explain how these anomalies could be caused by methyl mercury.

The answers to these questions are at the back of the book.

21

The Integumentary System

The integumentary system consists of the skin and its appendages: sweat glands, nails, hair, sebaceous glands, and arrector muscles of hairs (L. *arrector pili*). The system also includes the mammary glands and teeth.

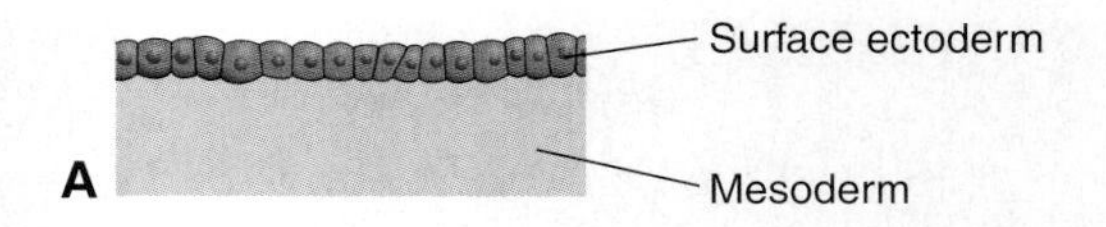

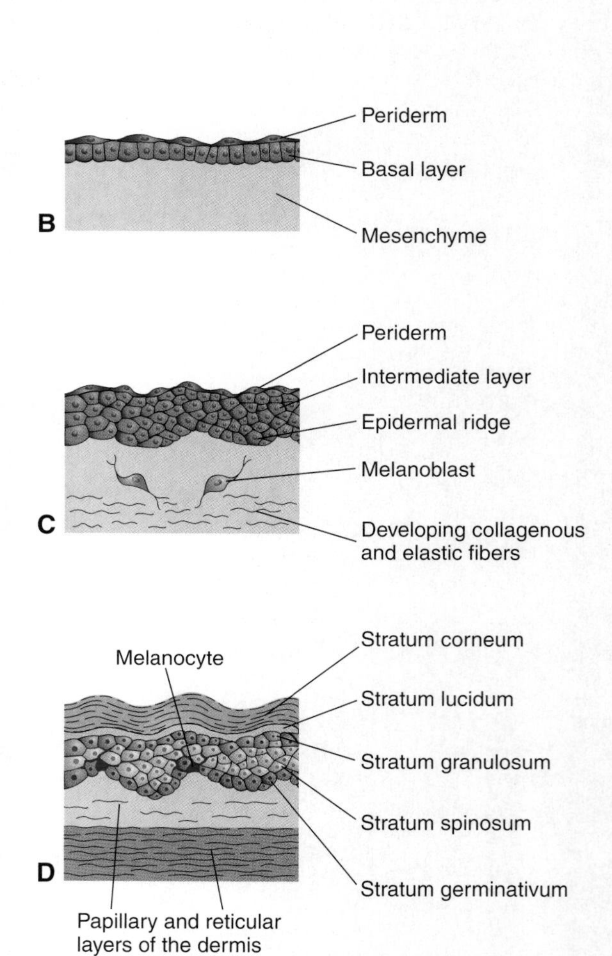

Figure 21 - 1. Successive stages of skin development. *A,* Four weeks. *B,* Seven weeks. *C,* Eleven weeks. The cells of the periderm continually undergo keratinization and desquamation. Exfoliated peridermal cells form part of the vernix caseosa. *D,* Newborn infant. Note the position of the melanocytes in the basal layer of the epidermis and that their processes extend between the epidermal cells to supply them with melanin.

Development of the Skin

The skin, one of the largest structures of the body, is a complex organ system that forms a protective covering for the body. The skin consists of two layers that are derived from two different germ layers (Fig. 21 - 1): ectoderm and mesoderm.

- The **epidermis** is a superficial epithelial tissue that is derived from **surface ectoderm**.
- The **dermis** is a deeper layer composed of dense, irregularly arranged connective tissue that is derived from **mesoderm**.

Ectodermal (epidermal)/mesenchymal (dermal) interactions involve mutual inductive mechanisms. Skin structures vary from one part of the body to another. For example, the skin of the eyelids is thin and soft and has fine hairs, whereas the skin of the eyebrows is thick and has coarse hairs. The embryonic skin at 4 to 5 weeks of development consists of a single layer of surface ectoderm overlying the mesenchyme (Fig. 21 - 1).

Epidermis

During the first and second trimesters, epidermal growth occurs in stages, the result of which is an increase in epidermal thickness. The primordium of the epidermis is the layer of surface ectodermal cells (Fig. 21 - 1*A*). These cells proliferate and form a layer of squamous epithelium, the **periderm**, and a basal (germinative) layer (Fig. 21 - 1*B*). The cells of the periderm continually undergo keratinization and desquamation and are replaced by cells arising from the **basal layer**. The exfoliated peridermal cells form part of a white greasy substance — the **vernix caseosa** — that covers the fetal skin. Later, the vernix (L., varnish) contains sebum, the secretion from sebaceous glands in the skin. The vernix protects the developing skin from constant exposure to amniotic fluid containing urine during the fetal period. In addition, the vernix facilitates birth of the fetus because of its slippery nature.

The basal layer of the epidermis becomes the **stratum germinativum** (Fig. 21 - 1*D*), which produces new cells that are displaced into the layers superficial to it. By 11 weeks, cells from the stratum germinativum have formed an **intermediate layer** (Fig. 21 - 1*C*). Replacement of peridermal cells continues until about the 21st week; thereafter, the periderm disappears and the **stratum corneum** forms (Fig. 21 - 1*D*). Proliferation of cells in the stratum germinativum also forms **epidermal ridges**, which extend into the developing dermis. These ridges begin to appear in embryos of 10 weeks and are permanently established by the 17th week. The epidermal ridges produce grooves on the surface of the palms of the hands and the soles of the feet, including the digits. The pattern that develops is determined genetically and constitutes the basis for examining fingerprints in criminal investigations and medical genetics. **Dermatoglyphics** is the study of the configurations of the characteristic ridge patterns. Abnormal chromosome complements affect the development of ridge patterns; for example, infants with Down syndrome have distinctive ridge patterns on their hands and feet that are of diagnostic value (see Chapter 9).

Late in the embryonic period, **neural crest cells** migrate into the mesenchyme in the developing dermis and differentiate into **melanoblasts** (Fig. 21 - 1*B* and *C*). Later, these cells migrate to the dermoepidermal junction and differentiate into **melanocytes** (Fig. 21 - 1*D*). The differentiation of melanoblasts into melanocytes involves the formation of pigment granules. *Wnt signaling* regulates this process. In white races, the cell bodies of melanocytes are usually confined to the basal layers of the epidermis; however, their dendritic processes extend between the epidermal cells. Only a few melanin-containing cells are normally present in the dermis. The melanocytes begin producing **melanin** (Gr. *melas,* black) before birth and distribute it to the epidermal cells. After birth, increased amounts of melanin are produced in response to ultraviolet light. The relative content of melanin in the melanocytes accounts for the different colors of skin. *Molecular studies* suggest that MSH cell surface receptor and melanosomal P-protein are genes that determine the degree of pigmentation by regulating tyrosinase levels and activities.

Dermis

The dermis develops from mesenchyme, which is derived from the mesoderm underlying the surface ectoderm (Fig. 21 - 1*A* and *B*). Most of the mesenchyme that differentiates into the connective tissue of the dermis originates from the somatic layer of lateral mesoderm, but some of it is derived from the dermatomes of the somites (see Chapter 16). By 11 weeks, the mesenchymal cells have begun to produce collagenous and elastic connective tissue fibers (Fig. 21 - 1*D*). As the **epidermal ridges** form, the dermis projects into the epidermis, forming **dermal ridges**. Capillary loops develop in some of the dermal ridges and provide nourishment for the epidermis. Sensory nerve endings form in others. The developing afferent nerve fibers apparently play an important role in the spatial and temporal sequence of dermal ridge formation. The development of the *dermatomal pattern of innervation of the skin* is described in Chapter 18. The blood vessels in the dermis begin as simple, endothelium-lined structures that differentiate from mesenchyme. As the skin grows, new capillaries grow out from the primordial vessels. Some capillaries acquire muscular coats through differentiation of myoblasts developing in the surrounding mesenchyme and become arterioles and arteries. Other capillaries, through which a return flow of blood is established, acquire muscular coats and become venules and veins. As new blood vessels form, some transitory ones disappear. By the end of the first trimester, the major vascular organization of the fetal dermis is established.

Glands of the Skin

Two kinds of glands, sebaceous and sweat glands, are derived from the epidermis and grow into the dermis. Mammary glands develop in a similar manner.

Sebaceous Glands

Most sebaceous glands develop as buds from the sides of developing **epidermal root sheaths** of hair follicles (Fig. 21 - 2). The glandular buds grow into the surrounding embryonic connective tissue and branch to form the primordia of alveoli and their associated ducts. The central cells of the alveoli break down, forming an oily secretion called **sebum**; this is released into the hair follicle and passes to the surface of the skin, where it mixes with desquamated peridermal cells to form **vernix caseosa**. Sebaceous glands, independent of hair follicles (e.g., in the glans penis and labia minora), develop in a similar manner to buds from the epidermis.

Sweat Glands

Eccrine sweat glands develop as epidermal downgrowths into the underlying mesenchyme (Fig. 21 - 2). As the bud elongates, its end coils to form the primordium of the secretory part of the gland. The epithelial attachment of the developing gland to the epidermis forms the primordium of the duct. The central cells of the primordial ducts degenerate, forming a lumen. The peripheral cells of the secretory part of the gland differentiate into myoepithelial and **secretory cells** (Fig. 21 - 2). The **myoepithelial cells** are thought to be specialized smooth muscle cells that assist in expelling sweat from the glands. Eccrine sweat glands begin to function shortly after birth. The **apocrine sweat glands** develop from downgrowths of the stratum germinativum of the epidermis that give rise to hair follicles. As a result, the ducts of these glands open, not onto the skin surface as do ordinary sweat glands, but into the upper part of hair follicles superficial to the openings of the sebaceous glands. They begin to secrete during puberty.

Development of Hair

Hairs begin to develop during the 9th to 12th weeks, but they do not become easily recognizable until about the 23rd week (Fig. 21 - 2). Expression of sonic hedgehog (Shh), an intercellular signaling molecule, has been

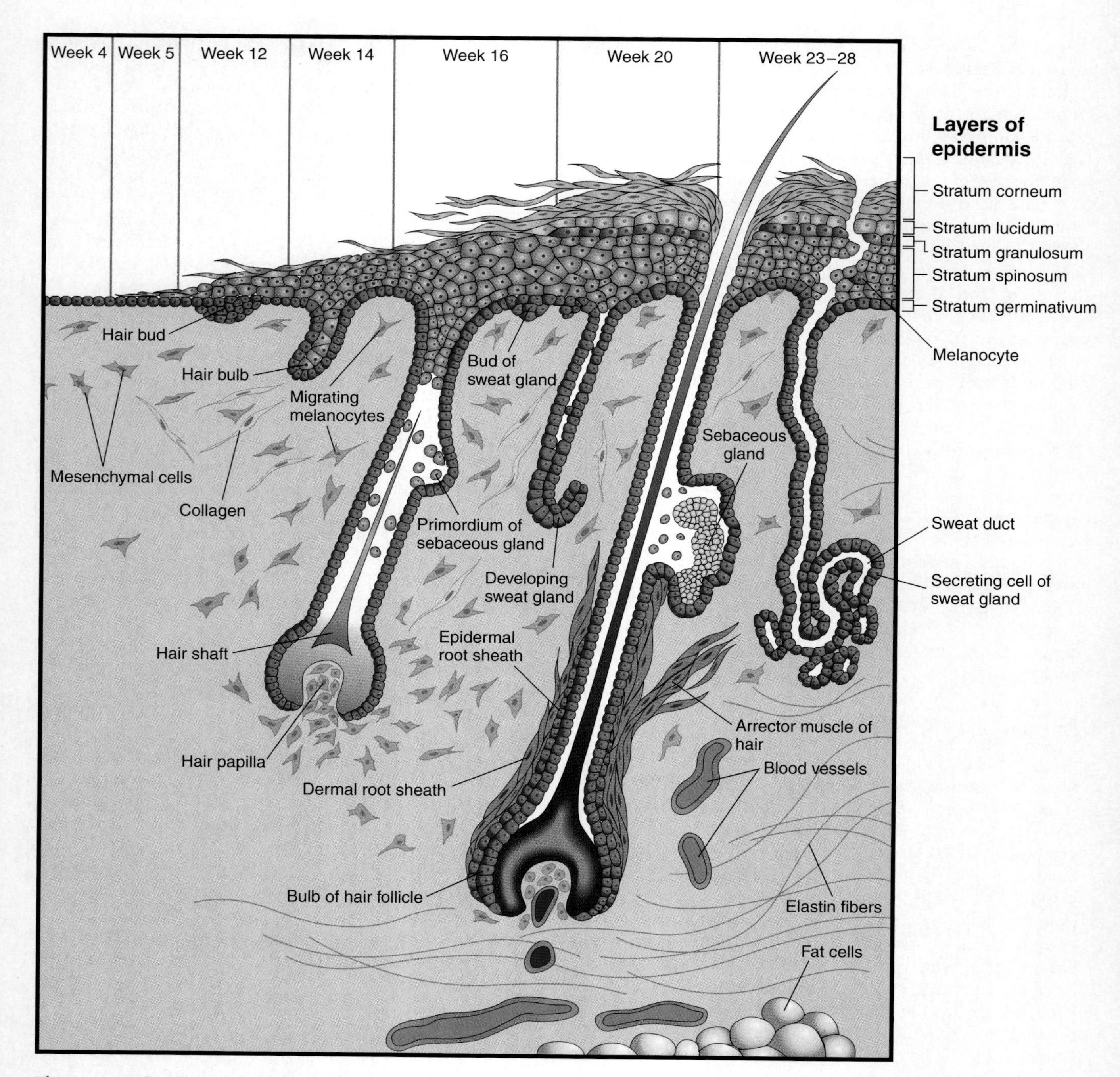

Figure 21 – 2. Successive stages in the development of a hair and its associated sebaceous gland and arrector muscle of hair. Note also the successive stages in the development of a sweat gland.

implicated in hair and follicle development. Hairs are first recognizable on the eyebrows, upper lip, and chin. A hair follicle begins as a proliferation of the stratum germinativum of the epidermis and extends into the underlying dermis. The **hair bud** soon becomes club-shaped, forming a **hair bulb**. The epithelial cells of the hair bulb constitute the **germinal matrix**, which later produces the hair. The hair bulb is soon invaginated by a small mesenchymal **hair papilla** (see Fig. 21 – 2). The peripheral cells of the developing hair follicle form the

epidermal root sheath, and the surrounding mesenchymal cells differentiate into the **dermal root sheath**. As cells in the germinal matrix proliferate, they are pushed toward the surface, where they keratinize to form the **hair shaft**. The hair grows through the epidermis on the eyebrows and upper lip by the end of the 12th week of development.

The first hairs that appear — **lanugo** (L. downy hair) — are fine, soft, and lightly pigmented. Lanugo begins to appear toward the end of the 12th week and is plentiful by 17 to 20 weeks. These hairs help to hold the vernix on the skin. Lanugo is replaced during the perinatal period by coarser hairs that persists over most of the body except in the axillary and pubic regions, where it is replaced at puberty by even coarser terminal hairs. In males similar coarse hairs also appear on the face and often on the chest. **Melanoblasts** migrate into the hair bulbs and differentiate into **melanocytes**. The melanin produced by these cells is transferred to the hair-forming cells in the germinal matrix several weeks before birth. The relative content of melanin accounts for different hair colors. **Arrector muscles of hairs**, small bundles of smooth muscle fibers, differentiate from the mesenchyme surrounding the hair follicle and attach to the dermal root sheath and the papillary layer of the dermis (see Fig. 21 – 2). The arrector muscles are poorly developed in the hairs of the axilla and in certain parts of the face. The hairs forming the eyebrows and the cilia forming the eyelashes have no arrector muscles.

Disorders of Keratinization

Ichthyosis (Gr. *ichthys,* fish) is a general term that is applied to a group of disorders resulting from excessive keratinization (Fig. 21 – 3). The skin is characterized by dryness and fish skin-like scaling, which may involve the entire body surface. A **harlequin fetus** results from a rare keratinizing disorder that is inherited as an autosomal recessive trait. The skin is markedly thickened, ridged, and cracked. Affected infants have a grotesque appearance, and most of them die during the first week of life. A **collodion baby** is covered at birth by a thick, taut membrane that resembles collodion or parchment. This membrane cracks with the first respiratory efforts and begins to fall off in large sheets. Complete shedding of membranes may take several weeks, occasionally leaving normal-appearing skin.

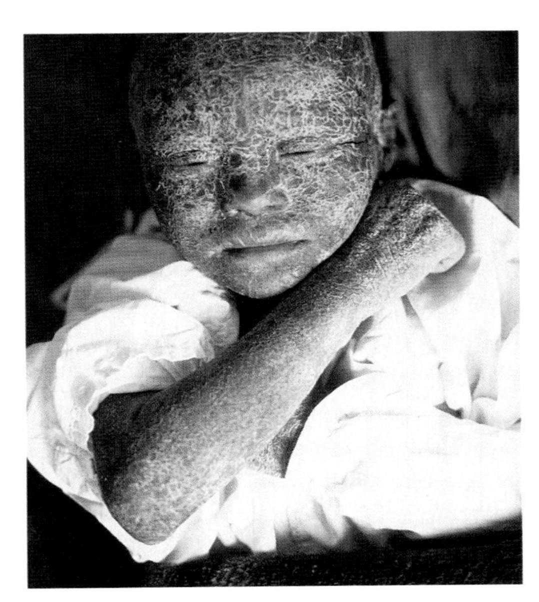

Figure 21 - 3. Photograph of a child with epidermolytic hyperkeratosis. This condition is characterized by severe hyperkeratosis from the time of birth. It has an autosomal dominant inheritance pattern. (Courtesy of Dr. Joao Carlos Fernandes Rodrigues, Servico de Dermatologia, Hospital de Desterro, Lisbon, Portugal.)

Angiomas of Skin

The vascular anomaly known as angioma of the skin is a developmental defect in which some transitory and/or surplus primordial blood or lymphatic vessels persist. These defects are called **angiomas**, even though they may not be true tumors. Those composed of blood vessels may be mainly arterial, venous, or cavernous, but they are often of a mixed type. Angiomas composed of lymphatics are called cystic lymphangiomas or **cystic hygromas** (see Chapter 15). True angiomas are benign tumors of endothelial cells, usually composed of solid or hollow cords; the hollow cords contain blood. Various terms are used to describe angiomatous anomalies ("birthmarks"). **Nevus flammeus** denotes a flat, pink or red, flamelike blotch that often appears on the posterior surface of the neck. A portwine stain or **hemangioma** is a larger and darker angioma than nevus flammeus and is nearly always anterior or lateral on the face and/or neck.

Albinism

In *generalized albinism,* which is an autosomal recessive trait, the skin, hair, and retina lack pigment; however, the iris usually shows some pigmentation. Albinism occurs when the melanocytes fail to produce melanin because of the lack of the enzyme tyrosinase. In *localized albinism* — piebaldism — an autosomal dominant trait, there is a lack of melanin in patches of skin and/or hair.

Development of the Nails

Toenails and fingernails begin to develop at the tips of the digits at about 10 weeks (Fig. 21 – 4). Development of fingernails precedes that of toenails by about 4 weeks.

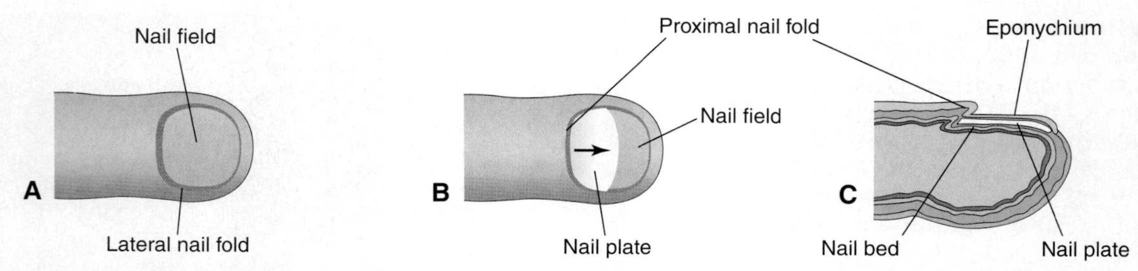

Figure 21 - 4. Successive stages in the development of a fingernail. *A,* The first indication of a nail is a thickening of the epidermis, the nail field, at the tip of the finger. *B,* As the nail plate develops, it slowly grows toward the tip of the finger. *C,* The fingernail normally reaches the end of the digit by 32 weeks.

The primordia of nails appear as thickened areas or fields of epidermis at the tip of each digit. Later these **nail fields** migrate onto the dorsal surface (Fig. 21 - 4*A*), carrying their innervation from the ventral surface. The nail fields are surrounded laterally and proximally by folds of epidermis called **nail folds**. Cells from the proximal nail fold grow over the nail field and keratinize to form the **nail plate** (Fig. 21 - 4*B*). At first, the developing nail is covered by superficial layers of epidermis called **eponychium** (Fig. 21 - 4*C*). This later degenerates, exposing the nail except at its base, where it persists as the **cuticle**. The fingernails reach the fingertips by about 32 weeks; the toenails reach the toetips by about 36 weeks.

Deformed Nails

Deformed nails occur occasionally and may be a manifestation of a generalized skin disease or systemic disease. A number of congenital diseases (e.g., psoriasis) are associated with nail defects.

Development of the Mammary Glands

Mammary glands are modified and highly specialized types of sweat glands. **Mammary buds** begin to develop during the sixth week of development as solid downgrowths of the epidermis into the underlying mesenchyme (Fig. 21 - 5*C*). These changes occur in response to an inductive influence from the mesenchyme. The mammary buds develop from **mammary crests** (ridges), which are thickened strips of ectoderm extending from the axillary to the inguinal regions (Fig. 21 - 5*A*). The mammary crests appear during the fourth week but normally persist in humans only in the pectoral area, where the breasts develop (Fig. 21 - 5*B*). Each primary **mammary bud** soon gives rise to several secondary mammary buds that develop into **lactiferous ducts** and their branches (Fig. 21 - 5*D* and *E*). Canalization of these buds is induced by maternal sex hormones entering the fetal circulation. This process continues until late gestation, and by term, 15 to 20 lactiferous ducts have been formed. The fibrous connective tissue and fat of the mammary gland develop from the surrounding mesenchyme.

During the late fetal period, the epidermis at the site of origin of the primordial mammary gland becomes depressed, forming a shallow **mammary pit** (Fig. 21 - 5*E*). The nipples are poorly formed and depressed in newborn infants. After birth, the nipples usually rise from the mammary pits. The rudimentary mammary glands of newborn males and females are identical and are often enlarged. Some secretion, often called "witch's milk," may be produced. These transitory changes are caused by maternal hormones passing through the placental membrane into the fetal circulation. At birth, the main lactiferous ducts are formed; the mammary glands remain underdeveloped until puberty. The mammary glands develop similarly and are of the same structure in both sexes. In females, the glands enlarge rapidly during puberty, mainly because of fat and other connective tissue development. Growth of the duct system also occurs because of the raised levels of circulating estrogens.

Gynecomastia

The rudimentary mammary glands in males normally undergo no postnatal development. Gynecomastia (Gr. *gyne,* woman, and *mastos,* breast) refers to excessive development of the male mammary tissue. It occurs in most newborn males because of stimulation of the glands by maternal sex hormones. This effect disappears in a few weeks. During midpuberty, about two thirds of boys develop varying degrees of hyperplasia of the breasts. About 80% of males with Klinefelter syndrome have gynecomastia (see Chapter 9).

Aplasia of the Breast

The breasts of a postpubertal female often differ somewhat in size. Marked differences are regarded as anomalies because both glands are exposed to the same hormones at puberty. In these cases, there is often associated rudimentary development of muscles, usually the pectoralis major (see Chapter 17).

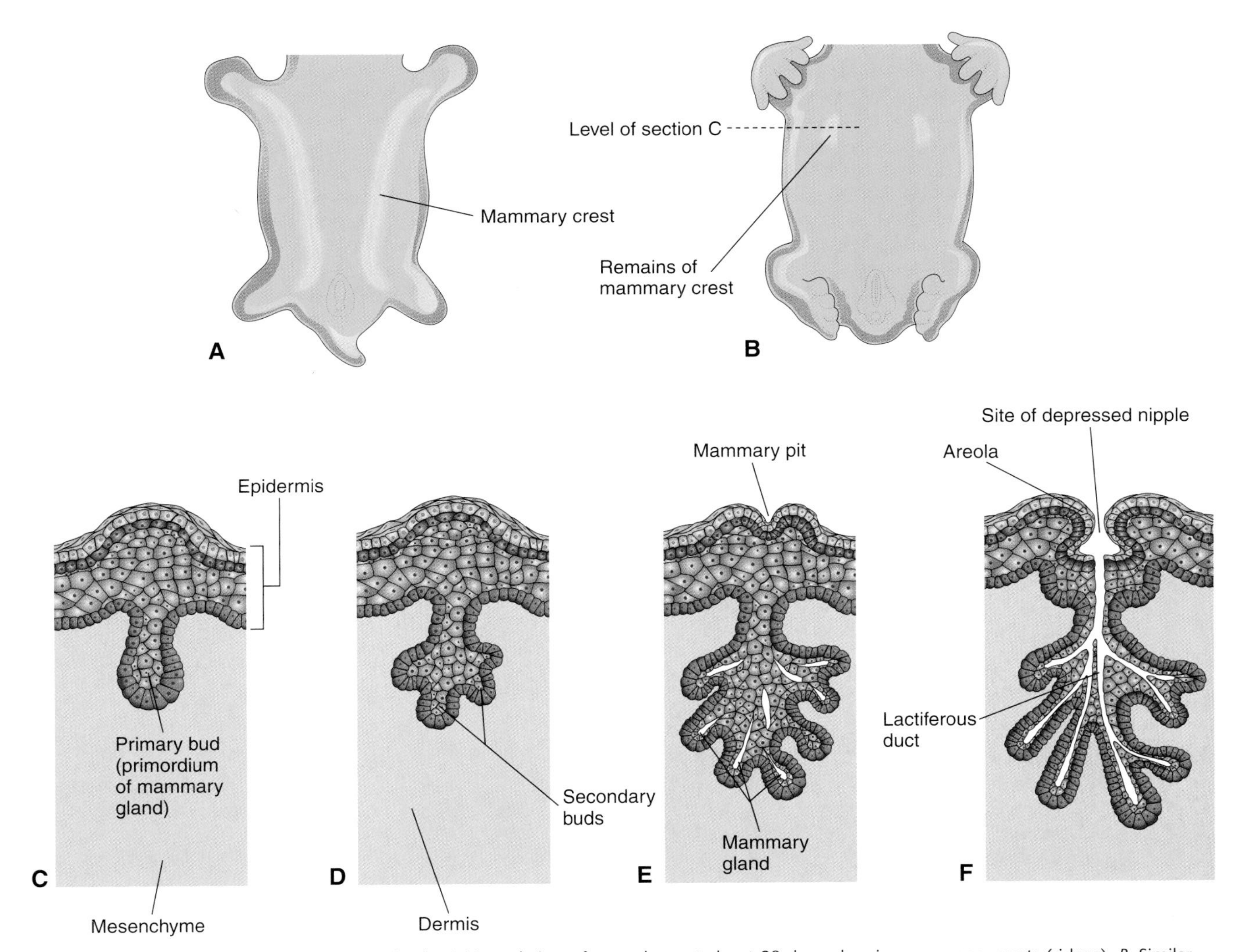

Figure 21 - 5. Development of mammary glands. *A,* Ventral view of an embryo at about 28 days, showing mammary crests (ridges). *B,* Similar view at 6 weeks, showing the remains of these crests. *C,* Transverse section of a mammary crest at the site of a developing mammary gland. *D* to *F,* Similar sections showing successive stages of breast development between the 12th week and birth.

Supernumerary Breasts and Nipples

An extra breast **(polymastia)** or nipple **(polythelia)** occurs in about 1% of the female population and is an inheritable condition. An extra breast or nipple usually develops just inferior to the normal breast. **Supernumerary nipples** are also relatively common in males; often, they are mistaken for moles. Less commonly, supernumerary breasts or nipples appear in the axillary or abdominal regions of females. In these positions, the nipples or breasts arise from extra mammary buds that develop along the mammary crests. They usually become obvious in women when pregnancy occurs. About one third of affected persons have two extra nipples or breasts.

Development of the Teeth

Two sets of teeth normally develop: the primary dentition or **deciduous teeth** and the secondary dentition or **permanent teeth**. Teeth develop from:

- oral ectoderm
- mesoderm
- neural crest cells

The enamel is derived from ectoderm of the oral cavity; all other tissues differentiate from the surrounding mesenchyme derived from mesoderm and neural crest cells. Experimental evidence suggests that neural crest cells are imprinted with morphogenetic information before or shortly after they migrate from the **neural crest**. The expression of homeobox MSX and Dlx genes in migrating neural crest cells, as well as in the ectoderm and mesenchyme, plays a critical role in initiation of tooth development. As the mandible and maxilla grow to accommodate the developing teeth, the shape of the face changes. **Odontogenesis** (tooth development) is initiated by the inductive influence of neural crest mesenchyme on the overlying ectoderm. Tooth development is a continuous process; however, it is usually divided into stages for descriptive purposes on the basis of the appearance of the developing tooth. Not all teeth begin to develop at the same time. The first tooth buds appear in the anterior mandibular region; later tooth development occurs in the anterior maxillary region and then progresses posteriorly in both jaws. Tooth development continues for years after birth (Table 21 - 1). The first indication of tooth development occurs in the sixth week of development as a thickening of the oral epithelium, a derivative of the surface ectoderm. These U-shaped bands, called **dental laminae**, follow the curves of the primordial jaws (Figs. 21 - 6*A* and 21 - 7*A*).

Table 21 - 1. **The Order and Usual Time of Eruption of Teeth and the Time of Shedding of Deciduous Teeth**

Tooth	Usual Eruption Time	Shedding Time
Deciduous		
Medial incisor	6 - 8 mos	6 - 7 yr
Lateral incisor	8 - 10 mos	7 - 8 yr
Canine	16 - 20 mos	10 - 12 yr
First molar	12 - 16 mos	9 - 11 yr
Second molar	20 - 24 mos	10 - 12 yr
Permanent*		
Medial incisor	7 - 8 yr	
Lateral incisor	8 - 9 yr	
Canine	10 - 12 yr	
First premolar	10 - 11 yr	
Second premolar	11 - 12 yr	
First molar	6 - 7 yr	
Second molar	12 yr	
Third molar	13 - 25 yr	

(Modified from Moore KL, Dalley AF: *Clinically Oriented Anatomy,* 4th ed. Baltimore, Williams & Wilkins, 1999).
*The permanent teeth are not shed. If they are not properly cared for or disease of the gingiva develops, they may fall out or have to be extracted.

Bud Stage of Tooth Development

Each dental lamina develops 10 centers of proliferation from which **tooth buds** grow into the underlying mesenchyme (Figs. 21 - 6*B* and 21 - 7*B*). These tooth buds develop into the first or **deciduous teeth**, which are so named because they are shed during childhood (Table 21 - 1). There are 10 tooth buds in each jaw, one for each deciduous tooth. The tooth buds for the **permanent teeth** that have deciduous predecessors begin to appear at about 10 weeks from deep continuations of the dental laminae (Fig. 21 - 7*D*). They develop lingual (toward the tongue) to the **deciduous tooth buds**. The permanent molars that have no deciduous predecessors develop as buds from posterior extensions of the dental laminae. The tooth buds for the permanent teeth appear at different times, mostly during the fetal period. The buds for the second and third permanent molars develop after birth.

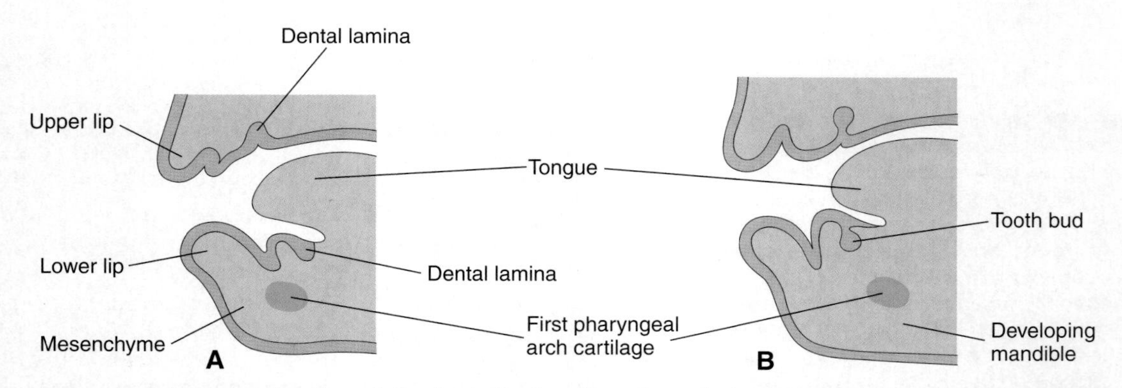

Figure 21 - 6. Sagittal sections through the developing jaws illustrating early development of the teeth. *A,* Early in the sixth week of development, the dental laminae are evident. *B,* Later in the sixth week, tooth buds arise from the laminae.

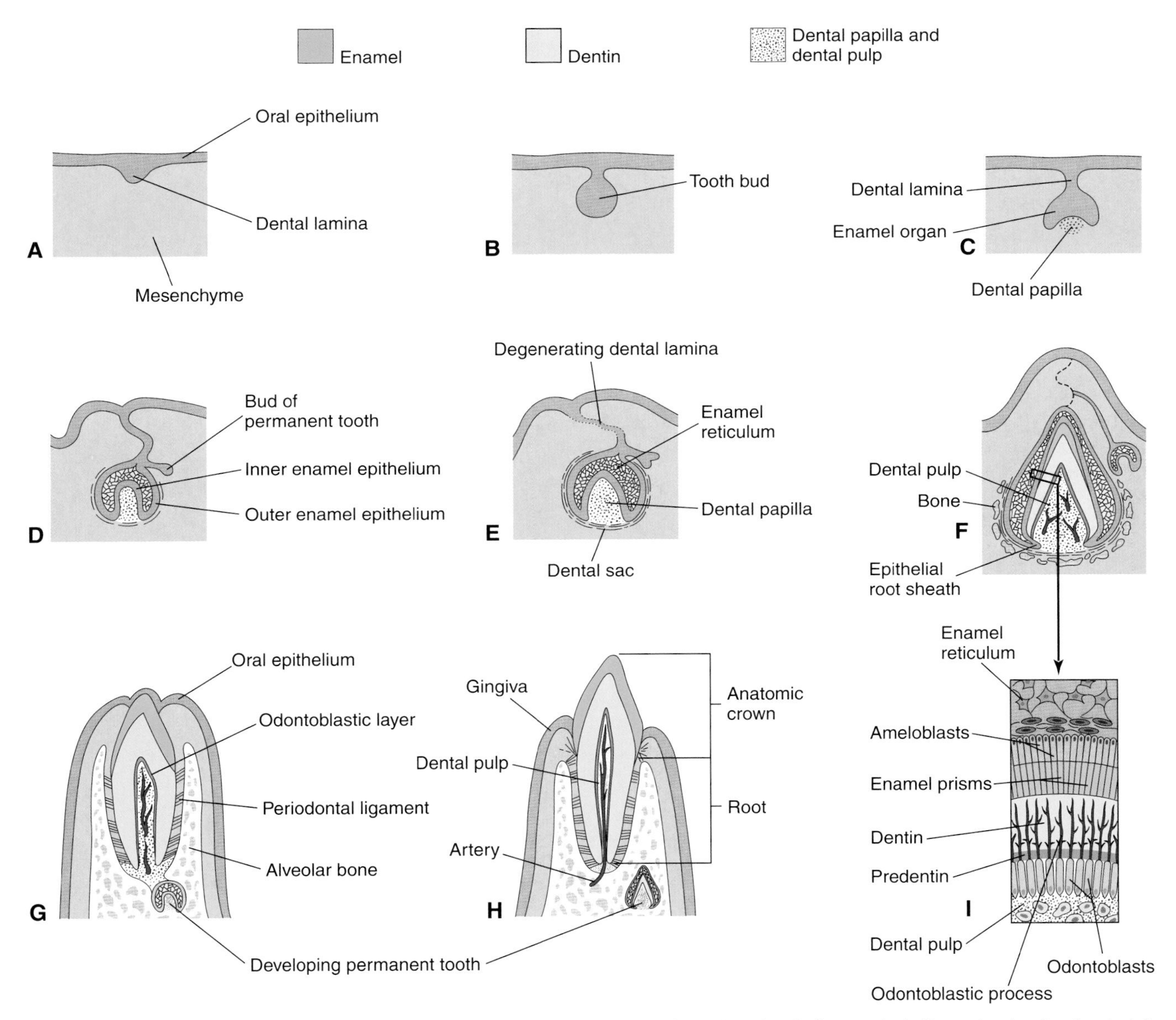

Figure 21 - 7. Sagittal sections illustrating successive stages in the development and eruption of an incisor tooth. *A,* Six weeks, showing the dental lamina. *B,* Seven weeks, showing the tooth bud developing from the dental lamina. *C,* Eight weeks, showing the cap stage of tooth development. *D,* Ten weeks, showing the early bell stage of a deciduous tooth and the bud stage of a permanent tooth. *E,* Fourteen weeks, showing the advanced bell stage of tooth development. Note that the connection (dental lamina) of the tooth to the oral epithelium is degenerating. *F,* Twenty-eight weeks, showing the enamel and dentin layers. *G,* Six months postnatal, showing early tooth eruption. *H,* Eighteen months postnatal, showing a fully erupted deciduous incisor tooth. The permanent incisor tooth now has a well-developed crown. *I,* Section through a developing tooth showing ameloblasts (enamel producers) and odontoblasts (dentin producers).

Cap Stage of Tooth Development

As each tooth bud is invaginated by mesenchyme — the primordium of the dental papilla — the bud becomes cap-shaped (Fig. 21 - 7*C*). The ectodermal part of the developing tooth, the **enamel organ** (dental organ), eventually produces enamel. The internal part of each cap-shaped tooth, the **dental papilla**, is the primordium of the dental pulp. Together, the dental papilla and enamel organ form the **tooth germ** (primordial tooth). The outer cell layer of the enamel organ is the **outer enamel epithelium**, whereas the inner cell layer lining the "cap" is the **inner enamel epithelium** (see Fig.

21 – 7*D*). The central core of loosely arranged cells between the layers of enamel epithelium is the **enamel (stellate) reticulum.** As the enamel organ and dental papilla of the tooth develop, the mesenchyme surrounding the developing tooth condenses to form the **dental sac,** a vascularized capsular structure (see Fig. 21 – 7*E*). The dental sac is the primordium of the cement (L. cementum) and periodontal ligament. The **cement** is the bonelike rigid connective tissue covering the root of the tooth. The **periodontal ligament** is the fibrous connective tissue that surrounds the root of the tooth, separating it from and attaching it to the alveolar bone (see Fig. 21 – 7*G*).

Bell Stage of Tooth Development

As the enamel organ differentiates, the developing tooth becomes bell-shaped (see Figs. 21 – 7*D* and 21 – 8). The mesenchymal cells in the dental papilla adjacent to the inner enamel epithelium differentiate into **odontoblasts,** which produce **predentin** and deposit it adjacent to the epithelium. Later, the predentin calcifies and becomes **dentin.** As the dentin thickens, the odontoblasts regress toward the center of the dental papilla; however, their cytoplasmic processes — **odontoblastic processes** or Tomes processes — remain embedded in the dentin (see Fig. 21 – 7*F* and *I*). The yellowish enamel is the second hardest tissue in the body. It overlies the brittle dentin, the hardest tissue in the body, and protects it from being fractured. Cells of the inner enamel epithelium differentiate into **ameloblasts**, which produce enamel in the form of prisms (rods) over the dentin. As the **enamel** increases, the ameloblasts regress toward the outer enamel epithelium. Enamel and dentin formation begins at the cusp (tip) of the tooth and progresses toward the future root. The **root of the tooth** begins to develop after dentin and enamel formation is well advanced. The inner and outer enamel epithelia come together in the neck region of the tooth where they form a fold, the **epithelial root sheath** (see Fig. 21 – 7*F*). This sheath grows into the mesenchyme and initiates root formation. The odontoblasts adjacent to the epithelial root sheath form dentin that is continuous with that of the crown. As the dentin increases, it reduces the pulp cavity to a narrow **root canal** through which the vessels and nerves pass. The inner cells of the dental sac differentiate into **cementoblasts**, which produce cement that is restricted to the root. Cement is deposited over the dentin of the root and meets the enamel at the neck of the tooth **(cementoenamel junction)**.

As the teeth develop and the jaws ossify, the outer cells of the dental sac also become active in bone formation. Each tooth soon becomes surrounded by bone, except over its crown. The tooth is held in its **alveolus** (bony socket) by the strong **periodontal ligament**, a derivative of the dental sac (see Fig. 21 – 7*G* and *H*). Some fibers of this ligament are embedded in the cement; other fibers are embedded in the bony wall of the alveolus. The periodontal ligament is located between the cement of the root and the bony alveolus.

Figure 21 – 8. Photomicrograph (× 17) of the crown and neck of a tooth. Observe the enamel (E), dentin (D), dental pulp (P), and odontoblasts (O). (From Gartner LP, Hiatt JL: *Color Textbook of Histology,* 2nd ed. Philadelphia, WB Saunders, 2001.)

Tooth Eruption

As the teeth develop, they begin a continuous slow movement toward the oral cavity (see Fig. 21 – 7*F* and *G*). The mandibular teeth usually erupt before the maxillary teeth, and girls' teeth usually erupt sooner than boys' teeth. A child's dentition contains **20 deciduous teeth**. As the root of the tooth grows, its crown gradually erupts through the oral epithelium. The part of the oral mucosa around the erupted crown becomes the **gingiva** (gum). Usually, eruption of the deciduous teeth occurs between 6 and 24 months after birth (see Table 21 – 1). The mandibular medial or **central incisors** usually erupt 6 to 8 months after birth, but this process may not begin until 12 or 13 months in some children. Despite this, all 20 deciduous teeth are usually present by the end of the second year in healthy children. Delayed eruption of all teeth may indicate a systemic or nutritional disturbance, such as hypopituitarism or hypothyroidism.

The **permanent teeth** develop in a manner similar to that described for deciduous teeth. As a permanent tooth grows, the root of the corresponding deciduous tooth is gradually resorbed by **osteoclasts**. Consequently, when the deciduous tooth is shed, it consists only of the crown and the uppermost part of the root. The permanent teeth usually begin to erupt during the sixth year and continue to appear until early adulthood (Fig. 21 – 9; see also Table 21 – 1). The shape of the face is affected by the development of the paranasal sinuses and the growth of the maxilla and mandible to accommodate the teeth (see Chapter 11). It is the lengthening of the **alveolar processes** (bony sockets supporting the teeth) that results in the increase in the depth of the face during childhood.

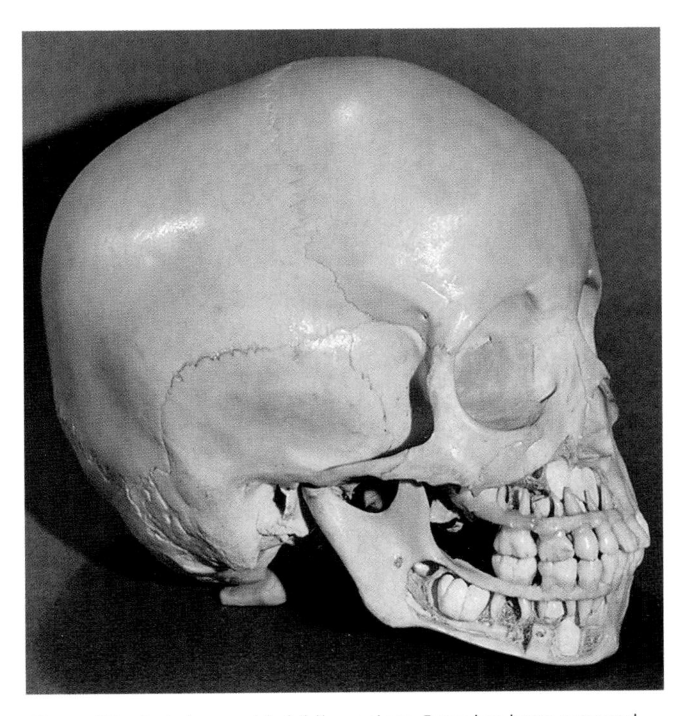

Figure 21 – 9. A 4-year-old child's cranium. Bone has been removed from the jaws to show the relation of the developing permanent teeth to the erupted deciduous teeth.

Natal Teeth

Natal teeth are those that are erupted at birth (L. *natus,* birth). There are usually two in the position of the mandibular incisors. Natal teeth, which are observed in about 1 in 2000 newborn infants, may produce maternal discomfort during breastfeeding. In addition, the infant's tongue may be lacerated, or the teeth may detach and be aspirated; for these reasons, natal teeth are sometimes extracted.

Enamel Hypoplasia

Defective enamel formation causes pits or fissures, or both, in the enamel (Fig. 21 – 10). These defects result from temporary disturbances of enamel formation. Various factors may injure ameloblasts (the enamel builders), such as nutritional deficiency, tetracycline therapy, and infectious diseases, such as measles. **Rickets** during the critical period of permanent tooth development (6 to 12 weeks) is the most common known cause of enamel hypoplasia. Rickets is a disease affecting children who are deficient in vitamin D.

Variations of Tooth Shape

Abnormally shaped teeth are relatively common (see Fig. 21 – 10*A* to *G*). Occasionally, there are spherical masses of enamel — **enamel pearls** — attached to the tooth. They are formed by aberrant groups of ameloblasts. In other children, the maxillary lateral incisor teeth may have a slender, tapering shape (peg-shaped incisors). **Congenital syphilis** affects the differentiation of the permanent teeth, resulting in screwdriver-shaped incisors, with central notches in their incisive edges.

Numerical Abnormalities

One or more supernumerary teeth may develop, or the normal number of teeth may fail to form (see Fig. 21 – 10*H* and *I*). **Supernumerary teeth** usually develop in the area of the maxillary incisors and disrupt the position and eruption of normal teeth. The extra teeth commonly erupt posterior to the normal ones. In **partial anodontia**, one or more teeth are absent. Congenital absence of one or more teeth is often a familial trait. In **total anodontia**, no teeth develop; this very rare condition is usually associated with *congenital ectodermal dysplasia* (disorders involving tissues that are ectodermal in origin).

Abnormally Sized Teeth

Disturbances during the differentiation of teeth may result in gross alterations of dental morphology, such as *macrodontia* (large teeth) and *microdontia* (small teeth).

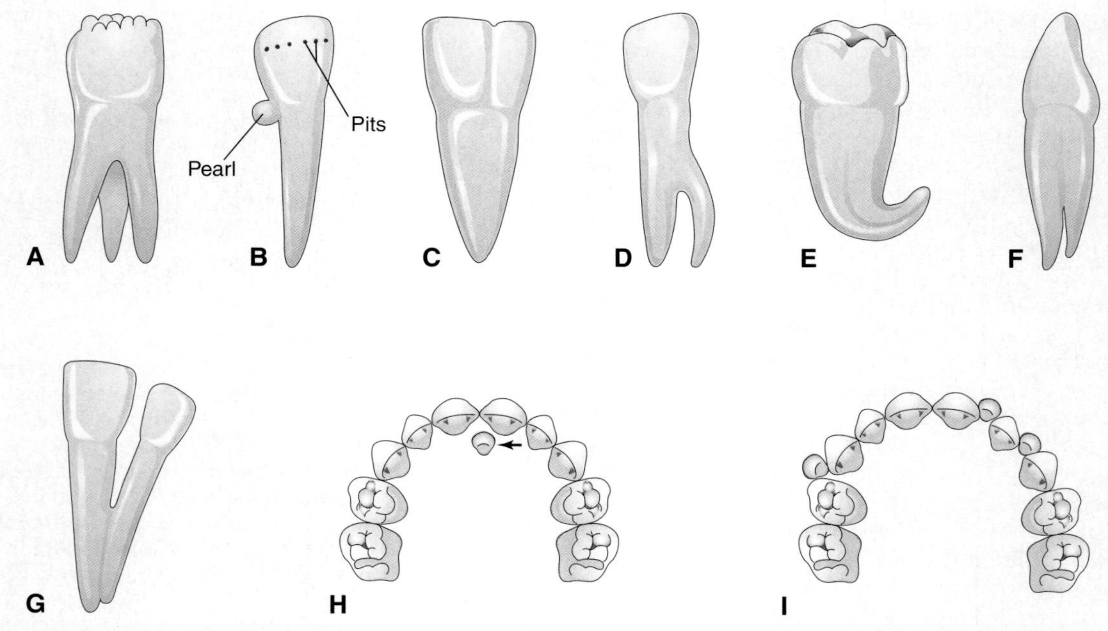

Figure 21 – 10. Anomalies of teeth. *A,* Irregular raspberry-like crown. *B,* Enamel pearl and pits. *C,* Incisor tooth with a double crown. *D,* Abnormal division of root. *E,* Distorted root. *F,* Branched root. *G,* Fused roots. *H,* Hyperdontia with a supernumerary incisor tooth in the anterior region of the palate (*arrow*). *I,* Hyperdontia with 13 deciduous teeth in the maxilla (upper jaw) instead of the normal 10.

Fused Teeth

Occasionally a tooth bud divides or two buds partially fuse to form fused teeth (Fig. 21 – 10*C* and *G*). This condition is commonly observed in the mandibular incisors of the deciduous dentition. "Twinning" of teeth results from division of the tooth bud. In some cases, the permanent tooth does not form; this suggests that the deciduous and permanent tooth primordia fused to form the deciduous tooth.

Dentigerous Cyst

A cyst containing an unerupted tooth may develop in a mandible, maxilla, or maxillary sinus. The dentigerous (tooth-bearing) cyst develops because of cystic degeneration of the enamel reticulum of the enamel organ of an unerupted tooth. Most cysts are deeply situated in the jaw and are associated with misplaced or malformed secondary teeth that have failed to erupt.

Amelogenesis Imperfecta

In amelogenesis imperfecta, the tooth enamel is soft and friable because of hypocalcification, and the teeth are yellow to brown in color (Fig. 21 – 11*A*). The teeth are covered with only a thin layer of abnormally formed enamel through which the yellow underlying dentin is visible. This gives a darkened appearance to the teeth. This autosomal dominant condition affects about 1 in 7000 to 8000 children depending on the population studied.

Dentinogenesis Imperfecta

Dentinogenesis imperfecta is relatively common in white children (Fig. 21 – 11*B*). In affected children, the teeth are brown to gray-blue, with an opalescent sheen, caused by failure of the odontoblasts to differentiate normally, producing poorly calcified dentin. Both deciduous and permanent teeth are usually involved. The enamel tends to wear down rapidly, exposing the dentin. This anomaly is inherited as an autosomal dominant trait.

Discolored Teeth

Foreign substances incorporated into the developing enamel and dentin discolor the teeth. The hemolysis (liberation of hemoglobin) associated with hemolytic disease of the newborn (see Chapter 8) may produce blue to black discoloration of the teeth. *All tetracyclines are extensively incorporated into the teeth.* The critical period of risk is from about 14 weeks of fetal life to the 10th postnatal month for deciduous teeth, and from about 14 weeks of fetal life to the 16th postnatal year for permanent teeth. Tetracyclines produce brownish-yellow discoloration (mottling) and enamel hypoplasia because they interfere with the metabolic processes of the ameloblasts. The enamel is completely formed on all but the third molars by about 8 years of age. For this reason, *tetracyclines should not be administered to pregnant women or children younger than 8 years of age.*

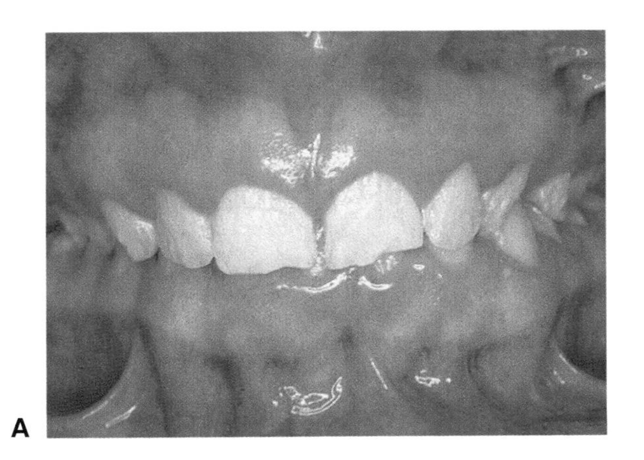

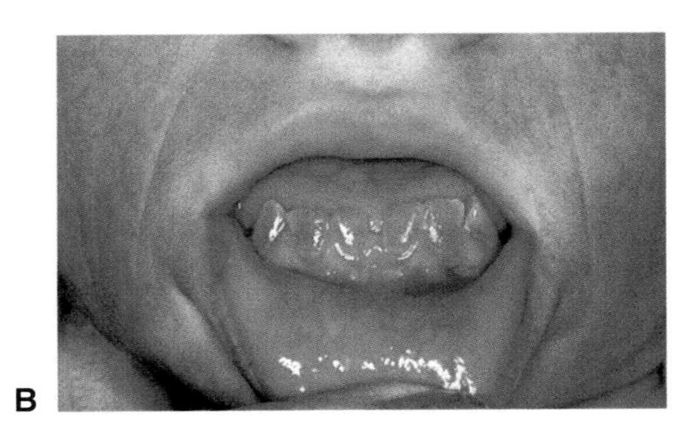

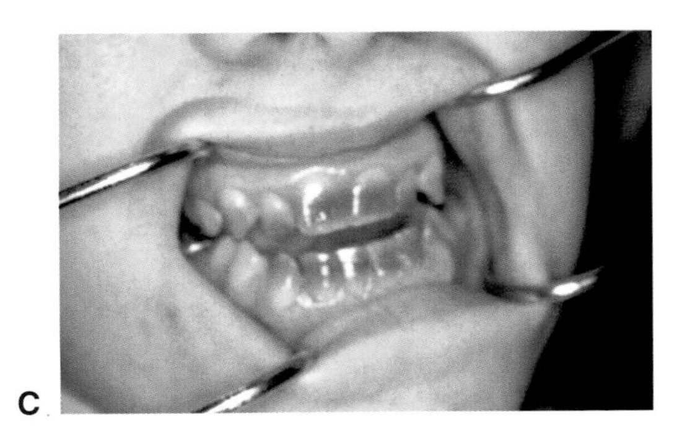

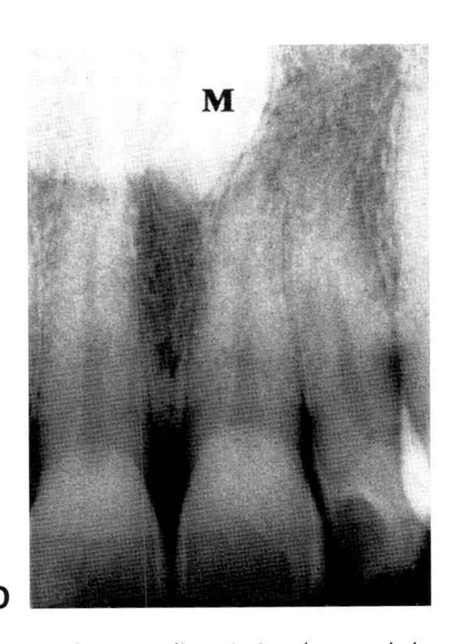

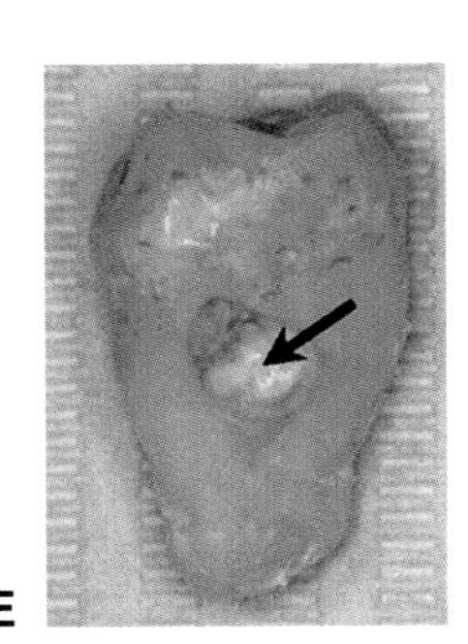

Figure 21 - 11. Photographs showing some common tooth anomalies. *A,* Amelogenesis imperfecta. *B,* Dentinogenesis imperfecta. *C,* Tetracycline-stained teeth. *D,* A midline supernumerary tooth (M, mesiodens), located near the root apex of the central incisor. The prevalence of supernumerary teeth is 1 to 3%. *E,* Molar tooth with an enamel pearl (*arrow*). *(A,* Courtesy of Dr. Blaine Cleghorn; *B to D,* Courtesy of Dr. Steve Ahing, Faculty of Dentistry, University of Manitoba, Winnipeg, Manitoba, Canada.)

Summary of the Integumentary System

The skin and its appendages develop from ectoderm, mesoderm, and neural crest cells. The epidermis is derived from surface ectoderm. Melanocytes are derived from **neural crest cells** that migrate into the epidermis. Castoff cells from the epidermis mix with secretions of sebaceous glands to form **vernix caseosa**, a whitish, greasy coating for the skin. This fatty substance protects the epidermis, probably making it more waterproof, and facilitates birth because of its slippery quality.

Hairs develop from downgrowths of the epidermis into the dermis. By about 20 weeks, the fetus is completely covered with fine, downy hairs called **lanugo**. These hairs are shed by birth or shortly thereafter and are replaced by coarser hairs. Most **sebaceous glands** develop as outgrowths from the sides of hair follicles; however, some glands develop as downgrowths of the epidermis into the dermis. **Sweat glands** also develop from epidermal downgrowths into the dermis. **Mammary glands** develop in a similar manner.

Congenital anomalies of the skin are mainly **disorders of keratinization** (ichthyosis) and pigmentation (albinism). Abnormal blood vessel development results in various types of **angioma**. Nails may be absent or malformed. Hair may be absent or excessive. Supernumerary breasts (polymastia) or nipples (polythelia) are relatively common.

Teeth develop from ectoderm, mesoderm, and neural crest cells. The enamel is produced by **ameloblasts**, which are derived from the oral ectoderm; all other dental tissues develop from mesenchyme, derived from mesoderm and neural crest cells. The common **congenital anomalies of teeth** are defective formation of enamel and dentin, abnormalities in shape, and variations in number and position. **Tetracyclines** are extensively incorporated into the enamel and dentin of developing teeth, producing brownish-yellow discoloration and hypoplasia of the enamel. Tetracyclines should not be prescribed for pregnant women or children under 8 years of age.

Clinically Oriented Questions

1. A baby was reportedly born without skin. Is this possible? If so, could such a baby survive?
2. A dark-skinned person presented with patches of white skin on his face, chest, and limbs. He even had a white forelock. What is this condition called and what is its developmental basis? Is there any treatment for these skin defects?
3. Some male babies have enlarged breasts at birth. Is this an indication of abnormal sex development? Some boys develop breasts during puberty. Are they intersexes?
4. A girl developed a breast in her axilla during puberty. She also had extra nipples on her chest and upper abdomen. What is the embryological basis for these anomalies?
5. A baby was born with two teeth. Would these be normal teeth? Is this a common occurrence? Are they usually extracted?

The answers to these questions are at the back of the book.

References and Suggested Reading

Chapter 1

Butler H, Juurlink BHJ: *An Atlas for Staging Mammalian and Chick Embryos*. Boca Raton, CRC Press, Inc, 1987.
Gasser R: *Atlas of Human Embryos*. Hagerstown, Harper & Row, 1975.
Gasser R: A DVD-ROM database of human embryo serial sections. New Orleans, LSU Health Sciences Center, 2002.
Jasny BR, Kennedy D (eds): The human genome. *Science* 291:1153, 2001.
Meyer AW: *The Rise of Embryology*. California, Stanford University Press, 1939.
Moore KL, Persaud TVN, Shiota K: *Color Atlas of Clinical Embryology*, 2nd ed. Philadelphia, WB Saunders, 2000.
Needham J: *A History of Embryology*, 2nd ed. Cambridge, Cambridge University Press, 1959.
Nussbaum RL, McInnes RR, Willard HF: *Thompson & Thompson Genetics in Medicine*, 6th ed. Philadelphia, WB Saunders, 2001.
O'Rahilly R, Müller F: *Developmental Stages in Human Embryos*. Washington, DC, Carnegie Institution of Washington, 1987.
Streeter GL: Developmental horizons in human embryos. Description of age group XI, 13 to 20 somites, and age group XII, 21 to 29 somites. *Contrib Embryol Carnegie Inst* 30:211, 1942.

Chapter 2

Bissonette F, Lapensée L: Investigating and treating infertility. How far have we come? Part I. *Can J Diag* 18:65, 2001.
Cooke HJ, Hargreave T, Elliott DJ: Understanding the genes involved in spermatogenesis: a progress report. *Fertil Steril* 69:989, 1998.
Dill-Macky MJ, Atri M: Ovarian sonography. *In* Callen PW (ed): *Ultrasonography in Obstetrics and Gynecology*, 4th ed. Philadelphia, WB Saunders, 2000.
Hillier SG: Gonadotropic control of ovarian follicular growth and development. *Mol Cell Endocrinol* 179:39, 2001.
Kubiak JZ, Johnson M: Human infertility, reproductive cloning and nuclear transfer: a confusion of meanings. *BioEssays* 23:359, 2001.
Oehninger S, Hodgen GD: Hypothalamic-pituitary-ovary-uterine axis. *In* Copeland LJ, Jarrell J (eds): *Textbook of Gynecology*, 2nd ed. Philadelphia, WB Saunders, 2000.
Weremowicz S, Sandstrom DJ, Morton CC et al: Fluorescence *in situ* hybridization (FISH) for rapid detection of aneuploidy: experience in 911 prenatal cases. *Prenat Diagn* 21:262, 2001.

Chapter 3

Allen CA, Green DPL: The mammalian acrosome reaction: gateway to sperm fusion with the oocyte. *BioEssays* 19:241, 1997.
Burmeister L, Palermo GD, Rosenwaks Z: IVF: the new era. *Int J Fertil* 46:137, 2001.
Eisenbach M, Tur-Kaspa I: Do eggs attract spermatozoa? *BioEssays* 21:203, 1999.
Hardy K, Wright C, Rice S et al: Future developments in assisted reproduction in humans. *Reproduction* 123:171, 2002.
Horne AW, White JO, Lalani E: The endometrium and embryo implantation. *Br Med J* 321:1301, 2000.
Kubiak JZ, Johnson M: Human infertility, reproductive cloning and nuclear transfer: a confusion of meanings. *BioEssays* 23:359, 2001.
Rock J, Hertig AT: The human conceptus during the first two weeks of gestation. *Am J Obstet Gynecol* 55:6, 1948.
Strom CM, Strom S, Levine E et al: Obstetric outcomes in 102 pregnancies after preimplantation genetic diagnosis. *Am J Obstet Gynecol* 182:1629, 2000.
Wilton L: Preimplantation genetic diagnosis for aneuploidy screening in early human embryos: a review. *Prenat Diagn* 22:512, 2002.

Chapter 4

Bianchi DW, Wilkins-Haug LE, Enders AC, Hay ED: Origin of extraembryonic mesoderm in experimental animals: relevance to chorionic mosaicism in humans. *Am J Med Genet* 46:542, 1993.
Cadkin AV, McAlpin J: The decidua - chorionic sac. A reliable sonographic indicator of intrauterine pregnancy prior to detection of a fetal pole. *J Ultrasound Med* 3:539, 1984.
Enders AC, King, BF: Formation and differentiation of extraembryonic mesoderm in the rhesus monkey. *Am J Anat* 181:327, 1988.
Hertig AT, Rock J: Two human ova of the pre-villous stage, having a developmental age of about eight and nine days, respectively. *Contrib Embryol Carnegie Inst* 33:169, 1949.
Levine D: Ectopic pregnancy. *In* Callen PW (ed): *Ultrasonography in Obstetrics and Gynecology*, 4th ed. Philadelphia, WB Saunders, 2000.
Lipscomb GH: Ectopic pregnancy. *In* Copeland LJ, Jarrell JF (eds): *Textbook of Gynecology*, 4th ed. Philadelphia, WB Saunders, 2000.
Luckett WP: Origin and differentiation of the yolk sac and extraembryonic mesoderm in presomite human and rhesus monkey embryos. *Am J Anat* 152:59, 1978.
Nussbaum RL, McInnes RR, Willard HF: *Thompson & Thompson Genetics in Medicine*, 6th ed. Philadelphia, WB Saunders, 2001.

Chapter 5

Bronner-Fraser M, Sternberg PW: Pattern formation and developmental mechanisms. The cell biological basis of inductive signaling. *Curr Opin Gen Develop* 10:347, 2000.
Christiansen JH, Coles EG, Wilkinson DG: Molecular control of neural crest formation, migration and differentiation. *Curr Opin Cell Biol* 12:719, 2000.
Cooke J: Vertebrate left and right: finally a cascade, but first a flow. *BioEssays* 21:537, 1999.
Dale KJ, Pourquié O: A clock-work somite. *BioEssays* 22:72, 2000.
Flake AW: The fetus with sacrococcygeal teratoma. *In* Harrison MR, Evans MI, Adzick NS, Holzgrev W (eds): *The Unborn Patient: The Art and Science of Fetal Therapy*, 3rd ed. Philadelphia, WB Saunders, 2001.

Fulop V, Mok SC, Gati I, Berkowitz RS: Recent advances in molecular biology of gestational trophoblastic diseases. *J Reprod Med* 47:369, 2002.
Roman BL, Weinstein BM: Building the vertebrate vasculature: research is going swimmingly. *BioEssays* 22:882, 2000.
Sabourin LA, Rudnicki MA: The molecular regulation of myogenesis. *Clin Genet* 57:16, 2000.
Seckl MJ, Fisher RA, Salerno G et al: Choriocarcinoma and partial hydatidiform moles. *Lancet* 356:36, 2000.
Smith JL, Schoenwolf GC: Neurulation: coming to closure. *TINS* 20:510, 1997.

Chapter 6

Barnea ER, Hustin J, Jauniaux E (eds): *The First Twelve Weeks of Gestation*. Berlin, Springer-Verlag, 1992.
Dickey RP, Gasser RF: Computer analysis of the human embryo growth curve: differences between published ultrasound findings on living embryos in utero and data on fixed specimens. *Anat Rec* 237:400, 1993.
Dickey RP, Gasser RF: Ultrasound evidence for variability in the size and development of normal human embryos before the tenth post-insemination week after assisted reproductive technologies. *Hum Reprod* 8:331, 1993.
Filly RA, Crane JP: Routine obstetric sonography. *J Ultrasound Med* 21:713, 2002.
Hay JC, Persaud TVN: Normal embryonic and fetal development. *In* Reece EA, Hobbins JC (eds): *Medicine of the Fetus and Mother*, 2nd ed. Philadelphia, Lippincott-Raven, 1999.
Iffy L, Shepard TH, Jakobovits A et al: The rate of growth in young human embryos of Streeter's horizons XIII and XXIII. *Acta Anat* 66:178, 1967.
Jirasek JE (ed): *An Atlas of the Human Embryo*. New York, Parthenon, 2000.
Kalousek DK, Fitch N, Paradice BA: *Pathology of the Human Embryo and Previable Fetus: An Atlas*. New York, Springer-Verlag, 1990.
Laing FC, Frates MC: Ultrasound evaluation during the first trimester of pregnancy. *In* Callen PW (ed): *Ultrasonography in Obstetrics and Gynecology*, 4th ed. Philadelphia, WB Saunders, 2000.
Moore KL, Persaud TVN, Shiota K: *Color Atlas of Clinical Embryology*, 2nd ed. Philadelphia, WB Saunders, 2000.
Nishimura H, Tanimura T, Semba R, Uwabe C: Normal development of early human embryos: observation of 90 specimens at Carnegie stages 7 to 13. *Teratology* 10:1, 1974.
O'Rahilly R, Müller F: *Developmental Stages in Human Embryos*. Washington, Carnegie Institute of Washington, 1987.
Schats R, Van Os HC, Jansen CAM, Wladimiroff JW: The crown-rump length in early human pregnancy: a reappraisal. *Br J Obstet Gynaecol* 98:460, 1991.
Shih J-C, Jaffe R, Hsieh F-J: Three-dimensional ultrasonography in early pregnancy. *Semin Perinatol* 25:3, 2001.

Chapter 7

ACOG (ACOG Practice Bulletin): Intrauterine growth retardation. *Int J Gynaecol Obstet* 72:85, 2001.
England MA: *Life Before Birth*, 2nd ed. London, Mosby-Wolfe, 1996.
Evans JA, Hamerton JL: Limb defects and chorionic villus sampling. *Lancet* 347:484, 1996.
Hay W Jr, Catz CS, Grave GD, Yaffe SY: Workshop summary: fetal growth: its regulation and disorders. *Pediatrics* 99:585, 1997.
Harrison MR, Evans MI, Adzick NS et al (eds): *The Unborn Patient*, 3rd ed. Philadelphia, WB Saunders, 2001.
Hinrichsen KV (ed): *Human embryologie*. Berlin, Springer-Verlag, 1990.
Kalousek DK, Fitch N, Paradice BA: *Pathology of the Human Embryo and Previable Fetus. An Atlas*. New York, Springer-Verlag, 1990.
Nussbaum RL, McInnes RR, Willard HF: *Thompson & Thompson Genetics in Medicine*, 6th ed. Philadelphia, WB Saunders, 2001.
Resnik R: Intrauterine growth restriction. *Obstet Gynecol* 99:490, 2002.
Sawin SW, Morgan MA: Dating of pregnancy by trimesters: a review and reappraisal. *Obstet Gynecol Surv* 51:261, 1996.
Shiota K: Development and intrauterine fate of normal and abnormal human conceptuses. *Congen Anom* 31:67, 1991.
Vohr BR (ed): Outcome of the very low-birth weight infant. *Clin Perinatol* 27:1, 2000.

Chapter 8

Battaglia FC: Fetoplacental perfusion and transfer of nutrients. *In* Reece EA, Hobbins JC (eds): *Medicine of the Fetus and Mother*, 2nd ed. Philadelphia, Lippincott-Raven, 1999.
Benirschke K, Kaufmann P: *Pathology of the Human Placenta*, 4th ed. New York, Springer-Verlag, 2000.
Berghella V, Kaufmann M: Natural history of twin-twin transfusion syndrome. *J Reprod Med* 46:480, 2001.
Collins JH: Umbilical cord accidents: human studies. *Seminar Perinatol* 26:79, 2002.
Cross JC: Formation of the placenta and extraembryonic membranes. *Ann NY Acad Sci* 857:23, 1998.
Dockery P, Bermingham J, Jenkins D: Structure-function relations in the human placenta. *Biochem Soc Trans* 28:202, 2000.
Doubilet PM, Benson CB: Ultrasound evaluation of amniotic fluid. *In* Callen PW (ed): *Ultrasonography in Obstetrics and Gynecology*, 4th ed. Philadelphia, WB Saunders, 2000.
Filly RA: Ultrasound evaluation during the first trimester. *In* Callen PW (ed): *Ultrasonography in Obstetrics and Gynecology*, 4th ed. Philadelphia, WB Saunders, 2000.
Foidart J-M, Hustin J, Dubois M, Schaaps J-P: The human placenta becomes haemochorial at the 13th week of pregnancy. *Int J Dev Biol* 36:451, 1992.
Fox H: The placenta, membranes and umbilical cord. *In* Chamberlain G (ed): *Turnbull's Obstetrics*, 2nd ed. Edinburgh, Churchill Livingstone, 1995.
Kraemer K, Noerr B: Placental transfer of drugs. *J Obstet Gynecol Neonatal Nurs (Neonatal Network)* 16:65, 1997.
Lala PK: Similarities between immunoregulation in pregnancy and malignancy: the role of prostaglandin E2 (Editorial). *Am J Reprod Immunol* 20:147, 1990.
Lewis SH, Perrin E: *Pathology of the Placenta*, 2nd ed. New York, Churchill Livingstone, 1998.

Loke YW, King A: Immunological aspects of human implantation. *J Reprod Fertil* (Suppl) 55:83, 2000.
Lysiak JJ, Lala PK: In situ localization and characterization of bone marrow-derived cells in the decidua of normal murine pregnancy. *Biol Reprod* 47:603, 1992.
Morgan BP, Holmes CH: Immunology of reproduction: protecting the placenta. *Current Biol* 10:R381, 2000.
Schmidt W: *The Amniotic Fluid Compartment: The Fetal Habitat*. Berlin, Springer, 1992.
Smith R (ed): *The Endocrinology of Parturition. Basic Science and Clinical Application*. Basel, Karger, 2001.
Spencer R: Theoretical and analytical embryology of conjoined twins: Part I: Embryogenesis. *Clin Anat* 13:36, 2000a.
Spencer R: Theoretical and analytical embryology of conjoined twins: Part I: Adjustments to union. *Clin Anat* 13:97, 2000b.
Townsend RR: Ultrasound evaluation of the placenta and umbilical cord. *In* Callen PW (ed): *Ultrasonography in Obstetrics and Gynecology*, 4th ed. Philadelphia, WB Saunders, 2000.

Chapter 9

Autti-Ramo I: Fetal alcohol syndrome — a multifaceted condition. *Dev Med Child Neurol* 44:141, 2002.
Briggs GG, Freeman RK, Yaffe SJ: *Drugs in Pregnancy and Lactation*, 5th ed. Baltimore, Williams & Wilkins, 2000.
Carvalho MHB, Brizot ML, Lopes LM et al: Detection of fetal structural abnormalities at the 11-14 week ultrasound scan. *Prenat Diagn* 22:1, 2002.
Chodirker BN, Chudley AE, Reed MH, Persaud TVN: Possible prenatal hydantoin effect in a child born to a nonepileptic mother. *Am J Med Genet* 27:373, 1987.
Chudley AE, Hagerman RJ: The fragile X syndrome. *J Pediatr* 110:821, 1987.
Cohen MM Jr: *The Child with Multiple Birth Defects*, 2nd ed. New York, Oxford University Press, 1997.
Gilbert GL: Infections in pregnant women. *MJA* 176:229, 2002.
Gregg NM: Congenital cataract following German measles in the mother. *Trans Ophthalmol Soc Aust* 3:35, 1941.
Hall JG: Chromosomal clinical abnormalities. *In* Behrman RE, Kliegman RM, Arvin AM (eds): *Nelson Textbook of Pediatrics*, 15th ed. Philadelphia, WB Saunders, 1996.
Jones KL: *Smith's Recognizable Patterns of Human Malformation*, 5th ed. Philadelphia, WB Saunders, 1997.
Karthikeyan K, Thappa DM: Early congenital syphilis in the new millennium. *Pediatr Dermatol* 19:275, 2002.
Kendrick JS, Merritt RK: Women and smoking: an update for the 1990s. *Am J Obstet Gynecol* 175:528, 1996.
Kliegman RM: Teratogens. *In* Behrman RE, Kliegman RM, Arvin AM (eds): *Nelson Textbook of Pediatrics*, 15th ed. Philadelphia, WB Saunders, 1996.
Levy HL, Guldberg P, Guttler F et al: Congenital heart disease in maternal phenylketonuria: report from the maternal PKU collaborative study. *Pediatr Res* 49:636, 2001.
Medicodes' Hospital and Payer: International Classification of Diseases, 9th Revision. *Clinical Modification*, 4th ed, vols 1–3. Salt Lake City, Medicode, Inc., 1995.
Nussbaum RL, McInnes RR, Willard HF: *Thompson & Thompson Genetics in Medicine*, 6th ed. Philadelphia, WB Saunders, 2001.
Persaud TVN: *Environmental Causes of Human Birth Defects*. Springfield, Charles C Thomas, 1990.
Remington JS, Klein JO (eds): *Infectious Diseases of the Fetus and Newborn Infant*, 4th ed. Philadelphia, WB Saunders, 1995.
Shepard TH: *Catalog of Teratogenic Agents*, 10th ed. Baltimore, The Johns Hopkins University Press, 2001.
Spranger J, Benirschke K, Hall JG et al: Errors of morphogenesis: concepts and terms. *J Pediatr* 100:160, 1982.
Tyrala EE: The infant of the diabetic mother. *Obstet Gynecol Clin North Am* 23:221, 1996.

Chapter 10

Braby J: Current and emerging treatment for congenital diaphragmatic hernia. *Neonatal Network* 20:5, 2001.
Harrison MR: The fetus with a diaphragmatic hernia. *In* Harrison MR, Evans MI, Adzick NS, Holzgreve W (eds): *The Unborn Patient: The Art and Science of Fetal Therapy*, 3rd ed. Philadelphia, WB Saunders, 2001.
Lally KP et al: (The Congenital Diaphragmatic Hernia Study Group): Estimating disease severity of congenital diaphragmatic hernia in the first five minutes of life. *J Pediatr Surg* 36:141, 2001.
Moore KL, Dalley AF: *Clinically Oriented Anatomy*, 4th ed. Baltimore, Williams & Wilkins, 1999.
Quah BS, Hashim I, Simpson H: Bochdalek diaphragmatic hernia presenting with acute gastric dilatation. *J Pediatr Surg* 34:512, 1999.
Schlembach D, Zenker M, Trautmann U et al: Deletion 15q24-26 in prenatally detected diaphragmatic hernia: increasing evidence of a candidate region for diaphragmatic development. *Prenat Diagn* 21:289, 2001.
Wells LJ: Development of the human diaphragm and pleural sacs. *Contr Embryol Carneg Instn* 35:107, 1954.

Chapter 11

Benacerraf BR: Ultrasound evaluation of the fetal face. *In* Callen PW (ed): *Ultrasonography in Obstetrics and Gynecology*, 4th ed. Philadelphia, WB Saunders, 2000.
Fisher DA, Polk DH: Development of the thyroid. *Bailliere's Clin Endocrin Metabol* 3:627, 1989.
Garg V, Yamagishi C, Hu T et al: *Tbx1*, a DiGeorge syndrome candidate gene, is regulated by sonic hedgehog during pharyngeal arch development. *Dev Biol* 235:62, 2001.
Gorlin RJ, Cohen MM Jr, Levin LS: *Syndromes of the Head and Neck*, 3rd ed. New York, Oxford University Press, 1990.
Hall BK: *The Neural Crest in Development and Evolution*. New York, Springer-Verlag, 1999.
Hinrichsen K: The early development of morphology and patterns of the face in the human embryo. *Adv Anat Embryol Cell Biol* 98:1–79, 1985.
Jones KL: *Smith's Recognizable Patterns of Human Malformation*, 5th ed. Philadelphia, WB Saunders, 1997.
Lele SM, Lele MS, Anderson VM: The thymus in infancy and childhood. *Chest Surg Clin North Am* 11:233, 2001.

Moore KL, Dalley AF: *Clinically Oriented Anatomy*, 4th ed. Baltimore, Williams & Wilkins, 1999.
Nishimura Y: Embryological study of nasal cavity development in human embryos with reference to congenital nostril atresia. *Acta Anat* 147:140, 1993.
Noden DM: Vertebrate craniofacial development: novel approaches and new dilemmas. *Curr Opin Genet Dev* 2:576, 1992.
Pleifer G (ed): *Craniofacial Abnormalities and Clefts of the Lip, Alveolus and Palate*. New York, Greorg Thieme Verlag, 1991.
Sandham A, Nelson R: Embryology of the middle third of the face. *Early Hum Dev* 10:313, 1985.
Sperber GH: *Craniofacial Development*. Hamilton, BC Decker, 2001.
Vermeij-Keers C: Craniofacial embryology and morphogenesis: normal and abnormal. *In* Stricker M, Van der Meulen JC, Raphael B, Mazzola R (eds): *Craniofacial Malformations*. Edinburgh, Churchill Livingstone, 1990.
Zalel Y, Gamzu R, Mashiach S et al: The development of the fetal thymus: an *in utero* sonographic evaluation. *Prenat Diagn* 22:114, 2002.

Chapter 12

Adzick NS: The fetus with a lung mass. *In* Harrison MR, Evans MI, Adzick NS, Holzgreve W (eds): *The Unborn Patient*, 3rd ed. Philadelphia, WB Saunders, 2001.
Beech DJ, Sibbons PD, Howard CV et al: Terminal bronchiolar duct ending number does not increase post-natally in normal infants. *Early Hum Dev* 59:193, 2000.
Bratu I, Flageole H, Chen M-F et al: The multiple facets of pulmonary sequestration. *J Pediatr Surg* 36:784, 2001.
Evans JA, Greenberg CR, Erdile L: Tracheal agenesis revisited: analysis of associated anomalies. *Am J Med Genet* 82:415, 1999.
Halliday H: Prevention and management of respiratory distress syndrome. *In* Levitt G et al (eds): *Practical Perinatal Care: The Baby Under 1000 Grams*. Oxford, Butterworth Heinemann, 1999.
Hume R, Richard K, Kaptein E et al: Thyroid metabolism and the developing human lung. *Biol Neonate* 80(Suppl 1):18, 2001.
Krummel TM: Congenital malformations of the lower respiratory tract. *In* Chernick V (ed): *Kendig's Disorders of the Respiratory Tract in Children*, 6th ed. Philadelphia, WB Saunders, 1998.
O'Rahilly R, Boyden E: The timing and sequence of events in the development of the human respiratory system during the embryonic period proper. *Z Anat Entwicklungsgesch* 141:237, 1973.
Perl AKT, Whitsett JA: Molecular mechanisms controlling lung morphogenesis. *Clin Genet* 56:14, 1999.
Spencer C, Neales K: Antenatal corticosteroids to prevent neonatal respiratory distress syndrome. *BMJ* 320:325, 2000.
Warburton D, Schwarz M, Tefft D et al: The molecular basis of lung morphogenesis. *Mech Dev* 92:55, 2000.
Wells LJ, Boyden EA: The development of the bronchopulmonary segments in human embryos of horizons XVII and XIX. *Am J Anat* 95:163, 1954.
Whitsett JA, Wert SE: Molecular determinants of lung development. *In* Chernick V (ed): *Kendig's Disorders of the Respiratory Tract in Children*, 6th ed. Philadelphia, WB Saunders, 1998.

Chapter 13

Abu-Judeh HH, Methratta S, Ybasco A et al: Congenital colonic stenosis. *South Med J* 94:344, 2001.
Balistreri WF: Liver and biliary atresia. *In* Behrman RE, Kliegman RM, Arvin AM (eds): *Nelson Textbook of Pediatrics*, 15th ed. Philadelphia, WB Saunders, 1996.
Beck F, Tata F, Chawengsaksophak K: Homeobox genes and gut development. *BioEssays* 22:431, 2000.
Hill LM: Ultrasound of fetal gastrointestinal tract. *In* Callen PW (ed): *Ultrasonography in Obstetrics and Gynecology*, 4th ed. Philadelphia, WB Saunders, 2000.
Martucciello G, Ceccherini I, Lerone M et al: Pathogenesis of Hirschsprung's disease. *J Pediatr Surg* 35:1017, 2000.
Ramalho-Santos M, Melton DA, McMahon AP: Hedgehog signals regulate multiple aspects of gastrointestinal development. *Development* 127:2763, 2000.
Sparey C, Robson SC: Oesophageal atresia. *Prenat Diagn* 20:251, 2001.
Ulshen M: Stomach and intestines. *In* Behrman RE, Kliegman RM, Arvin AM (eds): *Nelson Textbook of Pediatrics*, 15th ed. Philadelphia, WB Saunders, 1996.
Zona JZ: Umbilical anomalies. *In* Raffensperger JG (ed): *Swenson's Pediatric Surgery*, 5th ed. Norwalk, Appleton & Lange, 1990.

Chapter 14

American Academy of Pediatrics: Evaluation of the newborn with developmental anomalies of the external genitalia. *Pediatrics* 106:138, 2000.
Belman AB: Hypospadias update. *Urology* 49:166, 1997.
Byrne J, Nussbaum-Blask A, Taylor WS et al: Prevalence of Müllerian duct anomalies detected by ultrasound. *Am J Med Genet* 94:9, 2000.
DiGeorge AM: Hermaphroditism. *In* Behrman RE, Kliegman RM, Arvin AM (eds): *Nelson Textbook of Pediatrics*, 15th ed. Philadelphia, WB Saunders, 1996.
Filly RA, Feldstein VA: Ultrasound evaluation of the genitourinary system. *In* Callen PW (ed): *Ultrasonography in Obstetrics and Gynecology*, 4th. Philadelphia, WB Saunders, 2000.
Gill FT: Umbilical hernia, inguinal hernias, and hydroceles in children: diagnostic clues for optimal patient management. *J Pediatr Health Care* 12:231, 1998.
Habert R, Lejeune H, Saez JM: Origin, differentiation and regulation of fetal and adult Leydig cells. *Mol Cell Endocrinol* 179:47, 2001.
Martinez-Frias ML, Bermejo E, Rodriguez-Pinilla E et al: Exstrophy of the cloaca and exstrophy of the bladder: two different expressions of a primary field defect. *Am J Med J* 99:261, 2001.
Neri G, Opitz J: Syndromal (and nonsyndromal) forms of male pseudohermaphroditism. *Am J Med Genet (Semin Med Genet)* 89:201, 1999.
Persaud TVN: Embryology of the female genital tract and gonads. *In* Copeland LJ, Jarrell J (eds): *Textbook of Gynecology*, 2nd ed. Philadelphia, WB Saunders, 2000.
Piscione TD, Rosenblum ND: The malformed kidney: how gene mutations perturb developmental pathways. *Frontiers in Fetal Health* 2:14, 2000.

Ward CJ, Hogan MC, Rossetti S et al: The gene mutated in autosomal recessive polycystic kidney disease encodes a large, receptor-like protein. *Nature Genet* 30:259, 2002.
Witschi E: Migration of the germ cells of human embryos from the yolk sac to the primitive gonadal folds. *Contr Embryol Carneg Instn* 32:67, 1948.
Woolf AS: A molecular and genetic view of human renal and urinary tract malformations. *Kidney Int* 58:500, 2000.

Chapter 15

Amato JJ, Douglas WI, Desai U et al: Ectopia cordis. *Chest Surg Clin North Am* 10:297, 2000.
Anderson RH, Webb S, Brown NA: Clinical anatomy of the atrial septum with reference to its developmental components. *Clin Anat* 12:362, 1999.
Belmont JW: Recent progress in the molecular genetics of congenital heart defects. *Clin Genet* 54:11, 1998.
Bernstein E: The cardiovascular system. *In* Behrman RE, Kliegman RM, Arvin AM (eds): *Nelson Textbook of Pediatrics*, 15th ed. Philadelphia, WB Saunders, 1996.
Cohen MS: Fetal diagnosis and management of congenital heart disease. *Clin Perinatol* 28:22, 2001.
Conway SJ, Kruzynska-Frejtag A, Kneer PL et al: What cardiovascular defect does my prenatal mouse mutant have, and why? *Genesis* 2002 (in press).
Harvey RP, Rosenthal N: *Heart Development*. Orlando, Academic Press, 1999.
Harvey RP: Seeking a regulatory roadmap for heart morphogenesis. *Cell Develop Biol* 10:99, 1999.
Le Douarin NM, Kalcheim C: *The Neural Crest*, 2nd ed. Cambridge, Cambridge University Press, 1999.
Moore KL, Dalley AF: *Clinically Oriented Anatomy*, 4th ed. Baltimore, Williams & Wilkins, 1999.
Olson EN: The path to the heart and the road not taken. *Science* 291:2327, 2001.
Roman BL, Weinstein BM: Building the vertebrate vasculature: research is going swimmingly. *BioEssays* 22:882, 2000.
Sansoucie DA, Cavaliere TA: Transition from fetal to extrauterine circulation. *Neonatal Network* 16:5, 1997.
Silverman NH, Schmidt KG: Ultrasound evaluation of the fetal heart. *In* Callen PW (ed): *Ultrasonography in Obstetrics and Gynecology*, 4th ed. Philadelphia, WB Saunders, 2000.
Srivastava D, Olson EN: A genetic blueprint for cardiac development. *Nature* 407:221, 2000.
Westmoreland D: Critical congenital cardiac defects in the newborn. *J Perinat Neonat Nurs* 12:67, 1999.
Yutzey KE, Kirby ML: Wherefore heart thou? Embryonic origins of cardiogenic mesoderm. *Dev Dyn* 223:307, 2002

Chapter 16

Brooks CGD, de Vries BBA: Skeletal dysplasias. *Arch Dis Child* 79:285, 1998.
Budorick NE: The fetal musculoskeletal system. *In* Callen PW (ed): *Ultrasonography in Obstetrics and Gynecology*, 4th ed. Philadelphia, WB Saunders, 2000.
Cohen MM Jr, MacLean RE (eds): *Craniosynostosis: Diagnosis, Evaluation, and Management*, 2nd ed. New York, Oxford University Press, 2000.
Erlebacher A, Filvaroff EH, Gitelman SE et al: Toward a molecular understanding of skeletal development. *Cell* 80:371, 1995.
Hall BK, Miyake T: All for one and one for all: condensation and the initiation of skeletal development. *BioEssays* 22:138, 2000.
Long F, Schipani E, Asahara H et al: The CREB family of activators is required for endochondral bone development. *Development* 128:541, 2001.
Marsh J: Set down in bone. *BioEssays* 22:402, 2000.
Moore KL, Dalley AF: *Clinically Oriented Anatomy*, 4th ed. Baltimore, Williams & Wilkins, 1999.
O'Rahilly R, Müller F, Meyer DB: The human vertebral column at the end of the embryonic period proper. 3. The thoracolumbar region. *J Anat* 168:81, 1990a.
O'Rahilly R, Müller F, Meyer DB: The human vertebral column at the end of the embryonic period proper. 4. The sacrococcygeal region. *J Anat* 168:95, 1990b.
Sadler TW: Embryology of the sternum. *Chest Surg Clin North Am* 10:237, 2000.
Sperber GH: *Craniofacial Development*. Hamilton, BC Decker, 2001.

Chapter 17

Arnold HH, Braun T: Genetics of muscle determination and development. *Curr Top Dev Biol* 48:129, 2000.
Budorick NE: The fetal musculoskeletal system. *In* Callen PW (ed): *Ultrasonography in Obstetrics and Gynecology*, 4th ed. Philadelphia, WB Saunders, 2000.
Cheng JCY, Tang SP, Chen MWN et al: The clinical presentation and outcome of treatment of congenital muscular torticollis in infants — a study of 1,086 cases. *J Pediatr Surg* 35:1091, 2000.
Gasser RF: The development of the facial muscles in man. *Am J Anat* 120:357, 1967.
Kablar B, Krastel K, Ying C et al: Myogenic determination occurs independently in somites and limb buds. *Dev Biol* 206:219, 1999.
Noden DM: Vertebrate craniofacial development — the relation between ontogenetic process and morphological outcome. *Brain Behav Evol* 38:190, 1991.
Nussbaum RL, McInnes RR, Willard HF: *Thompson & Thompson Genetics in Medicine*, 6th ed. Philadelphia, WB Saunders, 2001.
O'Rahilly R, Gardner E: The timing and sequence of events in the development of the limbs of the human embryo. *Anat Embryol* 148:1, 1975.
Ordahl CP, Williams BA: Knowing chops from chuck: roasting MyoD redundancy. *BioEssays* 20:357, 1998.
Sabourin LA, Rudnicki MA: The molecular regulation of myogenesis. *Clin Genet* 57:16, 2000.

Chapter 18

Ambler CA, Nowicki JL, Burke AC et al: Assembly of trunk and limb blood vessels involves extensive migration and vasculogenesis of somite-derived angioblasts. *Dev Biol* 234:352, 2001.

Brook WJ, Diaz-Benjumea FJ, Cohen SM: Organizing spatial pattern in limb development. *Ann Rev Cell Develop Biol* 12:161, 1996.

Budorick NE: The fetal musculoskeletal system. *In* Callen PW (ed): *Ultrasonography in Obstetrics and Gynecology*, 4th ed. Philadelphia, WB Saunders, 2000.

Cohn MJ, Patel K, Krumlauf R et al: *HOX* 9 genes and vertebrate limb specification. *Nature* 387:97, 1997.

Hinrichsen KV, Jacob HJ, Jacob M et al: Principles of ontogenesis of leg and foot in man. *Ann Anat* 176:121, 1994.

Marini JC, Gerber NL: Osteogenesis imperfecta. *JAMA* 277:746, 1997.

Martin GR: The roles of FGFs in the early development of vertebrate limbs. *Genes Dev* 12:1571, 1998.

Moore KL, Dalley AF: *Clinically Oriented Anatomy*, 4th ed. Baltimore, Williams & Wilkins, 1999.

Muragaki Y, Mundlos S, Upton J, Olsen BR: Altered growth and branching patterns in synpolydactyly caused by mutations in HOXD13. *Science* 272:548, 1996.

Nussbaum RL, McInnes RR, Willard HF: *Thompson & Thompson Genetics in Medicine*, 6th ed. Philadelphia, WB Saunders, 2001.

O'Rahilly R, Müller F: *Developmental Stages in Human Embryos*. Washington, Carnegie Institution of Washington, 1987.

Riddle RD, Tabin CJ: How limbs develop. *Sci Am* 280:74, 1999.

Slack J: *Essential Developmental Biology*. Oxford, Blackwell Science, 2001.

Van Heest AE: Congenital disorders of the hand and upper extremity. *Pediatr Clin North Am* 43:1113, 1996.

Van Allen MI: Structural anomalies resulting from vascular disruption. *Pediatr Clin North Am* 39:255, 1992.

Watson S: The pinciples of management of congenital anomalies of the upper limb. *Arch Dis Child* 83:10, 2000.

Chapter 19

Evans OB, Hutchins JB: Development of the nervous system. *In* Haines DE (ed): *Fundamental Neuroscience*, 2nd ed. New York, Churchill Livingstone, 2002.

Harland R: Neural induction. *Curr Opin Genet Develop* 10:357, 2000.

Jobe AH: Fetal surgery for myelomeningocele. *N Engl J Med* 347:230, 2002.

LeDouarin N, Kalcheim C: *The Neural Crest*, 2nd ed. Cambridge University Press, 1999.

Moore KL, Persaud TVN: *The Developing Human: Clinically Oriented Embryology*, 7th ed. Philadelphia, WB Saunders, 2003.

Naftel JP, Hardy SGP: Visceral motor pathways. *In* Haines DE (ed): *Fundamental Neuroscience*, 2nd ed. New York, Churchill Livingstone, 2002.

Nakatsu T, Uwabe C, Shiota K: Neural tube closure in humans initiates at multiple sites: evidence from human embryos and implications for the pathogenesis of neural tube defects. *Anat Embryol* 201:455, 2000.

O'Rahilly R, Müller F: *Embryonic Human Brain. An Atlas of Developmental Stages*, 2nd ed. New York, Wiley-Liss, 1999.

O'Rahilly R, Müller F: The two sites of fusion of the neural folds and the two neuropores in the human embryo. *Teratology* 65:162, 2002.

Otake M, Schull WJ: *In utero* exposure to A-bomb radiation and mental retardation: a reassessment. *Br J Radiol* 52:409, 1984.

Pilu G, Falco P, Perolo A et al: Ultrasound evaluation of the fetal neural axis. *In* Callen PW (ed): *Ultrasonography in Obstetrics and Gynecology*, 4th ed. Philadelphia, WB Saunders, 2000.

Pooh RK, Pooh KH: Transvaginal 3D and doppler ultrasonography of the fetal brain. *Sem Perinatology* 25:38, 2001.

Sanes DH, Reh TA, Harris WA: *Development of the Nervous System*. New York, Academic Press, 2000.

Uher BF, Golden JA: Neuronal migration defects of the cerebral cortex: a destination debacle. *Clin Genet* 58:16, 2000.

Chapter 20

Barishak YR: *Embryology of the Eye and Its Adnexa*, 2nd ed. Basel, Karger, 2001.

Bauer PW, MacDonald CB, Melhem ER: Congenital inner ear malformation. *Am J Otology* 19:669, 1998.

Carlson BM: *Human Embryology and Developmental Biology*, 2nd ed. St. Louis, Mosby-Year Book, 1999.

Eagleson GW, Johnson-Meeter LJ, Frideres J: Effects of retinoic acid upon eye field morphogenesis and differentiation. *Dev Dyn* 221:350, 2001.

Mallo M: Formation of the middle ear: recent progress on the developmental and molecular mechanisms. *Dev Biol* 231:410, 2001.

Mathers PH, Grinberg A, Mahon KA, Jamrich M: The Rx homeobox gene is essential for vertebrate eye development. *Nature* 387:603, 1997.

McAvoy JW, Chamberlain CG, delongh RV et al: Lens development. *Eye* 13:425, 1999.

Nelson L: Disorders of the eye. *In* Behrman RE, Kliegman RM, Arvin AM (eds): *Nelson Textbook of Pediatrics*, 15th ed. Philadelphia, WB Saunders, 1996.

O'Rahilly R: The prenatal development of the human eye. *Exp Eye Res* 21:93, 1975.

Reardon W, Mueller RF: Inherited deafness in childhood — the genetic revolution unmasks the clinical challenge. *Arch Dis Child* 82:319, 2000.

Sellheyer K: Development of the choroid and related structures. *Eye* 4:255, 1990.

Twefik TL, Der Kaloustian VM (eds): *Congenital Anomalies of the Ear, Nose, and Throat*. Oxford, Oxford University Press, 1996.

Wawersik S, Maas RL: Vertebrate eye development as modeled in Drosophila. *Hum Mol Genet* 12:917, 2000.

Wright KW: Embryology and eye development. *In* Wright KW (ed): *Textbook of Ophthalmology*. Baltimore, Williams & Wilkins, 1997.

Chapter 21

Cobourne MT: The genetic control of early odontogenesis. *Brit J Orthodontics* 26:21, 1999.

Darmstadt GL, Lane AT: The skin. *In* Berhman RE, Kliegman RM, Arvin AM (eds): *Nelson Textbook of Pediatrics*, 15th ed. Philadelphia, WB Saunders, 1996.

Eichenfield LF, Frieden IJ, Esterly NB: *Textbook of Neonatal Dermatology*. Philadelphia, WB Saunders, 2001.

Johnsen DC: The oral cavity. *In* Behrman RE, Kliegman RM, Arvin AM (eds): *Nelson Textbook of Pediatrics*, 15th ed. Philadelphia, WB Saunders, 1996.

Johnson CL, Holbrook KA: Development of human embryonic and fetal dermal vasculature. *J Invest Dermatol* 93(Suppl):105, 1989.

Kishimoto J, Burgeson RE, Morgan BA: Wnt signaling maintains the hair-inducing activity of the dermal papilla. *Genes Dev* 14:1181, 2000.

LeDouran N, Kalcheim C: *The Neural Crest*, 2nd ed. Cambridge, Cambridge University Press, 1999.

Levy ML (ed): Pediatric Dermatology. *Ped Clin NA* 47:No. 4, 2000.

Moore SJ, Munger BL: The early ontogeny of the afferent nerves and papillary ridges in human digital glabrous skin. *Dev Brain Res* 48:119, 1989.

Müller M, Jasmin JR, Monteil RA, Loubiere R: Embryology of the hair follicle. *Early Hum Dev* 26:59, 1999.

Osborne MP: Breast anatomy and development. *In* Harris JR (ed): *Diseases of the Breast*, 2nd ed. Philadelphia, Lippincott Williams & Wilkins, 2000.

Sharpe PT: Homeobox genes in initiation and shape of tooth during development in mammalian embryos. *In* Teaford MF, Smith MM, Ferguson MW (eds): *Development, Function and Evolution of Tooth*. Cambridge, Cambridge University Press, 2000.

Sinclair RD, Banfield CC, Dawber RPR: *Handbook of Diseases of the Hair and Scalp*. London, Blackwell, 1999.

Sperber GH: *Craniofacial Development*. Hamilton, BC Decker, 2001.

Taylor G, Lehrer MS, Jensen PJ et al: Involvement of follicular stem cells in forming not only the follicle but also the epidermis. *Cell* 102:451, 2000.

Ten Cate AR: Development of the tooth. *In* Ten Cate AR (ed): *Oral Histology. Development, Structure, and Function*, 5th ed. St. Louis, CV Mosby, 1998.

Watts A, Addy MA: Tooth discolouration and staining: a review of the literature. *Br Dent J* 190:309, 2001.

Winter GB: Anomalies of tooth formation and eruption. In Welbury RR (ed): *Paediatric Dentistry*, 2nd ed. Oxford, Oxford University Press, 2001.

Answers to Clinically Oriented Questions

Chapter 1

1. You should not attempt to reproduce the timetable of development. It is presented as an overview of human development before birth. Neither should you try to memorize the criteria for the stages (e.g., that stage 3 begins on day 4). These stages are used by embryologists and clinicians when they are describing embryos in detail. You should, however, be able to describe human development to laypersons, and the sketches in the timetables are helpful when explaining embryonic and fetal development to them.
2. The term *conceptus* refers to the embryo and its membranes (amnion, chorion, yolk sac, and allantois). A conceptus refers to the products of conception, that is, everything that develops from the zygote. The *embryo* is the embryonic part of the conceptus.
3. Everyone, especially those in the health sciences, should know about conception, contraception, and how people develop, both normally and abnormally. Health professionals are expected to give intelligent answers to the questions people ask, such as, When does the baby's heart start to beat? When does it look like a human being? When does it move its limbs?
4. Animal and human embryos look similar for the first few weeks; for example, they both have pharyngeal arches and tail-like caudal eminences. After the seventh week, human embryos do not resemble animal embryos, mainly because the head has a human appearance and the caudal eminence disappears at the end of the embryonic period. The head of the human embryo is larger than in animal embryos.
5. Physicians date pregnancies from the last normal menstrual period (LNMP) because this date is usually remembered by women. It is not usually possible to detect the precise time of ovulation and fertilization (when development begins); however, tests and ultrasound imaging can be done to detect when ovulation is likely to occur and when pregnancy has occurred. These tests are not routinely performed because of the costs involved. When dating pregnancies using LNMP, physicians are aware that the age of the developing human is about 2 weeks less than the "menstrual or gestational age," and they base their decisions, such as determining the embryo's vulnerability to drugs, on estimated fertilization age.

Chapter 2

1. The hymen usually ruptures during the perinatal period, forming the vaginal orifice. This opening usually enlarges during childhood as the result of physical activity. Contrary to popular myth, the rupture of this mucous membrane or the absence of bleeding secondary to tearing during initial intercourse is not necessarily an indication of the loss of virginity.
2. The term *erection* is rarely used when referring to the sexual excitement of a female; however, the clitoris — homologous to the penis — enlarges ("erects") when it is stimulated and the female is sexually aroused. It is a highly sensitive female sex organ.
3. Pregnant women do not menstruate even though there may be some bleeding at the usual time of menstruation. This blood leaks from the intervillous space because of partial separation of the placenta. Because there is no shedding of the endometrium, this is not menstrual fluid. In unusual cases, periodic bleeding may occur every month during pregnancy; again, this is loss of blood from the placenta.
4. It depends on when she forgot to take the pill. If it was at midcycle, ovulation might occur and pregnancy could result. The taking of two pills the next day would not prevent ovulation.
5. *Coitus interruptus* refers to withdrawal of the penis from the vagina before ejaculation occurs. This method depends on the self-discipline of the couple to part before climax of the male (i.e., ejaculation). Not only is this difficult to do, but it is neither reliable nor psychologically acceptable. Often, a few sperms are expelled from the penis with the secretions of the auxiliary sex glands (e.g., the seminal glands) before ejaculation occurs. One of these sperms may fertilize the oocyte.
6. *Spermatogenesis* refers to the complete process of sperm formation. *Spermiogenesis* is the transformation of a spermatid into a sperm. Therefore, spermiogenesis is the final stage of spermatogenesis.
7. An IUD may inhibit the capacitation of sperms and their transport through the uterus to the fertilization site in the uterine tube; in this case, it would be a *contraceptive device.* More likely, the IUD produces endometrial changes that present a hostile environment for the blastocyst; as a result, the blastocyst does not implant. In this case, the IUD would be a *contraimplantation device* that results in death of the embryo when it is a week or so old.
8. *Menopause* is the permanent cessation of the menses — the periodic physiological hemorrhage from the uterus at approximately 4-week intervals. *Climateric* is the period of psychological changes occurring in the transition to menopause.

Chapter 3

1. The ovarian and endometrial (menstrual) cycles cease between 47 and 55 years of age, with the average age being 48 years. Menopause results from the gradual cessation of gonadotropin production by the pituitary gland; however, it does not mean that the ovaries have exhausted their supply of oocytes. The risk of Down syndrome and other trisomies is increased in

the children of women who are 35 years of age or older (see Chapter 9). *Spermatogenesis also decreases after the age of 45 years,* and the number of nonviable and abnormal sperms increases. Nevertheless, sperm production continues until old age. The risk of producing abnormal gametes is much less common in men than in women; however, older men may accumulate mutations that the child might inherit. Mutations may produce congenital anomalies (see Chapter 9).

2. Considerable research on new contraceptive methods is being conducted, including the development of contraceptive pills for men. This research includes experimental work on nonhormonal prevention of spermatogenesis and stimulation of immune responses to sperms. Arresting the development of millions of sperms on a continuous basis is much more difficult than arresting the monthly development of a single oocyte monthly.
3. It is not known whether polar bodies are ever fertilized; however, it has been suggested that dispermic chimeras result from fusion of a fertilized oocyte with a fertilized polar body. *Chimeras* are rare individuals who are composed of a mixture of cells from two zygotes. More likely, dispermic chimeras result from the fusion of dizygotic (DZ) twin zygotes early in development. DZ twins are derived from two zygotes. If a polar body were fertilized and remained separate from the normal zygote, it could form an embryo.
4. The most common cause of spontaneous abortions during the first week of development is chromosomal abnormalities, such as those resulting from nondisjunction (see Chapter 2). Failure of production of an adequate amount of hCG by the blastocyst to maintain the corpus luteum in the ovary could also result in an early spontaneous abortion.
5. Yes, it is possible; however, this phenomenon is extremely rare. The term *superfecundation* indicates fertilization (during separate acts of coitus) of two or more oocytes that are ovulated at approximately the same time. The possibility of this process occurring in humans cannot be discounted because evidence exists from DZ (nonidentical) twins belonging to different blood groups, which cannot be accounted for in any other way.
6. The differences are slight. *Conception* means to become pregnant. *Fertilization* occurs when a sperm fuses with an oocyte; when this occurs, conception takes place. *Impregnation* means to make pregnant (as in a male impregnating a female).
7. Essentially, they do mean the same thing. *Mitosis* is the usual process of cell reproduction that results in the formation of daughter cells of the zygote. *Cleavage* is the series of mitotic cell divisions of the zygote. This process results in the formation of daughter cells, or *blastomeres*. The expressions "cleavage division" and "mitotic division" mean the same when referring to the dividing zygote.
8. The nutritional requirements of the dividing zygote are not great. The blastomeres may be nourished partly by the sparse yolk granules in these cells; however, the nutrients are derived mainly from the secretions of the uterine tubes and later, from the uterine glands.
9. Yes. One of the blastomeres could be removed, and the Y chromosome could be identified by staining the cell with quinacrine mustard (see Chapter 7). Blastomeres of a female embryo lack a fluorescent body (Y chromosome). This technique could be made available to couples with a family history of sex-linked genetic diseases (e.g., hemophilia or muscular dystrophy), and to women who have already given birth to a child with such a disease and are reluctant to have more children. In these cases, only female embryos developing in vitro would be transferred to the uterus.

Chapter 4

1. "Implantation bleeding" refers to the loss of small amounts of blood from the implantation site of a blastocyst that occurs a few days after the expected time of menstruation. Persons unfamiliar with this possible occurrence may interpret the bleeding as a light menstrual flow; in such cases, they would give the physician the wrong date for their LNMP. This blood is not menstrual fluid; it is blood from the intervillous space of the developing placenta. The loss of blood could also result from the rupture of chorionic arteries or veins, or both (see Chapter 8).
2. Drugs or other agents may cause abortion of the embryo, but they do not cause congenital anomalies if taken during the first 2 weeks of development. A drug or other agent either damages all the embryonic cells, killing the embryo, or it injures only a few cells and the embryo recovers to develop normally.
3. The insertion of an *IUD* usually prevents implantation of a blastocyst in the uterus; however, it does not prevent a sperm from entering the uterine tube and fertilizing an oocyte if one is present. Because the endometrium is hostile to implantation, a blastocyst could develop and implant in the uterine tube (i.e., an ectopic tubal pregnancy).
4. Abdominal pregnancies are very uncommon. Although such a pregnancy can result from primary implantation of a blastocyst in the abdomen, in most cases, it is believed to result from the ectopic implantation of a blastocyst that spontaneously aborted from the uterine tube and entered the peritoneal cavity. The risk of severe maternal bleeding and fetal mortality is

high in cases of abdominal pregnancy. However, if the diagnosis is made late in pregnancy and the patient (mother) is free of symptoms, the pregnancy may be allowed to continue until the viability of the fetus is ensured (e.g., 32 weeks), at which time it would be delivered by cesarean section.

5. Yes, it can occur but it is very rare. An intrauterine and ectopic tubal pregnancy is most common.

Chapter 5

1. Yes they do, if they become pregnant soon after they stop taking the pills. It takes from 1 to 3 months for normal menstrual cycles to occur. If pregnancy occurs before this time, a spontaneous abortion may occur a week or so after the first missed menstrual period. Most such embryos have been found to have severe chromosomal abnormalities. For this reason, most physicians recommend that other contraceptive techniques be used for 2 to 3 months after cessation of birth control pills in order to allow normal menstrual cycles to occur.
2. "Menstrual extraction" or aspiration refers to suction or vacuum curettage of the uterus, usually within 5 to 8 weeks after a missed menstrual period. Menstrual extraction or aspiration is often a euphemistic term for an early therapeutic or elective abortion. The conceptus is evacuated using an electrically powered vacuum source.
3. Yes, certain drugs can produce congenital anomalies if administered during the third week after the LNMP (see Chapter 9). Antineoplastic agents (antitumor drugs) can produce severe skeletal and neural tube defects in the embryo, such as acrania and meroanencephaly (partial absence of brain), if administered during the third week.
4. Yes, risks to the mother aged 40 years or older and her embryo are increased. Increased maternal age is a predisposing factor to certain medical conditions (e.g., kidney disorders and hypertension). Preeclampsia, a hypertensive disorder of pregnancy characterized by increased blood pressure and edema, for example, occurs more frequently in older pregnant women than in younger ones. Advanced maternal age is also associated with a significantly increased risk to the embryo/fetus. Most common are birth defects associated with chromosomal abnormalities, such as Down syndrome and trisomy 13 (see Chapter 9); however, women older than 40 may have normal children.

Chapter 6

1. Early in the eighth week, embryos look different from 9-week fetuses because of their webbed toes and stubby, tail-like caudal eminences; however, by the end of the eighth week, embryos and early fetuses appear similar. The name change is made to indicate that a new phase of development (rapid growth and differentiation) has begun and that the most critical period of development is completed.
2. The question of when an embryo becomes a human being is difficult to answer because views are affected by religious and personal views. The scientific answer is that the embryo has human potential, and no other, from the time of fertilization because of its human chromosome constitution.
3. No it cannot. During the embryonic period, more similarities than differences exist in the external genitalia (see Chapter 14). It is impossible to tell by ultrasound examination whether the primordial sexual organ (genital tubercle at 5 weeks and phallus at 7 weeks) will become a penis or a clitoris. Sexual differences are not clear until the early fetal period (10th to 12th week). Sex chromatin patterns and chromosomal analysis (fluorescence in situ hybridization, or *FISH*) of embryonic cells obtained during amniocentesis can reveal the chromosomal sex of the embryo (see Chapter 7).

Chapter 7

1. Ultrasound examinations have shown that mature embryos (8 weeks) and young fetuses (9 weeks) show spontaneous movements, such as twitching of the trunk and limbs. Although the fetus begins to move its back and limbs during the 12th week, the mother cannot feel her baby move until the 16th to 20th week. Women who have had several children (multigravida) usually detect this movement, called *quickening,* sooner than women who are pregnant for the first time because they know what fetal movements feel like. Quickening is the first time a woman recognizes fetal movement, which is often perceived as a faint flutter (quivering motion).
2. Folic acid supplementation before conception and during early pregnancy is effective in reducing the incidence of NTDs (e.g., spina bifida). It has been shown that the risk of having a child with NTD is significantly lower when a vitamin supplement containing 400 µg of folic acid is consumed daily. However, no consensus exists that vitamins are helpful in preventing these defects in most at-risk pregnancies.
3. There is no risk of damaging the fetus during amniocentesis when ultrasonography is used to locate the position of the fetus so that the needle will not injure it. The risk of inducing an abortion is slight (about 0.5%). Maternal or fetal infection is an unlikely complication when the procedure is performed by a trained person using modern techniques.

Chapter 8

1. A stillbirth is the birth of an infant that was dead prior to delivery and weighs at least 500 gm and is at least 20 weeks old. A stillborn shows no evidence of life. The incidence of having a stillborn infant is about three times greater among mothers older than 40 years of age than among women in their 20s. It is true that more male fetuses are stillborn than females. The reason for this is unknown.
2. Sometimes the umbilical cord is abnormally long, and wraps around part of the fetus, such as the neck or a limb. This "cord accident" obstructs blood flow in the umbilical vein to the fetus and in the umbilical arteries from the fetus to the placenta. If the fetus does not receive sufficient oxygen and nutrients, it dies. A true knot in the umbilical cord, formed when the fetus passes through a loop in the cord, also obstructs blood flow through the cord. *Prolapse of the umbilical cord* may also be referred to as a "cord accident." This occurs when the cord prolapses into the cervix at the level of a presenting part (often the head). This creates pressure on the cord and prevents the fetus from receiving adequate oxygen. Entanglement of the cord around the fetus can also cause congenital defects (e.g., absence of a forearm).
3. Most "over-the-counter" pregnancy tests are based on the presence of hCG. These tests are capable of detecting the relatively large amounts of hCG in the woman's urine. Such tests are positive a short time (a week or so) after the first missed period. hCG is produced by the syncytiotrophoblast of the chorion. These tests usually give an accurate diagnosis of pregnancy; however, a physician should be consulted as soon as possible because some tumors (*choriocarcinomas*) also produce this hormone.
4. The "bag of waters" is a colloquial term for the amniotic sac containing amniotic fluid (largely composed of water). Sometimes the chorionic and amniotic sacs rupture before labor begins, allowing fluid to escape. *Premature rupture of the membranes (PROM) is the most common event leading to premature labor (birth).* PROM may complicate the birth process; however, it is not a "dry birth." Sometimes sterile saline is infused into the uterus by way of a catheter — *amnioinfusion* — to alleviate fetal distress. A "dry birth" is due to a low volume of amniotic fluid. PROM may also allow a vaginal infection to spread to the fetus.
5. *Fetal distress* is synonymous with *fetal hypoxia, indicating decreased oxygenation* to the fetus resulting from a general decrease of the maternal oxygen content of the blood, decreased oxygen-carrying capacity, or diminished blood flow. Fetal distress exists when the fetal heart rate falls below 100 beats per minute. *Pressure on the umbilical cord* causes fetal distress secondary to impairment of blood supply to the fetus in about 1 in 200 deliveries. In these cases, the fetal body compresses the umbilical cord as it passes through the cervix and vagina. *Fetal stress* results from hypoxia and cardiac anomalies.
6. Yes, this statement is true for DZ twins but not for MZ twins. DZ twinning is an autosomal recessive trait that is carried by the daughters of mothers of twins; hence, *DZ twinning is hereditary.* MZ twinning, on the other hand, is a random occurrence that is not genetically controlled.

Chapter 9

1. No evidence indicates that the occasional use of aspirin in *recommended therapeutic dosages* is harmful during pregnancy; however, large doses at subtoxic levels (e.g., for rheumatoid arthritis) have not been proven to be harmless to the embryo and fetus. All pregnant women should discuss the use of over-the-counter medications with their physicians.
2. A woman who is addicted to a habit-forming drug (e.g., heroin) and takes it during pregnancy is almost certain to give birth to a child who shows signs of drug addiction. The fetus's chances of survival until birth, however, are not good; mortality and premature birth rates are high among fetuses of drug-addicted mothers.
3. All drugs prescribed in North America are tested for teratogenicity before they are marketed. The thalidomide tragedy, however, clearly demonstrates the need for improved methods for detecting potential human teratogens. Thalidomide has not been found to be teratogenic in pregnant mice and rats; however, it is a potent teratogen in humans during the fourth to sixth weeks of pregnancy. Because it is unethical to test the effects of drugs on embryos that are to be aborted, no way exists to prevent some human teratogens from being marketed. Human teratologic evaluation depends on retrospective epidemiologic studies and the reports of astute physicians. This is the way thalidomide teratogenicity was detected. Most new drugs contain a disclaimer in the accompanying package insert, such as, "This drug has not been proven safe for pregnant women." Some drugs may be used if, in the opinion of the physician, the potential benefits outweigh the possible hazards. All known teratogenic drugs that may be taken by a pregnant woman are available only through prescription by a physician.
4. Cigarette smoking during pregnancy is harmful to the embryo and fetus. Its most adverse effect is *intrauterine growth retardation.* Women who stop smoking during the first half of pregnancy have infants with birth weights closer to those of nonsmokers. Decreased placental blood flow, thought to be a nicotine-mediated effect, is believed to cause decreased intrauterine blood flow. No conclusive evidence exists that mater-

nal smoking causes congenital anomalies. The growth of the fetus of a woman who smokes but does not inhale is still endangered because nicotine, carbon monoxide, and other harmful substances are absorbed into the maternal bloodstream through the mucous membranes of the mouth and throat, as well as through the lungs. These substances are then transferred to the embryo/fetus through the placenta. Hence, refraining from inhaling smoke is safest; however in any case, smoking in any manner during pregnancy is not advisable.

5. Ample evidence indicates that most drugs do not cause congenital anomalies in human embryos; however, a pregnant woman should take only drugs that are essential and recommended by her physician. A pregnant woman with a severe lower respiratory infection, for example, would be unwise to refuse drugs recommended by her doctor to cure her illness; her health and that of her embryo or fetus could be endangered by the infection. Most drugs, including sulfonamides, meclizine, penicillin, antihistamines, and Bendectin, are considered safe drugs. Similarly, local anesthetic agents, dead vaccines, and salicylates (e.g., aspirin) in low doses are not known to cause congenital anomalies.

Chapter 10

1. Yes it is. When a baby is born with a congenital diaphragmatic hernia (CDH), part of its stomach and liver may be in its thorax (chest); however, this is uncommon. Usually, the abnormally placed viscera are intestines. The viscera enter the thorax through a posterolateral defect in the diaphragm, usually on the left side.
2. Yes it can. A baby born with CDH may survive; however, the mortality rate is high (about 76%). Treatment must be given immediately. A feeding tube is inserted into the stomach, and the air and gastric contents are aspirated with continuous suction. The displaced viscera are replaced into the abdominal cavity, and the defect in the diaphragm is surgically repaired. Infants with large diaphragmatic hernias who are operated on within 24 hours after birth have survival rates of 40 to 70%. CDH can be repaired before birth; however, this intervention carries considerable risk to the fetus and mother.
3. It depends upon the degree of herniation of the abdominal viscera. With a moderate hernia, the lungs may be mature but small. With a severe degree of herniation, lung development is impaired. Most babies with CDH die, but not because of the defect in the diaphragm or viscera in the thorax (chest); they die because the lung on the affected side is hypoplastic (underdeveloped).
4. Yes, it is possible to have a small CDH and not be aware of it. Some small diaphragmatic hernias may remain asymptomatic into adulthood and may be discovered only during a routine radiographic or ultrasound examination of the thorax. The lung on the affected side would probably develop normally because there would be little or no pressure on the lung during prenatal development.

Chapter 11

1. "Harelip" is the old, obsolete term for cleft lip. The term was adopted because the hare (a mammal resembling a large rabbit) has a divided upper lip. It is not an accurate comparison, however, because the cleft in the hare's lip is in the median part of the upper lip, whereas most human clefts are lateral to the median plane. The clinical name for "harelip" is cleft lip.
2. No, both statements are inaccurate. All embryos have grooves in their upper lips where the maxillary prominences meet the merged medial nasal prominences; however, normal embryos do not have cleft lips. When lip development is abnormal, the tissue in the floor of the labial groove breaks down, forming a cleft lip.
3. The risk in this case is the same as for the general population: about 1 per 1000.
4. Although environmental factors may be involved, it is reasonable to assume that the son's cleft lip and cleft palate were hereditary and recessive in their expression. This would mean that his father also carried a concealed gene for cleft lip and that his family was equally responsible for the son's anomalies.
5. Minor anomalies of the auricle of the external ear are common, and usually they are of no serious medical or cosmetic consequence. About 14% of newborn infants have minor morphological abnormalities; less than 1% of them have other defects. The child's abnormal ears could be considered branchial anomalies because the external ears develop from six small auricular hillocks (swellings) of the first two pairs of pharyngeal arches; however, such minor abnormalities of ear shape would not normally be classified in this way.

Chapter 12

1. The fetus cannot breathe before birth because the airways and primordial alveoli are distended with liquid. The fetal lungs do not function as organs of gas exchange; however, breathing movements are practiced by the fetus. Rapid, irregular respiratory movements occur during the terminal stages of pregnancy. The lungs must develop in such a way that

they can assume their breathing role as soon as the baby is born. Intra-alveolar fluid is rapidly replaced by air after birth.

2. The stimuli that initiate breathing at birth are multiple. "Slapping the buttocks" used to be a common physical stimulus; however, this action is usually unnecessary. Under normal circumstances, the infant's breathing begins promptly, which suggests that it is a reflex response to the sensory stimuli of exposure to air and touching. The changes in blood gases after interruption of the placental circulation are also important in stimulating breathing, such as the fall in oxygen tension and pH and the rise in Pco_2.
3. *Hyaline membrane disease* (*HMD*), a common cause of the respiratory distress syndrome (RDS), occurs after the onset of breathing in infants with immature lungs and a *deficiency of pulmonary surfactant*. The incidence of RDS is about 1% of all live births, and it is the leading cause of death in newborn infants. It occurs mainly in infants who are born prematurely. HMD is caused by environmental factors (mainly, surfactant deficiency).
4. A 22-week fetus is viable and, if born prematurely and given special care in a neonatal intensive care unit, may survive. Chances of survival, however, are poor for infants who weigh less than 600 gm because the lungs are immature and incapable of adequate alveolar-capillary gas exchange. Furthermore, the fetus's brain is not usually differentiated sufficiently to permit regular respiration.

Chapter 13

1. Undoubtedly, the baby had *congenital hypertrophic pyloric stenosis*, a diffuse hypertrophy (enlargement) and hyperplasia of smooth muscle in the pyloric part of the stomach. This condition produces a hard mass ("tumor"); however, it is not a true tumor. It is a benign enlargement and is definitely not a malignant tumor. The muscular enlargement causes narrowing of the exit canal (pyloric canal). In response to the outflow obstruction and vigorous peristalsis, the vomiting is projectile, as in the case of the baby described. Surgical relief of the pyloric obstruction is the usual treatment. The cause of pyloric stenosis is not known; however, it is thought to have a *multifactorial inheritance*; that is, genetic and environmental factors are probably involved.
2. It is true that infants with Down syndrome have an increased incidence of *duodenal atresia*. They are also more likely to have an *imperforate anus* and other congenital defects (e.g., *atrial septal defects*). These anomalies are likely caused by the abnormal chromosome constitution of the infants (i.e., three instead of two chromosomes 21). The atresia can be corrected surgically by bypassing the pyloric obstruction, a procedure called *duodenoduodenostomy*.
3. In very uncommon cases, when the intestines return to the abdomen, they rotate in a clockwise direction rather than in the usual counterclockwise manner. As a result, the cecum and appendix are located on the left side, a condition called *situs inversus abdominis*. A left-sided cecum and appendix can also result from a *mobile cecum*. If the cecum is not fixed to the posterior abdominal wall during the fetal period, the cecum and appendix are freely movable and could migrate to the left side.
4. Undoubtedly, the individual described had an *ileal* (*Meckel*) *diverticulum*, a fingerlike outpouching of the ileum. This common anomaly is sometimes referred to as a "second appendix," which is a misnomer. An ileal diverticulum produces symptoms that are similar to those produced by appendicitis. It is also possible that the person had a *duplication of the cecum*, which would result in two appendices.
5. Hirschsprung disease, or *congenital megacolon* (Gr. *megas*, big), is the most common cause of obstruction of the descending colon in newborn infants. The cause of the condition is *failure of migration of neural crest cells into the wall of the intestine*. As these cells form neurons, there is a deficiency of the nerve cells that innervate the muscular wall of the bowel. When the wall collapses, obstruction occurs and constipation results.
6. No, she was telling the truth. If the baby had an *umbilicoileal fistula*, the abnormal canal connecting the ileum and umbilicus could permit the passage of the contents of the ileum to the umbilicus. This occurrence would be an important diagnostic clue to the presence of this canal. The fistula results from the persistence of the intra-abdominal part the yolk stalk.

Chapter 14

1. Most people with a horseshoe kidney have no urinary problems. These abnormal kidneys are usually discovered at autopsy, during diagnostic imaging, or in the dissecting room. Nothing needs to be done with the abnormal kidney unless the person develops an uncontrolled infection of the urinary tract. In some cases, the urologist may divide the kidney into two parts, and fix them in positions that do not result in urinary stagnation.
2. His developing kidneys probably fused during the sixth to eighth weeks as they "migrated" from the pelvis. The fused kidneys then ascended toward the normal position on one side or the other. Usually no problems are associated with fused kidneys; however, surgeons have to be conscious of the possibility of this condition and recognize it for what it is. This abnormality is called crossed renal ectopia.

3. Some true hermaphrodites marry; however, most of them do not. Affected individuals have both ovarian and testicular tissue. Although spermatogenesis is uncommon, ovulation is not. Pregnancy and childbirth have been observed in a few patients; however, this is very unusual.
4. By 48 hours after birth, a definite gender assignment can be made in most cases. The parents are told that their infant's genital development is incomplete and that tests are needed to determine whether the baby is a boy or girl. They are usually advised against announcing their infant's birth to their friends until the appropriate sex has been assigned. The buccal smear test for the identification of sex chromatin is done as soon as possible. Chromatin-positive cells — those with sex chromatin in their nuclei — almost always indicate a female. Chromatin-negative cells usually indicate a male; however, study of the baby's chromosomes may be required before sex can be assigned. Hormone studies may also be required.
5. Virilization (masculinization) of the female fetus resulting from *congenital adrenal hyperplasia* (CAH) is the most common cause of ambiguous external genitalia. In other cases, androgens enter the fetal circulation following maternal ingestion of androgenic hormones. In unusual cases, the hormones are produced by a tumor on one of the mother's suprarenal glands (see text). Partial or complete fusion of the urogenital folds or labioscrotal swellings is the result of exposure to androgens prior to the 12th week of development. Clitoral enlargement occurs after this; however, androgens do not cause sexual ambiguity because the other external genitalia are fully formed by this time.

Chapter 15

1. Heart murmurs are sounds transmitted to the thoracic wall from turbulence of blood in the heart or great arteries. Loud murmurs often represent *stenosis (narrowing) of one of the semilunar valves* (aortic or pulmonary valve). A ventricular septal defect or a *patent oval foramen* (foramen ovale) may also produce a murmur.
2. Congenital heart defects are common. They occur in 6 to 8 of every 1000 newborn infants and represent about 10% of all congenital anomalies. *Ventricular septal defects* (*VSDs*) are the most common type of heart anomaly. They occur more frequently in males than females. The reason for this is unknown.
3. The cause of most congenital anomalies of the cardiovascular system is unknown. In about 8% of children with heart disease, a genetic basis is clear. Most of these anomalies are associated with obvious chromosomal abnormalities (e.g., trisomy 21) and deletion of parts of chromosomes. *Down syndrome* is associated with congenital heart disease in 50% of cases. The maternal ingestion of drugs, such as antimetabolites and Coumadin (an anticoagulant), has been shown to be associated with a high incidence of cardiac defects. Evidence suggests that heavy consumption of alcohol during pregnancy may cause heart defects; however, it is impossible to say whether the excessive use of alcohol by the woman caused her baby's heart condition.
4. Several viral infections are associated with congenital cardiac defects; however, only *rubella virus* is known to cause cardiovascular disease (e.g., patent ductus arteriosus). *Measles* is a general term that is used for two different viral diseases. *Rubeola* (common measles) does not cause cardiovascular defects; however, *rubella* (German measles) does. *Rubella virus vaccine* is available and is effective in preventing the development of rubella infection in a woman who has not had the disease and is planning to have a baby. It will subsequently prevent the rubella syndrome from developing in her baby as well. Because of the potential hazard of the vaccine to the embryo, the vaccine is given only if there is assurance that there is no likelihood of pregnancy for the next 2 months.
5. This anomaly is called *transposition of the great arteries* (TGA) because the positions of the great vessels (aorta and pulmonary trunk) are reversed. Survival after birth depends on mixing between the pulmonary and systemic circulations (e.g., through an *ASD — patent oval foramen*). TGA occurs in slightly more than 1 per 5000 live births and is more common in male infants than in female infants (by almost 2:1). Most infants with this severe cardiac anomaly die during the first months of life; however, corrective surgery can be done in those who survive for several months. Initially, an ASD may be created to increase mixing between the systemic and pulmonary circulations. Later, an arterial switch operation (reversing the aorta and pulmonary trunk) can be performed; however, more commonly, a baffle is inserted in the atrium to divert systemic venous blood through the mitral valve, left ventricle, and pulmonary artery to the lungs, and pulmonary venous blood through the tricuspid valve, right ventricle, and aorta. This physiologically corrects the circulation.
6. Very likely, the one twin has *dextrocardia,* which usually is of no clinical significance. The heart is simply displaced to the right. In the condition of the individual described, the heart presents a mirror image of the normal cardiac structure. This occurs during the fourth week of development when the heart tube rotates to the left rather than to the right. Dextrocardia is a relatively common anomaly in MZ twins.

Chapter 16

1. The most common congenital anomaly of the vertebral column is *spina bifida occulta.* This defect of the vertebral arch of the first sacral and/or last lumbar vertebra is present in about 10% of people. The defect can also involve the cervical and thoracic vertebrae. The spinal cord and nerves are usually normal, and neurological symptoms are usually absent. Spina bifida occulta does not cause back problems in most people; occasionally, however, it may be associated with neurological or musculoskeletal disturbances.
2. A rib associated with the seventh cervical vertebra is of clinical importance because it may compress the subclavian artery and/or brachial plexus, producing symptoms of artery and nerve compression. These ribs develop from the costal processes of the seventh cervical vertebra. Lumbar ribs are common and have a similar embryological basis; that is, they result from development of the costal processes of the first lumbar vertebra.
3. A hemivertebra can produce a lateral curvature of the vertebral column (*scoliosis*). A hemivertebra is composed of one half of a body, a pedicle, and a lamina. This anomaly results when mesenchymal cells from the sclerotomes on one side fail to form the primordium of half of a vertebra. As a result, more growth centers are found on one side of the vertebral column; this imbalance causes the vertebral column to bend laterally.
4. *Craniosynostosis* indicates premature closure of one or more of the cranial sutures. This developmental abnormality results in cranial malformations. *Scaphocephaly* or *dolichocephaly* — a long, narrow cranium — results from premature closure of the sagittal suture. This type of craniosynostosis accounts for about 50% of the cases of premature closure of cranial sutures.
5. The features of Klippel-Feil syndrome are short neck, low hairline, and restricted neck movements. In most cases, the number of cervical vertebra is fewer than normal.

Chapter 17

1. The *prune-belly syndrome* results from partial or complete absence of the abdominal musculature. Usually, the abdominal wall is thin. This syndrome is usually associated with malformations of the urinary tract, especially the urinary bladder (e.g., *exstrophy*).
2. Absence of the sternocostal part of the left pectoralis major muscle is usually the cause of an abnormally low nipple and areola. Despite its numerous and important actions, absence of all or part of the pectoralis major muscle usually causes no disability. The actions of other muscles associated with the shoulder joint compensate for the partial absence of this muscle.
3. The girl has a prominent sternocleidomastoid muscle (SCM). The SCM attaches the mastoid process to the clavicle and sternum; hence, continued growth of the side of the neck results in tilting and rotation of the head. This relatively common condition — *congenital torticollis* (wryneck) — may occur because of injury to the muscle during birth. Stretching and tearing of some muscle fibers may have occurred during delivery, resulting in bleeding into the muscle. Over several weeks, necrosis of some fibers occurs, and the blood is replaced by fibrous tissue. This results in shortening of the muscle and pulling of the child's head to one side. If the condition is not corrected, the shortened muscle also could distort the shape of the face on the affected side.
4. The young athlete probably had an accessory soleus muscle. It is present in about 6% of people. This anomaly probably results from splitting of the primordium of the soleus muscle into two parts.

Chapter 18

1. The ingestion of drugs did not cause the child's short limbs. The infant has a skeletal disorder known as *achondroplasia.* This type of short-limbed dwarfism has an incidence of 1 in 10,000 and shows an autosomal dominant inheritance. About 80% of affected infants are born of normal parents, and presumably, the condition results from fresh mutations (changes of genetic material) in the parents' germ cells. Most achondroplastic people have normal intelligence and lead normal lives within their physical capabilities. If the parents of an achondroplastic child have more children, the risk of having another child with this condition is slightly higher than the general population risk; however, the risk for the achondroplastic person's own children is 50%.
2. *Brachydactyly* is an autosomal dominant trait; that is, it is determined by a dominant gene. If the woman (likely bb) marries the brachydactylous man (likely Bb), the risk is 50% for a brachydactylous child and 50% for a normal child. It would be best for her to discuss her obvious concern with a medical geneticist.
3. Bendectin, an antinauseant, does not produce limb defects in human embryos. Several epidemiologic studies have failed to show an increased risk of birth defects after exposure to Bendectin or its separate ingredients during early pregnancy. In the case described, the mother took the drug more than 3 weeks after the end of the critical period of limb development (24 to 36 days after fertilization). Most limb reduction defects have a genetic basis.
4. *Cutaneous syndactyly* is the most common type of limb anomaly. It varies from cutaneous webbing between the digits to *synostosis* (union of phalanges,

the bones of the digits). This anomaly occurs when separate digital rays fail to form in the fifth week or when the webbing between the developing digits fails to break down. Simple cutaneous syndactyl is easy to correct surgically. The absence of the sternal head of the pectoralis major caused the nipple to be lower than the other one.

5. The most common type of clubfoot is *talipes equinovarus,* occurring in about 1 of every 1000 newborn infants. In this deformity, the soles of the feet are turned medially and the feet are sharply plantar flexed. The feet are fixed in the tiptoe position, resembling the foot of a horse (L. *equinus,* horse).

Chapter 19

1. Neural tube defects (NTDs) are hereditary. Meroanencephaly (anencephaly) and spina bifida cystica have a multifactorial inheritance; that is, both genetic and environmental factors are involved. Nutritional factors may be implicated. After the birth of one child with an NTD, the risk of a subsequent child having an NTD is divided about equally between the two defects. The recurrence risk in the United Kingdom, where NTDs are common (7.6 per 1000 in South Wales and 8.6 per 1000 in Northern Ireland), is about 1 in 25. It is probably about 1 in 50 in North America. NTDs can be detected prenatally by a combination of ultrasound scanning and measurement of alpha-fetoprotein levels in amniotic fluid.
2. The condition described is *hydranencephaly,* an extremely rare anomaly. Most of both cerebral hemispheres are reduced to membranous sacs that contain CSF. Absence of cerebral hemispheres can result from different developmental disturbances. The condition most likely results from vascular occlusion of both internal carotid arteries secondary to a severe intrauterine infection. In some cases, hydranencephaly appears to be a severe type of intrauterine hydrocephalus. These infants usually do not survive longer than 3 months.
3. Mental retardation and growth retardation are the most serious aspects of the *fetal alcohol syndrome.* Average IQ scores in affected children are 60 to 70. It has been estimated that the incidence of mental retardation resulting from heavy drinking during pregnancy may be as high as 1 in every 400 live births. Heavy drinkers are those who consume five or more drinks on one occasion, with a consistent daily average of 45 mL of absolute alcohol. At present, no safe threshold for alcohol consumption during pregnancy is known. Physicians recommend complete abstinence from alcohol during pregnancy.
4. No conclusive evidence indicates that maternal smoking affects the mental development of a fetus; however, cigarette smoking compromises oxygen supply to the fetus because blood flow to the placenta is decreased during smoking. Because it is well established that heavy maternal smoking seriously affects physical growth of the fetus and is a major cause of IUGR, it is not wise for mothers to smoke during pregnancy. The reduced oxygen supply to the brain could affect fetal intellectual development, even though the effect may be undetectable. Abstinence gives the fetus the best chance for normal development.
5. Most laypeople use the designation "spina bifida" in a general way. They are unaware that the common type, *spina bifida occulta,* is usually clinically insignificant. It is an isolated finding in up to 20% of radiographically examined vertebral columns. Most people are unaware that they have this vertebral defect, and most physicians would not tell them about it because it produces no symptoms unless it is associated with a NTD or abnormality of the spinal nerve roots. The various types of *spina bifida cystica* are of clinical significance. Meningomyelocele is a more severe defect than meningocele because neural tissue is included in the lesion. Because of this, the function of abdominal and limb muscles may be affected. Meningoceles are usually covered with skin, and motor function in the limbs is usually normal unless associated developmental defects of the spinal cord or brain are present. Management of infants with spina bifida cystica is complex and involves several medical and surgical specialties. Spinal meningocele is easier to correct surgically than spinal meningomyelocele, and the prognosis is also better.

Chapter 20

1. The chance of significant damage to the embryo/fetus after a rubella infection depends primarily on the timing of the viral infection. In cases of primary maternal infection during the first trimester of pregnancy, the overall risk of embryonic/fetal infection is about 20%. It is estimated that about 50% of such pregnancies end in spontaneous abortion, stillbirth, or congenital anomalies (deafness, cataract, glaucoma, and mental retardation). When infection occurs at the end of the first trimester, the probability of congenital anomalies is only slightly higher than for an uncomplicated pregnancy. Certain infections occurring late in the first trimester, however, may result in severe eye infections (e.g., chorioretinitis), which may affect visual development. *Deafness is the most common manifestation of late fetal rubella infection* (i.e., infection during the second and third trimesters). If a pregnant woman is exposed to rubella, an antibody test can be performed. If she is determined to

be immune, she can be reassured that her embryo/fetus will not be affected by the virus. Preventive measures are essential for the protection of the embryo. It is especially important that girls obtain immunity to rubella (e.g., by active immunization) before they reach childbearing age.

2. The purposeful exposure of young girls to rubella (German measles) is not recommended by physicians. Although complications resulting from such infections are uncommon, neuritis and arthritis (inflammation of nerves and joints, respectively) occasionally occur. *Encephalitis* (inflammation of the brain) occurs in about 1 in 6000 cases; furthermore, the rubella infection is often subclinical (difficult to detect) and yet represents a risk to pregnant women. There is a chance of injury to embryos because the danger period is greatest when the eyes and ears are developing. This occurs early enough in pregnancy that some women might be unaware that they are pregnant. A much better way of providing immunization against rubella is the administration of live-virus vaccine. This is given to children older than 15 months of age and to nonpregnant post-pubertal females who can be reasonably relied upon not to become pregnant within 3 months of immunization.
3. *Congenital syphilis* ("fetal syphilis") results from the transplacental transmission of the microorganism *Treponema pallidum*. The transfer of this microorganism from untreated pregnant women may occur throughout pregnancy; however, it usually takes place during the last trimester. Deafness and tooth deformities commonly develop in these children. These anomalies can be prevented by treating the mother early in pregnancy. The microorganism that causes syphilis is very sensitive to penicillin, an antibiotic that does not harm the fetus.
4. Several viruses in the herpes virus family can cause fetal blindness and deafness during infancy. *Cytomegalovirus* can cross the placenta, be transmitted to the infant during birth, and be passed to the baby in breast milk. *Herpes simplex viruses* (usually type 2 or genital herpes) are usually transmitted just before or during birth. The chances for normal development of infected infants are not good. Some infants develop microcephaly, seizures, deafness, and blindness.
5. *Methyl mercury is teratogenic in human embryos,* especially to the developing brain. Because the eyes and internal ears develop as outgrowths from the brain, it is understandable that their development is also affected. Besides the methyl mercury that passes from the mother to the embryo/fetus through the placenta, the newborn infant may receive additional methyl mercury from breast milk. Sources of methyl mercury include fish from contaminated water, flour made from methyl mercury-treated seed grain, and meat from animals raised on contaminated food.

Chapter 21

1. Congenital absence of the skin is very uncommon. Patches of skin may be absent, most often from the scalp, but patches of skin may also be missing from the trunk and limbs. Affected infants usually survive because healing of the lesions is uneventful and takes 1 to 2 months. A hairless scar persists. The cause of congenital absence of hair, termed *aplasia cutis congenita,* is usually unknown. Most cases are sporadic; however, several well-documented pedigrees demonstrate autosomal dominant transmission of this skin defect.
2. The white patches of skin on a dark-skinned person result from *partial albinism* (piebaldism). This defect, which also affects light-skinned persons, is a heritable disorder transmitted by an autosomal dominant gene. Ultrastructural studies show an absence of melanocytes in the depigmented areas of skin. Presumably, the cause is a genetic defect in the differentiation of melanoblasts. These skin and hair defects are not amenable to treatment; however, they can be covered with cosmetics and hair dyes.
3. The breasts, including the mammary glands within them, of males and females are similar at birth. Breast enlargement in a newborn infant is common and results from stimulation by maternal hormones that enter the infant's blood through the placenta. Therefore, enlarged breasts are a normal occurrence in male infants and do not indicate abnormal sex development. Similarly, physiological *pubertal gynecomastia* occurs in some males during their early teens as a result of decreased levels of testosterone. The breast enlargement is usually transitory. *Familial gynecomastia* is an X-linked or autosomal dominant sex-linked trait. Gynecomastia also occurs in about 50% of males with *Klinefelter syndrome* (described in Chapter 9). These boys and men are not intersexes because their external and internal genitalia are normal except for their testes, which are small because of the degeneration of seminiferous tubules.
4. An extra breast (polymastia) or nipple (polythelia) is common. The axillary breast may enlarge during puberty, or it may not be noticed until pregnancy occurs. The embryological basis of extra breasts and nipples is the presence of mammary crests (ridges) that extend from the axillary to the inguinal regions. Usually, only one pair of breasts develops; however, breasts can develop anywhere along the mammary crests. The extra breast or nipple is usually just superior or inferior to the normal breast. An axillary breast or nipple is very uncommon.

5. Teeth present at birth are natal teeth (L. *natalis,* to be born). A more appropriate term would be *congenital teeth* (L. *congenitus,* born with). *Natal teeth,* which are teeth that are erupted at birth, are observed in about 1 in 2000 newborn infants. Usually, two mandibular medial (central) incisors are present. The presence of natal teeth usually suggests that early eruption of other teeth may occur. Because natal teeth may detach and be aspirated into the lungs, the natal teeth are sometimes extracted. Often, they fall out on their own; however, there is a danger that they may be aspirated.

Index

Page numbers followed by an italic *f* or *t* denote figures or tables, respectively.

A

abdominal circumference, fetal, 79
abdominal pregnancy, 43, 43*f*
abdominal wall defects, 217, 219*f*
 ventral, alpha-fetoprotein assay for detection of, 86
abducent nerve, formation of, 367–368, 368*f*
abortion
 definition of, 2, 43
 spontaneous, 35, 43
  sporadic and recurrent, 43
abortion pill, 44
accessory auricular hillocks, 385, 385*f*
accessory hepatic ducts, 208
accessory muscles, 327
accessory nerve, formation of, 368*f*, 369
accessory pancreatic duct, 210, 211*f*, 226
accessory pancreatic tissue, 210
accessory placenta, 100, 102, 103*f*
accessory renal arteries and veins, 232, 237*f*
accessory ribs, 316
accessory spleen, 212
accessory thyroid tissue, 168*f*
ACE inhibitors, as teratogens, 134
acetylsalicylic acid, fetal effects of, 134
achondroplasia, 126, 126*f*, 319–320, 320*f*
acoustic meatus, external, 69, 73*f*, 160*f*, 162, 172*f*–173*f*, 174, 186, 380, 383*f*, 384, 386
 absence of, 385, 385*f*
 atresia of, 385
acquired immunodeficiency syndrome (AIDS), fetal effects of, 136
acrania, 317, 317*f*
acromegaly, 320
acrosin, and fertilization, 24, 28
acrosome, 14*f*, 17
acrosome reaction, 24, 29*f*
ACTH (adrenocorticotropin), and labor, 99
active transport, placental, 98
activin(s), and pancreatic development, 209
adenocarcinoma, diethylstilbestrol exposure and, 133
adenohypophysis, 360*t*, 360*f*, 361
adenoids (pharyngeal tonsils), 159, 302
adipose tissue, fetal, 80, 83, 87–88
adrenal glands, development of. *See* suprarenal glands.
adrenal hyperplasia, congenital (CAH), 243, 244*f*, 255, 255*f*, 261
adrenocorticotropin (ACTH), and labor, 99
adrenogenital syndrome, 243, 244*f*
AFP. *See* alpha-fetoprotein assay
afterbirth, 90, 100–105, 102*f*
aganglionic megacolon, congenital, 224, 224*f*, 370
age
 bone, 319
 conceptional, 78
 embryonic, estimation of, 55, 68*t*, 74, 76*f*
 fertilization
  definition of, 2
  estimation of, 74, 78*t*
 fetal, estimation of, 78*t*, 78–79
 gestational, 28
  estimation of, 74, 75*f*, 78–79, 90
  ultrasound assessment of, 75*f*, 78–79
 maternal, and chromosomal abnormalities, 17, 120, 123*t*
 menstrual, 78
AIDS, fetal effects of, 136
ala orbitalis, 313
alar plate, 344–346, 350*f*, 353–354, 356*f*–357*f*
albinism, 391, 399
 generalized, 391
 localized, 391
alcohol abuse
 and congenital anomalies, 128*t*, 132, 132*f*
 and fetal growth, 83, 85
 and mental retardation, 132
allantois, 50–53, 51*f*, 110, 110*f*, 114, 222, 223*f*, 239, 240*f*
 embryonic folding and, 62, 63*f*–64*f*, 75–76
allograft, placenta as, 103–105
alpha-fetoprotein assay, 86, 108
 amniotic fluid, 86, 352
 for neural tube defect detection, 86, 108, 352
 maternal serum, 352
alveolar cells
 type I, 197
 type II, 197
alveolar ducts, 196*f*
alveolar period, of lung maturation, 196*f*, 197–199
alveolar processes, of teeth, 397
alveolocapillary membrane, 196*f*, 197, 199
alveolus (alveoli), pulmonary
 definition of, 197
 development of, 196*f*, 197–198
alveolus of teeth, 396
ambiguous genitalia, 253
amelia, 336
ameloblasts, 395*f*, 396, 400
amelogenesis imperfecta, 398, 399*f*
amino acids
 in fetal metabolism and growth, 83
 transplacental transport of, 97*f*
aminopterin, as teratogen, 128*t*, 134
amnioblasts, 39
amniocentesis, 108
 diagnostic, 85–86, 87*f*
amniochorionic membrane, 91*f*, 94*f*–95*f*, 95–96
 rupture of, 95–96
amnion, 39, 39*f*, 105–108, 106*f*–107*f*, 114
 blood supply to, 265*f*
 number in twin pregnancies, 113*t*
amniotic band disruption complex, 108
amniotic band syndrome, 108, 109*f*, 137
amniotic cavity, formation of, 39*f*, 39–41, 44
amniotic fluid, 39, 39*f*, 105–108
 circulation of, 95*f*, 106–108
 composition of, 106
 exchange of, 106
 fetal swallowing of, 106, 202
 significance of, 108
 volume, disorders of, 108 (*See also* oligohydramnios; polyhydramnios)
 water content of, 106
amniotic sac, 105, 106*f*–107*f*
ampulla of uterine tube, 10, 12*f*
 fertilization in, 28
ampullae of semicircular ducts, 382
anal agenesis, with fistula, 224–225, 225*f*
anal canal, development of, 222, 224*f*
anal membrane, 222, 223*f*, 226, 254*f*
anal pit, 202, 202*f*, 222, 223*f*, 226
anal stenosis, 225, 225*f*
anaphase lagging, 123
anatomical position, descriptive terms for, 7, 8*f*
androgen(s)
 and masculinization of female fetus, 132, 132*f*, 255
 and testes development, 247
 as teratogens, 128*t*, 132*f*, 132–133
androgen insensitivity syndrome, 255
androstenedione, 247
anencephaly, 53, 317. *See also* meroanencephaly.
aneuploidy, 119
Angelman syndrome, 124, 127
angioblastic cords, 264, 264*f*
angioblasts, 56
angiogenesis, 56, 265
angiogenesis factor, and ovarian follicles, 19
angiomas of skin, 391, 399
angiotensin-converting enzyme (ACE) inhibitors, as teratogens, 134
ankyloglossia, 170, 171*f*
anodontia
 partial, 397
 total, 397
anomalies, congenital. *See* congenital anomalies and birth defects
anoperineal fistula, 224, 225*f*
anorectal agenesis, 225*f*, 226
 with fistula, 225*f*, 226
anorectal anomalies, 222–227
anorectal atresia, 224*f*–225*f*, 224–226
anovulation, 21
anovulatory menstrual cycle, 23
anoxia, fetal, umbilical cord problems and, 105
anterior, as descriptive term, 8*f*
anterior commissure, 361, 363*f*
antibiotics, as teratogens, 133
antibodies, maternal, transplacental transport of, 97*f*, 98
anticoagulants, as teratogens, 133
anticonvulsants, as teratogens, 128*t*, 133, 352
antimüllerian hormone (AMH), 247
antinauseants, as teratogens, 134
antineoplastic agents, as teratogens, 134
antithyroid drugs, as teratogens, 134

antrum, 19
anular epiphyses, 312, 312*f*
anular ligament, failed differentiation of, 384
anular pancreas, 210, 212*f*, 226
anulus fibrosus, 311, 311*f*
anus
agenesis of, 224–225, 225*f*
development of, 222, 224*f*, 226, 252, 254*f*
ectopic, 224–225
imperforate, 224, 224*f*–225*f*
membranous atresia of, 224*f*–225*f*, 224–226
aorta, 337*f*
coarctation of, 293, 295*f*
juxtaductal, 293
postductal, 293, 295*f*
preductal, 293, 295*f*
dorsal, 152–154, 265*f*, 268
transformation and adult derivatives of, 292*f*
formation of, 283*f*, 284
renal blood supply from, 232, 236*f*
semilunar valves of, 284, 285*f*
aortic arches, 152–154, 155*f*–156*f*, 265*f*, 268, 291*f*
1st (first) pair of, derivatives of, 291
2nd (second) pair of, derivatives of, 291
3rd (third) pair of, derivatives of, 291–293, 292*f*
4th (fourth) pair of, derivatives of, 292*f*, 293
5th (fifth) pair of, 293
6th (sixth) pair of, derivatives of, 292*f*, 293, 294*f*
circulation through, 271, 272*f*
congenital anomalies of, 293–295
derivatives of, 154, 291–293, 292*f*, 304
double, 293, 296*f*
right, 293, 297*f*
with retroesophageal component, 293
without retroesophageal component, 293
aortic atresia, 289
aortic sac, 265*f*, 268, 272*f*, 291, 291*f*, 304
transformation and adult derivatives of, 292*f*
aortic stenosis, 289, 290*f*
aortic vestibule, 281*f*, 284
aorticopulmonary septum, 281*f*, 282–284, 283*f*
aorticoventricular junction, 285*f*
apical ectodermal ridge, 318*f*, 330, 335*f*, 340
apocrine sweat glands, 389
apoptosis
and limb development, 330, 340
endometrial, and implantation, 38
apoptosis-inducing ligands, and placental immunoprotection, 104
appendicular skeleton, development of, 317–320, 318*f*–319*f*
appendix, of intestine
development of, 212, 214*f*, 215, 217*f*
pelvic, 215
retrocecal, 215
retrocolic, 215
subhepatic, 218, 220*f*
appendix of epididymis, 248*f*
appendix vesiculosa, 248*f*, 250*t*
applied embryology, 7
aqueductal stenosis, congenital, 364
aqueous chambers of eye, development of, 377*f*–379*f*, 379
arachnoid mater, 348, 352*f*
arachnoid villi, 356
arched collecting tubule, 230, 234*f*, 235*f*
archicerebellum, 357*f*
areola, 393*f*
Arnold-Chiari malformation, 366, 367*f*
arrector muscles of hair (arrector pili muscles), 390*f*, 391
arteriocapillary networks, 57–58, 96*f*
arteriocapillary-venous system, in chorionic villi, 96, 96*f*
artery (arteries)
aorta (*See* aorta; aortic arches)
axial, primary, 334, 337*f*
brachial, 334, 337*f*
brachiocephalic, 292*f*
carotid
common, 291, 292*f*, 294*f*
external, 174, 291, 294*f*
internal, 174, 291–293, 292*f*
celiac trunk, 141, 142*f*, 202*f*, 206, 213*f*, 268
chorionic, 96
deep, of thigh (profunda fermoris), 334, 337*f*
endometrial, 12*f*, 23, 39–40, 94, 94*f*–95*f*, 97
femoral, 337*f*
fibular, 337*f*
foregut, 141, 142*f*
great, transposition of, 288, 290*f*
hindgut, 141, 142*f*
hyaloid, 372, 373*f*, 375*f*, 377*f*, 378
persistence of, 378
iliac, 337*f*
common, 268, 337*f*
external, 337*f*
internal, 268, 300*f*–301*f*
intercostal, 268, 310
interosseous, 337*f*
common, 334, 337*f*
intersegmental, 268, 292*f*, 310, 311*f*, 334, 337*f*
dorsal, 265*f*, 268
ischiadic, 337*f*
lumbar, 268
maxillary, 291
median, 337*f*
mesenteric
inferior, 141, 142*f*, 202*f*, 222, 268
superior, 141, 142*f*, 202*f*, 206, 212, 214*f*, 222, 268
midgut, 141, 142*f*
plantar, 337*f*
popliteal, 337*f*
pudendal, 222
pulmonary
left, 292*f*, 293
right, 292*f*, 293
radial, 334, 337*f*
rectal, 222
inferior, 222
superior, 222
renal, 232, 236*f*
accessory, 232, 237*f*
retinal, central, 372, 375*f*, 377*f*
sacral, lateral, 268
spiral endometrial, 12*f*, 23, 39–40, 94, 94*f*–95*f*, 97
stapedial, 291
subclavian, 292*f*
right, 292*f*, 293, 294*f*
anomalous, 295, 298*f*
retroesophageal, 295
tibial, 337*f*
ulnar, 334, 337*f*
umbilical, 53, 96, 265*f*, 300*f*, 337*f*
absence of (single umbilical artery, SUA), 105, 106*f*, 234
adult derivatives of, 299, 301*f*
constriction of, 299
fate of, 268
vertebral, 268
vesical, superior, 268, 299
vitelline, 265*f*, 337*f*
fate of, 268
arytenoid cartilage, formation of, 157*t*
arytenoid swellings, 190, 192*f*
asphyxia, intrauterine, and surfactant production, 198
aspirin, fetal effects of, 134
astroblasts, histogenesis of, 344, 349*f*
astrocytes, histogenesis of, 344, 349*f*
atrial septal defects, 284–288, 286*f*–287*f*, 290*f*
atrioventricular bundle, 282*f*, 284
atrioventricular canal
circulation of blood through, 271, 272*f*
development of, 272*f*, 273*f*, 274*f*–275*f*
partitioning of, 271–276, 273*f*
atrioventricular node, 282*f*, 284
atrioventricular septum, 274*f*–275*f*
atrioventricular valves, development of, 282*f*, 284, 285*f*
atrium (atria)
left, formation of, 277, 279*f*
primordial, 267*f*, 269*f*, 270*f*, 272*f*, 273*f*–278*f*, 284, 304
partitioning of, 271–276, 273*f*, 274*f*–275*f*, 277*f*
auditory ossicles, 380, 383, 386
auditory (pharyngotympanic) tube, 159, 186, 383, 383*f*
Auerbach plexus, 224
auricle (cardiac), 276, 278*f*
auricle (ear), 69, 73*f*, 74, 74*f*, 174, 380, 384, 384*f*
congenital anomalies of, 385, 385*f*
auricular appendages, 385, 385*f*
auricular fistulas, 385
auricular hillocks, 69, 73*f*, 162, 174, 384, 384*f*, 386
accessory, 385, 385*f*
auricular sinuses and cysts, 162, 163*f*

auricularis muscle, formation of, 158*f*
autonomic ganglia, 366–367
autonomic nervous system, 344
  development of, 369–370
autosomal recessive inheritance, 126
autosomes
  disorders of, 119
  trisomy of, 120, 122*t*
axial artery, primary, 334, 337*f*
axial dysraphic disorders, 316, 317*f*
axial skeleton
  congenital anomalies of, 316–317, 316*f*–317*f*
  development of, 55, 310–316, 311*f*, 320
axons, 332
azygos vein, 265–268, 266*f*, 267*f*

B

balanced translocation carriers, 123
Bartholin glands, 248*f*, 251
basal body temperature, ovulation and, 21
basal layer
  in skin development, 388
  of endometrium, 10, 12*f*
basal plate, 313, 344–348, 350*f*, 353–354, 356*f*–357*f*, 367
battledore placenta, 105
bell stage, of tooth development, 396
Bendectin, 134
benzodiazepine derivatives, as teratogens, 134
betamethasone, and fetal lung maturity, 198
bicornuate uterus, 256, 257*f*
  with rudimentary horn, 256, 257*f*
bifid nose, 185*f*, 186
bifid ureter, 236, 238*f*
bile duct, 206*f*, 208
bile formation, 207
biliary apparatus, development of, 202, 206*f*, 207–208, 226
biliary atresia, 208
bilirubin, transplacental transport of, 97*f*, 98
binge drinking, during pregnancy, 132
biparental inheritance, 31
biparietal diameter (BPD), 79
birth(s)
  multiple, 110–114 (*See also* twin(s))
    and fetal growth, 85, 85*f*
    placenta and fetal membranes in, 111–114, 113*f*
  process of, 99–105, 101*f*
birth defects, 117–138. *See also* congenital anomalies
  and infant deaths, 118
  causes of, 118, 118*f*
  chromosomal abnormalities and, 118, 118*f*, 119–126
  drug/chemical dosage and, 129–131
  environmental factors in, 118, 118*f*, 127–138
  genetic factors in, 118, 118*f*, 119–127, 138
  genetic mutation and, 118, 118*f*, 126–127
  multifactorial inheritance and, 118, 118*f*, 137, 138*f*
birth weight
  cigarette smoking and, 131
  extremely low, 78
  low, 78, 83, 131
birthmarks, 391
bladder. *See* gallbladder; urinary bladder
blastocyst(s)
  definition of, 2
  formation of, 31–35, 33*f*
  implantation of, 31–33, 34*f*, 38*f*–40*f*, 38–41, 43–44
    inhibition of, 44
    sites of, 42–43, 43*f*
blastocystic cavity, 31, 33*f*, 35
blastomeres, 31, 33*f*, 35
blink-startle response, 80
blood, development of, 50–53, 56, 57*f*, 109–110
blood cells, development of, 56
blood flow. *See* circulation of blood
blood islands, 56, 57*f*, 60
blood vessels, development of, 56, 57*f*, 60, 265*f*, 265–269, 266*f*–267*f*
blood-air barrier, 197
BMPs. *See* bone morphogenetic proteins
body cavities. *See also specific body cavities*
  development of, 139–150, 140*f*–144*f*
  embryonic, 140–145
    division of, 141–145
bone(s)
  development of, 306–308, 320
    intracartilaginous ossification and, 306–308, 309*f*
    intramembranous ossification and, 306, 308*f*
  endochondral formation of, 306, 308*f*, 320
  histogenesis of, 306
  intramembranous formation of, 306, 320
  limb
    development of, 308, 317–319, 318*f*–319*f*, 330
    ossification of, 74, 308, 318–319, 330
  shaft of, formation of, 306
bone age, 319
bone marrow, development of, 306–308
bone matrix, 306
bone morphogenetic proteins (BMPs)
  and bone development, 306
  and cardiovascular development, 276, 282
  and limb development, 330
  and nervous system development, 344
bony labyrinth of internal ear, 382*f*, 382–383, 386
BPD (biparietal diameter), 79
brachial artery, 334, 337*f*
brachiocephalic artery, 292*f*
brachiocephalic vein, left, 265, 266*f*–267*f*
brachycephaly, 317*f*
brachydactyly, 338
bradykinin, and ductus arteriosus closure, 299
brain
  congenital anomalies of, 362–366
    major, incidence of, 131*t*
  development of, 53, 69*f*–72*f*, 344, 347*f*, 352–362, 370
    critical period of development of, 129
  protective case of (neurocranium), development of, 313, 320
brain flexures, 347*f*, 353, 356*f*
brain stem developmental anomaly, and sudden infant death syndrome, 284
brain vesicles, 352–353, 358*f*, 361, 362*f*
  primary, 352, 355*f*
  secondary, 352–353, 355*f*
branch villi, 58, 94–96, 95*f*, 96*f*
branchial anomalies, 162, 163*f*–165*f*, 186
branchial apparatus, 152. *See* pharyngeal apparatus
branchial cysts, 162, 163*f*–165*f*, 186
branchial fistula, 162, 163*f*–164*f*, 186
branchial sinuses, 162, 163*f*–164*f*, 186
  external, 162, 163*f*–164*f*
  internal, 162, 163*f*
branchial vestiges, 162, 163*f*
breasts
  aplasia of, 392
  congenital anomalies of, 392–393
  development of, 392, 393*f*
  excessive development in males (gynecomastia), 392
  supernumerary, 393, 399
breathing movements, fetal, 198
brevicollis, 316
broad ligaments, 250, 253*f*
bronchi
  development of, 141–142, 192*f*, 193–199, 194*f*–195*f*
  main, 194–195, 195*f*, 199
  secondary or stem, 195, 195*f*, 199
  segmental, 195, 199
bronchial buds, 141–142, 144*f*, 192*f*–194*f*, 193–194, 199
bronchioles, 195–196, 196*f*
bronchopulmonary segment, 195, 199
brown fat, fetal, 80
buccinator muscle, formation of, 158*f*
bud stage, of tooth development, 394, 394*f*–395*f*
bulbar conjunctiva, 380
bulbar ridges, 280, 281*f*, 283*f*
bulbourethral gland, 11*f*, 13, 24, 25*f*, 248*f*, 249, 250*t*
bulboventricular groove, 272*f*
bulboventricular loop, 271
bulbus cordis, 267*f*, 269*f*, 271, 304
  circulation through, 271, 272*f*
  partitioning of, 280–284, 281*f*, 283*f*
bundle branches, 282*f*, 284
busulfan, as teratogen, 128*t*

C

C cells, 161
caffeine consumption, during pregnancy, 131–132
CAH. *See* congenital adrenal hyperplasia
calcitonin, C cell production of, 161

calcium, and bone development, 308
calices
development of, 230, 234*f*
major, 230, 234*f*
minor, 230, 234*f*
calvaria, 313, 320
defects of, 317, 317*f*, 362–364
canalicular period, of lung maturation, 196*f*, 196–197, 199
cancer
diethylstilbestrol exposure and, 133
treatment of, teratogens in, 134
capacitation of sperms, 24, 29*f*
carbon dioxide, transplacental transport of, 97*f*, 98
carbon monoxide, transplacental transport of, 98
carboxyhemoglobin, cigarette smoking and, 131
cardiac jelly, 268, 269*f*, 270*f*, 272*f*
cardiac muscle, development of, 327
cardiac muscle fibers, 327
cardiac myoblasts, 327
cardiac valves, development of, 282*f*, 284, 285*f*
cardinal veins, 272*f*, 304, 337*f*
anterior, 265, 266*f*–267*f*
common, 143–144, 143*f*–144*f*, 264–265, 266*f*–267*f*
development of, 264–265, 266*f*–267*f*
posterior, 266*f*–267*f*
cardiogenic area, 49, 54*f*, 56
cardiovascular system, 263–304. *See also specific cardiovascular structures*
development of, 263–304
early, 56, 57*f*–58*f*, 264–268
primordial, 60
Carnegie Embryonic Staging System, 68*t*, 75
carotid arteries
common, 291, 292*f*, 294*f*
external, 174, 291, 294*f*
internal, 174, 291–293, 292*f*
carpus, 318*f*, 319*f*, 335*f*
cartilage
bone development in, 306–308, 309*f*
development of, 306–308
histogenesis of, 306
hyaline, 306, 309*f*, 310, 318
hypophysial, 313
parachordal, 313
pharyngeal arch, derivatives of, 154–158, 157*f*, 157*t* (*See also specific pharyngeal arches*)
cartilage matrix, 309*f*
cartilaginous joints, 308–310, 310*f*, 320
cartilaginous neurocranium, 313
cartilaginous otic capsules, 382, 382*f*–383*f*
cartilaginous stage, of vertebral development, 312*f*
cartilaginous viscerocranium, 313–314
cataracts, congenital, 379, 386
rubella virus and, 135, 136*f*
cauda equina, 350
caudal, as descriptive term, 8*f*
caudal eminence, 67, 69*f*, 72*f*, 73*f*
caudal fold, 62, 64*f*
caudal neuropore, 69*f*–71*f*, 344, 345*f*, 346*f*
defect in closure of, 353*f*
caudal ridges, 143
caudate nucleus, 361, 363*f*–364*f*
cavities, body. *See also specific body cavities*
development of, 139–150, 140*f*–144*f*
embryonic, 140–145
division of, 141–145
CDH. *See* congenital diaphragmatic hernia
cecal diverticulum, 212, 214*f*, 215, 217*f*
cecum
development of, 212, 214*f*, 215, 217*f*
subhepatic, 218, 220*f*
celiac trunk (artery), 141, 142*f*, 202*f*, 206, 213*f*, 268
cell cultures, fetal, 86
cell death, programmed. *See* apoptosis
cement (cementum), dental, 396
cementoblasts, 396
cementoenamel junction, 396
centers of growth, 171
central artery and vein of retina, 372, 375*f*, 377*f*
central canal of spinal cord, 344, 350*f*, 370
central incisors, 397
central nervous system, 344, 370. *See also* brain; spinal cord
histogenesis of cells in, 349*f*
primordium of, 49*f*, 50, 51*f*–52*f*, 53, 60
central tendon of diaphragm
development of, 62, 145*f*
primordial, 141, 143*f*, 145–146
centrum, 310, 312, 312*f*
cerebellar cortex, 356
cerebellum, development of, 353–356, 356*f*–358*f*, 370
cerebral aqueduct, 356–357, 357*f*–358*f*
cerebral commissures, 361–362, 363*f*
anterior, 363*f*
hippocampal, 363*f*
posterior, 363*f*
cerebral hemispheres, development of, 357–358, 358*f*–360*f*, 361–362, 363*f*–453*f*, 365*f*, 370
cerebral peduncles, 357, 358*f*
cerebral vesicles, 352–353, 358*f*, 361, 362*f*
primary, 352, 355*f*
secondary, 352–353, 355*f*
cerebrospinal fluid (CSF)
excessive, in hydrocephalus, 364, 366*f*
formation of, 348, 357*f*
cervical canal, 10
cervical flexure, 347*f*, 353, 356*f*
cervical myotomes, 325, 325*f*, 326*f*, 330
cervical ribs, 316, 316*f*
cervical sinus, 67, 73*f*, 152, 154*f*, 156*f*, 162
persistent remnant of, 163*f*–165*f*
cervical somites, 147*f*
cervix, 10, 11*f*–12*f*
CHAOS (congenital high airway obstruction syndrome), 190
cheeks, development of, 174
chemicals, as teratogens, 128*t*, 135
dosage and, 129–131
Chiari malformation, 366, 367*f*
chickenpox, fetal effects of, 136
childbirth, 99–105, 101*f*
chimeric models, 7
CHL. *See* crown-heel length
choanae, 176–177, 178*f*
primordial, 176, 178*f*
chondrification centers, 306, 312*f*, 330
chondroblasts, 306, 309*f*
chondrocranium, 313
chondrocyte, 309*f*
chondrogenesis, 306
chordae tendineae, 280, 282*f*
chordee, 255
chordomas, 311
choriocarcinomas, 58, 103
chorion, 40*f*, 41
blood supply to, 265*f*
number in twin pregnancies, 113*t*
smooth, 90, 91*f*–94*f*
villous, 90, 91*f*, 93, 95*f*
chorionic arteries, 96
chorionic cavity, 41, 44
ultrasound assessment of, 75*f*
chorionic fluid, 90, 91*f*
chorionic plate, 94, 94*f*, 95*f*, 96*f*, 97
chorionic sac
development of, 40*f*, 41, 42*f*, 90, 91*f*–93*f*
diameter of, ultrasound measurement of, 41, 74, 75*f*, 90–93
fusion with decidua, 95*f*
chorionic vessels, 92*f*, 103
chorionic villi, 90, 91*f*–96*f*, 93–94, 95*f*
arteriocapillary-venous system in, 96, 96*f*
branch (terminal), 58, 94–96, 95*f*, 96*f*
development of, 41, 42*f*, 57–58, 59*f*, 60
primary, 40*f*, 41, 42*f*, 57–58, 60
secondary, 57–58, 59*f*, 60
stem (tertiary), 57–58, 59*f*, 60, 94–95, 95*f*–96*f*
chorionic villus sampling, 86, 87*f*
diagnostic value of, 86
choroid, development of, 374*f*, 377*f*–378*f*, 380, 386
choroid fissure, 361, 364*f*
choroid plexus, 363*f*
development of, 356, 356*f*–357*f*, 361
chromaffin cells, 367
chromaffin system, 367
chromatid, 14
chromatophores, 376
chromosomal abnormalities, 118, 118*f*, 119–126
and spontaneous abortion, 35
detection of, 85–86
in gametes, 17–19, 18*f*
maternal age and, 17, 120, 123*t*
numerical, 119–123
structural, 123–126, 125*f*
chromosome(s)
breakage, 123
crossing over of, 14, 31
deletion of, 123–124, 125*f*
double chromatid, 14, 16*f*
duplication of, 125*f*, 125–126

haploid number of, 13–14
homologous, 13, 17, 119
in gametogenesis, 13–14
inversion of, 125*f*, 126
microdeletions and microduplications of, 124
nondisjunction of, 17, 18*f*, 43, 119, 120*f*
ring, 123–124, 125*f*
sex (*See* sex chromosomes)
single chromatid, 14, 16*f*
translocation of, 123, 125*f*
chyle cistern, 302, 303*f*
cigarette smoking, fetal effects of, 83–85, 85*f*, 131
ciliary body, development of, 376, 377*f*, 386
ciliary muscle, 376
ciliary processes, 376, 380
circulation of amniotic fluid, 95*f*, 106–108
circulation of blood
fetal, 295–296, 300*f*
neonatal, 295–299, 301*f*
transitional, 296–299
placental, 95*f*–96*f*, 96–99, 271
fetal, 95*f*–96*f*, 96
impaired, and fetal growth, 85
maternal, 95*f*, 97
through primordial heart, 271–276, 272*f*
uteroplacental
impaired, and fetal growth, 85
primordial, 39–40, 44, 56, 58*f*
circumvallate papillae, 170*f*
cisterna chyli, 302, 303*f*
cleavage of zygote, 31, 33*f*, 35, 36*f*
definition of, 2
cleft face, 185*f*, 185–186
cleft foot, 338, 338*f*
cleft hand, 338, 338*f*
cleft lip and palate, 181*f*–185*f*, 181–186
anterior, 181, 182*f*
bilateral cleft lip, 183, 183*f*
complete, 182*f*, 183
embryological basis of, 177*f*, 182*f*, 183–184, 184*f*
involving upper lip, 181*f*, 182–183
median cleft of lower lip, 183, 185*f*
median cleft of upper lip, 183, 185*f*
of anterior and posterior palate, 182*f*, 184
of anterior (primary) palate, 182*f*, 183
of posterior (secondary) palate, 182*f*, 183–184
posterior, 181–182, 182*f*
unilateral cleft lip, 181*f*, 183, 183*f*–184*f*
cleft uvula, 182*f*, 183
cleft vertebral column, 316, 317*f*
climacteric, 24
clinical embryology, 7
clitoris, 11*f*, 13, 13*f*
development of, 250*t*, 253, 254*f*, 261
cloaca, 62, 64*f*, 75, 202*f*, 222, 223*f*, 226
partitioning of, 222, 223*f*
cloacal membrane, 49, 49*f*, 51*f*–52*f*, 62, 64*f*, 75–76, 202, 202*f*, 222, 223*f*
cloning, 7
closing plug, 39*f*, 40
clubfoot, 137, 339, 340*f*
c-met, and muscle development, 324*f*, 326
coarctation of aorta, 293, 295*f*
juxtaductal, 293
postductal, 293, 295*f*
preductal, 293, 295*f*
cocaine use
and birth defects, 128*t*, 135
and fetal growth, 85
cochlea, membranous, 382, 382*f*
cochlear duct, 381–382, 382*f*
cochlear ganglion, 382, 382*f*
cochlear nerve, 369
cochlear nuclei, 369
coelom. *See also specific body cavities*
extraembryonic, 40, 40*f*, 44, 140, 140*f*–141*f*
intraembryonic
development of, 54*f*, 56, 60, 140, 140*f*, 149
embryonic folding and, 62, 63*f*–64*f*, 140, 141*f*
coelomic spaces, 54*f*, 56
collecting tubules, 230, 234*f*, 250*t*
collodion baby, 391
coloboma
of eyelid (palpebral coloboma), 380
of iris, 376, 377*f*
of retina, 374
colon
congenitally enlarged (megacolon), 224, 224*f*
development of, 212–215, 214*f*, 220
left-sided (nonrotation of midgut), 217, 220*f*
positioning of, 215, 216*f*
color flow Doppler ultrasound, of umbilical cord, 105
commissures
cerebral, 361–362, 363*f*
anterior, 361, 363*f*
hippocampal, 361, 363*f*
posterior, 363*f*
labial, 253
common placenta, in twin pregnancy, 111
compact layer, of endometrium, 10, 12*f*
compaction, 31
complement regulatory proteins, and placental immunoprotection, 104
computed tomography, for fetal assessment, 86
conceptional age, 78
conceptus, definition of, 2
conchae, 176, 178*f*
condensation, and bone development, 306
conduction system of heart, 282*f*, 284
abnormalities of, 284
cones (retinal), 372
congenital adrenal hyperplasia (CAH), 243, 244*f*, 255, 255*f*, 261
congenital anomalies, 117–138
causes of, 118, 118*f*
chromosomal abnormalities and, 118, 118*f*, 119–126
clinically significant types of, 118
definition of, 118
environmental factors in, 118, 118*f*, 127–138
genetic factors in, 118, 119–127, 138
genetic mutation and, 118, 118*f*, 126–127
multifactorial inheritance and, 118, 118*f*, 137, 138*f*
neurulation abnormalities and, 53
of anus and rectum, 222–227
of aortic arches, 293–295
of bladder, 241*f*–242*f*
of brain, 362–366
of duodenum, 207, 207*f*–208*f*, 226
of ear, 384*f*, 384–386
of esophagus, 190–193, 199, 202–203
of eye, 135, 136*f*, 374, 376, 378–380, 386
of face, 181*f*–185*f*, 181–186
of genital system, 247, 253–256, 255*f*–256*f*
of head and neck, 152, 162–166, 163*f*–166*f*, 186
of heart and great arteries, 284–289, 304
of hindgut, 222–227
of inguinal canals, 259–260
of kidneys and ureters, 234–237, 237*f*–239*f*
of larynx, 190
of limbs, 334–340, 338*f*–340*f*
causes of, 340
major, incidence of, 131*t*
terminology for, 335–336
thalidomide and, 118, 134, 135*f*, 335, 338*f*, 340
of lip and palate, 181–185, 181*f*–185*f*
of liver, 208
of lungs, 198
of lymphatic system, 303
of mammary glands, 392
of midgut, 215–220, 226
of muscles, 327, 328*f*
of nails, 392, 399
of pancreas, 210, 212*f*, 226
of pharyngeal apparatus, 152, 162–166
of skeleton, 316–317, 316*f*–317*f*, 319–320
of skin, 391, 391*f*, 399
of spinal cord, 350–352
of spleen, 212
of stomach, 203
of teeth, 397–398, 398*f*–399*f*, 400
of thyroid gland, 166*f*, 167, 168*f*–169*f*, 186
of tongue, 170, 171*f*
of trachea, 190–193, 199
of urinary bladder, 239–241, 241*f*–242*f*, 261
of uterus, 256, 257*f*, 261
of vena cava, 268
study of (teratology), 118
congenital diaphragmatic hernia (CDH), 147–149, 148*f*–149*f*
and pulmonary hypoplasia, 147, 148*f*, 198
prenatal diagnosis of, 147
congenital heart defects (CHDs), 284–289, 304
congenital high airway obstruction syndrome (CHAOS), 190
congenital rubella syndrome, 135

conjoined twins, 113, 116*f*
conjunctiva, 380, 386
  bulbar, 380
  palpebral, 380
conjunctival sac, 380
connecting stalk, 62, 63*f*
connective tissue framework, 96*f*
contiguous gene syndromes, 124
contraceptives, oral
  and ovulation, 21
  fetal effects of, 132–133
conus arteriosus, 281*f*, 284
copula, 169, 170*f*
cornea, development of, 372, 377*f*–379*f*, 380, 386
corneal endothelium, 380
corneal epithelium, 380
corniculate cartilage, formation of, 157*t*
corona radiata, 15*f*, 17
  passage of sperm through, 28, 29*f*–30*f*
coronal plane, 8*f*
coronary sinus, 267*f*, 276–277, 278*f*
corpora cavernosa clitoridis, 250*t*
corpora cavernosa penis, 250*t*, 252
corpus albicans, 21
corpus callosum, 361–362, 363*f*
corpus luteum, 21, 22*f*, 39
  involution and degeneration of, 21
  of menstruation, 21
  of pregnancy, 21
corpus spongiosum penis, 250*t*, 252
corpus striatum, 361–362, 363*f*
cortex
  cerebellar, 356
  gonadal, 244, 246*f*, 250*t*
  suprarenal, 243, 244*f*, 250*t*
corticosteroids, fetal effects of, 134
corticotropin-releasing hormone (CRH), and labor, 99
cortisol, and labor, 99
costal processes, 311, 312*f*, 313
costodiaphragmatic recesses, 146, 146*f*
costovertebral joints, 312*f*, 313
cotyledons, 94, 102*f*, 103
  retention of, and uterine hemorrhage, 102
coxsackie virus, transplacental transport of, 99
cranial, as descriptive term, 8*f*
cranial fold, 62, 64*f*
cranial ganglion, 355
cranial meningocele, 363
cranial nerves
  formation of, 157*t*, 366–369, 368*f*
  innervation of pharyngeal arches by, 158–159
  somatic efferent, 367–368, 368*f*
  special sensory, 368*f*, 369
  special visceral afferent components of, 158
  special visceral efferent (branchial) components of, 158
cranial ridges, 143
cranial vault (calvaria), 313, 320
  defects of, 317, 317*f*
craniofacial anomalies, benzodiazepine derivatives and, 134
craniopharyngiomas, 361
craniosynostosis, 317, 317*f*
cranium
  congenital anomalies of, 317, 317*f*, 362–363, 365*f*
  development of, 310, 313–316, 320
  fetal, molding of, 313
  newborn, 314*f*, 315
  postnatal growth of, 315–316
cranium bifidum, 362–363, 365*f*
cremasteric muscle and fascia, 258*f*, 259
cretinism, 134, 320
CRH (corticotropin-releasing hormone), and labor, 99
cri du chat syndrome, 123
"crib death," 284
cribriform plate, 369
cricoid cartilage, formation of, 157*t*
cricothyroid muscle, formation of, 157*t*
crista dividens, 296
crista terminalis, 276, 278*f*
cristae ampullares, 382
critical periods, of development, 129, 130*f*
CRL. *See* crown-rump length
crossed renal ectopia, 235, 238*f*
crossing over, of chromosomes, 14, 31
crown-heel length (CHL), 75, 76*f*, 79
crown-rump length (CRL), 74–75, 75*f*–76*f*, 78*t*, 78–79
  at full term, 83
crura of diaphragm, 145*f*, 146
cryptorchidism, 259, 259*f*, 261
CSF. *See* cerebrospinal fluid.
cumulus oophorus, 19
cuneate nuclei, 353, 356*f*
cuneiform cartilage, formation of, 157*t*
cutaneous nerve area, 334
cutaneous syndactyly, 338–339, 339*f*
cuticle, 392
CVS (chorionic villus sampling), 86, 87*f*
cyanosis/cyanotic heart disease, 284, 288
cyst(s)
  auricular, 162, 163*f*
  branchial, 162, 163*f*–165*f*, 186
  dentigerous, 398
  Gartner duct, 248*f*
  lingual, congenital, 167*f*, 170
  thyroglossal duct, 167, 167*f*, 168*f*, 186
  urachal, 239, 241*f*
cystic duct, 206*f*, 207
cystic duplication of intestines, 220
cystic hygroma, 303, 304, 391
cystic kidney disease, 237
cytogenetics, molecular, 124–125
cytomegalovirus
  as teratogen, 128*t*, 136
  transplacental transport of, 97*f*, 99
cytotrophoblast, 33, 34*f*, 38, 38*f*–39*f*, 43, 96*f*, 97
cytotrophoblastic shell, 58, 59*f*, 60, 93–94, 95*f*

D

deaf-mutism, 384
deafness
  congenital, 135, 384, 386
  streptomycin and, 133
decidua, 90
  regions of, 90, 91*f*
decidua basalis, 90, 91*f*, 93, 94*f*, 95*f*
decidua capsularis, 90, 91*f*, 94
decidua parietalis, 90, 91*f*, 93*f*, 94*f*
decidual cells, 38, 40, 90
decidual reaction, 40, 44, 90
deciduous teeth
  development of, 316, 393–394
  eruption of, 396–397, 397*f*
  shedding of, order and usual time of, 394*t*
deep artery of thigh, 334, 337*f*
deep inguinal ring, 256
deferentectomy, 26
deformation, 118
deletion, chromosomal, 123–124, 125*f*
delivery, 99–105, 101*f*
  expected date of (EDD), 83
dental lamina, 394, 394*f*–396*f*
dental papilla, 395, 395*f*
dental sac, 395*f*, 396
dentate nucleus, 356, 357*f*
dentigerous cyst, 398
dentin, 395*f*–396*f*, 396
dentinogenesis imperfecta, 398, 399*f*
dermal ridges, 389
dermal root sheath, 390*f*
dermatoglyphics, 389
dermatomal patterns
  of limb development, 336*f*
  of skin innervation, 389
dermatome, 332
dermis, development of, 388–389
dermomyotome, 306, 307*f*
desquamation, 388, 388*f*
detached retina, congenital, 374
dextrocardia, 284, 286*f*
  isolated, 284
  with situs inversus, 284
diabetes mellitus, fetal effects of, 134
diaphragm
  central tendon of
    development of, 62, 145*f*
    primordial, 141, 143*f*, 145–146
  congenital absence of, 327
  crura of, 145*f*, 146
  defects of, 147–148
    posterolateral, 147, 148*f*–149*f*
  development of, 145–147, 145*f*–147*f*, 150
    muscular ingrowth from lateral body walls, 145*f*–146*f*, 146
  eventration of, 147–148, 148*f*
  innervation of, 146–147
  positional changes of, 146–147, 147*f*
  primordial, 146
diaphragmatic atresia, 218
diaphragmatic hernia, congenital, 147–149, 148*f*–149*f*

diaphragmatic hernia, congenital—cont'd
and pulmonary hypoplasia, 147, 148*f*, 198
prenatal diagnosis of, 147
diaphysial-epiphysial junction, 308
diaphysis, 306–308, 309*f*, 318
diazepam, use during pregnancy, 134
diencephalon, 347*f*, 352, 355*f*, 357–361, 359*f*, 370
and pituitary development, 359*f*–360*f*, 361
diethylstilbestrol, as teratogen, 128*t*, 133
differentiation, and bone development, 306
digastric muscle, formation of, 157*t*, 158*f*
DiGeorge syndrome, 165
digestive system, development of, 201–227
digit(s)
congenital anomalies of, 338–339, 339*f*
development of, 67–69, 71, 73*f*, 330, 331*f*–334*f*
supernumerary, 338, 339*f*
webbing of, 338–339, 339*f*
digital rays, 69, 73*f*, 318*f*, 330, 333*f*–334*f*, 340
notches between, 71, 73*f*, 330, 333*f*, 340
dilation stage of labor, 100, 101*f*
dilator pupillae muscle, 327, 376, 386
diploid number, 13
direction, anatomical, descriptive terms for, 7, 8*f*
discoid kidney, 235, 238*f*
discoid placenta, 100, 102*f*
discordant twins, 119
dispermy, 28
disruption, 118
distal tongue buds, 169, 170*f*
distress, fetal, 87
diverticulum
cecal, 212, 214*f*, 215, 217*f*
hepatic, 206*f*, 207, 209*f*
hypophysial, 359–361, 360*f*
ileal, 109–110, 220, 221*f*, 226
laryngotracheal, 190, 191*f*, 192*f*, 198
metanephric, 230, 233*f*, 234*f*, 260
branching of, 230, 236*f*
congenital disorders of, 234, 238*f*
neurohypophysial, 361
respiratory, 191*f*
dizygotic twins, 111, 112*f*, 114
placental vascular anastomosis with, 111
dolichocephaly, 317, 317*f*
Dolly (cloned sheep), 7
dominantly inherited congenital anomalies, 126
Doppler ultrasound, color flow, of umbilical cord, 105
dorsal, as descriptive term, 8*f*
dorsal gray columns, 346
dorsal (gray) horns, 346, 367
dorsal mesentery, 140–141, 141*f*–142*f*
of esophagus, diaphragm development from, 145*f*, 145–146, 150
dorsal mesogastrium, 203, 204*f*–205*f*, 213*f*
dorsal primary ramus, 324, 325*f*
dorsal root of spinal nerve, 350*f*, 367
dorsal septum, 346
double bubble sign, of duodenal atresia, 207, 208*f*
double chromatid chromosomes, 14, 16*f*
Down syndrome, 17, 120, 122*t*, 123
and intrauterine growth retardation, 85
duodenal atresia with, 207
incidence of, 123*t*
maternal age and, 17, 120, 123*t*
phenotype of, 119, 121*f*
skin development in, 389
tongue anomalies in, 170
drug(s). *See also specific drugs*
as teratogens, 128*t*, 131–135
dosage and, 129–131
transplacental transport of, 97*f*, 98–99
drug abuse, fetal effects of, 85, 135
duct(s)
alveolar, 196*f*
bile, 206*f*, 208
cochlear, 381–382, 382*f*
cystic, 206*f*, 207
ejaculatory, 11*f*, 13, 25*f*, 239, 249, 250*t*
endolymphatic, 381, 382*f*
Gartner, 248*f*, 250*t*. Duct of epoophoron
cysts of, 248*f*
genital
development of, 248*f*, 249–252, 251*f*, 253*f*
female, 248*f*, 249–251, 253*f*
male, 248*f*, 249
embryonic, vestigial structures derived from, 251–252
hepatic, accessory, 208
lactiferous, 392, 393*f*
lymphatic
development of, 302, 303*f*
right, 302, 303*f*
mesonephric, 230, 232*f*–234*f*, 239, 240*f*
adult derivatives and vestigial remains of, 230, 250*t*, 251–252
and male genital system, 246*f*, 247, 248*f*, 249, 251*f*, 261
remnants of, 248*f*
nasolacrimal, 174
atresia of, 174
nephrogenic, 233*f*
of epoophoron, 250*t*
pancreatic
accessory, 210, 211*f*, 226
main, 209–210, 211*f*
paramesonephric, 245*f*
adult derivatives and vestigial remains of, 250*t*, 251–252
and female genital system, 246*f*, 247–251, 248*f*, 251*f*, 261
remnants of, 248*f*
suppression, in male development, 247, 261
semicircular, 382, 382*f*
submandibular, 171
thoracic, development of, 302, 303*f*
thyroglossal, 166, 166*f*
cysts and sinuses of, 167, 167*f*, 168*f*, 186
ductuli efferentes, 250*t*. Efferent ductules
ductus arteriosus, 292*f*, 293, 294*f*, 296, 300*f*
adult derivatives of, 299, 302*f*
closure/constriction, 296–299, 302*f*
patent, 288, 299, 302*f*, 304
ductus deferens, 11*f*, 13, 24, 25*f*, 248*f*, 249, 250*t*, 259
ductus epididymis, 246*f*, 247, 248*f*, 249
ductus reuniens, 382, 382*f*
ductus venosus, 265, 267*f*, 295–296, 300*f*
adult derivatives of, 299
sphincter mechanism of, 296
duodenal atresia, 207, 207*f*, 226
ultrasound of, 207, 208*f*
duodenal obstruction, 217
duodenal papilla, minor, 210
duodenal stenosis, 207, 207*f*
duodenum
congenital anomalies of, 207, 207*f*–208*f*, 226
development of, 202, 202*f*, 206*f*, 206–207, 212, 226
positioning of, 215, 216*f*
duplication, chromosomal, 125*f*, 125–126
dura mater, 348, 352*f*
dwarfism, 126, 126*f*, 319–320, 320*f*
dysplasia, 118

E

ear, 380–386
congenital anomalies of, 384*f*, 384–386
development of, 159, 160*f*, 380–386
external, 380
development of, 69, 73*f*, 74, 74*f*, 172*f*–173*f*, 174, 186, 383*f*–384*f*, 384, 386
internal, 381
development of, 67, 69*f*, 381*f*–382*f*, 381–383, 386
low-set, 385–386
middle, 380
development of, 154–158, 157*t*, 157*f*, 313, 383*f*, 383–384, 386
eardrum. *See* tympanic membrane
earlobe, 384
early pregnancy factor, 28, 46
eccrine sweat glands, 389
ectoderm, embryonic, 41*f*, 46, 47*f*, 48, 60
derivatives of, 46, 65*f*, 380, 386, 388
ectodermal dysplasia, congenital, 397
ectopia cordis, 284
ectopic anus, 224–225
ectopic kidney, 234–235, 238*f*
ectopic parathyroid gland, 165, 166*f*
ectopic pregnancies, 42–43, 43*f*
ectopic testes, 259, 259*f*
interstitial, 259
ectopic thyroid gland, 166*f*, 167, 168*f*, 186
ectopic ureter, 237, 239*f*
ectromelia, 336

Edwards syndrome (trisomy 18), 120, 121*f*, 122*t*, 124, 385–386
and intrauterine growth retardation, 85
efferent ductules, 230, 247, 248*f*
egg. *See* oocyte(s)
eighth week, 71–74, 74*f*
ejaculate, 24
ejaculatory duct, 11*f*, 13, 25*f*, 239, 249, 250*t*
elastic cartilage, 306
electrolytes, transplacental transport of, 97*f*, 98
electromagnetic fields, fetal effects of, 137
eleventh week, 81*f*
embryo(s)
definition of, 2
development of (*See* embryonic development)
folding of, 62, 63*f*–64*f*, 75–76, 141*f*
and cardiovascular system, 62, 64*f*, 268, 271*f*
lateral, 62, 63*f*, 75
longitudinal, 62, 64*f*
genotype of, and effect of teratogens, 131
heartbeat of, ultrasound detection of, 56, 58*f*, 280
implantation of, 31–33, 34*f*, 38*f*–40*f*, 38–41, 43–44
measurement of, 68*t*, 75*f*–76*f*
nutrient transfer to, 38
spontaneous abortion of, 35, 43
ultrasound of, 74–75, 75*f*–76*f*
embryo transfer, 31, 32*f*
embryoblast, 31, 33*f*, 35
embryological timetables, 129, 130*f*
embryology
applied, 7
clinical, 7
importance of and advances in, 2–7
terminology in, 2
embryonic age, estimation of, 55, 68*t*, 74, 76*f*
embryonic development. *See also specific anatomy and processes*
1st (first) week, 27–35, 36*f*
2nd (second) week, 37–44
3rd (third) week, 45–60
4th (fourth) week, 67, 69*f*–71*f*, 69*f*–72*f*
5th (fifth) week, 67, 73*f*
6th (sixth) week, 67–71, 73*f*
7th (seventh) week, 71
8th (eighth) week, 71–74, 74*f*
control of, 66–67
inductive tissue interactions in, 66–67
molecular biology of, 7
stages of, criteria for estimating, 68*t*
teratogens and, 129, 130*f* (*See also specific teratogens*)
timetable of, 129, 130*f*
embryonic disc
bilaminar, formation of, 39*f*, 39–40, 44
formation of, 39*f*, 39–41, 44
trilaminar, 46, 47*f*
folding of, 62, 63*f*–64*f*, 75–76
embryonic period, 6*f*
embryonic pole, 34, 34*f*, 38
embryonic tissues, origin of, 41*f*
embryotroph, 39
enamel, tooth, development of, 393–396, 395*f*–396*f*, 400
abnormal, 397, 398*f*–399*f*, 400
enamel epithelium, 395*f*, 395–396
inner, 395–396
outer, 395
enamel hypoplasia, 397, 398*f*
enamel organ, 395*f*
enamel pearls, 397, 398*f*–399*f*
enamel (stellate) reticulum, 395*f*, 396
endocardial cushions, 272*f*, 276, 281*f*
defects of, 287*f*, 288
endocardial heart tubes, 56, 60, 264, 268, 269*f*, 270*f*–271*f*, 303–304
endocardium, 269*f*, 270*f*
endochondral bone formation, 306, 308*f*, 320
endocrine synthesis, in placenta, 99
endocrine system
in climacteric, 24
parathyroid glands
abnormal number of, 166
congenital absence of, 166
congenital anomalies of, 165–166, 166*f*
development of, 160*f*–161*f*, 161, 186
ectopic, 165, 166*f*
inferior, 161
superior, 161
pineal gland (body), 358, 359*f*
pituitary gland
and female reproductive cycles, 19, 20*f*
derivation and terminology of, 360*t*
development of, 359–361, 360*f*
glandular part of (adenohypophysis), 360*t*, 360*f*, 361
nervous part of (neurohypophysis), 360*t*, 360*f*, 361
suprarenal glands, development of, 236*f*, 243, 244*f*
thyroid gland
accessory tissue of, 168*f*
congenital anomalies of, 166*f*, 167, 168*f*–169*f*, 186
development of, 166*f*, 166–167, 186
ectopic, 166*f*, 167, 168*f*, 186
isthmus of, 166
lingual tissue of, 167, 168*f*
sublingual, 167, 169*f*
endoderm, embryonic, 41*f*, 46, 47*f*, 48, 60
derivatives of, 46, 65*f*
endolymphatic duct, 381, 382*f*
endolymphatic sac, 381, 382*f*
endometrial (spiral) arteries, 12*f*, 23, 39–40, 94, 94*f*–95*f*, 97
endometrial menstrual cycle, 22*f*, 22–24
anovulatory, 23
duration of, 23
ischemic phase of, 23
luteal (secretory) phase of, 10, 23
menstrual phase of, 23
pregnancy phase of, 24
proliferative phase of, 23
endometrial veins, 39–40, 94*f*–95*f*
endometrium, 10, 11*f*–12*f*
as mirror of ovarian cycle, 23
blastocyst implantation in, 31–33, 34*f*, 38*f*–40*f*, 38–41, 43–44
gravid, 90
layers of, 10, 12*f*
endothelium of fetal capillaries, 96*f*, 97
endovaginal ultrasound, 75
environmental chemicals, as teratogens, 128*t*, 135
dosage and, 129–131
environmental factors, and birth defects, 118, 118*f*, 127–138
epaxial division, of myotomes, 324, 325*f*, 326*f*
derivatives of, 324–325
ependyma, 344
ependymal cells, 344
epiblast, 39, 39*f*, 41*f*, 44
and gastrulation, 46–48, 60
epicardium, 268, 269*f*, 270*f*
epidermal ridges, 388*f*, 388–389
epidermal root sheath, 389–391
epidermis, development of, 388*f*, 388–389
epididymis, 11*f*, 13, 17, 25*f*, 248*f*, 250*t*
appendix of, 248*f*
development of, 246*f*, 248*f*, 249
epigastric hernia, congenital, 148–149
epiglottis, development of, 158, 190, 192*f*, 198–199
epiphyses, 308, 309*f*, 319
anular, 312, 312*f*
epiphysial cartilage plates (growth plates), 308, 309*f*, 319
epispadias, 241, 242*f*, 261
epithalamic sulcus, 358, 359*f*
epithalamus, 358, 359*f*
epithelial root sheath, 396
eponychium, 392, 392*f*
epoophoron, 248*f*, 250*t*
duct of, 250*t*
equatorial zone of lens, 377*f*, 378
erectile tissue, of penis, 11*f*, 13
errors of metabolism, 86
erythroblastosis fetalis (hemolytic disease of newborn), 99
intrauterine fetal transfusion for, 86
erythrocytic mosaicism, 111
erythropoiesis, fetal, 80, 83
esophageal atresia, 108, 202
tracheoesophageal fistula with, 190–193, 194*f*, 199, 202
esophageal muscles, formation of, 157*t*
esophageal stenosis, 203
esophagus
congenital anomalies of, 190–193, 199, 202–203
development of, 190, 191*f*, 192*f*, 198, 202, 202*f*, 226

esophagus—cont'd
dorsal mesentery of, diaphragm development from, 145*f*, 145–146, 150
recanalization, failure of, 202
esterases, and fertilization, 28
estrogen
and female genital development, 253
and labor, 99–100
and mammary development, 392
and menstrual cycle, 23
and ovarian cycle, 19, 20*f*
production of
ovarian, 10
placental, 99
ethisterone, avoidance in pregnancy, 132
excretory system. *See* urinary system
exocoelomic cavity, 38*f*, 39
exocoelomic membrane, 38*f*, 39
expected date of confinement (EDC), 78, 83
expected date of delivery (EDD), 83
expulsion stage of labor, 100, 101*f*
exstrophy of bladder, 241, 241*f*–242*f*, 261
external os of uterus, 10, 12*f*
extraembryonic coelom, 40, 40*f*, 44, 140, 140*f*–141*f*
extraembryonic mesoderm, 38*f*–41*f*, 39, 44
somatic, 40*f*, 41, 56
splanchnic, 40*f*, 41, 56
extrahepatic biliary atresia, 208
extravillous trophoblast, 103
extremely low birth weight, 78
eye(s)
anterior chamber of, 377*f*, 379, 379*f*
congenital anomalies of, 135, 136*f*, 374, 376, 378–380, 386
development of, 53, 67, 69, 69*f*, 74*f*, 171, 172*f*–173*f*, 357, 372–380, 373*f*–379*f*, 385–386
inductive tissue interaction and, 66, 372
movement, fetal, 80
muscles, 325*f*, 326, 326*f*
posterior chamber of, 377*f*, 379
eyelid(s)
congenital anomalies of, 380, 386
development of, 74, 74*f*, 83, 87, 173*f*, 377*f*–379*f*, 380

F

face
congenital anomalies of, 181*f*–185*f*, 181–186
development of, 152, 154*f*–156*f*, 171–176, 172*f*–178*f*
postnatal growth of, 316
prenatal smallness of, 176
skeleton of (viscerocranium), development of, 313–315, 320
facial clefts, 185*f*, 185–186
facial expression, muscles of
formation of, 157*t*, 175–176
innervation of, 368–369
facial muscles, development of, 326*f*
facial nerve, formation of, 157*t*, 158, 366, 368*f*, 368–369
facial primordia, 171, 172*f*–173*f*
facilitated diffusion, placental transport via, 98
falciform ligament, 208, 209*f*, 210*f*, 213*f*
fallopian tubes. *See* uterine tubes
false knots, in umbilical cord, 105
false ribs, 313
fat, fetal, 87–88
brown, 80
white, 83
female anatomy, 10–13, 244–251, 253, 261. *See also specific anatomical entities*
female fetus, masculinization of, 98, 132, 132*f*, 243, 244*f*, 255, 255*f*, 261
female pronucleus, 28, 30*f*, 35
female pseudohermaphroditism, 255, 255*f*, 261
female reproductive cycles, 19–24, 20*f*–22*f*
female reproductive organs, 10–13, 11*f*–13*f*
femoral artery, 337*f*
femur, 74, 319*f*
femur length, fetal, 78–79
fertilization, 13, 24, 25*f*, 28–31, 29*f*–30*f*, 36*f*
in vitro, 31, 32*f*
phases of, 28–31, 29*f*–30*f*
results of, 31
site of, 28
fertilization age
definition of, 2
estimation of, 74, 78*t*
fertilized oocyte, 15*f*, 17
fetal age, estimation of, 78*t*, 78–79
fetal alcohol effects, 132
fetal alcohol syndrome, 132, 132*f*
fetal distress, 87
fetal growth, factors influencing, 83–85
fetal hydantoin syndrome, 131, 133, 133*f*
fetal membranes, 89–115
after birth, 100–105, 102*f*
development of, 91*f*
functions and activities of, 90
multiple pregnancies and, 111–114, 113*t*
premature rupture of, 96, 108
fetal movement, 80
fetal period, 6*f*, 77–88
highlights of, 79–83
nine to twelve weeks, 79–80, 79*f*–81*f*
seventeen to twenty weeks, 80, 82*f*
thirteen to sixteen weeks, 80, 82*f*
thirty to thirty-four weeks, 83
thirty-five to thirty-eight weeks, 83, 84*f*–85*f*
twenty-one to twenty-five weeks, 80–83
twenty-six to twenty-nine weeks, 83
fetal surgery, 88
fetal weight, 78*t*, 79
fetomaternal junction, 93–94, 95*f*
fetomaternal organ, placenta as, 90
fetoplacental blood flow, 95*f*–96*f*, 96
impaired, and fetal growth, 85
fetus
as unborn patient, 85
body proportions of, 79–80, 80*f*, 81*f*
computed tomography of, 86
definition of, 2
harlequin, 391
magnetic resonance imaging of, 86
measurements and characteristics of, 78*t*, 79
status of, assessment of, 85–88
ultrasound of, 78–79, 79*f*
viability of, 2, 78
fibrinoid material, 96*f*, 98
fibroblast growth factor(s)
and cardiovascular development, 282
and limb development, 330
and lung development, 197–198
and nervous system development, 344
and pancreatic development, 209
fibroblast growth factor receptor 3 gene, mutation, and achondroplasia, 126
fibrocartilage, 306
fibrous joints, 308, 310*f*, 320
fibrous pericardium, 143, 144*f*
fibula, 319*f*, 332
fibular artery, 337*f*
fifth week, 67, 73*f*
filiform papillae, 169
filum terminale (terminal filum), 350, 352*f*
fimbriae, of uterine tube, 22*f*, 24
finger(s), development of, 67–69, 71, 73*f*, 330, 331*f*–334*f*
fingernails, development of, 80, 391–392, 392*f*, 399
first arch syndrome, 162–165, 165*f*, 384–385
first meiotic division, 13–14, 16*f*
in oogenesis, 15*f*, 17
in spermatogenesis, 14, 15*f*
first trimester, 79
first week, 27–35, 36*f*
FISH (fluorescent in situ hybridization), 86, 125
fissure(s)
choroid, 361, 364*f*
retinal, 372, 373*f*, 375*f*
defects in closure of, 374, 376, 386
ventral median, 348
fistula(s)
anal agenesis with, 224–225, 225*f*
anoperineal, 224, 225*f*
auricular, 385
branchial, 162, 163*f*–164*f*, 186
lingual, 167*f*, 170
perineal, 225*f*
rectocloacal, 225*f*
rectourethral, 225*f*, 226
rectovaginal, 225*f*, 226
rectovesical, 226
rectovestibular, 226
tracheoesophageal, 190–193, 194*f*, 199, 202

fistula(s)—cont'd
umbilicoileal, 220, 221f
urachal, 241, 241f
flexures, brain, 347f, 353, 356f
floating ribs, 313
flocculonodular lobe, 357f
fluorescent in situ hybridization (FISH), 86, 125
folding of embryo, 62, 63f–64f, 75–76, 141f
and cardiovascular system, 62, 64f, 268, 271f
lateral, 62, 63f, 75
longitudinal, 62, 64f
foliate papillae, 169
folic acid antagonists, as teratogens, 134
folic acid supplements, and neural tube defects, 352
follicles, ovarian. *See* ovarian follicles
follicle-stimulating hormone (FSH)
and ovarian cycle, 19, 20f, 22f
and reproductive cycles, 19–22, 20f, 22f
oral contraceptives and, 23
ovulation induction and, 21
follicular fluid, 19, 22f
fontanelles, 314f
foot
cleft, 338, 338f
congenital anomalies of, 338–339, 338f–340f
development of, 330, 331f, 333f, 340
foot length, fetal, 78t, 79
foot plate, 73f, 330, 334f
foramen cecum of tongue, 160f–161f, 167–169, 170f
foramen magnum, 313
foramen of Bochdalek, 147
foramen of Magendie, 356
foramen of Morgagni, herniation through, 149
foramen ovale. *See* oval foramen
foramen primum, 273f, 274f–275f, 276
fate of, 273f, 274f–275f
patent, 287f, 288
foramen secundum, 273f, 274f–275f, 276
foramina of Luschka, 356
forebrain, 67, 69f, 71f–72f, 347f, 352, 355f, 357–361, 359f, 370
and eye development, 357, 372, 373f, 386
defects of, 364–366
division into telencephalon and diencephalon, 352
foregut, 62, 64f, 75, 143f, 202
arteries of, 141, 142f
derivatives of, 202–210, 226
foreskin (prepuce), 11f, 243f, 252
fossa ovalis (oval fossa), 299
fourth week, 67, 69f–72f
fragile X syndrome, 126, 127f
fraternal (dizygotic) twins, 111, 112f, 114
placental vascular anastomosis with, 111
frenulum of upper lip, 176, 180f
frontal plane, 8f
frontal sinuses, 178
frontalis muscle, formation of, 158f
frontonasal prominence, 171, 172f–177f, 176
fructose, in semen, 24
FSH. *See* follicle-stimulating hormone
functional layer, of endometrium, 10, 12f
fungiform papillae, 169
fused kidney, 235
fused ribs, 316
fused teeth, 398, 398f

G

gallbladder, development of, 206f, 207–208
gamete(s), 10, 13, 14f. *See also* oocyte(s); sperms
chromosomally abnormal, 17
male versus female, 14f, 17
transportation of, 22f, 24, 25f
viability of, 26
gametogenesis, 13–19, 15f
abnormal, 17–19, 18f
ganglion (ganglia)
autonomic, 366–367
cochlear, 382, 382f
cranial, 355
geniculate, 369
preaortic, 369
spinal, development of, 348, 350f–351f, 366–367
spiral, 382f
sympathetic, 243, 351f, 367, 369
trigeminal, 368
Gartner duct, 248f, 250t
Gartner duct cysts, 248f
gases, transplacental transport of, 83, 97f, 98
gastroschisis, 86, 148–149, 217, 219f
gastrosplenic ligament, 213f
gastrulation, 46–53, 47f, 58–60
gene(s)
housekeeping, 127
inactivation of, 119
specialty, 127
gene mapping, 125
gene mutation, and birth defects, 118, 118f, 126–127
general somatic afferent, 354, 356f
general somatic efferent, 354, 356f
general visceral afferent, 354, 356f–357f
general visceral efferent, 354, 356f–357f
genetic disorders, preimplantation diagnosis of, 35
genetic imprinting, 124
geniculate ganglion, 369
genital ducts
development of, 248f, 249–252, 251f
female, 248f, 249–251, 253f
male, 248f, 249
embryonic, vestigial structures derived from, 251–252
genital glands
auxiliary, in females, 251
development of
female, 248f, 249–251, 253f
male, 248f, 249
genital system, 230
congenital anomalies of, 247, 253–256, 255f–256f
development of, 231f, 243–253, 261
indifferent state of, 244, 251f, 261
genital tubercle, 242f, 252, 254f
genitalia
ambiguous, 253
external
development of, 252–253, 254f
female, 253, 254f, 261
male, 252, 254f
visualization, for sex determination, 253
female, masculinization of, 243, 244f, 255, 255f
genome sequencing, 127
genomic imprinting, 127
genotype of embryo, and effects of teratogens, 131
germ cells, 13. *See also* gamete(s)
primordial, 109, 244–245, 245f, 247, 261
germ layers
derivatives of, 62–66, 65f, 75
formation of, 46–53, 47f
German measles. *See* rubella
germinal matrix, 390
gestational age, 28
estimation of, 74, 75f, 78–79, 90
ultrasound assessment of, 75f, 78–79
gestational sac. *See* chorionic sac
gestational time units, 78–79, 79t
gestational trophoblastic disease, 103
gigantism, 320
gingivae (gums), development of, 175–176, 180f, 395f, 396–397
glandular plate, 241, 243f, 252, 254f
glans clitoris, 250t, 254f
glans penis, 11f, 243f, 250t, 252, 254f
glanular hypospadias, 255, 256f
glaucoma, congenital, 379, 386
rubella virus and, 135, 136f
glial-derived neurotrophic factor, and renal development, 236f
glioblasts, histogenesis of, 344, 349f
glomerular capsule, 233f, 235f
glomerulus, 230, 233f, 235f
glossopharyngeal nerve
formation of, 157t, 158, 366, 368f, 369
innervation of tongue by, 170
glottis, primordial, 190
glucagon, pancreatic secretion of, 210
glucocorticoids, and fetal lung maturity, 198
glucose
in fetal metabolism and growth, 83
transplacental transport of, 98
GnRH (gonadotropin-releasing hormone)
and female reproductive cycles, 19, 20f, 22f
oral contraceptives and, 23
goiter, congenital, thyroid drugs and, 134
gonad(s), 10. *See also* ovaries; testes
congenital anomalies of, 247

gonad(s), 10. *See also* ovaries; testes—cont'd
development of, 233*f*, 244–249, 245*f*–246*f*, 248*f*, 261
indifferent, 244–245, 245*f*, 247
adult derivatives and vestigial remains of, 250*t*
gonadal ridge, 230, 244, 245*f*–246*f*
gonadal veins, 266*f*
gonadotropin(s), for ovulation induction, 21
gonadotropin-releasing hormone (GnRH)
and female reproductive cycles, 19, 20*f*, 22*f*
oral contraceptives and, 23
gracile nuclei, 353, 356*f*
gravid endometrium, 90
gray columns, 346
gray communicating ramus, 369
gray horns, 346, 367
great arteries, transposition of, 288, 290*f*
great vessels, formation of, 56
greater cornu, formation of, 157*f*, 157*t*, 158
greater curvature of stomach, 203, 204*f*–205*f*
greater omentum, 203, 204*f*–205*f*
greater vestibular glands, 248*f*
development of, 250*t*, 251
greatest length (GL), 74, 76*f*
growth
centers of, 171
fetal, factors influencing, 83–85, 85*f*
growth and differentiation factor 5 (Gdf5), and bone development, 306
growth factor(s)
and bone development, 306
embryonic, and gastrulation, 48
growth hormone, in fetal metabolism and growth, 83
growth plates, 308, 309*f*, 319
growth spurt, pubertal, 10
gubernaculum, 248*f*, 250*t*, 256, 258*f*, 259
gubernaculum testis, 250*t*, 258*f*
gums, development of, 175–176, 180*f*, 395*f*, 396–397
gut
abnormal fixation of, 215, 226
malrotation of, 215, 226
primordial, 62, 64*f*, 75–76, 109, 141, 202, 202*f*, 226 (*See also* foregut; hindgut; midgut)
gynecomastia, 392
gyri, cerebral, 362, 365*f*

H

hair
development of, 80, 389–391, 390*f*, 399
abnormal, 391, 391*f*
lanugo, 80, 83, 87, 391, 399
hair bud, 390
hair bulb, 390, 390*f*
hair follicle, 389–391, 390*f*
hair papilla, 390
hair shaft, 390*f*, 391
hand(s)
cleft, 338, 338*f*
congenital anomalies of, 338–339, 338*f*–339*f*
development of, 67–69, 71, 73*f*, 318*f*, 330, 331*f*–334*f*, 340
hand plates, 67–69, 71, 73*f*, 330, 333*f*–334*f*
haploid number, 13–14
hard palate, 179, 179*f*–180*f*
cleft of, 181–183
harlequin fetus, 391
haversian systems, 306
hCACTH (human chorionic corticotropin), placental synthesis of, 99
hCG. *See* human chorionic gonadotropin
hCS (human chorionic somatomammotropin), placental synthesis of, 99
hCT (human chorionic thyrotropin), placental synthesis of, 99
head
congenital anomalies of, 152, 162–166, 163*f*–166*f*, 186
development of, 151–187
embryonic, growth of, 67
fetal
circumference of, 79
growth of, 79–80, 81*f*, 87
measurement of, 78–79
head fold, 62, 63*f*–64*f*, 75–76, 141*f*
and cardiovascular system, 62, 64*f*, 268, 271*f*
and intraembryonic coelom, 62, 63*f*–64*f*
heart
conducting system of, 282*f*, 284
abnormalities of, 284
congenital anomalies of, 284–289, 304
major, incidence of, 131*t*
development of
critical period for, 304
early, 264*f*, 264–268, 267*f*, 303–304
further, 268–284, 304
functional start of, 264
position of, head fold and, 62, 64*f*, 271*f*
primordial, 49, 56, 57*f*, 60, 143*f*, 143–144, 264*f*, 268–284
circulation through, 271–276, 272*f*
partitioning of, 271–284, 273*f*, 274*f*–275*f*, 277*f*, 304
veins associated with, development of, 264–268, 265*f*–267*f*
heart murmurs, 289
heart prominence, 69, 73*f*, 172*f*
heart rate, fetal, monitoring of, 85, 87
heart tubes, 56, 60, 264, 268, 269*f*, 270*f*–271*f*, 303–304
heartbeat
beginning of, 264, 303
ultrasound of, 56, 58*f*, 280
hemangioblasts, 56, 60
hemangioma, 391
hematogenesis, 56
hematoma, and placental separation, 100
hematopoiesis
fetal, 83
hepatic, 207
hematopoietic tissue, hepatic, 207
hemiazygos vein, 265–268, 266*f*
hemimelia, 336
hemivertebra, 316, 316*f*
hemolytic disease of newborn, 99, 398
intrauterine fetal transfusion for, 86
hemopoietic cells, of bone marrow, 308
heparin, use during pregnancy, 133
hepatic cords, 206*f*, 207, 226, 264
hepatic diverticulum, 206*f*, 207, 209*f*
hepatic ducts, accessory, 208
hepatic segment, of inferior vena cava, 266*f*, 268
hepatic sinusoids, 207, 264, 296
hepatic veins, 264, 266*f*, 300*f*–301*f*
hepatoduodenal ligament, 208, 210*f*
hepatogastric ligament, 208, 210*f*, 213*f*
hermaphroditism, 253–255, 261
true, 255, 261
hernia
diaphragmatic, congenital, 147–149, 148*f*–149*f*
and pulmonary hypoplasia, 147, 148*f*, 198
prenatal diagnosis of, 147
epigastric, congenital, 148–149
inguinal, congenital, 260, 260*f*
internal, 218, 220*f*
midgut, physiological, reduction of, 212
parasternal, 149
retrosternal, 149
umbilical, 71, 215–217
physiological, 212, 215*f*
heroin, fetal effects of, 135
herpes simplex virus, as teratogen, 128*t*, 136
herpes zoster virus, as teratogen, 129*t*, 136
high-resolution chromosome banding, 124
hindbrain, 347*f*, 352–356, 355*f*, 357*f*, 370
division into metencephalon and myelencephalon, 352–353
hindgut, 62, 64*f*, 75, 202, 202*f*
arteries of, 141, 142*f*
congenital anomalies of, 222–227
derivatives of, 222, 226
hip, congenital dislocation of, 137, 339–340
hippocampal commissure, 361, 363*f*
Hirschsprung disease, 224, 224*f*, 370
HIV infection, fetal effects of, 136
Hofbauer cells, 96*f*
holoprosencephaly, 364–366
homeobox (HOX) genes, 7, 127
and gut development, 202
and limb development, 330
and nervous system development, 344
and prostate development, 249
and skeletal development, 306, 318
and spleen development, 210
homologous chromosomes (homologs), 13, 17, 119
hormones
placental synthesis of, 99
transplacental transport of, 97*f*, 98
horseshoe kidney, 235, 239*f*

housekeeping genes, 127
HOX genes. *See* homeobox genes
hPL (human placental lactogen), placental synthesis of, 99
human chorionic corticotropin (hCACTH), placental synthesis of, 99
human chorionic gonadotropin (hCG), 38–39, 42, 46
  and male genital development, 249
  placental synthesis of, 99
human chorionic somatomammotropin (hCS), placental synthesis of, 99
human chorionic thyrotropin (hCT), placental synthesis of, 99
human genome, 127
Human Genome Project, 127
human immunodeficiency virus (HIV) infection, fetal effects of, 136
human leukocyte antigens, and placental immunoprotection, 103–104
human parvovirus B19, as teratogen, 128*t*
human placental lactogen (hPL), placental synthesis of, 99
humerus, 318*f*, 319*f*, 335*f*
hyaline cartilage, 306, 309*f*, 310, 318
hyaline membrane disease, 198
hyaloid artery, 372, 373*f*, 375*f*, 377*f*, 378
  persistence of, 378
hyaloid canal, 377*f*, 378
hyaloid vein, 372, 373*f*, 375*f*
hyaluronidase, and fertilization, 24
hydatid (of Morgagni), 248*f*, 250*t*
hydatidiform mole, 58
hydramnios, 108
  esophageal atresia/tracheoesophageal fistula and, 193, 202
  with meroanencephaly, 364
hydrocele, 260, 260*f*
  of spermatic cord, 260, 260*f*
  of testis, 260*f*, 260–261
hydrocephalus, 364, 366*f*, 370
  nonobstructive or communicating, 364, 366, 370
  obstructive or noncommunicating, 364, 370
hydronephrosis, 232
hygroma, cystic, 303, 304, 391
hymen, 10, 13*f*
  development of, 248*f*, 250*t*, 251, 251*f*, 254*f*
  imperforate, 256
hyoid arch. *See* pharyngeal arches, 2nd (second)
hyoid bone, formation of, 152, 157*f*, 157*t*, 158, 314
hypaxial division, of myotomes, 324, 325*f*, 326*f*
  derivatives of, 325–326
hyperdiploid, 119
hyperdontia, 398*f*
hyperpituitarism, 320
hypertrophic pyloric stenosis, congenital, 203
hypoblast, 34*f*, 34–35, 39, 39*f*, 41*f*, 44
hypobranchial eminence, 158, 170*f*
hypodiploid, 119
hypogastric vein, 266*f*
hypoglossal nerve
  formation of, 367–368, 368*f*
  innervation of tongue by, 169–170
hypoglossal nerve roots, 367–368
hypoglossal nucleus, 367
hypoglycemic drugs, fetal effects of, 134
hypoparathyroidism, congenital, 165
hypopharnygeal eminence, 158, 169, 170*f*, 190, 192*f*
hypophysial cartilage, 313
hypophysial diverticulum (pouch), 359–361, 360*f*
hypophysial portal system, 19
hypospadias, 132, 255, 256*f*, 261
hypothalamic sulcus, 358, 359*f*
hypothalamus
  and female reproductive cycles, 19, 20*f*
  development of, 358, 359*f*
hypothyroidism, 320
hypoxia, fetal
  cigarette smoking and, 131
  ductus arteriosus closure in, 299

I

ichthyosis, 391, 391*f*, 399
identical (monozygotic) twins, 111, 113*f*, 114, 115*f*
  conjoined, 113, 116f
ileal diverticulum, 109–110, 220, 221*f*, 226
ileum, development of, 212–215, 216*f*
iliac arteries, 337*f*
  common, 268, 337*f*
  external, 337*f*
  internal, 268, 300*f*–301*f*
iliac lymph sacs, 303*f*
iliac veins
  common, 266*f*
  external, 266*f*
  internal, 266*f*
ilium, 319*f*
illicit drugs, fetal effects of, 85, 135
immature infants, 78
immunogobulins, transplacental transport of, 97*f*, 98
immunoprotection, of placenta, 103–105
imperforate anus, 224, 224*f*–225*f*
imperforate hymen, 256
implantation, of blastocyst, 31–33, 34*f*, 38*f*–40*f*, 38–41, 43–44
  inhibition of, 44
  sites of, 42–43, 43*f*
implantation bleeding, 43, 83
in vitro fertilization, 31, 32*f*
inborn errors of metabolism, 86
incisive canals, 179–181
incisive fossa, 179*f*, 179–181, 183
incisive papilla, 181, 233*f*
incisor teeth, 179, 179*f*, 395*f*, 397
incus, 380, 383, 383*f*, 386
  formation of, 154, 157*f*, 157*t*, 313
indifferent gonad, 244–245, 245*f*, 247
  adult derivatives and vestigial remains of, 250*t*
indifferent state of sex development, 244, 251*f*, 261
indoleamine 2,3 deoxygenase, and placental immunoprotection, 104–105
inductions, 66–67, 372
infants
  circulation in, 295–299, 301*f*
    transitional, 296–299
  immature, 78
  lungs of, 198
  newborn, 100
  premature, 78
    ductus arteriosus closure in, 299
    patent ductus arteriosus in, 299
infection, fetal, 99
infectious agents
  as teratogens, 128*t*–129*t*, 135–137
  transplacental transport of, 97*f*, 99
inferior, as descriptive term, 8*f*
inferior colliculi, 357, 358*f*
inferior mesenteric artery, 141, 142*f*, 202*f*, 222, 268
inferior mesenteric lymph nodes, 222
inferior parathyroid gland, 161
inferior recess of omental bursa, 203
inferior rectal artery, 222
inferior rectal nerve, 222
inferior rectal vein, 222
inferior vena cava, 295–296, 300*f*– 301*f*
  development of, 265–268, 266*f*–267*f*, 304
  hepatic segment of, 266*f*, 268
  interruption of abdominal course, 268
  postrenal segment of, 266*f*, 268
  prerenal segment of, 266*f*, 268
  renal segment of, 266*f*, 268
  valves of, 276–277, 278*f*
infertility
  male, 25
  treatment of, 31, 32*f*
infracardiac bursa, 203
infundibular stem, 360*t*, 360*f*, 361
infundibular stenosis, 288, 290*f*
infundibulum, 360*f*, 361
inguinal canals
  congenital anomalies of, 259–260
  development of, 248*f*, 256–259, 258*f*
inguinal hernia, congenital, 260, 260*f*
inguinal lymph nodes, superficial, 222
inguinal ring
  deep, 256
  superficial, 256
inner cell mass, 31, 33*f*, 35
insula, 362
insulin
  in fetal metabolism and growth, 83
  pancreatic secretion of, 210
insulin therapy, fetal effects of, 134
insulin-like growth factors, in fetal metabolism and growth, 83
integumentary system, 387–400
intercalated discs, 327

intercostal arteries, 268, 310
interleukin-10, and placental immunoprotection, 104
intermaxillary segment, 173*f*, 175, 177*f*, 179
intermediate layer, in skin development, 388, 388*f*
intermediate mesoderm, 52*f*, 53, 54*f*
- derivatives of, 65*f*
- metanephric mass of, 230, 234*f*, 235*f*, 261

intermediate zone, 344, 362
intermediolateral cell column, 369
internal capsule, 361, 363*f*
internal hernia, 218, 220*f*
internal os of uterus, 10, 12*f*
interosseous artery, 337*f*
- common, 334, 337*f*

intersegmental arteries, 268, 292*f*, 310, 311*f*, 334, 337*f*
- dorsal, 265*f*, 268

intersexuality, 253–255, 261
interstitial cells (of Leydig), 246*f*, 247, 261
interthalamic adhesion, 359*f*
interventricular foramen, 280, 280*f*–281*f*, 361, 362*f*–364*f*
- closure of, 280

interventricular septum
- membranous part of, 280, 281*f*–282*f*
- muscular part of, 277–280, 281*f*
- primordial, 273*f*, 277–280, 280*f*

intervertebral disc, 310, 311*f*
intervillous space, 40, 91*f*, 94–95, 97
- maternal blood in, 94*f*

intestine(s)
- atresia of, 218–219
- congenital anomalies of, 215–220
- development of, 79–80, 212–215, 214*f*
- duplication of, 220
  - cystic, 220
  - tubular, 220
- fixation of, 215, 216*f*
  - abnormal, 215, 226
- herniation of, 215–218
- return to abdomen, 81, 214*f*, 215, 216*f*
- small, development of, 212–215, 214*f*
- stenosis of, 218–219
- volvulus of, 217–218, 220*f*

intracartilaginous ossification,306–308, 309*f*
intracytoplasmic sperm injection, 31
intraembryonic coelom
- development of, 54*f*, 56, 60, 140, 140*f*, 149
- embryonic folding and, 62, 63*f*–64*f*, 140, 141*f*

intraembryonic ectoderm. *See* ectoderm, embryonic
intraembryonic endoderm. *See* endoderm, embryonic
intraembryonic mesoderm. *See* mesoderm, embryonic
intramembranous bone formation, 306
intramembranous ossification, 306, 308*f*
intraocular tension, 379
intraretinal space, 372, 373*f*–374*f*, 376*f*–378*f*
intrauterine device (IUD), and implantation, 44
intrauterine fetal transfusion, 86
intrauterine growth retardation, 78, 83–85
- cigarette smoking and, 83–85, 131
- cytomegalovirus infection and, 136
- definition of, 84
- genetic factors and, 85
- placental circulation and, 85
- social drugs and, 85
- triploidy and, 28, 123

inversion, chromosomal, 125*f*, 126
iodides, as teratogens, 134
iodine, maternal deficiency of, fetal effects of, 134
ionizing radiation, as teratogen, 129*t*, 137, 380
iris
- coloboma of, 376, 377*f*
- color of, 376
- development of, 372, 376, 378*f*–379*f*, 386

ischemic phase, of menstrual cycle, 23
ischiadic artery, 337*f*
isochromosomes, 125*f*, 126
isotretinoin, as teratogen, 128*t*, 134
IUD (intrauterine device), and implantation, 44
IUGR. *See* intrauterine growth retardation

J

jaundice, 208
jaws
- development of, 171, 173*f*, 176
- postnatal growth of, 316
- primordium of, 152, 153*f*–154*f*

jejunum, development of, 212–215, 216*f*
joint(s)
- cartilaginous, 308–310, 310*f*, 320
- costovertebral, 312*f*, 313
- development of, 308–310, 310*f*, 320
- fibrous, 308, 310*f*, 320
- neurocentral, 312, 312*f*
- synovial, 308–310, 310*f*, 320, 332

jugular lymph sacs, 303*f*
jugular vein, 266*f*

K

keratinization, 388, 388*f*
- disorders of, 391, 391*f*, 399

keratinocyte growth factor, and lung development, 197–198
kidney(s)
- blood supply, changes in, 232, 236*f*
- congenital anomalies of, 234–237, 237*f*–239*f*, 261
  - major, incidence of, 131*t*
- cystic disease of, 237
- development of, 67, 73*f*, 230–232, 232*f*–235*f*, 260–261
  - molecular studies of, 230, 236*f*
- discoid (pancake), 235, 238*f*
- double, 236, 238*f*
- ectopic, 234–235, 238*f*
- fused, 235, 238*f*
- horseshoe, 235, 239*f*
- interim, 67, 230, 232*f*, 233*f*
- lobulation of, 230–232
- malrotation of, 234, 238*f*
- pelvic, 235, 238*f*
- permanent, development of, 230–232, 233*f*, 234*f*
- positional changes of, 232, 236*f*, 261
- supernumerary, 235–237, 238*f*–239*f*

killer-inhibitory receptors, and placental immunoprotection, 103
Klinefelter syndrome, 124*t*, 124*f*
Klippel-Feil syndrome, 316
Kupffer cells, hepatic, 207

L

labia majora, 11*f*, 13, 13*f*, 248*f*, 250*t*, 253, 261
labia minora, 11*f*, 13, 13*f*, 250*t*, 253, 261
labial commissures, 253
labial groove, persistent, 183, 184*f*
labiogingival groove, 176, 180*f*
labiogingival lamina, 176, 180*f*
labioscrotal folds, 253
labioscrotal swellings, 252, 254*f*, 256
- adult derivatives and vestigial remains of, 250*t*

labor, 99–100
- definition of, 99
- drugs for managing, transplacental transport of, 99
- factors triggering, 99
- stages of, 100

lacrimal glands, development of, 380, 386
lacrimal sac, 174
lactiferous ducts, 392, 393*f*
lacunae, 23, 38*f*, 38–39, 43
lacunar network, 23, 39*f*, 40, 94
lamina terminalis, 361, 362*f*–364*f*
lanugo, 80, 83, 87, 391, 399
laryngeal atresia, 190
laryngeal cartilages, 192*f*
- formation of, 157*t*, 157*f*, 158, 190, 314

laryngeal inlet, 190, 192*f*
- primordial, 190, 192f

laryngeal lumen, temporary occlusion of, 190
laryngeal muscles, 190
laryngeal nerves, 158–159, 190, 369
- recurrent, 293, 294*f*, 369
- superior, 369

laryngeal ventricles, 190
laryngotracheal diverticulum, 190, 191*f*, 192*f*, 198
laryngotracheal groove, 170*f*, 190, 191*f*, 192*f*, 198
laryngotracheal tube, 190, 192*f*–193*f*, 198
larynx
- congenital anomalies of, 190
- development of, 152, 190, 192*f*, 198–199
- intrinsic muscles of, formation of, 157*t*
- recanalization of, 190

last normal menstrual period (LNMP), 28, 46, 74, 78, 83
- and expected date of delivery, 83

lateral, as descriptive term, 8*f*

lateral aperture, 356
lateral folding, of embryo, 62, 63*f*, 75
lateral horns of spinal cord, 346
lateral mesoderm, 52*f*, 53, 54*f*
derivatives of, 65*f*
lateral nasal prominence, 172*f*, 174, 174*f*–177*f*, 176
lateral palatine process, 178*f*, 180*f*–181*f*, 181
lead, as teratogen, 135
lens, development of, 66–67, 69*f*, 372, 373*f*–374*f*, 378, 378*f*–379*f*, 385–386
lens capsule, 378
lens epithelium, subcapsular, 377*f*, 378
lens fibers, 379*f*
primary, 378
secondary, 378
lens pit, 73*f*, 372, 373*f*–374*f*, 385
lens placodes, 67, 69*f*, 73*f*, 155*f*, 172*f*, 372, 374*f*, 385
invaginated, 374*f*
lens vesicle, 372, 373*f*, 378, 385
lentiform nucleus, 361, 363*f*–364*f*
leptomeninges, 348
lesser cornu, formation of, 157*t*
lesser curvature of stomach, 203
lesser omentum, 203, 208, 209*f*, 210*f*, 211*f*
lesser sac of peritoneum. *See* omental bursa
levator palpebrae superioris muscle, abnormal development of, 380
levator veli palatini, formation of, 157*t*
Lewis, Edward B., 7
Leydig cells, 246*f*, 247, 261
LH. *See* luteinizing hormone
ligament(s)
anular, failed differentiation of, 384
broad, 250, 253*f*
falciform, 208, 209*f*, 210*f*, 213*f*
gastrosplenic, 213*f*
hepatoduodenal, 208, 210*f*
hepatogastric, 208, 210*f*, 213*f*
of malleus, anterior, 154, 157*f*, 157*t*
ovarian, 250*t*, 259
periodontal, 395*f*, 396
round
of liver, 299, 301*f*
of uterus, 248*f*, 250*t*, 259
sphenomandibular, 154, 157*t*
splenorenal, 213*f*
stylohyoid, formation of, 157*f*, 157*t*, 158
umbilical
medial, 239, 268, 299, 301*f*
median, 50–53, 90, 110, 110*f*, 239
ligamentum arteriosum, 292*f*, 293, 294*f*, 299, 302*f*
ligamentum teres, 299, 301*f*
ligamentum venosum, 299, 301*f*
limb(s), 329–341
blood supply to, 334, 337*f*
bones
development of, 308, 317–319, 318*f*–319*f*, 330
ossification of, 74, 308, 318–319, 330
congenital anomalies of, 334–340, 338*f*–340*f*
causes of, 340
major, incidence of, 131*t*
terminology for, 335–336
thalidomide and, 118, 134, 135*f*, 335, 338*f*, 340
cutaneous innervation of, 332–334
development of, 67, 68*t*, 69*f*, 153*f*–154*f*, 318*f*, 329–341, 331*f*–334*f*
critical period of development, 129, 334
dermatomal patterns of, 336*f*
early stages of, 330, 331*f*–334*f*
final stages of, 330, 335*f*–336*f*
molecular control of, 330
innervation of, 367
muscles, 325*f*, 326–327, 340
rotation of, 332, 336*f*, 340
limb buds, 330, 340
lower, 69*f*, 73*f*, 154*f*, 330, 331*f*, 334*f*, 340
upper, 68*t*, 69*f*, 73*f*, 153*f*, 330, 331*f*–334*f*, 340
limbus fossae ovalis, 299
lingual cysts and fistulas, congenital, 167*f*, 170
lingual papillae, 169
lingual septum, 169
lingual thyroid tissue, 167, 168*f*
lingual tonsil, 302
lip(s)
cleft, 181*f*–185*f*, 181–186
bilateral, 183, 183*f*
involving upper lip, 181*f*, 182–183
median
of lower lip, 183, 185*f*
of upper lip, 183, 185*f*
unilateral, 181*f*, 183, 183*f*
development of, 171, 173*f*, 176, 177*f*, 180*f*
philtrum of, 173*f*, 175
upper, frenulum of, 176, 180*f*
lithium carbonate, as teratogen, 128*t*, 134
liver
congenital anomalies of, 208
development of, 202, 206*f*, 207–208, 209*f*–210*f*, 213*f*, 226
positioning of, 213*f*
primordium of, 202*f*, 207
visceral peritoneum of, 208, 209*f*
LNMP. *See* last normal menstrual period
lobster-claw anomalies, 338, 338*f*
long bones
development of, 318, 318*f*
osteogenesis of, 330
longitudinal folding, of embryo, 62, 63*f*
loop of Henle. *See* nephron loop.
low-birth weight, 78, 83
cigarette smoking and, 131
lower limb buds, 67, 69*f*, 73*f*, 154*f*, 330, 331*f*, 334*f*, 340
lower respiratory organs. *See also* bronchi; larynx; lung(s); trachea
development of, 152, 189–199
lumbar arteries, 268
lumbar myotomes, 325*f*, 326*f*, 330
lumbar rib, 316
lung(s)
aeration at birth, 198
clearance of fluid from, at birth, 198
congenital anomalies of, 198
development of, 80–83, 141–144, 192*f*, 193–199, 194*f*–196*f*
glucocorticoids and, 198
molecular studies of, 197–198
normal, important factors for, 198
oligohydramnios and, 198
primordial, 145*f*
hypoplasia, 198
congenital diaphragmatic hernia and, 147, 148*f*, 198
maturation of, 195–199, 196*f*
alveolar period (late fetal period to childhood) of, 196*f*, 197–199
canalicular period (16 to 25 weeks) of, 196*f*, 196–197, 199
pseudoglandular period (5 to 17 weeks) of, 195, 196*f*, 199
terminal saccular period (24 weeks to birth) of, 196*f*, 197, 199
neonatal, 198
lung bud, 190, 192*f*, 193, 199
luteal (secretory) phase, of menstrual cycle, 10, 23
luteinizing hormone
and corpus luteum, 21
and ovarian cycle, 19–21, 20*f*, 22*f*
and ovulation, 19–21, 20*f*, 22*f*
and reproductive cycles, 19–22, 20*f*, 22*f*
oral contraceptives and, 23
ovulation induction and, 21
luteinizing hormone surge, 19–21, 22*f*
lymph ducts
development of, 302, 303*f*
right, 302, 303*f*
lymph nodes
development of, 302, 304
inferior mesenteric, 222
superficial inguinal, 222
lymph sacs
development of, 299–302, 303*f*, 304
iliac, 302, 303*f*
jugular, 302, 303*f*
retroperitoneal, 302, 303*f*
lymph sinuses, 302
lymphatic nodules, 161
lymphatic system
congenital anomalies of, 303
development of, 299–302, 303*f*, 304
lymphedema, congenital, 303
lymphoblasts, 302
lymphocytes, development of, 302
lysergic acid diethylamide (LSD), as teratogen, 135

M

macrodontia, 397
macroglial cells (macroglia), histogenesis of, 344, 349*f*, 370

macrostomia, 185*f*, 186
magnetic resonance imaging, for fetal assessment, 86
major histocompatibility complexes, and placental immunoprotection, 103
male anatomy, 10, 13, 244–247, 249, 252, 261. *See also specific anatomical entities*
male pronucleus, 28, 30*f*, 35
male pseudohermaphroditism, 255, 261
male reproductive organs, 11*f*, 13
malformation, 118. *See also* congenital anomalies
malleus, 380, 383, 383*f*, 386
  anterior ligament of, 154, 157*f*, 157*t*
  formation of, 154, 157*t*, 313
malnutrition, maternal, and fetal growth, 84, 85*f*
mammary buds, 393*f*
mammary crests, 392, 393*f*
mammary glands
  congenital anomalies of, 392
  development of, 392, 393*f*, 399
    postnatal/pubertal, 392
mammary pit, 392, 393*f*
mammary ridges, 392, 393*f*
mammillary bodies, 358, 359*f*
mandible, formation of, 319*f*
mandibular arch. *See* pharyngeal arches, 1st (first)
mandibular nerve, formation of, 158
mandibular prominences, 152, 153*f*–154*f*, 171, 172*f*–175*f*, 176
mandibulofacial dysostosis, 165
manubrium, 313
marginal zone, 344, 362
marijuana, fetal effects of, 85
masculinization of female fetus, 98, 132, 132*f*, 243, 244*f*, 255, 255*f*, 261
masculinizing hormones, 249
masseter muscle, formation of, 158*f*
mastication, muscles of, formation of, 157*t*, 158, 176
mastoid antrum, 159, 186, 383
mastoid processes, 383
maternal age, and chromosomal abnormalities, 17, 120, 123*t*
maternal factors, as teratogens, 137
matrix-mediated interaction, 67
mature oocyte, 15*f*, 17
mature sperm, 15*f*, 17
maxilla
  development of, 67, 177*f*, 179*f*, 181*f*
  formation of, 319*f*
  intermaxillary segment of, 173*f*, 175, 177*f*, 179
  premaxillary part of, 175, 177*f*, 179, 179*f*
maxillary arteries, 291
maxillary nerve, formation of, 158
maxillary prominences, 67, 152, 153*f*–154*f*, 171, 172*f*–177*f*, 174–176, 314–315
maxillary sinuses, 178
measles
  German or three-day (*See* rubella)
  transplacental transport of virus, 99
meatal plug, 383*f*, 384
mechanical factors, as teratogens, 137
Meckel cartilage (first arch cartilage), 154
Meckel (ileal) diverticulum, 109–110, 220, 221*f*, 226
meconium, 108, 208
medial, as descriptive term, 8*f*
medial nasal prominence, 172*f*–177*f*, 174–176
medial umbilical ligament, 239, 268, 299, 301*f*
median aperture, 356
median artery, 337*f*
median eminence, 360*t*, 360*f*, 361
median palatine process, 178*f*, 179, 180*f*–181*f*
median plane, 8*f*
median sulcus of tongue, 170*f*
median tongue bud, 169, 170*f*
median umbilical ligament, 50–53, 90, 110, 110*f*, 239
mediastinum, primordial, 143*f*, 144
medulla
  gonadal, 244
  of suprarenal glands, 243, 244*f*
medulla oblongata, development of, 353–354, 356*f*–358*f*, 370
medulla of suprarenal glands, 250*t*
medullary center, 362
medullary cone, 350, 352*f*
megacolon, congenital, 224, 224*f*, 370
meiosis, 13–19, 16*f*
  significance of, 14
meiotic division
  first, 13–14, 16*f*
    in oogenesis, 15*f*, 17
    in spermatogenesis, 14, 15*f*
  second, 14, 16*f*
    in spermatogenesis, 14–17, 15*f*
    of oogenesis, 15*f*, 17, 28, 29*f*
meiotic nondisjunction of chromosomes, 119. *See also* nondisjunction of chromosomes
Meissner plexus, 224
melanin, 389, 391
melanoblast, 367, 388*f*, 389, 391
melanocytes, 367, 388*f*, 389, 390*f*, 391, 399
membranes. *See also specific membranes*
  fetal, 89–115
    after birth, 100–105, 102*f*
    development of, 91*f*
    functions and activities of, 90
    multiple pregnancies and, 111–114, 113*t*
    premature rupture of, 96, 108
membranous atresia of anus, 224*f*–225*f*, 224–226
membranous cochlea, 382*f*
membranous labyrinth of internal ear, 381–383, 382*f*–383*f*, 386
membranous neurocranium, 313
membranous viscerocranium, 314–315
menarche, 10
meninges, spinal
  anomalies of, 353*f*
  development of, 348, 352*f*
meningocele, 353*f*, 363
meningoencephalocele, 363, 365*f*
meningohydroencephalocele, 363, 365*f*
meningomyelocele, 353*f*–354*f*
meninx, primordial, 348
menopause, 21, 24
menses, 23
menstrual age, 78
menstrual cycle, 22*f*, 22–24
  anovulatory, 23
  duration of, 23
  ischemic phase of, 23
  luteal (secretory) phase of, 10, 23
  menstrual phase of, 23
  pregnancy phase of, 24
  proliferative phase of, 23
menstrual period
  first (menarche), 10
  last normal (LNMP), 28, 46, 74, 83
    and expected date of delivery, 83
menstruation, 23
  cessation of, and pregnancy, 46
  corpus luteum of, 21
mental retardation
  alcohol and, 132
  cerebral anomalies and, 364
mercury, as teratogen, 128*t*, 135
meroanencephaly, 53, 317, 317*f*, 352, 355*f*, 363–364
meromelia, 134, 135*f*, 336, 340
mesencephalon. *See* midbrain
mesenchymal cells, 344, 349*f*
mesenchyme, 48, 49*f*, 344, 380
  and skeletal development, 306
mesenteric artery
  inferior, 141, 142*f*, 202*f*, 222
  superior, 141, 142*f*, 202*f*, 206, 212, 214*f*, 222, 268
mesenteric lymph nodes, inferior, 222
mesentery (mesenteries), 140–141
  definition of, 140
  development of, 141*f*–142*f*
  dorsal, 140–141
    of esophagus, diaphragm development from, 145*f*, 145–146, 150
  of stomach, 203, 204*f*–205*f*
  ventral, 206*f*, 208, 209*f*–210*f*, 213*f*
mesoblast, 48, 48*f*
mesocardium, dorsal, 271
mesoderm
  embryonic, 41*f*, 46, 47*f*, 48, 60
    derivatives of, 46, 65*f*, 306, 372, 388
  extraembryonic, 38*f*–41*f*, 39, 44
    somatic, 40*f*, 41, 56
    splanchnic, 40*f*, 41, 56
  intermediate, 52*f*, 53, 54*f*
    derivatives of, 65*f*
    metanephric mass of, 230, 234*f*, 235*f*, 261
  lateral, 52*f*, 53, 54*f*
    derivatives of, 65*f*
  paraxial, 52*f*, 53–56, 54*f*, 306, 344
    derivatives of, 65*f*, 307*f*, 344

mesodermal cells, and skeletal development, 306
mesogastrium
dorsal, 203, 204f–205f
ventral, 203, 204f–205f
mesonephric ducts, 230, 232f–234f, 239, 240f
adult derivatives and vestigial remains of, 230, 250t, 251–252
and male genital system, 246f, 247, 248f, 249, 251f, 261
remnants
in females, 248f
in males, 248f
mesonephric ridges, 67, 73f
mesonephric tubules, 230
adult derivatives and vestigial remains of, 230, 250t
mesonephric vesicles, 233f
mesonephroi, 67, 230, 232f, 233f, 260
mesorchium, 246f, 247
mesovarium, 246f, 249
metabolism, errors of, 86
metacarpals, 318f, 319f, 335f
metamorphosis, in spermatogenesis, 17
metanephric diverticulum, 230, 233f, 234f, 260
branching of, 230, 236f
congenital disorders of, 234, 238f
metanephric mass of intermediate mesoderm, 230, 234f, 235f, 261
metanephric tubules, 230, 235f
metanephric vesicles, 230, 235f
metanephrogenic blastema, 230, 234f
metanephroi, 230–232, 232f, 233f, 234f, 260–261
metatarsals, 319f
metencephalon, 347f, 352–356, 355f, 357f
methadone, fetal effects of, 135
methotrexate, as teratogen, 128t, 134
methylmercury, as teratogen, 128t, 135
microcephaly, 317
microdeletions, 124
microdontia, 397
microduplications, 124
microglial cells (microglia), histogenesis of, 344, 349f, 370
micrognathia, 165
microstomia, 185f, 186
microtia, 385, 385f
midbrain, 73f, 347f, 352–353, 355f, 356–357, 357f–359f, 370
midbrain flexure, 353, 356f
midgut, 62, 63f, 76, 202, 202f
arteries of, 141, 142f
congenital anomalies of, 215–220, 226
derivatives of, 212–215, 226
mixed rotation and volvulus, 217, 220f
nonrotation of, 217, 220f
return to abdomen, 212, 214f
reversed rotation of, 218, 220f
volvulus of, 218, 220f
midgut hernia, physiological, reduction of, 212
midgut loop (umbilical intestinal loop), 210f, 214f.
rotation of, 212, 214f
Minamata disease, 135
mitosis, 66
mitral valve, development of, 282f, 284
mittelschmerz, and ovulation, 21
Mohr syndrome, 183
molding of fetal cranium, 313
molecular biology, of human development, 7
molecular cytogenetics, 124–125
molecular studies
of lung development, 197–198
of muscle development, 324, 324f, 326
of renal development, 236f
monochorionic-monoamniotic twin placenta, 112–113
monosomy, 19, 119
monozygotic twins, 111, 113f, 114, 115f
conjoined, 113, 116f
mons pubis, 253
morning-after pills, 44
morphogenesis, 46, 66
morula, 31, 33f, 35
definition of, 2
mosaicism, 120, 123
erythrocytic, 111
motor axons, 332
mouth
congenital anomalies of, 181–186 (*See also* cleft lip and palate)
primordial (stomodeum), 62, 64f, 152, 153f–155f, 159, 171, 172f, 202, 202f
pituitary gland development from ectodermal roof of, 359–361, 360f
mucous plug, 91f, 94f
müllerian inhibiting substance (MIS), 247, 261
müllerian (sinual) tubercle, 249, 251, 251f
multicystic dysplastic kidney disease, 237
multifactorial inheritance, 118, 118f, 137, 138f, 185–186, 284, 339–340
multigravidas, 100
multiple pregnancy, 110–114. *See also* twin(s)
and fetal growth, 85, 85f
placenta and fetal membranes in, 111–114, 113f
muscle(s), 323–328. *See also specific muscles*
accessory, 327
cardiac, development of, 327
congenital anomalies of, 327, 328f
molecular control of, 324, 324f, 326
skeletal, development of, 324–327, 325f–326f
smooth, development of, 327
variations in, 327
muscle fibers, 324
cardiac, 327
muscular system, development of, 323–328
myelencephalon, 347f, 352–354, 355f, 356f
myelin sheaths, 332, 350
myelination
of optic nerve fibers, 374
of spinal nerve fibers, 350
myeloschisis, 353f–354f
Myf-5, and muscle development, 324, 324f
mylohoid muscle, formation of, 157t, 158f
myoblasts, 306, 307f, 324, 330
cardiac, 327
myocardium
development of, 268, 269f, 270f, 304
primordial, 268, 304
MyoD, 324, 324f
myoepithelial cells, 389
myofilaments, 324, 327
myogenesis, 324
myogenic regulatory factors, 324
myometrium, 10, 11f–12f
myotomes, 306, 307f, 324–327, 325f, 326f, 330
epaxial division of, 324, 325f, 326f
hypaxial divisions of, 324, 325f, 326, 326f
occipital (postotic), 326
preotic, 326, 326f
myotubes, 324

N

Nägele rule, 83
nail(s)
absence of, 399
deformed, 392, 399
development of, 80, 391–392, 392f
nail fields, 392, 392f
nail folds, 392, 392f
nail plate, 392, 392f
narcotics, fetal effects of, 135
nasal capsules, 313
nasal cavities, development of, 152, 175f, 176–178, 178f
nasal conchae, 176, 178f
nasal pits, 73f, 172f, 174–176, 174f–177f
nasal placodes, 73f, 153f, 154f, 155f, 172f, 174, 175f, 176
nasal prominences
lateral, 172f, 174, 174f–177f, 176
medial, 172f–177f, 174–176
nasal sacs, primordial, 176, 178f
nasal septum, 178f, 179, 180f
nasolacrimal duct, 174
atresia of, 174
nasolacrimal groove, 73f, 172f–173f, 174, 176f, 177f
nasopalatine canal, 179
natal teeth, 397
natural killer cells, killer-inhibitory receptors in, and placental immunoprotection, 103
nausea medication, as teratogen, 134
neck
congenital anomalies of, 152, 162–166, 163f–166f, 186
development of, 151–187
neocerebellum, 357f
neonate, 100
circulation in, 295–299, 301f
transitional, 296–299
lungs of, 198
nephrogenic cord, 230, 231f–234f
nephrogenic duct, 233f
nephron(s), development of, 230–232, 235f

nephron loop, 230, 235*f*
nerve(s)
abducent, formation of, 367–368, 368*f*
accessory, formation of, 368*f*, 369
cochlear, 369
cranial
formation of, 157*t*, 366–369, 368*f*
innervation of pharyngeal arches by, 158–159
somatic efferent, 367–368, 368*f*
special sensory, 368*f*, 369
special visceral afferent components of, 158
special visceral efferent (branchial) components of, 158
facial, formation of, 157*t*, 158, 366, 368*f*, 368–369
glossopharyngeal
formation of, 157*t*, 158, 366, 368*f*, 369
innervation of tongue by, 170
hypoglossal
formation of, 367–368, 368*f*
innervation of tongue by, 169–170
laryngeal, 158–159, 190, 369
recurrent, 293, 294*f*, 369
superior, 369
mandibular, formation of, 158
maxillary, formation of, 158
oculomotor
abnormal development of, 380
formation of, 367–368, 368*f*
olfactory, 177–178, 178*f*, 369
formation of, 368*f*
optic
development of, 369, 372, 375*f*–376*f*
fibers, myelination of, 374
phrenic, 146
rectal, inferior, 222
sensory, special, formation of, 368*f*
spinal
development of, 324, 350*f*, 351*f*, 366–367
dorsal root of, 350*f*, 367
dorsal roots of, 350*f*, 367
myelination of fibers, 350
ventral roots of, 346–348, 350*f*, 367
trigeminal
formation of, 157*t*, 158, 366, 368, 368*f*
innervation of tongue by, 170
trochlear, formation of, 367–368, 368*f*
vagus
formation of, 157*t*, 158–159, 366, 368*f*, 369
innervation of esophagus by, 202
innervation of tongue by, 170
laryngeal branches of, 157*t*, 158–159, 190, 293, 294*f*, 369
superior laryngeal branch of, formation of, 157*t*
vestibular, 369
vestibulocochlear, formation of, 366, 368*f*, 369
visceral, 366
vomeronasal, 178*f*
nervous system, 343–370
autonomic, 344
development of, 369–370
central, 344, 377 (*See also* brain; spinal cord)
histogenesis of cells in, 349*f*
primordium of, 49*f*, 50, 51*f*–52*f*, 53, 60
components of, 344
origin of, 66, 344, 345*f*, 346*f*
parasympathetic, development of, 370
peripheral, 344
development of, 366–369
sympathetic, development of, 369
neural canal, 344, 346*f*, 370
neural crest
derivatives of, 324, 351*f*
development of, 53, 55*f*, 60, 345*f*
maldevelopment of, 186
neural crest cells, 53
and cardiovascular development, 280–282
and eye development, 372, 380
and facial development, 171, 186
and laryngeal development, 190, 199
and limb development, 332
and nervous system development, 66, 366–367, 370
and pharyngeal arch development, 152, 186
and skeletal development, 306
and skin development, 389, 399
and spinal development, 348, 350, 351*f*
and suprarenal gland development, 243, 244*f*
and thymic development, 161
and tooth development, 393–394, 400
neural folds, 49*f*, 52*f*, 53, 54*f*–55*f*, 60, 345*f*–346*f*, 370
neural groove, 49*f*, 52*f*, 53, 54*f*–55*f*, 60, 345*f*, 370
neural plate, 49*f*, 50, 51*f*–52*f*
and origin of nervous system, 344, 345*f*
formation of, 52*f*, 53
neural retina, 372, 376*f*–379*f*
neural tube, 53
and brain development, 352
and origin of nervous system, 344, 345*f*, 370
and spinal cord development, 344, 347*f*–348*f*
development of, 53, 54*f*–55*f*, 60, 69*f*–70*f*, 344, 345*f*–346*f*
in axial skeletal development, 310, 311*f*
nonfusion of, 350 (*See also* neural tube defects)
neural tube defects, 350–352, 353*f*–355*f*, 362–364, 370
causes of, 352, 362
detection of, alpha-fetoprotein assay for, 86, 108, 352
ultrasound of, 355*f*
neuraminidase, and fertilization, 28
neurenteric canal, 50, 52*f*
neuroblasts, 344
histogenesis of, 349*f*
neurocentral joints, 312, 312*f*
neurocranium
cartilaginous, 313
development of, 313, 320
membranous, 313
neuroectoderm, 53
derivatives of, 53, 65*f*, 372, 376, 386
neuroepithelial cells, 344, 349*f*
neuroepithelium, 372, 376*f*
neuroglial cells, histogenesis of, 344, 349*f*
neurohypophysial diverticulum, 361
neurohypophysis, 360*t*, 360*f*, 361
neurolemma (Schwann) cells, 332, 350, 351*f*, 366
neurolemmal sheaths, 332
neurons, development of, 344, 349*f*, 351*f*
neuropores, 69*f*–71*f*, 344
caudal, 69*f*–71*f*, 344, 345*f*, 346*f*
defect in closure of, 353*f*
closure of, 344, 346*f*, 357
rostral, 69*f*–71*f*, 153*f*, 344, 346*f*, 357
defect in closure of, 362–364
neurulation, 49*f*, 52*f*, 53, 54*f*–55*f*, 344
abnormal, congenital anomalies resulting from, 53
nevus flammeus, 391
newborn infant, 100
nicotine, fetal effects of, 131
ninth week, 79–80, 79*f*–80*f*
nipples
development of, 392, 393*f*
supernumerary, 393, 399
nodal factors, 284
nondisjunction of chromosomes, 17, 18*f*, 43, 119, 120*f*
norethisterone, avoidance in pregnancy, 132
nose. *See also entries at* nasal
bifid, 185*f*, 186
congenital absence of, 186
development of, 171–176, 172*f*, 175*f*
nostril, single, 185*f*, 186
Notch signaling pathway, and somite development, 56
notochord, 48–50, 51*f*–52*f*
and neural plate/tube formation, 53
and origin of nervous system, 344, 345*f*, 370
development of, 49*f*, 49–50, 51*f*–52*f*
in axial skeleton development, 310–311, 311*f*
remnant of, and chordoma, 311
notochordal canal, 48–49, 51*f*, 60
notochordal plate, 50, 52*f*, 60
notochordal process, 41*f*, 47*f*, 48–50, 49*f*, 51*f*–52*f*
NTDs. *See* neural tube defects
nucleus pulposus, 311, 311*f*
numerical chromosomal abnormalities, 119–123
Nüsslein-Volhard, Christiane, 7

nutrients
transfer to embryo, 38, 109
transplacental transport of, 83, 97*f*, 98

O

oblique facial cleft, 185, 185*f*
oblique muscle, external, 326*f*
oblique vein, 266*f*, 267*f*
obstructive uropathy, and amniotic fluid disorders, 108
occipital bone, 319*f*
occipital myotomes, 326
occipitalis muscle, formation of, 158*f*
ocular anomalies, 135, 136*f*, 374, 376, 378–380, 386
ocular muscles, 325*f*–326*f*, 326
oculomotor nerve
abnormal development of, 380
formation of, 367–368, 368*f*
odontoblastic processes, 395*f*, 396
odontoblasts, 395*f*–396*f*, 396
odontogenesis, 394
olfactory bulb, 177–178, 178*f*
olfactory epithelium, 177, 178*f*
olfactory nerve, 177–178, 178*f*, 369
formation of, 368*f*
olfactory receptor cells, 177
oligodendroblasts, histogenesis of, 344, 349*f*
oligodendrocytes, 350
histogenesis of, 344, 349*f*
oligohydramnios, 108
and deformation, 137
and lung development, 198
bilateral renal agenesis and, 234
olivary nucleus, 354, 356*f*
omental bursa, 203, 204*f*–205*f*, 213*f*, 214*f*
inferior recess of, 203
superior recess of, 203
omental foramen, 203, 204*f*–205*f*
omphalocele, 86, 215, 218*f*, 226
omphaloenteric duct. *See* yolk stalk.
oocyte(s), 2, 10, 13, 14*f*–15*f*, 17, 35
abnormal, 17–19
definition of, 2
fertilized, 15*f*, 17
plasma membrane, fusion with sperm plasma membrane, 28, 29*f*–30*f*
primary, 15*f*, 17
second meiotic division of, 15*f*, 17, 28, 29*f*
secondary (mature), 15*f*, 17
transport of, 22*f*, 24
versus sperm, 14*f*, 17
viability of, 26
oogenesis, 13, 15*f*, 17
abnormal, 17–19, 18*f*
oogonia, 17, 246*f*, 247–249
ophthalmic (hyaloid) artery, 373*f*, 375*f*
optic chiasm, 362, 363*f*
optic cup, 372, 373*f*–374*f*, 376, 377*f*, 385
optic fissure. *See* retinal fissure.
optic groove, 372, 373*f*, 385
optic nerve
development of, 369, 372, 375*f*–376*f*
fibers, myelination of, 374
optic stalk, 372, 373*f*–375*f*
optic vesicle, 66, 154*f*, 171, 347*f*, 357, 372, 373*f*, 385
oral contraceptives
and ovulation, 21, 23
fetal effects of, 132–133
orbicularis oculi muscle, formation of, 158*f*, 380
orbicularis oris muscle
formation of, 158*f*
loss of continuity, with cleft anomalies, 183
organogenetic period, 61–76
teratogens and, 129, 130*f*, 138
oronasal membrane, 176, 178*f*
oropharyngeal membrane, 48–49, 49*f*, 52*f*, 62, 63*f*–64*f*, 75, 152, 153*f*, 202
osseous syndactyly, 339
ossification, 80, 306–308, 320
intracartilaginous, 306–308, 309*f*
intramembranous, 306, 308*f*
of appendicular skeleton, 319*f*
of cranium, 313
of limb bones, 74, 308, 318–319, 330
of pharyngeal arch cartilages, 154–158, 157*f*
of vertebral column, 312*f*, 312–313
primary centers of, 79–80, 306–308, 309*f*, 312, 312*f*, 318, 319*f*, 330
secondary centers of, 308, 309*f*, 312, 312*f*, 318–319, 319*f*
osteoblasts, 306
osteoclasts, 306, 397
osteocytes, 306, 308*f*
osteogenesis, 306
of long bones, 330
osteoid tissue, 306
ostium primum defects, 288
ostium secundum defect, 287*f*, 288
otic capsules, 313, 382
cartilaginous, 382, 382*f*–383*f*
otic pits, 67, 69*f*, 381, 381*f*
otic placodes, 153*f*, 380, 381*f*, 386
otic vesicles, 381, 381*f*–383*f*, 386
saccular part of, 381, 382*f*, 386
utricular part of, 381, 382*f*, 386
outer cell mass, 31
oval foramen, 274*f*–275*f*, 276, 277*f*, 296–299, 300*f*–301*f*
adult derivatives of, 299
closure at birth, 299, 301*f*
patent, 284–288, 286*f*–287*f*, 304
oval fossa, 299
ovarian cortex, 246*f*, 250*t*
ovarian cycle, 19–21, 20*f*–22*f*, 36*f*
ovarian follicles
development of, 19, 20*f*–21*f*, 250*t*
mature, 19
primary, 19, 20*f*
primordial, 21*f*, 246*f*, 247–249
reproductive cycles and, 19, 20*f*–21*f*
secondary, 19, 20*f*–21*f*
ovarian follicular fluid, 19, 22*f*
ovarian ligament, 250*t*, 259
ovarian vein, 266*f*
ovaries, 10, 11*f*–12*f*
and reproductive cycles, 19, 20*f*
descent of, 259
development of, 233*f*, 244–249, 245*f*–246*f*, 248*f*, 250*t*, 253*f*, 261
ovulation, 10, 19–21, 20*f*–22*f*
cessation of (anovulation), 21, 23
induction of, 21
luteinizing hormone surge and, 19–21, 22*f*
mittelschmerz and, 21
oral contraceptives and, 21, 23
ovum. *See* oocyte
oxazepam, use during pregnancy, 134
oxycephaly, 317, 317*f*
oxygen
and ductus arteriosus closure, 299
transplacental transport of, 97*f*, 98
oxytocin, and labor, 99

P

palate
cleft, 181*f*–185*f*, 181–186
complete, 182*f*, 183
embryological basis of, 177*f*, 182*f*, 183–184
of anterior and posterior palate, 182*f*, 184
of anterior (primary) palate, 182*f*, 183
of posterior (secondary) palate, 182*f*, 183–184
development of, 175, 178*f*, 178–181, 179*f*–181*f*
critical period of development, 178
hard, 179, 179*f*–180*f*
primary, 178*f*, 179, 180*f*
secondary, 175, 178*f*–180*f*, 179
soft, 179, 179*f*–180*f*
palatine process (shelf)
lateral, 178*f*, 179, 180*f*–181*f*
median, 178*f*, 179, 180*f*–181*f*
palatine raphe, 179, 180*f*
palatine tonsil, 159, 160*f*, 186, 302
palatogenesis, 178
paleocerebellum, 357*f*
palpebral coloboma, 380
palpebral conjunctiva, 380
pancake (discoid) kidney, 235, 238*f*
pancreas
accessory tissue of, 210
anular, 210, 212*f*, 226
body of, 211*f*
congenital anomalies of, 210, 212*f*, 226
development of, 202, 206*f*, 208–210, 210*f*, 211*f*, 213*f*, 226
fetal, secretion by, 83
head of, 211*f*
positioning of, 213*f*, 215, 216*f*
tail of, 211*f*
pancreatic buds, 206*f*, 208–209, 210*f*, 211*f*, 226

pancreatic buds—cont'd
dorsal, 208–209, 210*f*, 211*f*, 226
ventral, 208–209, 211*f*, 226
pancreatic ducts
accessory, 210, 211*f*, 226
main, 209–210, 211*f*
pancreatic tissue, accessory, 210
papilla (papillae)
circumvallate, 170*f*
dental, 395, 395*f*
duodenal, minor, 210
filiform, 169
foliate, 169
fungiform, 169
hair, 390
incisive, 181, 233*f*
lingual, 169
of tongue, development of, 169–170, 170*f*
vallate, 169–170
papillary muscles, 280, 282*f*
paracentric inversion, 125*f*, 126
parachordal cartilage, 313
paradidymis, 248*f*, 250*t*
parafollicular cells, 161
paraganglia, 367
paralingual sulcus, 171
paramesonephric duct, 245*f*
adult derivatives and vestigial remains of, 250*t*, 251–252
and female genital system, 246*f*, 247–251, 248*f*, 251*f*, 261
anomalies of, 256
remnants
in females, 248*f*
in males, 248*f*
suppression, in male development, 247, 261
parametrium, 251
paranasal sinuses
development of, 178, 178*f*
in neonatal and postnatal development, 178, 316
parasitic twin, 116*f*
parasternal hernia, 149
parasympathetic nervous system, development of, 370
parathyroid glands
abnormal number of, 166
congenital absence of, 166
congenital anomalies of, 165–166, 166*f*
development of, 160*f*–161*f*, 161, 186
ectopic, 165, 166*f*
inferior, 161
superior, 161
paraurethral glands, 248*f*, 250*t*, 251
paraxial mesoderm, 52*f*, 53–56, 54*f*, 306
derivatives of, 65*f*, 307*f*, 344
parietal bone, 319*f*
parietal pleura, 194*f*, 195
paroophoron, 248*f*, 250*t*
parotid glands, development of, 171
pars distalis, 360*t*, 361
pars intermedia, 360*t*, 360*f*, 361
pars nervosa, 360*t*, 360*f*, 361
pars tuberalis, 360*t*, 360*f*, 361
parturition, 99–105, 101*f*
parvovirus B19, as teratogen, 128*t*
Patau syndrome (trisomy 13), 120, 122*f*, 122*t*, 185, 386
patent ductus arteriosus, 288, 299, 302*f*, 304
patent foramen primum, 287*f*, 288
patent oval foramen, 284–288, 286*f*–287*f*, 304
pectinate line, 222, 224*f*
pectoral girdle, development of, 317–319
pectoralis major muscle, congenital anomalies of, 327
pelvic appendix, 215
pelvic girdle, development of, 317–319
pelvic kidney, 235, 238*f*
pelvis, renal
development of, 230, 234*f*, 250*t*
duplication of, 235–237
penicillin, use during pregnancy, 133
penile hypospadias, 255
penile raphe, 252, 254*f*
penis, 10, 11*f*, 13
congenital anomalies of, 241*f*–242*f*, 255–256, 256*f*
development of, 242*f*, 243*f*, 250*t*, 252, 254*f*
erectile tissue of, 11*f*, 13
penoscrotal hypospadias, 255
pentasomy, 123
percutaneous umbilical cord blood sampling (PUBS), 86, 105
pericardial cavity, 54*f*, 56
development of, 140, 143–144, 149, 264*f*, 269*f*, 270*f*–271*f*, 271
pericardial sinus, transverse, 271
pericardioperitoneal canals, 54*f*, 140–144, 142*f*–144*f*, 145*f*, 149
pericardium
fibrous, 143, 144*f*, 147
visceral (endocardium), 268, 269*f*, 270*f*
pericentric inversion, 126
perichondrium, 308, 309*f*
periderm, 388, 388*f*
perilymph, 383
perilymphatic space, 382–383, 383*f*
perimetrium, 10, 11*f*–12*f*
perinatal medicine, 85
perinatal morbidity, cigarette smoking and, 85
perinatology, 85
perineal body, 222
perineal fistula, 225*f*
perineal hypospadias, 255
periodontal ligament, 395*f*, 396
periosteum, 308, 309*f*
peripheral nervous system, 344
development of, 366–369
peristalsis, of uterine tube, 24
peritoneal cavity, 54*f*, 56
development of, 140–144, 149
peritoneum, lesser sac of. *See* omental bursa
peromelia, 336
phalanges, 318*f*, 335*f*
phallus, 252, 254*f*
adult derivatives and vestigial remains of, 250*t*
pharyngeal apparatus, 151–187
and developing respiratory system, 191*f*
components of, 152, 153*f*
congenital anomalies of, 152, 162–166, 163*f*–166*f*
pharyngeal arches, 67, 69*f*, 71*f*–72*f*, 73*f*, 152–159, 153*f*–154*f*, 186
1st (first), 67, 69*f*, 71*f*–72*f*, 152, 153*f*
cartilage, derivatives of, 154–158, 157*f*, 157*t*, 170*f*, 313, 383*f*, 383–384
muscles, derivatives of, 157*t*, 158, 158*f*
2nd (second), 67, 69*f*, 72*f*, 152, 153*f*–154*f*, 313
cartilage, derivatives of, 157*f*, 157*t*, 158, 170*f*, 383*f*, 384
fate of, 154*f*, 156*f*
muscles, derivatives of, 157*t*, 158*f*
3rd (third), 69*f*, 72*f*, 154*f*, 314
cartilage, derivatives of, 157*t*, 157*f*, 158, 170*f*
muscles, derivatives of, 157*t*, 158*f*
4th (fourth), 69*f*, 72*f*
cartilage, derivatives of, 157*t*, 157*f*, 158, 170*f*, 314
muscles, derivatives of, 157*t*, 158*f*
6th (sixth)
cartilage, derivatives of, 157*t*, 158, 170*f*, 314
muscles, derivatives of, 157*t*, 157*f*, 158*f*
and tongue development, 170*f*
arteries of, 152–154 (*See also* aortic arches)
cartilages of, derivatives of, 154–158, 157*t*, 158*f*
components of, 152–154, 155*f*
fate of, 152–159, 154*f*–156*f*
muscles, 325*f*, 326
derivatives of, 157*t*, 158, 158*f*
nerves
derivatives of, 157*t*, 158–159
formation of, 367–369, 368*f*
ossification of, 154–158, 157*f*
pharyngeal grooves (clefts), 152, 153*f*, 156*f*, 160*f*, 161–162, 186
pharyngeal hypophysis, 361
pharyngeal membranes, 153*f*, 155*f*, 159, 161, 186
pharyngeal pouches, 153*f*, 155*f*–156*f*, 159–161, 186
1st (first), derivatives of, 159, 160*f*
2nd (second), derivatives of, 159–161, 160*f*–161*f*
3rd (third), derivatives of, 160*f*–161*f*, 161
4th (fourth), derivatives of, 160*f*–161*f*, 161
5th (fifth), 161
pharyngeal tonsils, 159, 302
pharyngotympanic tube, 159, 186, 383, 383*f*

pharynx
  constrictor muscles of, formation of, 157*t*
  development of, 152, 202, 226
  primordial, 153*f*–156*f*, 159, 190, 202
phenobarbital, use during pregnancy, 133
phenotype, 119
phenylketonuria (PKU), fetal effects of, 137
phenytoin, as teratogen, 128*t*, 133, 133*f*
philtrum of lip, 173*f*, 175
phocomelia, 336
phosphorous, and bone development, 308
phrenic nerves, innervation of diaphragm by, 146
pia mater, 348, 356, 357*f*
piebaldism, 391
Pierre Robin syndrome, 165
pigmentation of skin, 389, 391, 399
pineal gland (body), 358, 359*f*
pinocytosis, placental transport via, 98
pituitary gland
  and female reproductive cycles, 19, 20*f*
  derivation and terminology of, 360*t*
  development of, 359–361, 360*f*
  glandular part of (adenohypophysis), 360*t*, 360*f*, 361
  nervous part of (neurohypophysis), 360*t*, 360*f*, 361
PKU (phenylketonuria), fetal effects of, 137
placenta, 90–99, 114
  abnormalities of, 103, 104*f*
    and fetal growth, 85
  accessory, 100, 102, 103*f*
  after birth, 100–105, 102*f*
  anastomosis of blood vessels in, twin pregnancy and, 111
  as allograft, 103–105
  battledore, 105
  common, in twin pregnancy, 111, 113*t*
  development of, 40, 59*f*, 90–93, 91*f*–96*f*
  discoid, 100, 102*f*
  endocrine synthesis and secretion in, 99
  examination of, 100–103
  fetal part of, 90, 91*f*, 93, 94*f*, 107*f*, 114
  fetal surface of, 95*f*, 102*f*–136*f*, 103
  fetomaternal junction of, 95*f*
  functions and activities of, 90, 98, 114
  immunoprotection of, 103–105
  maternal part of, 90, 91*f*, 93, 94*f*, 114
  maternal surface of, 95*f*, 102*f*, 103, 103*f*
  metabolism of, 98
  monochorionic-monoamniotic twin, 112–113
  multiple pregnancies and, 111–114, 113*t*
  retention of, and uterine hemorrhage, 102
  shape of, 91*f*, 94
    variations in, 100–102, 102*f*
  transport across, 83, 90, 97*f*, 98–99
    active, 98
    by facilitated diffusion, 98
    by pinocytosis, 98
    by simple diffusion, 98
    mechanisms of, 98
    of drugs and drug metabolites, 97*f*, 99
    of electrolytes, 97*f*, 98
    of gases, 83, 97*f*, 98
    of hormones, 97*f*, 98
    of infectious agents, 97*f*, 99
    of maternal antibodies, 97*f*, 98
    of nutrients, 83, 97*f*, 98
    of waste products, 97*f*, 99
  ultrasound of, 90
  umbilical cord attachment to, 102*f*–136*f*, 103, 105
placenta accreta, 103, 104*f*
placenta percreta, 103, 104*f*
placenta previa, 42, 103, 104*f*
placental barrier. *See* placental membrane
placental circulation, 95*f*–96*f*, 96–99, 271
  fetal, 95*f*–96*f*, 96
  impaired, and fetal growth, 85
  maternal, 95*f*, 97
placental membrane, 96, 96*f*–97*f*, 97–98, 114
  transfer across, 83, 90, 97*f*, 98–99
  vasculosyncytial, 97
placental septa, 94, 95*f*, 102*f*, 103
placental stage, of labor, 100, 101*f*
plagiocephaly, 317
planes, anatomical, descriptive terms for, 7, 8*f*
plantar artery, 337*f*
plasma membranes, sperm and oocyte, fusion of, 28, 29*f*–30*f*
platysma muscle, formation of, 158*f*
pleura
  parietal, 194*f*, 195
  visceral, 194*f*, 195
pleural cavities, 56
  development of, 140, 143–144
  primordial, 143
pleuropericardial folds, 143
pleuropericardial membranes, 143–144, 143*f*–144*f*, 149–150
pleuroperitoneal folds, 143–144
pleuroperitoneal membranes, 144, 145*f*
  diaphragm development from, 145*f*, 145–146, 150
pleuroperitoneal openings, closure of, 144–145, 145*f*
pneumocytes, 197
pneumonitis, tracheoesophageal fistula and, 193
pneumothorax, with congenital diaphragmatic hernia, 147
Poland syndrome, 327
polar body
  first, 15*f*, 17, 28, 29*f*–30*f*
  second, 15*f*, 17, 28, 30*f*
poliomyelitis virus, transplacental transport of, 99
polychlorinated biphenyls, as teratogens, 128*t*, 135
polycystic kidney disease, 237
polycythemia, twin-twin transfusion syndrome and, 111, 114*f*
polydactyly, 338, 339*f*
polyhydramnios, 108
  esophageal atresia/tracheoesophageal fistula and, 193, 202
  with meroanencephaly, 364
polymastia, 393, 399
polypeptides, in fetal metabolism and growth, 83
polyploidy, 119
polythelia, 393, 399
pons, development of, 353–356, 356*f*–358*f*, 370
pontine flexure, 347*f*, 353–354, 356*f*
popliteal artery, 337*f*
porta hepatis, 208
portal system, 304
portal vein, development of, 264, 267*f*, 300*f*–301*f*
port-wine stain, 391
position, anatomical, descriptive terms for, 7, 8*f*
posterior, as descriptive term, 8*f*
posterior commissure, 363*f*
postotic myotomes, 326
Prader-Willi syndrome, 124, 127
preaortic ganglia, 369
preauricular sinuses, 385, 385*f*
prechordal plate, 40*f*, 41, 44, 47*f*–51*f*, 48, 60
predentin, 395*f*, 396
pregnancy
  corpus luteum of, 21
  drug use in, 85, 128*t*, 135
  ectopic, 42–43, 43*f*
  multiple, 110–114 (*See also* twin(s))
    and fetal growth, 85, 85*f*
    placenta and fetal membranes in, 111–114, 113*f*
  ultrasound detection of, 46, 46*f*
  uterine growth in, 99, 100*f*
pregnancy phase, of menstrual cycle, 24
pregnancy testing, 38–39, 46
premature by date, 83
premature by weight, 83
premature infants, 78, 83
  ductus arteriosus closure in, 299
  patent ductus arteriosus in, 299
premature rupture of fetal membranes, 96, 108
prenatal development, timetable of, 3*f*–5*f*
preotic myotomes, 326, 326*f*
prepuce (foreskin), 11*f*, 243*f*, 252
primigravidas, 100
primitive groove, 46–48, 47*f*–48*f*
primitive node, 46, 47*f*–49*f*, 60
primitive pit, 46–48, 47*f*–48*f*
primitive streak, 41*f*, 46–53, 47*f*–49*f*, 58–60
  remnants, and sacrococcygeal teratoma, 48, 50*f*
  tail fold and, 62, 64*f*
primordium, definition of, 2
probe patent oval foramen, 284–288, 286*f*

processus vaginalis, 256, 258*f*, 259
 persistent, 260*f*, 260–261
proctodeum, 202, 202*f*, 222, 223*f*, 226
profunda fermoris artery, 334, 337*f*
progenitor cells. *See* stem cells
progesterone
 and menstrual cycle, 23
 and ovarian cycle, 19, 20*f*
 production of
  ovarian, 10
  placental, 99
progestogens/progestins
 as teratogens, 128*t*, 132–133
 transplacental transport of, 98
proliferative phase, of menstrual cycle, 23
pronephroi, 230, 232*f*, 260
pronucleus, female and male, formation of, 28, 30*f*, 35
prosencephalic organizing center, 171
prosencephalon. *See* forebrain
prospective approach, in proving teratogenicity, 131
prostaglandin(s)
 and ductus arteriosus closure, 299
 and labor, 99–100
 and placental immunoprotection, 104
 in semen, 24
prostate, 11*f*, 13, 24, 25*f*
 development of, 248*f*, 249, 250*t*, 252*f*
prostatic utricle, 248*f*, 250*t*
pseudoglandular period, of lung maturation, 195, 196*f*, 199
pseudohermaphroditism
 female, 255, 255*f*, 261
 male, 255, 261
ptosis, congenital, 380
pubertal growth spurt, 10
puberty, 10
 definition of, 10
 female reproductive cycles in, 19
 mammary development in, 392
 oogenesis in, 17
 spermatogenesis in, 14
PUBS (percutaneous umbilical cord blood sampling), 86
pudendal artery, 222
pudendal vein, 222
pulmonary artery
 left, 292*f*, 293
 right, 292*f*, 293
pulmonary atelectasis, 327
pulmonary hypoplasia, 198
 congenital diaphragmatic hernia and, 147, 148*f*, 198
pulmonary trunk, 280, 281*f*, 283*f*, 284
 semilunar valves of, 284, 285*f*
pulmonary valve stenosis, 288, 290*f*
pulmonary vein, 300*f*–301*f*
 primordial, 277, 279*f*
pupillary membrane, 377*f*, 378
Purkinje fibers, 327
pyloric stenosis, congenital hypertrophic, 203
pyramidal lobe, 167
pyramids, 354, 356*f*

Q

quadruplets, 111, 114
quickening, 80
quintuplets, 111, 114

R

rachischisis, 316, 317*f*
radial artery, 334, 337*f*
radiation, as teratogen, 129*t*, 137, 380
radius, 318*f*, 319*f*, 332, 335*f*
 congenital absence of, 338
ramus
 dorsal primary, 324, 325*f*
 gray communicating, 369
 ventral primary, 325*f*
 white communicating, 351*f*, 369
Rathke pouch (hypophysial diverticulum), 359–361, 360*f*
reciprocal translocation, 125*f*
recombinant DNA technology, 7
recovery stage, of labor, 100
rectal artery, 222
 inferior, 222
 superior, 222
rectal atresia, 225*f*, 226
rectal nerve, inferior, 222
rectal veins, 222
 inferior, 222
 superior, 222
rectocloacal fistula, 225*f*
rectourethral fistula, 225*f*, 226
rectouterine pouch, 250, 253*f*
rectovaginal fistula, 225*f*, 226
rectovesical fistula, 226
rectovestibular fistula, 226
rectum
 agenesis of, 225*f*, 226
 development of, 222, 223*f*–224*f*, 226
rectus abdominis muscle, 325*f*, 326*f*
recurrent laryngeal nerve, 157*t*, 294*f*, 369
Reichert cartilage, 158
renal agenesis, 108, 234
 bilateral, 234
 unilateral, 234, 238*f*
renal arteries, 232, 236*f*
 accessory, 232, 237*f*
renal corpuscle, 230, 235*f*
renal pelvis
 development of, 230, 234*f*, 250*t*
 duplication of, 235–237
renal veins
 accessory, 232, 237*f*
 development of, 266*f*
reproduction, 9–26
reproductive cycles, female, 19–24, 20*f*–22*f*
reproductive organs, 10–13. *See also specific organs*
 female, 10–13, 11*f*–13*f*
 male, 10, 11*f*, 13
reproductive system. *See* genital system
respiratory bronchioles, 195–196, 196*f*
respiratory distress syndrome, 198–199
respiratory diverticulum, 191*f*
respiratory primordium, 190, 191*f*
respiratory system, development of, 189–199, 191*f*, 192*f*, 202
Ret proto-oncogene, and congenital megacolon, 224
rete ovarii, 246*f*, 247, 250*t*
rete testis, 246*f*, 247, 250*t*
retina
 central artery and vein of, 372, 375*f*, 377*f*
 coloboma of, 374
 congenital detachment of, 374
 development of, 53, 357, 372, 373*f*–379*f*, 385–386
 neural, 372, 376*f*–379*f*
retinal fissures, 372, 373*f*, 375*f*
 defects in closure of, 374, 376, 386
retinal pigment epithelium, 69, 372, 374*f*, 376*f*–379*f*
retinoic acid
 endogenous, and embryonic development, 7
 exogenous, as teratogen, 128*t*, 134
retrocecal appendix, 215
retrocolic appendix, 215
retroesophageal subclavian artery, 295
retroperitoneal lymph sacs, 303*f*
retrospective approach, in proving teratogenicity, 131
retrosternal hernia, 149
Rh disease, 99
rhombencephalic organizing center, 171
rhombencephalon. *See* hindbrain
ribs
 accessory, 316
 cervical, 316, 316*f*
 congenital anomalies of, 316, 316*f*
 development of, 310, 312*f*, 313, 319*f*, 320
 false, 313
 floating, 313
 fused, 316
 lumbar, 316
 true, 313
rickets, 308, 397
rima glottis, 170*f*
ring chromosome, 123–124, 125*f*
Robertsonian translocation, 125*f*
Robin morphogenetic complex, 165
rods (retinal), 372
root canal, 396
rostral neuropore, 69*f*–71*f*, 153*f*, 344, 346*f*, 357
 defect in closure of, 362–364
round ligament of liver, 299, 301*f*
round ligaments of uterus, 248*f*, 250*t*, 259
RU486, 44
rubella virus
 as teratogen, 118, 128*t*, 135, 136*f*
  in ear development, 384, 386
  in eye development, 118, 135, 136*f*, 380
  in heart development, 284
 transplacental transport of, 97*f*, 99

S

saccule, 196*f*, 197, 199, 381, 382*f*, 386
sacral arteries, lateral, 268
sacral dimple, 317*f*
sacral vein, median, 266*f*
sacrococcygeal teratoma, 48, 50*f*
sagittal plane, 8*f*
salicylates, fetal effects of, 134
salivary glands, development of, 171
satellite cells, 351*f*, 366
scala tympani, 382*f*, 383
scala vestibuli, 382*f*, 383
scalp hair patterning, 80
scalp vascular plexus, 71, 74, 74*f*
scaphocephaly, 317, 317*f*
scapula, 318*f*, 319*f*, 335*f*
Schwann cells, 332, 350, 351*f*, 366
sclera, development of, 374*f*, 377*f*–378*f*, 380, 386
scleral venous sinus, 377*f*, 379
sclerotomes, 306, 307*f*, 310, 311*f*, 320
scoliosis, 316
scrotal raphe, 252, 254*f*
scrotum, 11*f*
  development of, 248*f*, 250*t*, 252, 254*f*
"seal limbs," 134
sebaceous glands, 389, 390*f*, 399
sebum, 389
second meiotic division, 14, 16*f*
  in spermatogenesis, 14–17, 15*f*
  of oocyte, 15*f*, 17, 28, 29*f*
second trimester, 79
second week, 37–44
secondary sex characteristics, 10
secretory cells, 389
secretory (luteal) phase, of menstrual cycle, 10, 23
semen, 11*f*, 13, 24–25, 25*f*, 249
semicircular canals, 382
semicircular ducts, 382, 382*f*
semilunar valves, 284, 285*f*, 290*f*
seminal colliculus, 250*t*, 251*f*
seminal glands (vesicles), 11*f*, 13, 24, 25*f*, 248*f*, 249
seminiferous cords, 246*f*, 247
seminiferous tubules, 11*f*, 25*f*, 247, 250*t*
  primordium of, 246*f*, 247
sensory axons, 332
sensory nerves, special, formation of, 368*f*, 369
septum pellucidum, 362
septum primum, 273*f*, 274*f*–275*f*, 276, 277*f*, 299
  defects of, 288
septum secundum, 273*f*, 274*f*–275*f*, 276, 277*f*, 299
  defects of, 288
septum transversum, 62, 64*f*, 141, 141*f*–143*f*, 207, 209*f*, 226, 264, 264*f*, 271*f*
  diaphragm development from, 62, 145*f*, 145–146, 150
septuplets, 114
Sertoli cells, 246*f*, 247, 261
seventeenth week, 80, 82*f*
seventh week, 71
sex characteristics, secondary, 10
sex chromatin, 119
sex chromosomes, 13–14, 245–247, 261
  disorders of, 119, 247
  in oocyte, 17
  in sperm, 15*f*, 17
  nondisjunction of, 18*f*
  trisomy of, 120–123, 124*f*, 124*t*
sex cords
  primary, 244–245, 245*f*, 247
  secondary, 247
sex determination, 15*f*, 17, 245–247, 246*f*, 261
  for fetal diagnostic testing, 35, 253–255
  visualization of fetal genitalia for, 253
sex organs, 10–13. *See also specific organs*
  female, 10–13, 11*f*–13*f*
  male, 10, 11*f*, 13
sextuplets, 114
shaft of bone, formation of, 306
SIDS (sudden infant death syndrome), 284
signaling pathways
  and embryonic development, 66–67
  and muscle development, 324, 324*f*
  and renal development, 236*f*
  and somite development, 56
Simonart band, 183
simple diffusion, placental transport via, 98
single chromatid chromosomes, 14, 16*f*
single umbilical artery, 105, 106*f*, 234
sinovaginal bulbs, 251, 253*f*
sinual (Müller) tubercle, adult derivatives and vestigial remains of, 249, 251, 251*f*
sinuatrial node, 282*f*, 284
sinuatrial valve, 271, 272*f*, 276–277
sinus(es)
  auricular, 162, 163*f*
  branchial, 162, 163*f*–164*f*, 186
    external, 162, 163*f*–164*f*
    internal, 162, 163*f*
  cervical, 67, 73*f*, 152, 154*f*, 156*f*, 162
    persistent remnant of, 163*f*–165*f*
  coronary, 267*f*, 276–277, 278*f*
  frontal, 178
  lymph, 302
  maxillary, 178
  paranasal
    development of, 178, 178*f*
    in neonatal and postnatal development, 178, 316
  pericardial, transverse, 271
  preauricular, 385, 385*f*
  scleral venous, 377*f*, 379
  sphenoid, 178
  thyroglossal duct, 167, 167*f*, 168*f*, 186
  tonsillar, 159, 160*f*–161*f*
  transverse pericardial, 271
  urachal, 239–241, 241*f*
  urogenital, 222, 223*f*, 226, 237–239, 240*f*, 248*f*, 251*f*, 260
    adult derivatives and vestigial remains of, 250*t*
    caudal phallic part of, 239, 240*f*, 241, 251*f*
    cranial vesical part of, 239, 240*f*
    pelvic part of, 239, 240*f*
sinus venarum, 276, 278*f*
sinus venosus, 264–265, 265*f*–267*f*, 268–271, 269*f*, 270*f*, 271, 272*f*, 304, 337*f*
  changes in, 276–277, 278*f*
  circulation in, 272*f*
  horns of, 267*f*, 271, 272*f*, 278*f*
sinus venosus atrial septal defects, 287*f*, 288
sinusoids, 40
  hepatic, 207, 264, 296
sixth week, 67–71, 73*f*
skeletal muscle, development of, 324–327, 325*f*–326*f*
skeletal system, development of, 305–321, 307*f*
  critical period of, 129
skeleton
  appendicular, development of, 317–320, 318*f*–319*f*
  axial
    congenital anomalies of, 316–317, 316*f*–317*f*
    development of, 55, 310–316, 311*f*, 320
  congenital anomalies of, 316–317, 316*f*–317*f*, 319–320
skin
  congenital anomalies of, 391, 391*f*, 399
  development of, 388*f*, 388–389, 399
  glands of, 389, 390*f*
skin tags, auricular, 385, 385*f*
skull. *See* cranium
small intestine, development of, 212–215, 214*f*
smoking, fetal effects of, 83–85, 85*f*, 131
smooth chorion, 90, 91f–94*f*
smooth muscle, development of, 327
social drugs, fetal effects of, 85, 135
soft palate, 179, 179*f*–180*f*
  cleft of, 181–183
soleus muscle, accessory, 327
somatic afferents, general and special, 354, 356*f*–357*f*
somatic efferent columns, 367
somatic efferent cranial nerves, 367–368, 368*f*
somatic efferents, general and special, 354, 356*f*–357*f*
somatic mesoderm, extraembryonic, 40*f*, 41, 56
somatopleure, 54*f*, 56
somatostatin-containing cells, pancreatic, 210
somites
  and embryonic age, 55
  and skeletal development, 55, 306, 307*f*, 310, 311*f*
  cervical, 147*f*
  development of, 53–56, 54*f*, 60, 67, 69*f*–72*f*, 73*f*
  myotomes of, 307*f*, 324–327, 325*f*

Sonic hedgehog
and hair development, 389
and limb development, 330
and nervous system development, 344
and pancreatic development, 209
and prostate development, 249
special somatic afferent, 354, 356*f*
special visceral afferent, 354, 356*f*
special visceral efferent, 354, 356*f*
special visceral efferent column, 368
specialty gene, 127
sperm, 2, 10, 13, 14*f*, 35
abnormal, 17–19
acrosome of, 14*f*, 17
acrosome reaction of, 24, 29*f*
capacitation of, 24, 29*f*
definition of, 2
ejaculation of, 24
head of, 14*f*, 17
intracytoplasmic injection of, 31
maturation of, 24–26, 29*f*
mature, 17
neck of, 14*f*, 17
plasma membrane, fusion with oocyte plasma membrane, 28, 29*f*–30*f*
sex chromosomes in, 15*f*, 17
tail of, 14*f*, 17
transport of, 24, 25*f*
versus oocyte, 14*f*, 17
viability of, 26
sperm counts, 25
spermatic cord, hydrocele of, 260, 260*f*
spermatic fascia
external, 258*f*, 259
internal, 258*f*, 259
spermatic vein, 266*f*
spermatids, 14–17
spermatocytes
primary, 14, 15*f*
secondary, 14, 15*f*
spermatogenesis, 13–17, 15*f*
abnormal, 17–19, 18*f*
spermatogonia, 14, 15*f*, 246*f*, 247
spermatozoon. *See* sperm
spermiogenesis, 14–17
sperm-oocyte interactions, 28
sphenoid sinuses, 178
sphenomandibular ligament, 154, 157*t*
sphincter pupillae muscle, 327, 376, 386
spina bifida, 316, 353*f*
with meningomyelocele, 352, 353*f*–354*f*
spina bifida cystica, 53, 118, 316, 352, 353*f*–355*f*
hydrocephalus with, 364
with meningocele, 352, 353*f*
spina bifida occulta, 316, 351, 353*f*–354*f*
spinal cord
central canal of, 344, 350*f*, 370
congenital anomalies of, 350–352
development of, 53, 344–350, 347*f*–348*f*, 350*f*–351*f*, 370
myelination of nerve fibers in, 350
positional changes of, 349–350, 352*f*
white matter of, 344
spinal ganglion, development of, 348, 350*f*–351*f*, 366–367
spinal ganglion cells, 351*f*
spinal meninges
anomalies of, 353*f*
development of, 348, 352*f*
spinal nerves
development of, 324, 350*f*, 351*f*, 366–367
dorsal roots of, 350*f*, 367
myelination of fibers, 350
ventral roots of, 346–348, 350*f*, 367
spiral endometrial arteries, 12*f*, 23, 39–40, 94, 94*f*–95*f*, 97
spiral ganglion, 382*f*
spiral organ (of Corti), 369, 381–382, 382*f*, 386
splanchnic mesoderm, extraembryonic, 40*f*, 41, 56
splanchnopleure, 56
spleen
accessory, 212
development of, 210, 213*f*, 302
fetal, hematopoiesis in, 83
splenorenal ligament, 213*f*
spongioblasts, histogenesis of, 344, 349*f*
spongy layer, of endometrium, 10, 12*f*
spongy urethra, 241–242, 243*f*, 252, 254*f*
spontaneous abortion, 35, 43
sporadic and recurrent, 43
squamous temporalis bone, formation of, 314–315
stapedial arteries, 291
stapedius muscle, formation of, 157*t*, 384
stapes, 380, 383, 383*f*, 386
congenital fixation of, 384
formation of, 157*t*, 157*f*, 158, 313
stem cells
and placental immunoprotection, 104
embryonic, 7
stem villi, 57–58, 59*f*, 60, 94–95, 95*f*–96*f*
sternal bars, 313
sternebrae, 313
sternocleidomastoid muscle
formation of, 158*f*
in congenital torticollis, 327
sternocostal hiatus, herniation through, 149
sternum, development of, 310, 313
steroid hormones
placental synthesis of, 99
unconjugated, transplacental transport of, 98
stigma, 19, 22*f*
stomach
congenital anomalies of, 203
development of, 202*f*, 202–203, 204*f*–205*f*, 213*f*
mesenteries of, 203, 204*f*–205*f*
positioning of, 213*f*, 216*f*
rotation of, 203, 204*f*–205*f*
stomodeum, 62, 64*f*, 152, 153*f*–155*f*, 159, 171, 172*f*, 202, 202*f*
pituitary gland development from ectodermal roof of, 359–361, 360*f*
straight collecting tubule, 234*f*, 235*f*
stratum corneum, 388, 388*f*, 390*f*
stratum germinativum, 388, 388*f*, 390*f*
stratum granulosum, 388*f*, 390*f*
stratum lucidum, 388*f*, 390*f*
stratum spinosum, 388*f*, 390*f*
streptomycin/streptomycin derivatives, as teratogens, 133
stromal cells, and placental immunoprotection, 104
structural chromosomal abnormalities, 123–126, 125*f*
stylohyoid ligament, formation of, 157*f*, 157*t*, 158
stylohyoid muscle, formation of, 157*t*, 158*f*
styloid process, formation of, 157*t*, 157*f*, 158, 313–314
stylopharyngeus muscle, formation of, 157*t*, 158*f*
subaortic stenosis, 289
subarachnoid space, 348, 356
subcapsular lens epithelium, 377*f*
subcardinal veins, 265, 266*f*
subclavian arteries, 292*f*
right, 292*f*, 293, 294*f*
anomalous, 295, 298*f*
retroesophageal, 295
subclavian veins, 266*f*
sublingual glands, development of, 171
sublingual thyroid gland, 167, 169*f*
submandibular duct, 171
submandibular glands, development of, 171
substantia nigra, 357, 358*f*
sudden infant death syndrome (SIDS), 284
sulci, cerebral, 362
sulcus limitans, 344, 350*f*, 353, 356*f*
sulcus terminalis, 276, 278*f*
superficial inguinal lymph nodes, 222
superficial inguinal ring, 256
superior, as descriptive term, 8*f*
superior colliculi, 357, 358*f*
superior laryngeal nerve, 369
formation of, 157*t*, 369
superior mesenteric artery, 141, 142*f*, 202*f*, 206, 212, 214*f*, 222, 268
superior parathyroid gland, 161
superior recess of omental bursa, 203
superior rectal artery, 222
superior rectal vein, 222
superior vena cava, 296, 300*f*–301*f*
development of, 265, 267*f*, 304
persistent left, 268
superior vesical arteries, 301*f*
supernumerary breasts and nipples, 393, 399
supernumerary digits, 338, 339*f*
supernumerary kidney, 235–237, 238*f*–239*f*
supernumerary parathyroid glands, 166
supernumerary teeth, 397, 398*f*–399*f*
supracardinal veins, 265, 266*f*
suprarenal glands, development of, 236*f*, 243, 244*f*
suprarenal veins, development of, 266*f*
surfactant, 80, 197, 199
deficiency of, 197–199
surfactant replacement therapy, 198

surgery, fetal, 88
sutures (cranial), 313, 314*f*
premature closure of, 317, 317*f*
sweat glands, 389, 390*f*, 399
apocrine, 389
eccrine, 389
sympathetic ganglion, 243, 351*f*, 367, 369
sympathetic nervous system, development of, 369
sympathetic trunk, 351*f*, 369
syncytial knots, 96*f*, 98
syncytiotrophoblast, 33–35, 34*f*, 38, 38*f*–39*f*, 43, 90, 96*f*, 97, 99
syndactyly, 327, 330, 338–339, 339*f*
synovial cavity, 310, 320
synovial joints, 308–310, 310*f*, 320, 332
syphilis
and birth defects, 128*t*, 136–137
and tooth development, 397
primary maternal infections, 136
secondary maternal infections, 136–137
transmission across placenta, 99

T

tail fold, 62, 63*f*–64*f*, 75–76
talipes (clubfoot), 137, 339, 340*f*
talipes equinovarus, 339
tarsus, 319*f*
taste buds, development of, 169–170
tectum, 357
teeth
abnormally sized, 397
congenital anomalies of, 397–398, 398*f*–399*f*, 400
deciduous, 316, 393–394, 396–397, 397*f*
shedding of, order and usual time of, 394*t*
development of, 129, 393–397, 394*f*–397*f*, 400
bell stage of, 396
bud stage of, 394, 394*f*–395*f*
cap stage of, 395*f*, 395–396
syphilis and, 397
tetracycline and, 129, 398, 399*f*, 400
discolored, 398, 399*f*
eruption of, 316, 396–397, 397*f*
order and usual time of, 394*t*
fused, 398, 398*f*
incisor, 179, 179*f*, 395*f*, 397
natal, 397
numerical abnormalities of, 397, 398*f*–399*f*
permanent, 316, 393–394, 397, 397*f*
root of, 396
shape of, variations in, 397, 398*f*
twinning of, 398
tegmentum, 357
tela choroidea, 356, 356*f*–357*f*
telencephalic vesicles, 357, 358*f*
telencephalon, 347*f*, 352, 355*f*, 357–358, 361
temporal bone, styloid process of, formation of, 157*t*, 157*f*, 158, 313–314
temporalis muscle, formation of, 158*f*
tendinous cords, 280, 282*f*
tensor tympani, 157*t*, 383–384
tensor veli palatini, formation of, 157*t*
teratogen(s), 62, 127–137, 128*t*–129*t*, 334–335
and critical periods of human development, 129, 130*f*
and organogenesis, 129, 130*f*, 138
behavioral, 135
definition of, 62
dose-response relationship for, 129–131
drugs as, 128*t*, 131–135
environmental chemicals as, 128*t*, 135
genotype and, 131
infectious agents as, 128*t*–129*t*, 135–137
known human, 131–137
mechanical factors as, 137
teratogenesis, basic principles in, 129–131
teratogenicity, proof of, 131
teratology, 118
definition of, 118
teratoma, sacrococcygeal, 48, 50*f*
terminal filum, 350, 352*f*
terminal groove, of tongue, 169, 170*f*
terminal saccular period, of lung maturation, 196*f*, 197, 199
terminal villi, 58, 94–96, 95*f*, 96*f*
terminology, embryological, 2
testes (testicles), 10, 11*f*, 13
descent of, 258*f*, 259
development of, 230, 233*f*, 244–247, 245*f*–246*f*, 248*f*, 250*t*, 261
ectopic, 259, 259*f*
interstitial, 259
hydrocele of, 260*f*, 260–261
undescended, 259, 259*f*, 261
testicular feminization syndrome, 255
testis-determining factor, 245–247, 246*f*, 261
testosterone
and male genital development, 247, 249, 252, 261
and masculinization of female fetus, 132
transplacental transport of, 98
tetracycline, as teratogen, 128*t*, 129, 133, 398, 399*f*, 400
tetralogy of Fallot, 289, 290*f*
tetraploidy, 123
tetrasomy, 123
thalamus, development of, 358, 359*f*
thalidomide, as teratogen, 118, 128*t*, 134, 135*f*, 335, 338*f*, 340
thalidomide syndrome, 134
theca externa, 19, 20*f*
theca folliculi, 19, 20*f*
theca interna, 19, 20*f*
third trimester, 79
third week of development, 45–60
thirty-sixth week of development, 84*f*
thoracic duct, development of, 302, 303*f*
thoracic myotomes, 325–326, 326*f*
thoracic wall, development, 143, 144*f*
thymic aplasia, congenital, 165
thymus
and lymphocyte development, 302
congenital anomalies of, 166*f*
development of, 160*f*–161*f*, 161, 186
thyroglossal duct, 166, 166*f*
thyroglossal duct cysts and sinuses, 167, 167*f*, 168*f*, 186
thyroid cartilage, formation of, 157*t*
thyroid drugs, as teratogens, 134
thyroid follicles, 167
thyroid gland
accessory tissue of, 168*f*
congenital anomalies of, 166*f*, 167, 168*f*–169*f*, 186
development of, 166*f*, 166–167, 186
ectopic, 166*f*, 167, 168*f*, 186
isthmus of, 166
lingual tissue of, 167, 168*f*
sublingual, 167, 169*f*
thyroid hypoplasia, 165
thyroid primordium, 166
thyroxine, and surfactant production, 198
tibia, 319*f*
tibial artery, 337*f*
time units, gestational, 78–79, 79*t*
timetable(s)
embryological, 129, 130*f*
of prenatal development, 3*f*–5*f*
toe buds, 330, 334*f*
toenails, development of, 80, 391–392, 392*f*, 399
Tomes (odontoblastic) processes, 395*f*, 396
tongue
congenital anomalies of, 170, 171*f*
development of, 160*f*–161*f*, 167–171, 170*f*
distal buds of, 169, 170*f*
median bud of, 169, 170*f*
midline groove (median sulcus) of, 169, 170*f*
muscles of
development of, 158*f*, 169, 326
innervation of, 170
nerve supply of, 170, 170*f*
papillae of, development of, 169–170, 170*f*
posterior third (pharyngeal part) of, 169–170, 170*f*
terminal groove of, 169, 170*f*
tongue-tie, 170, 171*f*
tonsil(s)
development of, 159, 302
lingual, 302
palatine, 159, 160*f*, 186, 302
pharyngeal, 159, 302
tubal, 302
tonsillar crypts, 159
tonsillar sinus, 159, 160*f*–161*f*
tooth. *See* teeth
tooth bud, 394, 395*f*
tooth germ, 395
torticollis, congenital, 327, 328*f*

Toxoplasma gondii (toxoplasmosis)
as teratogen, 128*t*, 136
fetal infection, 136
maternal infection, 136
transmission across placenta, 97*f*, 99
trabeculae carneae, 280, 282*f*
trabeculae cranii, 313
trachea
congenital anomalies of, 190–193, 199
development of, 190, 193*f*, 198–199
tracheal atresia, 193
tracheal bud, 190, 192*f*, 193, 199
tracheal stenosis, 193
tracheoesophageal fistula, 190–193, 194*f*, 199, 202
tracheoesophageal folds, 190, 192*f*
tracheoesophageal septum, 190, 192*f*, 198, 202
defective, 190–193, 199
tranquilizers, as teratogens, 134
transcription factors, 127
transferrin, transplacental transport of, 98
transforming growth factor-β
and bone development, 306
and cardiovascular development, 276, 284
and limb development, 330
and nervous system development, 344
and palate development, 179
and placental immunoprotection, 104
transfusion, intrauterine fetal, 86
transgenic mice, 7
translocation, chromosomal, 123, 125*f*
transport, across placenta, 90, 97*f*, 98–99
transposition of great arteries, 288, 290*f*
transvaginal ultrasound, 75
transverse plane, 8*f*
Treacher Collins syndrome, 165
Treponema pallidum
as teratogen, 128*t*, 136–137, 397
transmission across placenta, 99
tricuspid valve, development of, 282*f*, 284
trigeminal ganglion, 368
trigeminal nerve
formation of, 157*t*, 158, 366, 368, 368*f*
innervation of tongue by, 170
trigone region of bladder, 239
trigonocephaly, 317
trilaminar embryonic disc, 46, 47*f*
folding of, 62, 63*f*–64*f*, 75
trimester of pregnancy, 79
definition of, 2
trimethadione, as teratogen, 128*t*
triplets, 111, 114
triploidy, 28, 123
trisomy, 17–19
of autosomes, 120, 122*t*
of sex chromosomes, 120–123, 124*t*, 124*f*
trisomy 13, 120, 122*f*, 122*t*, 185, 386
trisomy 18, 120, 121*f*, 122*t*, 124, 385–386
and intrauterine growth retardation, 85
trisomy 21. *See* Down syndrome
trochlear nerve, formation of, 367–368, 368*f*
trophoblast, 31, 33*f*, 35. *See also* cytotrophoblast; syncytiotrophoblast
abnormal growth of, 58, 103
differentiation of, 38, 38*f*, 43
extravillous, 103
true knots, in umbilical cord, 105
true ribs, 313
truncal ridge, 283*f*
truncus arteriosus, 152–154, 267*f*, 268, 269*f*, 270*f*, 304
circulation through, 271, 272*f*
partitioning of, 280*f*, 280–284, 283*f*
persistent, 288, 289*f*
transformation and adult derivatives of, 292*f*
unequal division of, 288, 290*f*
tubal mucosal enzymes, 28
tubal pregnancy, 42–43, 43*f*
tubal tonsils, 302
tubotympanic recess, 160*f*, 383, 383*f*, 386
tunica albuginea, 246*f*, 247, 249
tunica vaginalis, 258*f*, 259
tunica vasculosa lentis, 377*f*, 378
Turner syndrome, 119, 121*f*, 124, 247
turricephaly, 317
twelfth week of development, 79–80
twentieth week of development, 80
twenty-fifth week of development, 80–83
twenty-ninth week of development, 83
twin(s), 110–114
and fetal membranes, 111–114, 113*t*
conjoined, 113, 116*f*
dizygotic, 111, 112*f*, 114
early death of, 113
fetal growth of, 85, 85*f*
monozygotic, 111, 113*f*, 114, 115*f*
discordant, 119
parasitic, 116*f*
placental vascular anastomosis with, 111
ultrasound of, 113
zygosity of, 111–113, 113*t*
twin-twin transfusion syndrome, 111, 114*f*
tympanic cavity, 159, 186, 383, 383*f*
tympanic membrane, 159, 160*f*, 161, 186, 380, 383*f*, 384

U

ulna, 318*f*, 319*f*, 332, 335*f*
ulnar artery, 334, 337*f*
ultimopharyngeal body, 160*f*–161*f*, 161
ultrasonic waves, fetal effects of, 137
ultrasound, 86, 87*f*
color flow Doppler, of umbilical cord, 105
for bone age determination, 319
for estimation of fetal/gestational age, 75*f*, 78–79
for fetal assessment, 86, 87*f*
for measuring chorionic sac diameter, 41, 74, 75*f*, 90–93
for sex determination, 253
of congenital diaphragmatic hernia, 149*f*
of duodenal atresia, 207, 208*f*
of ectopic pregnancy, 43*f*
of embryos, 74–75, 75*f*–76*f*
of fetus, 78–79, 79*f*
of heart and heartbeat, 56, 58*f*, 280
of neural tube defects, 355*f*
of omphalocele, 218*f*
of placenta, 90
of twin pregnancy, 113
pregnancy detected by, 46, 46*f*
safety of, 137
transvaginal, 75
umbilical artery(ies), 53, 96, 265*f*, 300*f*, 337*f*
absence of (single umbilical artery, SUA), 105, 106*f*, 234
adult derivatives of, 299, 301*f*
constriction of, 299
fate of, 268
umbilical cord, 62, 63*f*, 94*f*
attachment to placenta, 102*f*–136*f*, 103, 105
coiling of, 106*f*
color flow Doppler ultrasound of, 105
excessively long, 105
false knots in, 105
formation of, 73*f*, 76
looping around fetus, 105, 106*f*
prolapse of, 105
true knots in, 105
velamentous insertion of, 105, 105*f*
umbilical cord puncture, 86
umbilical hernia, 71, 215–217
physiological, 212, 215*f*
umbilical ligament
medial, 239, 268, 299, 301*f*
median, 50–53, 90, 110, 110*f*, 239
umbilical veins, 96, 269*f*, 272*f*, 295–296, 300*f*, 337*f*
adult derivatives of, 299, 301*f*
development of, 53, 264–265, 265*f*, 266*f*–267*f*
transformation of, 267*f*
umbilicoileal fistula, 220, 221*f*
unborn patient, 85
uncinate process, 209
unconjugated steroid hormones, transplacental transport of, 98
unicornuate uterus, 256, 257*f*
upper limb buds, 67, 68*t*, 69*f*, 73*f*, 153*f*, 330, 331*f*–334*f*, 340
urachus, 50–53, 110, 110*f*, 239, 240*f*
congenital anomalies of, 239–241, 241*f*
cysts of, 239, 241*f*
fistula of, 241, 241*f*
sinuses of, 239–241, 241*f*
urea, transplacental transport of, 97*f*, 99
ureter(s)
bifid, 236, 238*f*
congenital anomalies of, 234–237
development of, 230–232, 234*f*, 250*t*
ectopic, 237, 239*f*
ureteric bud, 233*f*, 234*f*
urethra
development of, 220, 226, 240*f*, 241–243, 243*f*, 250*t*

urethra—cont'd
male, 11*f*, 13, 25*f*, 241, 243*f*, 252
spongy, 241–242, 243*f*, 252, 254*f*
urethral glands, 251
urethral groove, 242*f*, 252, 254*f*
urethral plate, 252, 254*f*
uric acid, transplacental transport of, 97*f*, 99
urinary bladder
congenital anomalies of, 239–241, 241*f*–242*f*, 261
development of, 50–53, 220, 226, 237–239, 240*f*, 250*t*, 261
exstrophy of, 241, 241*f*–242*f*, 261
urinary system, 230–232
development of, 229–243, 231*f*, 260–261
urinary tract
duplications of, 235–237, 238*f*–239*f*
obstruction, and amniotic fluid disorders, 108
urine formation, fetal, 80, 230
uriniferous tubule, 230
urogenital folds, 252, 254*f*
adult derivatives and vestigial remains of, 250*t*
urogenital groove, 252, 254*f*
urogenital membrane, 222, 223*f*, 252, 254*f*
urogenital orifice, 252
urogenital ridge, 230, 231*f*, 233*f*
urogenital sinus, 222, 223*f*, 226, 237–239, 240*f*, 248*f*, 251*f*, 260
adult derivatives and vestigial remains of, 250*t*
caudal phallic part of, 239, 240*f*, 241, 251*f*
cranial vesical part of, 239, 240*f*
pelvic part of, 239, 240*f*
urogenital system
development of, 231*f*
embryonic structures in, adult derivatives and vestigial remains of, 250*t*
functional divisions of, 230
urorectal septum, 222, 223*f*, 226, 237, 240*f*
abnormal development of, 224
uterine fundus, 10, 12*f*
uterine hemorrhage, placental abnormalities and, 102
uterine tubes, 10, 12*f*
ampulla of, 10, 12*f*
fertilization in, 28
congenital anomalies of, 256
development of, 246*f*, 248*f*, 250, 250*t*, 261
transport of oocyte via, 22*f*, 24
uteroplacental circulation
impaired, and fetal growth, 85
primordial, 39–40, 44, 56, 58*f*
uterovaginal primordium, 249, 250, 251*f*
developmental arrest of, 256
uterus, 10, 12*f*
and reproductive cycles, 19, 20*f*
bicornuate, 256, 257*f*
with rudimentary horn, 256, 257*f*
body of, 10, 12*f*
congenital anomalies of, 256, 257*f*, 261
development of, 248*f*, 249–250, 250*t*, 253*f*, 261
abnormal, 256, 257*f*
double, 256, 257*f*, 261
growth in pregnancy, 99, 100*f*
horns of, 10, 12*f*
isthmus of, 10, 12*f*
septate, 257*f*
unicornuate, 256, 257*f*
utricle, prostatic, 248*f*, 250*t*
utricular part of otic vesicle, 381, 382*f*, 386
uvula, 179, 180*f*
cleft, 182*f*, 183

V

VACTERL syndrome, 132–133
vagina, 10, 11*f*–12*f*
adenocarcinoma of, diethylstilbestrol exposure and, 133
agenesis of, 256
congenital anomalies of, 256, 257*f*
development of, 248*f*, 250*t*, 250–251, 261
vaginal orifice, 10, 13, 13*f*, 251
vaginal plate, 248*f*, 251
abnormal development of, 256
vaginal process, 256, 258*f*, 259
persistent, 260*f*, 260–261
vagus nerve
formation of, 157*t*, 158–159, 366, 368*f*, 369
innervation of esophagus by, 202
innervation of tongue by, 170
laryngeal branches of, 157*t*, 158–159, 190, 293, 294*f*, 369
vallate papillae, 169–170
valproic acid, as teratogen, 128*t*, 133, 352
valve(s)
atrioventricular, development of, 282*f*, 284, 285*f*
cardiac, development of, 282*f*, 284, 285*f*
mitral, development of, 282*f*, 284
of inferior vena cava, 276–277, 278*f*
of oval foramen, 274*f*–275*f*, 301*f*
pulmonary, stenosis of, 288, 290*f*
semilunar, 284, 285*f*, 290*f*
sinuatrial, 271, 272*f*, 276–277
tricuspid, development of, 282*f*, 284
varicella virus
as teratogen, 129*t*, 136
transplacental transport of, 99
variola virus, transplacental transport of, 99
vas deferens. *See* ductus deferens
vascular accident, fetal, 219
vascular ring, with double aortic arch, 293, 296*f*
vascular structures, fetal, adult derivatives of, 299, 301*f*
vasculogenesis, 56
vasculosyncytial placental membrane, 97
vasectomy, 26
vein(s)
associated with heart, development of, 264–268, 265*f*–267*f*
azygos, 265–268, 266*f*, 267*f*
brachiocephalic, left, 265, 266*f*–267*f*
cardinal, 272*f*, 304, 337*f*
anterior, 265, 266*f*–267*f*
common, 143–144, 143*f*–144*f*, 264–265, 266*f*–267*f*
development of, 264–265, 266*f*–267*f*
posterior, 266*f*–267*f*
endometrial, 39–40, 94*f*–95*f*
gonadal, 266*f*
hemiazygos, 265–268, 266*f*
hepatic, 264, 266*f*, 300*f*–301*f*
hyaloid, 372, 373*f*, 375*f*
hypogastric, 266*f*
iliac
common, 266*f*
external, 266*f*
internal, 266*f*
jugular, 266*f*
oblique, 266*f*, 267*f*
ovarian, 266*f*
portal, development of, 264, 267*f*, 300*f*–301*f*
pudendal, 222
pulmonary, 300*f*–301*f*
primordial, 277, 279*f*
rectal, 222
inferior, 222
superior, 222
renal
accessory, 232, 237*f*
development of, 266*f*
retinal, central, 372, 375*f*, 377*f*
sacral, median, 266*f*
spermatic, 266*f*
subcardinal, 265, 266*f*
subclavian, 266*f*
supracardinal, 265, 266*f*
suprarenal, 266*f*
development of, 266*f*
umbilical, 96, 269*f*, 272*f*, 295–296, 300*f*, 337*f*
adult derivatives of, 299, 301*f*
development of, 53, 264–265, 265*f*, 266*f*–267*f*
transformation of, 267*f*
vena cava (*See* vena cava)
vitelline, 269*f*, 272*f*, 337*f*
development of, 265*f*, 266*f*–267*f*
velamentous insertion, of umbilical cord, 105, 105*f*
vena cava
congenital anomalies of, 268
inferior, 296, 300*f*–301*f*
development of, 265–268, 266*f*–267*f*, 304
interruption of abdominal course, 268
postrenal segment of, 266*f*, 268
prerenal segment of, 266*f*, 268
renal segment of, 266*f*, 268

vena cava—cont'd
valves of, 276–277, 278*f*
superior, 296, 300*f*–301*f*
development of, 265, 267*f*, 304
persistent left, 268
Venezuelan equine encephalitis virus, as teratogen, 128*t*
venous stasis, 23
ventral, as descriptive term, 8*f*
ventral abdominal wall defects (VWDs), detection of, alpha-fetoprotein assay for, 86
ventral axial line, 332–334
ventral horns, 346–348
ventral median fissure, 348
ventral median septum, 348
ventral mesentery, 206*f*, 208, 209*f*–210*f*, 213*f*
ventral mesogastrium, 203, 204*f*–205*f*
ventral primary ramus, 325*f*
ventral roots of spinal nerves, 346–348, 350*f*, 367
ventricles
cardiac
development of, 269*f*, 273*f*, 304
primordial, partitioning of, 277–280, 280*f*–282*f*
walls, cavitation of, 280
cerebral, 344
lateral, 357
third, 358
laryngeal, 190
ventricular septal defects, 288, 289*f*–290*f*
muscular, 288
ventricular system, cerebral, 344
ventricular zone, 344, 362
vernix caseosa, 80, 87, 388–389, 399
vertebrae, variation in number of, 313
vertebral arches, 312*f*, 316
nonfusion of, 351, 353*f* (*See also* spina bifida; spina bifida cystica; spina bifida occulta)
vertebral artery, 268
vertebral body, 312, 312*f*
vertebral column
congenital anomalies of, 316, 316*f*–317*f*
development of, 50, 55, 310–313, 311*f*, 319*f*, 320
bony stage of, 312*f*, 312–313
cartilaginous stage of, 311–312, 312*f*
vesical artery, superior, 268, 299
vesicle(s)
brain, 352–353, 358*f*, 361, 362*f*
primary, 352, 355*f*
secondary, 352–353, 355*f*
lens, 372, 373*f*, 378, 385
mesonephric, 233*f*
metanephric, 230, 235*f*
optic, 66, 154*f*, 171, 347*f*, 357, 372, 373*f*, 385
otic, 381, 381*f*–383*f*, 386
saccular part of, 381, 382*f*, 386
utricular part of, 381, 382*f*, 386
seminal, 11*f*, 13, 24, 25*f*
telencephalic, 357, 358*f*
vesicouterine pouch, 250, 253*f*
vesiculase, 24
vestibular folds, 190
vestibular nerve, 369
vestibular nuclei, 369
vestibule (female), 13, 252, 254*f*
vestibulocochlear nerve, formation of, 366, 368*f*, 369
vestibulocochlear organ, 381
viability
of fetuses, 2, 78
of gametes, 26
villous chorion, 90, 91*f*, 93, 93*f*, 95*f*
viruses
as teratogens, 128*t*–129*t*, 136
transplacental transport of, 97*f*, 99
visceral afferents, general and special, 354, 356*f*–357*f*
visceral efferents, general and special, 354, 356*f*–357*f*
visceral nerves, 366
visceral peritoneum of liver, 208, 209*f*
visceral pleura, 194*f*, 195
viscerocranium
cartilaginous, 313–314, 320
development of, 313–315
membranous, 314–315
visual acuity, 374
vitamin(s), transplacental transport of, 97*f*, 98
vitamin A, as teratogen, 134
vitamin D deficiency, 308
vitelline artery, 265*f*, 337*f*
fate of, 268
vitelline system, 304
vitelline veins, 269*f*, 272*f*, 337*f*
development of, 264, 265*f*, 266*f*–267*f*
vitreous body, 376*f*–377*f*, 378
vitreous humor, 378
primary, 378
secondary, 378
vocal folds (cords), 190
volvulus, 217–218, 220*f*
vomiting
projectile, with congenital hypertrophic pyloric stenosis, 203
with duodenal atresia, 207
vulva, 13

W

warfarin, as teratogen, 128*t*, 133
waste products, transplacental transport of, 97*f*, 99
water
in amniotic fluid, 106
transplacental transport of, 97*f*, 98
webbing of digits, 338–339, 339*f*
weight
birth
cigarette smoking and, 131
extremely low, 78
low, 78, 83, 131
fetal, 78*t*, 79
premature by, 83
Wharton jelly, 105
white communicating ramus, 351*f*, 369
white fat, fetal, 83
white matter (substance), of spinal cord, 344
Wieschaus, Eric. F., 7
Wilmut, Ian, 7
"witch's milk," 392
Wolffian ducts, 249. *See also* mesonephric ducts
wryneck (torticollis), 327, 328*f*

X

X chromosome, 13–14, 245–247, 261
inactivation of, 119
isochromosome of, 126
xiphoid process, 313
X-linked disorders, 126–127

Y

Y chromosome, 13–14, 245–247, 261
yolk sac, 91*f*, 108–110, 114
blood supply to, 265*f*, 271
endoderm of, 41*f*
fate of, 109–110
formation of, 38*f*, 39–41, 40*f*–41*f*
midgut (umbilical intestinal) loop and, 212
primary, 38*f*, 39
primordial gut and, 202, 202*f*, 226
secondary, 40, 40*f*
significance of, 109
yolk stalk, 62, 63*f*, 71, 76, 108–110
midgut loop and, 212, 214*f*
persistent remnants of, 109–110, 220, 221*f*

Z

zona fasciculata, 243, 244*f*
zona glomerulosa, 243, 244*f*
zona pellucida, 15*f*, 17, 31
penetration by sperm, 28, 29*f*–30*f*
shedding of, 31, 33*f*, 35, 43
zona reaction, 28
zona reticularis, 243, 244*f*
zone of polarizing acitivity (ZPA), 330
zygomatic bone, formation of, 314–315
zygosity, in twin pregnancies, 111–113, 113*t*
zygote, 2, 10, 28, 30*f*, 36*f*
cleavage of, 31, 33*f*, 35, 36*f*
definition of, 2
definition of, 2